Backarc Basins

Tectonics and Magmatism

Backarc Basins

Tectonics and Magmatism

Edited by

Brian Taylor

University of Hawaii at Manoa
Honolulu, Hawaii

Plenum Press • New York and London

Library of Congress Cataloging-in-Publication Data

Backarc basins : tectonics and magmatism / edited by Brian Taylor.
p. cm.
Includes bibliographical references and index.
ISBN 0-306-44937-4
1. Back-arc basins. 2. Magmatism. 3. Geology, Structural.
I. Taylor, Brian, 1953-
QE511.2.B33 1995
551.8--dc20 95-2949
CIP

ISBN 0-306-44937-4

A Division of Plenum Publishing Corporation
233 Spring Street, New York, N. Y. 10013

10 9 8 7 6 5 4 3 2 1

Printed in the United States of America

Contributors

Andrey A. Andreyev • Institute of Marine Geology and Geophysics, Yuzhno-Sakhalinsk, Sakhalin 693002, Russia

Jean-Marie Auzende • IFREMER/CB, 29280 Plouzané, France. *Present address*: ORSTOM, BPA5, Nouméa, New Caledonia

Peter F. Barker • British Antarctic Survey, National Environment Research Council, Cambridge CB3 0ET, England

Hervé Bellon • CNRS URA 1278-GDR "GEDO" 910, Université de Bretagne Occidentale, 29287 Brest, France

J. W. Cole • Department of Geology, University of Canterbury, Christchurch, New Zealand

Joseph Cotten • CNRS URA 1278-GDR "GEDO" 910, Université de Bretagne Occidentale, 29287 Brest, France

D. J. Darby • Institute of Geological and Nuclear Sciences, Lower Hutt, New Zealand

Kathleen A. Devaney • Department of Geological Sciences, University of Texas at El Paso, El Paso, Texas 79968

Jean-Philippe Eissen • Antenne ORSTOM-IFREMER/CB, 29280 Plouzané, France

Martin R. Fisk • College of Oceanic and Atmospheric Sciences, Oregon State University, Corvallis, Oregon 97331-5503

Patricia Fryer • Hawaii Institute of Geophysics and Planetology, School of Ocean and Earth Sciences and Technology, University of Hawaii at Manoa, Honolulu, Hawaii 96822

J. A. Gamble • Department of Geology, Research School of Earth Sciences, Victoria University of Wellington, Wellington, New Zealand

Martine Gérard • ORSTOM, 93143 Bondy, France

Helios S. Gnibidenko • Shirshov Institute of Oceanology, Russian Academy of Sciences, Moscow 117851, Russia. *Present address*: Geodynamics Research Institute, Texas A&M University, College Station, Texas 77843

Elena V. Gretskaya • Institute of Marine Geology and Geophysics, Yuzhno-Sakhalinsk, Sakhalin 693002, Russia

James W. Hawkins, Jr. • Geological Research Division, Scripps Institution of Oceanography, La Jolla, California 92093-0220

Thomas W. C. Hilde • Geodynamics Research Institute, Texas A&M University, College Station, Texas 77843

Shu-Kun Hsu • IFREMER, Centre de Brest, 29280 Plouzané, France

Jun-ichiro Ishibashi • Laboratory for Earthquake Chemistry, Faculty of Science, University of Tokyo, Bunkyo-ku, Tokyo 113, Japan

Jean-Louis Joron • Groupe des Sciences de la Terre, Laboratoire Pierre-Süe, CEN Saclay, 91191 Gif sur Yvette, France

Shigeru Kasuga • Hydrographic Department, Maritime Safety Agency, Chuo-ku, Tokyo 104, Japan

Randall A. Keller • College of Oceanic and Atmospheric Sciences, Oregon State University, Corvallis, Oregon 97331-5503

Kazuo Kobayashi • Japan Marine Science and Technology Center, Yokosuka 237, Japan

Lawrence A. Lawver • Institute for Geophysics, University of Texas at Austin, Austin, Texas 78759-8397

Char-Shine Liu • Institute of Oceanography, National Taiwan University, Taipei, Taiwan, Republic of China

Patrick Maillet • ORSTOM, Centre de Brest-GDR "GEDO" 910, 29280 Plouzané, France

Kathleen M. Marsaglia • Department of Geological Sciences, University of Texas at El Paso, El Paso, Texas 79968

Setsuya Nakada • Kyushu University, Fukuoka 812, Japan

Kyoko Okino • Hydrographic Department, Maritime Safety Agency, Chuo-ku, Tokyo 104, Japan

Bernard Pelletier • ORSTOM, Nouméa, New Caledonia

Alain Person • Laboratoire de Géologie des Bassins Sédimentaires, Université Pierre et Marie Curie, 75252 Paris, France

Richard C. Price • La Trobe University, Bundoora, Victoria 3083, Australia

Etienne Ruellan • CNRS Sophia-Antipolis, 06560 Valbonne, France

Chuen-Tien Shyu • Institute of Oceanography, National Taiwan University, Taipei, Taiwan, Republic of China

Jean-Claude Sibuet • IFREMER, Centre de Brest, 29280 Plouzané, France

T. A. Stern • Research School of Earth Sciences, Victoria University of Wellington, Wellington, New Zealand

Jorge A. Strelin • Departamento de Ciencias de la Tierra, Instituto Antártico Argentino, Buenos Aires, Argentina

Kensaku Tamaki • Ocean Research Institute, University of Tokyo, Nakano-ku, Tokyo 164, Japan

Tetsuro Urabe • Geological Survey of Japan, Tsukuba 305, Japan

I. C. Wright • New Zealand Oceanographic Institute, National Institute of Water and Atmospheric Research, Wellington, New Zealand

Preface

This is the first book to provide a comprehensive treatment of the tectonics and magmatism of backarc basins, from their initial rift stage to mature spreading. It focuses on the young backarc basins of the circum-Pacific, where the volcano-tectonic processes are best studied because they are still active. Twelve chapters describe the Taupo-Havre-Lau system, the North Fiji Basin, the New Hebrides, Mariana, Bransfield, and Okinawa troughs, the East Scotia and Japan Seas, and the Shikoku and Kuril Basins. Two chapters provide syntheses of hydrothermal activity related to arc-backarc magmatism and tectonic and magmatic controls on basin sedimentation. The authors of each chapter are the recognized experts for their area/topic.

The last ten years have seen fundamental changes in our understanding of the tectonics, volcanism, hydrothermal circulation, and sedimentation in backarc basins. This has resulted from exploration and detailed studies using swath bathymetry and sidescan acoustic imagery, multichannel seismics, manned submersibles and remotely operated vehicles, and deep ocean drilling. The studies in this volume summarize these new findings and use them to build improved models of the dynamic processes occurring in backarc basins. Below I preview some of the key themes woven throughout the chapters in this book, together with my own observations.

VOLCANO-TECTONIC SETTING

Backarc basins are situated above mantle exhibiting strong seismic wave attenuation, that is in turn above or beyond Wadati–Benioff seismic zones that trace the descent of oceanic lithosphere from bathymetric trenches into the mantle. Not all subduction zones have associated active arc and/or backarc volcanism, sometimes because the above wedge of mantle is missing and/or because there is compressional coupling between the subducting and overriding plates. The formation of backarc basins involves the splitting of volcanic arcs, initially forming rifts (elongate depressions bounded by faults) and subsequently backarc spreading centers. They may form within continental or intraoceanic arcs.

CAUSE AND LOCATION

The trench axes of all subduction systems with active backarc basins are observed to migrate seawards in a hotspot reference frame. Whether the sinking of the subducting plate beneath the overriding plate is a cause or an effect of backarc opening remains debated, but presently a majority favor the former. Depending on its other boundary conditions, the upper plate (being horizontally coupled by suction to the downgoing plate) is pulled seawards and stretched. Periodically the stretching is sufficient to pull the weakest part of the upper plate apart. This occurs where the plate is both hottest and has the thickest crust (the stronger mantle is thinnest), namely within about 50 kilometers of the volcanic front. Sometimes this is predominantly on the trench (forearc) side (e.g., Lau Basin), sometimes in the backarc (e.g., New Hebrides and Okinawa troughs), and sometimes variably on both sides along strike (e.g., Izu-Bonin-Mariana system).

INITIAL RIFTING

The formation of backarc basins involves extensive rifting of the preexisting arc massifs. The initial rift basins subside along a zigzag pattern (in plan view) of border faults, many of which dip at low angles (20–50°). Transfer zones that link opposing master faults and/or rift flank uplifts further subdivide the rifts into segments along strike. The rifts are often better developed between the arc volcanoes: magmatism, rather than stretching and subsidence, often fills the opening adjacent to arc volcanoes. Explosive, high-silica volcanism commonly reaches peak intensities during the initial rifting (as documented in the Japan, Izu-Bonin-Mariana, and Lau-Colville-New Zealand arcs). Nearby calderas shed voluminous pumiceous sediments into rapidly filled rift basins.

RIFT MAGMAS

Magmas intrude along faults into the rifted arc crust of the basins, sometimes producing lineated magnetic anomalies. The magmas may be derived from the suprasubduction zone mantle arc source and/or from pressure-release partial melting of the shallow mantle: both arc-like and mid-ocean ridge basalt (MORB)-like compositions are observed temporally and spatially juxtaposed. Nevertheless, slightly wet but otherwise MORB-like lavas erupt in even the earliest rifts, very proximal (~10 km) to arc volcanoes. Isotope data indicate that the mantle source region may be replenished by trench ward flow of aesthenosphere.

CONTINUED STRETCHING

As continued stretching widens the basins, the zone of greatest subsidence and intrusion migrates with the collapsing arc margin border faults. The result is an asymmetric basin in cross section, with a wider zone of stretched and intruded crust on the proto-remnant arc side whether the rifting initiated in the forearc or in the backarc. In the former case, there may be no active frontal arc, the proto-remnant arc may remain active during initial rifting, and half-graben depocenters within the rift basin may be largely isolated from margin sediments.

INITIAL SPREADING

After continued stretching and intrusion for several million years has produced a rift basin up to 200 kilometers wide, the shallow and deep sources of mantle partial melts become horizontally separated and an organized spreading center forms, fed by the shallow mantle. This evolution is spatially transgressive: just as rifting migrates laterally, spreading is localized in some areas first (commonly along a basin-bounding transform fault) and then propagates longitudinally into adjacent rifting areas. Eventually the suprasubduction zone arc magmas establish a new volcanic front along the rifted edge of the old one (the South Sandwich arc is actually built on backarc basin oceanic crust). However, in several but not all (Mariana) systems, arc activity declined markedly at the start of subsequent backarc spreading (or even terminated in the Izu-Bonin-Shikoku Basin case).

MATURE SPREADING

The recognition that backarc basins have an extensive rift stage prior to spreading has helped reconcile some of their initially perceived differences to mid-ocean spreading centers. The study of mature systems in the northern Lau, North Fiji, Manus, central Mariana, and East Scotia basins has confirmed first-order similarities in backarc spreading parameters (rates, morphotectonics, magnetization, crustal structure, and geochemistry) to those of mid-ocean ridges. Nevertheless there remain important differences, particularly in the diversity of magma, hydrothermal fluid, and sediment types, and possibly in the thermal and crustal structures, associated with their suprasubduction zone setting.

HYDROTHERMAL ACTIVITY

One sixth of the known sites of seafloor hydrothermal activity has been discovered in backarc basins. Silica, sulfate, and sulfide chimneys venting clear, white, and black solutions at temperatures of 200° to 400°C have been found on all the active backarc spreading centers studied (Lau, North Fiji, Manus, and Mariana). These fluids have low pH (2 to 5), have evidence of direct interaction with magma, and typically contain more zinc, barium, lead, cadmium, and arsenic than their mid-ocean ridge equivalents. Hydrothermal deposits discovered in backarc rifts such as the Okinawa Trough and Sumisu Rift are a modern analog of the Kuroko-type volcanic-hosted massive sulfide (VMS) deposits that formed in the middle Miocene Hokuroku Basin during the last phase of Japan Sea opening. Like all sites of major VMS ore bodies, and unlike mid-ocean ridge deposits, they occur in association with subduction-related volcanism including silica-rich lavas as well as basalt.

Acknowledgments

Diane Henderson of the SOEST Publications Group was of great assistance in coordinating, reformatting and copyediting the papers. I thank all the authors for contributing their work. The rigorous manuscript reviews by those listed below significantly improved the volume: Dallas Abbott, Hugh Bibby, Thomas Brocher, Glenn Brown, Anthony Crawford, Anthony Ewart, Fred Hochstaedter, Kevin Johnson, Lynn Johnson, Donna Jurdy, David Kemp, Loren Kroenke, Robert Larter, Lawrence Lawver, Neil Lundberg, Alex Malahoff, Fernando Martinez, James Natland, Richard Price, Ian Smith, Robert Stern, and Michael Underwood.

Brian Taylor

Contents

Chapter 2

The Southern Havre Trough: Geological Structure and Magma Petrogenesis of an Active Backarc Rift Complex

J. A. Gamble and I. C. Wright

Chapter 3

The Geology of the Lau Basin

James W. Hawkins, Jr.

Chapter 6

Geology of the Mariana Trough

Patricia Fryer

Chapter 8

Bransfield Strait, Antarctic Peninsula: Active Extension behind a Dead Arc

Lawrence A. Lawver, Randall A. Keller, Martin R. Fisk, and Jorge A. Strelin

Chapter 9

Structural and Kinematic Evolutions of the Okinawa Trough Backarc Basin

Jean-Claude Sibuet, Shu-Kun Hsu, Chuen-Tien Shyu, and Char-Shine Liu

Chapter 10

Shikoku Basin and Its Margins

Kazuo Kobayashi, Shigeru Kasuga, and Kyoko Okino

Chapter 11

Opening Tectonics of the Japan Sea

Kensaku Tamaki

Chapter 12

Kuril (South Okhotsk) Backarc Basin

Helios S. Gnibidenko, Thomas W. C. Hilde, Elena V. Gretskaya, and Andrey A. Andreyev

Chapter 13

Hydrothermal Activity Related to Arc-Backarc Magmatism in the Western Pacific

Jun-ichiro Ishibashi and Tetsuro Urabe

Chapter 14

Tectonic and Magmatic Controls on Backarc Basin Sedimentation: The Mariana Region Reexamined

Kathleen M. Marsaglia and Kathleen A. Devaney

Backarc Basins

Tectonics and Magmatism

1

Taupo Volcanic Zone and Central Volcanic Region
Backarc Structures of North Island, New Zealand

J. W. Cole, D. J. Darby, and T. A. Stern

ABSTRACT

Volcanism in North Island, New Zealand, is related to subduction within the Taupo–Hikurangi arc-trench system. The central volcanic region (CVR) is a wedge-shaped basin, approximately 4 m.y. old, which is defined by geophysical parameters. It corresponds to an area of low density, low-velocity volcanic rocks and anomalous seismic properties within the upper mantle. Seismic data also provide evidence for a 7.4–7.5 km/s layer at the unusually shallow depth of about 15 km beneath CVR; this material is inferred to be partially melted upper mantle. The youngest and active part of CVR is the Taupo volcanic zone (TVZ), the currently active backarc basin of the subduction system where volcanism is <2 m.y. old. TVZ is an area of high-convective-heat output (4×10^9 W). This output is equivalent to an average heat flow of about 800 mW/m^2, about 13 times greater than the continental norm and one of the highest on record for a backarc basin. Rates of extension in the currently active TVZ are about 7 mm/yr in the north and 18 mm/yr in the south, derived from shear strain rates varying between 0.18 and 0.5×10^{-6}/yr shear. Subsidence rates are 1–2 mm/yr. Focal mechanisms from recent small earthquakes are inconclusive, but most appear to be transcurrent. The larger 1987 Edgecumbe (M=6.3) earthquake, in the Bay of Plenty, had a normal focal mechanism. There are three volcanic assemblages in TVZ: high-Al basalt, basaltic andesite-andesite-dacite, and rhyolite-ignimbrite, in order of increasing volume. High-Al basalt and rhyolite occur in a bimodal assemblage associated with caldera structures in the extensional backarc basin, while the basaltic andesite-andesite-dacite assemblage occurs principally at the northern and southern ends of TVZ and in a narrow frontal arc along the eastern margin. The andesitic assemblage is undoubtedly slab related and was formed in a multistage process involving dehydration of the slab, anatexis of the

---

J. W. Cole • Department of Geology, University of Canterbury, Christchurch, New Zealand. *D. J. Darby* • Institute of Geological and Nuclear Sciences, Lower Hutt, New Zealand. *T. A. Stern* • Research School of Earth Sciences, Victoria University of Wellington, Wellington, New Zealand.

Backarc Basins: Tectonics and Magmatism, edited by Brian Taylor, Plenum Press, New York, 1995.

overlying mantle, fractionation of mineral phases, and minor crustal assimilation. The high-Al basalts are considered to be derived from a hot mantle wedge above the subduction zone with variation due to fractionation and minor assimilation. Origin of the rhyolites and ignimbrites is more controversial. They are most likely to be the result of either melting of earlier-formed igneous rocks or sedimentary/metamorphic rocks of a restricted composition from within the crust, perhaps with a minor upper mantle source component, or partial melting of a mantle-derived source with significant crustal contamination.

1. INTRODUCTION

The CVR/TVZ offers a rare opportunity to study backarc processes from land-based observations. Most backarc spreading areas are covered by ocean, which makes direct observations of geology, volcanism, and earthquake seismology difficult. In particular, reliable heat flow data from oceanic backarc basins are notoriously difficult to acquire (Sclater *et al.*, 1980).

The results of a variety of techniques are described in this study including seismic tomography, geodesy, geochemical analysis, and integrated heat flux. In this chapter three workers from what at times have been seemingly disparate subdisciplines of the earth sciences—geochemistry, geodesy, and geophysics—have attempted to produce a common synthesis of knowledge on the CVR/TVZ. During this synthesis one of the problems that arose was the differing inferences drawn from geophysics and geology about what the principal volcano-tectonic unit is in central North Island. This is a question we address but do not finally resolve. In general, however, we believe that effective progress will be made in learning about such as areas the CVR/TVZ by such a cross-disciplinary approach.

2. PLATE TECTONIC SETTING

New Zealand lies at the boundary between the Pacific and Australian plates, and in the northern part of New Zealand the plate boundary zone is dominated by subduction of oceanic crust of the Pacific plate beneath continental crust of the Australian plate (Fig. 1.1). The relative instantaneous pole of rotation for these two plates is sited at 62°S, 174°E by Chase (1978), and rotation is estimated to be 1.27°/m.y. As a result, subduction is oblique under North Island (Fig. 1.1) to form the Taupo–Hikurangi arc-trench system (Cole and Lewis, 1981). Subduction becomes progressively more oblique to the south until, in South Island, continental crust of the Chatham Rise is being obliquely obducted onto continental crust of the Australian plate in a zone of transpression about 10 km wide. Much of the movement in this zone occurs as reverse-dextral movement on the Alpine Fault (Fig. 1.1).

3. TAUPO–HIKURANGI ARC-TRENCH SYSTEM

The Taupo–Hikurangi arc-trench system extends from the Hikurangi Trough off the east coast of North Island, New Zealand, to the TVZ in central North Island (Fig. 1.1). The Hikurangi Trough is slightly offset in terms of bathymetry from the Kermadec Trench to the north (Carter, 1980) and merges southward into the Alpine Fault system.

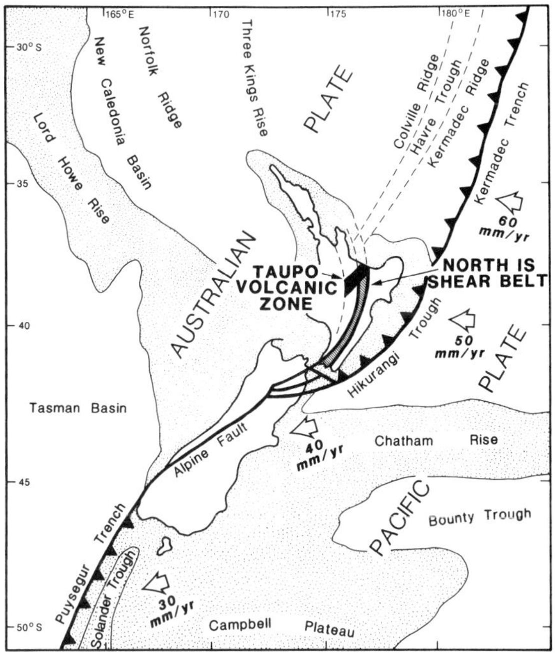

FIGURE 1.1. Location of Pacific-Australian plate boundary in the New Zealand region. Stippled area represents continental crust around N.Z. Arrows show motion of Pacific plate relative to Australian plate (after Cole, 1990). Rates are from Walcott (1987).

To the west of the trough is an accretionary prism, up to 150 km wide, which becomes progressively older from east to west (Lewis, 1980; Cole and Lewis, 1981). The western part of the accretionary prism is a frontal ridge of Upper Paleozoic–Mesozoic graywacke-argillite which is undergoing rapid uplift (e.g., Lamb and Vella, 1987) and is cut by a major zone of dextral strike-slip faults—the North Island shear belt (Fig. 1.1).

4. DISTINCTION BETWEEN CENTRAL VOLCANIC REGION AND TAUPO VOLCANIC ZONE

The CVR is a wedge-shaped basin of predominantly Quaternary rhyolitic volcanism. Calhaem (1973) noted that the northwestern boundary of CVR corresponded to the location of a number of low-potash andesites K-Ar dated at 4 Ma (although some of these rocks are now known to be altered, so the error limits of the dates are high). The TVZ is the eastern part of CVR and represents that portion (<2 Ma) that is currently volcanically active. The CVR and TVZ share a common eastern boundary with the present active volcanic front or arc.

The CVR is the principal volcanic basin structure defined by seismic and gravity parameters in central North Island (Fig. 1.2) and corresponds to an area of low-density, low-velocity volcanics that occupy both the eastern side of the Coromandel volcanic zone and TVZ. The western boundary of the TVZ within the CVR is equivocal and is drawn somewhat arbitrarily on Fig. 1.3 to correspond to (i) the westernmost limit of active faulting in central North Island, (ii) the apparent position of the active volcanic front at 2 Ma, and (iii) a marked change in structural trend in offshore Bay of Plenty. The CVR is associated with the latter stages of a volcanic arc that has been in existence for at least 22 m.y. (see

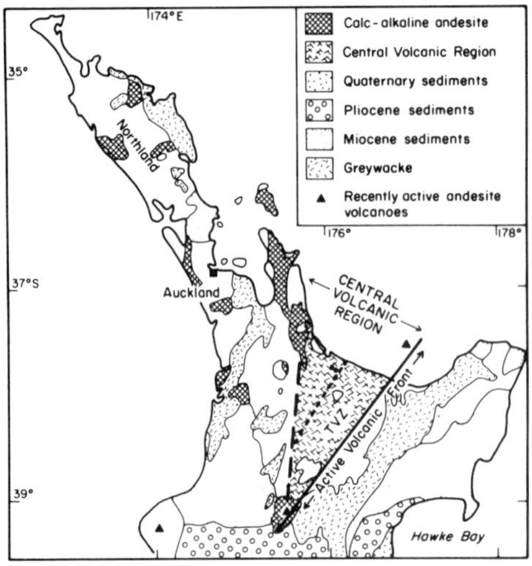

FIGURE 1.2. Map of the northern part of the North Island showing the location of the central volcanic region and the distribution of andesites, graywacke and principal sedimentary basins.

following). The TVZ corresponds to a change in orientation of the plate boundary resulting from opening of the Havre Trough in the past 4 m.y.

5. EARLIER VOLCANIC ARCS

Within Northland, outcrops of calc-alkaline andesite and basalt volcanism are found interspersed among older volcanics of an obducted ophiolite sequence (Brothers and Delaloye, 1982) and Permian-Triassic basement graywacke (Fig. 1.2). A number of workers have noted an apparent migration of the low-potash andesites southeastward through the

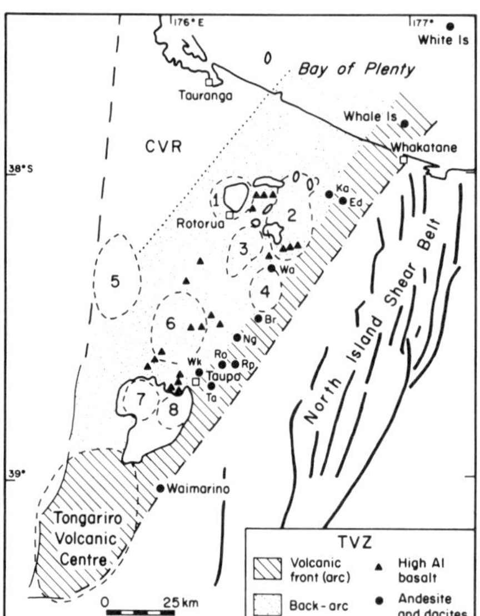

FIGURE 1.3. Map of the Taupo volcanic zone showing location of probable calderas (1, Rotorua; 2, Haroharo; 3, Kapenga; 4, Reporoa; 5, Mangakino; 6, Whakamaru; 7, Oruanui; 8, Taupo); high-Al basalts (triangles) and andesite-dacite localities (filled circles) mentioned in the text. (Ka, Kawerau; Ed, Edgecumbe; Wa, Waiotapu; Br, Broadlands; Ng, Ngatamariki; Ro, Rotokawa; Rp, Rolles Peak; Wk, Wairakei; Ta, Tauhara).

Northland-Auckland peninsula with time (e.g., Kear, 1959; Hatherton, 1969). With the availability of K-Ar dates of these andesites, various plate reconstructions have been developed that either take the plate boundary during the Tertiary as being adjacent and parallel to Northland (e.g., Calhaem, 1973; Cole and Lewis, 1981; Walcott; 1987, Cole 1990) or that a substantially shallower dipping Pacific plate extended beneath Northland from the current plate boundary (Brothers and Delaloye, 1982).

Regardless of what the exact orientation for the plate boundary was, two points should be noted. First, the Northland–Waikato area has been subjected to up to 22 m.y. of volcanism and, presumably, thermal weakening of the lithosphere; second, there has been a more rapid migration of the andesite arc across the CVR than within Northland.

Two distinct volcanic arcs can be recognized in Northland (Fig. 1.4): a western Waitakere arc comprising tholeiitic basalt to medium-K andesite erupted between 22 and 15 Ma (Hayward, 1979); geophysical anomalies and offshore drilling indicate that much of the arc is offshore. A second arc (Coromandel volcanic zone) extends from the eastern side of Northland through Great Barrier Island to Coromandel. The earliest activity was submarine with the eruption of medium-K andesites, comparable in age to those of the Waitakere arc, but volcanism continued in the Coromandel volcanic zone after the Waitakere arc ceased activity. About 10 Ma, rhyolitic volcanism began on the east side of Coromandel volcanic zone to form the Whitianga Group (Skinner, 1986), with numerous rhyolite domes and extensive ignimbrites, some associated with calderas.

6. OFFSHORE BAY OF PLENTY

The western boundary of the TVZ is perhaps best illustrated in offshore Bay of Plenty where a structure 40–50 km wide bounded by inward-facing normal fault zones is evident (Wright, 1992). The western margin of the extensional basin is marked by the Tauranga

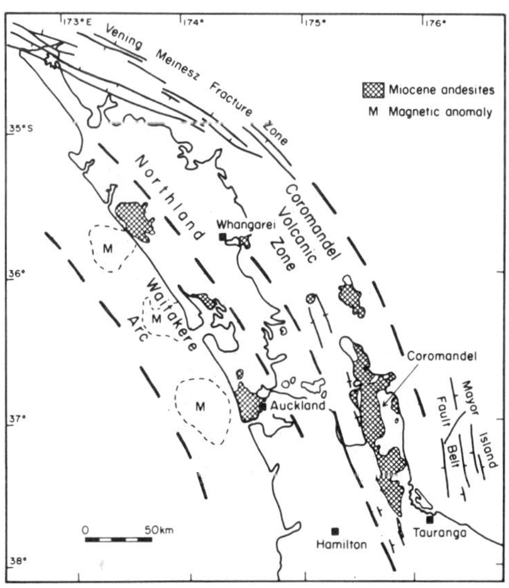

FIGURE 1.4. Map of Northland and Coromandel showing location of Miocene (22-12 Ma) volcanic arcs.

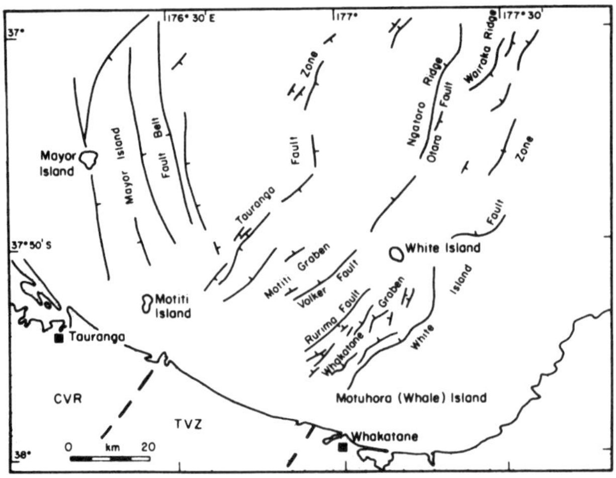

FIGURE 1.5. Faulting in offshore Bay of Plenty (after Wright, 1992).

fault zone and the eastern margin by the White Island fault zone (Fig. 1.5). Within the basin is a frontal nonvolcanic graben, a volcanic (arc) ridge, and a volcanic (back-arc) basin.

7. TAUPO VOLCANIC ZONE CHRONOLOGY

The commencement of volcanism in TVZ is difficult to establish. The first recognized activity that can be regarded as part of TVZ is the formation of the Mangakino caldera. This is sited at the intersection between the southern limit of the Coromandel volcanic zone and TVZ (Fig. 1.3). It was active from 1.6 to 0.9 Ma (Table I) and was the site of nine voluminous (>1100 km^3) ignimbrites erupted between 1.62 and 1.51 Ma (Ngaroma and ignimbrites B and C) and 1.23 to 0.91 Ma (Ongatiti, Ahuroa, Rocky Hill, Kaahu, Marshall, and Waioraka ignimbrites; Briggs *et al.*, 1993) together with two phreatomagmatic deposits (units D and E; Wilson, 1986).

Silicic volcanics of the central part of the TVZ are associated with seven main centers and associated calderas (e.g., Healy, 1962; Wilson *et al.*, 1984; Cole 1990; Nairn *et al.*, 1994) (Fig. 1.3). The earliest activity is probably now completely buried by later units. Wilson *et al.* (1984) recognized the Kapenga caldera as one of the oldest, and possibly the source of the 0.71 Ma Waiotapu ignimbrite (Table I). Activity then moved to the SW with the eruption of the Whakamaru Group ignimbrites ~0.35 Ma (Wilson *et al.*, 1986). These were probably erupted over a significant time span and filled parts of the earlier Mangakino and Kapenga structures. Since this time most rhyolitic activity has been from the Okataina, Reporoa, and Taupo calderas on the E side of the TVZ (Fig 1.3), but the Mamaku ignimbrite (~0.22 Ma) is likely to have come from the Rotorua caldera (Fig 1.3).

The last 0.20 m.y. of activity has been dominated by rhyolite dome formation, explosive eruptions, and the eruption of ignimbrites (e.g., Rotoiti, Oruanui, and Taupo) mainly from the Okataina or Taupo centers (Fig 1.3). Interspersed between these events has been the eruption of minor (<1% of total volume of TVZ volcanics) high-Al basalt, commonly from fissures.

The most complete sequence of andesitic lavas occurs in the Tongariro volcanic center at the southern end (Fig 1.3), where eruptive activity spans at least 0.25 m.y. (Hackett and

TABLE I
Stratigraphy of Ignimbrites from the Major Calderas, with Approximate Ages of Eruption[a]

Ma	Mangakino		Taupo/Whakamaru		Kapenga/Reporoa		Rotorua/Okataina	
			Taupo ig	0.0018			Rotoiti	0.065
			Oruanui ig	0.0226			Mamaku	0.22
					Kaingaroa	0.23		
0.2			Rangitaiki }	0.32	Pokai	?	Matahina	0.28
			Whakamaru }		Chimp	?		
			Te Kopia	0.34				
0.4								
					Waiotapu	0.71		
0.6								
	Marshall ig	0.91 ± 0.02						
0.8	Kaahu ig	0.92 ± 0.07						
	Rocky Hill ig	0.97 ± 0.02						
1.0								
	Aharoa ig	1.19 ± 0.03						
1.2								
	Ongatiti ig	1.23 ± 0.02						
1.4								
	Ignimbrite B	1.51 ± 0.02						
1.6	Ngaroma ig	1.60 ± 0.03						

[a]Data from Briggs *et al.*, 1993; Houghton *et al.*, 1995.

Houghton, 1989). Andesites have also been intersected by drilling at Rotokawa, Ngatamariki, Kawerau, and Waiotapu, and relationships to dated ignimbrite units suggest they may be up to 0.7 Ma.

8. CENTRAL VOLCANIC REGION STRUCTURE

Drilling for exploration and exploitation of geothermal power within the TVZ over the past 20 years has provided a great deal of stratigraphic data. About 10 holes have been drilled to depths of more than 2 km, and these show predominantly rhyolitic material in the form of ignimbrite sheets within the upper 1500 m or so, and then an increasing amount of dense volcanic rocks, particularly andesite, below 1500 m. A west-east cross section of drill-hole stratigraphy across the TVZ is given by Stern (1987). Mesozoic graywacke has been encountered at the base of some holes that are within 6 km of the eastern boundary of the TVZ, but no sedimentary basement rocks have been encountered in drill holes elsewhere within the TVZ. The deepest hole to date (at Ngatamariki) terminated in a quartz diorite at a depth of about 2.8 km (Wood, 1986), while other deep holes on the eastern side of the TVZ have terminated in andesite (e.g., Browne, 1971; Browne *et al.*, 1992).

9. GEOPHYSICAL DATA BEARING ON CRUSTAL AND UPPER MANTLE STRUCTURE, HEAT FLUX, AND KINEMATICS

Geophysical data from central North Island have provided a basis from which the deeper crust and upper mantle conditions can be interpreted. Together, these data indicate

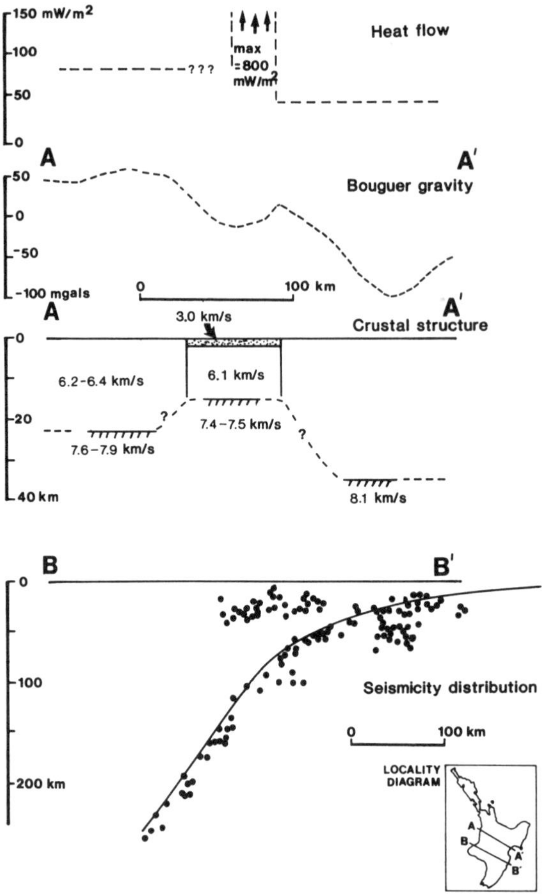

FIGURE 1.6. A cartoon cross-section of central North Island showing crustal and upper mantle structure and the inferred position of the subducted Pacific Plate (after Smith *et al.*, 1989). Seismic velocities from refraction surveys of Stern and Davey (1987) and Stern *et al.*, (1987).

an area of grossly anomalous crustal and upper mantle conditions. A cross section of central North Island, which includes the CVR, is shown in Fig 1.6. Information on this cross section will be discussed in the following sections.

9.1. Earthquake Seismology

Seismologists have known for some time that earthquake waves that traverse the upper mantle beneath central North Island are subjected to severe attenuation (Eiby, 1958; Mooney, 1970) and travel at anomalously slow velocities (Haines, 1979). For example, Haines's partition of the upper mantle beneath North Island shows an area beneath the CVR to have upper mantle P-wave velocities (Pn velocity) of 7.4 ± 0.1 km/s, while more normal Pn velocities of 8.1–8.5 km/s are found to the east.

Seismic intensity data have been inverted to determine the attenuation parameter Q (a measure of the amount of seismic energy lost as a wave passes through a medium; Fowler, 1990) in the mantle beneath North Island (Satake and Hashida, 1989). Figure 1.7 shows the results for Q in two depth ranges, 0–30 km and 30–80 km. Dominating both maps is a low-Q region that occupies the CVR and spreads out into the Coromandel volcanic zone.

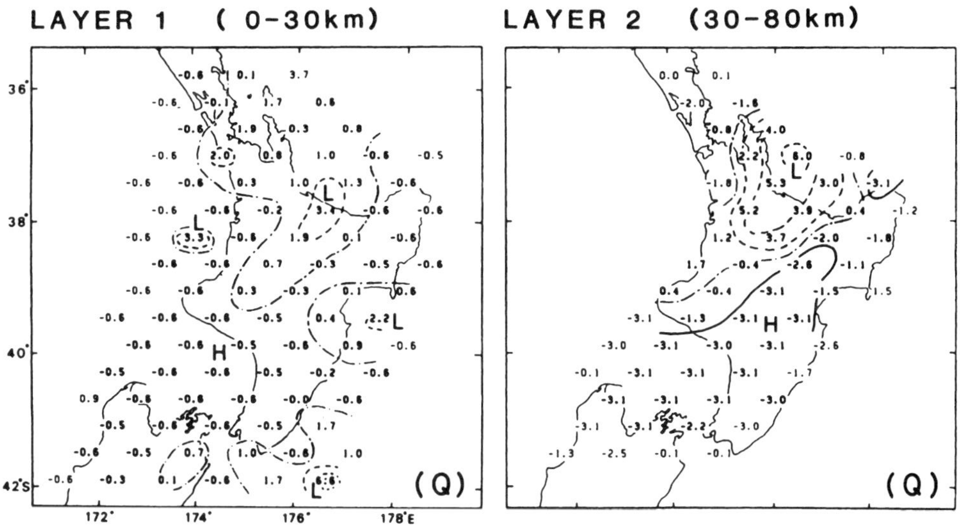

FIGURE 1.7. Attenuation structure (Q) determined by the inversion of felt intensities (Satake and Hashida, 1989). Q at two different depth ranges are shown. The numbers are deviations from an initial value of attenuation coefficient and are in units of 10^{-2} s^{-1}; negative values are weak attenuation (high Q) and positive values are strong attenuation (low Q).

This probably represents areas of pervasive partial melt in the lower crust and upper mantle (Haines, 1979).

9.2. Explosion Seismology

Seismic refraction, wide-angle reflection, and near-vertical reflection surveys conducted during the past 10 years have provided information on velocities and structures within the crust and depth to the upper mantle. Using shot points in the Bay of Plenty and Lake Taupo, Stern and Davey (1987) conducted a first-order crustal refraction survey, using analog microearthquake seismographs. They reported a crustal velocity model for the CVR that features a 3.0–6.1 km/s crust overlying a layer of velocity 7.4–7.5 km/s at a depth of 15±2 km (Fig 1.6). A recent offshore seismic reflection profile reported by Davey (1993) found a band of subhorizontal reflectors at 5–6 s of two-way travel time. For an average crustal velocity of, say, 5.9 km/s, these reflectors correspond to a depth range of 14–18 km and are thus broadly consistent with the top of the 7.4–7.5 km/s layer found in the refraction experiment of Stern and Davey (1987) described above.

9.3. Interpretation of the 7.4–7.5 km/s Layer

Seismic velocities of 7.4–7.5 km/s are ambiguous in that they may be interpreted as either lower crustal or upper mantle in origin. For example, velocities of 7.3–7.4 km/s have been detected at a depth of only 9 km beneath Fiordland, New Zealand (Davey and Broadbent, 1980), where they have been interpreted as uplifted lower crust. However, low-Pn velocities of 7.4–7.9 km/s are almost invariably associated with tectonically active areas of high heat flow (e.g., Eaton, 1984; KRISP working party, 1991). Therefore the 7.4–7.5

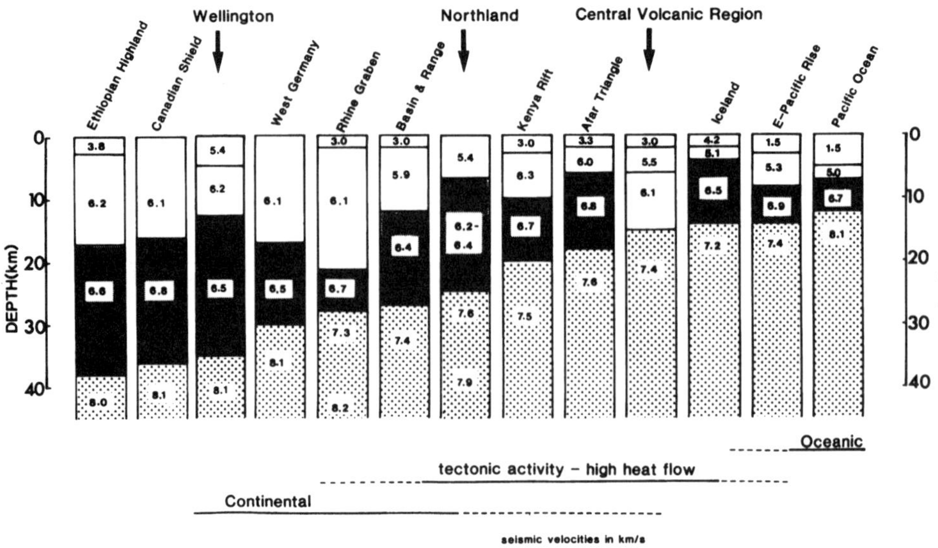

FIGURE 1.8. Crustal structures from different locations around the world illustrating a transition from continental-type crustal structure through tectonically active types to oceanic-type crust (after Garrick, 1968; Berckhemer *et al.*, 1975; KRISP working party, 1991; Stern *et al.*, 1987; Stern and Davey, 1987). Dark gray pattern represents lower crustal velocities of 6.4–7.0 km/s, and it is notable that this layer is not present in the CVR.

km/s layer beneath the CVR is likely to be the upper mantle, albeit in a partially molten state (see Black and Braille, 1982, for discussion on relationship between Pn velocities and heat flow). Figure 1.8 summarizes crustal structures from around the world and puts the proposed crustal structure model for the CVR in a global context.

9.4. Gravity Data

About 8000 gravity observations have been made within and adjacent to the CVR and these data are displayed on 1:250,000 maps of the Bouguer and isostatic gravity anomaly field (Reilly, 1972). An attempt to separate numerically the regional and residual components of the gravity field of the CVR was made by Stern (1979). Figure 1.9 shows the resulting residual anomalies. The principal feature of this map is that almost the whole CVR is dominated by a background residual low of −300 μN/kg, with some areas as low as −600 μN/kg, and other areas where the residual anomalies rise to zero (on the Bay of Plenty coast where a graywacke outcrop exists). Rogan (1982) has made a three-dimensional interpretation of the residual anomalies with a simple constant-density contrast between the volcanic rocks and basement. Interpretations by Stern (1986) have been made along three lines (location shown on Fig 1.9) that are controlled by seismic refraction data: one west-east line through the large gravity low at Mangakino (40 km northwest of Lake Taupo); a north-south line from the graywacke outcrop near the Bay of Plenty coast; and a west-east line through the gravity low just to the north of Lake Taupo. For the gravity lows near Lake Taupo and Mangakino, negative-mass anomalies are inferred deeper than the 4.9–5.5 km/s seismic basement. These anomalies may represent partially cooled rhyolitic intrusions or low-density pyroclastic deposits lying beneath higher-density ignimbrite units where the ignimbrite effectively constitutes basement (Stern, 1986).

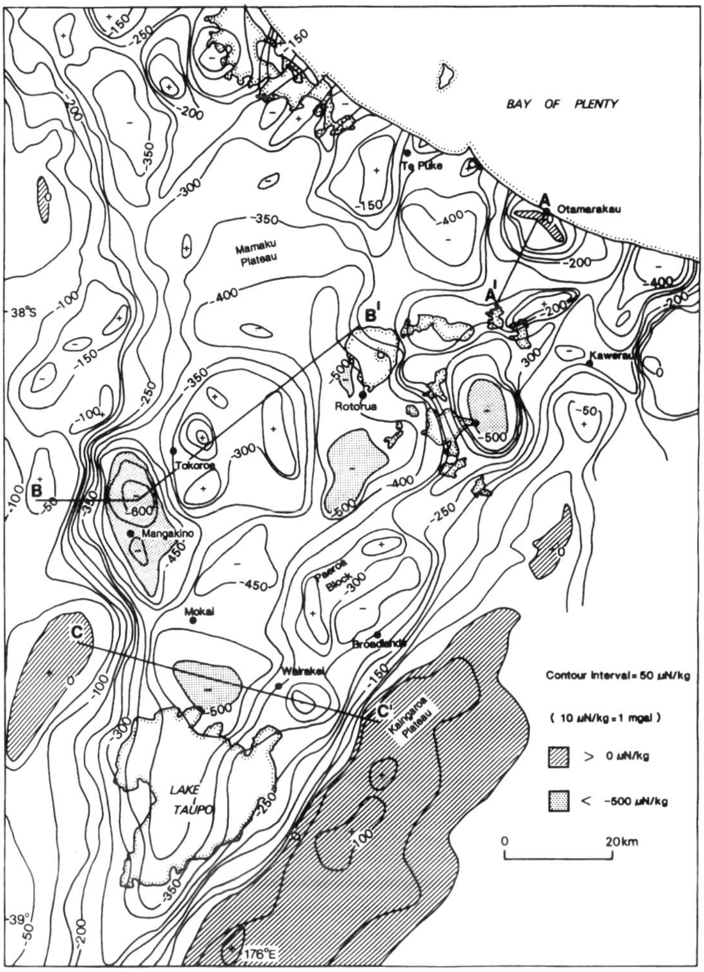

FIGURE 1.9. Residual gravity anomalies for the CVR (after Stern, 1979). The three profiles (A-A′, B-B′, and C-C′) are those interpreted by Stern (1986). Contour interval = 50 N/kg (10 N/kg = 1 mgal).

9.5. Heat Flux

Heat output from the CVR is almost totally expressed in the discharge of hot water and steam from hydrothermal systems within the TVZ, as shown in Fig 1.10 (Studt and Thompson, 1969). Natural heat output from the individual geothermal systems has been measured by a variety of methods (e.g., Dawson and Dickinson, 1970; Bibby *et al.*, 1984; Calhaem, 1973). Natural heat output data shown in Fig 1.10 are from Calhaem (1973) and Bibby *et al.* (1984) and sum to about 4×10^9 W. This figure is a minimum, as other "hidden" geothermal fields similar to Mokai (Bibby *et al.*, 1984) may yet be discovered.

Taking the recharge zone for these geothermal areas to be about 5000 km^2 (Stern, 1987), the equivalent heat flow is then about 800 mW/m^2, a value about 13 times greater than the continental norm and, as far as we are aware, one of the highest reported for any backarc basin. This may be because most backarc basins are oceanic and therefore covered by a shallow ocean that effectively masks the true heat output, thus resulting in an

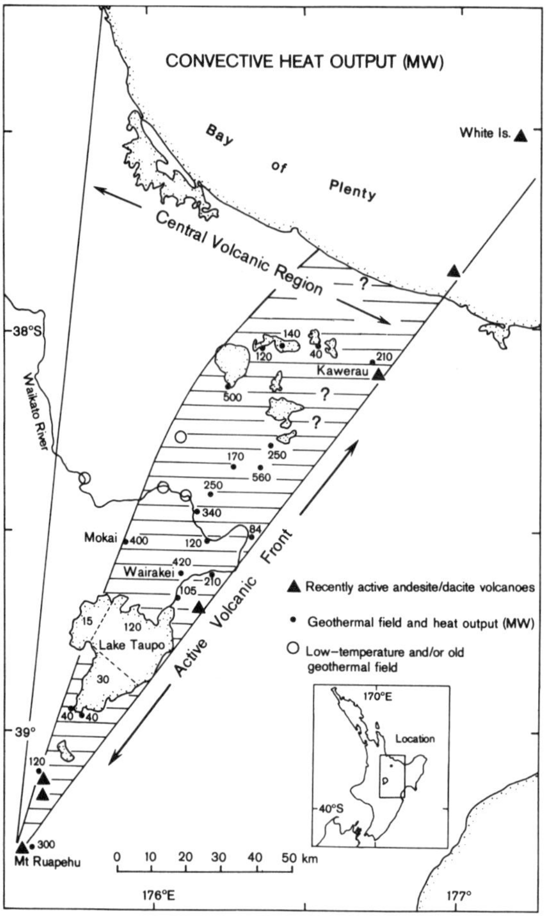

FIGURE 1.10. Distribution and heat output (in MW) of geothermal fields within the CVR (after Stern, 1987). Total natural heat output in shaded area from south end of Lake Taupo to the Bay of Plenty coast is about 4 x 10^9 W. Shaded region represents our best estimate of the region where convective heat transfer is occurring at present.

underestimate of the heat flow (Sclater *et al.*, 1980). Nevertheless, an average heat flow of 800 mW/m^2 is of the same magnitude as areas like Yellowstone caldera in the United States (Morgan *et al.*, 1977) and spreading centers like Iceland (Palmason and Saemundsson, 1974) and, in general, is indicative of a tectonic regime where a large-scale convective transfer of heat is taking place.

10. GEODETIC DATA

10.1. Large-Scale Horizontal Deformation

Five analyses of first-order triangulation have been undertaken to estimate large-scale deformation across North Island, particularly across the CVR and the TVZ (Sissons, 1979; Adams, 1984; Walcott, 1984, 1987; Crook and Hannah, 1989; Reilly, 1990), using the method of Bibby (1973, 1981, 1982).

A first-order North Island triangulation across the TVZ was established between 1923 and 1936 by the predecessor of the New Zealand Department of Survey and Land Information (DOSLI; Lee, 1978). A narrow belt of this first-order network was resurveyed and densified, with triangulation and trilateration completed by DOSLI in 1976 (Bevin *et al.*, 1984). Because of problems with scale calibration of the electronic distance meter (EDM) instruments used for the trilateration, Reilly (1990) used these data only as line ratios.

Sissons (1979) estimated 7 mm/yr secular extension across the TVZ from a comparison of these triangulation observations, partitioning the set of reobserved stations according to the boundaries of the TVZ (Fig 1.11). Shear rates were negligible for his networks BP1 and BP3 to the west and east of the TVZ. In contrast, the shear rate is highly significant for network BP2, which spans the TVZ. Apart from one station to the southwest, this network lies between the Rotorua–Tarawera region and the Bay of Plenty coast. The shear-rate component, relatively extensional at azimuth 130° and perpendicular to the dominant 040° strike of faults in the zone, is 0.18 ± 0.06 (1 s.e.) $\times 10^{-6}$/yr. Under the assumption that there is no length change parallel to the faults, this component would represent actual extension across the 40-km-wide zone of 7.2 ± 2.4 (1 s.e.) mm/yr.

Adams (1984) compared the 1920s survey with both the trilateration and triangulation of the 1976 survey. His results show a uniform strain rate across most of North Island, with a marked increase in rate, but little change in orientation, near the coast at East Cape. This study indicated a shear strain component corresponding to an extension perpendicular to the faulting trend, assuming there is no parallel length change, of typically 0.09 ± 0.03 (1 s.e.) $\times 10^{-6}$/yr. Careful reading of Sissons (1979) and Adams (1984) appears to preclude the possibility that the factor-of-2 difference between the two results is due to confusion between tensor and engineering definitions of shear magnitude, but this nevertheless still seems the most likely explanation.

Only triangulation measurements appear to have been used by Walcott (1984), who reported shear results of 0.15 ± 0.03 (1 s.e.) $\times 10^{-6}$/yr across the entire CVR, zero in the eastern Bay of Plenty, and a much larger value near East Cape. Reilly (1990) used both direction and distance measurements throughout New Zealand and determined horizontal

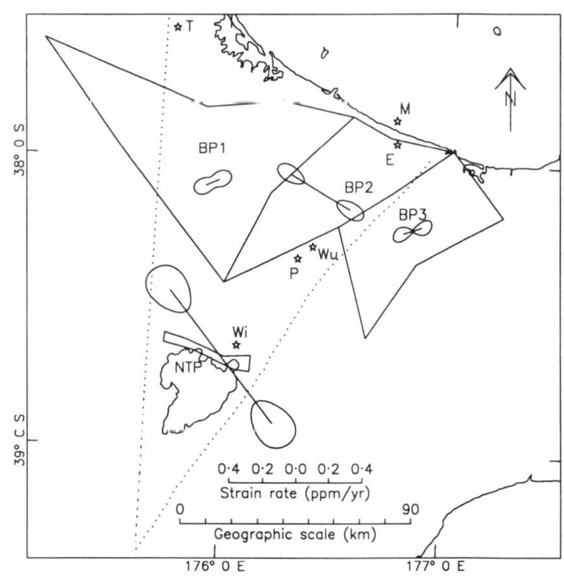

FIGURE 1.11. Shear strain results from Sissons (1979) (BP1, BP2, and BP3) and from Darby and Williams (1991) (NTP). The shear strain is plotted as a bar whose half-length is proportional to the maximum engineering shear and with the orientation of maximum relative extension. Confidence regions of one standard error for these values are shown. Lettered stars show locations of earthquakes referred to in the text: Wi, Wairakei; Wu, Waiotapu; M, Matata; E, Edgecumbe; T, Te Aroha; P, Paeroa.

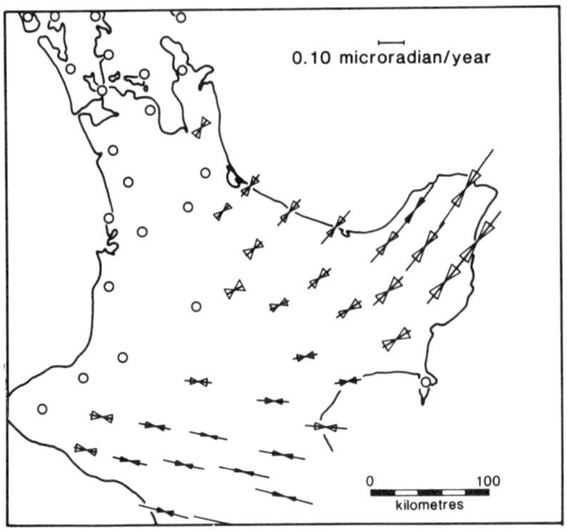

FIGURE 1.12. Shear strain results, redrafted from Reilly (1990). The bars represent the axis of maximum relative contraction, the length being proportional to the tensor shear (half the engineering shear). The sector symbols give the limits of two standard deviations in both magnitude and direction; circles denote results less than two standard deviations.

derivatives, up to the third order, of station horizontal velocities to describe the variation of strain (Fig 1.12). In his Raglan to East Cape transect, he found the rate of shear steadily increasing from zero in the west, to a maximum of 0.30±0.08 (1 s.e.) $\times 10^{-6}$/yr in the east with a rate in CVR of about 0.1×10^{-6}/yr, consistent with Walcott's estimate. Walcott (1987) considered this to be a typical smoothed value for the Bay of Plenty region; he multiplied it by the 120-km maximum width of the CVR over which it was estimated to obtain an extension rate of 12 mm/yr, but this value obviously depends strongly on the assigned width of CVR. Finally, Crook and Hannah (1988, 1989) estimated shear of 0.18±0.11 $\times 10^{-6}$/yr (1 s.e.), relatively extensional at azimuth 150°, for the period 1951–1977 by comparing one quadrilateral of the first-order network spanning the TVZ at the Bay of Plenty coast with lower-order triangulation. This result is very similar to that of Sissons (1979).

All these results are for networks across the northern part of the CVR/TVZ, within 40 to 70 km of the coast. There are no large-scale results farther south, but global positioning system (GPS) surveys of the TVZ were completed in 1990 and 1991 with a view to their comparison with triangulations of the 1920s, 1950s, and 1970s (Blick *et al.*, 1992) and analysis of these data is still in progress.

10.2. Intermediate-Scale Horizontal Deformation

Comprehensive second- and third-order triangulation was established by DOSLI between 1949 and 1955 in the Bay of Plenty, Rotorua, Taupo, and Raglan districts (Williams, 1989). A survey across the Taupo fault belt immediately north of Lake Taupo was completed in 1986 comprising triangulation, trilateration, and vertical angle measurements (Williams 1987). From these surveys, Darby and Williams (1991) estimated uniform shear of $0.5 \pm 0.1 \times 10^{-6}$/yr (1 s.e.), relatively extensional at azimuth 14±29° (1 s.e.) immediately to the north of Lake Taupo for the period 1950–1986 (Fig 1.11). Investigation of variable strain across the network showed a west-to-east difference in magnitude from 0.4×10^{-6}/yr to 1.0×10^{-6}/yr and in relatively extensional orientation from 102° to 170°. Integration of the uniform shear result across the 40-km width of the TVZ yields 18±5 mm/yr extension perpendicular to the faults.

10.3. Vertical Deformation

Lake-leveling observations on Lake Taupo have recorded vertical deformation from 1985 to the present (Otway and Sherburn, 1994) (Fig 1.13). The main results are a long-term subsidence centered on the Taupo fault belt and relatively-short-term variations most noticeable outside the fault belt. While the subsidence is undoubtedly related to the intermediate-scale extension discussed, Otway and Sherburn (1994) show there is no obvious relationship of the short-term variations with fluctuations in the low-level seismicity in the region. This is in contrast to the uplift preceding the Taupo earthquake swarms in 1983 and the subsidence during them (Otway, 1986).

10.4. Active Surface Faulting

In the late Pleistocene normal faulting was dominant and formed the Taupo fault belt from Ruapehu and Tongariro to Lake Taupo, the Paeroa Range, Tarawera, and Whakatane (Fig 1.14). North of Lake Taupo, this belt cuts across the trend of the older Hauraki graben that contains the mid-rift Kerepehi Fault (Fig 1.14). Faulting is dense southwest of the Okataina volcanic center but less so to the northeast. The belt is about 20 km wide with an overall strike of 040°, swinging to 055° in the northeast (Healy *et al.*, 1964).

The Taupo fault belt has been active in the late Quaternary with movements dated from pyroclastic deposits of known age (Nairn, 1976). Southwest of Okataina volcanic center, the faults have moved repeatedly during the last 50 k.y. with many having moved between 13 and 11 ka. Some displacements postdate the 1.85-ka Taupo eruption (Nairn and Hull, 1985). There were also historical surface ruptures of up to 3 m in the 1922 Taupo earthquake sequence (Grange, 1932; Sissons, 1979; Grindley and Hull, 1986) and of 50 mm in the 1983 Taupo earthquake swarms (Otway, 1986). Northeast of Okataina volcanic center the faults displace the eroded surface of Rotoiti breccia (50 ka), whereas others displace the 9-ka Rotoma ash (Nairn, 1981). Ruptures of up to 2.7 m occurred with the 1987 Edgecumbe earthquake (Beanland *et al.*, 1990).

Rates of faulting are best known within the Whakatane graben, which is the expression of the TVZ at the coast between Matata and Whakatane. Nairn and Beanland (1989) infer vertical faulting rates exceed 1.9 mm/yr from the offset of 0.28 Ma Matahina ignimbrite and point out that this is consistent with their inference of 1–2 mm/yr subsidence within TVZ and 1 mm/yr uplift rates outside. Horizontal extension rates are poorly known. Beanland

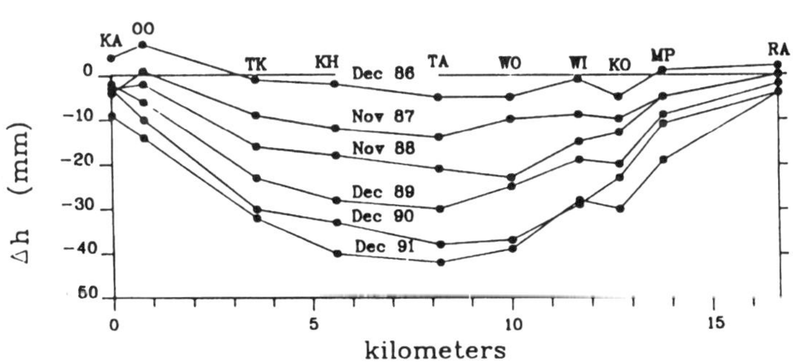

FIGURE 1.13. Profiles of apparent height change of Taupo fault belt stations relative to an origin station from 1986 to 1991 (after Otway and Sherburn, 1994); KA, Kawakawa; OO, Omoho; TK, Te Kauwae; KH, Kinloch; TA, Te Itarata; WO, Whakaipo; WI, Waikarariki; KO, Kaiapo; MP, Mine Point; RA, Rangatira.

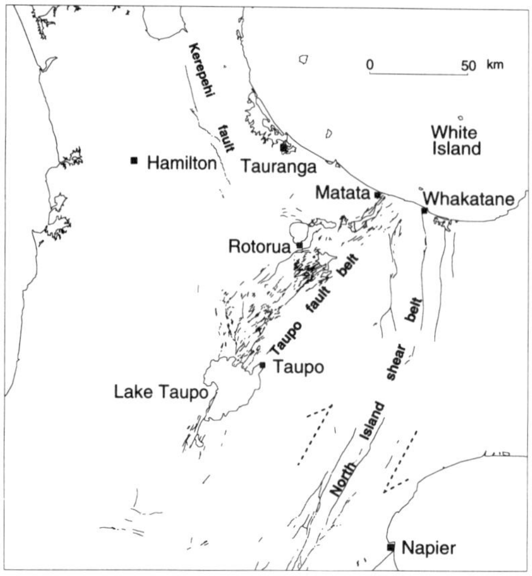

FIGURE 1.14. Active faulting in the central North Island. (Source: Institute of Geological and Nuclear Sciences 1:250,000 digital compilation).

et al. (1990) argue that if 2–3 mm/yr vertical displacement is assumed to occur at both the eastern and western margins on 45° dipping faults, then the horizontal extension would be 4–6 mm/yr. This rate would be a minimum because it does not include extension on smaller-scale structures within the graben.

11. FOCAL MECHANISMS

Seismological studies in the CVR and TVZ have not lessened the difficulty of interpreting focal mechanisms. Despite the preponderance of normal fault traces, for smaller-magnitude events normal faulting focal mechanisms have generally been absent, both for the less reliable composite kind associated with swarms and microearthquakes and the more reliable single events.

Transcurrent mechanisms have been more commonly observed: for example, NE dextral fault mechanisms in a microseismic survey of the Wairakei geothermal field (Hunt and Latter, 1982); possibly NE sinistral for the Waiotapu M=5.1 earthquake of 1983 (Smith *et al.*, 1984); both NE dextral and NE sinistral for the Taupo swarms, reaching M=3.7 and M=4.3, in 1983 (Webb *et al.*, 1986); and NE dextral with a normal component for the Matata M=5.4 earthquake of 1977 (Richardson, 1989).

The clearest, predominantly normal focal mechanism was for the 1987 Edgecumbe M=6.3 earthquake (Anderson and Webb, 1989). Other evidence is less compelling. The 1972 Te Aroha M=5.3 earthquake may have had a normal focal mechanism but with different mechanisms for the aftershocks (Adams *et al.*, 1972). The fault strike was possibly north but is very poorly constrained (T. Webb, personal communication, 1990). Also, small events following the 1982 June M=4.2 Paeroa earthquake had first motions consistent with a normal mechanism (Smith *et al.*, 1984). It is unfortunate that the only other available seismic studies, of the 1964–65 Taupo swarms that reached M=4.6 (Gibowicz, 1973) and of a microseismic survey from the Taupo fault belt to the Kaingaroa plateau (Evison *et al.*, 1976), were not able to provide focal mechanism solutions.

12. KINEMATICS

An unusually diverse collection of independent data bears upon crustal kinematics of the CVR. These data include the apparent migration of low-potash andesites across the CVR (Fig. 1.15a,b), geodetically determined strain rates (Sissons, 1979; Walcott, 1984; Darby and Williams, 1991), and paleomagnetic data from Tertiary sediments to the east of the CVR (Walcott, 1984; Wright and Walcott, 1986). These data all point to a surprisingly consistent, late Tertiary, kinematic evolution of central North Island (summarized in Stern, 1987). Included in this evolution are rotations of about 6°/m.y. over the past 4 m.y. and translations, or spreading, of 7–18 mm/yr over the same period. Both the rotation and translation contribute to the development of a wedge-shaped CVR. What makes this evolutionary solution particularly compelling is that the data sets previously described are totally independent and pertain to time scales from 20 m.y. to 30 yr. Similar evolutionary schemes for backarc extension, which also involve wedge-shaped rotations and translations, have been proposed elsewhere (Otofuji *et al.*, 1985).

Darby and Williams (1991) point out that while all the geodetic results are broadly consistent with each other, Sissons's, Adams's, Walcott's, and Reilly's results provide no distinction in rate of deformation between the narrow TVZ and the broader CVR. They generally found little change in orientation of the strain field across the CVR. Furthermore, they show no marked western boundary to the CVR, which may have been expected from Adams's and Walcott's network subdivisions if not from Reilly's polynomial smoothing. This may be due to the inability of the typically 40-km station spacing to resolve variations in strain at smaller distance scales. Darby and Williams's result of a significantly higher strain may be due to a local maximum of a regionally varying strain, or it may be influenced by systematic effects arising from the poor geometry of the network and the relatively large number of eccentric stations used in the later survey compared with the earlier survey. In general, it is unfortunate that strain rates are integrated by some investigators over different distances to derive widening rates and then become widely quoted without the associated distances.

13. MAGMA TYPES AND GENESIS

The TVZ is characterized by three distinct groups of eruptives: high-Al basalt, basaltic andesite-andesite-dacite, and rhyolite, in order of increasing volume. The high-Al basalt and rhyolite occur in a bimodal assemblage associated with caldera structures (see Fig. 1.3) in the backarc part of the TVZ, while the basaltic andesite-andesite-dacite assemblage occurs at the surface, particularly at the northern (White Island) and southern (Tongariro) ends of the TVZ and in a narrow frontal arc along the eastern margin (see Fig. 1.3).

13.1. High-Al Basalts

High-Al basalts have erupted sporadically at 15 locations in the TVZ (Houghton *et al.*, 1987) (see Fig. 1.3), ranging from single monogenetic cones (e.g., Johnsons Road) to fissure eruptions (e.g., Tarawera, 1886) and phreatomagmatic tuff rings (e.g., Kaiapo and Acacia Bay). The basalts range from aphyric to coarsely porphyritic (Cole, 1973) with phenocrysts of olivine + plagioclase ± clinopyroxene. A summary of mineralogy is given in Gamble *et al.* (1990). Chemically the basalts are high-alumina (HAB) as defined by Crawford *et al.*

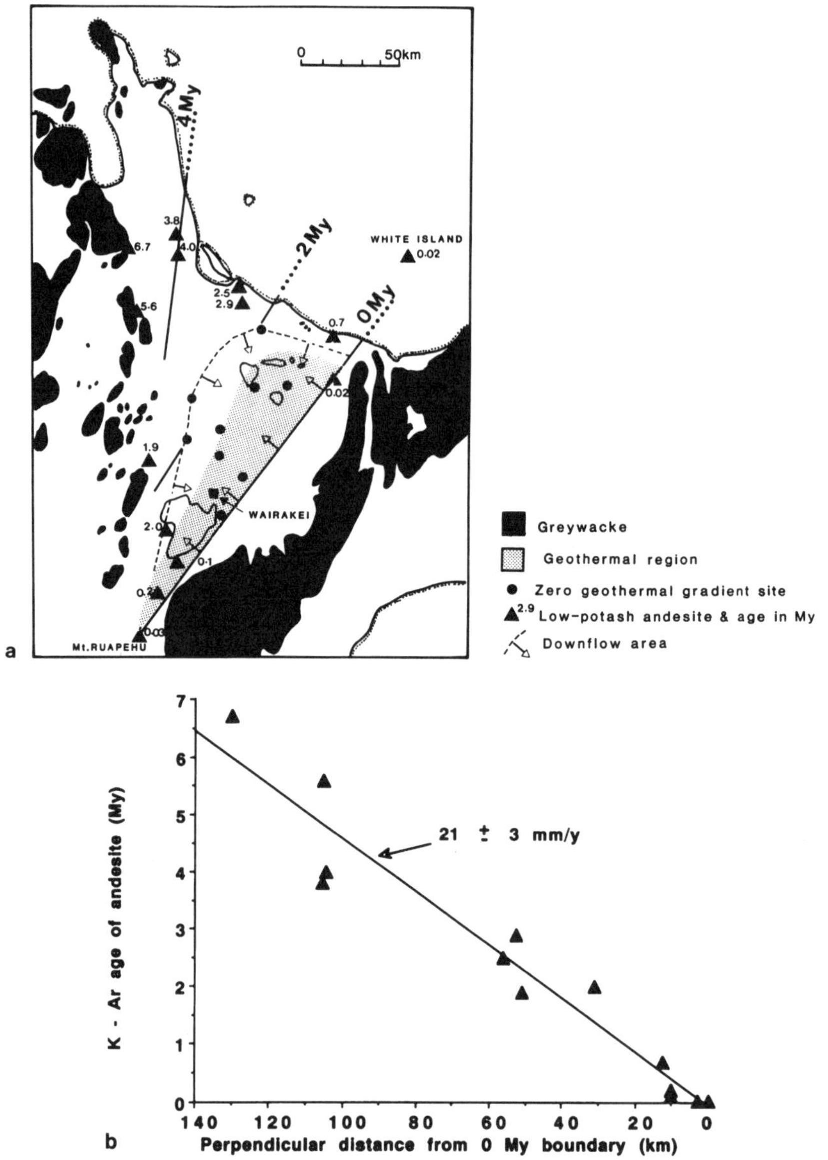

FIGURE 1.15. (a) Distribution of geothermal phenomena and low-potash andesites within and adjacent to the CVR. Solid lines indicate possible positions of the active volcanic front at times indicated in My. (b) Plot of K/Ar ages versus perpendicular distance from volcanic front (0 Ma boundary) for low-potash andesites (modified from Stern, 1987). The rate of 21 ± 3, mm/yr is the apparent average rate of migration in a southeasterly direction.

(1987) with 100 Mg/(Mg + Fe) values between 56 and 72, and CaO/Al_2O_3 ratios between 0.59 and 0.70. They have a low abundance of K_2O (Fig. 1.16) and high-field-strength elements, but show lower Ti/Zr, higher Ti/V and Ti/Sc ratios, and generally higher Zr abundance than MORB (Gamble *et al.*, 1993). They are light rare-earth enriched (Ce/Yb = 2–3) with flat heavy rare-earth patterns. Isotopically they define an almost linear array of decreasing Nd- and increasing Sr-isotopic ratios with Kakuki basalt ($^{87}Sr/^{86}Sr$ = 0.70388

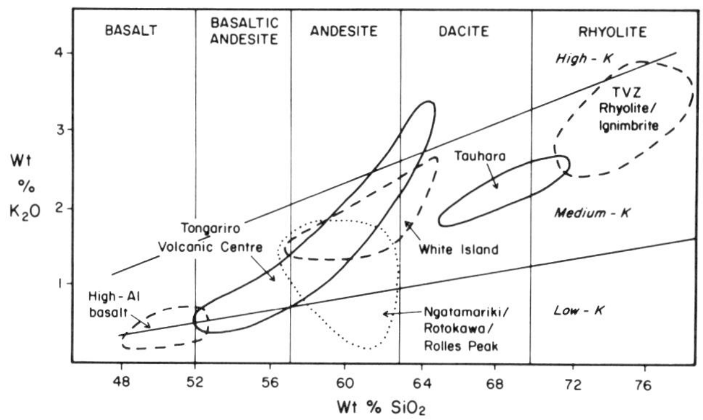

FIGURE 1.16. Weight percent K_2O versus weight percent SiO_2 for volcanic rocks of TVZ. Classification is from Le Maitre (1989), fields from Graham *et al.* (1992).

and εNd = +5.1) isotopically the most primitive (Fig. 1.17). Kakuki also has the least radiogenic Pb-isotope signature (Graham *et al.*, 1992), while Tarawera basalt has the least radiogenic signature (Fig. 1.18). The latter is, however, riddled with small rhyolitic inclusions, and despite best efforts to avoid these in analysis a crustal component is likely to be reflected in isotope values.

Gamble *et al.* (1993) conclude that the TVZ basalts contain both crustal and slab-derived subduction signatures. They consider that the slab component comes from a relatively fertile hot mantle and may well reflect the youth of the TVZ magma system. Ascent of the magmas was hindered by the low-density continental crust, and variation is caused largely by fractionation of olivine ± plagioclase, followed by clinopyroxene and Fe-Ti oxides in the more evolved lavas.

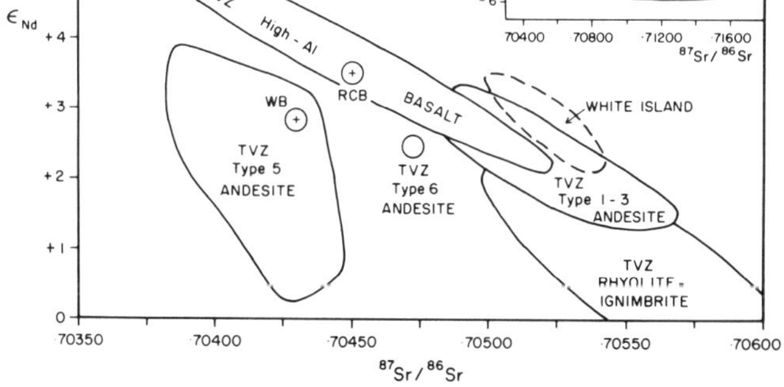

FIGURE 1.17. $^{87}Sr/^{86}Sr$ versus εNd for TVZ rocks (after Graham and Cole, 1991). "Mantle Array" from Nohda (1984). KB, Kakuki basalt; RB, Ruapehu basalt, RCB, Red Crater basalt; WB, Waimarino basalt.

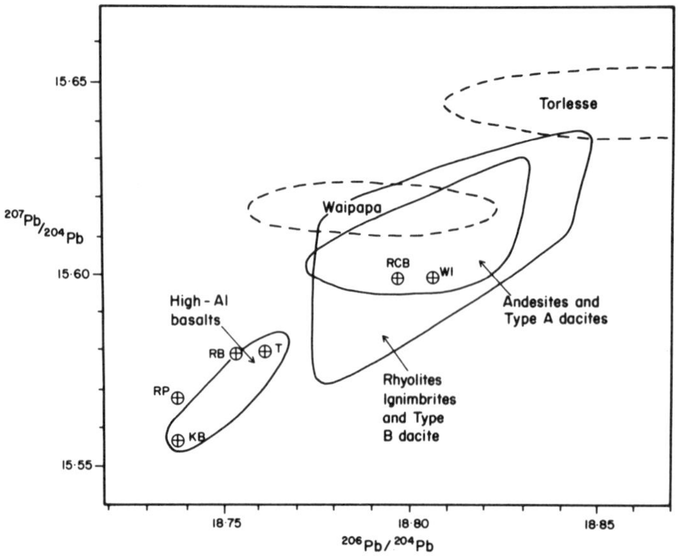

FIGURE 1.18. $^{207}Pb/^{204}Pb$ versus $^{206}Pb/^{204}Pb$ for TVZ rocks. Field for each volcanic rock type and Waipara/Torlesse metasediments from Graham *et al.*(1992). KB, Kakuki basalt; RB, Ruapehu basalt; T, Tarawera basalt; RCB, Red Crater basalt; RP Rolles Peak basaltic andesite; WI, White Island basaltic andesite.

13.2. Basaltic Andesite, Andesite, and Dacite

The most mafic members of this association occur on Ruapehu, Red Crater (Tongariro volcanic center), and Waimarino (SW of Lake Taupo). The latter is a magnesian quartz tholeiite which is porphyritic with 16% olivine (Fo_{90}) and 7% clinopyroxene and has high MgO, Cr, and Ni (Graham and Hackett, 1987). However, it has low ϵNd, high $^{87}Sr/^{86}Sr$ (Fig. 1.17), and moderately radiogenic $^{206}Pb/^{204}Pb$ (Fig. 1.18), reflecting crustal contamination. The two basaltic andesites from the Tongariro volcanic center have lower Al_2O_3 contents than HAB and contain plagioclase, orthopyroxene, and rare crustal xenoliths. The Tongariro basaltic andesite has comparable $^{87}Sr/^{86}Sr$ ratios to the HAB (Graham *et al.*, 1992), but Ruapehu basaltic andesite has higher ϵNd (Fig. 1.17) and Red Crater basaltic andesite higher $^{207}Pb/^{204}Pb$ (Fig. 1.18).

Most of the volume of the Tongariro volcanic center is composed of medium-K andesite (Fig. 1.16). Of the two massifs in the center, Ruapehu has been most extensively studied geochemically. Graham and Hackett (1987) identified six main types; a seventh type was added by Patterson and Graham (1988). These are characterized by differing phenocryst assemblages, major and trace element chemistries (Fig. 1.19) and $^{87}Sr/^{86}Sr$ ratios (Fig. 1.17). Andesites also occured in the 1977 eruption of White Island (Cole and Graham, 1989; Graham and Cole, 1991), where Mg-rich bombs and blocks were erupted (Fig. 1.19), at Kawerau (Browne, 1978), Waiotapu (Hedenquist, 1983), Broadlands (Browne, 1971; Wood, 1983), Rotokawa and Ngatamariki (Browne *et al.*, 1992), Rolles Peak (Graham and Worthington, 1988), and Wairakei (Grindley, 1965); see Fig. 1.3 for locations.

Dacites are volumetrically minor in the TVZ and are of two types (Reid and Cole, 1983). Only one type (Type A; Graham *et al.*, 1992) appears directly related to the basaltic

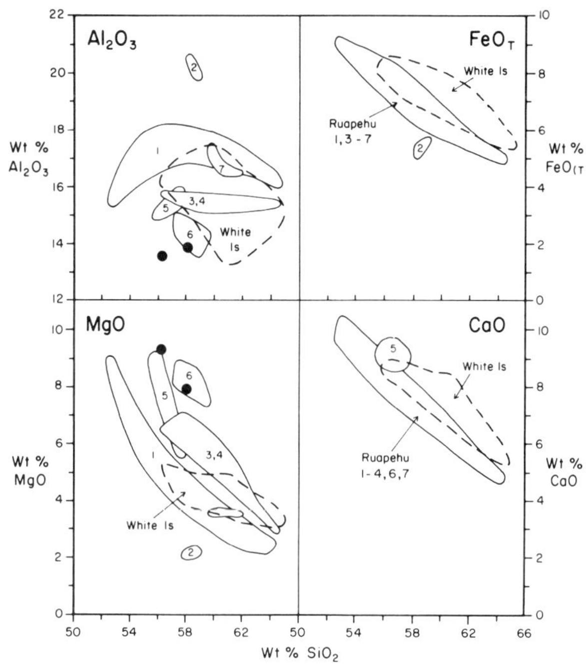

FIGURE 1.19. Major element plots for andesites of Tongariro volcanic center and White Island. Fields from Graham and Cole (1991). Solid circles are 1977 eruptives from White Island.

andesite-andesite-dacite association. Dacites of this type are medium K (Fig. 1.16) and occur in the Tongariro volcanic center and White Island. They are characterized by slightly higher $^{87}Sr/^{86}Sr$ and ϵNd than those of Tongariro (Fig. 1.17) but comparable $^{207}Pb/^{204}Pb$ (Fig. 1.18). The second type (Type B; Graham *et al.*, 1992) occurs at Tauhara, Broadlands, Waiotapu, Edgecumbe, and Whale Island (Fig. 1.3). These are most likely to be products of mixing between a Rolles Peak (type) andesite and rhyolite (Graham and Worthington, 1988).

The origin of the basaltic andesite-andesite-dacite (Type A) association has been discussed by a number of authors (Cole 1978, 1990; Graham and Hackett, 1987; Graham and Cole, 1991, Graham *et al.*, 1992; Browne *et al.*, 1992). Most agree that the assemblage was formed by a multistage process involving (1) subsolidus slab dehydration fluxing the overlying mantle wedge, (2) anatexis of the asthenosphere or subcontinental lithosphere of the mantle wedge to produce low-Al basalt, and (3) fractionation of plagioclase-orthopyroxene/olivine-augite-magnetite (POAM) mineral phases (15–55%) with minor subarc crustal assimilation (1–30%, particularly by Torlesse metasediment; Graham *et al.*, 1992) (Fig. 1.18) or crustal accumulation. Both Type 5 in the Tongariro volcanic center and the Rolles Peak andesite (with high Sr and lower $^{87}Sr/^{86}Sr$ ratios (Fig. 1.17) must have come from a different mantle source to the bulk of the lavas. Rolles Peak also has lower $^{207}Pb/^{204}Pb$ (Fig. 1.18).

13.3. Rhyolites and Ignimbrites

The silicic volcanics are voluminous (12,000 km^3; Cole, 1979), representing at least 80% of the total eruptives of the TVZ (Hochstein *et al.*, 1993). The rhyolites range from an anhydrous assemblage of plagioclase, pyroxene, magnetite, and ilmenite to a strongly porphyritic assemblage of quartz, plagioclase, calcic-hornblende, biotite, magnetite, and ilmenite. Details are given in Ewart *et al.* (1976) and Cole (1979). Chemically they range from 70% to 79% SiO_2 with 2–4% K_2O (Fig. 1.16) and $Na_2O > K_2O$ in most rocks. Isotopically there is a range of $^{87}Sr/^{86}Sr$ ratios from 0.70500 to 0.70650 (Fig. 1.17), whereas $^{206}Pb/^{204}Pb$ ratios (Fig. 1.18) are remarkably uniform (Graham *et al.*, 1992). There is no systematic difference between rhyolites and ignimbrites nor over time in the stratigraphic sequence.

The origin of the silicic volcanics is much more controversial. Ewart and Stipp (1968), Cole (1981), and Reid (1983) considered them a product of partial melting of subvolcanic basement, particularly Waipapa Group metasediments. This origin was consistent with general chemistry, $^{87}Sr/^{86}Sr$ ratios (Fig. 1.17), and $^{207}Pb/^{204}Pb$ ratios (Fig. 1.18). However, Blattner and Reid (1982) pointed out that such an origin is incompatible with low $\delta^{18}O$ values and Conrad *et al.* (1988) showed by experimental studies that the mildly peraluminous Waipapa metasediments were unlikely to produce the predominantly diopside-normative rhyolites. Graham *et al.* (1992) showed that the Pb-isotope data were consistent with a derivation from primary mantle-derived basaltic liquids by assimilation-fractional crystallization (AFC); see Fig. 1.18), but he had serious reservations because of bimodality and relative magma volumes. Most recently Briggs *et al.* (1993), studying the Mangakino center, have suggested a multiple origin including partial melting of a crustal source similar to western (Waipapa) graywacke or its metamorphic equivalent, melting of a plagioclase-rich plutonic or metamorphic source similar to anorthosite or trondjemite, and a minor upper mantle source component.

14. DISCUSSION

A question that has puzzled geophysicists for a long time is, what is the heat source for the TVZ (e.g., Evison *et al.*, 1976; Studt and Thompson, 1969)? If we make the reasonable assumption that the present-day natural heat output is representative, and not a local perturbation in time, and we take note of geological evidence (Grindley, 1965) that the present-day geothermal fields may have lifetimes of, say, 1 m.y., then a major source of heat is required.

Stern (1987) proposed one possible model whereby all the heat is provided by cooling igneous rocks within the TVZ. He found that for a spreading rate of at least 16 mm/yr, and with spreading being accompanied by the intrusion of new igneous rock, then a steady-state 4×10^9 W heat output can be maintained if the full 13 km (15 km less the top 2 km of volcanic ash, etc.) of crust is made of igneous rocks.

Other solutions are possible. For example, Hochstein *et al.* (1993) argue that about half the heat output may be supplied by heat released during plastic deformation. Another alternative may be heat from the mantle being transferred by meteoric water that convects to depths of 13 km or so and then efficiently transfers heat from the upper mantle to the surface.

In order to account for the preponderance of normal fault traces in the near absence of normal focal mechanisms, Smith and Webb (1986) proposed a crustal model consisting of a complex system of conjugate transcurrent faults joined by normal faults. The absence of surface transcurrent faults was attributed to the inability of the weak 2-km-thick pyroclastic surface layer to transmit transcurrent movement. Webb *et al.* (1986) added the provision that the normal faulting is aseismic, perhaps because of high fluid pressure reducing the normal stress, and noted that neither temporal changes in the stresses nor a chaotic stress distribution near the apex of the region of active widening could be ignored.

Darby and Williams (1991) countered these explanations by arguing that a lowered normal stress component over a region would affect normal and transcurrent faulting equally, that a peculiar strength anisotropy of the weak pyroclastic layer would be required to favor dip-parallel over strike-parallel shear failure, and that normal faulting did occur during the 1987 Edgecumbe (M=6.3) earthquake, albeit after the papers of Smith and Webb (1986) and Webb *et al.* (1986) had appeared.

In consideration of smaller earthquakes down to M=5, there is general agreement with the findings of Smith and Webb (1986) and Otway and Sherburn (1994) that there does not appear to be a simple relationship between the lower-magnitude seismicity data and surface deformation as manifested by faulting or by geodetic measurement. However, the largest historical earthquakes in the TVZ (viz. 1922 Taupo and 1987 Edgecumbe) had major components of normal faulting consistent with geological expectations and the geodetic results reviewed here.

Kinematic data suggest that the CVR has been spreading for at least 4 m.y., with most of the recent evolution in the TVZ. Shear rates are negligible in the older part of the CVR (BP1 in Fig. 1.11) but highly significant in TVZ, leading Cole (1990) to suggest that the TVZ may be transtensional with current orientations of the fault belts controlled by a combination of dextral shear in the basement-northeastern continuation of the North Island shear belt beneath the TVZ and back-arc extension. This possibility is, however, only likely to be resolved with greater GPS station density and information on both dilational and rotational components of strain tensor.

Origin of the voluminous ignimbrites is a major problem in TVZ. It seems likely that most may be a result of remelting earlier formed igneous rocks (e.g., Graham *et al.*, 1992) or sedimentary/metamorphic rocks of a restricted composition together with a small mantle component (e.g., Briggs *et al.*, 1993). Some may be a result of fractional crystallization of a mantle source with significant crustal contamination (e.g., McCulloch *et al.*, 1994). Cole (1990) noted that rhyolitic centers only occur within a restricted area (between 38° and 39°S) in the TVZ (see Fig. 1.3), while back-arc extension is clearly taking place throughout the zone. It may thus be that it is only in this area where suitable sources occur that would partially melt to form rhyolitic magmas. Cole (1990) has suggested that if these are igneous rocks they may be a SE extension of the Coromandel volcanic zone.

Some of these possibilities should be explored by a combined geophysical-geochemical study of subvolcanic features in the TVZ. In particular, the advent of a new generation of digital seismological equipment with enhanced dynamic range should provide us with a higher resolution of the crust and upper mantle than has been possible in the past.

15. SUMMARY AND CONCLUSIONS

1. The CVR is the principal volcanic basin structure defined by seismic and gravity parameters in central North Island, and the TVZ represents the eastern part of the CVR that is currently volcanically and tectonically active.

2. Activity in the TVZ began with eruptions from the Mangakino caldera about 1.6 m.y. ago, while activity in the main part of the TVZ has taken place within the past 0.7 m.y.
3. Earthquake seismology indicates that the mantle and lower crust beneath the CVR are highly attenuative and is probably in a state of partial melt. A combination of earthquake and explosion seismology data indicates a crustal thickness of 15±2 km for much of the CVR.
4. CVR is dominated by low gravity (<300 μN/kg) with areas −600 μN/kg representing, in part, cooled or partially cooled rhyolitic magma. It has an average heat flow of 800 mW/m^2, about 13 times greater than the continental norm and one of the highest on record for a backarc basin. This may be due to high rates of extension.
5. Rates of extension in the currently active TVZ vary from about 7 mm/yr with a shear of $0.18 \pm 0.11 \times 10^{-6}$yr (1 s.e.) in the Bay of Plenty to about 18 mm/yr and a shear of 0.5±0.1 (1 s.e.) in the Taupo fault belt. Vertical faulting rates indicate 1–2 mm/yr subsidence.
6. Focal mechanisms have been difficult to interpret. Most appear to have a large transcurrent component, although the 1987 Edgecumbe earthquake was normal.
7. The TVZ volcanism is characterized by three distinct types of eruptives: high-Al basalt; basaltic andesite, andesite, and dacite; and rhyolite and ignimbrites, in order of increasing volume.
8. The high-Al basalts come from a relatively fertile hot mantle, with magmas rising into the crust, fractionating, and assimilating small amounts of crust.
9. The basaltic andesite-andesite-dacite assemblage, of which most of the near-surface volume is in the Tongariro volcanic center, was formed by a multistage process involving subsolidus slab dehydration fluxing the overlying mantle wedge, anatexis of the mantle wedge, fractionation, and minor crustal assimilation.
10. The rhyolites and ignimbrites form the greatest volume (~12,000 km^3) of the TVZ volcanics. Their origin is debatable, but they may have a multiple origin, including partial melting of crustal rocks with a small mantle component or fractionation of a mantle source with crustal contamination.
11. The simplest explanation of the extreme heat output is by the cooling of magma and the transfer of heat to the surface geothermal fields by meteoric water. It is not clear, however, at what depth this heat transfer takes place.

REFERENCES

Adams, D. A. 1984. Some methods of analysis of geodetic data and their applications to the measurement of crustal deformation, M.Sc. thesis, Victoria University of Wellington.

Adams, R. D., Muir, M. G., and Kean, R. J. 1972. Te Aroha earthquake, 9 January 1972, *Bull. N.Z. Nat. Soc. Earthquake Eng.* **5**(2)**:**54–58.

Anderson, H., and Webb, T. 1989. The rupture process of the 1987 Edgecumbe earthquake, New Zealand, *N.Z. J. Geol. Geophys.* **32**(1)**:**43–52.

Beanland, S., Blick, G. H., and Darby, D. J. 1990. Normal faulting in a back arc basin: geological and geodetic characteristics of the 1987 Edgecumbe earthquake, New Zealand, *J. Geophys. Res.* **95**(B4)**:**4693–4707.

Berckhemer, H., Baier, B., Bartelsen, H., Behle, A., Burkhardt, A., Gebrande, H., Makris, J., Menzel, H., Miller, H., and Vees, R. 1975. Deep seismic soundings in the Afar region and on the highland of Ethiopia, in *Afar Depression of Ethiopia,* Vol. 1 (A. Pilger and A. Roster, eds.), pp. 89–107, E. Schweizer, Velagbuchhandl, Stuttgart.

Bevin, A. J., Otway, P. M., and Wood, P. R. 1984. Geodetic monitoring of crustal deformation in New Zealand, in *An Introduction to Recent Crustal Movements of New Zealand* (R. I. Walcott, ed.), pp. 13–60, Royal Society of New Zealand Misc. Ser. 7.

Bibby, H. M. 1973. The reduction of geodetic survey data for the detection of earth deformation, DSIR Geophysics Division Report 84.

Bibby, H. M. 1981. Geodetically determined strain across the southern end of the Tonga-Kermadec-Hikurangi subduction zone, *Geophys. J. R. Astron. Soc.* **66:**513–533.

Bibby, H. M. 1982. Unbiased estimate of strain from triangulation data using the method of simultaneous reduction, *Tectonophysics* **82**(1–2)**:**161–174.

Bibby, H. M., Dawson, G. B., Rayner, H. H., Stagpoole, V. M., and Graham, D. J. 1984. The structure of the Mokai geothermal field based on geophysical observations, *J. Volcanol. Geotherm. Res.* **20:**1–20.

Black, P. R., and Braille, L. W. 1982. Pn velocity and cooling of continental lithosphere, *J. Geophys. Res.* **87:**10557–10563.

Blattner, P., and Reid, F. W. 1982. The origin of the lavas and ignimbrites of the Taupo volcanic zone, New Zealand in the light of oxygen isotope data, *Geochim. Cosmochim. Acta* **46:**1417–1429.

Blick, G. H., Darby, D. J., Meertens, C. M., Otway, P. M., Perin, B., Rocken, C., and Scott, B. J. 1992. Crustal deformation surveys in the Taupo Volcanic Zone and central North Island using GPS measurements, *N.Z. Surveyor* **33**(281)**:**299–312.

Briggs, R. M., Gifford, M. G., Moyle, A. R., Taylor, S. R., Norman, M. D., Houghton, B. F., and Wilson, C. J. N. 1993. Geochemical zoning and eruptive mixing in ignimbrites from Mangakino caldera, Taupo volcanic zone, New Zealand, *J. Volcanol. Geotherm. Res.* **56:**175–203.

Brothers, R. N., and Delaloye, M. 1982. Obducted ophiolites of North Island, New Zealand: origin, age, emplacement and tectonic implications for Tertiary and Quaternary volcanicity, *N.Z. J. Geol. Geophys.* **25:**257–274.

Browne, P. R. L. 1971. Petrological logs of drillholes, Broadlands Geothermal Field, *N.Z. Geol. Surv. Rep.* 52.

Browne, P. R. L. 1978. Petrological logs of drillholes, Broadlands Geothermal Field, *N.Z. Geol. Surv. Rep.* 84.

Browne, P. R. L., Graham, I. J., Parker, R. J., and Wood, C. P. 1992. Subsurface andesite lavas and plutonic rocks in the Rotokawa and Ngatamariki geothermal systems, Taupo volcanic zone, New Zealand, *J. Volcanol. Geotherm. Res.* **51:**199–215.

Calhaem, I. M. 1973. Heat flow measurements under some lakes in North Island, New Zealand, unpublished Ph.D. thesis, Victoria University of Wellington, Wellington.

Carter, L. 1980. *New Zealand Region Bathymetry* 1:6,000,000, 2nd ed., New Zealand Oceanographic Inst. Chart. Misc. Series 15.

Chase, C. G. 1978. Plate kinematics: the Americas, East Africa and the rest of the world, *Earth Planet Sci. Lett.* **37:**355–348.

Cole, J. W. 1973. High-alumina basalts of Taupo volcanic zone, New Zealand, *Lithos* **6:**53–64.

Cole, J. W. 1978. Andesites of the Tongariro volcanic center, North Island, New Zealand, *J. Volcanol. Geotherm. Res.* **3:**121–153.

Cole, J. W. 1979. Structure, petrology and genesis of Cenozoic volcanism, Taupo volcanic zone, New Zealand—a review, *N.Z. J. Geol. Geophys.* **22:**631–657.

Cole, J. W. 1981. Genesis of lavas of the Taupo volcanic zone, North Island, New Zealand, *J. Volcanol. Geotherm. Res.* **10:**317–337.

Cole, J. W. 1990. Structural control and origin of volcanism in the Taupo volcanic zone, New Zealand, *Bull. Volcanol.* **52:**445–459.

Cole, J. W., and Graham, I.J. 1989. Petrology of strombolian and phreatomagmatic ejecta from the 1976–82, White Island eruption sequence, in The 1976–82 Eruption Sequence at White Island Volcano (Whakaari), Bay of Plenty, New Zealand, *N.Z. Geol. Surv. Bull.* **103:**61–68.

Cole, J. W., and Lewis, K. B. 1981. Evolution of the Taupo–Hikurangi subduction system, *Tectonophysics* **72:** 1–21.

Conrad, W. K., Nicholls, I. A., and Wall, V. J. 1988. Water-saturated and under-saturated melting of metaluminous and peraluminous crustal compositions at 10 kb: Evidence for the origin of silicic magmas in the Taupo volcanic zone, New Zealand and other occurrences, *J. Petrol.* **29:**765–803.

Crawford, A. J., Falloon, T. J., and Eggins, S. 1987. The origin of island arc high-alumina basalts, *Contrib. Mineral. Petrol.* **97:**417–430.

Crook, C. N., and Hannah, J. 1988. Regional horizontal deformation associated with the March 2, 1987, Edgecumbe earthquake, New Zealand, *Geophys. Res. Lett.* **15**(4)**:**361–364.

Crook, C. N., and Hannah, J. 1989. Regional horizontal deformation associated with the 1987 Edgecumbe earthquake, Bay of Plenty, New Zealand—an introduction, *N.Z. J. Geol. Geophys.* **32**(1)**:**93–98.

Darby D. J., and Williams, R. O. 1991. A new geodetic estimate of deformation in the central volcanic region of the North Island, New Zealand, *N.Z. J. Geol. Geophys.* **34:**127–136.

Davey, F. J. 1993. Crustal seismic reflection measurements over a continental backarc basin, North Island, New

Zealand, in Programme & Abstracts, Geophysical Symposium, N.Z. Geophysical Society, Victoria University of Wellington, abstract.

Davey, F. J., and Broadbent, M. 1980. Seismic refraction measurements in Fiordland, southwest New Zealand, *N.Z. J. Geol. Geophys.* **23**:395–406.

Dawson, G. B., and Dickinson, D. J. 1970. Heat flow studies in thermal areas of the North Island of New Zealand, *Geothermics Spec. Issue* **2**:466–473.

Eaton, G. P. 1984. The Miocene Great Basin of western North America as an extending backarc region, *Tectonophysics* **102**:275–295.

Eiby, G. A. 1958. The structure of New Zealand from seismic evidence, *Geol. Rundsch.* **47**:647–662.

Evison, F. F., Robinson, R., and Arabasz, W. J. 1976. Microearthquakes, geothermal activity, and structure, central North Island, New Zealand, *N.Z. J. Geol. Geophys.* **19**(5):625–637.

Ewart, A., Hildreth, W., and Carmichael, I. S. E. 1976. Quaternary acid magma in New Zealand, *Contrib. Mineral. Petrol.* **51**:1–27.

Ewart, A., and Stipp, J. J. 1968. Petrogenesis of the volcanic rocks of the central North Island, New Zealand as indicated by a study of $^{87}Sr/^{82}Sr$ ratios and Sr, Rb, U, and Th abundances, *Geochim. Cosmochim. Acta* **32**:699–736.

Fowler, C. M. R. 1990. *The Solid Earth—an Introduction to Solid Earth Geophysics*, Cambridge University Press.

Gamble, J. A., Smith, I. E. M., Graham, I. J., Kokelaar, B. P., Cole, J. W., Houghton, B. F., and Wilson, C. J. N. 1990. The petrology, phase relations and tectonic setting of basalts from the Taupo Volcanic Zone, New Zealand and the Kermadec Island arc-Havre Trough, S.W. Pacific, *J. Volcanol. Geotherm. Res.* **43**:253–270.

Gamble, J. A., Smith, I. E. M., McCulloch, M. T., Graham, I. J., and Kokelaar, B. P. 1993. The geochemistry and petrogenesis of basalts from the Taupo volcanic zone and Kermadec Island arc, S.W. Pacific, *J. Volcanol. Geotherm. Res.* **54**:265–290.

Garrick, R. A. 1968. A reinterpretation of the Wellington crustal refraction profile (letter), *N.Z. J. Geol. Geophys.* **11**:1280–1294.

Gibowicz, S. J. 1973. Variation of frequency-magnitude relationship during Taupo earthquake swarm of 1964–65, *N.Z. J. Geol. Geophys.* **16**(1):18–51.

Graham, I. J., and Cole, J. W. 1991. Petrogenesis of andesites and dacites of White Island Volcano, Bay of Plenty, New Zealand in the light of new geochemical and isotopic data, *N.Z. J. Geol. Geophys.* **103**:650–661.

Graham, I. J., Gulson, B. L., Hedenquist, J. W., and Mizon, K. 1992. Petrogenesis of Late Cenozoic volcanic rocks from the Taupo volcanic zone, New Zealand, in the light of new lead isotope data, *Geochim. Cosmochim. Acta* **56**:2797–2819.

Graham, I. J., and Hackett, W. R. 1987. Petrology of calc-alkaline lavas from Ruapehu Volcano and associated vents, Taupo volcanic zone, New Zealand, *J. Petrol.* **28**:531–567.

Graham, I. J., and Worthington, T. I. 1988. Petrogenesis of Tauhara Dacite (Taupo volcanic zone, New Zealand)—Evidence for mixing between high-alumina andesite and rhyolite, *J. Volcanol. Geotherm. Res.* **35**:279–294.

Grange, L. I. 1932. Taupo earthquakes, 1922. Rents and faults formed during earthquake of 1922, in Taupo district, *N.Z. J. Sci. Technol.* **14**(3):139–141.

Grindley, G. W. 1965. The geology structure and exploitation of the Wairakei geothermal field, Taupo New Zealand, *N.Z. Geol. Surv. Bull.* no. 75.

Grindley, G. W., and Hull, A. G. 1986. Historical Taupo earthquakes and earth deformation, in *Recent Crustal Movements of the Pacific Region*, (W. I. Reilly; B. E. Harford, eds.), *R. Soc. N.Z. Bull.* **24**:173–186.

Hackett, W. R., and Houghton, B. F. 1989. A facies model for a Quaternary andesitic composite volcano: Ruapehu, New Zealand, *Bull. Volcanol.* **51**:51–68.

Haines, A. J. 1979. Seismic wave velocities in the uppermost mantle beneath New Zealand, *N.Z. J. Geol. Geophys.* **22**:245–257.

Hatherton, T. 1969. The geophysical significance of calc-alkaline andesites in New Zealand, *N.Z. J. Geol. Geophys.* **12**:436–459.

Hayward, B., 1979, Eruptive history of the Early to Mid-Miocene Waitakere volcanic arc and paleogeography of the Waitemata Basin, northern New Zealand, *J. R. Soc. N.Z.* **9**:297–320.

Healy, J. 1962 Structure and volcanism in the Taupo volcanic zone, New Zealand, in *Crust of the Pacific Basin*, pp. 151–157, Geophys. Monogr. Ser., Vol. 6, American Geophysical Union, Washington, DC.

Healy, J., Schofield, J. C., and Thompson, B. N. 1964. *Sheet 5, Rotorua* (1st ed.), *Geological Map of New Zealand 1:250,000*, New Zealand Department of Scientific and Industrial Research.

Hedenquist, J. W. 1983. Waiotapu, New Zealand: the geochemical evolution and mineralisation of an active hydrothermal system, Ph.D. thesis, University of Auckland.

Hochstein, M. P., Smith, I. E. M., Regenauer-Lieb, K., and Ehara, S. 1993. Geochemistry and heat transfer processes in Quaternary rhyolitic systems of the Taupo volcanic zone, New Zealand, *Tectonophysics* **223:**213–237.

Houghton, B. F., Wilson, C. J. N., Lloyd, E. F., Gamble, J. A., and Kokelaar, B. P. 1987. A catalogue of basaltic deposits within the central Taupo volcanic zone, *N.Z. Geol. Survey Rec.* **18:**95–101.

Houghton, B. F., Wilson, C. J. N., McWilliams, M. D., Lanphere, M. A., Weaver, S. D., Briggs, R. M. and Pringle, M. S. 1995. Chronology and dynamics of large silicic magmatic system: Central Taupo Volcanic Zone, New Zealand, *Geology.*

Hunt, T. M., and Latter, J. H. 1982. A survey of seismic activity near Wairakei geothermal field, New Zealand, *J. Volcanol. Geotherm. Res.* **14**(3–4)**:**319–334.

Kear, D. 1959. Stratigraphy of New Zealand's Cenozoic volcanism northwest of the volcanic belt, *N.Z. Geol. Geophys.* **2:**578–589.

KRISP working party. 1991. Large scale variation in lithsopheric structure along and across the Kenya Rift, *Nature* **354:**223–227.

Lamb, S. H., and Vella, P. 1987. The last million years of deformation in part of the New Zealand plate boundary zone, *J. Struct. Geol.* **9:**877–891.

Le Maitre, R. W. (ed.). 1989. *A Classification of Igneous Rocks and Glossary of Terms*, Blackwell, Oxford.

Lee, L. P. 1978. *First-Order Geodetic Triangulation of New Zealand 1909–49 and 1973–74*, Department of Lands and Survey Technical Ser. No. 1.

Lewis, K. B. 1980. Quaternary sedimentation on the Hikurangi oblique-subduction and transform margin, New Zealand, in *Sedimentation on Oblique Slip Mobile Zones* (P. F. Ballance and H. G. Reading, eds.), Int. Assoc. Sedimentology. Spec. Publ. **4:**171–189.

McCulloch, M. T., Kyser, T. K., Woodhead, J., and Kinsley, L. 1994. Pb-Sr-Nd-O isotopic constraints on the origin of rhyolites from the Taupo volcanic zone of New Zealand: evidence for assimilation followed by fractionation of basalt, *Contrib. Mineral. Petrol.* **115:**303–312.

Mooney, H. M. 1970. Upper mantle inhomogeneity beneath New Zealand: seismic evidence, *J. Geophys. Res.* **75:**285–309.

Morgan, P., Blackwell, D. D., and Spafford, R. E. 1977. Heat flow measurements in Yellowstone lake and the thermal structure of Yellowstone caldera, *J. Geophys. Res.* **82:**3719–3732.

Nairn, I. A. 1976. Late Quaternary faulting in the Taupo volcanic zone, in *Excursion Guide No. 55A and 55C* (S. Nathan, comp.), 25th Int. Geological Congr.

Nairn, I. A. 1981. Some studies of the geology, volcanic history, and geothermal resources of the Okataina volcanic center, Taupo volcanic zone, New Zealand, Ph.D. thesis, Victoria University of Wellington.

Nairn, I. A., and Beanland, S. 1989. Geological setting of the 1987 Edgecumbe earthquake, New Zealand, *N.Z. J. Geol. Geophys.* **32**(1)**:**1–13.

Nairn, I. A., and Hull, A. G. 1985. Post-1800 years B.P. displacements on the Paeroa fault zone, Taupo volcanic zone, DSIR, *N.Z. Geol. Surv. Rec.* **8:**135–142.

Nairn, I. A., Wood, C. P., and Bailey, R. A. 1994. The Reporoa caldera, Taupo volcanic zone: source of the Kaingaroa ignimbrites, *Bull. Volcanol.* **56:**529–537.

Nohda, A. 1984. Classification of island arcs by Nd-Sr isotopic data, *Geochem. J.* **18:**1–19.

Otofuji, Y., Matsuda, T., and Nohda, S. 1985. Opening mode of the Japan Sea inferred from the paleomagnetism of the Japan arc, *Nature* **317:**603–604.

Otway, P. M. 1986. Vertical deformation associated with the Taupo earthquake swarm, June 1983, *R. Soc. N.Z. Bull.* **24:**187–200.

Otway, P. M., and Sherburn, S. Vertical deformation and shallow seismicity around Lake Taupo, New Zealand, 1985–90. *N.Z. J. Geol. Geophys.* **37:**195–200.

Palmason, G., and Saemundsson, K. 1974. Iceland in relation to the mid-Atlantic ridge. *Annu. Rev. Earth Planet. Sci.* **2:**25–50.

Patterson, D. B., and Graham, I. J. 1988. Petrogenesis of andesitic lavas from Mangatepopo valley and Upper Tama Lake, Tongariro volcanic center, New Zealand, *J. Volcanol. Geotherm. Res.* **35:**17–29.

Reid, F. W. 1983. Origin of the rhyolitic rocks of the Taupo volcanic zone, *J. Volcanol. Geotherm. Res.* **15:** 315–338.

Reid, F. W. and Cole, J. W. 1983. Origin of dacites of the Taupo volcanic zone, New Zealand, *J. Volcan. Geotherm. Res.* **18:**191–214.

Reilly, W. I. 1972. The New Zealand gravity map series, *N.Z. Geol. Geophys.* **15:**3–15.

Reilly, W. I. 1990. Horizontal crustal deformation on the Hikurangi margin, *N.Z. J. Geol. Geophys.* **33**(2)**:**393–400.

Richardson, W. P. 1989. The Matata earthquake of 1977 May 31: a recent event near Edgecumbe, Bay of Plenty, New Zealand, *N.Z. J. Geol. Geophys.* **32**(1)**:**17–30.
Rogan, M. 1982. A geophysical study of the Taupo volcanic zone, New Zealand, *J. Geophys. Res.* **87:**4073–4088.
Satake, K., and Hashida, T. 1989. Three dimensional attenuation structure beneath the North Island, New Zealand, *Tectonophysics* **159:**181–194.
Sclater, J. G. Jaupart, C., and Galson, D. 1980. The heat flow through oceanic and continental crust and heat loss of the earth, *Rev. Geophys. Space Phys.* **18:**269–312.
Sissons, B. A. 1979. The horizontal kinematics of the North Island of New Zealand, Ph.D. thesis, Victoria University of Wellington.
Skinner, D. N. B. 1986. Neogene volcanism of the Hauraki volcanic region, in *Late Cenozoic Volcanism in New Zealand* (I. E. M. Smith, ed.), *R. Soc. N.Z. Bull.* **23:**21–47.
Smith, E. G. C., Scott, B. J., and Latter, J. H. 1984. The Waiotapu earthquake of 1983, December 14, *Bull. N.Z. Nat. Soc. Earthquake Eng.* **17**(4)**:**272–279.
Smith, E. G. C., Stern, T. A., and Reyners, M. 1989. Subduction and backarc activity at the Hikurangi convergent margin, New Zealand, *Pure Appl. Geophys.* **129:**385–409.
Smith, E. G. C., and Webb, T. H. 1986. The seismicity and related deformation of the central volcanic region, North Island, New Zealand, in *Late Cenozoic Volcanism in New Zealand* (I. E. M. Smith, ed.), *R. Soc. N.Z. Bull.* **23:**112–133.
Stern, T. A. 1986. Geophysical studies of the upper crust within the central volcanic region, New Zealand, in *Late Cenozoic Volcanism in New Zealand,* (I. E. M. Smith, ed.), *R. Soc. N.Z. Bull.* **23:**92–111.
Stern, T. A. 1987. Asymmetric backarc spreading, heatflux and structure associated with the central volcanic region of New Zealand, *Earth Planet. Sci. Lett.* **85:**265–276.
Stern, T. A. 1979. Regional and residual gravity fields, central North Island, New Zealand, *N.Z. J. Geol. Geophys.* **22:**479–485.
Stern, T. A., and Davey, F. J. 1987. A seismic investigation of crustal and upper mantle structure within the central volcanic region of New Zealand, *N.Z. J. Geol. Geophys.* **30:**217–231.
Stern, T. A., Smith, E. G. C., Davey, F. J., and Muirhead, K. H. 1987. Crust and upper mantle structure of the northwestern North Island, New Zealand, *Geophys. J. R. Astronom. Soc.* **91:**913–936.
Studt, F. E., and Thompson, G. E. K. 1969. Geothermal heat flow in the North Island of New Zealand, *N.Z. J. Geol. Geophys.* **12:**673–683.
Walcott, R. I. 1984. The kinematics of the plate boundary zone through New Zealand: a comparison of short- and long-term deformations, *Geophys. J. R. Astronom. Soc.* **79:**613–633.
Walcott, R. I. 1987. Geodetic strain and the deformational history of the North Island of New Zealand during the late Cainozoic, *Phil. Trans. R. Soc. London A* **321:**163–181.
Webb, T. H., Ferris, B. G., and Harris, J. S. 1986. The Lake Taupo, New Zealand, earthquake swarms of 1983, *N.Z. J. Geol. Geophys.* **29**(4)**:**377–389.
Williams, R. O. 1987. The June 1986, Survey of the North Taupo regional monitoring pattern, Department of Scientific and Industrial Research, N.Z. Geol. Surv. Rep. EDS111.
Williams, R. O. 1989. Horizontal angle survey data, 1949–1954, acquired for studies of contemporary deformation of the Taupo belt, *N.Z. Geol. Surv. Rep.* EDS122.
Wilson, C. J. N. 1986. Reconnaissance stratigraphy and volcanology of ignimbrites from Mangakino Volcano, in *Late Cenozoic Volcanism in New Zealand* (I. E. M. Smith, ed.), *R. Soc. N.Z. Bull.* **23:**179–193.
Wilson, C. J. N., Houghton, B. F., and Lloyd, E. F. 1986. Volcanic history and evolution of the Maroa-Taupo area, central North Island, in *Late Cenozoic Volcanism in New Zealand* (I. E. M. Smith, ed.), *R. Soc. N.Z. Bull.* **23:**194–223.
Wilson, C. J. N., Rogan, A. M., and Smith, I. E. M. 1984. Caldera volcanoes of the Taupo volcanic zone, New Zealand, *J. Geophys. Res.* **89:**8463–8484.
Wood, C. P. 1983. Petrological logs of drillholes Bl 26, to Br 40: Broadlands Geothermal Field, *N.Z. Geol. Surv. Rep.* **108**.
Wood, C. P. 1986. Stratigraphy and petrology of N.M.4—Ngatamariki geothermal field. Unpublished report of the N.Z. Geol. Survey, DSIR. Lower Hutt, New Zealand.
Wright, I. C. 1992. Shallow structure and active tectonism of an offshore continental backarc spreading system: the Taupo volcanic zone New Zealand, *Mar. Geol.* **103:**287–309.
Wright, I. C., and Walcott, R. I. 1986. Large tectonic rotation of part of New Zealand in the last 5 Ma, *Earth Planet. Sci. Lett.* **80:**348–352.

2

The Southern Havre Trough
Geological Structure and Magma Petrogenesis of an Active Backarc Rift Complex

J. A. Gamble and I. C. Wright

ABSTRACT

Associated with Pacific–Australian plate convergence, the Havre Trough–Kermadec arc-backarc system (SW Pacific) is an archetypal example of an "island arc" plate boundary. Swath data from the southern Havre Trough between 33–34°S and 35°30′–37°S reveal that within an active, young backarc basin highly complex and heterogeneous tectonic and magmatic processes can be associated with a tectonically "simple" subduction system. The compressed across arc-backarc length scale would appear to be a significant control on the complexity of the tectonism and magmatism. A near-contiguous, segmented axial rift system, flanked by marginal rifts and subbasins, and horst ridges form the dominant backarc morphology. Arc edifices of the present frontal arc lie within the "backarc region," sited 15–25 km west of the Kermadec frontal ridge. At its southern extremity, the axial rift system is propagating southward toward New Zealand, penetrating the continental-oceanic arc crustal boundary. Contrary to previous models, the entire Havre Trough is now interpreted to be undergoing backarc rifting prior to true spreading, with the attendant tectonic fabric dextrally oblique to the bounding Colville and Kermadec ridges. "Pseudo-linear" magnetic anomalies are interpreted to result from the generally irregular spatial and temporal emplacement of sheeted lava and dike intrusives between older arc basement. Significant magma source heterogeneity is observed both along and orthogonal to the frontal Kermadec arc. Basalts from the northern and southern sectors of the Kermadec arc define two distinctive isotopic arrays, with the latter overlapping with primitive basalts from the onshore Taupo Volcanic Zone. Across the arc-backarc system, lavas within the axial rift system have a marked spatial asymmetry and heterogeneity, with mid-ocean ridge basalts (MORB)-like and arc basalts being separated by as little as 15 km.

J. A. Gamble • Department of Geology, Research School of Earth Sciences, Victoria University of Wellington, Wellington, New Zealand. *I. C. Wright* • New Zealand Oceanographic Institute, National Institute of Water and Atmospheric Research, Wellington, New Zealand.

Backarc Basins: Tectonics and Magmatism, edited by Brian Taylor, Plenum Press, New York, 1995.

1. INTRODUCTION

The Havre Trough–Kermadec arc-backarc system associated with Pacific–Australian plate convergence (Fig. 2.1), along with its contiguous northern and southern extensions (the Lau Basin and Taupo Volcanic Zone (TVZ)), is an archetypal example of an "island arc" plate boundary. Indeed the Havre Trough–Kermadec system formed an original example of a convergent plate boundary from which the concept of backarc basin evolution was initially developed over 20 years ago (e.g., Karig, 1970a, 1974; Packham and Falvey, 1971). In spite of its early recognition as a young, active backarc basin the nature and spatial distribution of present-day tectonic and magmatic processes within the Havre Trough are proving now to be not well known. Varying interpretations of Havre Trough backarc tectonism and magmatism, based on limited data, have been proposed (e.g., Karig, 1970a; Malahoff *et al.*, 1982; Taylor and Karner, 1983). Segments of the Havre Trough backarc system, now relatively well studied, reveal that tectonic and magmatic processes are highly complex and variable, even at length scales of ~15 km. Such heterogeneity is significant within a backarc basin linked to a tectonically "simple" subduction system where the plate boundary is linear, convergence is more or less orthogonal to it, and subduction polarity has not, if ever, reversed from the present westward directed convergence within the last 30–40 m.y. The implications of these observations are important as we unravel from the rock record the processes of volcanism and extension within, and tectonic settings of, backarc arc systems.

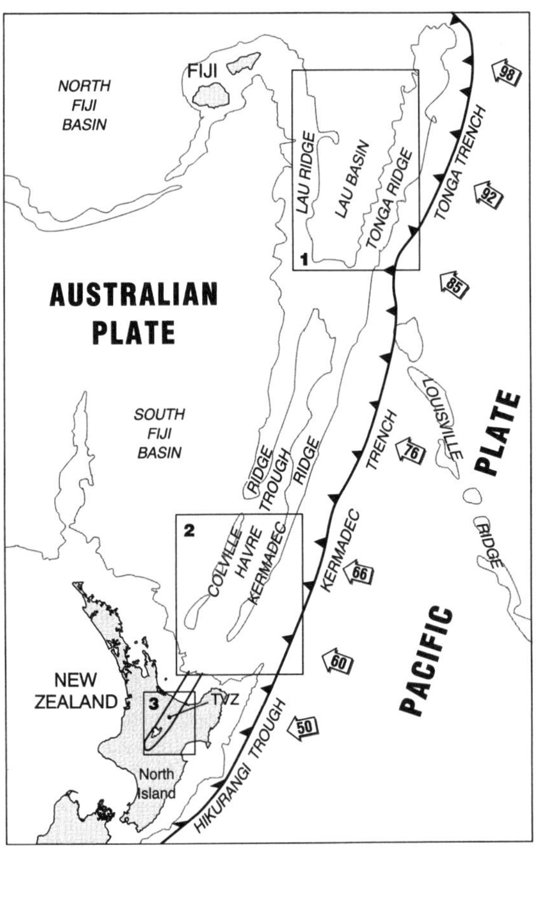

FIGURE 2.1. Australian-Pacific "island arc" plate boundary and associated contiguous Lau-Havre-Taupo backarc system. Boxes 1, 2, and 3 refer to sections of this backarc system described by, respectively, Hawkins (Chapter 3 this volume), this chapter, and Cole *et al.* (Chapter 1 this volume). Open arrows give the rate of relative plate convergence in mm yr^{-1} from Minster and Jordan (1978).

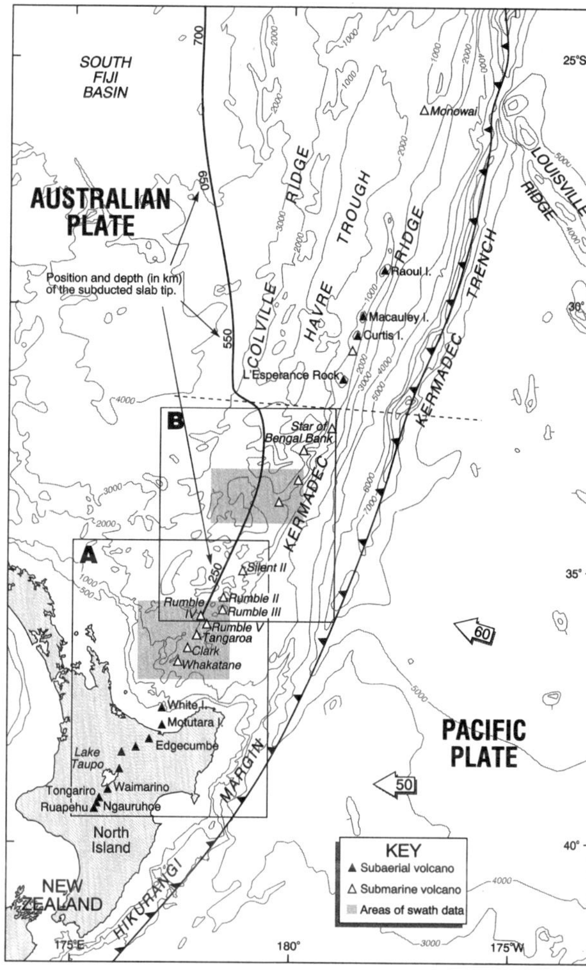

FIGURE 2.2. Regional bathymetry (in meters) and tectonic structure of the Kermadec-Hikurangi system and associated Havre-Taupo backarc system. Position and depth of the subducted slab are from Pelletier and Dupont (1990). Dashed line is the inferred position at about 32°S of changes in the morphology of the Kermadec subduction system (see text for discussion). Box outlines A and B show locations of Figs. 2.3 and 2.9. Shaded inserts show the two sectors of the backarc system mapped with swath data.

Within the last six years two sectors of the southern Havre Trough (SHT) have been relatively well surveyed with modern swath data (Fig. 2.2): a segment between 33° and 34°S (Caress, 1991) and the southernmost sector of the Havre Trough between 35°30′ and 37°S (Seafloor Surveys International, 1990, 1993; Wright, 1990; Wright *et al.*, 1990; Blackmore and Wright, 1995). Both sectors appear to display a range of typical and atypical tectonic morphologies for at least the SHT. The southernmost sector is particularly important because (1) it in part forms the transition from an *oceanic* backarc-arc system into the *continental* Taupo Volcanic Zone of North Island, New Zealand (Karig, 1970b; Gamble *et al.* 1993b; Cole *et al.*, Chapter 1 this volume), and (2) seafloor photographs and geochemical data are available from this segment (Gamble *et al.*, 1993b, 1994; Wright, 1993a, 1994) from which to characterize SHT volcanism and hence relate the concomitant processes of tectonism and magmatism. This chapter provides a synopsis of these new data from the SHT, so as to document the complexity and heterogeneity of tectonic and magmatic processes within an active, young, tectonically "simple," backarc basin.

2. REGIONAL STRUCTURE AND TECTONICS

2.1. General

The regional structure and tectonics of the submarine Lau and Havre segments of the nearly 2700-km-long contiguous Tonga-to-New Zealand backarc system, although displaying variations along strike, are typical of active backarc basins. The Lau–Havre system is characterized by variable but relatively high heat flow (Sclater *et al.*, 1972; Watanabe *et al.*, 1977), extremely complex seafloor morphology (Caress, 1991; Wright, 1993a), minimal or absent sediment cover (Wright, 1993b), heterogeneous and spatially dispersed volcanism (Gamble *et al.*, 1993b; Hawkins *et al.*, 1994), and extensive shallow seismicity (Sykes *et al.*, 1969; Pelletier and Louat, 1989). The first-order variation in the backarc structure, along the strike of the Lau–Havre system, is essentially a function of distance along the plate boundary from the Australian–Pacific rotation pole located at 62°S, 174°E. Increasing distance from the rotation pole northward along the plate boundary naturally coincides with an increase in the rate of plate convergence (Fig. 2.1). Between 35°S and 20°S the rate increases from about 53 to 90 mm yr^{-1} (Minster and Jordan, 1978; de Mets *et al.*, 1990); estimates of the rate do vary by some 15%, depending on whether Tonga–Kermadec earthquake-slip vectors are incorporated into the plate motion models. This northward increase in the rate of convergence is also matched by a general northward increase in the rate of backarc opening. The latter is shown by present-day spreading occurring in the Lau Basin to the north (Parson and Hawkins, 1994) and rifting to the south within the Havre Trough (Wright, 1993a).

Pelletier (1990) and Pelletier and Dupont (1990) have, using reconnaissance data, recognized clear differences in the morphology, depth, and position of the subduction trench and backarc region, north and south of 32°S, which they relate to an increase in the length of the subducted slab north of 32°S (Fig. 2.2). Thus, along the northern Havre-Kermadec sector, tectonic erosion of the inner subduction trench wall, a relatively simple, thickly sedimented backarc region as shallow as 2500 m coincides with subducted slab seismic activity to a depth of 550–650 km. To the south, accretion along the inner subduction trench wall, and a structurally complex backarc region as deep as 3000 to 4000 m, coincides with subducted slab seismic activity to a depth of 250 km. Limited swath data substantiate that the changes in structural morphology, at least in the backarc region (Caress, 1991), occur near 32°S.

2.2. Southern Havre–Kermadec Backarc-Arc System

The SHT (south of 32°S) is, on average, 120 km wide between the 2000-m isobaths of the parallel and flanking Colville and Kermadec ridges. Normal faults mark the boundaries between the trough and the inner margins of both ridges (Karig, 1970a; Caress, 1991; Wright, 1994). Both ridges are characterized by basalt-andesitic arc volcanism (Cole, 1986; Gamble *et al.*, 1990) and are interpreted to have been a single magmatic arc prior to opening of the Havre Trough (Karig, 1970a; Wright, 1993a). Present-day arc volcanism, including eruptions at submarine and subaerial volcanoes, generally occurs within the vicinity of the Kermadec Ridge (Fig. 2.2; Cole, 1986; Dupont, 1988), although the precise position of the volcanic front varies along the plate boundary (Wright, 1994). Volcanism, however, may persist at the southern end of the Colville Ridge (Wright *et al.*, 1990; Seafloor Surveys International, 1993).

The intratrough tectonic fabric comprises a series of dextral en echelon basement ridges and rift grabens that are generally 20–30° oblique to the trend of the bounding Colville and Kermadec ridges (Wright *et al.*, 1990; Caress, 1991). Slip vectors from shallow seismic events within the southern Havre Trough have near identical oblique azimuths (Pelletier and Louat, 1989). At its southern extremity, the SHT is marked by the 3300-m-deep Ngatoro rift graben system (Fig. 2.3), which in part forms a major bathymetric reentrant impinging on the northeast New Zealand continental slope (Cole and Lewis, 1981; Lewis and Pantin, 1984).

2.3. Taupo–Havre Continental-Oceanic Transition

The transition between the oceanic and continental segments, recognizable from bathymetric, magnetic anomaly, and petrologic data forms a near-linear boundary along the northeast New Zealand continental margin (Fig. 2.3; Wright, 1993a). Marked changes in both the amplitude and spatial fabric of magnetic anomalies occur at the boundary (Davey and Robinson, 1978; Malahoff *et al.*, 1982; Lewis and Pantin, 1984; Wright, 1992). Magnetic and gravity modeling indicate that the Raukumara Plain to the northeast of the crustal boundary is underlain by oceanic crust (Gillies and Davey, 1986). Similarly, Gamble *et al.* (1990, 1993b, 1994) have documented significant differences in the compositional range, mineralogy, and petrologic nature of backarc rift and arc-related volcanic rocks across this

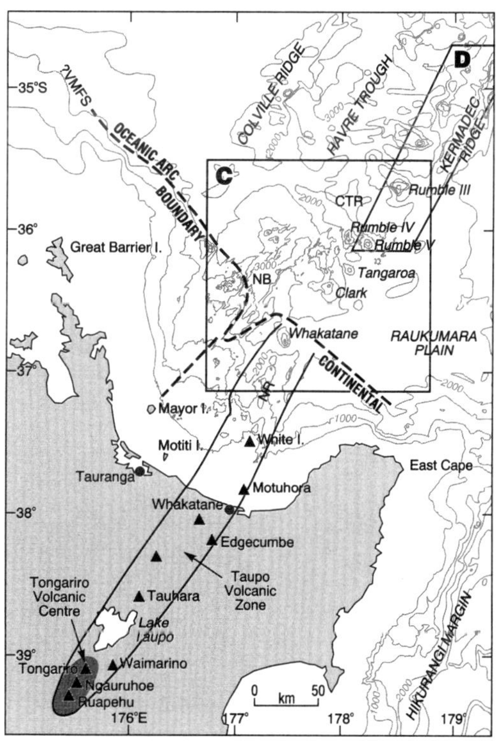

FIGURE 2.3. Synoptic bathymetry (meters) and structure of the southern Kermadec-Havre Trough region. Box outlines C and D show locations of Figs. 2.4 and 2.8. NB, Ngatoro Basin; NR, Ngatoro Ridge; VMFS, Vening Meinesz fracture zone.

boundary. These relations have been interpreted to reflect the different crustal settings of the volcanism.

Two features of the oceanic-continental transition are pertinent to the regional structure and tectonism of the SHT. First, this crustal boundary is remarkably distinct and linear (Fig. 2.3) and appears to coincide with the inferred position of the Vening Meinesz fracture zone (van der Linden, 1967). Hence, the present oceanic arc–continental boundary may be an inherited feature resulting from transform motion along the outer edge of the New Zealand continental margin during Oligocene backarc spreading of the South Fiji Basin (Davey, 1982). Second, the sole deviation from the linearity of the crustal boundary is the tectonic reentrant coinciding with the southern Ngatoro graben basin (Fig. 2.3). This reentrant is interpreted to be attentuated continental crust, with morphological and geochemical characteristics transitional between normal continental crust and oceanic arc crust (Gamble *et al.*, 1993b, 1994), and marks the southern limit of SHT rifting that is propagating into the New Zealand continental margin (Wright, 1993a).

The continental segment of the backarc (the TVZ) extends southward for some 370 km into North Island, New Zealand. The axes of oceanic SHT and continental TVZ backarc extension are not continuous (Fig. 2.3), having a left lateral offset of 45–50 km (Lewis and Pantin, 1984; Wright *et al.*, 1990). A series of dextrally oblique and en echelon faults form the surficial expression of this offset (Wright, 1992). The offshore TVZ comprises a 40–45-km-wide volcano-tectonic structure consisting of a volcanic arc ridge (the Ngatoro Ridge) flanked by rift grabens (Wright, 1992). Volcanism within the offshore TVZ is widely distributed and has a range of compositions, including silicic dome complexes (Gamble *et al.*, 1993b).

3. STRUCTURE AND MORPHOLOGY OF THE 35°30′S–37°S SECTOR

3.1. General

The 35°30′S–37°S sector comprises a morphologically heterogeneous 50–65-km-wide zone of complexly arranged basement ridges and troughs, split by an axial rift system (Fig. 2.4; Wright, 1993a). The characteristic morphology of the rift system is a series of segmented half-grabens bounded by master normal faults. Each half-graben typically has the constituent elements of (1) an outer rift graben with associated outer escarpments, (2) a series of generally downstepping fault block ridges and terraces, and (3) an inner rift graben (Fig. 2.5).

3.2. Outer Rift Graben

The outer rift graben forms a 15–25-km-wide, segmented and en echelon rift system that trends 045°. The morphology of the rift system and shallow seismic reflection data (Figs. 2.4a and 2.5) suggest that the opposing outer graben escarpments comprise a parallel pair of inward-facing master and antithetic normal faults. These master and antithetic fault escarpments typically have relief of 800–1000 m and 150–200 m, respectively. In plan, the outer escarpments have an en echelon and zigzag arrangement with both the master and antithetic fault segments trending either 030° or 060° (Fig. 2.6). Generally, the fault segments are right-stepping, with both fault azimuths appearing to have similar strike lengths.

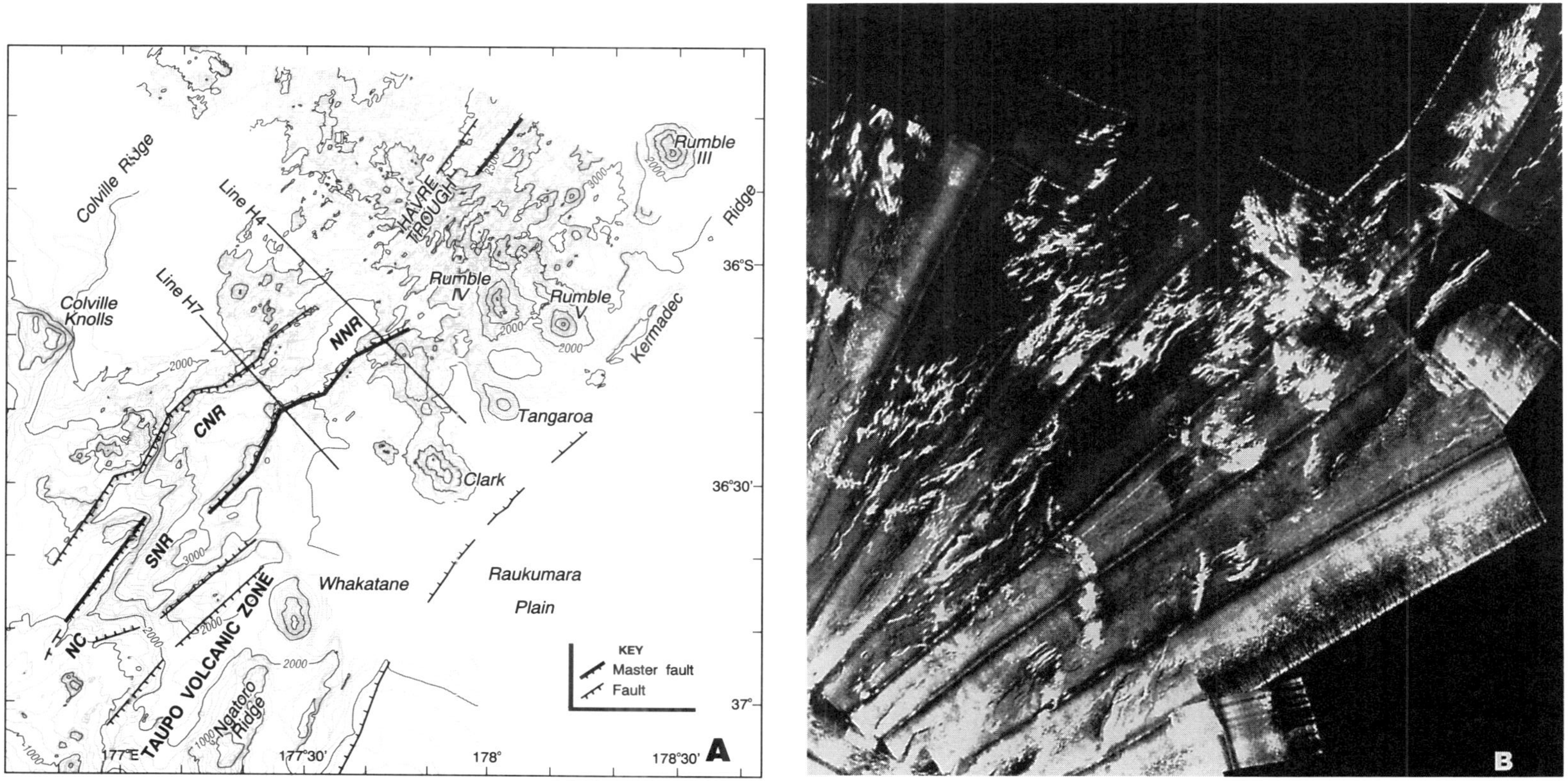

FIGURE 2.4. (a) Bathymetry (100-m contour interval) of the southern Kermadec Ridge and Havre Trough, and offshore Taupo Volcanic Zone region (after Blackmore and Wright, 1995). SNR, CNR, NNR, and NC refer to the southern, central and northern Ngatoro rifts, and Ngatoro Canyon, respectively. Lines H4 and H7 locate single-channel seismic reflection profiles shown in Fig. 2.5. (b) GLORIA mosaic of the southern Kermadec Ridge and Havre Trough, and offshore Taupo Volcanic Zone region. Bright zones are areas of high acoustic reflectivity.

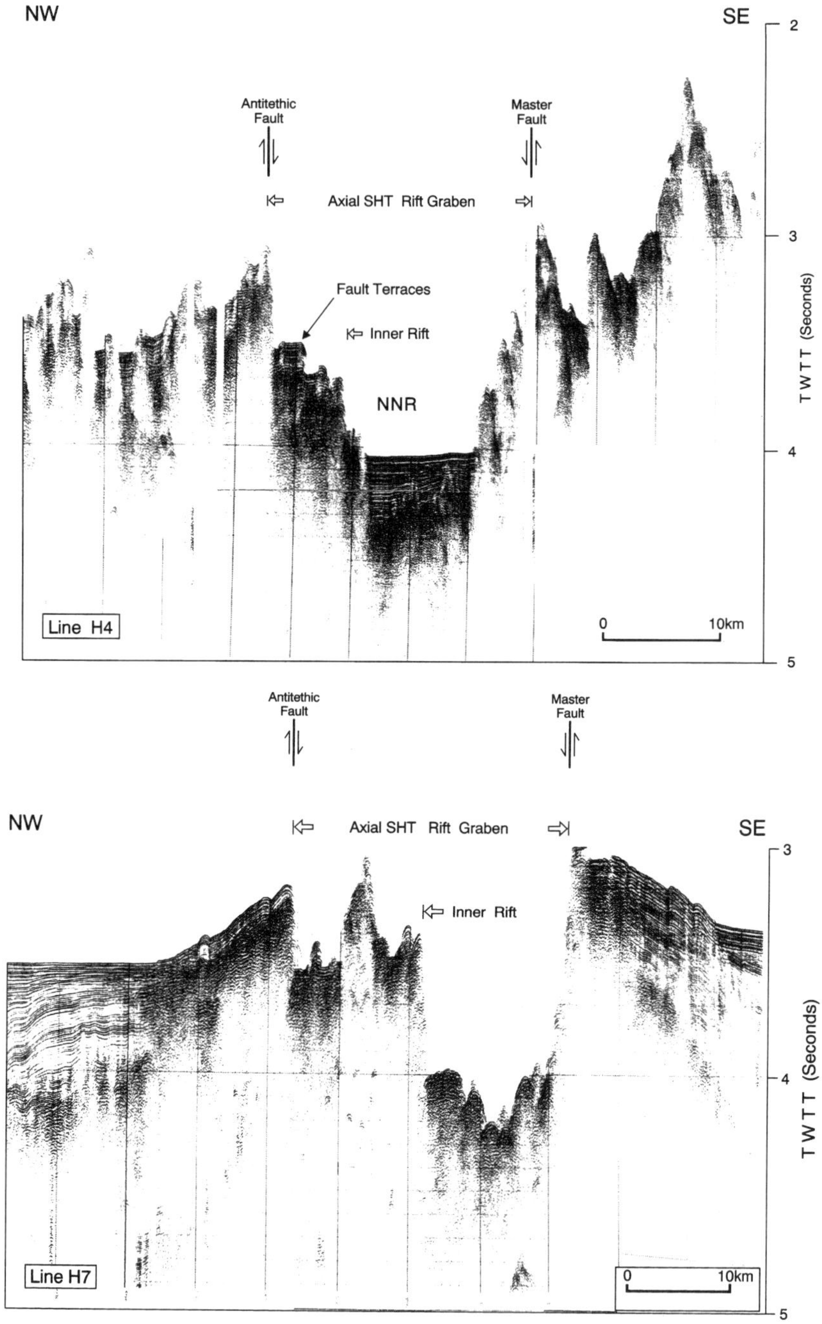

FIGURE 2.5. Single-channel air-gun seismic reflection profiles showing the tectonic morphology of the Ngatoro rift system, southern Havre Trough (after Wright, 1993a). TWTT=two-way travel time (s).

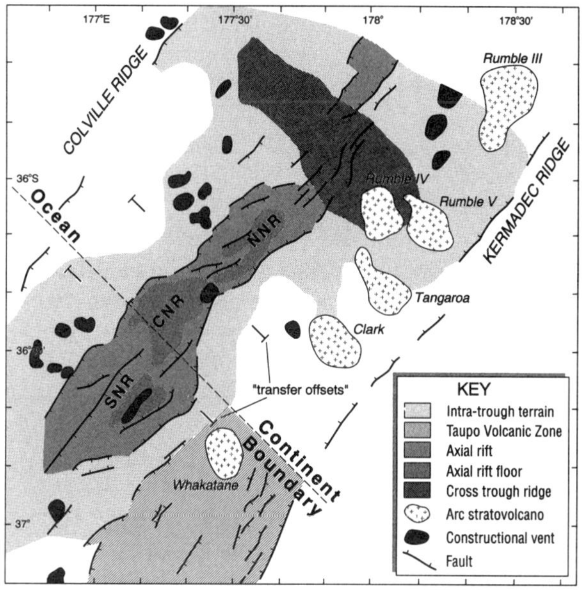

FIGURE 2.6. Synoptic geological-tectonic interpretation of the Ngatoro rift system, southern Havre Trough, showing the rift graben, graben fill sediments, faults, volcanic vents, and arc stratovolcanoes (after Wright, 1993a). SNR, CNR, and NNR, refer to the southern, central and northern Ngatoro rifts, respectively.

Between the outer antithetic fault escarpment and the inner rift, fault block terraces and ridges form a heterogeneous structure of downstepping normal faults interspersed with elongate ridges and depressions (Figs. 2.4a and 2.5). Faults bounding individual blocks comprise linear or near-linear scarps with relief of 50–100 m and are longitudinally continuous over distances of at least 6–10 km. Fault terraces are back-tilted and infilled to form a descending series of ponded sediment dams.

The structural fabric of the SHT rift system is disrupted by "transfer offsets," which partition the half-grabens into 25–30-km-long, partially linked contiguous rift segments (Fig. 2.6). These transfer offsets disrupt the spatial arrangement of both the outer and inner rift grabens. The surficial morphology of the rifts suggests that the asymmetry of the half-grabens varies between rift segments along the strike of the system, alternating at the transfer offsets.

3.3. Inner Rift Graben

At the southern extremity of the SHT rift system, the three inner rift grabens (the southern, central, and northern Ngatoro rifts) are each 24–26 km long and 10–12 km wide. At the "transfer offsets" the inner rift grabens are connected by narrow (2–5 km wide) "interrift corridors" that consist of bathymetric sills or steps with relief of 100–200 m coinciding with sediment-free basement highs.

Bathymetric and geologic long-range inclined asdic (GLORIA) data (Figs. 2.4 and 2.6) highlight the pronounced V-shape of the southern Ngatoro rift basin, which penetrates the continental edge of New Zealand by some 25 km (Wright, 1993a). The progressive southwestward narrowing of the rift appears to be accommodated by the antithetic faults altering their orientation, while the opposing master fault retains its linearity. At the head of the graben, the rift escarpments progressively converge to become a series of parallel normal faults that subsequently merge with the tectonically controlled, 500-m-wide Ngatoro canyon.

Along the length of the southern rift floor is a 3-km-wide, 150-m-high volcanic ridge capped with conical knolls (Blackmore and Wright, 1995). This ridge, comprising in part glassy basalts (Gamble *et al.*, 1993b) and largely devoid of sediment cover, forms the main recent constructional volcanic feature within at least the three segments of the Ngatoro rift system (Wright, 1993a).

Landward of the propagating southern Ngatoro rift, a narrow linear structural "furrow" comprising inward-facing active normal faults is identifable from single-channel seismic reflection data. This structural furrow (Fig. 2.3) is inferred to form a line of incipient continental rifting (Cole and Lewis, 1981; Wright, 1992), extending southwestward to the basalt-pantellerite volcano of Mayor Island (Ewart, 1968; Cole, 1978; Houghton *et al.*, 1992). Farther southwestward in North Island, New Zealand, across the projected line of rifting, geodetic data indicate considerable extension around 4 mm yr^{-1}

The central and northern Ngatoro rifts (Fig. 2.6) form the main, generally flat-floored and partially sediment infilled, grabens of the Ngatoro segment of the SHT system. Although the morphology of these rifts appears typical of SHT rifts in general (Blackmore and Wright, 1995), the significant sediment infilling of up to 600 m (Wright, 1993b) is not. The latter is interpreted to be a function of sediment supply, due to a high flux of terrigenous detritus from the New Zealand landmass via the Ngatoro Canyon system (Wright *et al.*, 1990). Isopachs of graben fill clearly define an underlying structural basement comprising elongate ridges and troughs with relief of 200 m (Wright, 1993a) that are interpreted to mark the position of subsurface synthetic and antithetic faults. Most of the extension within these rifts appears to be accommodated by the bounding master and antithetic faults. Some active deformation does, however, occur on the intrarift synthetic and antithetic faults, as evinced by the development of seafloor scarps on the sediment-flooded rift floor (Seafloor Surveys International, 1990).

3.4. Rift Volcanism

Recent volcanism within the Ngatoro rift system is generally not associated with youthful constructional terrains on the rift graben floor. The major exception is the axial ridge of the southern rift (Fig. 2.6). Likewise, farther to the north (between 35°35′ and 35°50′S) constructional volcanic features within the half-graben rift (Fig. 2.4a) are minor or absent (Seafloor Surveys International, 1993). Sites of "neovolcanism" occur along both the master and antithetic fault escarpments of at least the Ngatoro rifts. Photographic and rock dredge samples show that the rift escarpment volcanism comprises lobate and ropy pillow flows of basalt with fresh 4–5-mm-thick glassy rinds (Gamble *et al.*, 1993b; Wright, 1993a).

3.5. Rift Flank Structure and Volcanism

Flanking the axial rift system, the intratrough region encompasses a heterogeneous terrain of isolated knolls and seamounts and elongate ridges and depressions. The latter have axes parallel or subparallel to the axial rift system (Fig. 2.4). Apparent morphological differences occur between the two terrains flanking the rift system (Wright, 1993a), although these have yet to be quantified. Distant from the effects of hemipelagic sedimentation derived from New Zealand, significant areas of the flanking terrains consist of exposed basement rock (Caress, 1991; Seafloor Surveys International, 1993). Side-scan imagery shows clearly that the flanking margins, at least those of the 35°35′ to 35°50′S rift segment, comprise major fields of extrusive volcanism (Seafloor Surveys International, 1993), which are possibly active. Like the structural fabric, this extrusive volcanism also appears to vary between the respective margins flanking the rift, being more extensive on the flank nearest to the volcanic front and subduction trench.

3.6. Seismicity

Regional compilations of New Zealand shallow earthquakes during the period 1956–1987 (Hatherton, 1980; Reyners, 1989) show that the southern extremity of the SHT rift system is a seismically active region (Fig. 2.7). As noted by Reyners (1989), most of this seismicity occurred during two earthquake swarms in 1974 and 1984–1985. The site of the earlier swarm, lying along the western margin of the southern Ngatoro rift, is clearly associated with active extensional tectonism of the rift system (Wright, 1993a). Further, the concentration of the 1974 swarm along the western margin of the southern rift is consistent with the interpreted position of the master fault on the western flank for that rift segment. To the north of 36°30′S the apparent decrease in seismicity largely reflects the limit of onshore instrument coverage, although the seismicity that is recorded is almost entirely restricted to the axial rift system. Elsewhere intratrough shallow seismicity (e.g., Pelletier and Louat, 1989) cannot be readily related to the tectonic structure of the backarc, because ocean bottom seismometers have yet to be deployed within the Havre Trough.

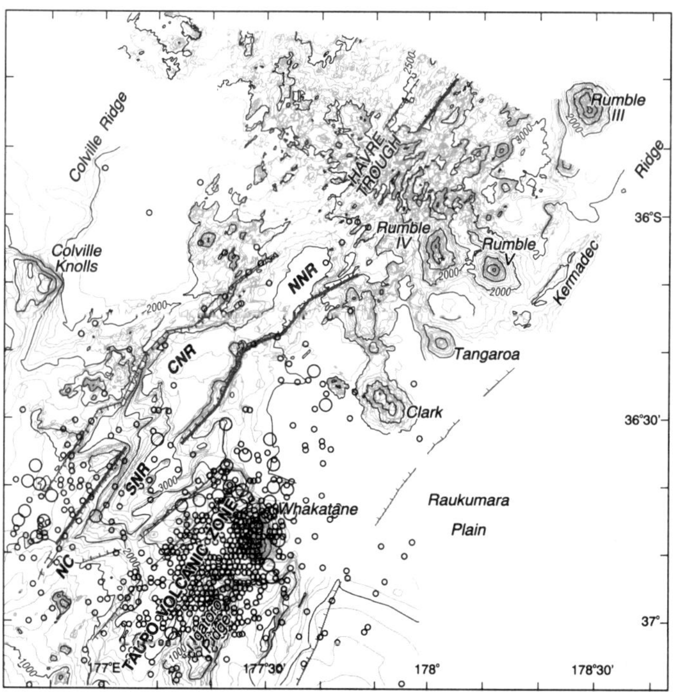

FIGURE 2.7. Compilation of shallow seismicity (M_L 4.0 and depth <40 km) lying within the limits of the New Zealand Seismograph Network for the period January 1964-September 1991 within the Ngatoro rift system (after Wright, 1993a).

The 1984–1985 swarm was more extensive than the 1974 swarm (Reyners, 1989) and coincides with the northern terminus of the offshore TVZ (Fig. 2.7; Wright, 1993a). Seismological evidence indicates that the 1984–1985 events are the result of normal and strike-slip faulting with a consistent T-axis striking northeast-southwest, parallel with SHT structure.

3.7. Heat Flow

Although the precise magnitude of heat flow within submarine backarc basins is difficult to establish (Sclater *et al.*, 1980), reconnaissance measurements within and near the Ngatoro rift system reveal a variable but generally high heat flow (Whiteford, 1990, 1992). Outside the rift graben mean values from both flanks lie within the range of 106–114 mW m^{-2}, whereas within the central Ngatoro rift segment the mean value is 86 mW m^{-2}. Heat flow within the onshore TVZ segment of the backarc basin is about 800 mW m^{-2} (Stern, 1987; Cole *et al.*, Chapter 1 this volume).

3.8. Volcanic Front

Like the northern Kermadec Ridge where active subaerial island volcanoes (e.g., Raoul, Macauley, and Curtis) lie along the ridge axis (Fig. 2.2), it has been generally

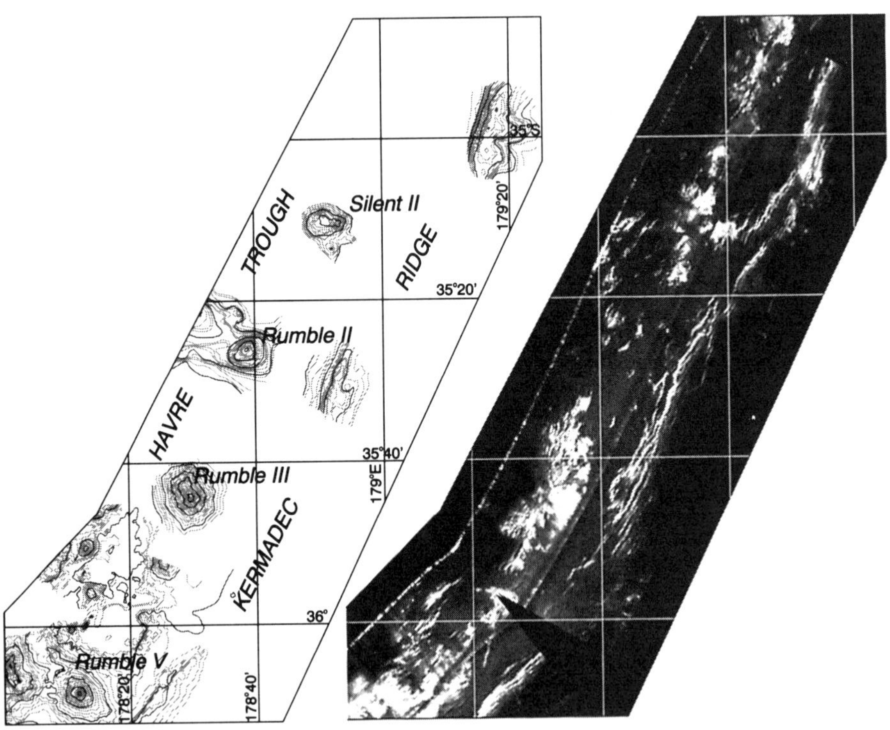

FIGURE 2.8. Bathymetry (100-m contour interval) and GLORIA sonograph of the southern Kermadec Ridge-Havre Trough (35°–36°S) (after Wright, 1994). Signal polarity is the same as Fig. 2.4.

assumed that the southern Kermadec volcanic front coincides with the frontal ridge (e.g., Dupont, 1988). Within the 35°–37°S sector, the Kermadec volcanic front is defined by seven discrete submarine arc volcanoes (Figs. 2.4 and 2.8): Rumble II, III, IV, V, Silent II, Tangaroa, and Clark (Kibblewhite and Denham, 1967; Wright, 1994). To the south the volcanic front continues with Whakatane volcano sited at the northern extremity of the TVZ (Wright, 1992; Blackmore and Wright, 1995). All seven of the edifices of the volcanic front lie within the eastern confines of the Havre Trough, some 15–25 km west of the southern Kermadec frontal ridge axis (Wright, 1994), and 35–45 km east of the axial backarc rift graben system (Wright, 1993a). Along the same latitudinal sector of the subduction system, constructional arc volcanoes are not identified on the Kermadec Ridge. The temporal history of southern Kermadec arc volcanism is largely unknown; however, very limited data suggest that arc volcanism associated with the present 35–37°S sector of the arc front extends back to least 0.77 Ma (Wright, 1994). These arc volcanoes continue to be the sites of recent volcanism with fresh lavas devoid of sediment cover observed on most volcanoes (Gamble *et al.*, 1993b; Blackmore and Wright, 1995), and modern submarine eruptions were recorded from Rumble III (Latter and Hall, 1986). Hence, the late Quaternary to modern southern Kermadec volcanic front lies within the "backarc" region, west of the Kermadec frontal ridge, and thus postdates the early phases of backarc widening.

4. STRUCTURE AND MORPHOLOGY OF THE 33°S–34°S SECTOR

4.1. Rift System

Data from the only other swath-mapped sector (between 33° and 34°S) within the Havre Trough (Fig. 2.9) indicate that the axial, en echelon rift graben morphology recognizable between 35°40′ and 37°S is the characteristic structural fabric for at least the SHT. A series of short, segmented en echelon grabens, which are oblique by some 20° to the bounding arc ridges (Fig. 2.9), are clearly identifiable from GLORIA and Sea Beam data (Caress, 1991). Like the segment to the south these grabens are 15–35 km in length, 6–10 km wide, have depths of 3500–4000 m, and are generally 500-1000 m deeper than the outer rift flanks. As with the 35°40′–37°S sector the morphology of the grabens is dominated by extensional faulting with no obvious constructional volcanic features within the area mapped by Sea Beam. The individual rift segments are typically offset by 14–30 km and linked, like those to the south, by "transfer offsets" comprising 100–300-m-high, but narrow, topographic sills. Sparse single-channel seismic reflection data (Caress, 1991) indicate the rifts are probably asymmetric half-grabens like those to the south. The only apparent difference between the two swath-mapped sectors is the degree of sediment infilling of the rift grabens, which is interpreted to be a function of sediment supply (Wright, 1993a).

4.2. Volcanic Front

As in the southern sector, the arc stratovolcanoes of the 33–34°S sector lies within the "backarc" region, sited some 25 km west of the Kermadec Ridge (Wright, 1994). GLORIA (Caress, 1991) and conventional bathymetric data (Wright, 1994) show that the volcanic front is marked by three major (as yet unnamed) volcanoes (Fig. 2.9), each with a constructional volume around 100 km^3. Similarly, these data show no evidence of late Quaternary constructional arc stratovolcanoes on the attendant section of the Kermadec frontal ridge.

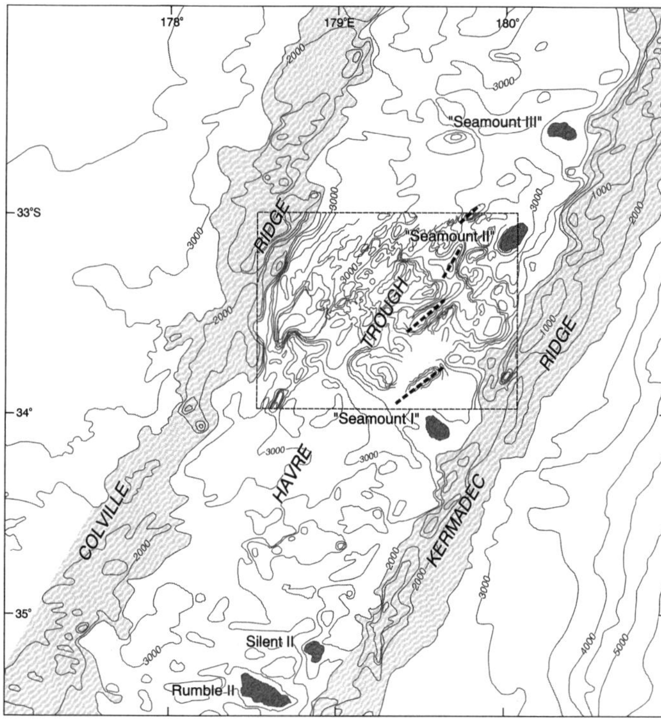

FIGURE 2.9. Bathymetry (at 500-m interval) of the Kermadec Ridge, Havre Trough, and Colville Ridge between 32°S and 35°S based on swath data (Caress, 1991) and conventional soundings (Wright, 1994). Insert box and dashed lines mark, respectively, the swath survey area and positions of the four grabens described by Caress (1991).

5. TECTONISM OF THE SOUTHERN HAVRE TROUGH

5.1. Rifting Model

The Havre Trough, until recently, has been almost exclusively interpreted as a site of intraoceanic spreading with the accretion of new oceanic crust along a spreading center system (Malahoff *et al.*, 1982; Caress, 1991; Wright, 1992, 1993b). More recently Wright (1993a) has interpreted the shallow structure and surficial morphology of, at least, the SHT as evidence of rifting, with its concomitant attenuation by rift block development (Fig. 2.10). In particular, the recognition of a segmented axial rift system, with its constituent asymmetric half-grabens along an apparently near-contiguous 440-km section of the SHT (between at least 33°S and 37°S), is considered to be convincing evidence for rifting rather than spreading. The shallow structure and surficial morphology of the SHT are almost identical to that documented within the Izu–Bonin backarc system (Taylor *et al.*, 1990, 1991). Multichannel crustal seismic reflection studies and Ocean Drilling Program (ODP) drill-hole data (Taylor *et al.*, 1991; Klaus *et al.*, 1992; Taylor, 1992) provide conclusive evidence for rifting within the Izu–Bonin system. The absence of ODP drill-hole data and crustal seismic reflection or refraction data from the Havre Trough means the nature of the inferred rift structure and associated magmatism at subcrustal depths is unknown. However, by analogy with the Izu–Bonin and western Lau Basin backarc basins (Hawkins *et al.*, 1994; Hawkins, Chapter 3 this volume), it is inferred that variable volumes of pillow and

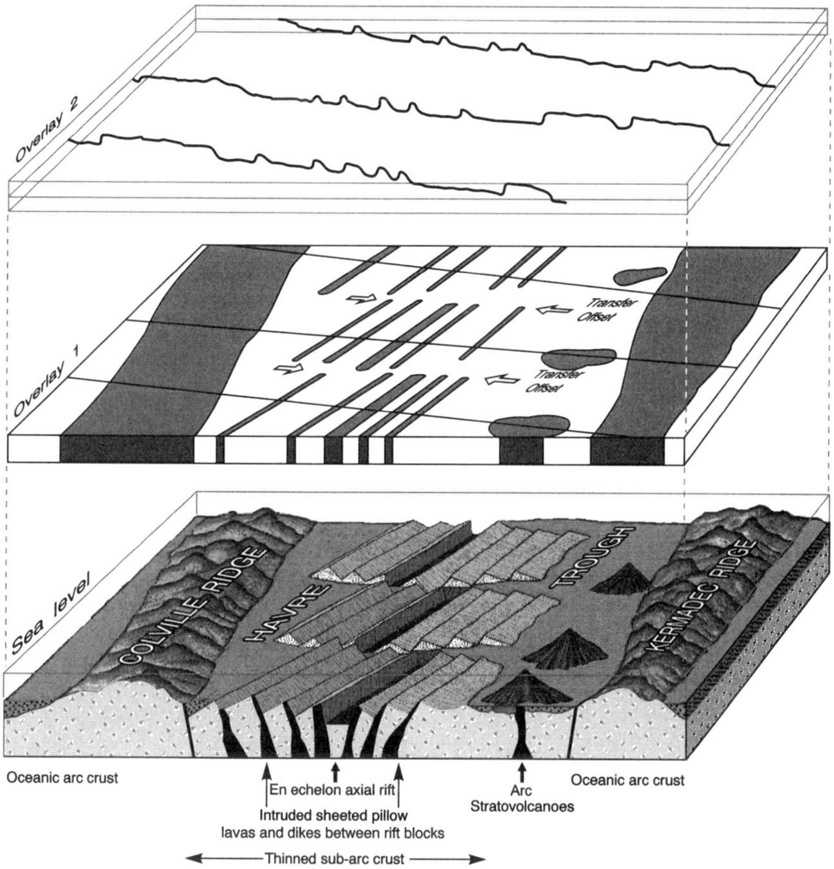

FIGURE 2.10. Schematic block model interpretation showing the southern Havre Trough rift morphology with attendant sheeted pillow lava and dike intrusives. Overlay 1 shows a schematic magnetization model of the rifted structure with magnetic inter-block intrusives emplaced between rift blocks of older low-magnetization arc crust. Overlay 2 shows the resultant magnetic anomaly profiles along arbitrary survey lines which form "pseudo-linear magnetic anomalies" (after Wright, 1993a).

sheet lava flows and sheeted dikes are emplaced within thinned, pervasively faulted Havre Trough subarc crust (Fig. 2.10; Wright, 1993a).

Further evidence for rifting within the Havre Trough comes from the recognition that crustal spreading in the Lau Basin, along the Valu Fa spreading ridge, has propagated southward only as far as ~23°S (Parson *et al.*, 1990; Weidicke and Collier, 1993). Hence, crustal spreading has yet to propagate southward into the Havre Trough. The termination of crustal spreading within the southern region of the Lau Basin is consistent with a marked increase in the rate of backarc widening from about 8 to 80 mm yr^{-1} at this latitude (Pelletier and Louat, 1989). Thus, Lau Basin crustal spreading is propagating southward at 110 mm yr^{-1} (Parson and Hawkins, 1994) toward the Havre Trough, while 1000 km to the south, rifting in the Havre Trough is propagating into the already extensional TVZ of continental New Zealand. The extremely high heat flow within the onshore TVZ has been taken as evidence for true crustal spreading (Stern, 1987) but may be better explained as reflecting massive emplacement of arc and rift intrusives within continental crust.

A corollary of the rifting model is that the magnetic anomalies, identified within the SHT by aeromagnetic survey (Malahoff *et al.*, 1982), do not record the accretion of new backarc crust associated with spreading. The SHT magnetic anomaly sequence forms a variable and, in part, low-fidelity record, with both irregular and apparently symmetrical anomaly profiles. Much of the displacement and irregularity of the anomalies was originally attributed to the presence of transform faults (Malahoff *et al.*, 1982).

Alternatively, Wright (1993a) has interpreted the low-fidelity SHT magnetic anomaly sequence as recording the emplacement of magnetic sheeted pillow lavas and dike intrusives between older, less magnetized, crustal arc rift blocks (Fig. 2.10). Hence, the magnetic anomalies are interpreted as "pseudolinear magnetic anomalies," resulting from the generally irregular spatial and temporal emplacement of subcrustal intrusives between these rift block segments (which are <35 km in strike length) flanking the axial rift graben. The magnetic anomaly pattern is complicated further by the location of the volcanic arc stratovolcanoes within the "backarc" region (Fig. 2.10). The proposed diffuse, low-fidelity anomaly sequence resulting from this rifting model is argued by Wright (1993a) to be more consistent with the observed SHT anomaly data than with crustal spreading.

5.2. Age and Rate of Extension

Interpretation of the SHT magnetic anomalies as recording phenomena other than the accretion of oceanic crust necessitates that the age and rate of basin widening be resolved from data other than magnetic anomaly sequences. Previously the anomaly sequences have been interpreted as recording a whole spreading rate and age of basin opening as, respectively, either 54 mm yr^{-1} and ~2 Ma (Malahoff *et al.*, 1982) or 20 mm yr^{-1} and 4 Ma (Korsch and Wellman, 1988).

Three sets of independent proxy data are available, however, to constrain the timing of basin initiation and the subsequent rate of extension (Wright, 1993a). These data are (1) K-Ar dating of eastward-migrating North Island intracontinental arc volcanism (Stern, 1987), (2) onshore North Island geodetic retriangulation (e.g., Walcott, 1984; Darby and Williams, 1991), and (3) slip vectors of present-day plate motion (e.g., Pelletier and Louat, 1989). These data provide a first approximation for the timing of basin opening and subsequent rate of widening for the 35°–37°S segment of the Havre Trough, which Wright (1993a) has taken as 5 Ma and 15–20 mm yr^{-1},respectively. The estimate of 5 Ma for the initiation of SHT rifting is consistent with a 10–5 Ma age for the earliest phases of rifting within the Lau Basin to the north (Parson and Hawkins, 1994) and concordant with a model of southward-directed propagation of rifting and spreading along the Lau–Havre backarc system.

6. GEOCHEMISTRY

6.1. General

Geochemical data for volcanic rocks from the oceanic Kermadec arc have been reported by Brothers (1967) and Ewart *et al.* (1977), and more recently by Ewart and Hawkesworth (1987), Smith and Brothers (1988), and Gamble *et al.* (1990, 1993a, b, 1994). For the backarc region, geochemical data are sparse and presently confined to a pillow basalt from 34°28′S 178°52′E (Gamble *et al.*, 1990, 1993a) and a suite of pillow lavas from the eastern and western flanks of the Ngatoro rift system between 36° and 37°S (Gamble *et al.*, 1993b, 1994). For the continental setting of New Zealand, geochemical data on basalts and andesites are more numerous from the onshore TVZ (e.g., Cole, 1986; Graham and Hackett, 1987; Gamble *et al.*, 1990; Graham and Cole, 1991) and its northward offshore extension in the Bay of Plenty (e.g., Gamble *et al.*, 1993b, 1994). Gamble *et al.* (1993b, 1994) have commented on the contrasting relative abundances of volcanic rocks between the oceanic and offshore continental sectors, with the former dominated by basalt and basaltic andesite and the latter by a spectrum of compositions extending from basalt and andesite to dacite and rhyolite. In this section we summarize briefly the major geochemical features of basaltic volcanism along and orthogonal to the arc front in the "oceanic" and "continental" settings. Three main geochemical provinces are defined by (1) the oceanic Kermadec arc, (2) the oceanic Havre Trough backarc complex, and (3) the offshore continental arc (defined by the Ngatoro ridge–Whakatane volcano) and the continental backarc region. The following discussion concentrates on the basalts as they are the magma type common to all the provinces.

6.2. Kermadec Arc

Representative chemical analyses of basaltic rocks from the northern subaerial Kermadec arc volcanoes of Raoul, Macauley, Curtis, and L'Esperance and the southern submarine volcanoes of Rumble II, III, and IV at the southern termination of the oceanic segment are presented in Table I. The total alkalis + silica diagram (Fig. 2.11a; LeBas *et al.*, 1986) and plot of TiO_2 versus Na_2O (Fig. 2.11b) serve to identify some principal chemical differences between basalts from the two sectors. For example, Kermadec arc basalts show low total alkali contents (Fig. 2.11a) overlapping with the basalt fields from the offshore and onshore TVZ, which are very similar and extend to higher-alkali contents, particularly Na_2O (Fig. 2.11b). Incompatible multielement plots for representative Kermadec arc basalts (Fig. 2.12a,b), normalized to MORB (Pearce, 1983), bear the distinctive hallmark of subduction-related magmas with high-alkali earth and metal and depleted high-field-strength (HFS) element abundances relative to MORB.

The Sr- and Nd-isotope measurements on Kermadec arc basalts (Ewart and Hawkesworth, 1987; Gamble *et al.*, 1993a, 1994, unpublished data), shown on a standard covariation plot (Fig. 2.13), define two clearly discernible fields for the respective northern and southern segments of the Kermadec arc. The two fields have overlapping Nd isotopes but distinctive Sr isotopes. A single basalt from the flanks of the Rumble IV volcano is exceptional, having a Sr-isotope signature lower than other southern Kermadec basalts and being more characteristic of northern Kermadec lavas.

6.3. Southern Havre Trough–Ngatoro Rift System

Chemical analyses of basaltic rocks from the SHT are contained in Table II. On the total alkalis + silica diagram (Fig. 2.11a) basalts from the SHT show higher total alkalis, largely a reflection of their higher Na_2O, and define a vertical array at more or less constant SiO_2 content. Similarly, these rocks show higher TiO_2 and Na_2O contents (Fig. 2.11b) than basalts from the Kermadec arc and both the onshore and offshore segments of the TVZ. On incompatible multielement plots, basalts with >7% MgO (Fig. 2.12c) show subtle enrichments of the alkali earths and alkali metals (note the vertical scale), together with HFS elements Ce, P, and Zr relative to MORB.

A significant aspect of SHT backarc volcanism is the marked spatial asymmetry and heterogeneity of the erupted basalts, as evinced by the Ngatoro rift graben lavas (Gamble

TABLE I

Representative Chemical Analyses of Basalts from the Kermadec Arc[a]

Sample no.	AU23396	AU14782	AU37546	AU37548	AU14837
Field no.	KA 4	KA 2	KA 10	KA 11	KA 15
Locality	Raoul I	Herald I	Macauley I	Macauley I	L'Esperance
SiO_2	50.78	48.76	48.36	48.39	51.64
TiO_2	0.77	0.79	0.52	0.52	1.01
Al_2O_3	14.01	17.75	17.37	17.34	18.18
Fe_2O_3	1.35	1.39	1.20	1.20	1.57
FeO	9.01	9.24	8.98	7.98	10.46
MnO	0.25	0.19	0.17	0.18	0.24
MgO	11.98	7.33	8.06	8.01	4.97
CaO	9.64	11.44	14.19	14.15	9.27
Na_2O	1.32	1.75	1.25	1.23	1.33
K_2O	0.17	0.12	0.16	0.16	0.30
P_2O_5	0.12	0.08	0.16	0.04	0.06
LOI	0.50	1.28	0.04	0.12	0.47
Total	99.90	100.12	100.46	99.32	99.47
Mg#	0.72	0.61	0.67	0.66	0.48
Sc	37	47	44	44	39
V	275	355	300	297	388
Cr	25	90	155	151	17
Ni	28	41	50	54	19
Cu	130	150	120	121	47
Zn	88	92	65	66	109
Ga	16	16	15	15	17
Rb	4	2	2	2	8
Sr	173	161	194	193	220
Y	24	19	11	11	21
Zr	32	30	20	18	37
Nb	1	2	1	1	1
Ba	105	33	48	47	115
La	3	4.00	7.50	7.60	3.30
Ce	7	2.93	3.85	6.19	9.69
Pb	—	2	2	2	3
Th	—	0.5	0.8	0.84	0.4
U	—	0.3	0.45	0.46	0.2
$^{87}Sr/^{86}Sr$	0.703452 ± 14	0.703405 ± 15	0.703446 ± 14	0.703444 ± 20	0.703936 ± 18
$^{143}Nd/^{144}Nd$	0.513029 ± 8	0.513055 ± 10	0.513030 ± 6	0.513020 ± 6	0.512986 ± 5
eps Nd	+7.4	+7.8	+7.4	+7.2	+6.5

TABLE I
(*Continued*)

AU36982 KA 16 Rumble II	AU36986 KA 17 Rumble III	VUW11600 X162/1 Rumble IV	VUW11606 X168/1A Rumble IV	VUW11607 X168/1B Rumble IV	VUW11595 X169/1 Rumble IV
50.83	51.69	50.88	52.03	50.25	49.64
0.61	0.71	0.87	0.82	0.65	0.82
13.70	15.19	18.30	16.56	15.40	16.96
1.22	1.32	1.15	0.97	1.11	0.95
8.13	8.81	7.68	6.46	7.37	6.34
0.18	0.20	0.17	0.16	0.14	0.14
9.22	6.47	5.08	7.27	8.31	9.10
12.97	11.08	11.12	10.55	12.52	12.22
1.45	2.42	2.62	2.96	2.11	2.78
0.53	0.47	0.45	0.59	0.45	0.25
0.15	0.11	0.11	0.20	0.06	0.09
0.26	0.25	0.77	1.00	1.14	0.00
99.25	98.72	99.20	99.57	99.51	99.29
0.69	0.59	0.54	0.67	0.67	0.72
41	37	30	29	38	30
315	325	300	290	278	196
327	172	64	24	355	345
69	50	32	23	90	125
133	142	90	118	95	66
69	90	76	96	68	55
12	15	19	17	14	15
9	6	8	9	8	7
353	229	288	263	239	222
15	21	20	20	14	17
32	48	57	62	40	59
1	1	—	2	2	2
263	230	187	206	235	123
7.45	6.71	5	4	3	5
15.90	12.40	16	18	12	15
2	2	7	6	5	5
1.29	0.70	1	3	1	1
0.45	0.42	—	1	—	—
0.703966 ± 17	0.704090 ± 13	0.703976 ± 12	0.704251 ± 13	0.704270 ± 13	0.703348 ± 11
0.512919 ± 5	0.513032 ± 5	0.513000 ± 6	0.513014 ± 3	0.512989 ± 12	0.513050 ± 7
+5.2	+7.4	+7.1	+7.3	+6.8	+8.0

[a]All major element and trace element XRF Analyses were done at the Analytical Facility of Victoria University, Wellington (Gamble *et al.*, 1993a, b, 1994). Isotope analyses were done in the Research School of Earth Science, Australian National University. Details of the analytical methods used are in Gamble *et al.* (1993a). La, Ce, Th, and U values quoted at less than whole number significance were determined by INNA (data from Gamble *et al.*, 1993a). Dashes mean element concentration was less than detection limit

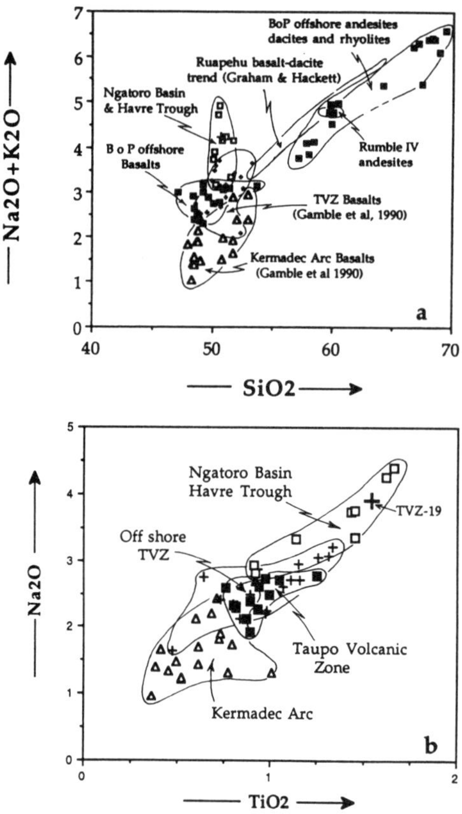

FIGURE 2.11. Offshore Bay of Plenty rocks compared to TVZ and Kermadecs. (a). Total alkali-silica (TAS diagram, Le Bas *et al.*, 1986) for volcanic rocks from TVZ, Kermadec arc, Havre Trough, Ngatoro Basin (SHT). The field of Ruapehu basalts-dacites (from Graham and Hackett, 1987) is shown for comparison. Filled squares: Bay of Plenty offshore basalts and andesites to rhyolites. Open squares: Ngatoro basin basalts. Large cross: Havre Trough basalt. Open triangle: Kermadec arc basalts. Small crosses: TVZ basalts. (b). Na_2O–TiO_2 variation diagram for basalts from TVZ, Kermadec arc, Ngatoro Basin, Havre Trough and offshore Bay of Plenty. Symbols as in (a), above.

et al., 1993b; Wright, 1993a). Basalts with higher TiO_2 and Na_2O contents are restricted to the western escarpment of the Ngatoro rift (that farthest away from the subduction and volcanic front), whereas basalts from the eastern escarpment are geochemically indistinguishable from the Kermadec arc lavas (Fig. 2.14). This heterogeneity occurs over a length scale of 15–20 km (the width of the rift graben) and appears to have a nearly perfect compositional separation in this best-surveyed segment of the SHT.

Sr- and Nd-isotope measurements of basalts from the SHT, including the western Ngatoro rift lavas (Gamble *et al.*, 1993a, 1994; Fig. 2.13), are similar both to many MORB basalts (Ito *et al.*, 1987) and backarc basalts (BABB) from the southern Lau Basin (Hergt and Hawkesworth, 1994). The western SHT rift basalts are appreciably less radiogenic (higher Nd- and lower Sr-isotope ratios) than basalts from the Kermadec arc or TVZ. The only exception is the basalt from the floor of the southern Ngatoro rift, which plots in the TVZ array, consistent with its transitional setting in the attenuated continental crust of the propagating reentrant.

6.4 Offshore TVZ Volcanic Front and Backarc

Basalts from the Ngatoro ridge–Whakatane volcano section of the offshore TVZ segment (Table III) are very similar to onshore TVZ basalts with considerable overlap (Fig. 2.11). Incompatible multielement plots show the distinctive signature of subduction-related magmas (Fig. 2.12d,e).

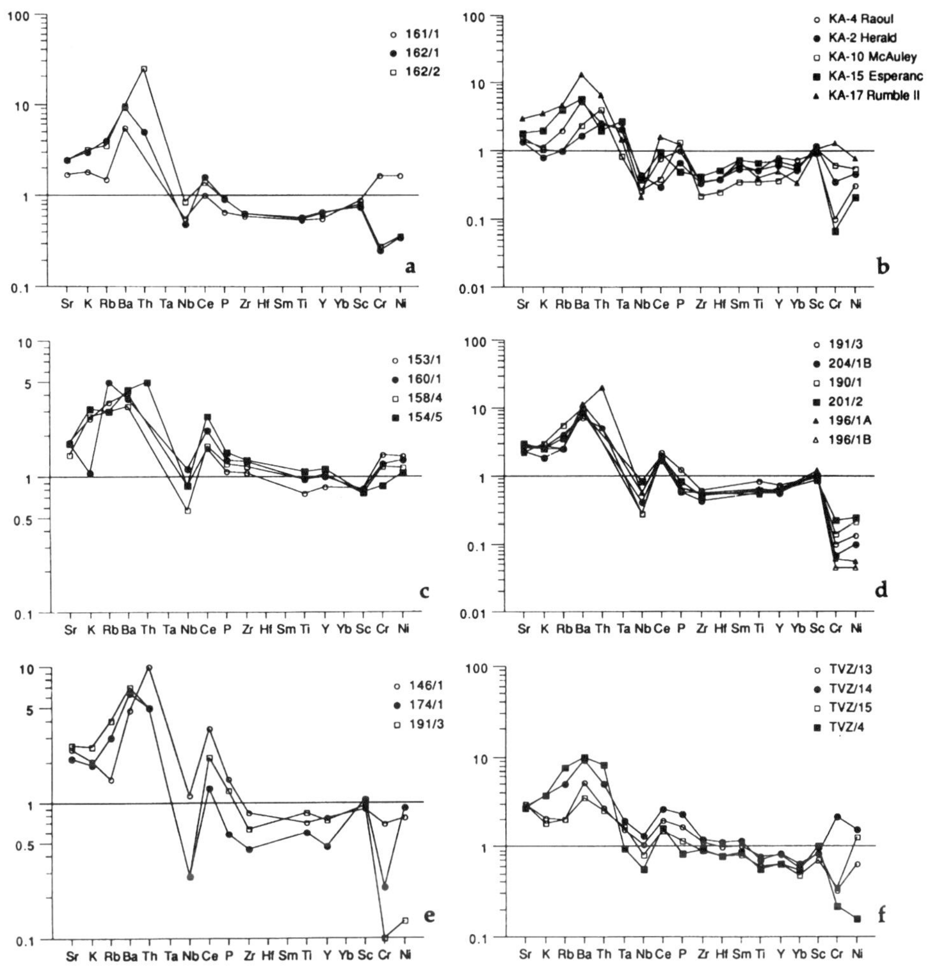

FIGURE 2.12. MORB normalized multielement diagrams for basalts from the Kermadec-New Zealand subduction system. Normalizing values from Pearce (1983). (a) Basalts from the vicinity of Rumble IV volcano at the southern end of the Kermadec arc. (b) Basalts from Raoul Island, Herald Island, Macauley Island, L'Esperance rock in the northern Kermadec Island arc and Rumble II toward the south. (c) Basalts from the Ngatoro Basin, note scale of vertical axis. (d) Basalts from Ngatoro Ridge area, offshore Bay of Plenty. (e) Basalts from Alderman Trough, offshore Bay of Plenty, note scale of vertical axis. (f) Basalts from Taupo Volcanic Zone, New Zealand. All analyses from Gamble *et al.* (1993a, 1994).

Isotope ratios of Sr and Nd are distinct from those of the Kermadec arc and the Havre Trough backarc rift system, showing higher-Sr isotopes and lower-Nd isotopes. Furthermore, for a given Nd-isotope ratio the offshore basalts are generally more radiogenic in Sr than the onshore TVZ or Taranaki basalts.

Further, andesites, dacites, and rhyolites are appreciably more abundant from the offshore TVZ region than from the oceanic sector to the north. Offshore andesites to rhyolites overlap with the onshore Ruapehu andesite-rhyolite lineage (Graham and Hackett, 1987) and are broadly similar to the general onshore TVZ andesite-dacite suite (Cole, 1979), with the exception that none of the felsic volcanics sampled in the offshore region approach the high SiO_2 contents (>75%) of younger (<20 ka) rhyolites from onshore TVZ.

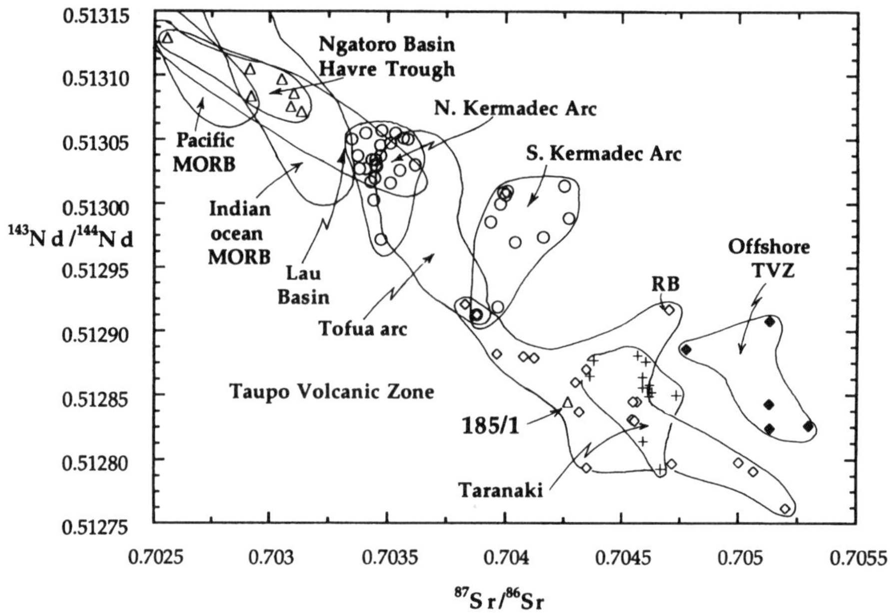

FIGURE 2.13. Sr- and Nd-isotope covariation diagram for basalts from the Tonga-New Zealand convergent plate boundary. Open triangles: Ngatoro and Havre Trough basalts, note how sample 185/1 falls in the field of TVZ basalts. Open circles: Kermadec arc basalts, subdivided into northern and southern Kermadecs. Open diamonds: TVZ basalts, RB is basalt from Ruapehu. Filled diamonds: basalts from Ngatoro Ridge, offshore Bay of Plenty. Crosses: Taranaki (Egmont) young basalts (Price *et al.*, 1992). Fields for Pacific MORB, Indian Ocean MORB, Lau Basin basalts, and Tofua arc lavas are from Ito *et al.* (1987), Klein *et al.* (1991), and Hergt and Hawkesworth (1994).

7. PETROGENESIS

7.1. General

Subduction zones mark the sites of recycling lithosphere into the deep earth and as such have been influential in the development of models of plate tectonics, crustal growth (Taylor and McLennan, 1985), and fluid transfer into the convecting mantle. Indeed, this latter process has been the focus of much recent geochemical work on volcanic arcs (e.g., Gill, 1981; Ellam and Hawkesworth, 1988; Morris *et al.*, 1990; McCulloch and Gamble, 1991; Hawkesworth *et al.*, 1993). Geochemical studies of volcanic rocks from the Tonga–Kermadec island arc system have been important in evolving models of arc magma petrogenesis (e.g., Ewart *et al.*, 1977; Ewart and Hawkesworth, 1987). Extension of this simple oceanic island arc system southward into the continental TVZ in New Zealand introduces the additional quanta of continental crust and further concomitant complexities to unravel (Wilson, 1989).

The coupling of backarc basins to volcanic arcs was first recognized in the margins of the Pacific Ocean basin (Karig, 1974). Petrological study of these regions of extension recognized the broadly MORB-like geochemical characteristics of the basaltic rocks (e.g.,

Hawkins, 1976; Saunders and Tarney, 1979, 1991; Hawkins and Melchior, 1985; Sinton and Fryer, 1987). More recent studies have furnished an abundance of new geochemical data from the western Pacific backarc basins (e.g., Hochstaedter *et al.*, 1990a, b; Loock *et al.*, 1990; Hergt and Hawkesworth, 1994; Hergt and Farley, 1994). From these data, models have begun to emerge linking both arc and backarc basin magma sources (e.g., Woodhead *et al.*, 1993; Gamble *et al.*, 1994) into a dynamic model of melt generation and extraction.

7.2. Taupo–Havre–Lau System

Whole rock and (where available) coexisting glass data for basalts from TVZ and the Kermadec arc (Gamble *et al.*, 1990, 1993a), Taranaki (Price *et al.*, 1992), offshore TVZ, southern Havre Trough (PPTUW/5) (Gamble *et al.*, 1990; 1993b, 1994), and the Lau Basin (Sunkel, 1990) are shown in Fig. 2.15 in a CaO—MgO—Al_2O_3—SiO_2 (CMAS) pseudoternary projection from plagioclase (Walker *et al.*, 1979). The majority of TVZ, Kermadec arc, and Lau Basin basalts cluster toward the evolved (hypersthene-quartz normative) end of the 1-atm cotectic as does the evolved basalt from the floor of the southern Ngatoro rift. Many of these lavas are strongly porphyritic, and their phase relations have been complicated by processes of mixing, assimilation, and storage in crustal magma chambers (cf. Gamble *et al.*, 1990). Young basalts from Egmont volcano (Price *et al.*, 1992) and basalts from the Havre Trough–Ngatoro Basin and offshore TVZ define parallel lineages from mildly nepheline normative compositions toward the 1-atm cotectic. Possible interpretations of these data are that they reflect either (a) different degrees of melting of comparable sources or (b) melting of progressively more refractory sources (higher normative olivine). In each case the primary magmas evolve into the 1-atm cotectic by olivine + plagioclase ± clinopyroxene fractionation. The implication here is that the trajectories are source related and that the bulk of basalts plotting in the shaded region (Fig. 2.15) have their primary characteristics obliterated by secondary modification effects such as crystal fractionation and mixing.

In an attempt to see through these effects Gamble *et al.* (1993a) used plots of incompatible HFS elements and ratios (e.g., Ti/Zr versus Zr); these elements are relatively insoluble in aqueous fluids or silicic melts, which are commonly thought to be slab derived, yet sensitive to mantle wedge source mineralogy and degree of melting. This approach was later amplified and extended to volcanic arc-backarc basin systems in the circum-Pacific region (Woodhead *et al.*, 1993). The observed systematic relationships between HFS elements were modeled in terms of multistage melting of a primary MORB source. It was concluded that ratios such as Ti/Zr and V/Ti in backarc basin basalts could be produced by partial melting of a source similar to that of MORB, but that many arc basalts required a source with higher Ti/Zr and V/Ti, which could be produced by prior melt extraction in the backarc basin regime. This model, moreover, was in keeping with current geodynamic models of mantle flow into the wedge corner region (e.g., Spiegelman and McKenzie, 1987).

The Sr- and Nd-isotope data provide additional clues to significant magma source heterogeneity both along and across the 3000 km of the Tonga–Kermadec–New Zealand arc system. Fields for arc and backarc basalts from the entire region, together with fields for Pacific and Indian Ocean MORB, are given in Fig. 2.13. We note the following points: (1) Basalts from the southern Havre Trough and western Ngatoro rift are broadly MORB-like, span the fields of Pacific and Indian Ocean MORB, and, in general, are similar to

TABLE II
Representative Chemical Analyses of Basalts from the Southern Havre Trough[a]

Sample no.	VUW11574	VUW11578	VUW11579	VUW11580	VUW11581
Field no.	X153/1	X154/1	X154/5	X158/1	X158/2
Locality	Ngatoro Basin	Ngatoro Basin	Ngatoro Basin	Ngatoro Basin	Ngatoro Basin
SiO_2	49.99	50.47	50.43	51.05	50.19
TiO_2	1.14	1.66	1.62	1.43	0.91
Al_2O_3	16.52	16.27	16.18	16.10	16.86
Fe_2O_3	1.02	1.19	1.19	1.14	0.93
FeO	6.76	7.90	7.95	7.58	6.23
MnO	0.15	0.18	0.18	0.16	0.13
MgO	8.06	6.89	7.06	6.92	7.66
CaO	10.60	9.31	9.36	9.45	12.61
Na_2O	3.33	4.39	4.25	3.73	2.93
K_2O	0.40	0.52	0.47	0.51	0.24
P_2O_5	0.13	0.18	0.18	0.15	0.07
LOI	1.19	0.28	0.50	1.13	0.58
Total	99.29	99.24	99.37	99.35	99.34
Mg#	68.00	60.90	61.30	61.90	68.70
Sc	33	29	31	30	34
V	234	286	282	257	209
Cr	363	209	217	293	280
Ni	128	90	96	100	86
Cu	52	43	45	49	74
Zn	67	79	78	81	62
Ga	17	20	20	19	17
Rb	7	7	6	7	4
Sr	217	213	210	173	190
Y	25	33	34	32	19
Zr	97	119	119	110	60
Nb	3	3	3	4	1
Ba	82	95	88	63	58
La	4	6	5	5	4
Ce	16	23	28	17	10
Pb	5	4	5	5	6
Th	1	—	1	—	1
U	—	—	—	—	—
$^{87}Sr/^{86}Sr$	0.703090 ± 14		0.703103 ± 14	0.702917 ± 10	
$^{143}ND/^{144}Nd$	0.513076 ± 7		0.513086 ± 8	0.513105 ± 7	
eps Nd	+8.5		+8.7	+9.1	

basalts from the Lau Basin to the north (Hergt and Hawkesworth, 1994). (2) Basalts from the northern and southern sectors of the Kermadec arc define two distinctive arrays distinguished by generally higher Sr in the southern Kermadec arc. The data field for the active Tofua arc of the Tonga–Lau Basin arc-backarc system (Ewart and Hawkesworth, 1987) spans both the northern and southern Kermadec fields. Perhaps a more detailed study of the isotope systematics of this arc segment would turn up distinctive fields as recognized along the Kermadecs. (3) Basalts from the southern Kermadec arc overlap with primitive

TABLE II
(*Continued*)

VUW11582 X158/3 Ngatoro Basin	VUW11583 X158/4 Ngatoro Basin	VUW11599 X160/1 Ngatoro Basin	VUW11616 X185/1 Indentor	VUW33441 PPTUW/5 Havre T
51.70	50.74	50.09	51.51	50.66
1.45	1.45	1.45	0.92	1.49
16.05	16.15	15.92	17.68	16.09
1.12	1.14	1.21	1.14	1.11
7.45	7.59	8.07	7.87	7.37
0.16	0.16	0.17	0.16	0.16
7.29	7.03	7.04	4.78	7.49
9.69	9.42	9.95	10.83	10.26
3.76	3.75	3.35	2.81	3.95
0.41	0.42	0.56	0.54	0.29
0.15	0.15	0.16	0.12	0.18
0.16	1.35	1.35	0.71	0.54
99.39	99.35	99.32	99.07	99.59
63.30	62.30	60.90	52.00	66.00
28	30	32	32	34
248	255	311	209	246
280	300	310	222	280
103	104	120	65	97
40	48	60	64	58
77	80	79	113	78
18	19	18	17	16
5	6	10	10	3
168	172	209	263	174
30	31	30	22	35
106	107	116	113	127
3	2	4	4	2
64	66	75	196	23
5	5	5	11	6.4
17	17	22	25	21.4
5	6	5	8	3
1	2	2	2	0.72
—	—	—	1	0.41
	0.702922 ± 15	0.703052 ± 10	0.704267 ± 14	0.702556 ± 30
	0.513084 ± 26	0.513097 ± 6	0.512845 ± 2	0.513129 ± 8
	+8.7	+9.0	+4.0	+9.3

[a]Analytical details are outlined in Table I. Details of sample locations are contained in Gamble *et al.* (1993a,b).

basalts from the TVZ, and the latter define an array toward higher-Sr and lower-Nd isotopes, which Gamble *et al.* (1993a) interpreted in terms of contamination due to interaction with New Zealand continental lithosphere. (4) Recent data (Gamble *et al.*, 1994) on basalts from offshore TVZ arc displaced to the right of the TVZ array, showing higher-Sr isotopes at equivalent Nd. These analyses are all for samples from the offshore arc front, delineated by the Ngatoro Ridge–Whakatane volcano lineament. Interestingly, the TVZ sample most closely approximating these data is a basalt from Ruapehu (RB in Fig. 2.15)

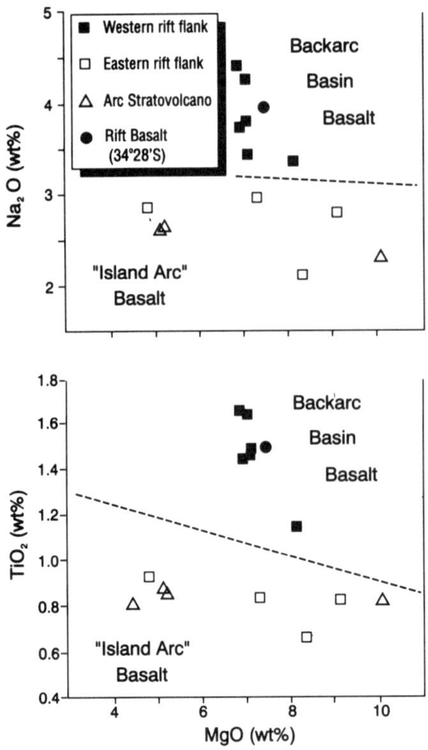

FIGURE 2.14. Variation diagram of TiO_2 and Na_2O plotted against MgO content for southern Havre Trough–Ngatoro rift basalts.

defining the volcanic front in TVZ. This observation led Gamble *et al.* (1994) to suggest that the source region for arc-front lavas might be particularly susceptible to fluxing by slab-derived fluids as a function of previous melt extraction events. Clearly this observation has important implications for the evolution of magma sources in arcs with coupled backarc basins and may be especially important in systems, such as TVZ and southern Havre Trough, where the subduction zone length scales have compressed the arc-backarc dimensions.

8. SYNTHESIS

Modern geological studies (e.g., Taylor *et al.*, 1992; Hawkins *et al.*, 1994) have shown that backarc basins are the sites of an extremely complex and heterogeneous range of extensional tectonic and magmatic processes. The SHT, as shown here, is no exception. Such heterogeneity occurring within a young rifting backarc basin associated with a tectonically "simple" subduction system is significant.

One of the more notable observations from the SHT is the compositional heterogeneity of the rift volcanism over short length scales (Gamble *et al.*, 1993b, 1994; Wright, 1993a). The emplacement of MORB-like or BABB along one rift escarpment while "arc"-

like basalts are emplaced only 15 km away along the opposite rift escarpment, apparently without transitional compositions, requires special tectonic and magmatic conditions. A two-stage process is envisaged whereby variable fluxing of variably depleted mantle wedge sources produces heterogeneous primary melts (Gamble *et al.*, 1994) which subsequently ascend and are emplaced into attenuated, subarc crust that is pervasively disrupted by rift block faults (Gamble *et al.*, 1993b). These faults apparently control the final emplacement of the lavas, with the concomitant proximity of markedly different basalt compositions. The geometry of the SHT subcrustal faults is, however, unknown. The interpretation that the surficial morphology of the rift system appears to comprise a series of asymmetric half-grabens (Fig. 2.4a), and crustal studies within other back-arc basins (e.g., Taylor *et al.*, 1991), would suggest that the crustal structure beneath the SHT rifts consists of low-angle detachment faults. However, as noted by Wright (1993a), high-angle normal faults would more easily penetrate the crust to tap heterogeneous magma melts and then act as conduits for ascending magma than would low-angle detachment faults. It is possible though that the SHT rift system has evolved to a stage where high-angle antithetic and synthetic faults are the dominant tectonic structure and they now control subarc rift magmatism. In the past, similar compositional heterogeneity recognized within the rock record would most likely be interpreted as separate "terranes" rather than as constituent margins of the same rift system.

The association of BABB within the rifting SHT, as in the backarc rifts of the Izu–Bonin system (Fryer *et al.*, 1990; Hochstaedter *et al.*, 1990a), and early history of the Lau Basin (Hawkins *et al.*, 1994) demonstrate that MORB-like magmatism can occur within prespreading rift grabens of young backarc basins. These data support the contention of Taylor and Karner (1983) and Fryer *et al.* (1990) that early models of the temporal evolution of backarc magmatism, whereby BABB occur only in mature spreading backarc basins, are not universally valid.

A further significant aspect of SHT tectonic and magmatic evolution is the position of the present southern Kermadec arc front. Two observations are noteworthy. The first is that the position of the present volcanic front sited within the "backarc" region (Figs. 2.8–2.10) indicates that the spatial relationship of the arc and backarc is nebulous (Wright, 1994). A similar indistinct spatial relationship of the arc front and regions of extension and subsidence occurs within the TVZ to the south (Gamble *et al.*, 1990; Wright, 1992). The second observation reiterates an earlier conclusion about fine-scale compositional heterogeneity of the arc-backarc volcanism. The Kermadec arc stratovolcanoes, and concomitant basaltic-andesite arc volcanism, lie within 30–40 km of the SHT rift graben erupting BABB (Wright, 1993a; Gamble *et al.*, 1993b). Again the implications of these observations for interpreting the tectonic settings and subduction polarity of backarc basins within the rock record are significant.

The essential observation from the SHT is that the influence of the geometry of subduction is fundamental. It would appear that spatial parameters of a subduction system (e.g., the volume of the underlying mantle wedge, distance between the rift axis and arc front, distance between the arc front and subduction front, rate of subduction, rate of backarc widening, rate of trench rollback, length and terminal depth of the subducted slab) are as important, or more important, than the temporal evolution of a backarc basin. When these parameters are shortened, as in a proximal arc-backarc system like the TVZ and SHT, their influence may be especially important.

TABLE III
Representative Chemical Analyses of Basalts from Offshore TVZ[a]

Sample no.	VUW11626	VUW11622	VUW11628	VUW11628	VUW11630	VUW11633
Field no.	X191/3	X190/1	X196/1A	X196/1B	X201/2	X204/1B
Locality	W. Alderman T	Ngatoro R.	Ngatoro R.	Ngatoro R.	Ngatoro R.	Ngatoro R.
SiO_2	49.14	50.55	49.57	49.96	48.41	48.41
TiO_2	1.25	0.80	0.99	0.89	0.93	0.87
Al_2O_3	18.18	17.77	19.15	19.10	20.01	19.23
Fe_2O_3	1.46	1.30	1.28	1.31	1.20	1.33
FeO	9.74	8.67	8.51	8.73	8.00	8.86
MnO	0.21	0.19	0.18	0.18	0.15	0.17
MgO	4.83	5.75	3.94	3.91	5.17	5.44
CaO	10.70	11.34	12.05	11.88	12.49	12.34
Na_2O	2.78	2.33	2.48	2.39	2.26	2.13
K_2O	0.39	0.45	0.41	0.37	0.37	0.27
P_2O_5	0.15	0.07	0.08	0.08	0.10	0.07
LOI	0.35	0.09	0.19	0.32	0.39	0.19
Total	99.18	99.31	98.83	99.12	99.48	99.31
Mg#	46.9	54.2	45.2	44.4	53.5	52.2
Sc	38	38	47	43	35	41
V	429	315	425	357	320	334
Cr	25	35	15	11	58	17
Ni	12	20	5	4	22	9
Cu	43	102	40	35	72	50
Zn	104	85	108	73	83	87
Ga	21	17	20	19	20	19
Rb	8	11	5	5	7	5
Sr	320	257	320	315	350	276
Y	22	20	19	18	19	17
Zr	57	52	51	50	47	40
Nb	1	1	2	2	3	2
Ba	140	200	229	221	165	168
La	6	4	6	5	7	3
Ce	22	18	20	18	19	17
Pb	5	7	5	6	6	6
Th	1	1	1	4	—	—
U	0.4	0.4	0.4	0.4	0.4	0.4
$^{87}Sr/^{86}Sr$		0.705128 ± 1	0.705128 ± 1	0.705132 ± 1		0.705297 ± 1
		2	4	0		4
$^{143}Nd/^{144}Nd$		0.512908 ± 9	0.512843 ± 7	0.512824 ± 3		0.512196 ± 7
eps Nd		+5.3	+4.0	+3.6		+3.7

TABLE III
(*Continued*)

VUW11613 X174/1 So Colville	VUW11585 X146/1 So Colville	VUW11613 X174/2 So Colville	VUW11627 X194/1 Raukumara Pl	VUW11652 A-002 Volcanolog	VUW11653 B-002 Volcanolog
48.29	47.05	49.20	53.62	51.33	51.05
0.94	1.05	0.97	0.76	0.81	0.81
16.54	17.89	16.26	17.85	17.90	18.05
1.36	1.20	1.40	1.08	1.10	1.13
9.09	8.06	9.36	7.18	7.34	7.54
0.44	0.20	0.30	0.16	0.15	0.15
6.37	7.51	6.52	4.98	4.97	5.02
11.51	11.66	11.70	9.64	10.35	10.39
2.62	2.71	2.73	2.59	2.28	2.30
0.29	0.31	0.27	0.56	0.82	0.80
0.07	0.18	0.06	0.11	0.25	0.26
1.30	1.61	0.21	0.61	1.34	1.48
98.82	99.25	98.98	99.14	99.64	99.98
56.5	62.4	55.4	55.3	55.7	54.3
43	36	42	31	35	37
333	292	345	264	257	265
58	173	57	45	41	41
83	69	53	16	11	10
139	46	130	35	26	18
102	99	99	79	85	83
18	17	19	18	20	19
6	3	5	11	27	26
259	299	258	252	305	300
14	23	16	19	23	22
40	75	38	76	94	93
1	4	0	3	4	4
126	95	133	279	312	293
2	10	6	6	10	11
13	35	12	15	25	24
7	8	5	7	9	7
1	2	1		5	5
0.3	0.3	0.4	0.2	0.2	0.2

[a]Analytical details are outlined in Table I. Details of sample locations are contained in Gamble *et al.* (1993a,b).

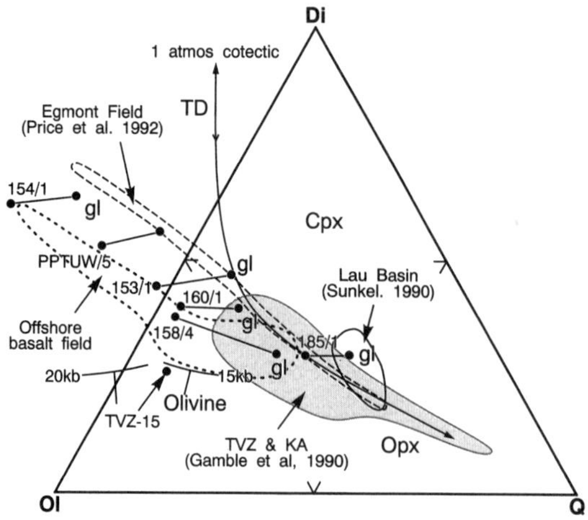

FIGURE 2.15. CaO—MgO—Al_2O_3—SiO_2 (CMAS) pseudo-ternary projection from plagioclase onto the plane Di-Ol-Qtz (Walker *et al.*, 1979) for basalts and coexisting glasses (where available) from the Kermadec–New Zealand arc-backarc system. Cotectics for 1 atm, 15 kb, and 20 kb are shown. TD = thermal divide on 1-atm cotectic. Fields for Egmont lavas (Price *et al.*, 1992), offshore TVZ, Lau Basin (Sunkel, 1990), and the majority of Kermadec arc and TVZ basalts are indicated.

Acknowledgments

Thanks are due to the officers and crew of the RV *Rapuhia*, RRS *Charles Darwin*, and RV *Akademik Lavrenteyev* for their forbearance while at sea. Special thanks are due to the Australian Overseas Telecommunications Commission (OTC) for allowing use of the PacRimEast SYS09 cable route survey data. ICW thanks the Royal Society (UK) and the Royal Society of New Zealand for financial assistance toward study leave at IOSDL (UK), where much of the manuscript was completed. I. Smith, J. Baker, R. Christie, and J. Woodhead provided much useful discussion. K. Majorhazi and P. Bennett produced the figures. This work was funded by New Zealand Foundation Research Science Technology contracts 93-VIC-8308 (JAG) and 93-WAR-32-384 (ICW).

REFERENCES

Blackmore, N. A., and Wright, I. C. 1995. *Southern Kermadec Volcanoes. 1:400,000*, New Zealand Oceanographic Institute, Miscellaneous Ser. No. 71, National Institute of Water and Atmospheric Research, Wellington.

Brothers, R. N. 1967. Andesite from Rumble II volcano, Kermadec Ridge, southwest Pacific, *Bull. Volcanol.* **31:**17–19.

Caress, D. W. 1991. Structural trends and backarc extension in the Havre Trough, *Geophys. Res. Lett.* **18:**853–856.

Cole, J. W. 1978. Tectonic setting of Mayor Island volcano (Note), *N.Z. J. Geol. Geophys.* **21:**645–647.

Cole, J. W. 1979. Structure, petrology and genesis of Cenozoic volcanism, Taupo volcanic zone, New Zealand—a review, *N.Z. J. Geol. Geophys.* **22:**631–657.

Cole, J. W. 1986. Distribution and tectonic setting of late Cenozoic volcanism in New Zealand. in: *Late Cenozoic Volcanism in New Zealand* (I. E. M. Smith, ed.), *Roy. Soc. N.Z. Bull.* **23:**7–20.

Cole, J. W., and Lewis, K. B. 1981. Evolution of the Taupo–Hikurangi subduction system, *Tectonophysics* **72:**1–21.

Darby, D. J., and Williams, R. O. 1991. A new geodetic estimate of deformation in the central volcanic region of the North Island, New Zealand, *N.Z. J. Geol. Geophys.* **34:**127–136.

Davey, F. J. 1982. The structure of the South Fiji Basin, *Tectonophysics* **87:**185–241.

Davey, F. J., and Robinson, A. G. 1978. Cook (lst ed.) *Magnetic Total Force Anomaly Map Oceanic Series 1:1,000,000*, New Zealand Department of Scientific and Industrial Research, Wellington.

de Mets, C., Gordon, R. G., and Stein, S. 1990. Current plate motions, *Geophys. J. Int.* **101:**425–478.

Dupont, J. 1988. The Tonga and Kermadec ridges, in *The Ocean Basins and Margins*, Vol. 7B, *The Pacific Ocean* (A. E. M. Nairn, F. G. Stehli, and S. Uyeda, eds.), pp. 375–409, Plenum Press, New York.

Ellam, R. M., and Hawkesworth, C. J. 1988. Elemental and isotope variations in subduction related basalts: evidence for a three component model, *Contrib. Mineral. Petrol.* **98:**72–80.

Ewart, A. 1968. Geochemistry of the pantellerites of Mayor Island, New Zealand, *Contrib. Mineral. Petrol.* **17:**116–140.

Ewart, A., Brothers, R. N. and Mateen, A. 1977. An outline of the geology and geochemistry and the possible petrogenetic evolution of the volcanic rocks of the Tonga–Kermadec–New Zealand island arc, *J. Volcanol. Geotherm. Res.* **2:**205–250.

Ewart, A., and Hawkesworth, C. J. 1987. The Pleistocene to Recent Tonga–Kermadec arc lavas: interpretation of new isotope and rare earth data in terms of a depleted source model, *J. Petrol.* **28:**495–530.

Fryer, P., Taylor, B., Langmuir, C. H., and Hochstaedter, A. G. 1990. Petrology and geochemistry of lavas from the Sumisu and Torishima backarc rifts, *Earth Planet. Sci. Lett.* **100:**161–178.

Gamble, J. A., Smith, I. E. M., Graham, I. J., Kokelaar, B. P., Cole, J. W., Houghton, B. F., and Wilson, C. J. N. 1990. The petrology, phase relations and tectonic setting of basalts from the Taupo volcanic zone, New Zealand, and the Kermadec Island arc–Havre Trough, S.W. Pacific, *J. Volcanol. Geotherm. Res.* **43:** 235–270.

Gamble, J. A., Smith, I. E. M., McCulloch, M. T., Graham, I. J., and Kokelaar, B. P., 1993a. The geochemistry and petrogenesis of basalts from the Taupo volcanic zone and Kermadec Island arc, S.W. Pacific, *J. Volcanol. Geotherm. Res.* **54:**265–290.

Gamble, J. A., Wright, I. C., and Baker, J. A. 1993b. Seafloor geology and petrology of the oceanic to continental transition zone of the Kermadec–Havre–Taupo Volcanic arc system, New Zealand, *N.Z. J. Geol. Geophys.* **36:**417–435.

Gamble, J. A., Wright, I. C., Woodhead, J. D., and McCulloch, M. T. 1994, Arc and backarc geochemistry in the southern Kermadec arc-Ngatoro Basin and offshore Taupo volcanic zone, in *Volcanism Associated with Extension at Consuming Plate Margins* (J. L. Smellie, ed.), pp. 193–212, Geol. Soc. Lond. Spec. Publ. 81.

Gill, J. W. 1981. *Orogenic Andesites and Plate Tectonics*, Berlin, Springer–Verlag.

Gillies, P. N., and Davey, F. J. 1986. Seismic reflection and refraction studies of the Raukumara forearc basin, New Zealand, *N.Z. J. Geol. Geophys.* **29:**391–403.

Graham, I. J., and Cole, J. W. 1991. Petrogenesis of andesites and dacites of White Island volcano, Bay of Plenty, New Zealand, in the light of new geochemical and isotopic data, *N.Z. J. Geol. Geophys.* **34:**303–315.

Graham, I. J., and Hackett, W. R. 1987. Petrology of calc-alkaline lavas from Ruapehu volcano and related vents, Taupo volcanic zone, New Zealand, *J. Petrol.* **28:**531–567.

Hatherton, T. H. 1980. Shallow seismicity in New Zealand 1956–75, *J. Roy. Soc. N.Z.* **10:**19–25.

Hawkesworth, C. J., Gallagher, K., Hergt, J. M., and McDermott, F. 1993. Trace element fractionation processes in the generation of island arc basalts, *Phil. Trans. R. Soc. London, A* **342:**179–191.

Hawkins, J. W. 1976. Petrology and geochemistry of basaltic rocks of the Lau Basin, *Earth Planet. Sci. Lett.* **28:**283–297.

Hawkins, J. W., and Melchior, J. T. 1985. Petrology of the Mariana Trough and Lau Basin basalts, *J. Geophys. Res.* **90:**11,431 11,468.

Hawkins, J. W., Parson, L. M., Allan, J., *et al.* 1994. Introduction to the Scientific Results of Leg 135: Lau Basin-Tonga Ridge Drilling Transect *Proc. ODP, Sci. Results*, 135, Ocean Drilling Program, College Station, TX, pp. 3–5.

Hergt, J. M. and Farley, K. 1994. Major, trace element, and isotope (Pb, Sr and Nd) variations in site 834 basalts: Implications for the initiation of backarc opening, in *Proc. ODP, Sci. Results*, 135 (J. W. Hawkins, L. M. Parson, J. Allan, *et al.*, eds.), pp. 471–485, Ocean Drilling Program, College Station, TX.

Hergt, J. M. and Hawkesworth, C. J. 1994. The Pb, Sr and Nd isotopic evolution of the Lau Basin: Implications for mantle dynamics during backarc opening, in *Proc. ODP, Sci. Results*, 135 (J. W. Hawkins, L. M. Parson, J. Allan *et al.*, eds.), pp. 505–517, Ocean Drilling Program, College Station, TX.

Hochstaedter, A. G., Gill, J. B., Kusakabe, M., Newman, S., Pringle, M., Taylor, B., and Fryer, P. 1990a. Volcanism in the Sumisu Rift. I: Major element, volatile, and stable isotope geochemistry, *Earth Planet. Sci. Lett.* **100:**179–194.

Hochstaedter, A. G., Gill, J. B., and Morris, J. D. 1990b. Volcanism in the Sumisu Rift. II: Subduction and non-subduction related components, *Earth Planet. Sci. Lett.* **100:**283–297.

Houghton, B. F., Weaver, S. D., Wilson, C. J. N., and Lanphere, M. A. 1992. Evolution of a Quaternary peralkaline volcano: Mayor Island, New Zealand, *J. Volcanol. Geotherm. Res.* **51:**217–236.

Ito, E., White, W. M., and Copel, C. 1987. The O, Sr, Nd, and Pb isotope composition of MORB, *Chem. Geol.* **62**:157–176.

Karig, D. E. 1970a. Ridges and basins of the Tonga-Kermadec Island arc system, *J. Geophys. Res.* **75**:239–254.

Karig, D. E. 1970b. Kermadec arc–New Zealand tectonic confluence, *N.Z. J. Geol. Geophys.* **13**:21–29.

Karig, D. E. 1974. Eolution of arc systems in the western Pacific, *Ann. Rev. Earth Planet. Sci.* **2**:51–75.

Kibblewhite, A. C., and Denham, R. N. 1967. The bathymetry and total magnetic field of the South Kermadec Ridge seamounts, *N.Z. J. Sci.* **10**:53–67.

Klaus, A., Taylor, B., Moore, G. F., Murakami, F., and Okamura, Y. 1992. Back-arc rifting in the Izu–Bonin island arc: Structural evolution of Hachijo and Aoga rifts, *The Island Arc* **1**:16–31.

Klein, E. M., Langmuir, C. H., and Staudigel, H. 1991. Geochemistry of basalts from the southeast Indian Ridge, *J. Geophys. Res.* **92**:8089–8115.

Korsch, R. J. and Wellman, H. W. 1988. The geological evolution of New Zealand and the New Zealand region, in *The Ocean Basins and Margins*, Vol. 7B, *The Pacific Ocean* (A. E. M. Nairn, F. G. Stehli, and S. Uyeda, eds.), pp. 411–482, Plenum Press, New York.

Latter, J. H., and Hall, L. 1986. Rumble III volcano, *Bull. Global Volcan. Network* **11**(7):15.

Le Bas, M. J., Le Maitre, R. W., Streckeisen, A., Zanettin, B. 1986. A classification of igneous rocks based upon the alkali-silica diagram, *J. Petrol.* **27**:745–750.

Lewis, K. B., and Pantin, H. M. 1984. Intersection of a marginal basin with a continent: Structure and sediments of the Bay of Plenty, New Zealand, in *Marginal Basin Geology: Volcanic and Associated Sedimentary and Tectonic Processes in Modern and Ancient Marginal Basins* (B. P. Kokelaar and M. F. Howells, eds.), pp. 121–135, Geol. Soc. Lond. Spec. Publ. 16.

Loock, G., McDonough, W. L., Goldstein, S. L. and Hofmann, A. W. 1990. Isotopic compositions of volcanic glasses from the Lau Basin, *Mar. Mining* **9**:235–245.

Malahoff, A., Feden, R. H., and Fleming, H. S. 1982. Magnetic anomalies and tectonic fabric of marginal basins north of New Zealand, *J. Geophys. Res.* **87**:4109–4125.

McCulloch, M. T. and Gamble, J. A. 1991. Geochemical and geodynamical constraints on subduction zone magmatism, *Earth Planet. Sci. Lett.* **102**:358–374.

Minster, J. B., and Jordan, T. H. 1978. Present day plate motions, *J. Geophys. Res.* **83**:5331–5354.

Morris, J. D., Leeman, W. P., and Tera, F. 1990. The subducted component in island arc lavas: Constraints from Be isotopes and B-Be systematics, *Nature* **344**:31–36.

Packham, G. H., and Falvey, D. A. 1971. An hypothesis for the formation of marginal seas in the western Pacific, *Tectonophysics* **11**:79–109.

Parson, L. M., and Hawkins, J. W. 1994. Two-stage ridge propagation and the geological history of the Lau backarc basin, in *Proc. ODP, Sci. Results*, 135 (J. W. Hawkins, L. M. Parson, J. Allan *et al.*, eds.), pp. 819–828, Ocean Drilling Program, College Station, TX.

Parson, L. M., Pearce, J. A., Murton, B. J., Hodkinson, R. A, Bloomer, S., Ernewein, M., Hugget, Q. J., Miller, S., Johnson, L., Rodda, P., and Helu, S. 1990. Role of ridge jumps and ridge propagation in the tectonic evolution of the Lau Basin backarc basin, southwest Pacific, *Geology* **18**:470–473.

Pearce, J. A. 1983. Role of sub-continental lithosphere in magma genesis at active continental margins, in *Continental Basalts and Mantle Xenoliths* (C. J. Hawkesworth and M. J. Norry, eds.), pp. 231–249, Shiva, Nantwich.

Pelletier, B. 1990. Tectonic erosion, accretion, backarc extension and slab length along the Kermadec subduction zone (Abstract), p. 65, *Fifth Circum-Pacific Energy and Mineral Resources Conf.*, Honolulu.

Pelletier, B., and Dupont, J. 1990. Tectonic erosion, accretion, extension arriere-arc et longuer du plan de subduction le long de la marge active des Kermadec, Pacifique Sud-Ouest, *C.R. Acad. Sci. Paris* **310**(II):11657–11664.

Pelletier, B., and Louat, R. 1989. Seismotectonics and present-day relative plate motions in the Tonga–Lau and Kermadec–Havre region, *Tectonophysics* **165**:237–250.

Price, R. C., McCulloch, M. T., Smith, I. E. M., and Stewart, R. B. 1992. Pb-Nd-Sr isotopic compositions and trace element characteristics of young volcanic rocks from Egmont volcano, New Zealand and comparisons with basalts and andesites from the Taupo volcanic zone, *Geochim. Cosmochim. Acta* **56**:941–953.

Reyners, M. 1989. New Zealand seismicity 1964–1987: An interpretation, *N.Z. J. Geol. Geophys.* **32**:307–315.

Saunders, A. D., and Tarney, J. 1979. The geochemistry of basalts from a backarc spreading centre in the east Scotia Sea, *Geochim. Cosmochim. Acta* **43**:555–572.

Saunders, A. D. and Tarney, J. 1991. Back-arc basins, in *Oceanic Basalts* (P. A. Floyd, ed.), pp. 219–263, Blackie, Glasgow and London.

Sclater, J. G., Hawkins, J. W., Mammerickx, J., and Chase, C. G. 1972. Crustal extension between the Tonga and Lau ridges: petrologic and geophysical evidence, *Geol. Soc. Am. Bull.* **83**:505–518.

Sclater, J. G., Jaupart, C., and Galson, D. 1980. The heat flow through oceanic and continental crust and heat loss of the earth, *Rev. Geophys. Space Phys.* **18**:261–312.

Seafloor Surveys International. 1990. PacRimEast Submarine Telecommunications Cable Route Survey, Auckland to Oahu, Vol. I (Charts 1–23), Seattle.

Seafloor Surveys International. 1993. PacRimEast Route Diversion Survey. Detailed Survey Charts of the Seafloor, Vol. II (Charts 21–40), Seattle.

Sinton, J., and Fryer, P. 1987. Mariana Trough lavas from 18°N: Implications for the origin of backarc basin basalts, *J. Geophys. Res.* **92**:12,782–802.

Smith, I. E. M., and Brothers, R. N. 1988. Petrology of Rumble seamounts, southern Kermadec Ridge, southwest Pacific, *Bull. Volcanol.* **50**:139–147.

Spiegelman, M., and McKenzie, D. P. 1987. Simple 2-D models for melt extraction at mid-ocean ridges and island arcs, *Earth Planet. Sci. Lett.* **83**:137–152.

Stern, T. A. 1987. Asymmetric backarc spreading, heat flux and structure associated with the central volcanic region of New Zealand, *Earth Planet. Sci. Lett.* **85**:265–276.

Sunkel, G. 1990. Origin of petrological and geochemical variations of Lau Basin lavas (SW Pacific), *Mar. Mining* **9**:205–234.

Sykes, L. R., Issacks, B., and Oliver, J. 1969. Spatial distribution of deep and shallow earthquakes of small magnitude in the Fiji–Tonga region, *Bull. Seismol. Soc. Am.* **59**:1093–1113.

Taylor, B. 1992. Rifting and volcanic-tectonic evolution of the Izu–Bonin–Mariana arc. in *Proc. ODP, Sci. Results*, 126 (B. Taylor, K. Fujioka, *et al.*, eds.), pp. 627–651, Ocean Drilling Program, College Station, TX.

Taylor, B., Brown, G. R., Fryer, P., Gill, J., Hochstaedter, F., Hotta, H., Langmuir, C., Leinen, M., Nishimura, A., and Urabe, T. 1990. Alvin-SeaBeam studies of the Sumisu rift, Izu–Bonin arc, *Earth Planet. Sci. Lett.* **100**:127–147.

Taylor, B., Fujioka, K., *et al.* 1992. *Proc. ODP, Sci. Results*, 126, Ocean Drilling Program, College Station, TX.

Taylor, B., and Karner, G. D. 1983. On the evolution of marginal basins, *Rev. Geophys. Space Phys.* **21**:1727–1741.

Taylor, B., Klaus, A., Brown, G. R., and Moore, G. F. 1991. Structural development of Sumisu Rift, Izu–Bonin arc, *J. Geophys. Res.* **96**:16,113–16,129.

Taylor, S. R., and McLennan S. M. 1985. *The Continental Crust: Its Composition and Evolution*, Blackwell Scientific, Oxford.

van der Linden, W. J. M. 1967. Structural relationships in the Tasman Sea and southwest Pacific Ocean, *N.Z. J. Geol. Geophys.* **10**:1250–1301.

Walcott, R. I. 1984. The kinematics of the plate boundary zone through New Zealand: A comparison of short and long-term deformations, *Geophys. J. Roy. Astron. Soc.* **79**:613–633.

Walker, D., Shibata, T., and De Long, S. E. 1979. Abyssal tholeiites from the Oceanographer Fracture Zone. II: phase equilibria amd mixing, *Contrib. Mineral. Petrol.* **70**:111–125.

Watanabe, T., Langseth, M. G., and Anderson, R. N., 1977. Heatflow in backarc basins of the western Pacific, in *Island Arcs, Deep Sea Trenches and Back-Arc Basins* (M. Talwani and W. C. Pitman III, eds.), Vol. 1, pp. 137 167, Maurice Ewing Series, American Geophysical Union, Washington, DC.

Weidicke, M., and Collier, J. 1993. Morphology of the Valu Fa spreading ridge in the southern Lau Basin, *J. Geophys. Res.* **98**:11,769–11,782.

Whiteford, P. C. 1990. Heat flow measurements in the Bay of Plenty, New Zealand, Report of the February 1988 Vulkanolog Cruise, Research Report No. 221, Geophysics Division, New Zealand Department of Scientific and Industrial Research, Wellington.

Whiteford, P. C. 1992. Heat flow measurements in and near the Ngatoro Basin, Bay of Plenty, New Zealand, Report of the February 1990 Vulkanolog Cruise, Research Report No. 240, Geophysics Division, New Zealand Department of Scientific and Industrial Research. Wellington.

Wilson, M. 1989. *Igneous Petrogenesis: A Global Tectonic Approach*, Unwin Hyman, London.

Woodhead, J. D., Eggins, S., and Gamble, J. A. 1993. High field strength and transition element systematics in island arc and back-arc basin basalts: Evidence for multi-stage melt extraction and ultra-depleted mantle wedge, *Earth Planet. Sci. Lett.* **114**:491–504.

Wright, I. C. 1990. *Bay of Plenty–Southern Havre Trough Physiography, 1:400,000*, New Zealand Oceanographic Institute Misc. Ser. No. 68, New Zealand Department of Scientific and Industrial Research, Wellington.

Wright, I. C. 1992. Shallow structure and active tectonism of an offshore continental spreading system: The Taupo volcanic zone, New Zealand, *Mar. Geol.* **103**:287–309.

Wright, I. C. 1993a. Pre-spread rifting and heterogeneous volcanism in the southern Havre Trough backarc basin, *Mar. Geol.* **113**:179–200.

Wright, I. C. 1993b. Southern Havre Trough–Bay of Plenty (New Zealand): Structure and seismic stratigraphy of an active backarc basin complex, in *South Pacific Sedimentary Basins*. 2. *Sedimentary Basins of the World* (P. F. Ballance, ed.), pp. 195–211, Elsevier, Amsterdam.

Wright, I. C. 1994. Nature and tectonic setting of the southern Kermadec submarine arc volcanoes: An overview, *Mar. Geol.* **118**:217–236.

Wright, I. C., Carter, L., and Lewis, K. B. 1990. GLORIA survey of the oceanic-continental transition of the Havre–Taupo backarc basin, *Geo-Marine Lett.* **10**:59–67.

3

The Geology of the Lau Basin

James W. Hawkins, Jr.

ABSTRACT

The Lau Basin comprises oceanic crust that separates the remnant Lau Ridge volcanic arc from the active Tofua arc. The basin is situated above mantle exhibiting strong seismic wave attenuation; it overlies the west dipping seismic zone of the Tonga Trench subduction system and is presently opening at rates as high as 1.6 cm/yr. Seaward rollback of the trench axis, coupled with upwelling high temperature mantle diapirs, is proposed as the main driving force to cause crustal extension in this region of oceanic plate convergence. The trapezoidal shape of the basin suggests that it has opened progressively from north to south; the trace of the Louisville seamount chain may help constrain the southern apex of the basin. The basin has opened in two tectonic styles. Initially, beginning at about 6 Ma, attenuation and rifting of the forearc or outer part of the Lau Ridge formed a series of half-grabens that were partly sedimented and received basaltic flows. This extension and magmatism overlapped with Lau Ridge arc volcanism. The second stage of opening (about 5.5–5 Ma) was promoted by a southward-propagating rift system that formed new crust by seafloor spreading. A second propagator (about 1.5 Ma) has overtaken the first and forms an overlapping ridge system. A small three-limbed spreading system in the northeastern basin forms a triple junction. A fourth spreading system is recognized in the northwestern part of the basin.

In spite of the suprasubduction zone (SSZ) setting and the proximity to volcanic arcs, the basin is dominated by mid-ocean ridge basalt (MORB)-like crust. Isotope data indicate that both Pacific and Indian MORB-source mantle have been involved; Pacific source formed some of the oldest crust, the Indian source feeds the modern spreading ridges. Helium-isotope data suggest a Samoan "plume" component is important in generating seamounts of the northern part of the basin. Lau Basin crust exhibits a general depletion in high-field-strength elements, and some samples show varied enrichment in large-ionic-radius lithophile elements, but the source must be like the MORB-source in most respects. However, mixing of MORB-like and arclike magmas has been important in the basin's history. Locally, crust is transitional to arc compositions but this occurs largely in areas of older crust or where rifting first encountered older crust/mantle. Heterogeneity of the SSZ

James W. Hawkins, Jr. • Geological Research Division, Scripps Institution of Oceanography, La Jolla, California 92093-0220.

Backarc Basins: Tectonics and Magmatism, edited by Brian Taylor, Plenum Press, New York, 1995.

mantle is indicated by the presence of intrabasinal arc-composition volcanoes erupted nearby to ridges where MORB crust was forming. The data suggest that a wave of arc-composition volcanism migrated across the basin as it opened; presently it forms the Tofua arc, which is a relatively young feature. The variety of mantle sources involved, as expressed in the range of chemical signatures in the crustal rocks, points to the complexity of backarc magmas. They reflect the complexity of SSZ mantle. No simple definition of backarc basin basalt is possible.

1. BACKARC BASINS

The western Pacific Ocean basin is rimmed by an array of island-arc systems and their related trenches. Alfred Wegener (1929) proposed that

> the island arcs, and particularly the eastern Asiatic ones, are marginal chains which were detached from continental masses, when the latter drifted westwards and remained fast in the old seafloor, which was solidified to great depths. Between the arcs and the continental margins later, still-liquid areas of seafloor were exposed as windows.

These "windows" of seafloor are the backarc basins. Most of them are relatively shallow (typically less than 3000 m deep) regions of ocean crust that are considerably younger than the seafloor being subducted at the adjacent oceanic trenches. They are situated between inactive, remnant volcanic arcs and the active volcanic arcs that form in response to the subduction of oceanic lithosphere. The Lau Basin (Fig. 3.1) is a well-studied, classic example.

This discussion summarizes the geology of the Lau Basin as an interim report on our understanding of its evolution. The discussion focuses mainly on the tectonic setting and the petrology of Lau Basin crust and draws on the wealth of information on these topics that has been acquired in the last 25 years. Major new insights have been gained as a result of geological long-range inclined asdic (GLORIA) imagery (Parson *et al.*, 1990) and data from six holes drilled in the western part of the Lau Basin on Ocean Drilling Project (ODP) Leg 135 (Parson *et al.*, 1992; Hawkins *et al.*, 1994).

Several of our new findings may be summarized as follows:

1. The Lau Basin has formed by two tectonic styles of opening. Initially the western part of the basin formed by crustal extension and rifting—not by seafloor spreading. The extension was accompanied by magmatism that partly filled rift basins.
2. The second phase of opening was by seafloor spreading. Spreading developed on propagating rifts that started from a transform fault boundary.
3. Basaltic magmatism in the rift basins and on the propagating rifts was contemporary with arc activity on the (present) remnant arc.
4. A triple junction is forming new crust in the northeastern part of the basin.
5. As the basin opened by extension, arc-composition seamounts formed within the basin close by to MORB-like eruptions.
6. The basin is in a suprasubduction zone setting, yet the new crust is dominated by MORB-like magma compositions. Both an Indian mantle MORB source and a Pacific mantle MORB source are recognized with isotope data.

These new views on the geological evolution of the Lau Basin are useful in understanding other (SSZ) convergent plate margins with arc-backarc basin systems. Our better understanding of the petrologic and tectonic complexity of these systems also gives insight

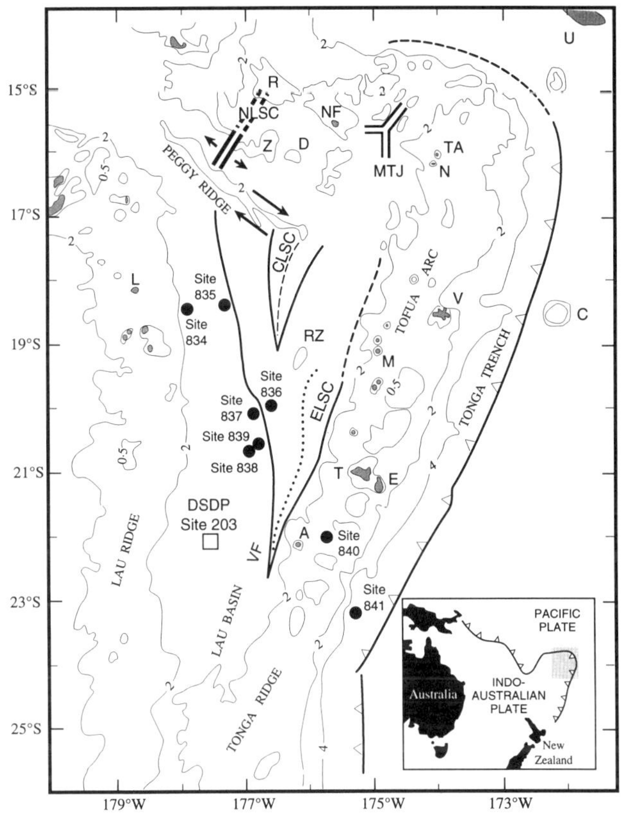

FIGURE 3.1. Lau Basin and Tonga Trench arc system showing locations of Leg 135 Sites 834–841 and Deep Sea Drilling Project (DSDP) Site 203. Major features shown include sites of modern volcanism in the Lau Basin: Mangatolu triple junction (MTJ) (Hawkins, 1989; Nilsson *et al.*, 1989; Nilsson, 1993); central Lau spreading center (CLSC), eastern Lau spreading center (ELSC), relay zone (RZ), and Valu Fa Ridge (VF) (Parson *et al.*, 1990; von Stackelberg and von Rad, 1990). Islands and shoals are Ata (A), Capricorn seamount (C), Donna seamount (D), 'Eua (E), Lakemba (L), Metis Shoal (M), Niuataputopu (N), Niuafo'ou (NF), Rochambeau Bank (R), Tongatapu (T), Tafahi (TA), Upolu, Western Samoa (U), Vava'u (V), Zephyr Shoal (Z). Contour interval in kilometers.

to the type of oceanic lithosphere we find preserved in ophiolites; namely, most of them probably come from SSZ settings.

Although backarc-arc-forearc-trench systems are situated in zones of lithosphere shortening, between plates with opposing relative motion, abundant evidence exists that these systems are loci of crustal extension and formation of new crust as illustrated in Fig. 3.2. Ocean Drilling Program (ODP) studies have provided good evidence that the "new" crust of backarc basins has formed contemporaneously with the volcanic island arcs that bound them (Hawkins *et al.*, 1991; Fryer and Pearce, 1992; Taylor, 1992; Taylor *et al.*, 1992; Hawkins *et al.*, 1994). Seafloor spreading has been an important factor in their evolution, but lithosphere extension and rifting has also been of major importance (Hawkins, 1994; Parson and Hawkins, 1994). Magmas that form the volcanic arcs and the crust of backarc basins form in SSZ settings as a result of fractional melting of upwelling mantle diapirs. These melts are largely derived from the mantle wedge lying above the subduction zone—

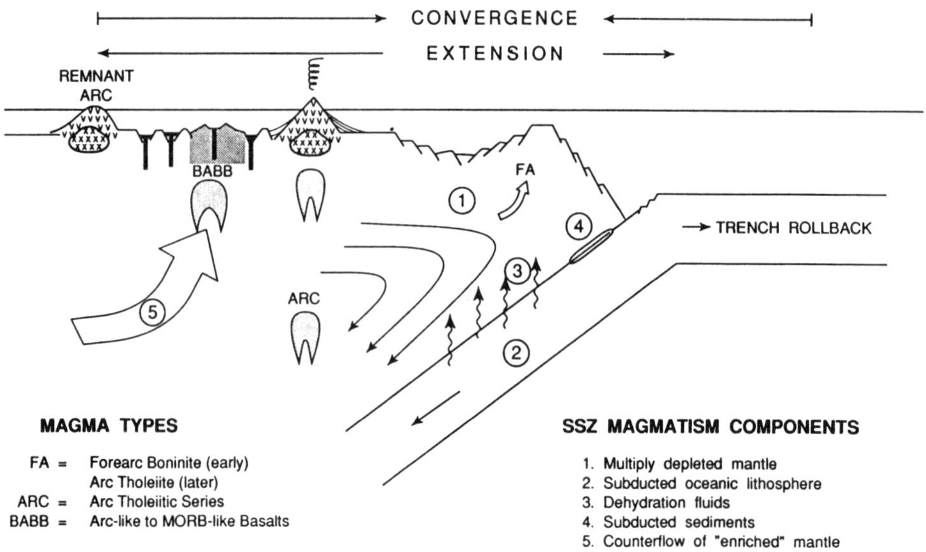

FIGURE 3.2. Schematic cross section of an intraoceanic convergent margin showing the remnant arc, backarc basin, active arc, and forearc (FA). Sites of magmatic activity are shown by arrows and shaded crescentic patterns. Numbers refer to potential components that may contribute to SSZ magmatism.

that is, the SSZ mantle (Pearce *et al.*, 1984). Additional contributions may come from the counterflow of mantle into the SSZ region and from the subducted lithosphere plate (e.g., Tatsumi *et al.*, 1986; Takazawa *et al.*, 1992). The result is new crust having varied enrichments in low-partition-coefficient elements relative to crust formed at oceanic spreading centers.

More than 40 years after Wegener's (1929) classic paper, several nearly contemporary papers by Karig (1970, 1971), Packham and Falvey (1971), Sleep and Toksoz (1971), and Moberly (1972) presented the first discussions of the geometry and tectonic setting of western Pacific backarc basin-arc-trench systems. The authors proposed that new crust had been formed in a zone of extension between an inactive remnant arc and an active volcanic island arc; a kinematic relation between the subduction process and the backarc basin extension was postulated, but neither the mechanism nor the evolution were well understood.

Various attempts to model and explain backarc extension have been made. These may be broadly separated into models invoking mantle diapirism, induced aesthenospheric convection, and global plate kinematics. Taylor and Karner (1983) discussed these general models and concluded that none were adequate. Chase (1978) proposed that motion of the upper plate away from the trench resulted in backarc extension, whereas Hynes and Mott (1985) attributed extension to seaward migration of the subducted plate. Sleep and Toksoz (1971) called on induced asthenospheric convection above the subduction zone. Karig (1971) proposed mantle diapirism triggered by the subducted slab. Rollback of the trench axis (Elsasser, 1971) is rejected by some authors (e.g., Hynes and Mott, 1985) but is used to explain the observations by others (e.g., Chase, 1978; Carlson and Melia, 1984).

The geologic evidence shows a relation between subduction, crustal extension, and the production of backarc magmas. The MORB-like nature of backarc magmagenesis requires mantle upwelling under the backarc basins. The upwelling and lithosphere extension are either a cause or the result of seaward rollback of the trench (Elsasser, 1971; Uyeda and Kanamori, 1979); the Tonga Trench may be migrating eastward at up to 10 cm/yr (Carlson and Melia, 1984). Furlong *et al.* (1982) proposed that an increase in subduction velocity may promote backarc spreading. Ribe (1989) presented a model showing how backarc spreading, regardless of how it started, would induce mantle flow above the subduction zone. He concluded that this helps account for the observed distribution of distinctive magma types.

In this discussion I will use a model in which crustal extension above the subduction zone is a consequence of trench rollback coupled with mantle diapirism induced by subduction. I propose that mantle counterflow above the subducting lithosphere plate and concurrent mantle upwelling probably are the main driving forces behind backarc basin evolution (e.g., Hawkins *et al.*, 1984). Unloading of the crust by extension and thinning promotes the rise and decompression melting of mantle diapirs (e.g., McKenzie and Bickel, 1988).

2. GEOLOGIC SETTING OF THE LAU BASIN

Intraoceanic convergent plate margins are dynamic systems that evolve by rifting, extension, and magmatism above the subduction zone. The geology of all parts of the system needs to be understood. The Lau Basin is but one element in the convergent margin system that includes the inactive Lau Ridge remnant volcanic arc and the Tonga Ridge (Figs. 3.1 and 3.2). A brief summary of the geology of these crustal segments follows.

2.1. Tonga Ridge

The Tonga Ridge extends for more than 1100 km; its southern end joins the Kermadec Ridge at the mutual intersection with the Louisville seamount chain. The Tonga Ridge comprises two belts of seamounts, shoals, atoll reefs, and islands that separate the 2- to 3-km-deep Lau Basin from the 10.5-km-deep Tonga Trench. The western belt is formed of seamounts, shoals, and volcanic islands that constitute the Tofua arc. The arc has been active historically, with many submarine eruptions having been reported within the last 50 years. Rocks of the Tofua arc constitute an arc tholeiitic series that ranges in composition from basalt to low-K rhyolite; the main magma types are basalt and basaltic andesite (Bryan *et al.*, 1972; Ewart and Bryan, 1972; Ewart *et al.*, 1977; Bryan, 1979a; Ewart, 1979; Gill, 1981; Ewart and Hawkesworth, 1987). The age of initiation of the Tofua arc is poorly constrained and may have varied along strike with oldest edifices occurring at the north end. For example, a 3-Ma age has been reported for Niuatopatapu Island by Tappin *et al.* (1994), but there is no evidence for similar ages at the south end. It is not likely that the present distribution of arc volcanoes existed prior to about 1 Ma (see discussion in Hawkins, 1994). There is a broad compositional similarity to the main phases of volcanism on the Lau Ridge, but, as we will discuss, the Tofua arc appears to have a more significant contribution of a subduction component than is seen on the Lau Ridge.

The eastern belt, or Tonga platform, comprises uplifted blocks of Tertiary platform carbonates that overlie a crystalline basement formed of middle Eocene to late Miocene arc-composition volcanic and plutonic rocks. Quaternary reef limestones form a cap on the platform (Scholl *et al.*, 1985). Depths increase from south to north on the Tonga platform.

Numerous cross faults with differential uplift give rise to three major morphologic segments. The northern segment is deeper (e.g., north of Vava'u depths range from 1000 to 1500 m), whereas the central and southern segments are from 500 to 1000 m deep, with many emergent banks and islands. The chain of active volcanoes of the Tofua arc is aligned along the crest of the northern segment, whereas it is offset to the west in the region south of Vava'u.

The oldest rocks on the platform are exposed on the island of 'Eua, where there are beach boulders of arc-tholeiitic-composition hypersthene gabbro which have been dated as 46 to 40 Ma (Duncan *et al.*, 1985). These are overlain by late-middle Eocene calcareous conglomerates and breccias (Cunningham and Anscombe, 1985). 'Eua also has arc tholeiitic andesitic lava flows that give ages of 33–31 Ma and andesitic to silicic andesite dikes with ages of 19–17 Ma (Duncan *et al.*, 1985; Hawkins and Falvey, 1985). These crystalline rocks are interpreted as part of the Lau Ridge that was rifted away as the Lau Basin opened. Drill core material (e.g., exploration wells, Exon *et al.*, 1985; and ODP Site 840, Parson *et al.* 1992), dredged samples (Exon *et al.*, 1985; Stevenson, 1985), and island exposures (Cunningham and Anscombe, 1985) show that the upper levels of the platform consist largely of gravity flow deposits formed of volcaniclastic turbidites and debris flows of sand and gravel. These were derived from a Miocene age volcanic arc; probably it was the Lau Ridge, but Cawood (1985) points out that volcanic rocks exposed on the Lau Ridge are not compositionally equivalent to the clastic rocks of equivalent age found on the Tonga Ridge. Along-strike compositional zonation may be an explanation for this, or the sediment sources may be submerged in the Lau Basin. The clastic rocks are interbedded with thin intervals of pelagic, or hemipelagic, nannofossil chalks and marlstones (Clift and Dixon, 1994; Ledbetter and Haggerty, 1994). The clastic rocks show an overall fining upward in the late Miocene prior to beginning of Lau Basin rifting. The late Pliocene-Pleistocene beds are a carbonate-rich sequence with minor interbeds of ash and volcanic sands (Cunningham and Anscombe, 1985; Exon *et al.*, 1985; Clift and Dixon, 1994).

Drill core data from ODP Site 841 on the Tonga Ridge (Fig. 3.1) show that the Miocene sediments are faulted against volcanic rocks comprising a low-K rhyolitic edifice formed of flows, welded and nonwelded tuffs, and tuff breccias that must have been erupted in shallow water or subaerially (Bloomer *et al.*, 1994). Radiometric dates (44 ± 2 Ma) and paleontologic ages for overlying sediments indicate a late-middle Eocene age (McDougall, 1994). Also found at Site 841 are arc tholeiitic basalt and basaltic andesite dikes or sills which have intruded distal facies of upper Miocene (foraminifer Zone N16 to Subzone N17a) volcaniclastic turbidites (Parson *et al.*, 1992). These intrusives, and the turbidites, were formed in the forearc to the Lau Ridge volcanic arc prior to opening of the Lau Basin.

2.2. Tonga Trench

The deepest levels of the Tonga Ridge are exposed on the wall of the Tonga Trench, where there is a cross section of oceanic crust that is largely formed of arc-related mafic and ultramafic rocks. Rocks dredged from the inner slope of the trench include highly depleted serpentinite, dunite, harzburgite, clinopyroxenite, gabbro, diabase, arc tholeiitic basalts, andesite, boninite, quartz diorite, tuffs, and volcanic breccia. Petrologic studies of this rock assemblage indicate that it is largely derived from a volcanic arc (Fisher and Engel, 1969; Hawkins *et al.*, 1972; Vallier *et al.*, 1985; Bloomer and Fisher, 1987; Hawkins, 1988). Some rocks indicative of the accretion of seafloor material have been dredged as well, but the bulk of the material suggests that the inner slope exposes the lower levels of an island arc as originally proposed by Fisher and Engel (1969). Bloomer and Fisher (1987) proposed a 4000-m stratigraphic reconstruction, based on depths of dredge hauls, that establishes an arclike crustal section capped by mafic and intermediate composition volcanic rocks.

2.3. Lau Ridge

The Lau Ridge remnant arc forms the western boundary of the Lau Basin. It comprises islands and atoll reefs as well as submerged conical features assumed to be volcanic seamounts; the emergent Lau Island Group forms a chain about 500 km long and up to 80 km wide. Barrier reefs, fringing reefs, or both, commonly surround the islands. The Lau Ridge and the submarine extension to the south, the Colville Ridge, form a 2400-km-long remnant arc. Lau Ridge volcanism began at least by mid-Miocene time, 14 Ma, and was active until early Pliocene time, 1.5–2.5 Ma (Gill, 1976; Cole *et al.*, 1985; Woodhall, 1985). Basaltic andesite and andesite, including both tholeiitic and calc-alkaline low- to high-K magma series, are the main rock types preserved on the Lau Ridge. Lesser amounts of basalt, dacite, and rhyolite are also present as well as intrusive rocks compositionally similar to the volcanic rocks (Woodhall, 1985). The Lau Ridge magma series differs from broadly similar rocks of the Tofua arc in having lower ratios of $^{87}Sr/^{86}Sr$ (e.g., 0.7030–0.7033 for the Lau Ridge (Gill, 1976), and 0.70361–0.70399 for the Tofua arc (Ewart and Hawkesworth, 1987)) that are interpreted as representing less of a subduction component in the source for Lau Ridge arc magmas.

2.4. Lau Basin—An Overview

The Lau Basin (Fig. 3.1) is a trapezoidal-shaped backarc basin that separates the inactive Lau Ridge remnant volcanic arc from the Tonga Ridge (Karig, 1970; Hawkins, 1974). The shape of the basin suggests that it has opened more widely at its north end, perhaps because opening began there earlier and has progressed southward in the basin. At present, the southern end apparently is hinged where the Louisville seamount chain intersects the Tonga Trench. Opening of the Lau Basin, as a consequence of crustal extension above the Tonga Trench subduction system, is the most recent (i.e., $<$6 Ma) event in a long sequence of crustal extension episodes in the southwestern Pacific that may be traced back to late Cretaceous time and the initial breakup of the eastern Australian continental margin (Kroenke, 1984; Hawkins, 1994).

The Lau Basin has many geologic characteristics that are typical of other western Pacific backarc basins, but it also has several features that may be unique. Like most backarc basins, the Lau Basin is situated above a well-defined Wadati–Benioff zone that marks the location of the subducted Pacific plate. The inclined seismic zone reaches depths on the order of 700 km to the west under the North Fiji Basin (Fig. 3.3; Isacks and Barazangi, 1977; Billington, 1980; Giardini and Woodhouse, 1984, 1986; Pelletier and Louat, 1989). Seafloor magnetic lineations and drill core data indicate that Cretaceous-age Pacific plate lithosphere is being subducted into the Tonga Trench beneath the Indo–Australian plate at rates estimated from 10.5 cm/yr (Minster and Jordan, 1978) to as much as 17.8 cm/yr (Pelletier and Louat, 1989), depending on the latitude. The seismic zone lies at a depth of about 140 km below the active volcanoes of the Tofua (Tonga) arc and is about 250 km under the central Lau spreading center (CLSC) and eastern Lau spreading center

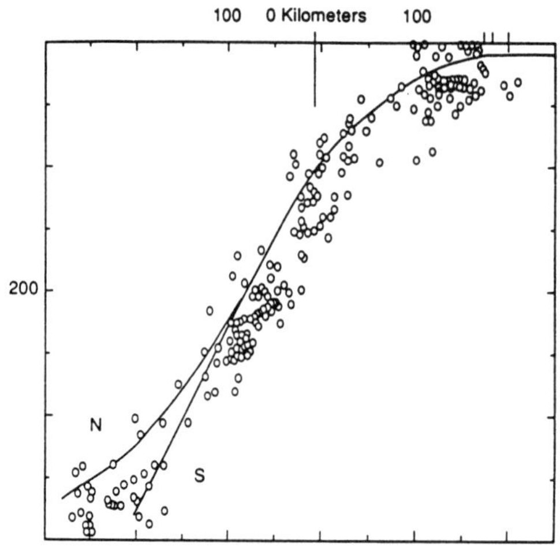

FIGURE 3.3. Composite cross section of earthquake hypocenters for the Tonga Trench subduction zone. The "0 kilometer" mark on the horizontal axis corresponds to the location of the Tofua arc. Locations of projections of the trench axis are shown by the vertical lines. Scale is in kilometers (Isacks and Barazangi, 1977).

(ELSC) (Isacks and Barazangi, 1977). The upper surface of the seismic zone has been modeled as an irregular and strongly curved surface (Billington, 1980). It terminates at its northern end where the Tonga Trench curves sharply to the west and the plate boundary becomes a transform fault boundary between the Tonga and New Hebrides–Vanuatu trenches. An inflection in the seismic zone at depths of 525 to 575 km, first identified by Billington (1980), was interpreted as due to imbrication of the subducted lithosphere by Louat and Dupont (1982). They estimated that the lithosphere at the depth of this inflection was subducted about 7 to 8 m.y. ago. This is close to the 6-Ma age estimated for beginning of opening of the Lau Basin (Parson *et al.*, 1992); the two events may be related. Interpretation of the configuration of the Tonga Trench seismic zone suggest a half-spoon shape sharply curved to the west at its north end and abruptly terminated along the west-trending part of the Tonga Trench (Fig. 3.4; Billington, 1980). A broad zone of strong seismic wave attenuation was recognized west of the Tonga Ridge beneath the Lau Basin by Barazangi and Isacks (1971), who attributed it to high temperature or partial melting in the mantle. The mantle under the basin has low Q (strong seismic-wave attenuation) and is inferred to be hotter than the surrounding mantle (Fig. 3.5; Barazangi and Isacks, 1971). The ridge axes of the modern spreading centers rise to depths as shallow as 2200 m, and the average basin depth is about 2500 to 3000 m (Hawkins, 1974); this is anomalous because it is the depth range for most of the ridge crest of mid-ocean ridge system as originally noted by Sclater *et al.*(1972). They interpreted this depth anomaly as being related to the relative youth of the basin and the likelihood of a broad region of upwelling hot mantle beneath it. The Lau Basin depths are in striking contrast to the Mariana Trough, where depths as great as 4500 m are common near the axial ridge and the crest of this ridge is about 3500 m deep (e.g., Hawkins *et al.*, 1990). A possible corollary of this is that the Mariana Trough lacks a similar large region of upwelling hot mantle.

Karig's proposal that the Lau Basin was floored with basalt was substantiated by the first geologic studies based on dredged samples (Hawkins *et al.*, 1970; Sclater *et al.*, 1972). These papers, and subsequent work by Hawkins (1974, 1976), established that there was young basaltic crust in the Lau Basin and that it showed considerable similarity to MORB. A range in basaltic types was found, but all were distinctly different from arc tholeiites. The modern spreading ridge system of the Lau Basin, comprising the central and eastern Lau

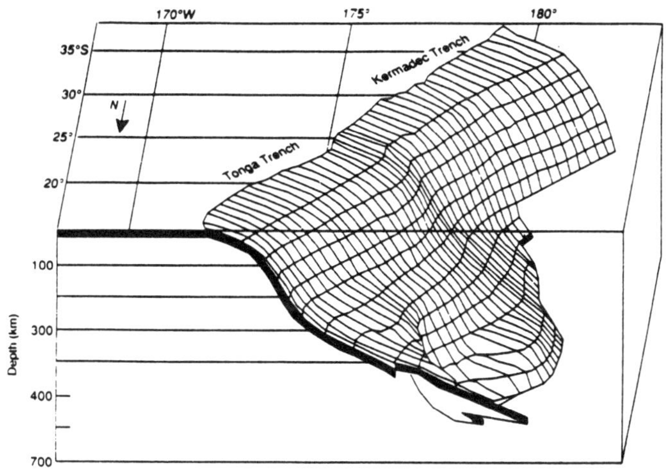

FIGURE 3.4. View, looking south, of a grid representing the upper surface of the Wadati–Benioff zone of the Tonga–Kermadec subduction zone. The projection is slightly distorted in that it does not take into account the Earth's sphericity (from Billington, 1980).

spreading centers, was discovered, and sampled, on the Scripps Institution of Oceanography *7-TOW* Expedition in 1970 (Hawkins *et al.*, 1970). Our single-beam profiler survey lines helped to delineate part of the ridge (Hawkins, 1974) and to recognize symmetric magnetic anomalies, but the anomalies proved difficult to trace throughout the basin (Lawver *et al.*, 1976; Lawver and Hawkins, 1978). Initially we all assumed that nearly the entire width of the Lau Basin had formed by seafloor spreading. Neither the overlapping spreading centers nor the wedge shape of the magnetic pattern and spreading system was recognized. Only the presence of fresh pillow lavas and the general trend of the ridge could be established. Additional data did not help solve the problem. The western part of the basin lacks a continuous magnetic fabric, and symmetric anomalies cannot be traced beyond the region near the axial ridges (e.g., Lawver *et al.*, 1976; Weissel, 1977; Lawver and Hawkins,

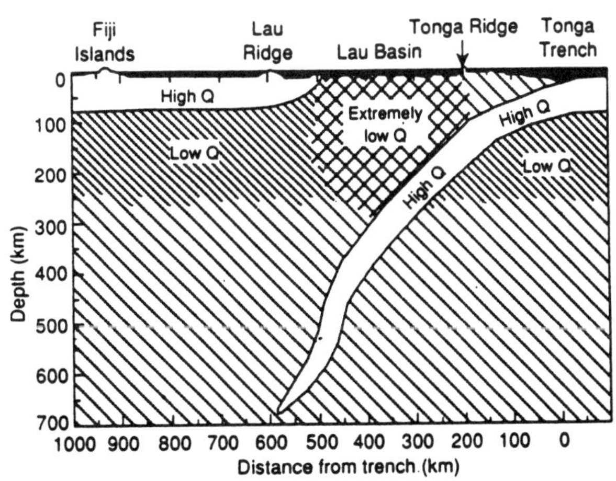

FIGURE 3.5 Cross section of the Fiji Islands to Tonga Trench region showing inferred and extrapolated extent of the region of high and low attenuation in the uppermost mantle. Note that the Lau Basin is underlain by mantle with extremely low Q (high attenuation) (from Barazangi and Isacks, 1971).

1978). Interpretations of the age of the earliest backarc crust in the Lau Basin varied from 5 to 10 Ma (Sclater *et al.*, 1972) to 2.5 to 3 Ma (Malahoff *et al.* 1982). Lawver *et al.* (1976) and Lawver and Hawkins (1978) concluded that true seafloor spreading had not been an important process throughout the entire history of the basin and proposed that some form of diffuse spreading on short ridge segments may have operated. Thus, it was clear early on that it was difficult to explain the entire evolution of the basin by seafloor spreading, although several attempts to do so (e.g., Weissel, 1977) appeared to give satisfactory models for some of the data.

Our understanding of the petrology of the basin's crust has undergone great change from an originally simple comparison to "normal" oceanic crust to the much more complex view now held. The earliest petrologic studies of Lau Basin crust focused on high-standing features that form the modern axial ridge system and the Peggy Ridge. Petrologic data for these ridges indicated a close similarity to MORB and led to the interpretation that petrogenetic process for backarc basins were similar to those forming oceanic crust (Hawkins *et al.*, 1970; Sclater *et al.*, 1972; Hawkins, 1974, 1976; Gill, 1976; Pineau *et al.*, 1976; Carlson *et al.*, 1978). More extensive sampling, including seamounts and high-standing older crust on the margins of the basin, showed that there were rocks having compositions transitional to arc chemistry and that the basin crust was zoned with borders more like arc compositions than MORB (Hawkins and Melchior, 1985). The compositional variability includes a wide range in rock types from basaltic to andesitic and multiple mantle source types have been inferred from a number of subsequent studies (e.g., Poreda, 1985; Jenner *et al.*, 1987; Volpe *et al.*, 1987, 1988; Hawkins 1989; Hawkins *et al.*, 1989; Nilsson *et al.*, 1989; Boespflug *et al.*, 1990; Davis *et al.*, 1990; Frenzel *et al.*, 1990; Loock *et al.*, 1990; Sunkel, 1990; Vallier *et al.*, 1991; Ernewein *et al.*, 1994). The most recent insights to the geology of the Lau Basin and its evolution have come from GLORIA imagery of much of the northern part of the basin (Parson *et al.*, 1990) and drilling on ODP Leg 135. The drill cores have given samples of hitherto inaccessible crust from the older, sediment-covered, western part of the basin and from sediment-covered crust near the modern axial ridges (Parson *et al.*, 1992; Hawkins *et al.*, 1994).

2.5. The New View of Lau Basin Evolution

Extensive bathymetric surveys with single-beam and multibeam profiling systems, seismic reflection profiling, shipboard and aeromagnetic surveys, and GLORIA imagery have given us a good understanding of both the regional bathymetric fabric and the relations between basin morphology and the magnetic patterns. Additional insight has come from drilling on ODP Leg 135. Together, these data have helped develop a new view of Lau Basin evolution that also has implications to other arc-backarc systems. The reason why the magnetic data have been so difficult to interpret is because they are not all part of a continuous symmetric system developed by seafloor spreading (Parson *et al.*, 1989, 1990, 1992; Parson and Hawkins, 1994).

Our new view of Lau Basin evolution is that it has opened in two stages each with its own tectonic style. Initially, the forearc to the Lau Ridge was stretched and attenuated by crustal extension and rifting. This formed a basin-range-type structure. Lavas ponded in the basins gave rise to magnetic anomalies, but the magnetic patterns do not form a continuous symmetric record because the lavas did not generate new crust at a "fixed" spreading center. Data from ODP Site 834 demonstrate that rifting started prior to 5.6 Ma; we estimate that it may have started by about 6 Ma (Hawkins, 1994; Hawkins and Allan, 1994). The early backarc crustal extension was contemporaneous with volcanism on the Lau Ridge (e.g., 4.5- to 2.5-Ma basalts of the Korobasaga Group). Whelan *et al.* (1985) correlate the Korobasaga Group with an "early rifting stage" that occurred late in the evolution of the Lau Ridge. The arc rifting and eruption of the Korobasaga Group, which was dominated by tholeiitic basalt, overlaps with beginning of crustal extension and basaltic volcanism in the western Lau Basin. This predates the time inferred for the beginning of seafloor spreading. Geometrically this process occurred in the *forearc*, and it is only because the modern Tofua arc subsequently formed to the seaward that the term *backarc basin* has any meaning.

The second stage of opening involved seafloor spreading that may have started at about 5.5 to 5 Ma. Parson and Hawkins (1994) propose that this spreading was initiated by southward propagation of a rift that started on the trace of the Peggy Ridge; this propagator formed the present ELSC. Thus, the age of crust formed by seafloor spreading is progressively younger to the south culminating at the Valu Fa Ridge at the south end of the ELSC. The wedge-shaped age pattern of new seafloor formed by the ELSC is further complicated by the development of a second propagator that cut through the older (ELSC) seafloor and formed the present CLSC (Parson *et al.*, 1989, 1990, 1992; Parson and Hawkins, 1994). The two spreading centers overlap and are separated by complex seafloor with traces of pseudofaults and an abandoned intermediate ridge or relay zone. Presently, the Lau Basin appears to be opening in an east-west direction by symmetric seafloor spreading on the well-defined, but segmented and offset, axial ridges aligned along longitude 176°30′ W.

The evolution of the Lau Basin has involved crustal extension and the partial dismembering of the Lau Ridge. The evidence suggests that the initial rift was on the outboard edge of the former arc or in the forearc (Hawkins, 1994). Other arc-backarc systems (e.g., Mariana) may have formed in a similar way although this evolutionary style is not necesarily true for all arcs. For example, the Sumisu rift has formed on the inboard side of the Izu arc (Taylor, 1992). There has been a long-standing controversy as to whether or not backarc and arc magmatism are synchronous. At least for the Lau Basin we have good evidence that they were synchronous although we have no good constraints as to relative volumetric importance. The 4.5-Ma to 2.5-Ma basalts of the Korobasaga Group erupted during the "early rifting stage" in the evolution of the Lau Ridge (Whelan *et al.*, 1985). The timing of this rifting and associated volcanism, which was dominated by tholeiitic basalt, broadly overlaps the beginning of crustal extension in the western Lau Basin (Hawkins, 1994). We also have evidence within the basin that a wave of arc composition volcanism migrated across it as the basin opened (Bednarz and Schmincke, 1994; Clift and Dixon, 1994). Presently, Tofua arc volcanism and backarc spreading are proceeding concurrently. It is also evident that arc and backarc magmatism operate contemporaneously in the modern Mariana arc and backarc.

3. LAU BASIN MORPHOLOGIC PROVINCES

3.1. Introduction

The Lau Basin may be subdivided into several morphologic-tectonic provinces that together constitute the extensional backarc basin (Fig. 3.1). The central part of the basin, where the ODP Leg 135 drill transect was located, comprises a western extensional basin-ridge province (here informally called WEB) and a triangular region of young crust that has formed by seafloor spreading. The latter area includes the two actively spreading ridges—the CLSC and the ELSC—and the relay zone (RZ) between them that collectively extend for about 600 km. These ridges are not centered in the Lau Basin but are offset toward the eastern side; much of the southern part of the ELSC lies within 50 km of the active Tofua arc. The Peggy Ridge (Fig. 3.1) separates the northern and southern parts of the Lau Basin. Peggy Ridge appears to be the location of a transform fault with right-slip displacement. The northern basin is not well sampled, but GLORIA imagery has helped define a probable spreading center, the northwestern Lau Spreading Center (NLSC), that trends northeasterly from the Peggy Ridge. A well-defined ridge-ridge-ridge triple junction has been recognized in the northeastern basin. Originally named Mangatolu triple junction (MTJ) by the discoverers (Hawkins, 1989; Hawkins *et al.*, 1989; Nilsson *et al.*, 1989), it is also called the Kings triple junction. Much of the northern basin is covered by seamounts. One robust feature forms the island of Niuafo'ou. Only three submerged features have been surveyed and sampled: Donna Seamount, Rochambeau Bank, and Zephyr Shoal. Each is petrologically distinct and has a different geologic history.

3.2. Eastern and Central Lau Spreading Centers

At present, new seafloor is being generated at three well-mapped areas of spreading (Fig. 3.1). These are the ELSC, CLSC, and MTJ. A fourth spreading center, NLSC, has been postulated but data for it are limited. The main site of seafloor spreading is on the well-defined axial ridge system that extends for more than 600 km in a north-south direction near 176°30′W. This ridge system is segmented in a morphotectonic sense, as well as having a petrologic segmentation, that allows further subdivision into secondary and tertiary segments on the scale of 10 s of km. It comprises the CLSC, ELSC and a RZ in the overlap region. Detailed charts and discussions of the morphology of parts of each ridge system are presented in Weidicke and Habler (1993) and Weidicke and Collier (1993).

3.2.1. Central and Eastern Lau Spreading Centers

The CLSC is shallower than 2300 m throughout its 190-km length, and it is flanked by 2800- to 2900-m-deep basins that parallel the ridge trend. The ridge crest morphology of the CLSC varies along axis. Parts of it have a narrow axial rift basin with flanking walls that rise 200 to 500 m above a low relief floor. Some areas are capped with small mounds and pinnacles. The southern end near 19°22′S is split by a narrow rift, and the ridge narrows and deepens and ends in an area of rough topography. The ELSC extends for about 400 km. Its crest has inward-facing scarps that define an axial rift valley. Small edifices are found within the valley, and at least one intrarift seamount has been split into two segments now on opposite sides of the axis. The floor of this rift shoals from about 3000 m at the dying northern end to about 2300 m near the southern end of the rifted part of the ridge at 21°S. Farther to the south, the ELSC axial crest lacks a rift and is capped by small elevated areas. The ELSC ends near 22°50′S on a segment of the ELSC known as the Valu Fa Ridge (VFR; Scholl *et al.*, 1985; Morton and Sleep, 1985). The morphology and segmentation of the VFR have been described by Weidicke and Collier (1993). The VFR lies only 30 to 40 km west of the trace of the Tofua arc. The VFR is petrologically distinct from the northern ELSC and is here treated as a separate major segment.

The ELSC began as a propagating rift at about 5.5 to 5 Ma and was followed by the CLSC at about 1.5 to 1.2 Ma (Parson and Hawkins, 1994). Both have propagated southward from the Peggy Ridge. The two ridges form an overlapping system near 19°22′S with the CLSC propagating southward at the expense of the ELSC. The southern end of the ELSC, the Valu Fa spreading center (Jenner *et al.*, 1987; Vallier *et al.*, 1991), appears to be propagating southward into older crust, but essentially nothing is known about the age or composition of that older crust. The rift tip of the propagator is near 23°S. Parson and Hawkins (1994) estimate that the rift tip of the ELSC has propagated southward at about 120 mm/yr.

Presently the northern Lau Basin may be opening at a rate of 120–145 mm/yr on a line between Lakemba, Fiji, and Vava'u as determined by global positioning system (GPS) measurements (Bevis *et al.*, 1993; F. W. Taylor, personal commun., 1994). The opening rate decreases southward toward the VFR. The complex magnetic fabric and the effects of propagating ridges make it difficult to estimate long-term average spreading rates. However, it is likely that similar rates of opening, 100–170 mm/yr, may have been in effect south of the Peggy Ridge since the initiation of spreading on the ELSC (Parson and Hawkins, 1994).

The most extensively studied part of the Lau Basin is along the axial ridge systems. The ridges comprise a range in rock types that is mainly basaltic with lesser amounts of fractionated types, including Fe-Ti-enriched basalt, "oceanic andesite," and low-K rhyolite. These are discussed in more detail below.

3.2.2. Mangatolu Triple Junction

The Mangatolu triple junction, in the northeastern part of the Lau Basin (Fig. 3.1) near 15°30′S, 175°45′W, comprises three limbs of a ridge-ridge-ridge triple junction (Hawkins *et al.*, 1989; Nilsson *et al.*, 1989; Nilsson, 1993). The MTJ is also known as the Kings triple junction (Falloon *et al.*, 1992). Three well-defined spreading axes have been defined for the MTJ by bathymetry, GLORIA imagery, magnetic anomalies, and dredged samples. The west-trending limb breaks up into at least three separate traces that have a poorly developed ridgelike cross section. The northeasterly trending limb intersects the line of volcanic edifices that constitutes the Tofua arc. A deep channel, or trough, cuts the Tonga Trench wall on the projection of the trace of this limb, but we have no data to support a direct relationship between the two. The southern axial limb has an overlap zone at about 15°53′S. The southern end is poorly defined by the available bathymetric data but can be traced as far south as 16°05′S.

The MTJ is in the widest part of the Lau Basin, more than 250 km wide, and has replaced older crust of uncertain age. This crust may include some of the oldest backarc basin crust, remnants of the forearc to the Lau Ridge, or trapped old ocean crust. Data for the seafloor are very limited, but, as discussed in a separate section, both arclike and MORB-like crust have been found.

3.2.3. Northwestern Lau Spreading Center

A northeasterly trending region of high sonar backscattering, near 177°W on the Peggy Ridge (Parson *et al.*, 1990; Parson and Tiffin, 1993; Parson and Hawkins, 1994) has been interpreted as a spreading center. The NLSC appears to trend northeasterly and may link the Peggy Ridge transform with the major structural discontinuity at the west trending segment of the Tonga Trench. Parson and Tiffin (1993) note that it is about 40 km wide where it intersects the Peggy Ridge and has the finely lineated GLORIA image typical of medium and slowly spreading axial ridges. The NLSC has been superposed on the crustal fabric of older sedimented rift basins. The trend of the NLSC is toward Rochambeau Bank, a prominent seamount described later. The interpretation of evidence from GLORIA imagery is supported by fresh basaltic samples from a few dredge sites near the Peggy Ridge and from seamounts and scarps on the probable trace of the spreading center (Hawkins, 1988). The NLSC cuts older crust that had a different origin.

3.3. Western Extensional Basin

The broad area south of Peggy Ridge, between the Lau Ridge and the active CLSC and ELSC, is characterized by narrow, partly sedimented, subbasins, elongated in a north to northeasterly direction. This part of the Lau Basin has rugged topography with up to 1500 m of relief in contrast to the smoother topography that flanks the two spreading centers. We described it as having basin and range topography to differentiate it from crust formed at the spreading centers (Parson *et al.*, 1992) and used the name western extensional basin (WEB) province. The basins are about 1–15 km wide and 10–85 km long. The subbasins are separated by ridges of thinly sedimented crust. Limited data from dredged samples show that these are mainly basalt, some of it moderately differentiated, having compositions transitional between MORB and arc material (Hawkins and Melchior, 1985). The age of the crust forming these ridges is not known, but several of the basins were drilled on ODP Leg 135 and we have good age data for these. There is a general age progression from about 6 Ma on the west to 0.6–0.8 Ma near the present axial ridges (Hawkins, *et al.*, 1994). The high-standing blocks between the subbasins are assumed to be older crust. The interpretation of Leg 135 data suggests that the WEB formed by crustal extension and rifting, not by seafloor spreading.

3.4. Peggy Ridge and the Northern Basin

The Peggy Ridge (Fig. 3.1) is a high-standing feature (1250 m) of uncertain origin. Parson and Tiffin (1993) noted that its upper surface is ribbed, irregular, characterized by clusters of linear peaks and troughs, and appears to be dominated by superimposed seamounts. The northeast side drops sharply into a 2500-m-deep valley. They interpret the steep northeast flank of the ridge as a fault. The northeast side of the deep valley rises abruptly about 1800 m to form a ridge subparallel to the Peggy Ridge. Presently Peggy Ridge marks the location of many shallow earthquakes that have right-lateral first motions (Eguchi, 1984; Hamburger and Isacks, 1988). The morphology of the ridge suggests to Parson and Tiffin (1993) that it may be a leaky transform fault. The ridge may have had a complex history, and all we can say with any degree of certainty is that its composition precludes it having been part of an island arc, and regional magnetic patterns make it unlikely that it was a spreading center (Sclater *et al.*, 1972; Hawkins, 1974, 1976). At present it marks the location of a transform fault accommodating differential motion between the CLSC and the postulated NLSC north of Peggy Ridge (Parson *et al.*, 1990; Parson and Tiffin, 1993; Parson and Hawkins, 1994). As discussed in a subsequent section, the Peggy Ridge may have played a major role in the development of the ELSC and CLSC rift propagators. Rocks from the Peggy Ridge are MORB-like basalts that resemble the rocks of the active ridges.

At least four distinct provinces may be recognized north of the Peggy Ridge. In addition to the MTJ and the NLSC, there is a broad region north of Peggy Ridge that Parson and Tiffin (1993) call a "sedimented block terrain." This includes an old ridge partially buried in sediment and a region of elevated topography that has several large seamounts and the island of Niuafo'ou. There are several areas of high heat flow (e.g., up to 4.22 HFU; Sclater *et al.*, 1972), and the recently active volcanic island of Niuafo'ou (Reay *et al.*, 1974; Taylor, 1991). Samples from the seamounts include a wide range of rock types such as N-MORB, E-MORB, OIB, and dacite.

South of the MTJ and west of the Tonga Ridge, the seafloor is characterized by north-south-trending ridges that separate sedimented basins. Parson and Tiffin (1993) call this the "ridge/interridge terrain." The narrow basins have relatively thick sediment fill similar to those in the WEB (e.g., more than 0.2-s two-way travel time, Hawkins, 1974). Parson and Tiffin (1993) suggest that this part of the basin may have had an origin by crustal extension and rifting like that which formed the western part of the Lau Basin.

4. PETROLOGIC DISCUSSION OF THE NEOVOLCANIC ZONES

4.1. SSZ Mantle Influence

Before discussing the petrogenesis of magmas that erupt on the axial ridges, a digression to comment on the tectonic setting is in order. Magmas of the Lau Basin neovolcanic zones form by partial melting of upwelling mantle in the SSZ tectonic environment (Fig. 3.2). The crust flooring subbasins of the WEB also came from this source. Current ideas about the nature of SSZ mantle are that it has been depleted by previous melting and selectively reenriched with a subduction component (e.g., Pearce *et al.*, 1984; Hawkins, 1994). It is not surprising that many backarc magmas carry the distinctive compositional signatures of this complex mantle source; that is, they show some affinity to arc magmas. This is expressed in the major element chemistry of *some* backarc basin basalt glasses in their relatively higher Al_2O_3 and Na_2O, but lower FeO* (all iron determined as FeO) and TiO_2, when compared to MORB glasses with equivalent levels of MgO. These characteristics were first pointed out by Fryer *et al.* (1981) for Mariana Trough basalts. They are found in most other backarc basins as well, but are by no means typical of all backarc basalts. Forearc magmas as well as those of island arcs also carry the SSZ signature. In addition to the different abundances of some major elements, there are several major distinctions in trace and minor constituents, differing from true MORB, that are typical of SSZ magma systems (e.g., Fryer *et al.*, 1981; Briqueu *et al.*, 1984; Hawkins and Melchior, 1985; Sinton and Fryer, 1987; Woodhead *et al.*, 1993). These are (1) a relative depletion in high-field-strength elements (HFSE), especially Nb and Ta. (2) The relatively higher Na is paralleled by a tendency for some, *but not all*, samples to be accompanied by variably higher K, Rb, Ba, and Sr relative to MORB; Ba in particular is relatively enriched. (3) The rare-earth element (REE) patterns vary from flat to MORB-like in form although for some of the least fractionated samples the low abundances indicate a depleted source. La is variable, but enrichment relative to MORB is common in many SSZ magmas (e.g., Gill, 1981). (4) SSZ magmas commonly are depleted relative to MORB in Cr, Ni, and Co.

(5) Backarc samples are generally more hydrous and more oxidized than otherwise comparable MORB. The distinctive differences from MORB in major elements, as well as some trace elements, are attributed to the more hydrous mantle source of the SSZ environment. Elevated P_{H_2O} delays the onset of plagioclase crystallization and leads to higher Al content of the melts. Higher oxygen fugacity promotes the crystallization of Fe-Ti oxides as well as Cr-spinel and causes the relative depletions in Fe, Ti, Cr of SSZ magmas. The field of olivine stability is expanded, and this causes the relative depletion of Ni in SSZ magmas. Clinopyroxene is an important liquidus phase and contributes to Co depletion.

Many, but not all, backarc basin basalts show varied effects of the SSZ influence described. The Lau Basin is no exception, as both MORB-like and arclike compositional varieties are found. The Mariana Trough axial ridge also displays both MORB-like and arc-like rock types (Hawkins *et al.*, 1990). This requires that the SSZ mantle must receive an input of MORB-source mantle as shown schematically on Fig. 3.2. In the Lau Basin there is a general pattern of MORB-like basalt being more common on the axial ridges and the transitional to arc compositions being more common on the eastern, western, and southern edges of the basin (Fig. 3.6; Hawkins and Melchior, 1985). The data for the edges of the basin are very limited, and this proposal still needs to be tested with many more samples than we have. However, the idea appears to be fairly well supported with the limited data we have. As we discussed, there are some deviations from this general pattern that probably are related to the extent or longevity of melt production in different areas as well as to source heterogeneity. The general inference for the Lau Basin is that, in spite of the SSZ setting, a MORB-source mantle has been the major contributor to the backarc crust, and there has been restricted mixing with arclike SSZ mantle sources. Some Lau Basin axial ridge lavas show an imprint of SSZ ("arc") compositional features but many do not. Compositions very close to MORB are found on all of the active ridges as well as in the drill core samples. Some of the earliest lavas to form new crust as the Lau Basin opened (ODP Site 834) are among the most MORB-like (Hergt and Farley, 1994; Hawkins and Allan, 1994). This requires massive upwelling and melting of MORB-source mantle that has overwhelmed the SSZ signature. This is discussed further in the section presenting a model for evolution.

4.2. Eastern and Central Lau Spreading Centers

4.2.1. Introduction

The combined ELSC and CLSC extend for about 600 km through the Lau Basin in a generally north-south direction culminating at the Valu Fa Ridge segment of the ELSC. There is an extensive data base for the ELSC and CLSC that offers insight to their evolution and the mantle sources for the melts that formed them (e.g., Hawkins, 1974, 1976, 1977, 1988, 1989; Hawkins and Melchior, 1985; Jenner *et al.*, 1987; Sunkel, 1990; Boespflug *et al.*, 1990; Vallier *et al.*, 1991; Falloon *et al.*, 1992; Ernewein *et al.*, 1994). Each spreading ridge carries its own distinctive isotopic and chemical signature, but all have a compositional stamp that indicates a predominantly MORB-like character with an SSZ overprint. A summary of representative data is in Table I and sample locations are in Fig. 3.6. The modern spreading axes all are situated above the Wadati Benioff zone which lies at about 250 km beneath the ELSC, CLSC, and MTJ, and about 300 to 350 km below the NLSC. As

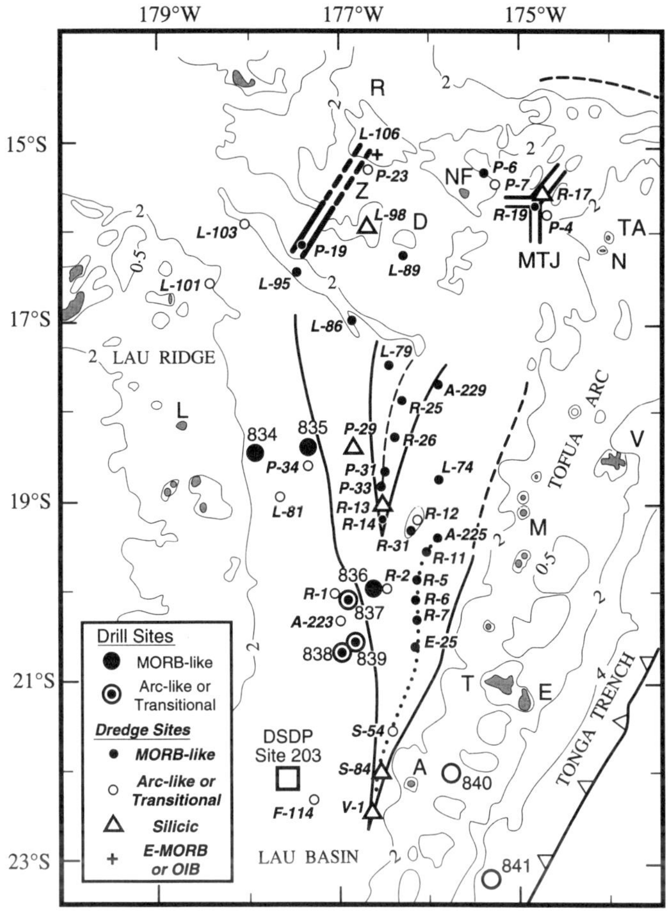

FIGURE 3.6. Locations of dredge sites referred to in text and on tables. Dredge site acronyms are A = ANT, L = 123 (both from Hawkins and Melchior, 1985); P = PPTU (Hawkins, 1988); R = RNDB (Hawkins, 1989); E = Ernewein *et al.* (1994); F = Frenzel *et al.* (1990); S = Sunkel (1990); V = Vallier *et al.* (1991). Numbers shown are the dredge site numbers.

a consequence, a major source of the melts must be in the SSZ mantle, but, as outlined, there is a general MORB-like character to all of the basaltic samples. Local variations in chemistry that point to variable affinity to arc-composition magmas are attributed to variable mixing with a subduction component. Proximity to the trace of the Tofua arc is suggested as an explanation for the compositional variation seen from north to south along the ELSC (Ernewein *et al.*, 1994).

ODP Site 836 was drilled on some of the oldest ELSC crust formed at this latitude, and the cores give data for early stages of ELSC magmagenesis on the propagating rift. Valu Fa Ridge samples also are early ELSC crust formed at the rift tip. They are discussed in a separate section because of their many distinctive features.

TABLE I

Representative Analyses, ELSC, CLSC, and Relay Zone

Location	ELSC								
Site	L-74	A-225	R-5	R-6	R-7	R-11	E-25	S-54	S-84
Sample	123	ANT	RNDB	RNDB	RNDB	RNDB	Darwin-3	Sonne	Sonne
	74-1 GL[a]	225-1 GL[a]	5-3 GL[a]	6-1 GL[a]	7-2 GL[a]	11-1 GL[a]	25-5-1	GC 54-1	KD 84-1
SiO_2	50.24	53.59	51.44	48.37	52.58	50.62	51.3	52.4	65.5
TiO_2	0.76	1.05	0.9	0.85	2.11	1.02	0.96	0.7	0.64
Al_2O_3	15.8	14.56	15.44	15.38	13.43	15.18	15.93	15.77	13.4
FeO*	9.13	11.55	9.39	7.95	14.76	9.54	9.28	8.07	6.69
MnO	0.21	0.21	0.19	0.17	0.21	0.18	0.17	0.16	0.16
MgO	7.39	5.27	8.02	6.21	3.99	7.51	6.5	7.78	1.34
CaO	13.37	10.22	12.87	11.01	8.55	12.83	11.29	12.5	4.54
Na_2O	2.1	2.34	1.91	2.09	3.24	1.9	2.11	1.79	4.2
K_2O	0.11	0.17	0.03	0.16	0.14	0.13	0.17	0.2	0.73
P_2O_5	0.09	0.11	0.06	0.11	0.2	0.12	0.08	0.08	0.14
Sum	99.2	99.07	100.26	99.21	99.21	99.04	97.79	99.45	97.34
Mg#	62.9	48.3	62.9	61.6	35.6	61.8	58.9	66.4	31.3
CaO/Al_2O_3	0.85	0.702	0.83	0.72	0.64	0.85	0.71	0.79	0.34
Trace elements (ppm)									
Cr	300	50	337			342	116	216	16
Ni	94	33	110	62	8	103	55	69	4
Co		90	43			43	35	38	12
V	282	359	307	277	502	338	376	258	54
Zr	40	45	53	50	127	57	47	36	141
Y	21	26	27	31	55	26	23	21	58
Nb	< 3		1	1	1	1	0.69	< 5	5
Hf			1.16			1.55	1.41		
Ta			< .05			0.08	0.04		
Rb	2	0.5	1	5	4	6	2.8	< 3	12
Ba	10.8	32.5	4	18	27	8	40.7	44	120
Sr	166	98	69	118	95	111	119	116	128

Location	ELSC						
Site	V-1	ODP 836 A	ODP 836 B	ODP 836 B	ELSC	ELSC	ELSC
Sample	Lee	Unit 3 RK	Unit 4 RK	Unit 5 RK	Mg# > 60	Mg# 50–59	Mg# < 50
	1–5	3H7/54-55	3R2/32-38	7R2/56-62	(Mean 20)	(Mean 7)	(Mean 2)
SiO_2	56.69	47.13	49.94	48.08	50.95	52.79	52.79
TiO_2	1.38	0.69	0.79	0.72	0.95	0.92	1.85
Al_2O_3	14.57	19.18	16.11	16.21	15.41	15.13	14.08
FeO*	11.51	8.08	8.27	8.13	9.36	10.47	13.62
MnO	0.22	0.16	0.12	0.14	0.18	0.19	0.21
MgO	3.73	8.88	7.68	9.07	7.52	6.52	4.28
CaO	7.65	13.03	14.85	13.78	12.64	11.47	8.9
Na_2O	3.36	1.9	1.96	2.33	2.03	1.92	3.03
K_2O	0.52	0.01	0.15	0.01	0.08	0.12	0.16
P_2O_5	0.16	0.14	0.06	0.14	0.08	0.08	0.19
Sum	99.2	99.2	99.93	98.59	99.22	99.61	99.09
Mg#	39.9	69.3	65.6	69.6	62.2	56.1	39.2
CaO/Al_2O_3	0.52	0.68	0.92	0.85	0.82	0.758	0.632
Trace elements							
Cr	4		318				
Ni	25	111	139	120			
Co							
V	350	216		157			
Zr	81	47	53	44			
Y	36	19	18	20			
Nb		2	1	2			
Hf	2.3						
Ta							
Rb	8.6	0.1	3	2			
Ba	96	17	14	19			
Sr	168	140	288	319			

(*continued*)

TABLE I
(Continued)

Location	CLSC							
Site	A-229	L-79	P-29	P-31	P-33	R-13	R-14	R-25
Sample	ANT	123	PPTU	PPTU	PPTU	RNDB	RNDB	RNDB
	229 GL	79-1 GL	29-3 GL	31-1 GL	33-1-GL	13-2 GL	14-1 GL	25-1 GL
SiO_2	48.11	49.33	72.96	49.51	50.45	60.41	51.28	50.96
TiO_2	0.88	1.94	0.34	1.64	1.93	1.62	2.39	1.05
Al_2O_3	17.35	13.93	12.08	16	14.11	12.93	12.63	14.65
FeO*	9.02	13.27	4.56	9.76	13.29	12.16	16.1	10.6
MnO	0.17	0.28	0.13	0.18	0.25	0.22	0.25	0.2
MgO	8.73	6.64	0.12	7.56	5.41	1.99	4.64	7.6
CaO	12.56	11.74	2.21	11.33	10.17	6.15	9.12	12.76
Na_2O	2.59	2.59	2.23	2.66	2.85	3.91	3.12	2.26
K_2O	0.03	0.09	1.18	0.42	0.08	0.42	0.14	0.03
P_2O_5	0.05	0.14	0.02	0.15	0.13	0.32	0.2	0.07
Sum	99.49	99.95	95.83	99.21	98.67	100.13	99.87	100.17
Mg#	67	51.2	5.4	61.4	45.5	25.2	37.1	59.5
CaO/Al_2O_3	0.72	0.84	0.18	0.71	0.72	0.48	0.72	0.87
Trace elements (ppm)			[b]					
Cr				198	20			303
Ni			3	68	37	7	18	116
Co				46	44			
V			13	345	455	147	469	332
Zr			201	68	128	441	165	45
Y			82	31	59	138	64	30
Nb			1	2	1			1
Hf				1.84	3.57			1.89
Ta				0.081	0.16			0.04
Rb			31	1	1	10	3	0.1
Ba			143	7	2	75	31	2
Sr			113	86	114	86	91	49

Location	CLSC					Relay Zone		
Site	CLSC	CLSC	CLSC	CLSC	CLSC	R-12	R-31	R-31
Sample	RNDB 26-2 GL	Mg# > 60 (Mean 5)	Mg# 50–59 (Mean 17)	Mg# 40–49 (Mean 10)	Mg# < 40 (Mean 7)	RNDB 12-6 GL	RNDB 31-1 GL	RNDB 31-4 GL
SiO_2	50.25	49.15	52.04	50.81	53.76	53	51.44	49.25
TiO_2	0.94	0.99	1.24	1.94	2.07	0.57	1.21	0.79
Al_2O_3	15.86	16.97	14.46	13.41	13.01	16.2	15.31	17.36
FeO*	9.79	9.19	10.7	14.1	14.44	7.95	10.61	8.65
MnO	0.19	0.18	0.2	0.26	0.24	0.14	0.22	0.17
MgO	8.25	8.48	6.71	5.63	3.82	6.89	6.28	9.1
CaO	13.06	12.14	11.37	10.21	8.34	12.28	11.6	12.82
Na_2O	2.05	2.57	2.53	2.88	3.27	1.54	2.54	2.2
K_2O	0.07	0.11	0.14	0.12	0.26	0.32	0.08	0.03
P_2O_5	0.08	0.07	0.1	0.16	0.28	0.1	0.07	0.03
Sum	100.53	99.88	99.67	99.55	99.56	99	99.37	100.41
Mg#	63.3	65.4	56.2	45	35.2	64	54.8	68.3
CaO/Al_2O_3	0.82	0.715	0.786	0.761	0.641	0.758	0.758	0.738
Trace elements (ppm)								
Cr	325							334
Ni	149					67	55	163
Co	43							44
V	323					271	326	212
Zr	51					66	83	68
Y	2929					17	31	24
Nb	2					1	3	3
Hf								1.14
Ta								<0.05
Rb	0.3					6	1	1
Ba	4					29	7	3
Sr	63					197	122	93

[a]GL = glass data. All other data are for aphyric rocks.
[b]Trace elements from aphyric rock.

4.2.2. Petrography—ELSC and CLSC

Rocks from the ELSC-CLSC ridge system are representative of nearly all of the rocks from the Lau Basin in terms of their petrography. They exhibit textures and mineralogy similar to rocks of mid-ocean ridges as summarized here.

4.2.2.1. Textures. Texturally, the dredged and cored samples from the ELSC-CLSC ridge system are mainly aphyric to sparsely phyric basalts. Highly phyric samples are relatively uncommon; those found are largely plagioclase phyric. Less common are olivine-plagioclase phyric rocks; olivine-phyric and clinopyroxene-plagioclase phyric samples are rare. Glass-rich vitrophyric chill margins are common as rinds on pillow fragments and probable sheet flows. Samples with chilled glass margins are common in the ODP Site 836 drill core samples and were useful in identifying five separate petrologic units and cooling units (Parson *et al.*, 1992). Essentially unaltered volcanic glass has been found on samples representing a wide range in composition and from rocks formed during the earliest stages of spreading on the ELSC. Some samples dredged from the VFR, discussed later, are almost entirely silica-rich glass (e.g., up to 57.5% SiO_2) with only a few microlites of plagioclase (Vallier *et al.*, 1991).

An important textural feature of all Lau Basin rocks is their highly vesicular nature. This is the only textural feature different from typical MORB samples. Vesicle contents ranging from 5% to 50% were observed in ODP drill core samples, and similar vesicularity occurs in samples dredged from ridge depths as great as 2600 m. Ernewein *et al.* (1994) contend that the ELSC samples are more vesicular (more enriched in volatiles) than CLSC rocks, but other data for the CLSC (Hawkins, 1976) do not support this. The high-silica glasses from the VFR have vesicle contents typically ranging from 10% to 25%, with some samples having up to 35%. Some vesicles are as large as 3 cm in length and spherical vesicles up to 1 cm in diameter are common (Vallier *et al.*, 1991). Glasses of comparable composition from the MTJ are equally vesicular; some samples have as much as 50% vesicles. The high vesicularity of all of the Lau Basin samples is indicative of the high volatile content of their parental magmas and may be typical of many backarc magma systems. For example, many Mariana Trough samples also are highly vesicular (Hawkins and Melchior, 1985; Hawkins *et al.*, 1990). High-water contents and high H_2O/CO_2 measured on backarc glasses (Garcia *et al.*, 1979; Muenow *et al.*, 1980; Newman, 1989) suggest that water was the main constituent forming the vesicles, but CO_2 and oxides of sulfur must have been important constituents as well (Nilsson, 1993; Nilsson and Peach, 1993; Farley, 1994).

4.2.2.2. Mineralogy. The mineralogy of the ELSC-CLSC axial ridge samples is typical of MORB assemblages. That is, they consist of Cr-spinel (SP), olivine (OL), plagioclase (PL), clinopyroxene (CPX), Fe and Ti oxides (OX), and sulfides of Fe and Cu-Fe (SU). This is the probable crystallization sequence as interpreted from petrographic study. Low-K rhyolitic samples, now off-axis from the CLSC, have minor quartz but lack alkali feldspar. These rocks may have formed close to the propagating rift tip of the CLSC.

Cr-spinel is common as an early-crystallized phase in relatively primitive basaltic magmas (i.e., least fractionated melts close in composition to the primary melts). Cr-spinel is useful in evaluating magma chemistry and evolution because the composition reflects parental melt chemistry, oxygen fugacity, pressure, and temperature during crystallization (e.g., Irvine, 1965, 1967; Fisk and Bence, 1980; Dick and Bullen, 1984; Allan, *et al.* 1988; Roeder and Reynolds, 1991; Allan, 1994). There is a good correlation between Mg# (where Mg# = (Mg/(Mg + Fe^{2+})) and Cr# (where Cr# = (Cr/(Cr + Al)) of Cr-spinel with Mg# of the melt and the forsterite (Fo) content of coexisting olivine (Hawkins and Allan, 1994). Hawkins and Melchior (1985) reported data for Cr-spinels that range from Cr# = 0.464 and Mg# = 0.735 to 0.295 and 0.706 respectively. Coexisting olivine ranges from $Fo_{89.3}$ to $Fo_{87.7}$. Sunkel (1990) reported Cr-spinel with Cr# = 0.54 and Mg# = 0.755 coexisting with olivine (Fo_{91}) in relatively primitive basalts. She also described basaltic andesites in which Cr-spinel with Cr# = 0.534 and Mg# = 0.613 coexisted with Fo_{86}. The abundance of Cr-spinel in Lau Basin samples may be a reflection of their relatively high initial fO_2 (e.g., Nilsson and Peach, 1993).

Plagioclase is the most abundant mineral in all samples and is present as phenocrysts, microphenocrysts, and as quench-textured microlites in vitrophyre. It also forms crystal aggregates with olivine, clinopyroxene, or both, and xenocrystic grains are common in some samples. The compositional range is typical of MORB (e.g., An_{60} to An_{85}). Some samples are extremely calcic and more typical of arc-composition magmas (e.g., up to An_{90}). In nearly all instances, these high-Ca grains are in rocks with compositions transitional between MORB and arc. It is important to note that plagioclase is present and abundant in all samples; thus, the elevated P_{H_2O} may have delayed onset of crystallization but did not completely suppress it.

Clinopyroxene is present in nearly all samples with Mg# <60. It occurs as a phenocryst phase as well as a groundmass constituent. Compositions range from diopside-salite to augite. The Fe/Mg ratio increases from least- to most-evolved rock compositions. Oxides are very minor constituents of all but the FeTi-enriched basalts. Both ilmenite and Ti-magnetite are present.

4.2.3. Petrology and Geochemistry—ELSC and CLSC

The general MORB-like character of crust formed at the active spreading centers is seen both in their mineralogy and chemistry. However, minor but distinctive chemical compositional features set some of the samples apart from N-MORB *sensu stricto*, and each ridge also has its own chemical characteristics. A summary of representative rock types dredged from the ELSC axial ridge, areas of ELSC crust sampled by drilling at ODP Site 836, and from the CLSC is in Table I and Figs. 3.7a–7d. Similar data for the Leg 135 drill sites are in Table II and Figs. 3.8a–8d for comparison. Trace element ranges, normalized to N-MORB, are in Fig. 3.9. Sample locations are in Fig. 3.6.

The chemical distinctions between MORB and the axial ridges samples are mainly due to subtle large-ionic-radius lithophile elements (LILE) enrichments and HFSE depletions that give variable similarity to arc tholeiitic magmas. This has been seen in other backarc basins (e.g., Mariana Trough; Hawkins and Melchior, 1985), and I consider it to be a reflection of the SSZ environment in which the magmas formed. The axial zone of the ELSC varies in composition along strike from N-MORB-like in the north to more arclike rocks in the south. This along-strike variation may in part be controlled by the increasing proximity to the trend of the Tofua arc and its SSZ mantle source as proposed by Ernewein *et al.* (1994). Sunkel (1990) describes samples from the axial ridge south of 21°50′S as being low-K tholeiitic basalts, depleted in HFSE, but enriched in LILE relative to N-MORB (i.e., showing more of the SSZ component). The southernmost part of the ELSC,

TABLE II
Representative Analyses, ODP Leg 135 Core Samples

Location Site Sample	WEB ODP 834 A Unit 1 12XCC/13-15	ODP 834 B Unit 7 mean (17)	ODP 834 B Unit 7 GL 29R1/07-12	ODP 834 B Unit 12 56R1/07-15	ODP 834 B Unit 13 GL[a] 59R2/16-24	ODP 835 Unit 1 GL[a] mean (27)	ODP 835 6R1/32-37
SiO_2	49.78	50.2	50.17	52.2	51.18	52.05	50.35
TiO_2	1.4	1.32	1.42	2.11	1.33	1.08	1.09
Al_2O_3	17.51	16.33	16.33	15.71	16.14	15.52	16.47
FeO*	8.47	8.78	8.67	12.41	9.08	9.99	8.46
MnO	0.15	0.17	0.2	0.2	0.19	0.19	0.16
MgO	5.06	7.73	7.73	3.4	6.1	6.12	7.67
CaO	12.71	11.88	11.87	8.58	11.41	11.14	12.9
Na_2O	3.08	3	3.08	3.87	2.98	2.35	1.82
K_2O	0.26	0.1	0.09	0.73	0.14	0.21	0.28
P_2O_5	0.15	0.12	0.11	0.21	0.1	0.12	0.11
Sum	99.48	99.63	99.67	100.78	98.66	98.77	100.23
Mg#	55	64.1	64.6	36	57.9	55.7	65
CaO/Al_2O_3	0.73	0.73	0.73	0.55	0.71	0.72	0.78
Trace elements (ppm)	[b]						
Cr	267			0			179
Ni	53		102	12			78
Co							
V			227				
Zr	120		102	146	79		44
Y	28		29	40	27		22
Nb	2		3	1	0.95		1
Hf							
Ta							
Rb	5		3	13	1.46		5
Ba	41		22	60	29		42
Sr	245		176	195	169		130

Location	WEB					
Site	ODP 837	ODP 837	ODP 838	ODP 839 A	ODP 839 B	ODP 839 B
Sample	Unit 1 mean (5)	Unit 1 4$1/36-41	Unit 1 GL 9H2/95-117	Unit 1 24X1/00-06	Unit 2 18R1/54-60	Unit 3 GL 22R1/19-23
SiO_2	55.16	55.23	52.95	53.03	54.92	53.21
TiO_2	1.34	1.33	0.92	0.65	0.9	0.7
Al_2O_3	15.34	15.34	15	15.06	16.08	15.9
FeO*	11.7	11.67	11.19	9.04	10.1	8.71
MnO	0.2	0.2	0.23	0.17	0.18	0.17
MgO	3.5	3.56	5.64	9	4.64	6.43
CaO	8.08	8.12	10.72	11.66	10.08	12.24
Na_2O	2.87	3.06	1.93	1.35	1.9	1.8
K_2O	0.69	0.75	0.35	0.23	0.56	0.32
P_2O_5	0.21	0.23	0.09	0.07	0.12	0.09
Sum	99.09	99.5	99.01	101.26	100.62	99.57
Mg#	38	38.5	50.8	67.1	48.5	60.2
CaO/Al_2O_3	0.53	0.53	0.96	0.77	0.63	0.77
Trace elements (ppm)						
Cr				618	7	
Ni		5		123	1	
Co						
V		314				
Zr		79		32	49	
Y		35		16	21	
Nb		5		1	0	
Hf						
Ta						
Rb		14		5	8	
Ba		80		78	85	
Sr		165		148	191	

[a]GL = glass data. All other data are for aphyric rocks.
[b]Major elements on glass trace elements, aphyric rock.

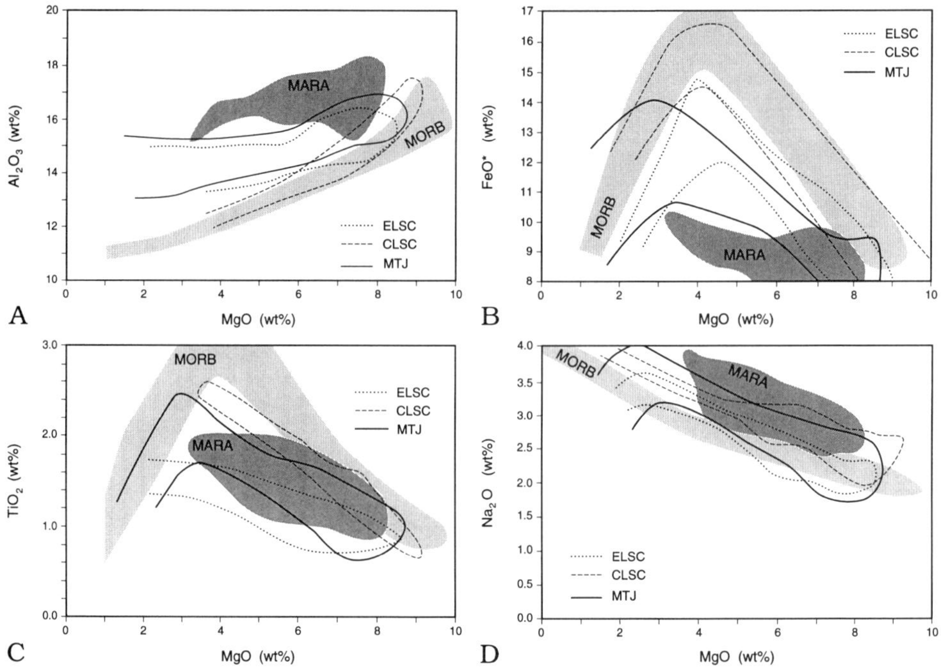

FIGURE 3.7. Variation diagrams showing major elements (A) Al_2O_3, (B) FeO*, (C) TiO_2, (D) Na_2O versus MgO. Fields shown are mid-ocean ridge basalt (MORB) (after Hochstaedter *et al.*, 1990a) and Mariana Trough glasses (MARA) (Hawkins *et al.*, 1990). Other fields are eastern Lau spreading center (ELSC), central Lau spreading center (CLSC), Mangatolu triple junction (MTJ). Data sources are discussed in the text.

the Valu Ridge, includes "oceanic andesite" (e.g., 55–60% SiO_2) having chemical and isotopic signatures resembling the nearby Tofua arc (Vallier *et al.*, 1991).

The ELSC began to form at about 5.5–5 Ma by rift propagation into older (pre–late Miocene) crust that may have been part of the forearc to the Lau Ridge or part of the arc (Fig. 3.1). Most of the ELSC data are from samples dredged on the neovolcanic zone of the ridge axis. Other data are from some of the older ELSC crust (about 0.6–0.8 Ma) that was drilled at ODP Site 836 and from dredge samples. Magmas that formed the northern part of the ELSC are mainly MORB-like tholeiitic basalt in terms of major and trace elements and element ratios (e.g., Bryan *et al.*, 1976; Bryan, 1979b; Viereck *et al.*, 1989). They range from relatively unfractionated (glasses have up to 8% MgO, Mg# 64, 0.9% TiO_2, <0.15% K_2O) to moderately fractionated (4–5% MgO and Mg# 43–50). There are lesser amounts of more strongly fractionated types, including Fe-Ti-enriched basalt (up to 2.1% TiO_2 and 14.8% FeO*). The range in composition of ELSC lavas is easily explained by low-pressure fractional crystallization of observed phenocryst assemblages.

ODP Site 836 is located close to the boundary between the WEB and probable ELSC crust. Five petrologic units were recognized. Units 1 and 2 (14–16 mbsf) are andesitic gravels probably derived from a nearby arc-composition seamount. The other three are MORB-like basaltic units which lie in the data fields for ELSC on Fig. 3.8; this supports the interpretation that they are genetically related to the ELSC and are some of the earliest ELSC crust formed at that latitude. Their trace element and REE abundances also resemble

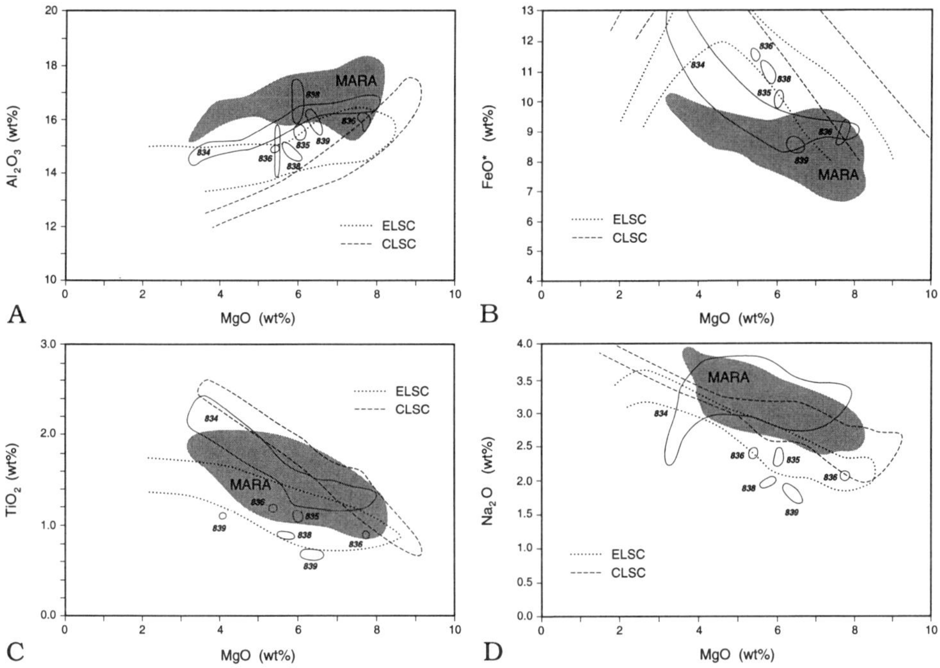

FIGURE 3.8. Variation diagrams showing major elements (A) Al_2O_3, (B) FeO*, (C) TiO_2, (D) Na_2O versus MgO. Numbered fields are for ODP Leg 135 drill sites (Hawkins, 1994). Other fields are as in Fig. 3.7.

the ELSC (Figs. 3.9a and 3.10a). Site 836 trace element data are discussed together with the ELSC data in a subsequent section.

Representative ELSC data are in Table I, and plots of major element composition versus MgO are shown in Fig. 3.7. ELSC samples with more than 7% MgO overlap the MORB and Mariana Trough fields for Al_2O_3. This may reflect higher f_{O_2}, which in turn may be due to higher water content in parental melts. Samples with lower MgO define fields that are shifted toward the Mariana Trough field for Al_2O_3 and lie parallel to the MORB field, but at a lower level, for FeO*. The TiO_2 field is subparallel to Mariana Trough but shows depletion. The Fe and Ti behavior also reflects the relatively higher water content than MORB. The behavior of Al, Fe, and Ti in fractionated samples resembles the (SSZ) trends seen in the Mariana Trough (e.g., Fryer *et al.*, 1981; Hawkins and Melchior, 1985; Hawkins, 1989; Hawkins *et al.*, 1990). The Na_2O field is MORB-like, and ELSC glasses lack the relatively higher-K content that distinguishes Mariana Trough (transitional to arc-type) basalts from MORB.

The ELSC magma chemistry appears to have remained fairly constant over at least 2 m.y., as shown by the similarity between axial ridge samples and dredge samples from the older part of the eastern limb of the ELSC (e.g., 123-74-1, Table I). An approximate age of about 2 Ma for this dredge site can be estimated from the model of Parson and Hawkins (1994). This relatively unfractionated sample (Mg#=62.9) has the Al, Fe, Ti characteristics of the Mariana Trough transitional basalts, although it has low K_2O. It is similar in most respects to the modern lavas of the axial zone. The record of ELSC magmatism from about 2 Ma to the present suggests a fairly uniform long-term composition for the magma

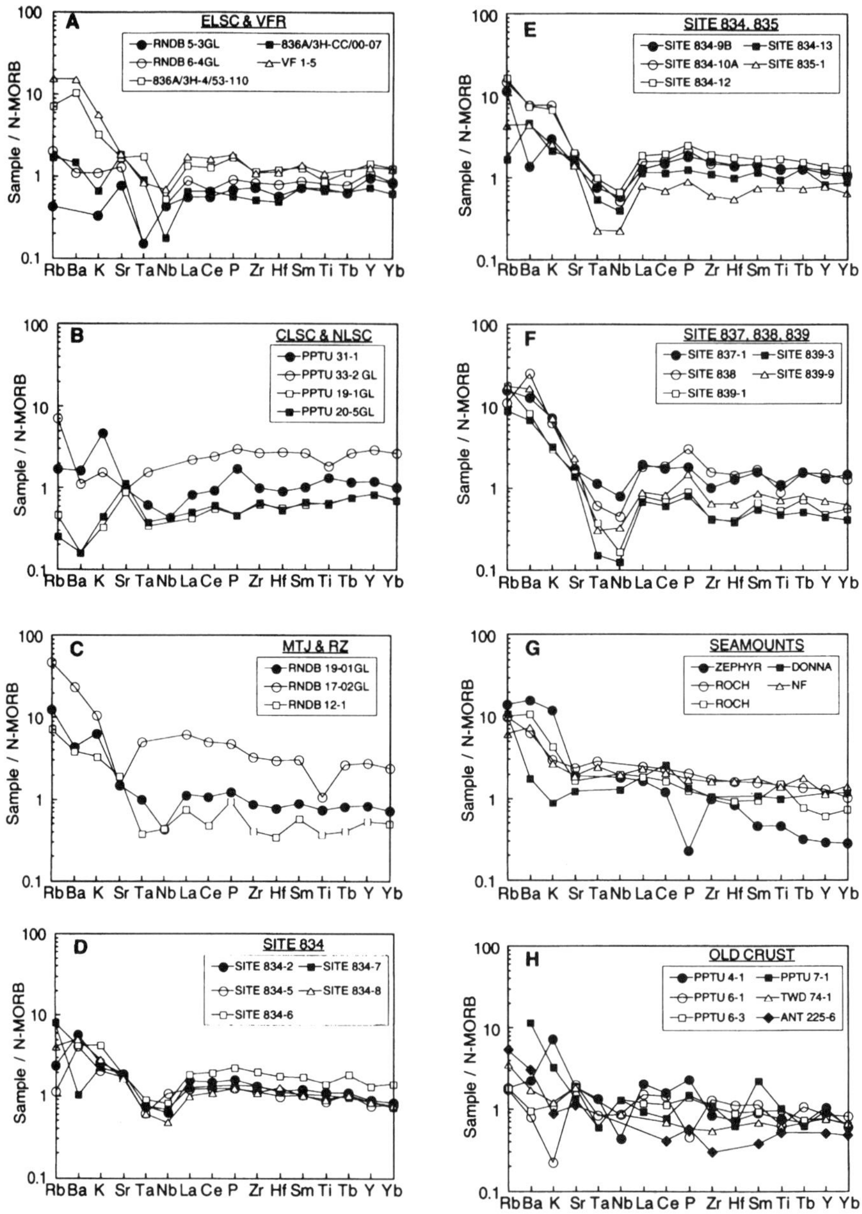

FIGURE 3.9. Trace elements normalized to N-MORB values of Sun and McDonough (1989). Note the relative depletion in Nb and Ta and enrichments in LILE that characterize SSZ magmas. (A) eastern Lau spreading center (ELSC) and Valu Fa ridge (VF). (B) central Lau spreading center (CLSC) and northwestern Lau spreading center (NLSC). (C) Mangatolu triple junction (MTJ) and relay zone (RZ) between ELSC and CLSC. (D,E,F) ODP drill sites (Hawkins *et al.*, Allan, 1994). (G) seamounts in northern basin, ROCH is Rochambeau Bank, and NF is Niuafo'ou Island. (H) PPTU samples are from northeastern Lau Basin (locations in Hawkins, 1988). TWD 74-1 is from eastern side of the CLSC and ANT 225-6 is from west side of the ELSC (locations in Hawkins, 1976; Hawkins and Melchior, 1985).

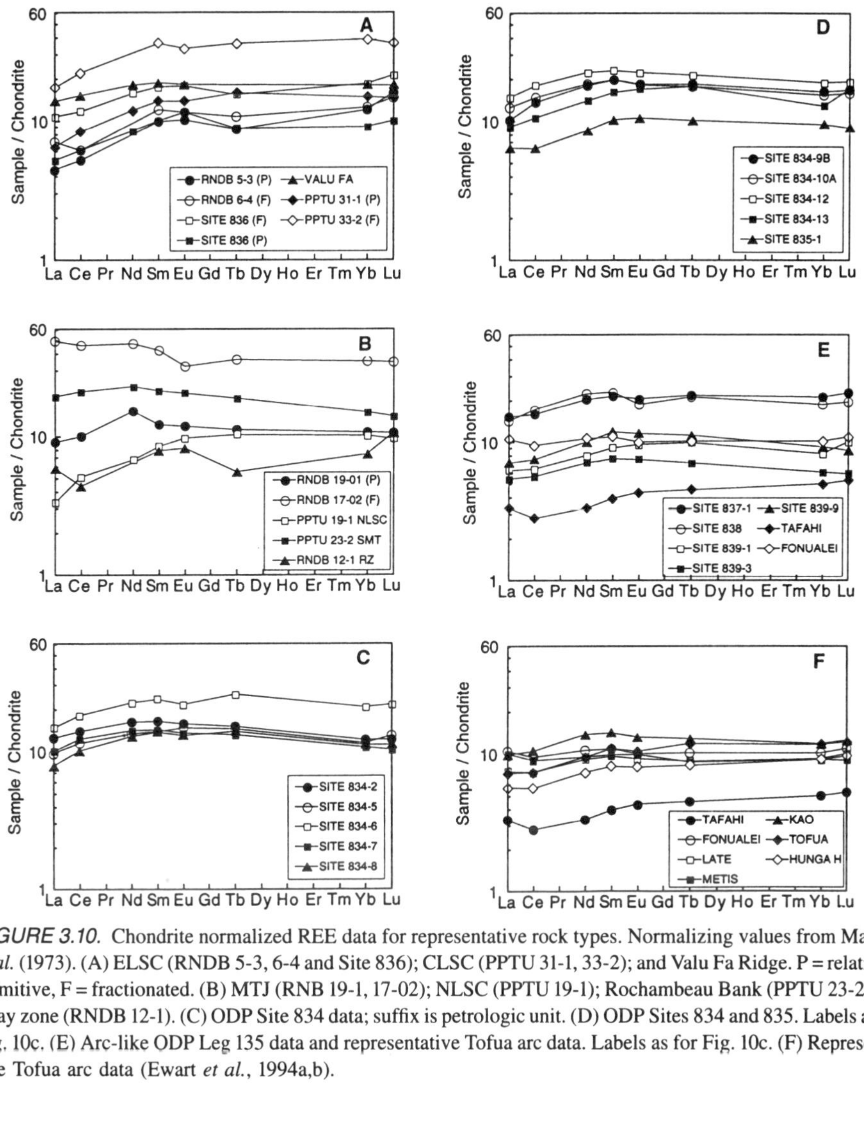

FIGURE 3.10. Chondrite normalized REE data for representative rock types. Normalizing values from Masuda *et al.* (1973). (A) ELSC (RNDB 5-3, 6-4 and Site 836); CLSC (PPTU 31-1, 33-2); and Valu Fa Ridge. P = relatively primitive, F = fractionated. (B) MTJ (RNB 19-1, 17-02); NLSC (PPTU 19-1); Rochambeau Bank (PPTU 23-2) and relay zone (RNDB 12-1). (C) ODP Site 834 data; suffix is petrologic unit. (D) ODP Sites 834 and 835. Labels as for Fig. 10c. (E) Arc-like ODP Leg 135 data and representative Tofua arc data. Labels as for Fig. 10c. (F) Representative Tofua arc data (Ewart *et al.*, 1994a,b).

sources. The main variability in ridge chemistry seems to be established near the rift tip or as the rift approaches the Tofua arc (e.g., Ernewein *et al.*, 1994).

The CLSC also formed by a propagating rift that penetrated crust formed by the ELSC (Fig. 3.1). It was initiated at about 1.5–1 Ma. In general, it comprises rock types like those for the ELSC and exhibits a similar range in composition. CLSC rocks have the chemical characteristics of MORB and are even closer to MORB in composition than are the ELSC rocks. Representative data are in Table I. CLSC rocks range from relatively unfractionated basalts, represented by glasses with up to 8.8% MgO and Mg# = 67.5, to more evolved Fe-Ti-enriched basalt glasses with 4.8% MgO, Mg# = 39.9, TiO = 2.31%, and FeO* = 14.7%.

Rocks from near the propagating rift tip include some low-K, high-silica andesites and low-K rhyolite (RNDB-13-2; PPTU 29-3, Table I) that probably are extreme fractionates formed as the rift penetrated older cold crust.

The CLSC magma chemistry appears to represent a marked change in the source chemistry from that which fed the ELSC. Sample ANT-229 (Table I) was collected from a prominent scarp near the inferred boundary between older crust generated by the ELSC and younger crust formed by the CLSC (Fig. 3.1). We can estimate the propagation rate for the CLSC from its length and the proposed age of 1.2 to 1.5 Ma for its initiation (Parson and Hawkins, 1994). The estimated crustal age at sample site ANT-229 is between 0.75 and 0.94 Ma. The dredged material could be either the youngest remnant of crust formed by the ELSC at this latitude or the oldest formed by the CLSC propagator. The glass data indicate a "primitive" Chr-Ol-Plag basalt with 8.73% MgO, Mg# = 67, and 0.88% TiO_2; K_2O is 0.03%. When plotted on the variation diagrams for oxides of Al, Fe, Ti, and Na, it is clear that it is part of the CLSC magma series and distinctly different from ELSC. Compare these data with those for sample 123-74-1, which was collected at essentially the same longitude but on crust that formed at the ELSC.

Representative CLSC data are in Table I, and plots of major element composition versus MgO are shown in Fig. 3.7. CLSC samples with more than 7% MgO plot in the MORB field for Al_2O_3 and thus are similar to the ELSC; the CLSC data follow the MORB trend at lower levels of MgO and define a distinctly different trend than for the ELSC. The field for Na_2O is displaced away from MORB and ELSC toward the Mariana Trough glasses. Both FeO* and TiO_2 define fields very close to those for MORB and suggest that f_{O_2} was lower than for the ELSC, and presumably the source was less hydrous. Samples with lower MgO define fields that are shifted toward the Mariana Trough field for Al_2O_3 and run parallel to the MORB field, but at a lower level, for FeO*. The TiO_2 field is subparallel to Mariana Trough but shows depletion, whereas the Na_2O field is MORB-like. The compositional range for the ELSC and CLSC may be explained in terms of low-pressure fractional crystallization of observed phenocryst assemblages. The spectrum of glass compositions for both the ELSC and CLSC lie close to the presumed low-pressure, anhydrous, multiply saturated cotectic as projected from PL in the OL-CPX-QTZ diagram (Hawkins and Allan, 1994), supporting the inference that fractionation may explain the compositional range.

Samples from the relay zone between the CLSC propagator and the dying north end of the ELSC have both transitional and MORB-like chemistry (Table I and Fig. 3.9c). Two samples with Mg#=68 are compared. Sample RNDB 12-6 shows the SSZ signature of having depletion in HFSE and elevated LILE (e.g., 0.3% K_2O, 24 ppm Ba), whereas RNDB 31-4 has HFSE somewhat depleted relative to MORB but lacks the elevated LILE (K_2O = 0.03%, 7 ppm Ba).

High volatile contents, mainly water, are a distinctive feature of the Mariana Trough and many Lau Basin samples (Garcia *et al.*, 1979; Hawkins and Melchior, 1985; Newman, 1989). This high level of volatiles is reflected in the high vesicularity of Lau Basin rocks. There are limited data for H_2O, CO_2, or S measured in ELSC or CLSC basaltic glasses; the range is 0.42–1.55%, 0–91 ppm, and 310–1880 ppm respectively (Nilsson, 1993; Nilsson and Peach, 1993). High volatile levels also are distinctive of samples from other backarc basins, for example, East Scotia Sea (Muenow *et al.*, 1980) and Sumisu rift, (Ikeda and Yuasa, 1989; Hochstaedter *et al.*, 1990a,b). Dick (1982) inferred high volatile contents as an explanation for the high vesicularity of Shikoku Basin rocks.

4.3. Valu Fa Ridge

4.3.1. Introduction

The southern end of the ELSC extends to at least 22°50′S, where it loses its morphologic identity in a zone of rough topography. The southernmost part of the ELSC, south of 21°20′S, is known as the Valu Fa Ridge (Morton and Sleep, 1985). The VFR is a well-defined, linear feature that extends for at least 165 km more or less parallel to the Tonga Ridge and approaches to about 40 km of the volcanic island of Ata in the Tofua arc. The VFR is propagating southward into older crust of uncertain composition. This crust may be equivalent to the rifted forearc terrane that forms the basin-range seafloor of the western Lau Basin.

The VFR rises about 800 m above the surrounding seafloor shoaling to depths of 1500 m. The ridge is narrow, about 10 km wide or less, lacks a crestal axial rift valley, and is capped by many small domelike edifices (Collier and Sinha, 1992; Wiedicke and Collier, 1993). It is offset about 3 km, in a left-lateral sense, near 22°12′S and there are several other minor offsets of the axis. Magnetic anomaly data indicate that the ridge has been spreading for at least 0.7 to 0.9 Ma, and the spreading rate is estimated to be about 6–7 cm/yr (Morton and Pohl, 1990). Multichannel seismic reflection data and seismic refraction lines give evidence for a strong reflection at about 3.5 km below seafloor (Morton and Sleep, 1985). Additional data indicate the depth as 3.2 ± 0.2 km below the seafloor (Collier and Sinha, 1992). The polarity of the reflection phase is shifted 180° as expected if this were the top of a low-velocity zone. It has been interpreted as the top of a flat-topped magma chamber about 2–3 km wide (Morton and Sleep, 1985). The magma chamber (lens?) probably extends along the ridge crest for at least 20 km between 22°10′S and 22°30′S (Collier and Sinha, 1992). These authors see a correlation between the width of the VFR crest and the width of the magma chamber relector. The offset in the ridge near 22°12′S is interpreted as an overlapping spreading center with a magma chamber under the overlap basin.

Hydrothermal circulation and metallogenesis on the VFR ridge crest is evidenced by high concentrations of particulate and dissolved Mn, by elevated ^{3}He, ferromanganese crusts up to 1 cm thick on apparently young fresh samples of lava dredged from the ridge crest, by sulfide mineralization, and by active hydrothermal vents. These features are dicussed in a separate section.

4.3.2. Petrography—VFR

The main rock types recovered from the VFR are highly vesicular, glass-rich blocky, tabular or bulbous fragments having an andesitic composition. Some samples have ropy surfaces and flow banding. Other types include andesitic breccias, ferromanganese crusts, and breccias of smectite and ferromanganese minerals. Photographs and imagery of the seafloor show "viscous domes," rare large pillows, thin sheets, and irregular rubble piles of lava fragment; some of the rubble piles are cemented by ferromanganese minerals and nontronite (Vallier *et al.*, 1991).

The samples are mainly vitrophyric or aphyric to sparsely phyric fine-grained rocks. Plagioclase is the main crystalline phase, but it is rare. It forms both microphenocrysts (An_{87-77}) and microlites (An_{62}-An_{38}). These have low K content (e.g., $Or_{0.1-0.4}$). Clinopyroxene is rare. It occurs as microphenocrysts; some are compositionally augite, others are subcalcic or pigeonitic. All are compositionally distinct from pyroxenes of the Tofua arc, including samples from Ata. The VFR pyroxenes all are Fe-rich with from 21 to 30% mol proportion Fs. Very minor amounts of Fe-Ti oxides form grains less than 0.1 mm in size.

4.3.3. Geochemistry—VFR

The glass samples analyzed are mainly low-K andesites ranging in composition from 54% to 60% SiO_2, 0.36–0.57% K_2O (Davis *et al.*, 1990; Vallier *et al.*, 1991). Representative analyses are in Table I. Sunkel (1990) reported the presence of three samples of high-silica dacitic glasses with from 65.5% to 69.8% SiO_2, 0.44% to 0.64% TiO_2, and 0.62% to 0.87% K_2O. Davis *et al.* (1990) included analyses of two basalt samples from the VFR that are moderately fractionated (MgO 5.66–6.07%). A plot of total alkalis versus silica (not shown) indicates that the suite is calcic. TiO_2, Na_2O, and FeO* all are relatively high when compared to Tofua arc data at comparable levels of MgO (Vallier *et al.*, 1991).

Trace element data show that the VFR samples are typical of other Lau Basin samples (Jenner *et al.*, 1987; Vallier *et al.*, 1991). That is, they have LILE enrichment and HFSE depletion, especially Ta and Nb, when compared to MORB, as seen in Fig. 3.9. However, these relative depletions and enrichments, while following the arc or SSZ pattern, are not as extreme. The chondrite normalized REE patterns (Fig. 3.10a) have a slightly convex upward (MORB-like) shape similar to the ELSC and CLSC rather than the flat pattern considered typical of arc tholeiites. The overall enrichment can be explained as the result of olivine and plagioclase fractionation. The shapes of the VFR patterns are similar to most of the Mariana Trough glasses and many of the Lau Basin samples (Hawkins, 1994; Hawkins and Melchior, 1985). However, note that the pattern also is similar to that for the Tofua arc volcano of Kao (Fig. 3.10f). Ratios of LILE to light rare-earth elements (LREE) for VFR, such as K/La, are enriched and resemble ratios for arc volcanics (Jenner *et al.*, 1987). The VFR data lie in the field defined by the Tofua arc on the Hf/3-Ta-Th diagram (Fig. 3.11).

4.3.4. Petrogenesis—VFR

The origin of this assemblage of intermediate composition rocks that forms the carapace of a very young segment of the ELSC is problematic; mixing of melts or sources, fractionation, and assimilation all may be involved. VFR rocks have many similarities to the Tofua arc, yet they also have isotopic and elemental characteristics of the ELSC and CLSC. Their major element chemistry resembles "oceanic andesites," like those found on the Galapagos Ridge (Sinton *et al.*, 1983) or on the CLSC, that are explained as the result of low-pressure crystal fractionation.

Vallier *et al.* (1991) summarize the elemental data as follows: A similarity to Tofua arc lavas is seen in the high-alkali metal and alkaline-earth values and ratios such as Rb/Cs and Ba/La. The depletions in HFSE are arclike, and the depletion in Nb gives high arclike ratios of Ba/Nb. An origin by differentiation of backarc basin basalt is suggested by the similarity to backarc signature shown in values for Sr/REE, Rb/Zr, and Ba/Zr lower than for most arcs. Higher values for Fe, Ti, and V than seen for arcs, as well as MORB-like Ti/V, are typical of backarc lavas. The arclike, SSZ signature would be an expected consequence of mixing of sources, or assimilation of older altered crust, at the tip of a propagating backarc rift into previously undisturbed SSZ lithosphere. This helps to explain the trace element and

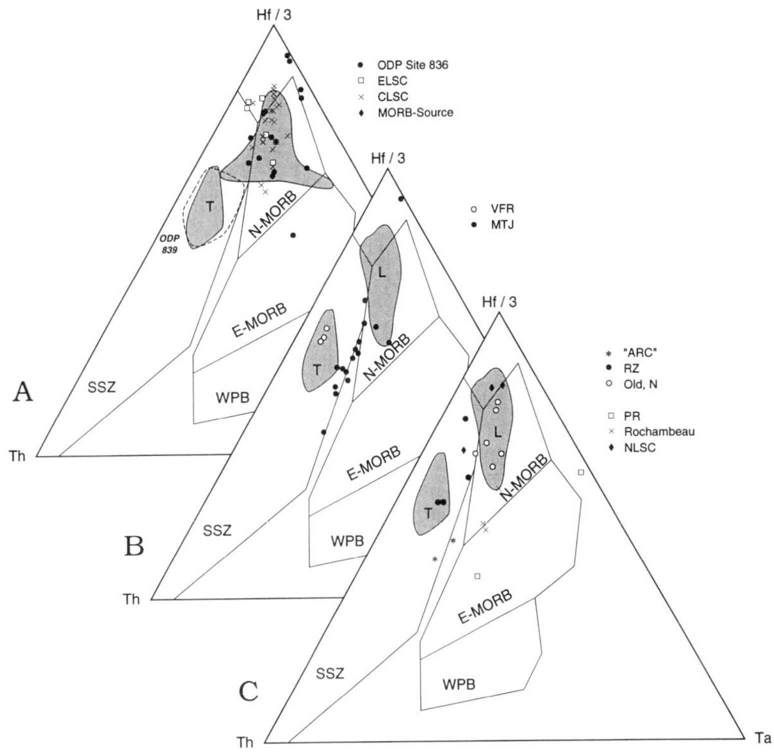

FIGURE 3.11. Discriminant diagram based on HFSE after Wood (1980) and Wood *et al.* (1981). WPB = within plate basalt. Other acronyms and labels as defined in the text. (A) Axial ridge system data. Shaded field is for Site 834 data (Hawkins, 1994). T = Tofua arc (Ewart *et al.*, 1994a,b). (B) Valu Fa Ridge and Mangatolu triple junction. Shaded field L=axial ridge from Fig. 11a, T = Tofua arc. (C) Old crust. ARC = seafloor southeast of MTJ; Old, N = crust northwest of MTJ; RZ = relay zone; PR = Peggy Ridge; NLSC = northwestern Lau spreading center.

REE patterns shown in Figs. 3.9a and 10a. Their similarity in Nd-Sr-isotope ratios to the Lau Basin axial ridge samples, and the Pb-isotopic similarity to Tofua arc lavas, suggest either mixing of sources or mixing of magmas. The isotope data are discussed in a separate section.

4.4. Mangatolu Triple Junction

4.4.1. Introduction

A northeasterly trending spreading axis was discovered in the northeastern Lau Basin, and samples were collected, on Scripps Institution of Oceanography PAPATUA Expedition, 1986. At one site we collected high-silica rocks from a small dome and recovered fragments of zinc sulfide-enriched hydrothermal vent chimneys—the first to be found in the western Pacific (Hawkins, 1986; Hawkins and Helu, 1986). Subsequently we discovered, charted, and sampled the other two limbs of this spreading system on SIO ROUNDABOUT expedition (Hawkins *et al.*, 1989; Nilsson *et al.*, 1989; Nilsson, 1993).

4.4.2. Petrography and Petrology

The samples include basaltic pillow fragments with glass-rich chill margins and blocky fragments of silica-rich vitrophyric material of broadly andesitic composition. The basaltic rocks are sparsely to moderately phyric comprising euhedral to subhedral phenocrysts of plagioclase, olivine, and clinopyroxene surrounded by glass, multiphase assemblages of spherulitic textured or finely crystalline material. Fe-Ti oxides are minor constituents and minute globules of Fe or Cu-Fe sulfides, 10 to 50 microns in diameter, are rare. The rocks are commonly vesicular, with vesicle contents as high as 20% and some vesicles as large as 5 mm in maximum dimension. As for other Lau Basin samples, a high content of volatiles is implied. Data for a limited number of samples have verified this (Nilsson and Peach, 1993).

The more evolved rocks have basaltic andesite and andesite bulk compositions. Nearly all of the samples are vitrophyres, with varied abundances of vesicles, and are sparsely phyric to nearly aphyric. Typically the dredged fragments are angular or block shaped, many have flow structures delineated by aligned microphenocrysts and trains of millimeter-sized vesicles. The sulfide-enriched chimney fragments comprise wurzite and sphalerite, with lesser amounts of chalcopyrite and pyrite, in a matrix of amorphous silica and barite (Hawkins, 1986).

Representative data for the range in rock types are in Table III and are summarized in Figs. 3.8–14. They share many of the characteristics of the VFR samples for major and trace elements, but in general are much more like the ELSC-CLSC lavas than like those of the Tofua arc. For example, on the Harker diagrams (Fig. 3.7), the trends for oxides of Al, Fe, Ti, and Na are MORB-like, although the field for Ti is intermediate to ELSC and CLSC; it coincides with the Mariana Trough field. The REE and the extended REE plots (Figs. 3.9 and 10) also resemble MORB as do some of the data shown on the Hf/3-Ta-Th diagram (Fig. 3.11). Samples with the highest Th levels lie close to, but not within, the SSZ field. Thus, even though lying close to the Tofua arc, and with one limb projecting into it, the MTJ samples are like those of the NLSC and CLSC. Nilsson (1993) concluded that major element variations can be explained by extensive low-pressure fractionation of plagioclase, olivine, clinopyroxene, and titanomagnetite. Despite the proximity to the Tofua arc, the MTJ lavas have been derived from a different mantle source.

5. *WESTERN EXTENSIONAL BASIN*

5.1. Introduction

The presently active spreading centers create new crust from mantle-derived melts with little evidence for a subduction component. The spreading centers have replaced older crust that formerly was part of the forearc to the Lau Ridge. The WEB has horsts of older crust separated by grabens partly filled with younger basalt. Some of the older crust of the WEB may be remnants of the Lau Ridge volcanic arc, some of it may be formed of basalts erupted in the early stages of extension of what had been the forearc of the Lau Ridge, and, possibly, there may be trapped fragments of old Pacific lithosphere. None of the latter has been identified. Part of the present Tonga Ridge is underlain by low-K rhyolites, dated at about 44 Ma, which formed in an arc setting, presumably an ancestral Lau Ridge. This was drilled at ODP Site 841 (Bloomer *et al.*, 1994). Other remnants of Eocene age arc crust, as well as Miocene-age volcanic units that probably were part of the Lau Ridge, are exposed

TABLE III

Representative Analyses, Peggy Ridge, NLSC and MTJ

Location	Peggy Ridge			NLSC		MTJ	
Site	L-95	L-86	L-103	P-19	P-20	R-17	R-19
Sample	123 95-12 GL	123 86-1	123 103-1 GL	PPTU 19-1 GL	PPTU 20-5 GL	RNDB 17-2 GL	RNDB 19-1 GL
SiO_2	50.42	49.66	52.84	49.17	48.19	59.98	50.89
TiO_2	0.34	2	1.75	0.82	0.79	1.37	0.94
Al_2O_3	15.75	14.65	15.37	17.15	17.41	15.16	15.93
FeO*	8.59	11.88	9.46	8.72	9.11	10.5	8.53
MnO	0.16	0.29	0.19	0.18	0.18	0.23	0.17
MgO	9.78	6.77	4.72	9.51	8.98	1.91	7.52
CaO	14.67	11.54	9.35	12.75	12.42	5.62	13.1
Na_2O	0.99	3.01	3.35	2.11	2.34	3.89	2.14
K_2O	0.01	0.16	1.01	0.03	0.04	0.97	0.21
P_2O_5	0.02	0.18	0.5	0.04	0.04	0.42	0.11
Sum	100.73	100.14	98.54	100.62	99.69	100.05	99.54
Mg#	70	53.9	50.5	69.1	66.9	27.2	64.4
CaO/Al_2O_3	0.93	0.79	0.61	0.74	0.71	0.37	0.82
Trace elements (ppm)							
Cr	670	700		741	470	39	309
Ni	271	380		312	221	6	90
Co	75			56	53	15	41
V	252			212	215	51	226
Zr	7			44	48	238	64
Y	11			20	21	78	24
Nb	3			1		9	1
Hf				1.15	1.08	6.09	1.59
Ta				0.045	0.045	0.65	0.133
Rb	0.153	2.9		0.3	0.1	27	7
Ba	1.16	30		1	1	147	27
Sr	12	109		79	100	135	133

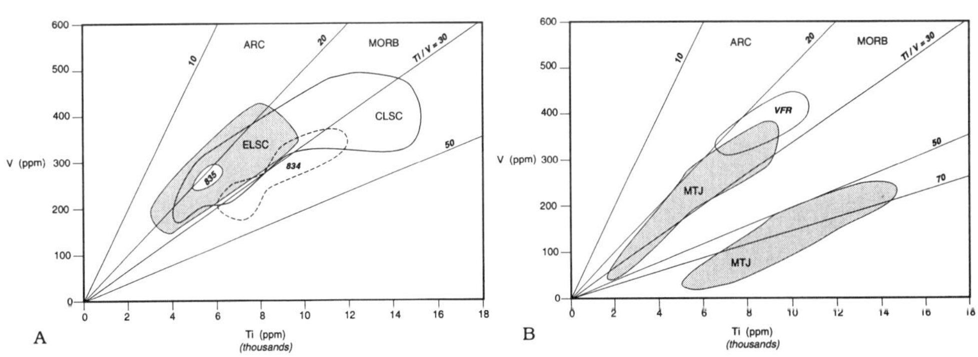

FIGURE 3.12. Ti-V plot. ARC and MORB fields from Shervais (1982). Field labels as for Figs. 3.10 and 3.11. ELSC field includes Site 836 data.

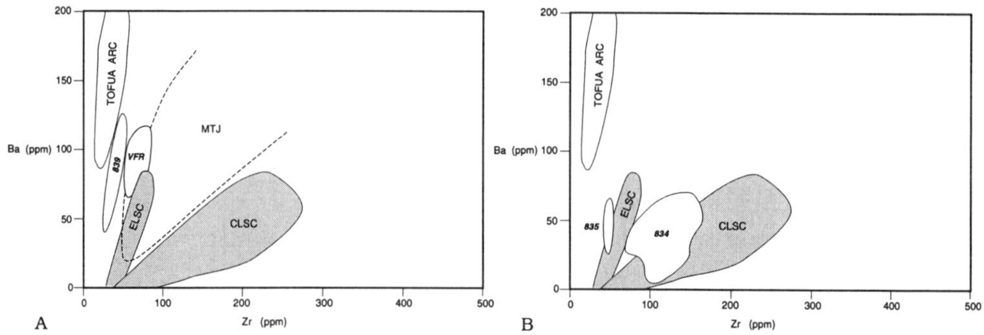

FIGURE 3.13. Ba-Zr plot. ELSC field includes data for Site 836.

on 'Eua in the present forearc (Cunningham and Anscombe, 1985). We know very little about the seafloor north and northeast of the Peggy Ridge, which may be comparable to the WEB. Dredged samples from several ridges and scarps in the WEB are compositionally transitional between MORB and arc (Hawkins and Melchior, 1985). Similar rock types were dredged on the eastern edge of the Lau Basin from similar morphologic features. This suggests that they were formerly part of the same petrologic/tectonic province.

The backarc crust of the WEB was sampled at six drill sites on ODP Leg 135. All but one of these sites were in subbasins that had opened during the early extension and rifting history of the basin. Three sites sampled MORB-like crust, and three were on arclike or transitional crust. One site (836) was on crust formed very early in the spreading history of the ELSC. A comprehensive discussion of the petrology of these drill sites is in Parson *et al.* (1992) and Hawkins *et al.* (1994). A brief summary of the drill site data and the dredged samples is presented in the following section.

5.2. Old Crust

There are large areas of seafloor that form ridges between the subbasins drilled in the WEB. For want of a better name I will call this "old crust." The name is also used for seafloor on the eastern side of the axial ridge system, in the northeastern part of the basin near the Mangatolu triple junction and for much of the region north of Peggy Ridge. Hawkins and Melchior (1985) proposed that the crust of the Lau Basin was compositionally

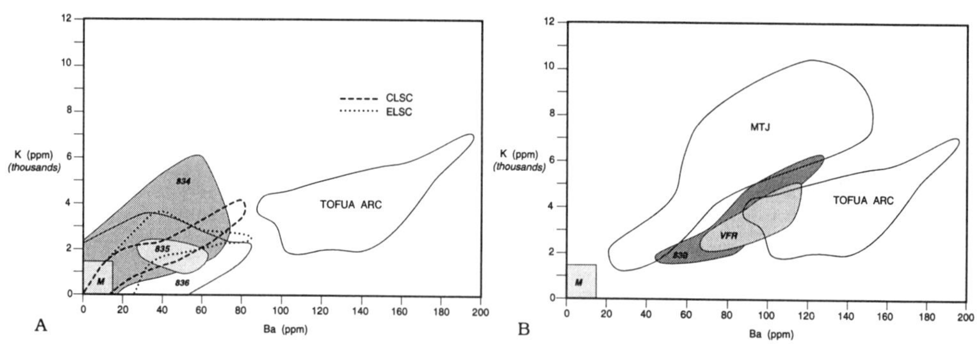

FIGURE 3.14. K-Ba plot. M is field for fresh N-MORB (Viereck *et al.*, 1989).

zoned from a MORB-like central area on the axial ridge systems to more arclike, or transitional to arc, composition crust on the outer margins. The "old crust" constitutes the outer margin. It is likely that this term includes a range in crustal ages; for most of it only a relative age can be assigned. It should be emphasized that the zoned pattern was based on analyses of glasses and fresh aphyric rocks and is not a reflection of alteration. The general pattern of the zonation, and locations of sample used to infer it, are in Fig. 3.6. Some representative analyses are in Table IV.

Examples of dredged samples from ridges in the WEB include RNDB 1-1 from a ridge crest west of Site 836, and RNDB 2-7, 2-8 (Table IV); the latter are from east of Site 836 but resemble older ridges to the west. DSDP Site 203 was drilled in a subbasin; it did not reach basement but recovered a small basalt cobble deep in the core. Glass in the vitrophyric rim has transitional chemistry (e.g., 52% SiO_2, 0.7% TiO_2, 0.28% K_2O in a sample with Mg#=66.1; Fryer *et al.*, 1981). It is likely that it came from the adjacent intrabasin ridge. There is but one sample from the southern end of the WEB, F-114 on Fig. 3.6, but it also shows the transitional chemistry of the postulated zonation. Rocks from dredge sites 123-81, PPTU 34 and ANT 223 (Table IV), are from crust which Hawkins and Melchior (1985) interpreted as transitional between MORB and arc. ANT-223 is from close to the western side of the ELSC crust about 100 km off the present axis and southwest of ODP Site 836. Sample 123-81 is from the WEB northwest of Site 837. Sample 123-101-4 is from the lower slopes of the Lau Ridge and represents either a part of the arc or some of the early formed backarc crust. Examples of old crust from the northeastern part of the basin, adjacent to the Mangatolu triple junction, are PPTU 4-1, 6-1 and 7-1. PPTU 4-1 and 7-1 have compositions transitional to arc chemistry, whereas PPTU 6-1 is more MORB-like.

Collectively these samples of "old crust" indicate that much of the earlier-formed crust in the opening of the Lau Basin was transitional in composition between MORB and arc, as originally proposed by Hawkins and Melchior (1985). However, there are problems with this simple model as evidenced by data discussed in the following section.

5.3. MORB-like Drill Sites

ODP Sites 834, 835, and 836 drilled rocks having MORB-like chemistry; representative data are in Table II. Sites 834 and 835 are in extensional basins, partly filled with sediments, and Site 836, discussed previously, is part of the earliest crust formed at the ELSC. It is in an extensional basin like Sites 834 and 835 that may represent a tectonic setting transitional from rifting to spreading. Age estimates from the paleontologic data indicate that Site 834 basalts range in age from about 5.5 Ma, or older, to 3.8 Ma. We proposed that magmatism may have started at about 6 Ma (Hawkins, 1994). It is important to note that the narrow subbasin at Site 834 was the locus of basalt eruption for more than 1.7 m.y. as it opened. The basaltic filling is interbedded with thin sedimentary layers. At Site 835, the age of the end of basâltic infilling is late Pliocene (3.5 Ma) as determined from the paleontologic age of overlying sediments. As for Site 834, we do not know when volcanism began. Site 836 basalts are intercalated with Pleistocene sediments. The basaltic crust is estimated to be from 0.8 to 0.64 m.y. old. The correlation of Site 836 basalts to the ELSC is based on the probable age of rift propagation past Site 836 (Parson and Hawkins, 1994) and the chemistry of Site 836 basalts.

Thirteen petrologic units were identified at Site 834 that range from a relatively "primitive" massive unit (unit 7) having Mg#=61–64 to fractionated units (e.g., unit 12) having Mg#=34–36 and SiO_2=54–56%. Identification of petrologic units was based on a combination of petrography, mineralogy, and chemistry as well as the presence of chilled margins and sedimentary interbeds. The sequence drilled shows cyclic eruptions of variably fractionated magmas and large variations in thickness of the units. The variations in thickness probably reflect variable volumes of production of parental melts. There is wide range in melt composition with depth and no correlation between depth and magma chemistry. For example, "primitive" unit 7 was encountered from 214–287.6 below seafloor (mbsf) and is early Pliocene or older, whereas "evolved" unit 12 was encountered from 363.5–407.9 mbsf. The deepest, oldest, unit sampled (unit 13) at 423 mbsf is a fractionated unit with Mg#=57–58 and about 51% SiO_2. One of the youngest units (unit 2), at 112 mbsf, has Mg#=52–55. It is clear that we did not drill to the bottom of the lava sequence into the earliest melts erupted at Site 834, but had encountered some relatively evolved rocks at the point where drilling had to be stopped. Fields for element abundances are in variation diagrams (Figs. 3.8–14). Trace element and isotope data and their bearing on petrogenesis are discussed in separate sections.

The cyclicity in magma chemistry and thickness of petrologic units indicates periodic eruption, refilling, and melt storage in near-surface magma chambers which experienced variable amounts of fractionation and mixing of melts. The variable thicknesses (volumes) of each petrologic unit may reflect variations in the volume of parental melt production. This in turn may be a response to nonuniform mantle upwelling. Several factors may have been involved, such as episodic increases in the flux of thermal energy to the mantle source, episodic pulses of trench rollback or rate of subduction, or perturbations in the convective overturn or mixing of SSZ mantle/MORB-source mantle.

Previous work (Hawkins and Melchior, 1985) suggested that the basin is compositionally zoned with rocks transitional to arc-composition forming zones on the east and west around the central part of the basin that has crust with more MORB-like composition. Figure 3.6 shows the location of samples that were used in making this interpretation. The work of Ernewein *et al.* (1994) adds to this picture by showing that the ELSC becomes more arclike from north to south, culminating at Valu Fa Ridge, which has a strong arclike signature. Site 834 was selected to sample some of the oldest crust in the basin and indeed the cores include the oldest Lau Basin crust yet identified. We expected that Site 834 lavas would be arclike because of their location and relative age. However, they include some of the most MORB-like rocks in the basin. The hypothesis was tested, and it appears to have failed. A similar relation between early rifting and MORB composition lavas has been documented at the Sumisu rift (Hochstaedter *et al.*, 1990a,b). What went wrong with this intuitively appealing idea?

Lavas cored at Site 834 formed over a time span of more than 1.8 m.y. by continued eruptions into a slowly opening basin. Relatively unfractionated basalts form thick petrologic units that have Na, Ti, and low-partition-coefficient trace element abundances, indicating a relatively small extent of source melting in comparison to other Lau Basin drill sites (Hawkins, 1994). The deepest samples recovered are fractionated rocks, which imply that we did not reach the base of new crust at Site 834. We have no idea of the chemistry of the first erupted lavas, but intricately zoned cores of plagioclase are interpreted as the result of mixing between arc and MORB-type melts (Bryan and Pearce, 1994). Strongly zoned Cr-spinels also support the idea of magma mixing (Allan, 1994). The Site 834 data do not necessarily refute the general idea of compositional zoning in the basin because rocks with transitional composition exist at the dredge sites on the intrabasin ridges (Fig. 3.6). The data for unit 13, the deepest petrologic unit sampled, also are instructive. The glass samples carry the depleted SSZ signature both in terms of LILE and HFSE. For example, sample 59R2/52-55 has Rb=1.46 ppm, Ba=29.4 ppm, Nb=0.95 ppm, Zr=79 ppm, and Hf=2.14 ppm (Hergt and Farley, 1994; Table II). The lack of K, Rb, Ba enrichment relative to MORB differs from the values we would have expected if these were the transitional basalts mapped elsewhere in the western Lau Basin. However, the Al, Fe, Ti, and Na values relative to MgO all would lie in the MARA (Mariana Trough basalt) field (Figs. 3.7 and 3.8) and are definitely not comparable to MORB but are like other backarc basin basalts. These samples must have come from depleted SSZ mantle that lacked the LILE enrichment. Was it a depleted MORB-source mantle? As discussed later, Hergt and Hawkesworth's (1994) isotopic data establish that the source had the Pacific MORB-source mantle isotopic composition.

An explanation for the lack of LILE enrichment may be that the SSZ mantle source for Site 834 escaped infiltration of subduction-derived fluids. A more likely answer is that the large volume of melt erupted to fill the basin at Site 834 may have extracted all of the LILE enriched SSZ component available and continued to melt depleted MORB-like mantle source material. For purposes of comparison with the other drill sites and the axial ridges, we need to consider that Site 834 lavas were erupted about 2 to 3 m.y. before the ELSC propagator tip passed by on the east, and that they are on a "flowline" relative to the present spreading direction of the CLSC (Parson and Hawkins, 1994). Site 834 samples are even more like the lavas from the modern CLSC than are some of the ELSC glasses. This is supported by the trace element abundances and ratios as well as the isotope chemistry. Site 834 glasses are, however, distinctive in having relatively high Na content when compared to either ELSC and CLSC glasses or to MORB. They resemble Mariana Trough glasses in having this relatively high Na_2O. Although they have a strong MORB element signature, data for least-fractionated samples (>6.5% MgO) overlap with the Mariana Trough fields and with the CLSC fields except for Al_2O_3 (Fig. 3.8). Differentiated samples (<6.5% MgO) define different trends relative to glasses from the axial ridge of the Mariana Trough. They lack the high Al_2O_3 and low FeO* and TiO_2 considered by some authors to be the backarc basin basalt signature.

The sodium anomaly presents a paradox. The thickness of some of the petrologic units indicates extensive melting of the source yet the high Na content implies a small extent of source melting. Site 834 magmas may have come from a source like the MORB source but relatively enriched in Na, or they may represent relatively small amounts of melting compared to typical MORB, or they may be the result of extensive interaction between magma and the conduits, or combinations of these processes may have occurred. Repeated eruption of melt batches, each formed by a small extent of melting of a source continually replenished with MORB-source mantle, could help explain the Na relations and would be consistent with the mantle counterflow and diapirism envisoned in Fig. 3.2.

Site 835 data are limited to one moderately fractionated basaltic petrologic unit. Like the Site 834 rocks, it is MORB-like in most respects, as seen in Figs. 3.8–3.10, and the data lie within the fields for ELSC rocks. An affinity to SSZ magmas is shown by the moderate LILE enrichment and the depletion in Nb and Ta (Fig. 3.9e). Comparison with the ELSC data is relevant because the ELSC propagator tip passed by to the east at about the time (3.6 Ma) that Site 835 lavas were erupted. The Na_2O and TiO_2 relative to MgO are lower than for Site 834, which may indicate that the source experienced a relatively larger extent of melting, or the source was more depleted in Na and Ti, or both. The Al_2O_3 and FeO* for

Site 835 lavas resemble the ELSC and show even less of the Mariana backarc basin signature. This displacement may reflect a less hydrous, less oxidizing, melt source than that for the Mariana Trough.

Data for Site 836 were presented in the discussion of ELSC crust. The two youngest units of the five cored at this site appear to have been derived from a nearby arc-composition seamount. Rocks dredged from a nearby ridge (e.g., RNDB 2-7, Table IV) may be representative of this source. The presence of this arc-composition material in close proximity to the spreading ridge points out the extreme small-scale heterogeneity of the backarc mantle melt sources.

Rocks from WEB drill Sites 834 and 835 resemble those from the active ridges in terms of major and trace element chemistry, which suggests a similar source. It is inferred that they are melting products from a MORB-source mantle with little or no chemical influence from a subduction component in spite of the SSZ setting in which they formed.

5.4. Arclike Drill Sites

Rocks at Sites 837, 838 and 839 are more arclike (Table II and Figs. 3.8–3.14) than those already described; some are closely similar to modern Tofua arc tholeiites (Ewart *et al.*, 1973, 1977, 1994a,b; Ewart and Hawkesworth, 1987; Hawkins and Allan, 1994). All three sites were drilled in small, narrow, partly sedimented basins, elongated in a north-easterly direction. The sites are all less than 90 km from the present axial zone of the ELSC and on what had been assumed to be backarc crust. Their crustal ages are estimated to be about 2 Ma, which corresponds closely to the time of nearest approach of the ELSC propagating tip. Their arclike composition was unexpected in view of their location, although some similar rocks had been dredged from nearby high-standing ridges (e.g., RNDB 1-1, 2-7, 2-8, Table IV).

Site 837 and 838 igneous rock samples are moderately differentiated arclike basalts and basaltic andesites. They are gravel clasts probably derived from a nearby seamount. No glasses were recovered at Site 837; bulk rock analyses have 54–56% SiO_2, and 3–3.4% MgO. These are markedly similar to VFR samples (Jenner *et al.*, 1987; Vallier *et al.*, 1991) and to island-arc tholeiites in general. FeO* and Al_2O_3 are like ELSC lavas, but Na_2O and TiO_2 resemble Mariana Trough data. Glasses from Site 838 comprise a bimodal suite with 52–53% SiO_2 and about 6% MgO as one mode and 69–73% SiO_2 and 0.2% MgO as the other. All are like Mariana Trough glasses for Al_2O_3 but more like the ELSC for FeO*, TiO_2, and Na_2O (Fig. 3.8). They resemble rocks from the relay zone between the ELSC and CLSC and other Lau Basin crust transitional between MORB and arc chemistry.

Site 839 lavas are among the more unusual samples collected by drilling on Leg 135. Some samples resemble those from Sites 837 and 838, whereas others are even more arclike. The samples include two-pyroxene basalts and basaltic andesite. Their SiO_2 ranges from 50–55% and MgO from 4.4% to 15%. The samples with more than 9% MgO are not liquid compositions as they have abundant phenocrysts of olivine and clinopyroxene as well as phenocrysts of chrome-spinel. Allan (1994) showed that the strongly zoned olivine and chrome-spinel phenocrysts support magma mixing as an explanation for some of the Site 839 lavas. Site 839 is considered to be an arc-composition edifice that formed within the Lau Basin at about 1.9 Ma. It was cut off and abandoned on the west side of the ELSC as the rift tip passed by.

The melts erupted at Sites 837, 838, and 839 were formed from mantle with a robust arc-source signature. If Site 839 lavas had been collected on the Tonga Ridge, they would have been considered to be part of the Tofua arc magma series. However, even though the arc signature is strong, the Site 839 samples lack the high Al_2O_3 seen in the Mariana Trough or in typical arc lavas. The Site 839 source may have been less hydrous than the Tofua arc SSZ source. Even though erupted synchronously with crust being formed at the nearby ELSC, rocks from these three drill sites are much less like MORB than is the ELSC crust. The mantle source must have remained heterogeneous, retaining arc-source SSZ domains, even while the MORB-source mantle influx was feeding melts to the ELSC propagating rift. Although unusual, this is not unique, as we found a young arc-composition volcanic edifice, aligned with other edifices having typical "backarc basin basalt" chemistry, on the crest of the axial ridge in the Mariana Trough (Hawkins *et al.*, 1990).

6. THE PEGGY RIDGE AND SEAMOUNTS OF THE NORTHERN BASIN

6.1. Introduction

The geology of the area north of Peggy Ridge is poorly known, although the morphology is well illustrated by sonar side-scan imagery made with GLORIA (Parson and Tiffin, 1993). The NLSC extends northeasterly from Peggy Ridge into older crust having numerous seamounts (Fig. 3.1). The morphology of this area suggests that there are numerous partly sedimented basins as well as some high-relief areas in which seamounts rise more than 3000 from nearby narrow basins (Hawkins, 1974). The region north of Peggy Ridge is poorly sampled and the data are mainly from seamounts. Each of the three major seamounts we have dredged and Niuafo'ou Island constitute distinctive magma types. This section presents a brief discussion of the area, although the data are few and it is not possible to develop an integrated picture of the regional geology.

6.2. Geology of Peggy Ridge

Peggy Ridge is a major feature that coincides with the trace of a fault zone having right-lateral displacement (Eguchi, 1984). This fault probably acts as a (leaky?) transform fault. It is not clear that the fault is related to the origin of the ridge, although a prior (?) phase of compression across the fault trace could be responsible for the ridge morphology. Although the shape of the ridge suggests that it may have formed as a spreading center, the magnetic anomaly patterns do not support this (Sclater *et al.*, 1972).

Samples from several places along its length (Fig. 3.6) indicate that it is formed of several basalt types (Sclater *et al.*, 1972; Hawkins, 1976, 1988, 1989). The westernmost end, near the Lau Ridge, has transitional basalt (e.g., 123-103-1, Table III) similar to that which forms much of the high areas of the western basin. These have element abundances and ratios suggesting an affinity to backarc basin basalts (BABB) or Mariana Trough-type lavas (Hawkins and Melchior, 1985). One site on the western part of the ridge yielded very primitive, highly depleted basalt that has many of the characteristics of a basaltic komatiite or a variety of boninite. Rocks and glasses from this dredge site (123-95-12, Table III; Hawkins and Melchior, 1985) represent extensive melting, up to 30%, of a depleted mantle source. Although having many of the chemical aspects of high Ca-boninites, the samples

TABLE IV
Representative Analyses, Old Crust and Seamounts

Location	Old Crust							
Site	A-223	F-114	L-81	L-101	P-4	P-6	P-7	P-34
Sample	ANT 223-1 GL	Sonne 114-GL	123 81-GL	123 101-4	PPTU 4-1 GL	PPTU 6-1 GL	PPTU 7-1 GL	PPTU 34-1 GL
SiO_2	50.89	50.88	51.57	55.97	49.72	48.22	50.08	49.18
TiO_2	1.32	0.83	1.11	1.04	0.96	0.79	1.28	0.79
Al_2O_3	17.1	18.84	16.03	14.45	16.17	17.46	15.48	15.58
FeO*	8.01	9.34	8.38	10.35	9.13	9.29	10.38	10.91
MnO	0.17	0.21	0.18	0.21	0.19	0.19	0.18	0.2
MgO	6.35	4.59	6.23	4.4	6.64	9.02	6.41	6.97
CaO	11.06	11.39	11.77	8.87	12.66	12.52	12.22	13.38
Na_2O	3.23	1.89	3.03	2.81	2.31	2.28	2.71	1.9
K_2O	0.38	0.46	0.43	0.61	0.66	0.02	0.3	0.17
P_2O_5	0.23	0.14	0.2	0.13	0.2	0.04	0.13	0.08
Sum	98.79	98.59	98.92	98.84	98.64	99.83	99.17	99.08
Mg#	61.9	50.1	60.9	46.6	59.8	66.6	55.9	56.7
CaO/Al_2O_3	0.647	0.6	0.73	0.61	0.78	0.72	0.79	0.86
Trace elements (ppm)				[a]				
Cr	640	74		510	681	177	222	
Ni	332	34		193	257	114	52	133
Co	225	31		85	52	50	32	
V	262	212		297	249	250	244	279
Zr	76	64		38	62	95	81	34
Y	26			29	25	24	26	21
Nb	2					2	1	
Hf					1.49	2.28	1.27	
Ta					0.175	0.11	0.078	
Rb	6			6	18	1	6	3
Ba	46.3	12		34.9	69	17	61	12
Sr	186	281		174	224	191	222	244

have relatively high Al_2O_3 (15–16%), which argues against that classification. However, CaO/Al_2O_3 is 0.9–1.0 which is comparable to many high-Ca boninites (Crawford *et al.*, 1989). The trace element data, especially REE, Cr, and Ni, offer a strong argument for a boninitic character (Hawkins and Melchior, 1985). Regardless of the name assigned, it is a rock type similar to early-formed melts in many SSZ settings and may represent some of first magma types erupted at the beginning of opening of the Lau Basin. I consider it unlikely that it formed by magma leakage on the transform fault because of the large amount of melting and unfractionated composition, implied by its chemistry.

The middle and eastern parts of the ridge are composed of MORB-like basalt (e.g., 123-86-1, Table III). These resemble some of the fractionated basalts from Site 834 (e.g., unit 13) and the axial ridges. Although they have relatively high TiO_2 (e.g., 2%), FeO* (e.g., 11–12%), and Na_2O (e.g., 3%), their calculated abundances at 8% MgO (after Klein and Langmuir, 1987) are about 1.6%, 9.7%, and 2.6% respectively. They may represent a relatively smaller extent of melting of a source similar to the source for Site 834 lavas.

TABLE IV
(*Continued*)

Location	Old Crust			Donna Smt.	Rochambeau Bank		Zephyr Sh.
Site	R-1	R-2	R-2	L-89	L-106	P-23	L-98
Sample	RNDB 1-1 RK	RNDB 2-7 GL	RNDB 2-8 GL	123 89-4 GL	123 106-1 GL	PPTU 23-2	123 98-4RK
SiO_2	49.33	55.86	56.43	50.53	48.77	48.88	62.99
TiO_2	0.51	1.2	1.25	1.26	1.6	1.89	0.47
Al_2O_3	11.89	14.95	14.14	14.3	16.09	14.94	14.34
FeO*	9.07	11.32	12.19	11.97	10.34	11.35	5.32
MnO	0.16	0.22	0.22	0.25	0.2	0.19	0.09
MgO	12.91	3.87	3.72	6.73	7.22	6.63	3.62
CaO	12.29	8.32	8.14	11.72	12.36	11.79	5.32
Na_2O	1.83	2.91	2.41	2.88	2.85	3.1	3.65
K_2O	0.35	0.24	0.29	0.08	0.32	0.26	0.88
P_2O_5	0.13	0.15	0.14	0.12	0.17	0.17	0.07
Sum	98.47	99.2	99.11	99.84	99.92	99.21	96.75
Mg#	72.8	41.2	38.5	53.5	58.9	54.5	58.2
CaO/Al_2O_3	1.03	0.557	0.576	0.82	0.77	0.79	0.37
Trace elements (ppm)		[b]	[b]				
Cr				150	255	238	
Ni	306	17	10	45	122	72	
Co				50	106	47	
V	197	435	404	334	287	355	
Zr	33	98	93	77	105	127	
Y	13	38	37	32	61	31	
Nb	3	1	2	3	5	7	
Hf						3.27	
Ta						0.375	
Rb	9	6	5	16	5.6	5	
Ba		55	55	11.1	66.6	40	
Sr	298	131	137	90	230	215	

[a]From aphyric sample 123-101-2.
[b]From aphyric rock samples. Other data for aphyric rock

The postulated NLSC abuts the Peggy Ridge and samples from the intersection are fresh glass-rich basalts also having MORB-like chemistry. We consider these to be part of the NLSC (e.g., PPTU-19-1, PPTU 20-5, Table III).

The ridge may be a submarine range of crustal blocks forced up by transpression on the northwest-trending fault. Partial support for this suggestion comes from the west-to-east variation in chemistry from arclike, to boninitic to MORB-like. The disposition of magma types collected from Peggy Ridge mimics the regional zonation seen in the western Lau Basin. The newly formed NLSC puts an overprint of young MORB composition lavas on the older crust.

6.3. Geology of Donna Seamount

Donna seamount (Fig. 3.1) is a very prominent high-standing feature that rises more than 2300 m above its surroundings to about 888 m deep. It is somewhat elongated in an east-west direction. The summit is highly irregular, and it may never have reached to wave base. The south flank drops into a deep basin from which diabasic/microgabbroic textured rocks were dredged (Hawkins, 1976). The morphology suggests that the seamount may be a very large uplifted, uptilted crustal block rather than a point-source magma leak. Rocks dredged from the southern and eastern flanks are mainly glass-rimmed pillow fragments comprising olivine-augite-labradorite basalt. Other dredged material includes minor amounts of breccia and tuffaceous sediment. Most of the breccias and sediment samples have Fe-Mn oxide-hydroxide rinds up to 1 cm thick. The age of this seamount is not known, but all indications are that it is one of the older features in the basin.

Rock mineralogy and chemistry (e.g., 123-89-4, Table IV) are typical of the MORB-like basalts found widely throughout the basin. The calculated Na and Fe at 8% MgO (2.4% and 9.9% respectively) are typical of other MORB-like basalts from this part of the basin. The REE pattern is typical for MORB (Hawkins and Melchior, 1985), and other trace elements also indicate derivation from a MORB-source mantle. There is no evidence that Donna seamount formed at the same time as the eruptions at Site 834, but the source composition and petrogenesis for the two sites were the same.

6.4. Geology of Rochambeau Bank

Rochambeau Bank is elongated in a northwesterly direction; it sits atop a broad 2000-m-deep platform and rises nearly 1500 above it. Parson and Tiffin (1993) noted that it lies along strike of the NLSC and suggest that the NLSC may extend at least to it. Rochambeau Bank offers several contrasts with Donna Seamount and with other samples from the postulated trace of the NLSC. Its form and the nature of the dredged samples suggest a point-source volcanic origin. Fresh glass-rinded plagioclase-rich pillow fragments or sheet flows have been collected. These are relatively more enriched in alkalis and HFS elements than most other Lau Basin rocks and are quite different from presumed NLSC samples (123-106-1, PPTU 23-2, Table IV). In some respects the major element chemistry is arclike, but the enriched HFSE suggest an ocean island basalt (OIB) affinity. Ti, Zr, Hf, Ta, Nb all are enriched relative to other Lau Basin rocks. Volpe *et al.* (1988) designated Rochambeau Bank samples as E-MORB on the basis of Nd- and Sr-isotope ratios. Poreda (1985) and Poreda and Craig (1993) reported $^3He/^4He$ of 11.0 and 14.1 $\times R_A$ (R_A = atmospheric ratio) for Rochambeau Bank samples and a ratio of 22.0 x R_A for a sample from the nearby seafloor. They propose that these ratios represent contributions from a Samoan plume component that has been entrained under the northern Lau Basin.

6.5. Geology of Niuafo'ou Island

The island of Niuafo'ou must represent a melting anomaly. It rises more than 1300 m above the surrounding seafloor, which itself is part of a broad pedestal, elongated in a northwesterly direction, and rising about 500 m above its surrounding seafloor. Its maximum elevation above sea level is 213 m. Although a Tongan island, it is not part of the Tofua arc; it is formed of MORB-like basalt similar to the nearby.

Niuafo'ou is a shield volcano, about 8 km in diameter at sea level and has a central collapse caldera about 5 km across. There have been 12 recorded eruptions in the last 170 years, the most recent being in the mid-1940s (Reay *et al.*, 1974; Taylor, 1991). The 1886 eruption was reported to have sent tephra to an estimated height of 2.3 km.

Surface exposures on the island include both lavas and tephra. Some tephra layers are as much as 6 m thick; both airfall and base surge deposits have been recognized (Taylor, 1991). Lavas range from glassy to porphyritic; plagioclase forms phenocrysts, and both olivine and plagioclase occur as microphenocrysts. The data for these samples (Reay *et al.*, 1974) indicate that the samples are moderately fractionated basalts: SiO_2=49–50%, MgO=5.1–7.2%, and TiO_2=1.4–1.75%. Reay *et al.* (1974) observed that the Ni content is low and resembles arc lavas and the Ba is high. However, the Ba enrichment is not unusual for Lau Basin lavas. The ratio Ti/V is 47 (from Reay *et al.*, 1974), which is at the high end of MORB-ratios (Fig. 3.8), and would plot close to a trend defined by seamounts (Shervais, 1982).

Isotope data (Sr, Nd, Pb) all show that Niuafo'ou lavas are distinctly different from Tofua arc samples. Ewart and Hawkesworth (1987) conclude that they are from a distinct magma source that has an OIB signature in spite of the general chemical similarity to MORB. Poreda and Craig (1993) presented $^3He/^4He$ data for gases collected in the crater lake Vai Kona on Niuafo'ou that have $7.8 \times R_A$. This MORB-like ratio is typical of the surrounding seafloor and contrasts with ratios up to $22 \times R_A$ (the Samoan "plume" component proposed by Poreda, 1985) found at Rochambeau Bank some 100 km to the northwest. The "plume" component and the OIB component must be decoupled.

6.6. Geology of Zephyr Shoal

Zephyr Shoal is a prominent feature with more than 1500 m relief that rises to water depths of less than 700 m. The top is capped with corals and reefal limestone (Sclater *et al.*, 1972; Hawkins, 1974, 1976, 1988). This seamount is anomalous in that it has arc composition rocks but is out in the middle of the northern Lau Basin. The rocks are hypersthene (Fs_{34})-augite-labradorite ($An_{70\text{-}55}$) phyric dacite (Hawkins, 1976). The groundmass is brown glass having low-K rhyolite chemistry (Hawkins, 1985). This glass composition is remarkably similar to glass from the 1967–1968 eruptions of Metis Shoal (Melson *et al.*, 1970). The bulk composition of the vitrophyres is dacitic with about 63% SiO_2, 0.5% TiO_2, and 0.9% K_2O (sample 123-98-4, Table IV; Hawkins, 1976).

The phenocryst assemblage could not have formed in equilibrium with the enclosing glass (melt); they probably are xenocrysts. The dacite chemistry is best explained as a hybrid formed by mixing between a highly fractionated liquid and "exotic" crystalline material. The extremely low REE patterns and general depletion in HFSE (Gill, 1976) suggest that Zephyr Shoal lavas may have been derived from a fractionated boninitic parental magma that mixed with crystal residue from other melts.

Although Zephyr Shoal samples are broadly similar in chemistry and mineralogy to Tofua arc lavas, it is unlikely that the two are related. Zephyr Shoal samples are much more sodic (Na_2O/K_2O about 4) than Tofua arc samples, which typically have a ratio of 2–3. Also, Zephyr Shoal has lower TiO_2 than otherwise comparable Tofua arc samples (Ewart *et al.*, 1973).

7. LAU BASIN PETROLOGIC EVOLUTION

7.1. Introduction

Mid-ocean ridge basalts form a distinct class of rocks that are relatively depleted in magmaphilic elements such as LILE and HFSE. It is inferred that they are derived from a previously melted, and relatively depleted, mantle source that gives them distinctive trace element abundances and ratios. These inferred depletions are further supported by isotopic signatures such as for the Sm-Nd, Rb-Sr, and U-Pb systems. The source for SSZ magmas has been depleted even more than the MORB source and is considered to be like the residue left after extraction of MORB (i.e., it is a multiply depleted mantle source). Another characteristic of many SSZ magmas is a relative enrichment in LILE and water. The subduction zone setting permits the reenrichment of the depleted mantle by fluids derived from the subducted ocean crust, by sediment involvement, either directly or through an intervening fluid phase, or by the addition of partial melts (e.g., plagiogranite) derived from the subducted amphibolitized ocean crust (Pearce *et al.*, 1992).

The early studies of backarc basin magma types emphasized their similarity to MORB (Hawkins *et al.*, 1970; Hart *et al.*, 1972; Sclater *et al.*, 1972), although the volatile and alkali-metal contents of some were recognized as being anomalously high (Fryer *et al.*, 1981). For the latter the term *backarc basin basalt* (BABB) was proposed. Some workers (e.g., Gill, 1976), pointed out that some samples were similar to arc magmas. The recently completed studies on ODP Leg 135 samples (Hawkins *et al.*, 1994) demonstrate the range in element and isotopic chemistry of Lau Basin rocks that span the range from MORB to arc composition.

A discussion of the petrologic evolution of the Lau Basin, and other backarc basins, needs to consider the nature of the multiple mantle and subduction-derived source materials present in the SSZ tectonic setting. In the section that follows, I discuss various element-element discriminants and isotope data that help delineate the nature of the Lau Basin crust and give insight to the magma sources.

7.2. Nature of the Magma Source

The SSZ mantle source for arc and backarc magmas has a complex chemical character that is the result of previous melting, addition of sediment, hydrous fluids, or melts from the subducted slab, and counterflow of "new" mantle into the source region. This new mantle may carry MORB-source or OIB-source chemical and isotopic signatures. It is well known that there is an extreme decoupling between the LILE and HFSE in SSZ magmas (e.g., Gill, 1981). In this section I discuss several of the chemical and isotopic tracers that help in constraining end members. The HFSE and REE perhaps offer the best constraints on depletion of the original mantle source relative to the MORB-source. Compatible elements (e.g., Cr, Ni, and V), give insight to melt/conduit reactions and to oxygen fugacity. The LILE are a possible tracer of slab-derived elements transported in fluid or assimilated with sediments. The isotope systems, especially Sr and the U-Pb series, give insight to mixing between different sources.

7.3. The Immobile Elements—HFSE and REE

The HFSE (e.g., Ti, Zr, Hf, Nb, Ta, Th) including the heavy rare-earth elements (HREE, Sm–Yb) are generally considered to be immobile under hydrous conditions up to the greenschist facies of metamorphism (Ludden and Thompson, 1979; Pearce and Norry, 1979). Inasmuch as the HFSE are not likely to be added to the mantle wedge by subduction-derived fluids, their abundances in IAV, and backarc basin crust, are determined by abundances in the SSZ mantle and the melting process. Moderate levels of alteration seem not to modify their relative abundances; thus, they are useful as tracers of the mantle source composition when least-fractionated samples are evaluated.

Island arc volcanic (IAV) magma series typically show depletions in HFSE relative to N-MORB (e.g., Arculus and Powell, 1986). Many IAVs show an overall relative depletion in HREE (Gill, 1981) but do not show progressively increasing depletion with atomic number as expected from garnet fractionation. A source depleted in all HFSE including the HREE, or a mechanism to retain all HREE uniformly in the mantle residue is required. The IAV are characterized by a marked relative depletion in Nb and Ta (Dupuy *et al.*, 1982; Briqueu *et al.*, 1984; Kelemen *et al.*, 1990) as seen on plots of trace element normalized to MORB (Fig. 3.9). The other HFSE show variable depletions; for some IAV series there is lesser depletion in Zr than in other arcs.

The general tendency for intraoceanic IAV series to be depleted in the HFSE has been explained by some authors as being due to previous depletion of their mantle source; that is, the mantle source is multiply depleted MORB-source mantle (Hawkins *et al.*, 1984; Woodhead *et al.*, 1993). Other explanations include retention of HFSE in a mantle mineral (e.g., a titanate) phase (Briqueu *et al.*, 1984; Ryerson and Watson, 1987), more extensive melting of MORB-source mantle (Dupuy *et al.*, 1982), and exchange reactions between melt and mantle enroute to the surface (Kelemen, 1986; Kelemen *et al.*, 1990). Kelemen *et al.* (1990) showed that melts derived from a relatively more fertile mantle may react with overlying depleted mantle and develop melts depleted in HFSE relative to MORB. This may be especially significant because long transit of small volumes of melts through large volumes of previously depleted mantle (e.g., through SSZ mantle) will develop strong depletions in HFSE found in IAV. The bulk distribution coefficients (D_0) for HFSE in depleted mantle will be controlled largely by the amount of residual pyroxene. D_0 for clinopyroxene-poor mantle increases for HFSE in the sequence Ta and Nb $<$ Zr $<$ Hf $<$ Ti.

Subduction-derived "dehydration fluids" would cause enrichments in the LILE but would have no effect on HFSE or HREE abundances in the mantle wedge because of immobility of these elements in aqueous fluids. However, partial melts of the subducted (amphibolitized) basaltic crust would be able to impregnate the SSZ mantle with the "immobile" elements (Pearce *et al.*, 1992). Consideration of distribution coefficients (e.g., Kelemen *et al.*, 1990) suggests that small amounts of melt would be enriched selectively in the relative order of LREE >Ta and Nb>Zr>Ti>Hf>HREE.

The discussion of HFSE partitioning in the genesis of IAV magmas applies to backarc magmagenesis as well. Plots of trace element abundances for Lau Basin basalts having Mg#>60, normalized to N-MORB (Fig. 3.9), show the characteristics of IAV. Niobium, in particular, is generally at or below x-ray fluorescence (XRF) detection limits of about 1–2 ppm. Nearly all Nb data obtained by ICP-MS or by emission spectrography are in the range of 0.4–1 ppm; Ta values are in the range 0.01–0.2 ppm (Ewart *et al.*, 1994a,b; Hergt and Hawkesworth, 1994). Tofua arc data show Nb = 0.5–1.4 ppm and Ta = 0.03–0.09 ppm (Ewart *et al.*, 1994a,b). Collectively, the data shown in Fig. 3.9 support a multiply depleted MORB-source mantle for the source of Lau Basin basalts. There is no indication of enrichment in Ti or Zr that might be due to impregnation of the SSZ mantle with melts from the subducted basaltic crust. We cannot rule out a possible role for these melts, but if they were a factor they did not impart a recognizable signal.

The REE data for the Lau Basin samples (Fig. 3.10) include both MORB-like patterns, with La/Sm_{ef} <1 and some with $La/Sm_{ef} \geqslant 1$. The variable LREE enrichment is typical of SSZ settings. The HREE are unremarkable in that they resemble MORB or show slight relative depletion. These depleted samples thus resemble low-K tholeiites of the Tofua arc (Gill, 1981; Ewart and Hawkesworth, 1987).

Thorium is generally considered as an HFSE, but it could be mobilized in subduction fluids. Data on the Hf/3-Ta-Th diagram (Fig. 3.11) indicate that nearly all of the Lau Basin samples have MORB-like relative abundances for these elements. Data for ELSC, CLSC, and Sites 834, 835, and 836 all lie in the MORB field. Many of the data are shifted toward Hf on these diagrams mainly because of the low levels for Ta and Th. Thorium is in very low concentrations, generally <0.2 ppm) and argues for a minimal contribution from sediment or sediment-derived fluids. This plot has been used to help evaluate the sources for ophiolites and is intended to help discriminate between SSZ sources and MORB sources. We see here that this is not a conclusive discriminant diagram inasmuch as the Lau Basin clearly is in an SSZ setting but the crust largely exhibits a MORB signature. The few data that plot in the SSZ field are either from old crust in the northeastern part of the basin, close to the Tofua arc; from the Valu Fa Ridge, where the ELSC is propagating into older (forearc?) crust; or from the relay zone separating the ELSC and CLSC. In each place the mantle source may retain fragments of the SSZ mantle that formed arclike melts. In contrast, the Mangatolu triple junction extends from the backarc basin into the Tofua arc. At the latitude of the MTJ, the Tofua arc may be as old as 3 Ma (Tappin *et al.*, 1994), and it has erupted depleted melts as seen on Tafahi Island (Ewart and Hawkesworth, 1987). Data for the MTJ trend across the MORB field toward the SSZ field in Fig. 3.11 but do not reach it. Possibly, the reason for the minimal arc signal in MTJ lavas, in spite of their location, is because the original "arc component" in the SSZ mantle wedge has long since been stripped out and has been replaced by MORB-source mantle.

Collectively, the HFSE data demonstrate that the Lau Basin mantle source is dominated by multiply depleted MORB-source mantle with local remnant domains of less depleted mantle or zones that have been replenished by counterflow with more fertile MORB-source mantle. There is little or no evidence for enrichment with partial melts of subducted crust. The HFSE shed no light on the role of infiltration of subduction-derived aqueous fluids.

7.4. Titanium and Vanadium

The covariation of Ti with V, and Ti/V ratios have been proposed as useful discriminants for recognizing the original tectonic setting of magma series now preserved in ophiolites (Shervais, 1982). The crystal/liquid partition coefficients for V (K_D) are sensitive to oxygen fugacity and range from >1 to <1; higher f_{O_2} causes K_D for V to decrease. Inasmuch as the bulk K_D for Ti is almost always <1, the relative enrichment or depletion of V relative to Ti is strongly influenced by f_{O_2}. Other factors include the original composition of the source, extent of melting, and effects of fractional crystallization. Ti-V data for the Lau Basin are shown in Fig. 3.12.

MORB-series magmas are characterized by Ti/V ranging from 20 to 50 and reflect derivation under conditions of relatively low f_{O_2}, whereas arc-series magmas have Ti/V ranging from 10 to 20, reflecting more oxidizing environments (Shervais, 1982). As noted by Shervais (1982), backarc basins display a range in Ti/V that includes both arclike and MORB-like ratios. This is well illustrated in the data for the Lau Basin (Fig. 3.12). For clarity, the arrays of data for each of the morphotectonic provinces are shown on separate plots.

Nearly all of the CLSC data lie within the MORB field. Some samples have Ti/V as low as 15, whereas Ti-rich samples (>12,000 ppm) have ratios between 30 and 50. The ELSC data lie on the plot between ratio lines of 15 and 30. The lower ratios straddle the boundary designated to separate MORB from arc. Thus it could be argued that the Ti-V relations indicate a transitional character for some CLSC and most of the ELSC samples. Data from VFR lie in the MORB field, even though many HFSE data give them an arc affinity (Figs. 3.9–3.11). Mangatolu triple junction data suggest that there may be two classes of magma types. Many of the data lie in the MORB field and overlap with Valu Fa data. There is also a broadly scattered field for high-Ti, low-V samples that extends into the field for alkali basalts (Shervais, 1982). Many of the low-V samples (e.g., <100 ppm) are from highly fractionated samples that likely lost both Ti and V as a result of magnetite fractionation, but additional V was lost to another phase. Fractionation of either clinopyroxene or amphibole or both would preferentially remove V relative to Ti.

Site 836 data lie within the ELSC field and are not shown separately. Sites 834 and 835 each define separate fields, although both are MORB-like.

7.5. Alkalis and Alkaline Earth Elements

A distinctive feature of N-MORB is their very low levels of K, Rb, Cs, Ba, and LREE (Kay *et al.*, 1970; Bryan *et al.*, 1976; Viereck *et al.*, 1989). N-MORB also have distinctive low ratios for $^{87}Sr/^{86}Sr$ (Hedge and Peterman, 1970; Hart *et al.*, 1974). Abundances of these trace elements (LILE), and isotope ratios, increase with low-temperature alteration. The K, Rb, and Cs increase much more readily than Ba, Sr, or LREE (Hart, 1969; Hart and Nawalk, 1970; Staudigel and Hart, 1983; Hart and Staudigel, 1989).

Unlike isotope ratios, abundances and ratios of trace elements can be modified by fractional crystallization. Inasmuch as the LILE are highly incompatible elements, one expects to find positive correlations between K, Rb, Cs, Ba and La, and SiO_2 or with HFS elements such as Zr. However, the difference in abundances of alkalis and alkaline earth elements between least-fractionated samples of IAV and MORB indicates that the IAV are relatively enriched in these elements. Equally important are the high ratios of Ba/La, alkalis/La, and an overabundance of Cs in IAV; these cannot be explained by crystal fractionation. They are considered to be signatures of IAV sources or signatures that arise during the melting process (White and Patchett, 1984). Similar chemical signatures are seen in many backarc basin basalts, especially those erupted on relatively youthful features such as the VFR (Jenner *et al.*, 1987; Vallier *et al.*, 1991) and the ELSC. Most BABB typically have abundances of K and Sr not much different from MORB. However, they are characterized by relatively high levels of Ba, and variable, but commonly high, Rb and La relative to MORB with comparable abundances of Mg. Many of the axial ridge glasses from the Mariana Trough and Lau Basin also are relatively enriched in Na (Fig. 3.7). These observations suggest that the backarc mantle source has been enriched selectively in a "subduction component."

Fractionation should cause a linear correlation between the LILE and elements such as Zr. Conversely, they should show scatter if they are the result of fluid transfer into the rock system. The LILE are highly mobile under hydrous low-temperature conditions of alteration and are likely constituents of fluids derived from dehydration of subducted crust (Tatsumi *et al.*, 1986; Tatsumi, 1989). Their presence in otherwise apparently unaltered glass in SSZ settings is strong evidence that they reflect a subduction fluid component in the melt source. For example, if Ba abundance correlates well with Zr, it can be argued that the Ba is a primary feature of the melt rather than an effect of alteration. Lau Basin data are shown in Fig. 3.12. We might then look for correlation between Ba and other cogeners such as K, Rb, and La to evaluate an enriched melt source rather than alteration to explain enrichments in this group of elements. The Lau Basin data for K and Ba are in Fig. 3.14.

Ewart and Hawkesworth (1987) note that the average K/Ba for Tonga (Tofua) arc lavas is about 30 and contrast this with values of 90 for fresh MORB and about 60 for altered MORB. They proposed that a subduction fluid component has been added to the source of the arc lavas. The range for all of the Lau Basin basalts (30–90) encompasses both MORB-source ratios and those for a subduction component. The most arclike ratios are from the VFR and parts of the ELSC and CLSC. This may be a reflection of their proximity to the Tofua arc. However, the Mangatolu triple junction actually cuts into the arc trend, yet it has ratios ranging mainly from 50 to 70. The highest K/Ba ratios (as high as MORB values of 90) are from the CLSC, from part of the Mangatolu triple junction, and areas of "old" crust. For comparison, Mariana Trough axial ridge glass samples have K/Ba ranging from 50 to 90 (Hawkins *et al.*, 1990).

Cesium is an especially sensitive tracer of alteration and is considered to be highly mobile in aqueous fluids (Tatsumi *et al.*, 1986). This mobility makes it potentially useful as an indication of a contribution to the SSZ mantle from subduction fluids. Cesium is enriched in IAV relative to MORB together with the other alkali metals. The limited data for Cs in Lau Basin samples require caution in attempting to determine how important subduction fluids have been in petrogenesis. Fresh, unaltered MORB has about 0.0086 ppm Cs, and Rb/Cs ranges from about 43–140 (e.g., Hart and Staudigel, 1989), whereas Aleutian IAV have 0.7–2 ppm Cs and Rb/Cs ranging from 24 to 30 (Morris and Hart, 1983). There are few data for the Tofua arc, but a Cs range of 0.04–0.8 ppm and Rb/Cs ranging from 10–95 is reported by Ewart *et al.* (1994a). The decrease in Rb/Cs of arc magmas can be attributed to input of sediment or sediment-derived fluids (e.g., Ben Othmman *et al.*, 1989). The Lau Basin data are limited, but most samples have MORB-like ratios. ODP Site 834 glasses have Cs abundances ranging from 0.01 to 0.36 ppm and Rb/Cs is about 50 (Ewart *et al.*, 1994a; Hergt and Hawkesworth, 1994). ODP Site 836 also has MORB-like values with Cs = 0.01–0.05 ppm and Rb/Cs = 90–180. The CLSC has $<$0.08–0.15 ppm Cs and Rb/Cs = 37–77. The VFR has some of the most arclike rocks; Cs data are very limited, but those available are MORB-like (Cs = 0.17–0.19 ppm and Rb/Cs = 39–45). More data may change the picture, but the data available indicate that introduction of Cs by subduction fluids or sediments has been very limited in importance.

7.6. Evidence for a Signature from Subducted Sediment

It is highly probable that at least some of the seafloor sediments delivered to the subduction zone are introduced to the mantle and influence the SSZ magma chemistry. Most of the studies that have addressed this problem have focused on island-arc volcanism, but the results are applicable to the entire SSZ system. The most useful tracers of sediment input to magmagenesis are abundances and ratios of the LILE, especially Ba and Sr, Cs (discussed above), the LREE, Th, Pb, B, Be, and isotope ratios such as $^{18}O/^{16}O$, $^{87}Sr/^{86}Sr$, $^{10}Be/^{9}Be$, and the Pb isotopes. Negative Ce anomalies, expressed as the Ce/Ce* ratio, are considered by some to be indicative of sediment input (e.g., Hole *et al.*, 1984). With the exception of Ce anomalies, nearly all of these element signatures may be explained equally well by transport in a fluid phase rather than by direct incorporation of sediment (see discussions in Ewart and Hawkesworth, 1987; Tatsumi, 1989; Woodhead, 1989). The argument for extensive *sediment* involvement seems weak and unnecessary, although it seems clear that dehydration fluids may play an important role. An effective way to evaluate this is with isotope data.

Meijer (1976) studied Mariana arc volcanoes and concluded that the Pb-isotope data permitted no more than 1–2% sediment input. He cautioned that the constraints depend critically on the assumption that the sediment data used are appropriate to what is actually incorporated. White and Patchett (1984) pointed out that other assumptions, such as using depleted MORB-source mantle or altered MORB as end members, are critical in the modeling. They showed that mixes of sediment and depleted MORB-source mantle were adequate to explain the observed Nd and Pb data for the Mariana arc, but the Sr isotope ratios derived from the modeling were lower than observed values.

Ewart and Hawkesworth (1987) showed that the Tonga (Tofua) arc and Kermadec arc volcanic series have Sr- and Nd-isotope data that lie in the mantle array. They noted that the Pb-isotope data for the Tonga–Kermadec arcs do not give unambiguous evidence for the incorporation of a subduction component, although the data allow it. Southern Tofua arc volcanoes may represent mixtures between MORB and about 2% sediment, whereas OIB sources could give rise to the Pb-isotope ratios observed throughout the region. They propose that the subduction component could have been dehydration fluids derived from subducted sediment.

8. DISCUSSION OF THE ISOTOPE DATA

8.1. Introduction

Isotope data provide important insights to the origin of magma systems; in particular, for studies of backarc basins they are important in identifying potential end members and for constraining possible mixing between them. There are extensive data for Sr and Nd isotopes for Lau Basin rocks; these are complemented by a somewhat lesser array of Pb-isotope data and some data for He and O. Here I will summarize some of the petrogenetic constraints inferred by other workers. Collectively the data show that multiple magma sources have been involved in magmagenesis. These include Pacific and Indian MORB-sources, "plume" or OIB sources, and a subduction component. Data for the Sr, Nd, and Pd isotopes are shown in Fig. 3.15–17.

The problems of magmagenesis in backarc basins are similar to those for IAV series. One of the distinctive signatures of IAV relative to MORB is their slightly more radiogenic $^{207}Pb/^{204}Pb$- and Sr-isotope ratios, an overabundance of Ba and Sr as well as alkalis, an enrichment in La, and a common, but not ubiquitous, depletion in Ce. Many of these features are seen in backarc basin basalts. Recognition of ^{10}Be in some arc magmas also has important implications for the role of subducted sediment in magmagenesis. To date there are no data indicating ^{10}Be in backarc lavas, but they seem to be an unlikely place to find it in view of the short half-life for ^{10}Be and the depth (transit time) to which the Be carrying phase would need to be subducted. IAV lavas tend to have more radiogenic Sr at a given level of $^{143}Nd/^{144}Nd$ than comparable MORB (e.g., DePaolo and Wasserburg, 1977), although White and Patchett (1984) argue that the field of MORB is broad enough to

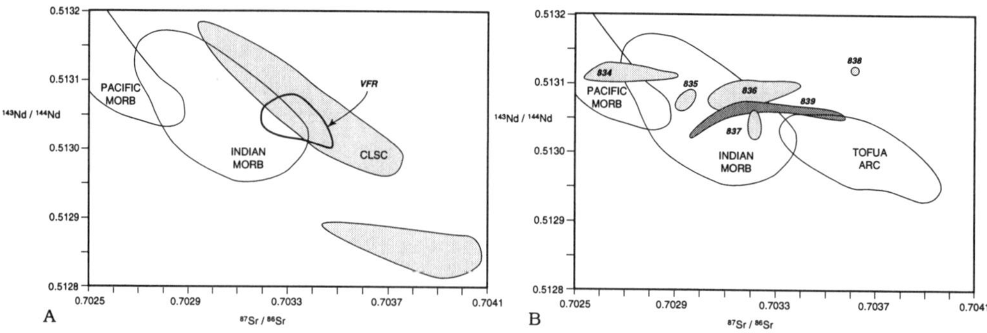

FIGURE 3.15. Nd-Sr-isotope data. Tofua arc field from Ewart and Hawkesworth (1987). Pacific and Indian MORB fields and ODP data from Hergt and Hawkesworth (1994). CLSC and ELSC fields from Volpe *et al.* (1988), Boespflug *et al.* (1990), and Loock *et al.* (1990).

include most of the IAV data. The IAV are, nonetheless, shifted to the right and at the lower Nd end of the MORB field. This characteristic is shared by many backarc basalts and is observed in Lau Basin samples (Volpe *et al.*, 1987, 1988, 1990; Loock *et al.*, 1990). Data from the ODP Leg 135 transect show that there are also less radiogenic, more MORB-like, basalts in the western Lau Basin (Hergt and Hawkesworth, 1994).

8.2. Isotope Data

It has long been recognized that the Lau Basin has anomalously high $^{87}Sr/^{86}Sr$ relative to MORB and to other western Pacific arc and backarc systems (Gill, 1976; Hawkins, 1976; Stern, 1982; Hawkins and Melchior, 1985; Volpe *et al.*, 1988; Loock *et al.*, 1990). Although most of the Sr-isotope ratios are higher than typical MORB values, the Nd-isotope ratios lie in the MORB field (Carlson *et al.*, 1978; Volpe *et al.*, 1988; Fig. 3.15). The isotope data, together with incompatible trace element abundances, have been used to suggest that several distinct mantle sources are required to explain the observed isotopic range of the active ridges and seamounts (e.g., Volpe *et al.*, 1988; Boespflug *et al.*, 1990; Loock *et al.*, 1990).

Volpe *et al.* (1988) showed that Nd- and Sr-isotope data for basalt glasses from the

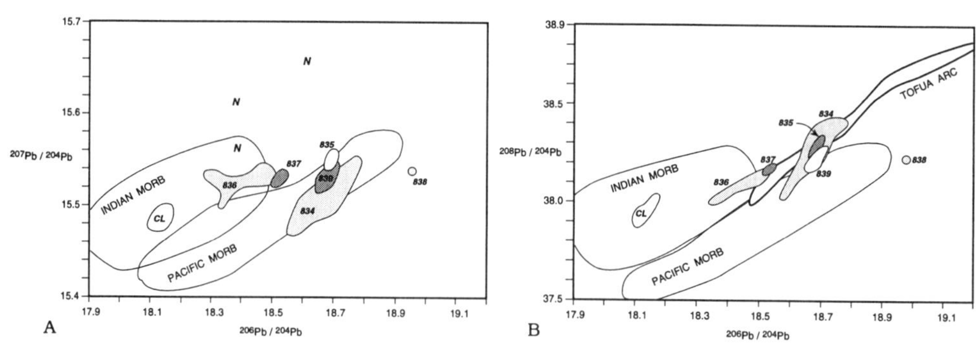

FIGURE 3.16. Pb-isotope data. N = Niuafo'ou. Sources as for Fig. 3.15.

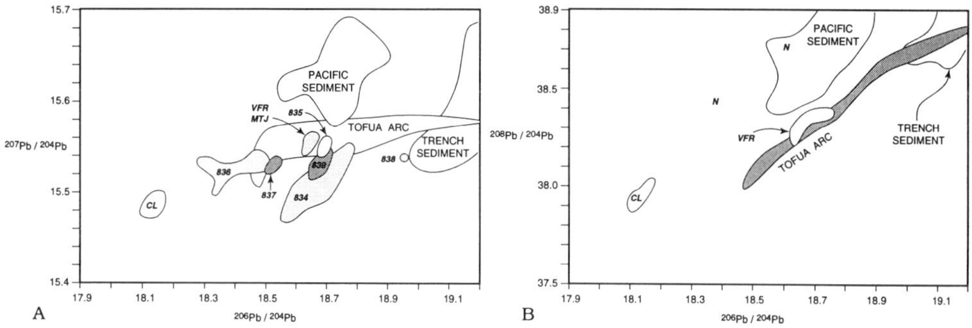

FIGURE 3.17. Pb-isotope data. Sources as for Fig. 3.15.

CLSC and seamounts in the northern Lau Basin require multiple mantle sources. They proposed (1) a MORB-like component depleted in incompatible elements, (2) a component similar to an E-MORB or OIB source that is enriched in incompatible elements, and (3) a distinct OIB-like component having high ^{87}Sr/^{86}Sr and low ^{143}Nd/^{144}Nd. The Nd-Sr relations are more like MORB from the Indian Ocean than to the East Pacific Rise, although all data are shifted to Sr-isotope ratios slightly higher than Indian MORB.

The Valu Fa Ridge (Jenner *et al.*, 1987; Vallier *et al.*, 1991) is the youngest segment of the ELSC rift zone where it is propagating into older oceanic crust at the southern end of the Lau Basin. The isotopic data give a mixed signal. The Sr- and Nd-isotopic data are similar to Lau Basin axial ridge samples, whereas the Pb-isotopic data trend toward arc values. Jenner *et al.* (1987) proposed that the trace element and isotopic data indicate a minor, but significant, contribution from the subducted slab. Vallier *et al.* (1991) conclude that VFR lavas are isotopically similar to the (Tofua arc) volcanic island of Ata that is 40 km to the east of the VFR. However, Ata is anomalous in the Tofua arc in having a MORB-like character. The VFR has ^{87}Sr/^{86}Sr ratios lower than most of the other Tofua arc volcanoes and thus resembles the backarc data. The VFR ratios for ^{143}Nd/^{144}Nd isotope data are higher than those of the Tofua arc, and they too are more like the backarc data. The high U/Th for VFR is arclike, as are the high ratios for ^{238}U/^{230}Th, ^{230}Th/^{232}Th, and ^{226}Ra/^{230}Th.

Loock *et al.* (1990) compared isotopic data for the CLSC, the Mangatolu triple junction in the northeastern Lau Basin, and Valu Fa Ridge. Their data further support the interpretation that glasses from much of the northern Lau Basin and CLSC have the Indian Ocean signature (^{87}Sr/^{86}Sr = 0.70309 to 0.70318 and ^{143}Nd/^{144}Nd = 0.51309 to 0.51312). The Mangatolu triple junction has ratios of 0.7040 and 0.51283, respectively, and the VFR has ratios of 0.70320 to 0.70339 and 0.51303 to 0.51306, respectively. Loock *et al.* (1990) and Boespflug *et al.* (1990) suggested that the arclike isotope data for the VFR could be explained by mixing between *Pacific* MORB and either a Tongan arc component or a flux of trace elements from the subducted slab. Hergt and Hawkesworth (1994) note that the VFR is the only part of the modern spreading ridge system in the Lau Basin that lacks an Indian MORB mantle source signature. Inasmuch as it is the southernmost and youngest part of the axial rift system, the influx of Indian-source mantle has not yet reached that far south.

Prior to drilling on ODP Leg 135, the best interpretation of the nature of Lau Basin mantle magma sources was that the axial ridges were dominated by Indian Ocean MORB, propagating rifts in the southern and northeastern parts of the basin have a strong signature of an arc component mixed with Pacific MORB, and the northern Lau Basin has an OIB component mixed with Indian MORB. The best evidence for each of these components is the Pb-isotope data, but it also is implicit in the Sr and Nd isotopes (Fig. 3.15; Hergt and Hawkesworth, 1994). These interpretations, based largely on samples from the modern spreading centers and other sites less than 1 m.y. old did not account for the sources important in the early history of the basin.

Drilling in the western part of the Lau Basin has given us a new view of the heterogeneity of Lau Basin mantle. One of the most significant results of Leg 135 is the isotope studies presented by Hergt and Hawkesworth (1994) that bear on mixing of multiple mantle components in the evolution of Lau Basin crust. They proposed that magmas erupted at the modern spreading centers are derived from a recent influx of mantle having the isotopic signature of the Indian plate mantle. The Indian source has displaced an older source having the isotopic signature of Pacific plate asthenosphere. The Pb data suggest two different mixing trends. These require an "arc" component and a component from either a Pacific MORB or an Indian MORB source. Hergt and Hawkesworth (1994) evaluate several possible explanations for the trace of the arc component. These are (1) magma mixing between upwelling mantle and arc magmas during the early stages of extension, (2) low but significant infiltration of the backarc magma source by slab-derived flux, and (3) contamination of the source by arc lithosphere during extension and upwelling. They propose that the "arclike" component has low Ce/Pb, low Nb/U, and relatively radiogenic Sr and Pb. The Pb composition of the "arc" end member is more radiogenic than most of the data from the Tofua arc but overlaps with Tafahi and Niuatoputapu, two of the northernmost Tofua arc volcanoes, and with old Tonga arc crust such as the Eocene rocks of 'Eua. No data are available from the Lau Ridge to make a comparison for Pb isotopes, but, arguably, 'Eua may be equivalent. The N-MORB component is defined by unit 7 at Site 834, which has relatively unradiogenic Pb and Sr like the Pacific N-MORB mantle sources. The trace element data for glasses from this unit have a smooth MORB-like normalized pattern (Hawkins, 1994; Hergt and Farley, 1994)

At Site 834, we cored basalts erupted in the earliest stage of Lau Basin extension. These basalts probably represent the isotopic composition of the "original" SSZ mantle. The Nd-isotope data are remarkably uniform (0.513109 to 0.513125) and are similar both to MORB and to the modern Lau spreading centers. The Sr data range from 0.702544 to 0.702958 (i.e., lower than most of the Lau Basin basalt). Although several samples are more radiogenic than MORB, they do not reach values typical of the modern Lau spreading centers (e.g., 0.7033). The Site 834 samples have the most MORB-like Nd and Sr values reported to date for the Lau Basin. Good correlations exist between Sr and Pb isotope compositions. The Pb-isotope data define linear trends indicating mixing that project into the field for *Pacific* MORB. This is quite remarkable in view of the widespread evidence elsewhere in the backarc basin for an Indian mantle source. Unit 7, at mid-depth in the hole, is the most MORB-like unit and has the least radiogenic Pb. The Pb-isotope ratios become progressively more radiogenic upward to the youngest flow studied (unit 2). Units 13 (deepest) to 8 also show a similar upward increase from less to more radiogenic; unit 8 and unit 2 are very similar. Units 7 and 8 offer the greatest contrast in Pb-isotope ratios. The variations with depth, and the apparent cyclicity of the variations, are interpreted as the result of mixing of different proportions of MORB-like and arclike melts (Hergt and Farley, 1994; Hergt and Hawkesworth, 1994). For example, these authors propose that the variation between unit 7, the most MORB-like sample, and units 2 and 8, the most radiogenic samples, may be explained by mixing between "essentially pristine MORB" and up to 50% of the "arc" endmember. The timing and nature of the mixing process are fairly well constrained and require mixing of mafic melts rather than melting of a heterogeneous source (Hergt and Farley, 1994).

Lavas from Sites 837 and 839 have a strong arc signature in terms of mineralogy and chemistry (e.g., they have two-pyroxene basaltic andesite and andesite). They also have arc isotopic characteristics (Figs. 3.15–17). Site 837 samples are similar to the modern Tofua arc, and Pb-isotope ratios lie on a trend between Indian mantle and arc. Site 839 Pb-isotope ratios are confined to a relatively small field, suggesting that they come from a well-mixed and homogeneous source. The Pb-isotope ratios lie on a trend between arc and the Pacific mantle and thus resemble Site 834 samples. The isotopic signatures for Site 839 resemble those for the modern Tofua arc volcanic island Ata as well as those for arc-derived intrusive dikes at forearc Site 841 (Bloomer *et al.*, 1994). The arc rocks of Site 839 probably formed at an arc edifice that was cut off and isolated from the Tonga Ridge by the southward propagation of the ELSC (Ewart *et al.*, 1994b; Hawkins, 1994).

Site 836 sampled crust formed early in the history of the ELSC. The propagating rift tip would have passed close to the site at about the time the Site 836 crust was formed (0.6 to 0.8 Ma; Parson and Hawkins, 1994). Probable igneous basement at Site 836 has a strong Indian mantle isotopic signature and resembles rocks of the present axial ridges. Samples from volcanic breccias or gravels are more radiogenic than the presumed basement and are similar to arclike Site 837 samples. The breccias probably came from a nearby seamount.

Collectively, the data of Hergt and Hawkesworth (1994), together with data of Loock *et al.* (1990), suggest that a remnant of "old Pacific" asthenosphere has been available under the easternmost part of the Lau–Tonga system for at least the past 6 m.y. This inference helps explain a Pacific MORB source as evidenced in the oldest drill site samples (Site 834). This depleted source mixed with an Indian MORB source and an arc component to form magmas of the modern spreading centers (e.g., CLSC, ELSC, and VFR). The remnant of Pacific mantle must have migrated eastward since about 5.5 Ma, perhaps in response to the eastward rollback of the Tonga Trench. Hergt and Hawkesworth (1994) suggest that the influx of Indian MORB-source mantle under the Lau Basin replaced Pacific MORB-source mantle within the past 5.5 m.y. They postulate that the influx was coupled with the southward propagation of the ELSC. The rift propagation may have been the result or the cause of clockwise rotation of the Tonga Ridge and its underlying mantle. They note that the complex mixture of Indian, Pacific, and arc sources at the Valu Fa rift tip may be explained as a consequence of this process. Another possibility is that the "Pacific" signature is derived from the subducted slab.

Mixing between MORB-source mantle and other components is not unique to the Lau Basin. For example the Mariana Trough may have both a MORB and an arc component (Volpe *et al.*, 1987, 1990). They interpreted the data as supporting a heterogeneous source rather than mixing of melts as proposed for the Lau Basin. The northern Mariana Trough narrows to a propagating rift tip penetrating the Mariana arc; mixing between MORB and either an arc component or a mantle metasomatized by fluid from the subduction zone (Stern *et al.* 1990). Nascent backarc basins behind the Shichito Ridge, in the Izu–Ogasawara arc, are interpreted as showing evidence for mixing between E-MORB and island arc *magmas* rather than derivation from heterogeneous magma sources (Ikeda and Yuasa, 1989). However, Hochstaedter *et al.* (1990a,b) invoke mixing relations between sources having E-MORB and a subduction component for the Sumisu Rift of the Izu–Ogasawara arc system.

The question of whether the isotopic heterogeneity of backarc magmas is caused by selective melting of separate domains comprising a heterogeneous source, or by mixing with injections of different batches of melt, may be closely tied to the dynamics of backarc opening. Injection of new batches of MORB melt to mix with another type of mantle or melt (e.g., SSZ mantle) could be a consequence of mantle counterflow, upwelling, and melting in the extensional region above the subduction zone. Magmas should evolve toward more MORB-like systems as the "other" component is melted, mixed with, and overwhelmed by the influx of MORB-source mantle. Alternatively, a passive dynamic response may be visualized in which heterogeneous mantle is progressively melted with the more fusible (enriched) domains being consumed first. This too would lead to crust dominated by less enriched rocks. The second sequence of events would lead to a linear progression in composition rather than cyclicity and would not necessarily have N-MORB crust as the evolutionary end point. The Lau Basin data suggest that the first model is a more likely explanation for Lau Basin evolution.

Helium-isotope data offer one of the more intriguing insights to mantle sources (Poreda, 1985; Poreda and Craig, 1993). The axial ridge system has MORB-like $^3He/^4He$ ratios of about 8–10 × atmospheric ratio (R_A). Seamounts and seafloor rocks in the northern part of the basin have R_A of up to 22. These are taken as evidence for a "Samoan plume" component in the mantle source that may be entering the SSZ region from the north or northwest. Other isotope data suggest an OIB component in these rocks. Oxygen-isotope data are limited to a collection of samples from the Peggy Ridge and the region around the CLSC (Pineau *et al.*, 1976). They all are similar to MORB values (e.g., $\delta^{18}O = +5.7$ to $+6$) and indicate equilibration with mantle source material.

9. HYDROTHERMAL ACTIVITY

9.1. Introduction

We have long known that the Lau Basin has many sites of high heat flow presumed to be related to hydrothermal activity and other areas with very low heat flow presumed to be in recharge zones (Sclater *et al.*, 1972). Considerable effort has been spent in sampling and mapping surface sediments in the basin to look for evidence of hydrothermal minerals and to understand their type, abundance, and distribution pattern. This has culminated in the discovery of several fields of active "black smoker" hydrothermal vents on the southern part of the ELSC. The presence of these vents and metalliferous deposits, on crust that is a potential progenitor to an ophiolite assemblage, has important implications for the origin of many economic deposits in continental settings.

9.2. Hydrothermal Sediments

The first reported occurrence of hydrothermal mineral deposits in the Lau Basin were of ferromanganese oxides in sediments (Berthine, 1974) and barite and opal in volcaniclastic sediments (Berthine and Keene, 1975) dredged from Peggy Ridge (dredge site 123-86, Fig. 3.1). Subsequently, there has been an intensive effort to map hydrothermal ferromanganese oxide (HTFM) mineral distribution in Lau Basin sediments (e.g., Cronan *et al.*, 1984; Hodkinson *et al.*, 1986; Hodkinson and Cronan, 1991, 1994). Sediment cores from ODP Leg 135 have given additional information about the regional and temporal distribution of HTFM minerals in nondetrital and carbonate-free sediments (Hodkinson and Cronan, 1994).

Hodkinson and Cronan (1991) recognized two hydrothermal phases in Lau Basin surface sediments: (1) an "oxide phase" with Mn, Fe, Co, Ni, and Cu, and (2) a "sulfide/weathered sulfide phase" with Fe, Cu, and Zn. The sulfide/weathered sulfide phase is restricted to sediments at the neovolcanic zone of actively spreading ridges, whereas the oxide phase is more widely dispersed throughout the basin. Using multivariate statistical analysis to evaluate the trace and minor element chemistry of the ODP drill core sediments, they were able to discriminate between a hydrothermal component and a weathered detrital component (Hodkinson and Cronan, 1994). The data were used to evaluate the relative abundances of HTFM minerals with depth in cores from the backarc basin drill sites. Only the oxide phase was found in the six backarc drill cores. The HTFM mineral concentrations give an integrated record of dispersal and fallout from hydrothermal plumes. In five of the six backarc drill cores the concentrations increase upward to a maximum value and then decrease. This pattern is different from that seen in sediments near mid-ocean ridges, where there is a concentration maximum at the interface between sediment and basalt and an upward decrease. Hodkinson and Cronan (1994) interpret these relations as indicating minimal input of HTFM minerals from the igneous crust of the backarc subbasins. They conclude that the HTFM concentration maximum at high levels in the core records hydrothermal activity on the ELSC or CLSC distant from the drill sites. They find a fairly good correlation with the stratigraphic age of the HTFM maximum in the cores and the probable time of passage of the propagating rift tip of the axial ridge. They note a time lag of from 0.4 to 0.7 Ma between passage of the rift tip and the age of the maximum HTFM signal in the cores. They explain the uphole decrease in HTFM as the result of increasing distance with time as the drill site location migrated away from the spreading ridge axis.

The record at Site 836 differed from that described above. Site 836 is on some of the early crust formed by the ELSC. There is a major HTFM mineral peak near the base of the sediment column at about 0.6 Ma. This is the approximate age of the igneous basement. There are several lesser spikes in the HTFM mineral concentrations at age increments of about 0.1 Ma. Even though this was drilled to igneous basement close to the former rise crest, no sulfide/weathered sulfide phases were found. All of the drill core data show many minor peaks in HTFM concentration in addition to the maximum. These are interpreted as being the reflection of variable hydrothermal activity at the ridge axis. The variable hydrothermal activity may be the result of repeated minor reorganization of the tectonic fabric of the ridge crest, which would lead to variations in crustal fractures and permeability.

9.3. Hydrothermal Vents, Crusts, and Massive Sulfides

Active hydrothermal vent systems and polymetallic sulfide deposits are known from many places on the global ocean ridge system and from several other backarc basins (e.g., Mariana Trough; Hawkins *et al.*, 1990). Evidence of hydrothermal circulation in the Lau Basin in the form of stained pillow surfaces and thin Mn-oxide crusts was photographed on the CLSC (von Stackelberg *et al.*, 1985); occurrences of hydrothermally formed nontronite and manganese oxide crusts (birnessite), and metallic sulfides (pyrite and chalcopyrite) impregnating volcanic rocks were dredged from the VFR (von Stackelberg *et al.*, 1985).

Three types of hydrothermal metal-rich crusts were identified on the VFR (von Stackelberg and von Rad, 1990): (1) ochre-colored Fe-rich crusts, (2) olive-colored Fe-rich crusts, and (3) silver-gray-colored Mn-rich crusts. The Fe-rich crusts are well-crystallized nontronite, whereas the Mn-rich crusts are birnessite. The authors interpret the growth of the sponge-like crusts as having been influenced by microorganisms.

The first evidence for hydrothermal vents impregnated with metallic sulfides in the Lau Basin was discovered on the east limb of the Mangatolu triple junction (Hawkins, 1986; Hawkins and Helu, 1986). Broken, cylindrical fragments of a sulfide-impregnated mass of barite and amorphous silica were dredged along with angular blocks of fresh, high-silica andesite vitrophyre. The sulfides are primarily wurtzite (ZnS) with minor amounts of sphalerite, chalcopyrite, and pyrite. The chimney fragments appeared to be unaltered, and it was assumed that they were from either an active vent system or one that had recently become inactive.

An extensive hydrothermal vent system, extending for about 100 km, has been found on the Valu Fa Ridge; four separate vent fields have been studied (von Stackelberg and von Rad, 1990). This field was explored and sampled with the deep-submersible *Nautile* (Fouquet *et al.*, 1991a,b). The sulfides occur mainly with basaltic andesite and rhyodacite. Four types of deposits were recognized, which the authors relate to the tectonic style of different segments of the VFR. These are (1) low-temperature hydrothermal Fe-oxides without visible hydrothermal activity; (2) low-temperature Mn-oxide crusts that cover higher-temperature sulfides, the low-temperature crusts being related to extensive hydrothermal discharges (40°C) through permeable andesitic rocks; (3) barite chimneys and masses of barite mixed with sulfides that were formed by medium- to high-temperature discharges and found in an inactive vent field; and (4) very high temperature black (320–400°C) and white (250–320°C) smokers. The chimneys are formed of Cu and Zn sulfides and barite. Fragments of altered andesite beneath the chimneys have stockwork mineralization with cm-thick veins of chalcopyrite.

The mineralogy of the VFR deposits consists of abundant barite, sphalerite, pyrite, chalcopyrite, and marcasite. Minor minerals include tennantite, galena, Pb-As sulfosalts (e.g., gratonite), and primary grains of native gold. The vent effluent has the lowest pH (2) ever measured in the ocean and the highest measured contents of Zn, Cd, As, and Pb (Fouquet *et al.*, 1991a).

The Valu Fa field is considered to be one of the most active on the seafloor and is distinctive both in the chemistry of the vent effluent and the mineralogy of the sulfide deposits (Fouquet *et al.*, 1991b). They describe it as being intermediate between typical mid-ocean ridge mineralization and the massive sulfide deposits of the Kuroko type. Von Stackelberg and von Rad (1990) speculate that the seafloor below the VFR may be impregnated with high-temperature sulfide minerals that would make the VFR comparable to a Cyprus-type ore deposit.

10. BACKARC BASIN SEDIMENTATION

Comprehensive discussions of Lau Basin sediment studies and their tectonic significance are presented in the scientific results volume for ODP Leg 135 (Hawkins *et al.*, 1994). Here I present a brief synthesis of data for volcanogenic sediments that is relevant to the magmatic evolution of the Lau Basin and the adjacent island arcs. One of the more important discoveries of ODP Leg 135 is the evidence for ephemeral arc-composition volcanism at numerous centers *within* the Lau Basin as it opened. Previously unrecognized, evidence for the existence of these intrabasinal volcanoes comes from the abundance of arc-composition detritus recovered in drill cores. Textures indicate minimum transport, and nearby intrabasinal arc constructs are inferred to be the sources. They were a major source of the volcaniclastic sediment deposited in the subbasins and were active synchronously with the nearby eruption of MORB-like lavas that formed much of the backarc basin crust (Bednarz and Schmincke, 1994; Clift and Dixon, 1994; Parson *et al.*, 1994).

The backarc basin drill sites, as well as the two drilled on the Tonga Ridge, display similar patterns of sedimentation. The oldest sediments recovered are late Miocene and the youngest are Holocene. The age of basal sediments differs at each site with a general pattern of younging eastward from Site 834, alongside the Lau Ridge, to Site 836 that was drilled on some of the oldest crust formed by the ELSC. Except for Site 836, the sediments primarily comprise a lower sequence of volcaniclastic turbidites interbedded with hemipelagic clayey nannofossil mixed sediments and nannofossil clays. These are overlain by an upper sequence of hemipelagic clayey nannofossil oozes, locally containing calcareous turbidites (Parson *et al.*, 1994; Rothwell *et al.*, 1994). These authors interpret the volcaniclastic turbidites as being derived from nearby arc fragments or constructs *within* the basin rather than from adjacent island arcs. Pliocene proximal volcaniclastic sediments are interpreted as having been derived from intrabasinal seamounts. The Pleistocene sediments are mainly hemipelagic. It is important to note that they view intrabasinal "arc" volcanism as a major source of the volcaniclastic material. Site 836, drilled on old ELSC crust, is distinctive in having only 20 m of sediment overlying presumed igneous basement. The lower half of this sediment sequence comprises more than 50% volcaniclastic material. The glass shards, representing a limited compositional range from 54.3% to 58.1% SiO_2, are best interpreted as the result of fractional crystallization of the MORB-like basaltic magmas that formed the basement at Site 836 and the nearby ELSC. Similar basaltic andesite has been dredged from the high-standing ridges near Site 836 (e.g., RNDB-02, Table IV).

The interpretation that the Pliocene age volcaniclastic beds were derived from intrabasinal seamount volcanoes is based on sedimentologic data. Rothwell *et al.* (1994) note that volcaniclastic rocks of the subbasin depocenters have delicate needlelike shards that could not have survived long-distance transport, and the glass shards are fresh. Neither the Lau Ridge nor the Tofua arc is considered a likely source. The chemistry of the glass shards indicates that nearly all have low-K arc tholeiitic compositions ranging from basaltic to rhyolitic. Some shards from Sites 834, 835, and 837 have medium-K arc chemistry. Glasses from Sites 834 and 835 span the full range from basaltic to rhyolitic, whereas Site 836 has a fairly unimodal composition (basaltic andesite, clustered around 55% SiO_2). Site 838 glasses also have a unimodal concentration of values ranging from 50% to 55% SiO_2. Bimodal suites are found at Sites 837 and 839 where basaltic andesites and rhyolites represent maxima. The Site 836 glasses are distinct in having trace element ratios similar to those of the nearby ELSC (e.g., MORB-like ratios for Ba/Zr of 0.9 to 1.4), in contrast to the much higher, arclike values (3.9 to 3.10), found at the other sites. In general, all of the trace element data for glasses from the backarc sites have broadly similar patterns when normalized to N-MORB. These patterns show the SSZ magma signature, namely moderate to strong enrichment in LILE, strong depletion in Nb, and minor depletion in other HFSE (Clift and Dixon, 1994; Rothwell *et al.*, 1994).

The provenance of the widespread volcaniclastic units is important to understanding the magmatic evolution of the Lau–Tonga system. The only epiclastic unit that can be tied to the Lau Ridge is a 3.3-Ma bed at Site 834 that closely matches the chemistry of the

Korobasaga volcanic group. Rothwell *et al.*(1994) conclude that it was unlikely that there was widespread or voluminous volcanism on the Lau Ridge after extension and rifting started in the Lau Basin. This idea is contradicted by the evidence that the Korobasaga Group was laid down on the Lau Ridge between 4.5 and 2.5 Ma (Whelan *et al.*, 1985) and was a potential source of sediment. Differences in eruptive style, sediment entrapment in the basin west of Site 834, a different pattern of dispersal, or a more limited locus of volcanism, all may have played a role in limiting volcaniclastic sedimentation in the subbasins sampled on Leg 135. Cole *et al.* (1985) and Woodhall (1985) note that the Korobasaga Group is limited to six islands, forming a 35-km-wide belt, in the northern 200 km of the Lau Group. This may account for its relative unimportance as a source of volcanic detritus for infilling the subbasins. The Tofua arc probably was not a robust system until about 1 Ma. It may have evolved over 3 m.y. by southward propagation of volcanic centers (Hawkins, 1994; Parson and Hawkins, 1994). There is an extensive record of Miocene volcanism on the Tonga Ridge. The early Miocene beds probably had a source on the Lau Ridge, but in latest Miocene time the most likely sources were small and ephemeral (?) intrabasinal eruptive sites (Clift and Dixon, 1994; Rothwell *et al.*, 1994). Zephyr Shoal (Table IV, Fig. 3.9g) is an example of such a feature. The low-K arc-series chemistry of many of the Tonga Ridge volcaniclastic units matches the compositions found in the sedimentary units at the backarc drill sites; this supports the idea of derivation from intrabasin volcanoes.

The age of opening and sedimentation of the basins appears to young eastward. The time of transition from deposition of volcaniclastic sediments to sediments dominated by clayey nannofossil ooze varies from site to site but also tends to young eastward. The eastward age progression in timing of the locally derived voluminous eruptions of arclike material suggests that there was a wave of arc volcanic activity that "tracked" the eastward retreat of the Tonga forearc as it followed the rollback of the trench. This is a previously unrecognized aspect of backarc basin evolution. Intrabasinal arc edifices erupted alongside magma leaks that form MORB-like backarc basin crust. The arc volcanic front, or zone, has been stabilized for about 1 Ma or less and constitutes the present Tofua arc. Well-dated volcaniclastic units help constrain the longevity of the intrabasinal arclike eruptive centers. For example, Site 835 records nearby activity from 3.5 to 2.9 or 3.0 Ma. At Site 837, proximal sources laid down deposits from 2.1 to 0.34 Ma; at Site 838, proximal sources formed volcaniclastic beds from 1.92 to 0.50 Ma; at Site 839, the range is from 1.73 to 0.37 Ma (Bednarz and Schmincke, 1994). I propose that the "intrabasinal arc" constructs had eruptive life spans around 1 to 2 m.y. before becoming inactive as the arc front migrated eastward.

11. REGIONAL GEOLOGIC HISTORY

Before presenting a synthesis of Lau Basin evolution, I will summarize the geologic history as we know it. Citations to appropriate references are given elsewhere in the text.

1. The Lau Ridge was an active volcanic island arc from at least 14 Ma to 2.5 Ma. The volcanism was in response to subduction at an ancestral Tonga Trench that may have had the same general orientation as the present one, but the trench axis probably lay 100 km or more to the west. Volcanic activity younger than 2.5 Ma appears to have been related to extensional stresses to the rear of the arc and backarc rather than to frontal arc volcanism.
2. The geologic history of the Lau Ridge may extend to late Eocene time, as suggested by the occurrence of Eocene arc volcanic rocks on Viti Levu, Fiji Islands, and

Eocene arc plutonic rocks on 'Eua, in the Tonga forearc. Leg 135 drilling at Site 841 also revealed pre-late Eocene and late Miocene arc series volcanic rocks on the Tonga Ridge. All of these rocks are interpreted as remnants of a former "Melanesian island arc" that included what are now parts of the Lau Ridge, Tonga Ridge, Fiji Platform, and New Hebrides–Vanuatu Ridge. The orientation of the trench related to this ancestral arc is not known, but it probably faced to the east or the northeast.

3. The Lau Basin began to open in latest Miocene time (about 6 Ma) as the eastern part of the Lau Ridge and its forearc were first extended and then rifted away to the east. Fragments of the Lau Ridge and its older basement are found on the Tonga Ridge.

4. There has been more or less continuous arc volcanism in the Lau–Tonga region since late Eocene time. Lau Ridge volcanism overlapped with the time of initial crustal extension that formed the Lau Basin, but during the last 2.5 Ma it has diminished and seafloor spreading in the Lau Basin has supplanted it. Within the last 1 m.y. the Tofua arc has become a major feature on the Tonga Ridge. The Tofua arc may have developed sequentially from north to south in consort with the rotational opening of the Lau Basin.

5. Lau Basin crust is largely floored by basaltic rocks having MORB mineralogy and many MORB-like chemical and isotopic characteristics. The central Lau and eastern Lau spreading centers (Fig. 3.1) are well-defined axial ridge systems formed mainly of N-MORB type basalts.

6. Although dominated by MORB-like chemistry, the crust of the Lau Basin, like other backarc basins, is heterogeneous. On a regional scale, the Lau Basin is compositionally zoned from older rocks having boninitic and "transitional to arc" compositions to younger MORB-like rocks on the spreading centers. However, this general pattern is contradicted by the rocks from Site 834. The basin drilled at Site 834, the oldest crust known in the basin, has MORB-like chemistry. Seamounts in the northern Lau Basin include rocks having an ocean island basalt composition component, and dacitic rocks of Zephyr Shoal may be silicic differentiates of boninite magma series. Rocks of the Valu Fa Ridge and the Mangatolu triple junction range from basaltic to basaltic andesite; the VFR has a strong arc signature and probably comes from a mixed source, whereas the MTJ has few arc characteristics. Sites 834, 835, and 836 (Fig. 3.1), although not on a true "flowline" in the sense of progressive accretion of new crust, sampled the evolution of magma types over a time span of about 5 m.y. The sites are located, respectively, on the oldest crust drilled (about 5.6 to 6 Ma, latest Miocene), intermediate age crust (about 3.4 Ma, late Pliocene), and the youngest crust drilled (about 0.64 to 0.8 Ma, middle Pleistocene). Site 836 lies about 50 km west of the axis of the ELSC, which is formed of MORB-like tholeiite. The ELSC becomes transitional from MORB to arc tholeiite toward the VFR. Sites 837, 838, and 839 form a cluster, about 125 km west of the active ELSC, on crust estimated to be 1.9, 2.0, and 2.1 Ma, respectively. Site 839 samples are arclike in chemistry in spite of their proximity to the ELSC.

12. SUMMARY AND A MODEL FOR LAU BASIN EVOLUTION

The evolution of the Lau Basin, and other backarc basins, must represent the kinematic response of the SSZ crust and upper mantle to oceanic plate subduction, mantle counterflow, and upwelling. This response is expressed as an extension of the suprasubduction zone lithosphere and the formation of new crust by emplacement of melts derived from the SSZ mantle. The magmatism derives from both the SSZ mantle and "new" mantle brought in by convective overturn. The tectonic setting and magmatic response are similar to that of extensional provinces in continental settings such as the Basin and Range Province of western North America, where crustal extension and thinning have been accompanied by voluminous volcanism. Backarc basin evolution may be a small-scale replica of the early stages of formation of the major ocean basins.

The opening of the Lau Basin has been accomplished by two modes of extension. Initially, rifting and extension formed basins separated by high-standing blocks; subsequently, seafloor spreading began and continues today. The rifting and extension phase resulted in formation of new crust within the subbasins. We do not know the deep structure of these subbasins, but it is likely that it is different from "true" deep-sea floor in that the new basaltic crust may overlie attenuated remnants of older crust (e.g., fragments of old arc or forearc). Extension was accompanied by extrusion of both arclike and mid-ocean-ridge-type magmas that have formed new crust. The proximity to active island arcs, and the similarity in many respects between arc and backarc magmas, suggests that there may be similarities in petrogenesis. Extreme rifting and extension ultimately separated all remnants of the older crust, and the potential void was replaced with new ocean crust that directly overlay its mantle source. Seafloor spreading began. The seafloor spreading was promoted by propagating rifts. Interpretations of the results of studies of the western Pacific backarc basins have given rise to a common model in which extension followed by rift propagation has played a key role in their evolution (Stern *et al.*, 1984; Parson *et al.*, 1990; Tamaki *et al.*, 1992; Fryer and Pearce, 1992; Parson *et al.*, 1992; Taylor, 1992; Parson and Hawkins, 1994).

A successful model for Lau Basin evolution has many problems to explain. These include an explanation for the occurrence of a spectrum of crustal rocks that range from MORB-like to arclike; the occurrence of both Pacific and Indian MORB-source mantle isotopic signatures as well as arc-source isotopic ratios; the cause for widespread extension at a convergent plate margin; the mechanism causing extension of the crust and its eventual rupture after which seafloor spreading becomes operative; and the apparent random location of the part of the remnant arc or forearc that becomes the site of extension and rifting.

The model presented in Fig. 3.18 is an updated version of earlier models by the author (Hawkins *et al.*, 1984; Hawkins and Melchior, 1985) and draws on more recent models of Stern *et al.* (1990), Parson *et al.* (1990), and Parson and Hawkins (1994). Stern and Bloomer (1992) proposed a similar interpretation of the early history of the Izu–Bonin–Mariana system and used it to explain the origin of ophiolite remnants in the California Coast Range. Interpretations of data from ODP drilling in the Sea of Japan, the Izu–Bonin system, and the Mariana arc all have led to broadly similar models (e.g., Fryer and Pearce, 1992; Pearce *et al.*, 1992; Tamaki *et al.*, 1992; Taylor, 1992). All data such as ages, rock and sediment chemistry, geophysical data, and the regional geology used in this Lau Basin model are from the numerous contributions cited elsewhere in the text in addition to extensive data collected by the author.

A discussion of the initial conditions in the southwestern Pacific at the beginning of Cenozoic time is beyond the scope of this chapter, but there is evidence for a vast zone of extension and upwelling along the 4400-km trace of the West Melanesian–Solomon–Vitiaz trench and arc system (see Kroenke, 1984, for an extensive discussion and Hawkins, 1994, for a summary). A similar tectonic/petrologic feature is required to explain the evolution of the Palau–Kyushu Ridge (e.g., Mattey *et al.*, 1980) and was used by Stern and Bloomer (1992) for the early stages of the Izu–Bonin–Mariana system. A dramatic change in rheology of the upper mantle, perhaps accompanied by a change in relative plate motion, was important in the genesis of these major geologic features. Mantle upwelling and crustal extension promoted the development of boninitic magma systems early in arc history and at relatively shallow mantle levels. This has been invoked for the Izu–Bonin arc (Pearce *et al.*, 1992; Stern and Bloomer, 1992) and for the Mariana forearc (Bloomer and Hawkins, 1987; Stern *et al.*, 1991). Boninite magmas were followed by, and perhaps overlapped with, eruption of arc tholeiitic series melts. There is a major problem in explaining the mismatch in the age of initiation of the western Pacific Eocene arc systems and the timing of the relative change in motion of the Pacific plate (about 43 Ma) as inferred from the age of the bend of the Emperor–Hawaii seamount chain. Evidence for volcanism at least 5 to 6 m.y. earlier is found on the Palau–Kyushu Ridge and in the Bonin Islands (Taylor, 1992, and references therein). Thus, it is difficult to relate *initiation* of arc volcanism to the change in motion even though plate convergence has been important subsequently. The boninitic and early arc tholeiitic stages of an arc require melting of a suitable mantle source at relatively shallow depths (e.g., Stern *et al.*, 1991; Pearce *et al.*, 1992; Stern and Bloomer; 1992); plate convergence may not be required, and, in fact, lithosphere extension could well have been the control on early boninitic magmatism. Admittedly this is speculative, but an early phase of extension could initiate the boninitic magmatism and set up crustal buoyancy differences that could be exploited by subsequent plate motion changes to cause subduction. A similar model was proposed by Stern and Bloomer (1992), and a less refined version was presented by Hawkins *et al.* (1984). Different, and equally plausible, models presented by Pearce *et al.* (1992) include subduction of a spreading ridge beneath young lithosphere. Whatever the mechanism for its initiation, the earliest stages of Eocene arc development in the northwestern Pacific resulted in an arc that was up to 300 km wide and 3000 km long formed of boninite and arc tholeiite (Taylor, 1992). A comparable arc developed concurrently in the southwestern Pacific.

Considering the present Papua–Vanuatu–Fiji–Lau–Tonga region, there is evidence for an Eocene Vitiaz arc or West Melanesian arc on the northeastern margin of the Indian plate (Colley and Hindle, 1984; Kroenke, 1984; Colley *et al.*, 1986, 1993; Gill, 1987; Wharton *et al.*, 1992) that may have been as much as 200 km wide and nearly 2000 km long. This ancestral arc is shown schematically in Fig. 3.18a; the 40- to 46-Ma rocks include arc tholeiitic gabbro and gabbro-norite, pillow basalts, low-K rhyolite, and boninite. Although these ages are close to the inferred age for the Emperor–Hawaii bend, they are hard to reconcile with a cause-effect relationship. We would not expect an instant response of magmatism to the beginning of subduction. It is probable that there was a delay of a few million years between the onset of magma separation and upwelling from a mantle source and the beginning of eruption. The silicic rocks are voluminous (?), highly differentiated melts, that must have had mafic precursors. A further delay is likely to allow the extreme fractionation. Whether these rocks are related to extension or are the result of subduction, they constitute some of the oldest crust of the West Melanesian or Vitiaz arc. Evidence for Oligocene arc magmatism is widespread and includes rocks on Fiji, 'Eua, and Vanuatu, as well as New Britain and New Ireland (Gill, 1987). These were no doubt a response to subduction after the change in Pacific plate motion. The Eocene–Oligocene arc is shown in Fig. 3.18b and in the following figures is labeled "E-O." Opening of the South Fiji Basin as a backarc system appears to have been a response to convergence at the Vitiaz Trench in Oligocene time, but for simplicity it is not shown on the figures.

The Miocene arc (Lau Ridge) shown in Fig. 3.18c probably faced the present Tonga Trench; although we have no control on its actual orientation, it is likely that there may have been some rotation. It is uncertain whether there was a break in arc volcanism in the early Miocene; neither outcrops nor drill site data offer an indication for continuity. The Eocene–

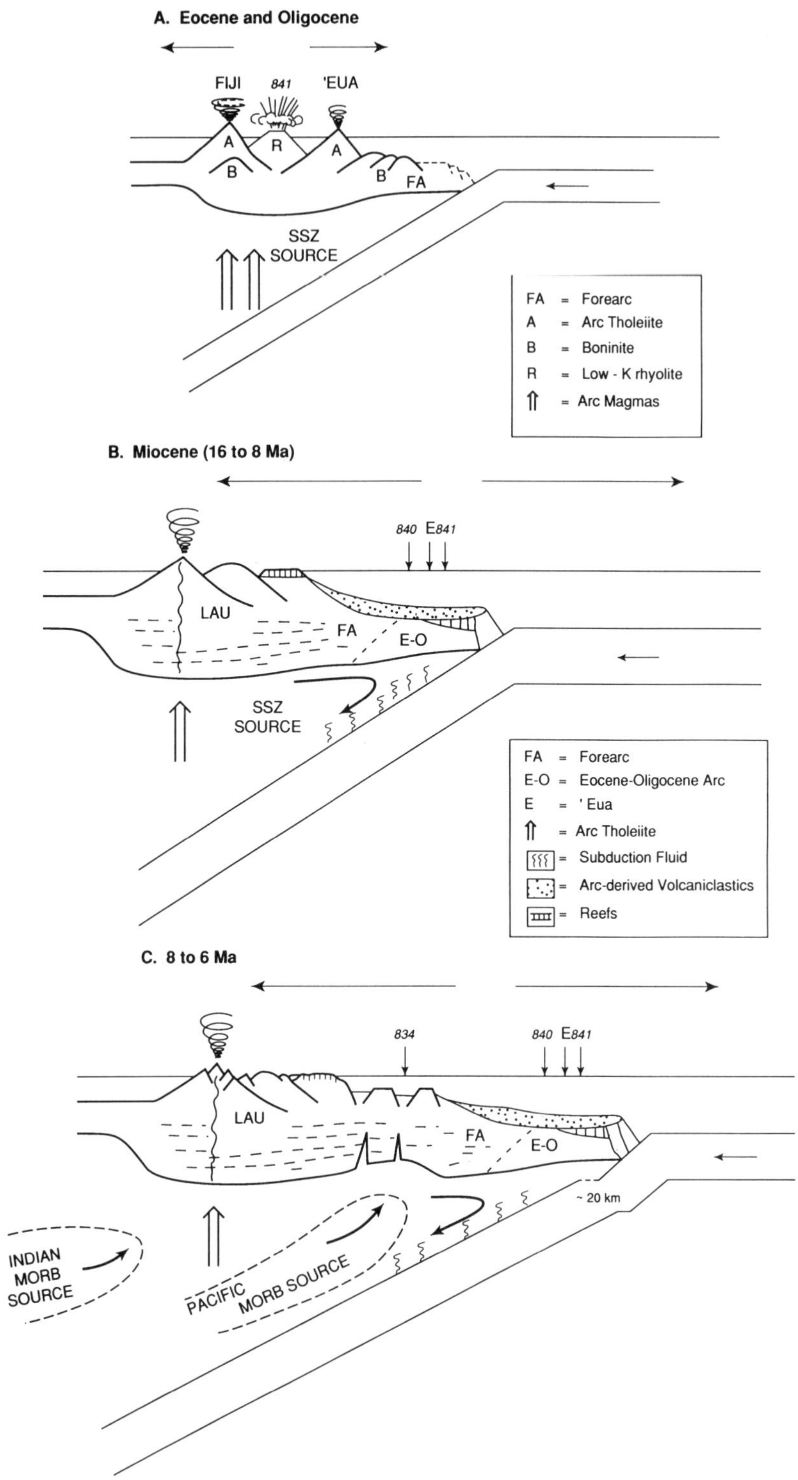
A. Eocene and Oligocene
FIJI
841
'EUA
A
R
A
B
B
FA
SSZ SOURCE
FA = Forearc
A = Arc Tholeiite
B = Boninite
R = Low - K rhyolite
= Arc Magmas
B. Miocene (16 to 8 Ma)
840 E841
LAU
FA
E-O
SSZ SOURCE
FA = Forearc
E-O = Eocene-Oligocene Arc
E = ' Eua
= Arc Tholeiite
= Subduction Fluid
= Arc-derived Volcaniclastics
= Reefs
C. 8 to 6 Ma
834
840 E841
LAU
FA
E-O
~ 20 km
INDIAN MORB SOURCE
PACIFIC MORB SOURCE

FIGURE 3.18.

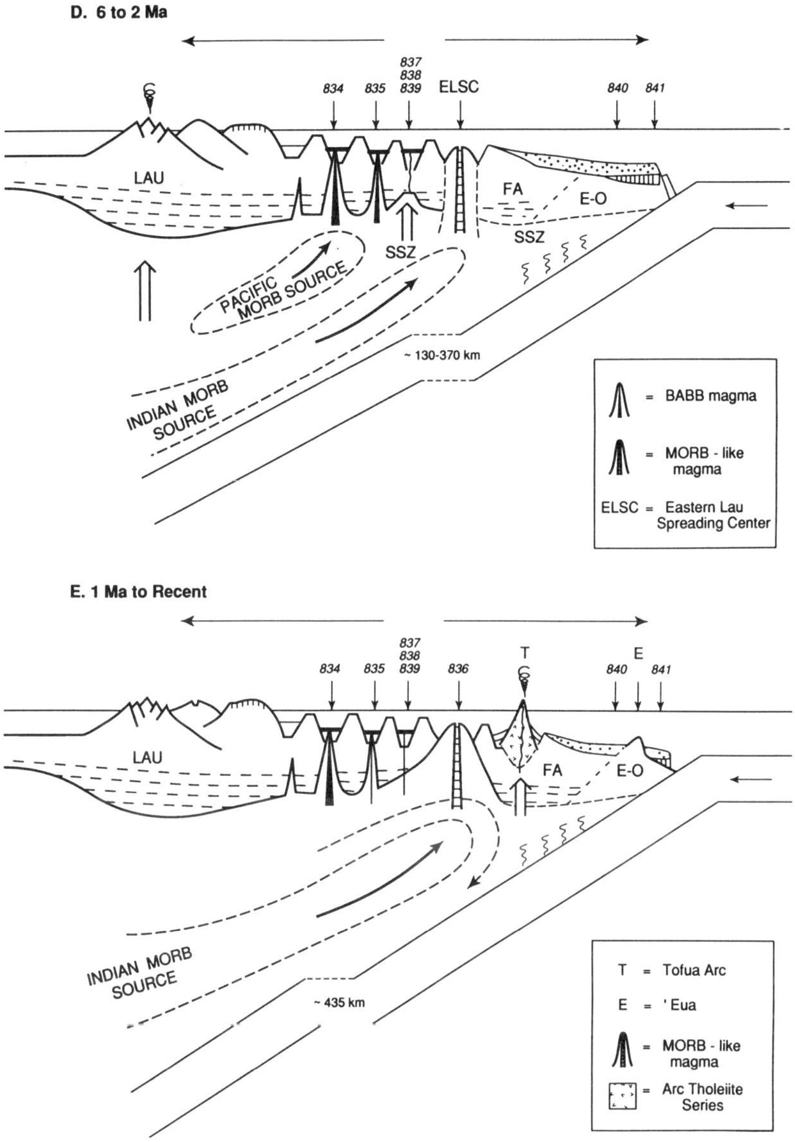

FIGURE 3.18. A model for the evolution of the Lau Basin. (A) An early phase of the Vitiaz or west Melanesian arc system forming arc rocks now found in the Fiji Islands and the Tonga platform. The rhyolitic rocks of Site 841 are postulated to be part of this system. Arc polarity is not known but is assumed to have faced north or northeasterly. (B) The Lau Ridge volcanic arc and possible locations of exposures on 'Eua and Sites 840 and 841. This view and subsequent views are projections onto latitude 20°S. (C) The beginning of crustal extension to form the WEB and rifting of the Lau arc. Magmatic activity has not yet filled the extensional basin, but mantle upwelling is setting the stage for it. Note the insertion of the flexure or imbrication in the subducted lithosphere. (D) Effects of extension, mantle upwelling, and magmatism in the backarc basin. The ELSC has been propagated southward and is fed by the Indian MORB source. Not shown are the ephemeral intrabasin arc-composition volcanoes that formed progressively from west to east. (E) In the present configuration the Tofua arc and the ELSC are both active. Not shown is the CLSC that is propagating southward. The vertical dimension in all sketches has been exaggerated to show details of crustal geology.

Oligocene arc probably formed part of the forearc to the Lau Ridge, as evidence from 'Eua and other Tongan islands indicates that Miocene dikes and volcaniclastic material were emplaced into, or deposited on, the older forearc rocks. Nearly all of the rocks of the forearc have an affinity to arc magmas, and this suggests that the underlying crust is largely arc material. Some evidence exists for older MORB-like crust under the forearc in the occurrence of basaltic clasts in the low-K rhyolite at Site 841 (Bloomer *et al.*, 1994). These are not unequivocal evidence for trapped older deep-sea floor inasmuch as MORB-like rocks are among the earliest magma types formed in arc rifting (e.g., Taylor *et al.*, 1992). MORB-like lavas in the Mariana forearc were reported by Johnson and Fryer (1990), who suggested that they may represent injection of magma during forearc rifting. I conclude that it is likely that much, or all, of the forearc is underlain by crust formed in an island arc system.

The SSZ source for the Lau Ridge volcanic arc is considered to be depleted harzburgite that had been metasomatized by fluid derived from subducted, altered ocean crust (e.g., Pearce *et al.*, 1984; Tatsumi *et al.*, 1986; Tatsumi, 1989) and possibly from subducted sediments. The initial depleted mantle source carried the isotope signature of the source of Pacific MORB. Late in the evolution of the Lau Ridge, at about 6 Ma, it began to be rifted, and at about the same time crustal extension began to form rift basins in the forearc to the Lau Ridge. An imbrication or a flexure of the subducted plate may have been formed at the trench at about 7 to 8 Ma, as shown schematically in Fig. 3.18d (see Billington, 1980, for a discussion of this important feature). This change in the subducted plate may have been either a cause or an effect of extension. The progressive deepening of the location of the flexure, presently at about 535 km deep, is shown in Fig. 3.18d and 18e. The opening of the Lau Basin included an extension and rifting phase followed by seafloor spreading. Extension and rifting formed the basins drilled at all of the backarc sites except for Site 836, which is related to the seafloor-spreading phase. The magmas at Site 834 appear to contradict the compositional zonation of the Lau Basin proposed by Hawkins and Melchior (1985) and shown by the distribution of sample types in Fig. 3.6. They should be transitional in composition like the samples dredged from nearby ridges. Some Site 834 petrologic units resemble Mariana Trough basalts in major elements and HFSE, but they lack LILE enrichment. A possible explanation may be that the long-term melting that partly filled the subbasin at Site 834 thoroughly depleted the source in the LILE-enriched SSZ component. Continued melting of the mantle tapped the depleted MORB-source component which dominated the later stages of melting. The rocks we drilled recorded only the products of the later melting.

Site 834 was fed mainly by a "Pacific" MORB source, as shown by Hergt and Hawkesworth (1994), whereas Sites 837 to 839 had an SSZ source with some mixture of the Pacific source. Site 836 magmas, and the modern spreading centers, were derived from an "Indian" MORB source, as originally proposed by Volpe *et al.* (1988). The two-dimensional sketch does not show the complexity of the probable mixing pattern. The influx of the Indian MORB source would have swept down the axis of the basin from the north in the wake of the propagating rift tip of the ELSC. The trace of a "Samoan plume component" in the northern part of the basin (Poreda, 1985) offers further support for this idea. The Pacific MORB source either was consumed, bypassed, or overwhelmed by the Indian source as the basin evolved. The final phase of evolution included the initiation and southward propagation of a second rift system at about 1 to 2 Ma that formed the CLSC. Mixing of MORB-like and arc magmas has been important throughout the history of opening of the basin (Hergt and Farley, 1994).

The development of a new volcanic arc (Tofua arc) on the outboard edge of the Lau Basin probably occurred within the past 3 m.y. and has propagated from north to south down the axis of the Tofua arc. A complexity not shown because of limitations of the scale of the sketches is the seaward march of the small, intrabasinal, arc composition volcanoes that contributed to the infilling of the rift basins. These may have been no less significant at any time than the scattering of arc volcanoes that constitute the present Tofua arc. Thus, the long-standing question of whether there was synchronous backarc and arc magmatism has an indefinite answer in the Lau Basin. Both phenomena have been operative more or less continuously *within* the basin as it opened. However, a well-defined arc on the Tonga Ridge probably did not appear until after about 6 m.y. of backarc spreading.

The Lau–Tonga system serves as a model for understanding relict convergent margin rocks found as terranes accreted to continental margins and as ophiolite series assemblages. It seems likely that eventually plate reorganization will lead to the amalgamation of the arc and backarc system of the western Pacific and their eventual accretion to a continent. There is ample evidence in areas of composite terranes such as the Philippines for amalgamation of arc-backarc-forearc systems like the Lau–Tonga system (e.g., Hawkins and Evans, 1983; Hawkins *et al.*, 1985). There is also general acceptance of the idea that many ophiolites are remnants of SSZ magma systems. In many ways the Leg 135 transect, together with the other western Pacific ODP studies, provides important new data and insight to understanding continental evolution through terrane accretion.

Acknowledgments

This synthesis summarizes results of many years of study of the Lau Basin and its surroundings by the author as well as the work of many others cited in the text. Within the last five years there have been a number of intensive studies with wide-beam echo sounding (SeaBeam), sonar imaging (GLORIA), submersibles, and drilling by the Ocean Drilling Program. Shipmates on ODP Leg 135 are acknowledged for their contributions toward a better understanding of Lau Basin geology. The ODP work in particular has brought together a wide range of studies that have given us new understanding of the complex history of the Lau Basin. This summary has drawn extensively on the collected works of the ODP and especially from the Scientific Results Volume for ODP Leg 135. The author acknowledges support received from the National Science Foundation, in particular from awards OCE 84-15472, OCE 87-16660, JOI JSC 5-89, and Texas A&M 20556. I thank Jo Griffith, Evelyn Hegemier, and Fred Florendo for making the illustrations. Fred Hochstaedter and Kevin Johnson are thanked for many helpful suggestions that have improved the original manuscript.

REFERENCES

Allan, J. F. 1994. Cr-spinel in depleted basalts from the Lau Basin backarc, ODP Leg 135; petrogenetic history from Mg-Fe crystal-liquid exchange, in *Proc. ODP, Sci. Results*, 135 (J. W. Hawkins, L. M. Parson, and J. F. Allan, *et al.*, eds.), pp. 565–584, Ocean Drilling Program, College Station, TX.

Allan, J. F., Sack, R. O., and Batiza, R. 1988. Cr-rich spinels as petrogenetic indicators: MORB-type lavas from the Lamont seamount chain, Eastern Pacific, *Am. Mineral.* **73**:741–753.

Arculus, R. J., and Powell, R. 1986. Source component mixing in the regions of arc magma generation, *J. Geophys. Res.* **91:**5913–5926.

Barazangi, M., and Isacks, B. 1971. Lateral variations of seismic-wave attenuation in the upper mantle above the inclined earthquake zone of the Tonga Island Arc: Deep anomaly in the upper mantle, *J. Geophys. Res.* **76:**8493–8516.

Bednarz, U., and Schmincke, H. U. 1994. Composition and origin of volcaniclastic sediments in the Lau Basin (southwest Pacific), Leg 135, (Sites 834–839), in *Proc. ODP, Sci. Results*, 135 (J. W. Hawkins, L. M. Parson, and J. F. Allan, *et al.*, eds.), pp. 51–74, Ocean Drilling Program, College Station, TX.

Ben Othman, D., White, W. M., and Patchett, J. 1989. The geochemistry of marine sediments, island arc magma genesis, and crust-mantle recycling, *Earth Planet. Sci. Lett.* **94:**1–21.

Berthine, K. K. 1974. Origin of Lau Basin Rise sediment, *Geochim. Cosmochim. Acta* **38:**629–640.

Berthine, K. K., and Keene, J. B. 1975. Submarine barite-opal rocks of hydrothermal origin, *Science* **188:**150–152.

Bevis, M., Taylor, F. W., Schutz, B. E., and Calmant, S. 1993. Geodetic observations of convergence and back-arc spreading in the SW Pacific (1988–1992), *Eos* **74:**60.

Billington, S. 1980. The morphology and tectonics of the subducted lithosphere in the Tonga–Kermadec–Fiji region from seismicity and focal mechanism solutions, Ph. D. dissertation, Cornell University, Ithaca, NY.

Bloomer, S. H., Ewart, A., Hergt, J. and Bryan, W. B. 1994. Geochemistry and origin of igneous rocks from the outer Tonga forearc, Site 841, in *Proc. ODP, Sci. Results*, 135, (J. W. Hawkins, L. M. Parson, and J. F. Allan *et al.*, eds.), pp. 625–646, Ocean Drilling Program, College Station, TX.

Bloomer, S. H., and Fisher, R. L. 1987. Petrology and geochemistry of igneous rocks from the Tonga Trench—a non-accreting plate boundary, *J. Geol.* **95:**469–495.

Bloomer, S. H., and Hawkins, J. W. 1987. Petrology and geochemistry of boninite series volcanic rocks from the Mariana Trench, *Contrib. Mineral. Petrol.* **97:**361–377.

Boespflug, X., Dosso, L., Bougault, H., and Joron, J.-L. 1990. Trace element and isotopic (Sr, Nd) geochemistry of volcanic rocks from the Lau Basin, *Geol. Jahrb.* **92:**503–516.

Briqueu, L., Bougault, H., and Joron, J.-L. 1984. Quantification of Nb, Ta, Ti, and V anomalies in magmas associated with subduction zones: Petrogenetic implications, *Earth Planet. Sci. Lett.* 68. 297–308.

Bryan, W. B. 1979a. Low K_2O dacite from the Tonga–Kermadec island arc: Petrography, chemistry, and petrogenesis, in *Trondhjemites, Dacites, and Related Rocks* (F. Barker, ed.), pp. 581–600, Elsevier, Amsterdam.

Bryan, W. B. 1979b. Regional variation and petrogenesis of basalt glasses from the FAMOUS area, Mid-Atlantic Ridge, *J. Petrol.* **20:**293–325.

Bryan, W. B., and Pearce, T. H. 1994. Plagioclase zoning in selected lavas from Holes 834B, 839B, and 841B, in *Proc. ODP, Sci. Results*, 135 (J. W. Hawkins, L. M. Parson, J. F. Allan, *et al.*, eds.), pp. 543–556, Ocean Drilling Program, College Station, TX.

Bryan, W. B., Stice, G. D., and Ewart, A. E. 1972. Geology, petrography, and geochemistry of the volcanic islands of Tonga, *J. Geophys. Res.* **77:**1566–1585.

Bryan, W. B., Thompson, G., Frey, F. A., and Dickey, J. S. 1976. Inferred geologic settings and differentiation in basalts from the Deep Sea Drilling Project, *J. Geophys. Res.* **81:**4285–4304.

Carlson, R. W., Macdougall, J. D. and Lugmair, G. W. 1978. Differential Sm/Nd evolution in oceanic basalts, *Geophys. Res. Lett.* **5:**229–232.

Carlson, R. L., and Melia, P. J. 1984. Subduction hinge migration, in *Geodynamics of Back-Arc Regions* (R. L. Carlson and K. Kobayashi, eds.), *Tectonophysics*, **102:**399–411.

Cawood, P. A. 1985. Petrography, phase chemistry, and provenance of volcanogenic debris from the southern Tonga Ridge: Implications for arc history and magmatism, in *Geology and Offshore Resources of Pacific Island Arcs—Tonga Region* (D. Scholl and T. Vallier, eds.), pp. 149–170, Circum-Pacific Council for Energy and Mineral Resources, Houston, TX.

Chase, C. G. 1978. Extension behind island arcs and motions relative to hot spots, *J. Geophys. Res.* **83:**5385–5387.

Clift, P., and Dixon, J. E. 1994. Variations in arc volcanism and sedimentation related to rifting of the Lau Basin (Southwest Pacific), in *Proc. ODP, Sci. Results*, 135 (J. W. Hawkins, L. M. Parson, J. F. Allan, *et al.*, eds.), pp. 23–50, Ocean Drilling Program, College Station, TX.

Cole, J. W., Gill, J. B., and Woodhall, D. 1985. Petrologic history of the Lau Ridge, Fiji, in *Geology and Offshore Resources of Pacific Island Arcs—Tonga Region* (D. Scholl and T. Vallier, eds.), pp. 379–414, Circum-Pacific Council for Energy and Mineral Resources, Houston, TX.

Colley, H., and Hindle, W. H. 1984. Volcano-tectonic evolution of Fiji and adjoining marginal seas, in *Marginal Basin Geology* (B. P. Kokelaar and M. F. Howells, eds.), pp. 151–162, Blackwell, Oxford.

Colley, H., Hindle, H. W., and Hathaway, B. 1986. Early arc and basinal igneous activity on Viti Levu, Fiji, IAVCEI Congress, New Zealand, p. 140 (Abstract).
Colley, H., Wharton, M. R., and Hathaway, B. 1993. Magmatism and initial arc formation in Fiji, in *Ancient Volcanism and Modern Analogues*, IAVCEI General Assembly, Canberra, Australia, p. 22 (Abstract).
Collier, J. S., and Sinha, M. C. 1992. Seismic mapping of a magma chamber beneath the Valu Fa Ridge, Lau Basin, *J. Geophys. Res.* **97**:14, 031–14, 053.
Crawford, A. J., Falloon, T. J., and Green, D. H. 1989. Classification, petrogenesis and tectonic setting of boninites, in *Boninites and Related Rocks* (A. J. Crawford, ed.), pp. 2–49, Unwin Hyman, London.
Cronan, D. S., Moorby, S. S., Glasby, G. P., Knedler, L., Thomson, J., and Hodkinson, R. A. 1984. Hydrothermal and volcaniclastic sedimentation on the Tonga-Kermadec Ridge and its adjacent marginal basins, *Geol. Soc. London Spec. Publ.* **16**:137–149.
Cunningham, J. K., and Anscombe, K. J. 1985. Geology of 'Eua and other islands, Kingdom of Tonga, in *Geology and Offshore Resources of Pacific Island Arcs—Tonga Region* (D. Scholl and T. Vallier, eds.), pp. 221–258, Circum-Pacific Council for Energy and Mineral Resources, Houston, TX.
Davis, A. S., Clague, D. A., and Morton, J. L. 1990. Volcanic glass compositions from two spreading centers in Lau Basin, southwest Pacific Ocean, *Geol. Jahrb.* **92**:481–501.
DePaolo, D. J., and Wasserburg, G. J. 1977. The sources of island arcs as indicated by Nd and Sr isotope studies, *Geophys. Res. Lett.* **4**:465–468.
Dick, H. J. B. 1982. The petrology of two back arc basins of the northern Philippine Sea, *Am. J. Sci.* **282**:644–700.
Dick, H. J. B., and Bullen, T. 1984. Chromian spinel as a petrogenetic indicator in abyssal and alpine-type peridotites and spatially associated lavas, *Contrib. Mineral. Petrol.* **86**:54–76.
Duncan, R. A., Vallier, T., and Falvey, D. A. 1985. Volcanic episodes on 'Eua, Tonga, in *Geology and Offshore Resources of Pacific Island Arcs—Tonga Region* (D. Scholl and T. Vallier, eds.), pp. 281–290, Circum-Pacific Council for Energy and Mineral Resources, Houston TX.
Dupuy, C., Dostal, J., Marcelot, G., Bougault, H., Joron, J. L., and Treuil, M. 1982. Geochemistry of basalts from central and southern New Hebrides arc: Implications for their source rock composition, *Earth Planet. Sci. Lett.* **60**:207–225.
Eguchi, T. 1984. Seismotectonics of the Fiji Plateau and Lau Basin, *Tectonophysics* **102**:17–32.
Elsasser, W. M. 1971. Sea-floor spreading as thermal convection, *J. Geophys. Res.* **76**:1101–1112.
Ernewein, M., Pearce, J. A., Bloomer, S. H., Parson, L. M., Murton, B. J., and Johnson, L. E. 1994. Geochemistry of Lau Basin volcanic rocks: Influence of ridge segmentation and arc proximity, in *Volcanism Associated with Extension at Consuming Plate Margins* (J. Smellie, ed.), Geol. Soc. Spec. Publ., London.
Ewart, A. 1979. A review of the mineralogy and chemistry of Tertiary—Recent dacitic, latitic, rhyolitic, and related salic volcanic rocks, in *Trondhjemites, Dacites, and Related Rocks* (F. Barker, ed.), pp. 13–122, Amsterdam Elsevier.
Ewart, A., Brothers, R. N., and Mateen, A. 1977. An outline of the geology and geochemistry, and the possible petrogenetic evolution of the volcanic rocks of the Tonga–Kermadec–New Zealand island arc, *J. Volcanol. Geotherm. Res.* **2**:205–250.
Ewart, A., and Bryan, W. B. 1972. Petrography and geochemistry of the igneous rocks from 'Eua, Tongan Islands, *Geol. Soc. Am. Bull.* **83**:3281–3298.
Ewart, A., Bryan, W. B., Chappell, B. W., and Rudnick, R. L. 1994a. Regional geochemistry of the Lau–Tonga arc and back-arc systems, in *Proc. ODP, Sci. Results*, 135 (J. W. Hawkins, L. M. Parson, J. F. Allan *et al.*, eds.), pp. 385–426. Ocean Drilling Program, College Station, TX.
Ewart, A., Bryan, W. B., and Gill, J. B. 1973. Mineralogy and geochemistry of the younger Tongan Islands, S. W. Pacific, *J. Petrol.* **14**:429–466.
Ewart, A., Hergt, J., and Hawkins, J. W. 1994b. Major, trace element and Pb, Sr, and Nd isotope geochemistry of Site 839 basalts and basaltic andesites: Implications for arc volcanism, in *Proc. ODP, Sci. Results*, 135 (J. W. Hawkins, L. M. Parson, J. F. Allan *et al.*, eds.), pp. 519–532. Ocean Drilling Program, College Station, TX.
Ewart, A. E., and Hawkesworth, C. J. 1987. Pleistocene to Recent Tonga– Kermadec arc lavas: Interpretation of new isotopic and rare earth data in terms of a depleted mantle source model, *J. Petrol.* **28**:495–530.
Exon, N. F., Herzer, R. H., and Cole, J. W. 1985. Mixed volcaniclastic and pelagic sedimentary rocks from the Cenozoic southern Tonga Platform and their implications for petroleum potential, in *Geology and Offshore Resources of Pacific Island Arcs—Tonga Region* (D. Scholl and T. Vallier, eds.), pp. 75–108, Circum-Pacific Council for Energy and Mineral Resources, Houston, TX.
Falloon, T. J., Malahoff, A., Zonenshain, L. P., and Bogdanov, Y. 1992. Petrology and geochemistry of back-arc basin basalts from Lau Basin spreading ridges at 15, 18, and 19°S, *Mineral. Petrol.* **47**:1–36.

Farley, K. N. 1994. Oxidation state and sulfur concentrations in Lau Basin basalts, in *Proc. ODP, Sci. Results*, 135 (J. W. Hawkins, L. M. Parson, J. F. Allan *et al.*, eds.), pp. 603–614, Ocean Drilling Program, College Station, TX.

Fisher, R. L., and Engel, C. G. 1969. Ultramafic and basaltic rocks dredged from the nearshore flank of the Tonga Trench, *Geol. Soc. Am. Bull.* **80**:1373–1378.

Fisk, M. R., and Bence, A. E. 1980. Experimental crystallization of chrome spinel in FAMOUS basalt 527-1-1, *Earth Planet. Sci. Lett.* **48**:113–121.

Fouquet, Y., von Stackelberg, U., Charlou, J. L., Donval, J. P., Foucher, J. P., Erzinger, J., Herzig, P., Muhe, R., Weidicke, M., Soakai, S., and Whitechurch, H. 1991a. Hydrothermal activity in the Lau back-arc basin: Sulfides and water chemistry, *Geology* **19**:303–306.

Fouquet, Y., von Stackelberg, U., Charlou, J. L., Donval, J. P., Erzinger, J., Foucher, J. P., Herzig, P., Muhe, R., Soakai, S., Weidicke, M., and Whitechurch, H. 1991b. Hydrothermal activity and metallogenesis in the Lau back-arc basin, *Nature* **348**:778–781.

Frenzel, G., Muhe, R., and Stoffers, P. 1990. Petrology of the volcanic rocks from the Lau Basin, southwest Pacific, *Geol. Jahrb.* **92**:395–479.

Fryer, P., and Pearce, J. A. 1992. Introduction to the scientific results of Leg 125, in *Proc. ODP, Sci. Results*, 125 (P. Fryer, J. A. Pearce, L. B. Stokking *et al.* eds.), pp. 3–14, Ocean Drilling Program, College Station, TX.

Fryer, P., Sinton, J., and Philpotts, J. A. 1981. Basaltic glasses from the Mariana Trough, in *Init. Repts. DSDP*, 60 (D. M. Hussong, S. Uyeda *et al.*, eds.), pp. 601–610, U. S. Govt. Printing Office, Washington, DC.

Furlong, K. P., Chapman, D. S., and Alfeld, P. W. 1982. Thermal modeling of the geometry of subduction with implications for the tectonics of the overriding plate, *J. Geophys. Res.* **87**:1786–1802.

Garcia, M., Liu, N. W. K., and Muenow, D. W. 1979. Volatiles from submarine volcanic rocks from the Mariana Island Arc and Trough, *Geochim. Cosmochim. Acta* **43**:305–312.

Giardini, D., and Woodhouse, J. H. 1984. Deep seismicity and modes of deformation in Tonga subduction zone, *Nature* **307**:505–509.

Giardini, D., and Woodhouse, J. H. 1986. Horizontal shear flow in the mantle beneath the Tonga arc, *Nature* **319**:551–555.

Gill, J. B. 1976. Composition and age of Lau Basin and ridge volcanic rocks: Implications for evolution of an interarc basin and remnant arc, *Geol. Soc. Am. Bull.* **87**:1384–1395.

Gill, J. B. 1981. *Orogenic Andesites*, Springer-Verlag, Berlin.

Gill, J. B. 1987. Early geochemical evolution of an oceanic island arc and backarc: Fiji and the South Fiji Basin, *J. Geol.* **95**:589–615.

Hamburger, M. W., and Isacks, B. L. 1988. Diffuse back-arc deformation in the southwestern Pacific, *Nature* **332**:599–604.

Hart, S. R. 1969. K, Rb, Cs contents and K/Rb, K/Cs ratios of fresh and altered submarine basalts, *Earth Planet. Sci. Lett.* **6**:295–303.

Hart, S. R., Erlank, A. J., and Kable, E. J. 1974. Sea floor basalt alteration: Some chemical and strontium isotope effects, *Contrib. Mineral. Petrol.* **44**:219–230.

Hart, S. R., Glassley, W. A., and Karig, D. E. 1972. Basalts and seafloor spreading behind the Mariana Island arc, *Earth Planet. Sci. Lett.* **15**:12–18.

Hart, S. R., and Nawalk, A. J. 1970. K, Rb, Cs, and Sr relationships in submarine basalts from the Puerto Rico Trench, *Geochim. Cosmochim. Acta* **34**:145–155.

Hart, S. R. and Staudigel, H. 1989. Isotopic characterization and identification of recycled components, in *Crust/Mantle Recycling at Convergence Zones* (S. R. Hart and L. Gulen, eds.), pp. 15–28, Kluwer, Boston.

Hawkins, J. W. 1974. Geology of the Lau Basin, a marginal sea behind the Tonga Arc, in *The Geology of Continental Margins* (C. A. Burk and C. L. Drake, eds.), pp. 505–520, New York, Springer-Verlag.

Hawkins, J. W. 1976. Petrology and geochemistry of basaltic rocks of the Lau Basin, *Earth Planet. Sci. Lett.* **28**:283–297.

Hawkins, J. W. 1977. Petrological and geochemical characteristics of marginal basin basalts, in *Island Arcs, Deep Sea Trenches, and Back Arc Basins* (M. Talwani and W. C. Pitman III, eds.), Vol. 1, pp. 355–365, Maurice Ewing Series, American Geophysical Union, Washington, DC.

Hawkins, J. W. 1985. Low-K rhyolitic pumice from the Tonga Ridge, in *Geology and Offshore Resources of Pacific Island Arcs—Tonga Region* (D. Scholl and T. Vallier, eds.), pp. 171–178, Circum-Pacific Council for Energy and Mineral Resources, Houston, TX.

Hawkins, J. W. 1986. "Black smoker" vent chimneys, Lau Basin, *Eos* **67**:430.

Hawkins, J. W. 1988. Cruise Report—PAPATUA expedition. Leg 04, R/V *Thomas Washington, SIO Ref. Ser.* 88–14.

Hawkins, J. W. 1989. Cruise Report—ROUNDABOUT expedition. Legs 14,15, R/V *Thomas Washington, SIO Ref. Ser.* 89–13.

Hawkins, J. W. 1994. Petrologic synthesis, ODP Leg 135—Lau Basin transect, in *Proc. ODP, Sci. Results*, 135 (J. W. Hawkins, L. M. Parson, J. F. Allan, *et al.*, eds.), pp. 879–908, Ocean Drilling Program, College Station, TX.

Hawkins, J. W., and Allan, J. F. 1994. Petrologic evolution of the Lau Basin, Sites 834–839, in *Proc. ODP, Sci. Results*, 135 (J. W. Hawkins, L. M. Parson, J. F. Allan, *et al.*, eds.), pp. 426–470, Ocean Drilling Program, College Station, TX.

Hawkins, J. W., and Evans, C. A. 1983. Geology of the Zambales Range, Luzon, Philippines: Ophiolite derived from an island-arc backarc basin pair, in *The Tectonic Evolution of Southeast Asian Seas and Islands*, Part 2 (D. Hayes, ed.), Vol. 27, pp. 124–138. Geophys. Monogr. Ser., American Geophysical Union, Washington, DC.

Hawkins, J. W., Bloomer, S. H., Evans, C. A, and Melchior, J. T. 1984. Evolution of intra-oceanic arc-trench systems, *Tectonophysics* **102:**174–205.

Hawkins, J. W., and Falvey, D. A. 1985. Petrology of andesitic dikes and flows from 'Eua, Tonga, in *Geology and Offshore Resources of Pacific Island Arcs—Tonga Region* (D. Scholl and T. Vallier), pp. 269–280, Circum-Pacific Council for Energy and Mineral Resources, Houston, TX.

Hawkins, J. W., Fisher, R. L., and Engel, C. G. 1972. Ultramafic and mafic rock suites exposed on the deep flanks of the Tonga Trench, *Geol. Soc. Am. Abs. Prog.* **4:**167–168 (Abstract).

Hawkins, J. W., and Helu, S. 1986. Polymetallic sulphide deposit from "black-smoker" chimney: Lau Basin, *Eos* **67:**378 (Abstract).

Hawkins, J. W., Lonsdale, P. F., Macdougall, J. D., and Volpe, A. M. 1990. Petrology of the axial ridge of the Mariana Trough backarc spreading center, *Earth Planet. Sci. Lett.* **100:**226–250.

Hawkins, J. W., and Melchior, J. T. 1985. Petrology of Mariana Trough and Lau Basin basalts, *J. Geophys. Res.* **90:**11431–11468.

Hawkins, J. W., Melchior, J. T., Florendo, F. F., and Nilsson, K. N. 1989. Evolution of backarc basin magmas and their mantle sources-examples from the Lau Basin and Mariana Trough, *Eos* **70:**1389.

Hawkins, J. W., Moore, G. F., Villamor, R., Evans, S., and Wright, E. 1985. Geology of the composite terranes of east and central Mindanao, in *Tectonostratigraphic Terranes of the Circum-Pacific Region* (D. G. Howell, ed.), pp. 437–463, Circum-Pacific Council for Energy and Mineral Resources, Earth Sci. Ser., Vol. 1, Houston, TX.

Hawkins, J. W., Parson, L. M., Allan, J. F., and Leg 135 Scientific Party. 1991. New insight to the evolution of arc-backarc systems, results of Ocean Drilling Program (ODP) Leg 135, Lau-Tonga transect, *Eos* **72:**541.

Hawkins, J. W., Parson, L. M., and Allan, J. F. *et al.* 1994. *Proc. ODP, Sci. Results*, 135, Ocean Drilling Program, College Station, TX.

Hawkins, J. W., Sclater, J. G., and Hohnhaus, G. 1970. Petrologic and geophysical characteristics of the Lau Basin Ridge: A spreading center behind the Tonga arc, *Geol. Soc. Am. Bull.* **2:**71, (Abstract).

Hedge, C. E., and Peterman, Z. E. 1970. The strontium isotopic composition of basalts from the Gorda and Juan de Fuca Rises, northeastern Pacific Ocean, *Contrib. Mineral. Petrol.* **27:**114–120.

Hergt, J. M. and Farley, K. N. 1994. Major, trace element, and isotope (Pb, Sr, and Nd) variations in Site 834 basalts: Implications for the initiation of backarc opening, in *Proc. ODP, Sci. Results*, 135 (J. W. Hawkins, L. M. Parson, J. F. Allan *et al.* eds.), pp. 471–486, Ocean Drilling Program, College Station, TX.

Hergt, J. M., and Hawkesworth, C. J. 1994. The Pb, Sr, and Nd isotopic evolution of the Lau Basin: Implications for mantle dynamics during the backarc opening, in *Proc. ODP, Sci. Results*, 135 (J. W. Hawkins, L. M. Parson, J. F. Allan *et al.*, eds.), pp. 505–518, Ocean Drilling Program, College Station, TX.

Hochstaedter, A. G., Gill, J. B., Kusakabe, M., Newman, S., Pringle, M., Taylor, B., and Fryer, P. 1990a. Volcanism in the Sumisu Rift. I: Major element, volatile, and stable isotope geochemistry, *Earth Planet Sci. Lett.* **100:**179–194.

Hochstaedter, A. G., Gill, J. B., and Morris, J. D. 1990b. Volcanism in the Sumisu Rift, II: Subduction and non-subduction related components, *Earth Planet. Sci. Lett.* **100:**195–209.

Hodkinson, R. A., and Cronan, D. S. 1991. Geochemistry of recent hydrothermal sediments in relation to tectonic environment in the Lau Basin, southwest Pacific, *Mar. Geol.* **98:**353–366.

Hodkinson, R. A., and Cronan, D. S. 1994. Variability in the hydrothermal component of the sedimentary sequence in the Lau back-arc basin: ODP Leg 135 Sites 834A, 835A, 836A, 837A, 838A and 839A, in *Proc. ODP, Sci. Results*, 135 (J. W. Hawkins, L. M. Parson, J. F. Allan, *et al.*, eds.), pp. 75–86, Ocean Drilling Program, College Station, TX.

Hodkinson, R. A., Cronan, D. S., Glasby, G. P., and Moorby, S. A. 1986. Geochemistry of marine sediments from the Lau Basin, *N. Z. J. Geol. Geophys.* **29:**335–344.

Hole, M. J., Saunders, A. D., Marriner, G. F., and Tarney, J. 1984. Subduction of pelagic sediments: Implications for the origin of Ce-anomalous basalts from the Mariana Islands, *J. Geol. Soc. London* **141:**453–472.

Hynes, A., and Mott, J. 1985. On the causes of back-arc spreading, *Geology*, **13:**387–389.

Ikeda, Y., and Yuasa, M. 1989. Volcanism in nascent back-arc basins behind the Shichito Ridge and adjacent areas in the Izu-Ogasawara arc, northwest Pacific: Evidence for mixing between E-type MORB and island arc magmas at the initiation of back-arc rifting, *Contrib. Mineral. Petrol.* **101:**377–393.

Irvine, T. N. 1965. Chromian spinel as a petrogenetic indicator. 1: Theory. *Can. J. Earth Sci.* **2:**648–672.

Irvine, T. N. 1967. Chromian spinel as a petrogenetic indicator, 2, Petrological applications, *Can. J. Earth Sci.* **4:**71–103.

Isacks, B. L., and Barazangi, M. 1977. Geometry of Benioff zones: Lateral segmentation and downwards bending of the subducted lithosphere, in *Island Arcs, Deep Sea Trenches, and Back Arc Basins* (M. Talwani and W. C. Pitman III, eds.), Vol. 1, pp. 94–114, Maurice Ewing Series, American Geophysical Union, Washington, DC.

Jenner, G. A., Cawood, P. A., Rautenschlein, M., and White, W. M. 1987. Composition of back-arc basin volcanics, Valu Fa Ridge, Lau Basin: Evidence for a slab-derived component in their mantle source, *J. Volcanol. Geotherm. Res.* **32:**209–222.

Johnson, L. E., and Fryer, P. 1990. The first evidence for MORB-like lavas from the outer Mariana forearc: Geochemistry, petrography and tectonic implications, *Earth Planet. Sci. Lett.* **100:**304–316.

Karig, D. E. 1970. Ridges and basins of the Tonga-Kermadec Island arc system, *J. Geophys. Res.* **75:**239–254.

Karig, D. E. 1971. Origin and development of marginal basins in the western Pacific, *J. Geophys. Res.* **76:**2542–2561.

Kay, R. W., Hubbard, N., and Gast, P. 1970. Chemical characteristics and origins of oceanic ridge volcanic rocks, *J. Geophys. Res.* **75:**238–254.

Kelemen, P. B. 1986. Assimilation of ultramafic rocks in subduction related magmatic arcs, *J. Geol.* **94:**829–843.

Kelemen, P. B., Johnson, K. T. M., Kinzler, R. J., and Irving, A. J. 1990. High-field strength element depletions in arc basalts due to mantle-magma interaction, *Nature* **345:**521–524.

Klein, E. M., and Langmuir, C. H. 1987. Global correlations of ocean ridge basalt chemistry with axial depth and crustal thickness, *J. Geophys. Res.* **92:**8089–8115.

Kroenke, L. 1984. *Cenozoic Development of the Southwest Pacific* CCOP/SOPAC, Tech. Bull., No. 6. U.N. Econ. Soc. Comm. Asia Pac., Suva, Fiji.

Lawver, L. A., and Hawkins, J. W. 1978. Diffuse magnetic anomalies in marginal basins: Their possible tectonic and petrologic significance, *Tectonophysics* **45:**323–339.

Lawver, L., Hawkins, J. W., and Sclater, J. G. 1976. Magnetic anomalies and crustal dilation in the Lau Basin, *Earth Planet. Sci. Lett.* **33:**27–35.

Ledbetter, J. K., and Haggerty, J. A. 1994. Late Miocene sedimentation history of the Tonga forearc at Site 840, in *Proc. ODP, Sci. Results*, 135 (J. W. Hawkins, L. M. Parson, J. F. Allan *et al.*, eds.), pp. 163–172, Ocean Drilling Program, College Station, TX.

Loock, G., McDonough, W. F., Goldstein, S. L. and Hoffman, A. W. 1990. Isotopic composition of volcanic glasses from the Lau Basin, *Mar. Mining* **9:**235–245.

Louat, R., and Dupont, J. 1982. Seismicite'de l'arc des Tonga-Kermadec, in *Equipe de Geologie-Geophysique du Centre ORSTOM de Noumea, contribution a' l'etude geodynamique du Sud-Ouest Pacifique*, pp. 299–317, Trav. Doc. 147, ORSTOM (Nouvelle Caledonie').

Ludden, J. N., and Thompson, G. 1979. An evaluation of the behavior of the rare earth elements during the weathering of sea-floor basalt, *Earth Planet. Sci. Lett.* **43:**85–92.

Malahoff, A., Feden, R. H., and Fleming, H. S. 1982. Magnetic anomalies and tectonic fabric of marginal basins north of New Zealand, *J. Geophys. Res.* **87:**4109–4125.

Masuda, A., Nakamura, N., and Tamaka, T. 1973. Fine structure of mutually normalized rare-earth element patterns of chondrites, *Geochim. Cosmochim. Acta.* **37:**239–248.

Mattey, D. P., Marsh, N. G. and Tarney, J. 1980. The geochemistry, mineralogy, and petrology of basalts from the west Philippine and Parece Vela Basins and from the Palau-Kyushu and West Mariana Ridges, DSDP Leg 59, in *Init. Repts. DSDP*, 59 (L. Kroenke and R. Scott, *et al.*, eds.), pp. 753–800, U. S. Govt. Printing Office, Washington, DC.

McDougall, I. 1994. Dating of rhyolitic glass in the Tonga forearc (Hole 841 B), in *Proc. ODP Sci. Results*. 135 (J. W. Hawkins, L. M. Parson, and J. F. Allan, eds.), pp. 923–924, Ocean Drilling Program, College Station, TX.

McKenzie, D., and Bickle, M. J. 1988. The volume and composition of melt generated by extension of the lithosphere, *J. Petrol.* **29:**625–679.

Meijer, A. 1976. Pb and Sr isotopic data bearing on the origin of volcanic rocks from the Mariana island-arc system, *Bull. Geol. Soc. Am.* **87:**1358–1369.

Melson, W. G., Jarosewich, E., and Lundquist, C. A. 1970. Volcanic eruption at Metis Shoal, Tonga, 1967–1968: Description and petrology, *Smithsonian Contrib. Earth. Sci.* **4:**1–18.

Minster, J. B., and Jordan, T. H. 1978. Present day plate motions, *J. Geophys. Res.* **83:**5331–5354.

Moberly, R. 1972. Origin of lithosphere behind island arcs with reference to the western Pacific, *Geol. Soc. Am. Mem.* **132:**35–55.

Morris, J. D., and Hart, S. R. 1983. Isotopic and incompatible element constraints on the genesis of island arc volcanics from Cold Bay and Amak Island, Aleutians, amd implications for mantle structure, *Geochim. Cosmochim. Acta.* **47:**2015-2-30.

Morton, J. and Sleep, N. H. 1985. Seismic reflections from a Lau Basin magma chamber, in *Geology and Offshore Resources of Pacific Island Arcs—Tonga Region* (D. Scholl and T. Vallier, eds.), pp. 441–453, Circum-Pacific Council for Energy and Mineral Resources, Houston, TX.

Morton, J., and Pohl, W. 1990. Magnetic anomaly identification in the Lau Basin and North Fiji Basin, southwest Pacific Ocean, *Geol. Jahrb. D.* **92:**93–108.

Muenow, D. W., Liu, N. W. K., Garcia, M. O., and Saunders, A. D. 1980. Volatiles in submarine volcanic rocks from the spreading axis of the East Scotia Sea backarc basin, *Earth Planet. Sci. Lett.* **47:**272–278.

Newman, S. 1989. Water and carbon dioxide contents in the basaltic glasses from the Mariana Trough, *Eos* **70:**1387.

Nilsson, K. 1993. Oxidation state, sulfur speciation, and sulfur concentration in basaltic magmas: Examples from Hess Deep and the Lau Basin, Ph. D. dissertation, Scripps Institution of Oceanography, University of California-San Diego, La Jolla.

Nilsson, K., Florendo, F., and Hawkins, J. W. 1989. Petrology of a nascent triple junction, northeastern Lau Basin, *Eos* **73:**1389.

Nilsson, K., and Peach, C. L. 1993. Sulfur speciation, oxidation state, and sulfur concentration in backarc magmas, *Geochim. Cosmochim. Acta* **57:**3807–3813.

Packham, G. H., and Falvey, D. A. 1971. An hypothesis for the formation of marginal seas in the western Pacific, *Tectonophysics* **11:**79–109.

Parson, L. M., and Hawkins, J. W. 1994. Two-stage ridge propagation and the geological history of the Lau backarc basin, in *Proc. ODP, Sci. Results*, 135 (J. W. Hawkins, L. M. Parson, J. F. Allan *et al.*, eds.), pp. 819–828, Ocean Drilling Program, College Station, TX.

Parson, L. M., Hawkins, J. W., Allan, J. F., *et al.* 1992. *Proc. ODP, Init. Repts.*, 135, Ocean Drilling Program, College Station, TX.

Parson, L. M., Pearce, J. A., Murton, B. J., and RRS *Charles Darwin* Scientific Party. 1990. Role of ridge jumps and ridge propagation in the tectonic evolution of the Lau back-arc basin, SW Pacific, *Geology* **18:**470–473.

Parson, L. M., Rothwell, R. G, and MacLeod, C. J. 1994. Tectonics and sedimentation in the Lau Basin, southwest Pacific, in *Proc. ODP, Sci. Results*, 135 (J. W. Hawkins, L. M. Parson, J. F. Allan, *et al.*, eds.), pp. 9–22, Ocean Drilling Program, College Station, TX.

Parson, L. M., and Tiffin, D. L. 1993. Northern Lau Basin: backarc extension at the leading edge of the Indo-Australian plate, *Geo-Marine Lett.* **13:**107–115.

Parson, L. M., *et al.* 1989. RRS *Charles Darwin* Cruise 33/88, 5 May–1 June, 1988, Geophysical and geological investigations of the Lau back-arc basin, SW Pacific. Inst. Oceanogr. Sci. Deacon Lab. Cruise Rep., No. 206.

Pearce, J. A., Lippard, S. J., and Roberts, S. 1984. Characteristics and tectonic significance of supra-subduction zone ophiolites, in *Geology of Marginal Basins* (P. Kokelaar and M. Howells, eds.), Geol. Soc. London. Spec. Publ. **16:**77–94.

Pearce, J. A., and Norry, M. J. 1979. Petrogenetic implications of Ti, Zr, Y, and Nb variations in volcanic rocks, *Contrib. Mineral. Petrol.* **69:**33–47.

Pearce, J. A., van der Laan, S. R., Arculus, R. J., Murton, B. J., Ishii, T., Peate, D. W., and Parkinson, I. J. 1992. Boninite and harzburgite from Leg 125 (Bonin–Mariana forearc): A case study of magma genesis during the initial stages of subduction, in *Proc. ODP, Sci. Results*, 125 (P. Fryer, J. A. Pearce, L. B. Stokking *et al.*, eds.), pp. 623–659, Ocean Drilling Program, College Station, TX.

Pelletier, B., and Louat, R. 1989. Seismotectonics and present day relative plate motions in the Tonga–Lau and Kermadec-Havre region, *Tectonophysics* **165:**237–250.

Pineau, F., Javoy, M., Hawkins, J., and Craig, H. 1976. Oxygen isotope variations in marginal basin and ocean ridge basalts, *Earth Planet. Sci. Lett.* **28**:299–307.

Poreda, R. 1985. Helium-3 and deuterium in backarc basin basalts: Lau Basin and Mariana Trough, *Earth Planet. Sci. Lett.* **76**:244–254.

Poreda, R., and Craig. H. 1993. He and Sr isotopes in the Lau Basin mantle: Depleted and primitive mantle components, *Earth Planet. Sci. Lett.* **119**:319–329.

Reay, A., Rooke, J. M., Wallace, R. C., and Whelan, P. 1974. Lavas from Niuafo'ou Island, Tonga, resemble ocean-floor basalts, *Geology* **2**:605–606.

Ribe, N. M. 1989. Mantle flow induced by back arc spreading, *Geophys. J. Int.* **98**:85–91.

Roeder, P. L., and Reynolds, I. 1991. Crystallization of chromite and chromium solubility in basaltic liquid, *J. Petrol.* **32**:909–934.

Rothwell, R. G., Bednarz, U., Boe, R., Clift, P., Hodkinson, R. A., Ledbetter, J. K., Pratt, C. E., and Soakai, S. 1994. Sedimentation and sedimentary processes in the Lau Basin: Results from Leg 135 of the Ocean Drilling Program, in *Proc. ODP, Sci. Results*, 135 (J. W. Hawkins, L. M. Parson, J. F. Allan, *et al.*, eds.), pp. 101–130, Ocean Drilling Program, College Station, TX.

Ryerson, F. J., and Watson, E. B. 1987. Rutile saturation in magmas; implications for Ti-Nb-Ta depletion in island arc basalts, *Earth Planet. Sci. Lett.* **86**:225–239.

Scholl, D. W., Vallier, T. L., and Packham, G. H. 1985. Framework geology and resource potential of southern Tonga Platform and adjacent terranes—a synthesis, in *Geology and Offshore Resources of Pacific Island Arcs—Tonga Region* (D. Scholl and T. Vallier, eds.), pp. 457–488, Circum-Pacific Council for Energy and Mineral Resources, Houston, TX.

Sclater, J. G., Hawkins, J. W., Mammerickx, J., and Chase, C. G. 1972. Crustal extension between the Tonga and Lau Ridges; petrologic and geophysical evidence, *Geol. Soc. Am. Bull.* **83**:505–518.

Shervais, J. W. 1982. Ti-V plots and the petrogenesis of modern and ophiolitic lavas, *Earth Planet. Sci. Lett.* **59**:101–118.

Sinton, J., and Fryer, P. 1987. Mariana Trough lavas from 18°N: Implications for the origin of back arc basin basalts, *J. Geophys. Res.* **92**:12782–12802.

Sinton, J. M., Wilson, D. S., Christie, D. M., Hey, R., and Delaney, J. R. 1983. Petrologic consequences of rift propagation on oceanic spreading ridges, *Earth Planet. Sci. Lett.* **62**:193–207.

Sleep, N., and Toksoz, M. N. 1971. Evolution of marginal basins, *Nature* **233**:548–550.

Staudigel, H., and Hart, S. R. 1983. Alteration of basaltic glass; mechanisms and significance for the ocean's crust-seawater budget, *Geochim. Cosmochim. Acta* **47**:337–350.

Stern, R. J. 1982. Strontium isotopes from circum-Pacific intraoceanic island arcs and marginal basins: Regional variations and implications for magmagenesis, *Geol. Soc. Am. Bull.* **93**:477–486.

Stern, R. J., and Bloomer, S. H. 1992. Subduction zone infancy: Examples from the Izu–Bonin–Mariana and Jurassic California arcs, *Geol. Soc. Am. Bull.* **104**:1621–1636.

Stern, R. J., Lin, P. N., Morris, J. D., Jackson, M. C., Fryer, P., Bloomer, S. H, and Ito, E. 1990. Enriched back-arc basin basalts from the northern Mariana Trough: Implications for the magmatic evolution of back-arc basins, *Earth Planet. Sci. Lett.* **100**:210–225.

Stern, R. J., Morris, J., Bloomer, S. H., and Hawkins, J. W. 1991. The source of the subduction component in convergent margin magmas: Trace element and radiogenic isotope evidence from Eocene boninites, Mariana forearc, *Geochim. Cosmochim. Acta* **55**:1467–1481.

Stern, R. J., Smoot, N. C., and Rubin, M. 1984. Unzipping of the Volcano Arc, Japan, *Tectonophysics* **102**:153–174.

Stevenson, A. J. 1985. Dredging and coring; southern Tonga platform, in *Geology and Offshore Resources of Pacific Island Arcs—Tonga Region* (D. Scholl and T. Vallier, eds.), pp. 31–38, Circum-Pacific Council for Energy and Mineral Resources, Houston, TX.

Sun, S.-S., and McDonough, W. F. 1989. Chemical and isotopic systematics of oceanic basalts: Implications for mantle composition and processes, in *Magmatism in the Ocean Basins* (A. D. Saunders and M. J. Norry, eds.), Geol. Soc. Spec. Publ., **42**:313–346, Blackwell, London.

Sunkel, G. 1990. Origin of petrological and geochemical variation of Lau Basin lavas (SW Pacific), *Mar. Mining* **9**:205–234.

Tamaki, K., Suyehiro, K., Allan, J., Ingle, J. C., and Pisciotto, K. A. 1992. Tectonic synthesis and implications of Japan Sea ODP drilling, in *Proc. ODP, Sci. Results*, 127/128, Pt. 2 (K. Tamaki, K. Suyehiro, J. Allan, M. McWilliams *et al.*, eds.), pp. 1333–1348, Ocean Drilling Program, College Station, TX.

Tappin, D. R., Bruns, T. R., and Geist, E. L. 1994. Rifting of the Tonga/Lau Ridge and formation of the Lau backarc basin: Evidence from Site 840 on the Tonga Ridge, in *Proc. ODP, Sci. Results*, 135 (J. W. Hawkins, L. M. Parson, J. F. Allan *et al.*, eds.), pp. 367–372, Ocean Drilling Program, College Station, TX.

Tatsumi, Y. 1989. Migration of fluid phases and genesis of basalt magmas in subduction zones, *J. Geophys. Res.* **94:**4697–4707.

Tatsumi, Y., Hamilton, D. L., and Nesbitt, R. W. 1986. Chemical characteristics of fluid phase released from a subducted lithosphere and origin of arc magmas: Evidence from high pressure experiments and natural rocks, *J. Volcanol. Geotherm. Res.* **29:**293–309.

Taylor, B. 1992. Rifting and the volcanic-tectonic evolution of the Izu-Bonin–Mariana Arc, in *Proc. ODP, Sci. Results*, 126 (B. Taylor and K. Fujioka, *et al.*, eds.), pp. 627–651, Ocean Drilling Program, College Station, TX.

Taylor, P. W. 1991. The geology and petrology of Niuafo'ou Island, Tonga: Subaerial volcanism in an active back-arc basin, M.Sc. thesis, Macquarie University.

Taylor, B., Fujioka, K., *et al.* 1992. *Proc. ODP, Sci. Results*, 126, Ocean Drilling Program, College Station, TX.

Taylor, B., and Karner, G. D. 1983. On the evolution of marginal basins, *Rev. Geophys. Space Phys.* **21:**1727–1741.

Takazawa, E., Frey, F. A., Shimizu, N., Obata, M., and Bodinier, J. L. 1992. Geochemical evidence for melt migration and reaction in the upper mantle, *Nature* **359:**55–58.

Uyeda, S., and Kanamori, H. 1979. Back-arc opening and the mode of subduction, *J. Geophys. Res.* **84:**1049–1061.

Vallier, T. L., Jenner, G. A., Frey, F. A., Gill, J. B., Volpe, A. M., Hawkins, J. W., Morris, J. D., Cawood, P. A., Morton, J. L., Scholl, D. W., Rautenschlein, M., White, W. M., Williams, R. W., Stevenson, A. J., and White, L. D. 1991. Subalkaline andesite from Valu Fa Ridge, a back-arc spreading center in the southern Lau Basin; petrogenesis, comparative chemistry, and tectonic implications, *Chem. Geol.* **91:**227–256.

Vallier, T. L., O'Conner, R. M., Scholl, D. W., Stevenson, A. J., and Quinterno, P. 1985. Petrology of rocks dredged from the landward slope of the Tonga Trench: Implications for middle Miocene volcanism and subsidence of the Tonga Ridge forearc, in *Geology and Offshore Resources of Pacific Island Arcs—Tonga Region* (D. Scholl and T. Vallier, eds.), pp., 109–120, Circum-Pacific Council for Energy and Mineral Resources, Houston, TX.

Viereck, L. G., Flower, M. F. J., Hertogen, J., Schmincke, H. U., and Jenner, G. A. 1989. The genesis and significance of N-MORB sub-types, *Contrib. Mineral. Petrol.* **102:**112–126.

Volpe, A. M., Macdougall, J. D., and Hawkins, J. W. 1987. Mariana Trough basalts (MTB): Trace element and Sr-Nd isotopic evidence for mixing between MORB-like and arc-like melts, *Earth Planet. Sci. Lett.* **82:**21–254.

Volpe, A. M., Macdougall, J. D., and Hawkins, J. W. 1988. Lau Basin basalts (LBB): Trace element and Sr-Nd isotopic evidence for heterogeneity in back-arc basin mantle, *Earth Planet. Sci. Lett.* **90:**174–186.

Volpe, A. M., Macdougall, J. D., Lugmair, G. W., Hawkins, J. W., and Lonsdale, P. F. 1990. Fine-scale isotopic variation in Mariana Trough basalts: Evidence for heterogeneity and a recycled component in backarc basin mantle, *Earth Planet. Sci. Lett.* **100:**251–264.

von Stackelberg, U., and von Rad, U. 1990. Geological evolution and hydrothermal activity in the Lau and North Fiji basins (Sonne cruise SO-35), *Geol. Jahrb.* *D***92:**629–660.

von Stackelberg, U., and Shipboard Party. 1985. Hydrothermal sulfide deposits in back-arc spreading centers in the southwest Pacific, *Bundes. Geowiss. Rohstoffe Circ.* **2:**2–14.

Wegener, A. 1929. *Die Entstehung der Kontinente und Ozeane* (translated by J. Biram, 1969, as *The Origin of Continents and Oceans*), Dover, New York.

Weidicke, M., and Collier, J. 1993. Morphology of the Valu Fa spreading ridge in the southern Lau Basin, *J. Geophys. Res.* **98:**11769–11782.

Weidicke, M., and Habler, W. 1993. Morphotectonic characteristics of a propagating spreading system in the northern Lau Basin, *J. Geophys. Res.* **98:**11, 783–797.

Weissel, J. K. 1977. Evolution of the Lau Basin by the growth of small plates, in *Island Arcs, Deep Sea Trenches, and Back Arc Basins* (M. Talwani and W. C. Pitman III, eds.), Vol. 1, pp. 429–436. Maurice Ewing Series, American Geophysical Union, Washington, DC.

Wharton, M. R., Hathaway, B., and Colley, H. 1992. Volcanism associated with extension in the Vitiaz island arc, Fiji, Symposium on Volcanism Associated with Extension at Consuming Plate Margins, Geological Society of London (Abstract).

Whelan, P. M., Gill, J. B., Kollman, E., Duncan, R., and Drake, R. E. 1985. Radiometric dating of magmatic stages in Fiji, in *Geology and Offshore Resources of Pacific Island Arcs—Tonga Region*, (D. Scholl and T. Vallier, eds.), pp. 415–440, Circum-Pacific Council for Energy and Mineral Resources, Houston, TX.

White, W. M., and Patchett, J. 1984. Hf-Nd-Sr isotopes and incompatible element abundances in island arcs: Implications for magma origins and crust-mantle evolution, *Earth Planet. Sci. Lett.* **67**:167–185.

Wood, D. A. 1980. The application of a Th-Hf-Ta diagram to problems of tectonomagmatic classification and to establishing the nature of crustal contamination of basaltic lavas of the British Tertiary volcanic province, *Earth Planet. Sci. Lett.* **50**:151–162.

Wood, D. A., Marsh, N. G., Tarney, J., Joron, J. L., Cotten, J., and Treuil, M. 1981. Geochemistry of igneous rocks recovered from a transect across the Mariana Trough, Arc, Fore-arc, and Trench, Sites 453 through 461, Deep Sea Drilling Project Leg 60, in *Init. Repts. DSDP*, 60 (D. M. Hussong and S. Uyeda, *et al.*, eds.), pp. 611–646, U. S. Govt. Printing Office, Washington, DC.

Woodhall, D., 1985, Geology of the Lau Ridge, in *Geology and Offshore Resources of Pacific Island Arcs—Tonga Region* (D. Scholl and T. Vallier, eds.), pp. 351–378, Circum-Pacific Council for Energy and Mineral Resources, Houston, TX.

Woodhead, J. D. 1989. Geochemistry of the Mariana arc (western Pacific) source compositions and processes, *Chem. Geol.* **76**:1–24.

Woodhead, J., Eggins, S., and Gamble, J. 1993. High field strength and transition element systematics in island arc and back-arc basin basalts: Evidence for multi-phase melt extraction and a depleted mantle wedge, *Earth Planet. Sci. Lett.* **114**:491–504.

4

The North Fiji Basin
Geology, Structure, and Geodynamic Evolution

Jean-Marie Auzende, Bernard Pelletier, and Jean-Philippe Eissen

ABSTRACT

As the result of intensive studies conducted by U.S., French, and Japanese scientific teams, the North Fiji Basin ridge, poorly known 10 years ago, is one of the most exhaustively investigated ridge axes of the world's oceans. Today, a ridge segment more than 800 km long and 100 km wide has been fully mapped with the Sea Beam and Furono echo sounders. This ridge axis shows four main segments characterized by the same morphostructural aspect and limits that characterize mid-oceanic ridges. Along the whole length of the axis, a water-column sample has been taken every 20 km and rock samples every 10 km. Different types of hydrothermal activity have been discovered and explored either during the *Nautile* cruise in 1989 or during the *Shinkai 6500* cruise in 1992. The most famous site is the "White Lady," located around 17°S; it is characterized by 285°C shimmering hot water, which is very poor in metallic elements, expelled through an anhydrite chimney. This water probably represents the low salinity end-member resulting from phase separation in the deep levels of the oceanic crust. Other active sites have been observed all along the axis showing different characteristics such as low-temperature diffusion zones. Even though some parts of the North Fiji Basin remain poorly investigated, the newly acquired data from the ridge axis and from the eastern and northwestern parts allow us to develop a new tectonic model of basin evolution since its creation 12 m.y. ago.

1. INTRODUCTION

The study of the North Fiji Basin (NFB) (Fig. 4.1) started between 20 and 25 years ago with focused efforts on selected areas and large scale profiling across the basin mainly conducted during U.S. cruises (Scripps Institution of Oceanography, Hawaii Institute of

Jean-Marie Auzende • IFREMER/CB, 29280 Plouzané, France. *Present address*: ORSTOM, BPA5, Nouméa, New Caledonia. *Bernard Pelletier* • ORSTOM, Nouméa, New Caledonia. *Jean-Phillippe Eissen* • Antenne ORSTOM-IFREMER/CB, 29280 Plouzané, France.

Backarc Basins: Tectonics and Magmatism, edited by Brian Taylor, Plenum Press, New York, 1995.

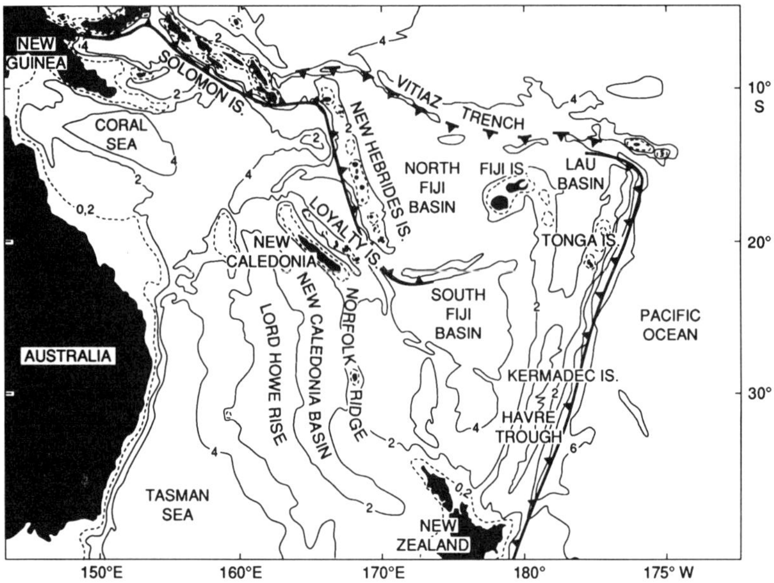

FIGURE 4.1. General geodynamic setting of the North Fiji Basin (NFB) in the SW Pacific.

Geophysics, and Woods Hole Oceanographic Institution) and French ORSTOM cruises. Since 1976, ORSTOM (Institut Français de Recherche Scientifique pour le Développement en Coopération) has conducted the EVA (Evolution des Arcs insulaires) program, which has been partly devoted to the study of the Lau and North Fiji basins. Fifteen cruises have been conducted in different areas, especially in the New Hebrides arc domain, in the southernmost part of the basin around the Matthew–Hunter zone, and more recently in the northwestern part of the basin.

The SEAPSO (Sea Beam Pacifique Sud Ouest) program in 1985 represented the first coordinated approach by French teams for backarc basin studies. As a part of this program, IFREMER (Institut Français de Recherche pour l'Exploitation de la Mer), ORSTOM, and INSU (Institut National des Sciences de l'Univers) jointly explored the NFB where the existence of an axial ridge was assumed but not demonstrated. One of the major results of the SEAPSO-Leg 3 cruise of the R.V. *Jean Charcot* was the partial mapping of an active accreting ridge in this marginal basin.

On the basis of the SEAPSO cruise results, Japanese and French scientists decided to undertake a joint project to study the rift systems of the western Pacific. This project, STARMER, coordinated by IFREMER in France and the Science and Technology Agency (STA) in Japan, was initiated in 1987. Its objective was a five-year interdisciplinary (geology, geochemistry, and geophysics) study of the NFB ridge. Since this date, seven cruises have been conducted, representing more than nine months at sea. Four cruises were dedicated to surface-ship surveys, including swath bathymetric mapping, geophysical profiling, water sampling, gravity coring, and dredging. The three others were diving cruises using the *Nautile* (June–July 1989), the new Japanese submersible *Shinkai 6500* (September–November 1991), and *Cyana* (December–January 1992).

Simultaneously, U.S., German, and SOPAC (South Pacific Applied Geoscience Commission) teams conducted SeaMARC, Sea Beam, and GLORIA (geologic long range inclined asdic) surveys on selected areas, especially in the northern and northeastern parts of the NFB. In total, more than 20 months of ship time have been spent in the basin in the past 10 years. The result of these efforts is an extensive multibeam bathymetric coverage of the central spreading axis, detailed rock and water sampling along the axis, and *in situ* observations by deep-tows and submersibles.

The NFB is one of the marginal basins located at the converging boundary between the Pacific and Australian major plates (Fig. 4.1). It lies between the New Hebrides arc to the west, the Fiji Platform to the east, the Vitiaz fossil subduction zone to the north, and the arcuate Matthew–Hunter zone to the south. Different models for opening of the NFB have been proposed (Chase, 1971; Gill and Gorton, 1973; Falvey, 1975; Dubois *et al.*, 1977; Malahoff *et al.*, 1982a; Auzende *et al.*, 1988b). These authors suggest that the opening of the basin started 10 m.y. ago after the locking of the Vitiaz subduction by the Ontong–Java Plateau and the reversal of its subduction polarity. This change of polarity involved the clockwise rotation of the New Hebrides arc and by secondary effect the anticlockwise rotation of the Fiji Platform. This first phase implies a NW-SE-trending spreading axis and N45° to N55° flowlines for the opening motion. The second phase resulted in the beginning of the collision of the New Hebrides arc with the d'Entrecasteaux and Loyalty Islands ridges (Daniel, 1982; Louat, 1982; Monzier *et al.*, 1984,1990), the change of the traction stresses to an E-W direction, and the emplacement of the N-S-trending spreading center in the central part of the basin. During this time N-S spreading occurred in the northern part of the basin and north of the Fiji Islands (Auzende *et al.*, 1988a,b).

1.1. Bathymetry and Structure

Bathymetric data concerning the NFB were rare before the SEAPSO III cruise of the R.V. *Jean Charcot* (December 1985) and the five surface cruises of the STARMER project. The only existing data were general and imprecise maps (Chase, 1971; Mammerickx *et al.*, 1971) or more detailed local surveys (Halunen, 1979; Kroenke *et al.*, 1987, 1991). The NFB was then called the North Fiji Plateau, and the main feature mapped was a 3000-m-deep flat area occupying the whole central part of the basin.

The bathymetric map shown in Fig. 4.2 (Mazé *et al.*, 1992, unpublished map) results from the compilation of all the available bathymetric data; it combines good-quality classical bathymetric surveys and recent multibeam surveys performed mainly within the SEAPSO and STARMER projects. This map especially uses the data from the maps of Monzier *et al.* (1984, 1991), Auzende *et al.* (1990b, 1992b) and Urabe *et al.* (1992). It shows the different physiographic zones characterizing the NFB. We have divided the basin into five physiographic and structural provinces: the western basin, the central spreading ridge, the eastern basin, the northwestern basin, and the northeastern basin.

1.1.1. The Western Basin

This domain is located south of the Hazel Holme Ridge (14°S) between the New Hebrides arc and 173°E (Fig. 4.2). It is still poorly known and contains the largest portion of old crust of the basin, including the initial NW-SE spreading center (Auzende *et al.*, 1988b; Pelletier *et al.*, 1993a). Although some topographic features exist, especially in its southern

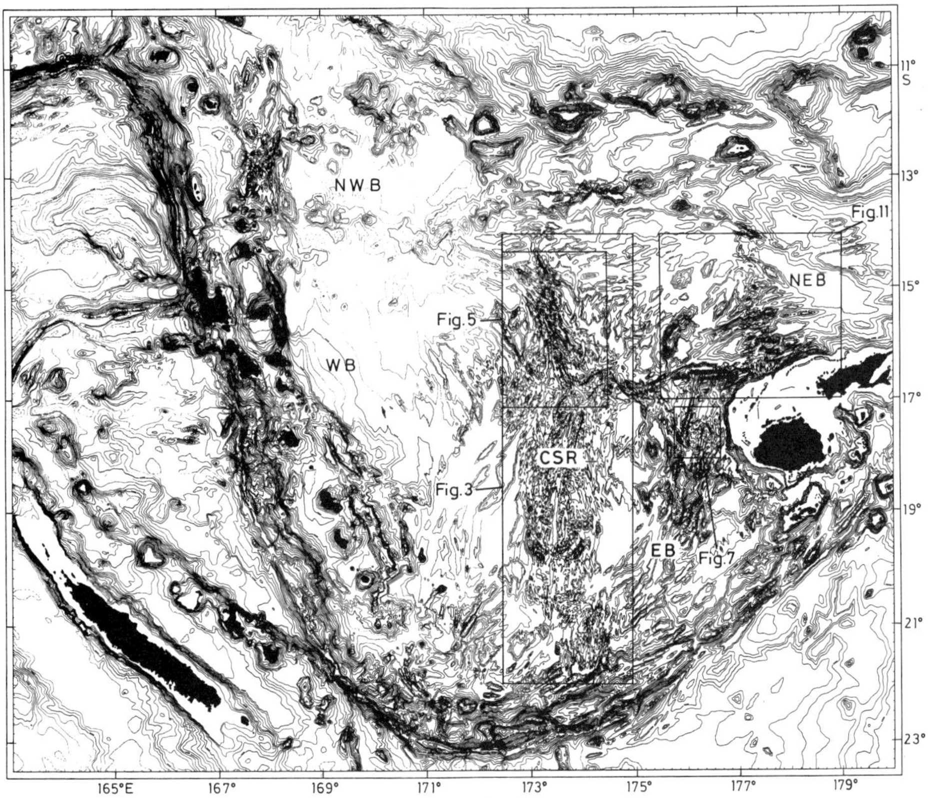

FIGURE 4.2. Bathymetric map of the NFB. This map results from a compilation (J.P. Mazé) of all multibeam data existing on the NFB (SEAPSO, Proligo, STARMER, Multipso cruises). The contour interval is 200 m. CSR: central spreading ridge; WB: western basin; E: eastern basin; NWB: northwestern basin; NEB: northeastern basin. Locations of Figs. 4.3, 4.5, 4.7, and 4.11 are shown.

part (south of 17°S), the western basin is mainly characterized by gently undulating and flat areas 3200–3300 m deep. The primitive oceanic crust topography is smoothed or obliterated by sedimentary cover which thickens toward the west from 0.1 s to more than 0.5 s of two-way travel time (Luyendyck *et al.*, 1974; Halunen, 1979; Kroenke *et al.*, 1994). The thickest sediments occur in the westernmost part of the basin along the New Hebrides arc, especially east of the central New Hebrides islands where the thickness reaches 1-s two-way travel time in seismic reflection profiles (Luyendyck *et al.*, 1974; Pelletier *et al.*, 1988).

From 17°20′S to 20°50′S, the westernmost part of the western basin is characterized by the N150° trending southern New Hebrides backarc troughs (Fig. 4.2) (Monzier *et al.*, 1984, 1991; Récy *et al.*, 1990; Price *et al.*, 1993). These troughs deepen southward from 2800 to 3600 m and are flanked eastward by a N150° volcanic ridge with an average depth of 1000 m and rising to 650 m above sea level at Futuna Island. A secondary N150°–170° trending volcanic ridge (600 to 1800 m deep) lies 30 km east of the Futuna Ridge. Isolated volcanoes exist farther east; one of them, the Constantine Bank, shoals to a depth of 104 m (Monzier *et al.*, 1984).

The morphology of the southern part of the western basin is still largely unknown but appears to be complex and characterized by NE-SW-trending structural features (Fig. 4.2)

(Monzier *et al.*, 1984,1991). Around 21°S, 172°E, two N40–45° trending ridges rising to a depth of 2000–2200 m and enclosing a 3600-m depression are the most prominent features and likely a major tectonic element of this southern area. Because NW-SE trending magnetic anomaly lineations exist immediately to the southeast, this paired trough-ridge is interpreted as a fracture zone.

Other important tectonic features are located at 17°30′S, 172°E in the central eastern part of the western basin, where NE-SW-trending, steep-sided, linear ridges rise to 1300 m and are commonly interrupted by 3000–3300-m depressions (Halunen, 1979). The ridge area, also characterized by absence of sediments, high heat flow, and presence of shallow earthquakes, has been interpreted as an active NE-SW spreading center (Halunen, 1979).

1.1.2. The Central Spreading Ridge

Locally recognized from bathymetric and magnetic profiles (Chase, 1971; Malahoff *et al.*, 1982a; Maillet *et al.*, 1986), the central spreading ridge (CSR) was for the first time partially multibeam-mapped during the SEAPSO III cruise of the R.V. *Jean Charcot* (December 1985) (Auzende *et al.*, 1986a,b, 1988a). Up to now within the French–Japanese joint STARMER project it has been mapped between 14°30′S and 21°40′S on more than 1° width by multibeam bathymetric full coverage (Auzende *et al.*, 1990b; Urabe *et al.*, 1992; Fig. 4.3). The ridge axis can be divided into four major segments, with lengths varying from 120 to more than 200 km, that are described below from south to north.

1.1.2.1. The Southernmost Segment. The southernmost segment is about 120 km long between 21°40′S and 20°30′S. It is characterized by a N-S trend and a complicated morphology of alternating ridges reaching 2500-m depth and depressions reaching below 3000 m. The atypical morphology and the lack of *in situ* observation make it difficult to locate precisely the present-day spreading axis on this segment. Only magnetic lineation analysis (Maillet *et al.*, 1989; Ruellan *et al.*, 1989) confirms the existence of an active spreading axis centered on 174°05′E. The transverse bathymetric profile of Fig. 4.4 illustrates this atypical morphology intermediate between fast and slow spreading ridges (Macdonald, 1983; Karson *et al.*, 1987).

1.1.2.2. The 310-km-Long North-South Segment. The 310-km-long north-south segment, between 21°S and 18°10′S, is apparently the simplest segment. Its linear ridge axis crosses a 2800-m-deep flat-topped dome cut in its center by a graben that is a few hundred meters wide and a few tens of meters deep. The width of the axial ridge, limited by the 3000-m isobath, is about 20 km. On both sides, the dome is flanked by symmetrical grabens more than 3000 m deep and trending N-S. The southern and northern tips of this segment exhibit V-shaped features interpreted as inner and outer pseudofaults of propagating rifts (de Alteris *et al.*, 1993; Ruellan *et al.*, 1994). In the middle part of the segment, around 19–19°40′S, symmetrical oblique volcanic lines crosscut the N-S oceanic structural grain. On the northwestern side of the area west of the inner pseudofault, the oceanic bottom shows successive arcuate ridges abutting the pseudofault. These peculiar features can be either interpreted as fossil overlapping spreading centers (OSC) (Ruellan *et al.*, 1994) or as failed rifts of a propagating rift system (de Alteris *et al.*, 1993).

1.1.2.3. The N15° Segment. The N15° segment is about 120 km long and shows spectacular along-strike morphological variation. In its southern part from 18°10′S to 17°55′S,

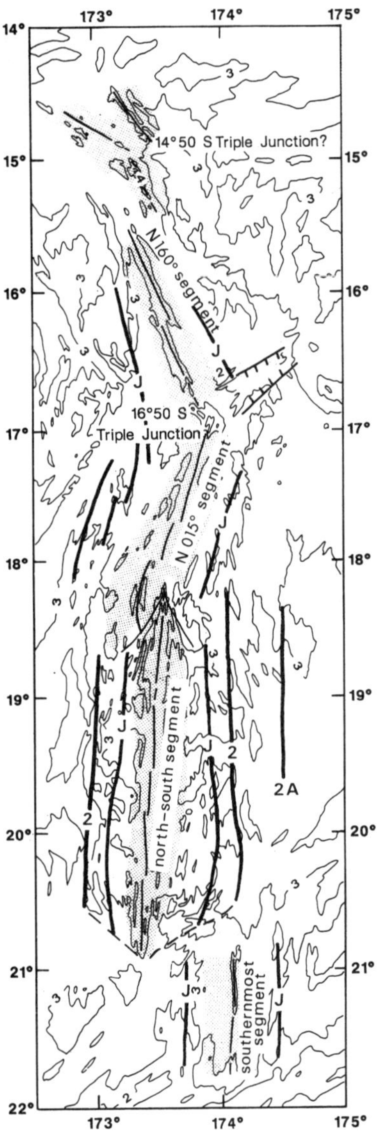

FIGURE 4.3. Simplified bathymetry (contour interval = 1000 m) and magnetic lineations (heavy lines) along the central spreading ridge from Auzende *et al.* (1990c) and de Alteris *et al.* (1993) (after Auzende *et al.*, 1994b). The present-day axis is shown with medium lines. In gray: anomaly 1; J: Jaramillo anomaly (1 Ma); 2: anomaly 2 (1.98 Ma); 2A: anomaly 2A (2.6–3.5 Ma). The ages of anomalies are after Cande and Kent (1992).

the present-day axis is not well defined and the accretion is distributed over numerous small volcanoes scattered over a wide area. This zone corresponds to the change of direction of the central spreading ridge from N-S to N15°. North of 17°55′S, the spreading axis is represented by a double ridge less than 2500 m high bounding a 2–3-km-wide, 200–300-m-deep graben. At the northern tip of the N15° segment, close to the 16°50′S triple junction, the ridge axis is located on the top of a shallow massif culminating at less than 1900 m deep and cut by a graben 500 m to 2 km wide and 200 m deep. This area is the site of the active hydrothermal vents explored by *Nautile* and *Shinkai 6500* (see following). The N15° axial ridge is flanked by curved grabens more than 3000 m deep interpreted by Ruellan *et al.* (1994) as fossil OSCs.

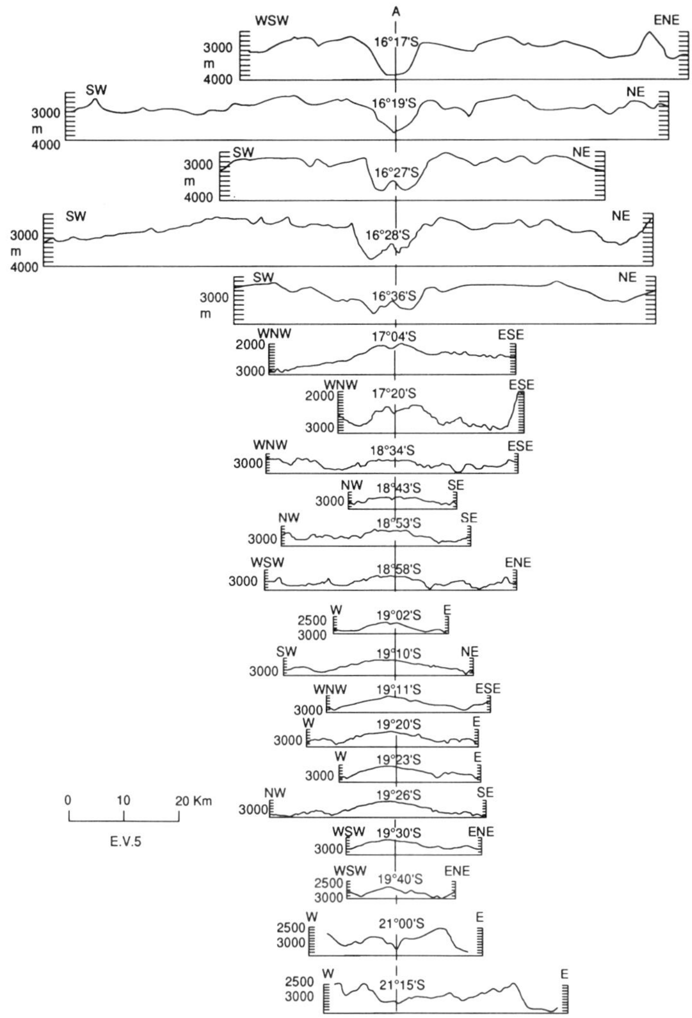

FIGURE 4.4. Transverse bathymetric profiles across the different segments of the central spreading ridge (16°17'S to 21°15'S). A: ridge axis.

1.1.2.4. The N160° Segment. The N160° segment is about 200 km long and can be described as three parts (Auzende *et al.*, 1991a, 1994a; Gracia-Mont, 1991, 1992; Jarvis *et al.*, 1993; Figs. 4.5 and 4.6): (1) *From 16°50S to 15°30S* the spreading axis is located in a 4000–4500-m-deep graben located between two subvertical walls. The average width of the graben is 8 km along the whole segment. In its axial part, the graben is cut by a 2–3-km-wide, 400–500-m-high ridge (Fig. 4.4). This morphology is very similar to those described at slow spreading (less than 40 mm/yr) ridges like the Mid-Atlantic Ridge (Kappel and Ryan, 1986; Karson *et al.*, 1987; Karson, 1990; Gente *et al.*, 1991). Around 16°10S, the remarkable linearity of the N160° axis is interrupted by a slight curve toward the north, offsetting the graben by about 4 km. On both sides the active domain is flanked by a large volcanic massif that shallows to less than 1700 m deep. The width of the volcanic massif

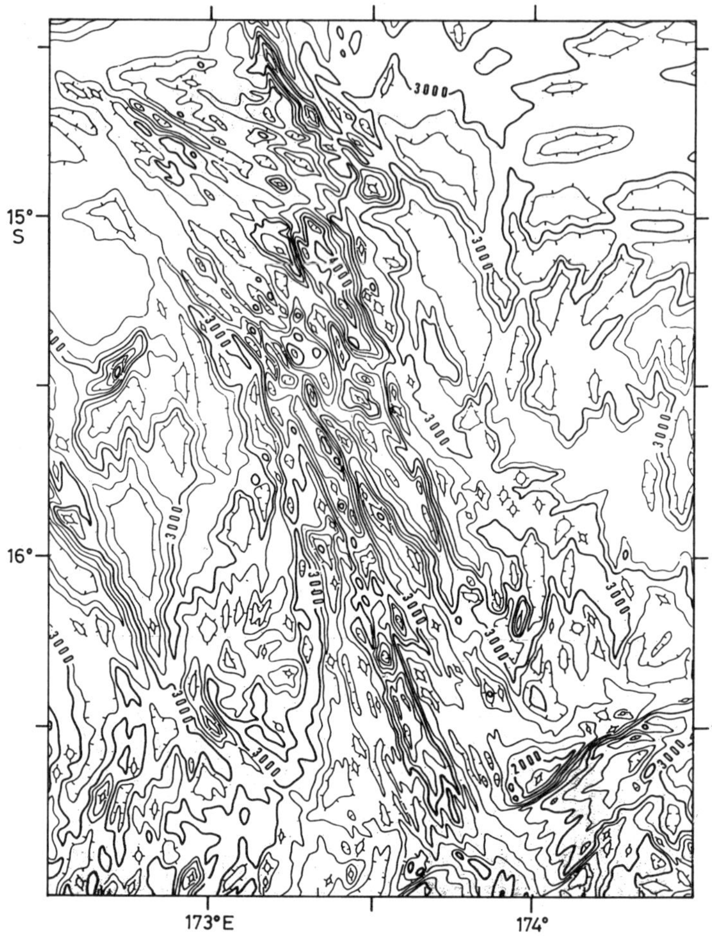

FIGURE 4.5. Bathymetry of the N160 segment and of the 14°50′S triple junction (enlargement of Fig. 4.2). The contour interval is 200 m.

decreases from 100 km in the south to a few kilometers in the north until it disappears north of 15°30′S. The magnetic data analysis allows an estimate of the beginning of the volcanic construction and the massif uplift at about 1 Ma. This uplift affects an older oceanic crust as demonstrated by the age of the sediments sampled from the northern edge of the N55° trending graben located east of the 16°50′S triple junction (Lagabrielle *et al.*, 1994). (2) *Between 15°30′S and 15°00′S* the accretion is distributed over a wide domain of two 60-km-long, 4000-m-deep en echelon grabens offsetting the axis by about 40 km to the northeast. Each of these grabens is made up of a succession of 10-km-long en echelon segments (Fig. 4.6). In this area, the magmatic supply appears to be concentrated along a narrow ridge that separates the grabens. For the past 1 m.y. the accretion seems to have been mainly amagmatic (Gracia-Mont, 1991, 1992). (3) *North of 15°00S* the spreading axis becomes more complex and is located within two distinct branches. The western one, trending N120°, is characterized by a 4-km-wide, 4000-m-deep graben that crosscuts an older oceanic crust showing different trends from N160° to N120°. The northern branch consists

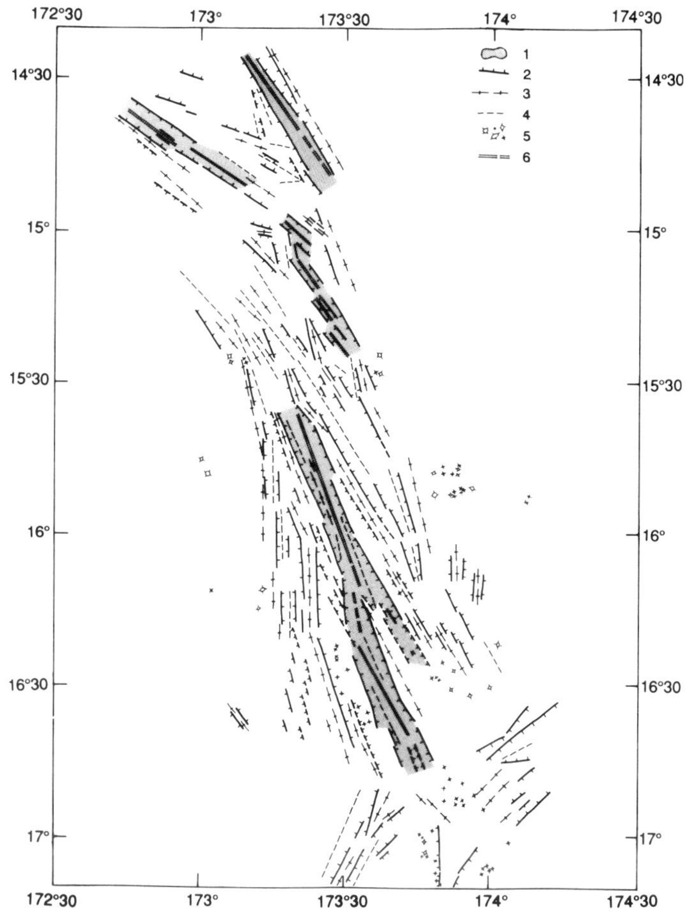

FIGURE 4.6. Structural map of the N160 segment and of the 14°50′S triple junction (after Auzende *et al.*, 1994a). 1: axial domain deduced from bathymetry and magnetic anomaly, 2: normal fault and scarps, 3: crests, 4: depressions, 5: isolated volcanoes, 6: ridge axis.

of a 2400-m-deep, N140°-trending ridge that connects with the N160° graben system at a 4000-m-deep depression at 14°50′S. This depression can be interpreted as a triple junction between the N160° and N120° grabens and the N140° ridge (Auzende *et al.*, 1994a).

1.1.3. The Eastern Basin

The eastern basin occupies the area between the Fiji Islands and the central spreading ridge. On the basis of different arguments, a spreading ridge was postulated west of Fiji (Sclater and Menard, 1967; Chase, 1971; Brocher and Holmes, 1985). Few recent data exist for this area, except a full coverage, one square-degree bathymetric survey located immediately west of Fiji (Fig. 4.7). This area is underlain by important shallow seismicity with strike-slip (Hamburger and Isacks, 1988, 1994) and normal-fault-type focal mechanism solutions (Louat and Pelletier 1989; Pelletier and Louat, 1989). Previously interpreted as a strike-slip deformation zone (Auzende *et al.*, 1986b), this area was recently reinterpreted in

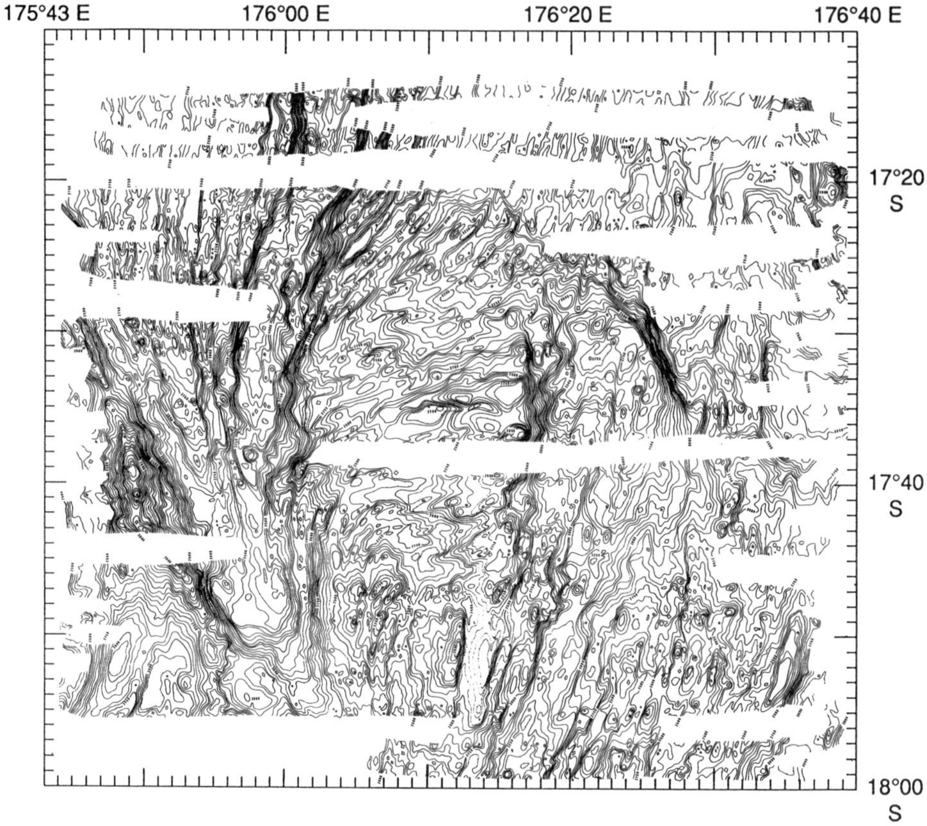

FIGURE 4.7. Sea Beam bathymetry of the West Fiji area (after Auzende *et al.*, 1988a). The contour interval is 50 m.

terms of a propagating rift (Auzende *et al.*, 1993, 1994b). The main morphostructural units of the area are represented in Fig. 4.8.

1.1.3.1. The Propagating Ridge (PR) and Its Tip (PT). The PR is located in the western part and comprises a central ridge bounded by two grabens. The eastern graben is well developed with a constant width of 10 km and depth of 4000 m. The western graben is narrower (2–3 km) and shallower (3000 m). The central ridge is 7 to 8 km wide and 2750 m deep. Its N-S to N05° trend changes to N155° at its southern tip. The width of the PR varies from 40 km in the north to 8 km in the south. South of 17°44′S, the ridge disappears and is replaced by a 3000 m deep flat area limited by converging faults. This flat area is similar to the "propagating tip" described by Hey *et al.* (1986) in the case of the 95.5°W propagating rift on the East Pacific Rise close to the Galapagos.

1.1.3.2. The Pseudofaults. The outer pseudofault (OPF) corresponds to a N155°-trending fault bounding the propagating system on the west. This fault separates the propagating from the "normal" oceanic bottom grain trending N15–20°. The inner pseudofault (IPF) is repeated north of 17°35′S (Fig. 4.8, dashed line, top center), with a N15° fault

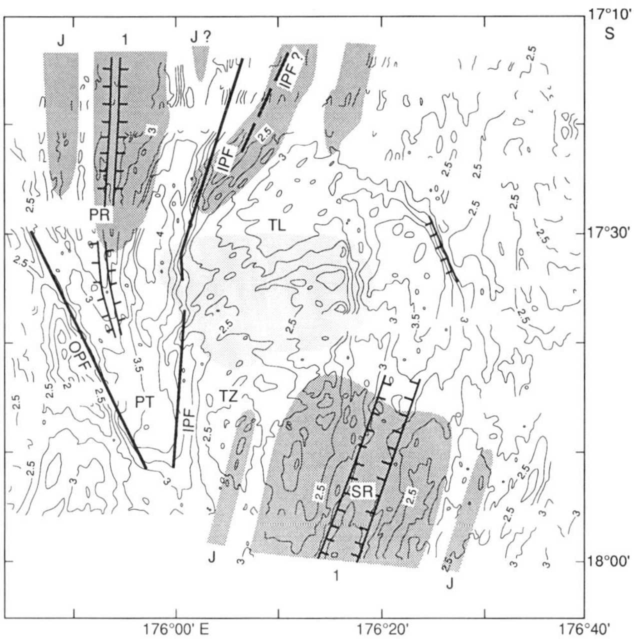

FIGURE 4.8. Structural sketch of the West Fiji Ridge (after Auzende *et al.*, 1993). PR axis: propagating axis; PT: propagating tip; OPF: outer pseudofault; IPF: inner pseudofault; SR: southern rift; TL: transferred lithosphere; TZ: transform zone. Main positive and negative magnetic anomaly lineations are underlined in gray and light gray respectively. Anomalies 1 and J are labeled.

converging toward the OPF with a 40° angle to the west and a more diffuse lineament trending N30° and converging with the OPF with a 55° angle to the east. South of 17°35′S, there is a unique fault converging with the OPF at a 20° angle.

1.1.3.3. The Southern Ridge (SR). This SR is located in the southeastern part of our survey (Fig. 4.8) and is characterized by a 3000-m-deep, few-kilometer-wide axial graben trending N10°. In the eastern part of the domain, the failed rift is represented by a succession of N10–N15° trending ridges abutting a N150° fault. These ridges are 2 to 3 km wide and 2750 m deep. They probably represent the old spreading axis abandoned during the southward propagation of the western active rift.

1.1.3.4. The Transform Zone (TZ) and the Transferred Lithosphere (TL). The junction between the propagating rift and the southern ridge is a wide zone of arcuate small ridges representing a transform zone as defined by Hey *et al.* (1986) on the East Pacific Rise. The TL is located north of the TZ and comprises fan-shaped structures deepening to 3000 m in the northern part.

The close structural similarities between this area and the Galapagos propagating rift confirm the interpretation of an active spreading axis. The spreading rate was estimated to be 30 mm/yr by Louat and Pelletier (1989) from seismicity analysis and kinematic reconstruction of the Lau and North Fiji basins. The width of the propagating rift area is consistent with this rate estimate (see following).

1.1.4. The Northwestern Basin

The northwestern basin, located north of 14°S and west of 174°E between the northern part of the New Hebrides island arc to the west and the Vitiaz Trench to the east, has been the subject of different speculations and interpretations, caused mainly by the lack of data. It was interpreted as (1) the result of a late Miocene fan-shaped opening along a median NW-SE-trending axis (Falvey, 1975; Malahoff *et al.*, 1982a; Auzende *et al.*, 1988b), (2) an old piece of Pacific plate (Chase, 1971; Luyendyck *et al.*, 1974), or (3) Australian plate (Halunen, 1979) trapped behind the Vitiaz paleotrench. Recently, the bathymetry and main morphostructural units of the whole northwestern part of the NFB have been recognized during the EVA 14 (1987) and Santa Cruz (1991) cruises of R.V. *Coriolis* and *Le Noroît*, respectively, and are described here (Pelletier *et al.*, 1988, 1993a,b) (Figs. 4.9 and 4.10).

1.1.4.1. The N-S-Trending Northern New Hebrides Backarc Troughs. The N-S-trending northern New Hebrides backarc troughs were previously known south of 12°S (Charvis and

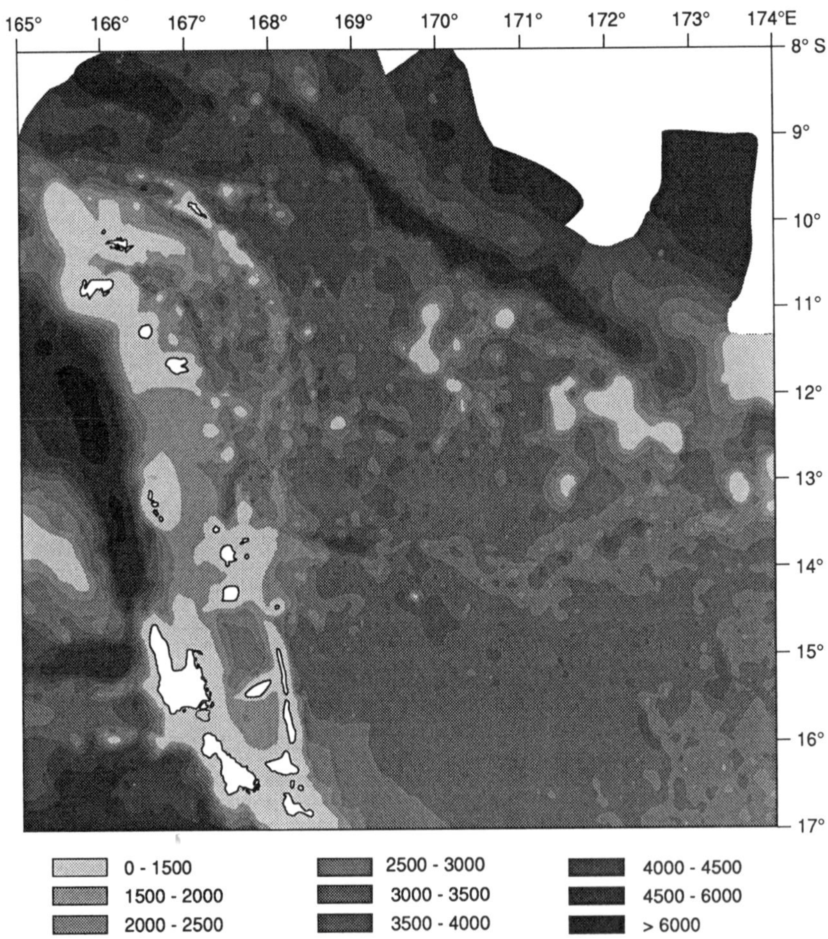

FIGURE 4.9. Bathymetric map of the northwestern basin.

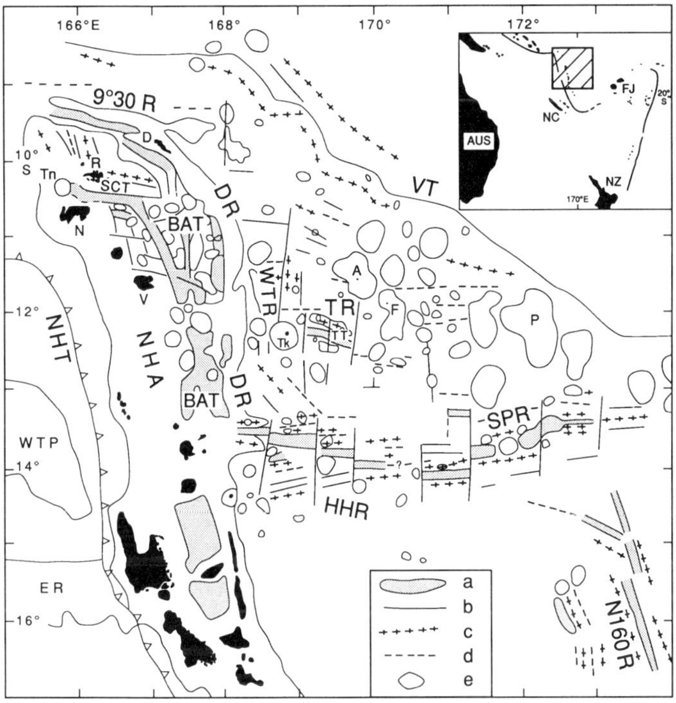

FIGURE 4.10. Structural map of the northwestern basin (after Pelletier *et al.*, 1993a). The studied area is shown in insert. AUS: Australia; NC: New Caledonia; FJ: Fiji; NZ: New Zealand. NHT: New Hebrides Trench; NHA: New Hebrides arc; VT: Vitiaz Trench; BAT: backarc troughs domain; 9°30 R: 9°30′S Ridge; DR: Duff Ridge; WTR: West Tikopia Ridge; TR: Tikopia Ridge; HHR: Hazel Holme Ridge; SPR: South Pandora Ridge; N160 R: N160 ridge of the central spreading ridge; TT: Tikopia Trough; SCT: Santa Cruz Trough; WTP: West Torres Plateau; ER: d'Entrecasteaux ridge. Islands and reefs are in black. D: Duff Islands; R: Reef Islands; Tn: Tinakula Island; N: Ndende Island; V: Vanikoro Island; Tk: Tikopia Island; A: Anuta Island; F: Fatutaka Island; P: Pandora Bank. (a) trough and depression; (b) structural trend and fracture; (c) structural high; (d) structural low; (e) volcanic high.

Pelletier, 1989; Récy *et al.*, 1990; Johnson *et al.*, 1993) and extend from 13°30′S to 9°30′S, east of the New Hebrides arc platform in the westernmost part of the basin. The troughs are formed by volcanic cones and ridges bounding N-S trending, 3000–3500-m-deep depressions, between a N-S- to NNW-SSE-trending major scarp to the west and Duff ridge to the east. They are filled by large volcanic highs at 12°S and 10°30′S and are 60 km wide south of 12°S and widen northward up to 100 km at 10°45′S. At 10°30′S the western part of the troughs abuts the N95° trending Santa Cruz trough, which cuts the arc platform.

1.1.4.2. The Duff Ridge. The Duff Ridge separates the backarc troughs from the central part of the northwestern basin. It is a continuous volcanic ridge more than 400 km long and 30 km wide, which strikes N-S and rises to around 1500 m south of 10°45′S. It trends NW-SE and shoals to 20 m farther north. The volcanic Duff Islands are located in the northernmost tip of the ridge. Duff Ridge is interpreted as a fossil volcanic line related to the New Hebrides subduction.

1.1.4.3. The Tikopia, West Tikopia, and 9°30′S Ridges. The Tikopia, West Tikopia, and 9°30′S ridges are a succession of orthogonal ridges, interpreted as spreading ridges or transform zones, lying in the axial part of the northwestern basin, which generally deepens to the north from 3200 to 4200 m. Tikopia ridge trends E-W and is located around 12°30′S between the Tikopia volcano and island at 168°45′E and the volcanoes located south of Fatutaka Island at 170°45′E. East of the large Tikopia volcano (3500 m high, 35 km wide at its base), the ridge is a 50-km-wide, N90°–100° trending elongated dome cut along strike by a 10-km-wide, 65-km-long trough (4200 m deep) bounded by 1000–1500-m-high scarps. Fresh basalts have been dredged on the northern wall of thc trough. West Tikopia Ridge is a N-S-trending, discontinuous ridge composed of aligned seamounts rising to 800 m west of the Tikopia volcano along 168°30′E from 12°20′S to 10°45′S. The 9°30 ridge is an almost continuous, N100° trending volcanic ridge from 165°45′E to 168°15′E north of the New Hebrides arc platform and the Duff Islands. The 9°30 ridge ends to the east at a large volcanic massif shoaling to 10 m in the northern prolongation of the West Tikopia Ridge.

Numerous volcanic highs, including *Anuta and Fatutaka Islands and Pandora Bank*, are located in the eastern part of the area between 10°30′S to 13°S and 169°E to 174°E , south of the Vitiaz paleotrench. These volcanic highs, previously considered as a part of the inactive Vitiaz arc, do not constitute a continuous chain but are isolated and occur as far as 240 km from the Vitiaz paleotrench. They constitute a series of massifs aligned on N-S lineaments. No volcanic arc lies immediately south of the Vitiaz paleotrench. However, a narrow ridge rising to about 2000 m between 12° and 11°S and a narrow and discontinuous swell (3200–3400 m deep) north of 10°30′S parallel the Vitiaz Trench (4500–6000 m deep), which shows a succession of NW-SE and E-W trending troughs.

The northwestern basin is separated from the rest of the basin by a complex series of E-W-trending ridges and troughs called the Hazel Holme fracture zone by Chase (1971). Right-lateral or left-lateral strike slip or even compressional motions have been proposed along this enigmatic feature (Luyendyck *et al.*, 1974; Halunen, 1979; Eguchi, 1984; Hamburger and Isacks, 1988,1994), which can be described as two parts.

1.1.4.4. The Hazel Holme Ridge, which corresponds to the western part of the Hazel Holme fracture zone of Chase (west of 171°E), was partly surveyed during the EVA 14 and Santa Cruz cruises and interpreted as an active extensional zone (Pelletier *et al.*, 1988, 1993a; Charvis and Pelletier, 1989; Louat and Pelletier, 1989). This seismically active feature is composed of several parallel narrow ridges and deep troughs trending N80–100° and extending over a maximum width of 120 km. The troughs are, on average, 3500–4000 m deep and are bounded by 500–1000-m-high scarps. The deepest trough (4500 m deep at about 169°E) is located in the axial part of the feature and appears to be right-laterally offset. West of 168°30′E the width of the ridge decreases, lateral troughs disappear, and a 2500–3500-m-deep E-W trough is bounded by symmetrical ridges rising to 1700 m. This trough ends at 168°10′E where it connects with the southern tip of the northern New Hebrides backarc troughs. The bathymetric profiles across the area exhibit a slow spreading ridge morphology.

1.1.4.5. The South Pandora Ridge, which corresponds to the eastern part of the Hazel Holme fracture zone of Chase (1971), was explored between 171°E and 173°30′E during a R.V. *Kana Keoki* reconnaissance cruise (1982) and near 174°E by a SeaMARC survey during the R.V. *Moana Wave* cruise (1987); it is interpreted as an E-W trending active slow spreading ridge within a broad transform domain (Kroenke *et al.*, 1991; Price and Kroenke, 1991). The ridge is 110 km wide and is cut by numerous transform offsets. It is composed of a series of E-W trending segments cut along strike by a central trough that is 10 to 20 km wide and 500 to 4500 m deep. The axial trough is flanked by 1000–2000-m-high scarps and is locally filled by large volcanoes rising in some places to sea level. Fresh pillow basalts recovered on the South Pandora Ridge have geochemical characteristics intermediate between mid-ocean ridge basalts (MORBs) and ocean island basalts (OIBs) (Price *et al.*, 1990; Price and Kroenke, 1991; Sinton *et al.*, 1994).

1.1.5. The Northeastern Basin

The northeastern basin is located east of 174°E and west of the northern Lau Basin (Fig. 4.2), between the Fiji Platform and the North Fiji fracture zone in the south and an imprecisely located lineament in the north that separates the young NFB oceanic crust from the old Pacific crust. This lineament is marked by a succession of deep troughs (Alexa Trough, Rotuma Trough, Horn Trough; Brocher, 1985) and connects the Vitiaz Trough to the west and the northern tip of the Tonga Trench to the east. Although some restricted areas have been recognized (see below), the overall structure of the northeastern basin is largely unknown. The northeastern basin is now the least-known part of the NFB.

On the basis of an aeromagnetic survey over the entire area, Cherkis (1980) reported E-W-trending magnetic lineations and proposed an active spreading center located near 14°30′S from 175° to 178°E and near 14°S from 178°E to 180°. The only area explored by surface ship is located north of Viti Levu Island between 176°30′E and 178°E and 13°30′S and 15°S (Halunen, 1979; Brocher, 1985). Halunen (1979) described a prominent NW-SE trending paired ridge-trough in the northwest part, more subdued WSW-ENE trending topography in the southwest, and uniform sediment cover about 0.08 s double-travel-time thick. On the basis of WNW-ESE magnetic lineations and bathymetric trends offset by an oblique NNW-SSE trough at 177°30′E and interpreted as a pseudofault, Brocher (1985) proposed a WNW-ESE inactive spreading center at 14°30′S and 14°S located respectively west and east of a fossil propagating rift at 177°30′E.

More recently, a multibeam bathymetric survey of a N-S elongated area at 177°–177°30′E from 13°50′S to 16°40′S was conducted during the R.V. *Sonne* cruise SO-35 (von Stackelberg *et al.*, 1985; von Stackelberg and von Rad, 1990). The different morphostructural units are described below from north to south (Fig. 4.11).

1. A 3500-m-deep, 10-km-wide steep-sided trough trends N120° at 14°05′S and is bounded by linear ridges rising to 2200 m. This trough, also described by Halunen (1979), corresponds to the northwestern extension of the pseudofaults of Brocher (1985) and is surprisingly interpreted by von Stackelberg and von Rad (1990) as a short narrow basin possibly belonging to the Vitiaz Trench lineament.

2. Two N120° trending narrow ridges, rising to 2400–2700 m and enclosing a 2900-m-deep depression, are observed at 14°30′S and are interpreted, following Brocher (1985), as an extinct spreading center. Although magnetic lineations are associated with these ridges, the unidentified anomalies do not allow an estimate of the age for the spreading. Altered ferrobasalts derived from a parental MORB magma and covered with hydrothermal Mn crusts were recovered from these ridges (Sinton *et al.*, 1985; Johnson and Sinton, 1990).

3. Volcanoes are present between 14°50′S and 15°15′S.

4. Two N70° trending prominent and parallel ridges exist at 15°40′S and 15°50′S. The northern one, called the Braemar Ridge, rises to 1600–1900 m deep and is bounded to the north by a 3600-m-deep depression. The southern ridge is 2000–2200 m deep and is

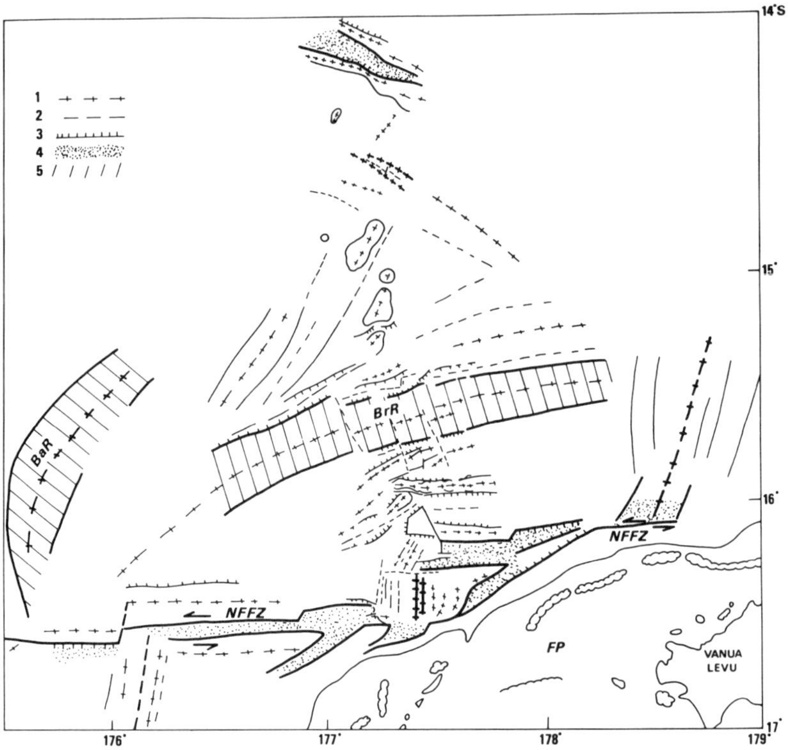

FIGURE 4.11. Structural sketch of a part of the northeastern basin based partly on the bathymetric map obtained during the R.V. *Sonne* cruise SO-35 (von Stackelberg *et al.*, 1985; von Stackelberg and von Rad, 1990), the GLORIA image (Jarvis *et al.*, 1994) and the map of Fig. 4.2. 1: structural high; 2: structural low; 3: scarp; 4: main trough; 5: main ridge. FP: Fiji Platform; NFFZ: north Fiji fracture zone; BrR: Braemar Ridge; BaR: Balmoral Ridge.

narrower than and separated from the Braemar Ridge by a 2900-m-deep depression. These two ridges are interpreted as uplifted faulted blocks of Pliocene volcanics and volcaniclastics derived from the fossil Fiji volcanic arc and deposited in a forearc basin between the Vanuatu–Fiji–Lau–Tonga arc and the Vitiaz Trench lineament (von Stackelberg and von Rad, 1990).

5. An E-W trending ridge lies at 15°55′S and appears to be made up of aligned volcanoes.

6. A 30–40-km-wide pull-apart basin with N-S structural trends, centered at 177°25′E between 16°10′S and 16°35′S, is developed between an E-W trending 2000-m-deep ridge and 3400–4000-m-deep troughs (von Stackelberg and von Rad, 1990). Fresh MORB and backarc basin basalts (BABB) lavas with hydrothermal sulfide mineralization have been recovered in the pull-part basin.

The topography of the area immediately north of the Fiji Platform is controlled by the active left-lateral north Fiji fracture zone. The zone extends from the northern tip of the Tonga Trench to the central part of the NFB and marks the present-day boundary between the Pacific and Australian plates. The fracture zone is mainly composed of E-W structural trends with some pull-apart basins (Louat and Pelletier, 1989; Pelletier and Louat, 1989; Johnson and Sinton, 1990; Hughes Clarke *et al.*, 1993; Jarvis *et al.* 1994). The fracture zone has been partly imaged by SeaMARC (Kroenke *et al.*, 1991) and GLORIA during the SOPAC cruise in 1990 (Hughes Clarke *et al.*, 1993; Jarvis *et al.*, 1994). A second pull-apart basin trending N45° was found at 15°30′S, 178°40′E. The ridges located in the northeastern basin north of the Fiji Platform, like the Braemar Ridge and the Balmoral Ridge and Balmoral Reef farther west, are interpreted by Jarvis *et al.* (1994) to be pieces of the Fiji Platform rifted away by successive spreading segments (pull-apart basins) during changes in the location of the north Fiji fracture zone.

2. MAGNETISM AND PALEOMAGNETISM

The first published data concerning the magnetism (Sclater and Menard, 1967; Chase, 1971; Luyendyck *et al.*, 1974; Halunen, 1979) emphasized the complexity of the basin. These authors developed various interpretations of the magnetic lineations with different locations of spreading centers being proposed in the NFB south of the Hazel Holme Ridge. The area located to the north of the Hazel Holme Ridge was considered to be part of the old Pacific or Australian plates.

Paleomagnetic results from Vanuatu and Viti Levu indicate a clockwise rotation of the New Hebrides arc of about 28° since 6 Ma (Falvey, 1978) and a counterclockwise rotation of 21° since 4 Ma (James and Falvey, 1978) and of 90° since 7 Ma (Malahoff *et al.*, 1982b) of the Fiji Platform. These rotations and the polarity reversal from the Vitiaz to the New Hebrides subduction since 10–8 Ma are the basis for the hypothesis of a NW-SE trending spreading center northwest of Fiji and an E-W spreading axis north of Fiji (Falvey, 1978; Malahoff *et al.*, 1982a,b).

An aeromagnetic survey carried out by the National Oceanic and Atmospheric Administration (NOAA) and the U.S. Naval Research Project in 1979 (Cherkis, 1980; Larue *et al.*, 1982; Malahoff *et al.*, 1982a, 1994) defined magnetic lineations in the NFB which have been differently interpreted by different authors. After reprocessing these aeromagnetic data, Auzende *et al.* (1988b) proposed a new interpretation of the magnetic lineations. In the central basin a very clear N-S axial lineation is bounded by two parallel lineations identified as the J (Jaramillo) (1 Ma) and 2 (1.9 Ma) anomalies (Cande and Kent, 1992). On the eastern limb of the axis, a possible 2A (2.6–3.5 Ma) anomaly can be distinguished but is not continuously identified on the western limb. In the western basin the magnetic pattern is dominated by NW-SE lineations crosscut by transverse N45° trending features interpreted as transform faults. The NW-SE lineations represent the trends of the initial stage of opening of the NFB between 10 and 3 Ma (Auzende *et al.*, 1988b) or 12? to 7 Ma (Pelletier *et al.*, 1993a). The magnetic pattern of the eastern basin is less well defined and shows indistinctly NW-SE trends mixed with N-S directions which are particularly dominant immediately west of the Fiji Islands. The northern and northeastern parts of the NFB are mainly characterized on the aeromagnetic survey by roughly E-W trending lineations that have not been dated.

In the northwestern basin recent data acquired during the EVA 14 and Santa Cruz cruises permit precise definition of the magnetic pattern (Pelletier *et al.*, 1988, 1993a,b). The magnetic map (Fig. 4.12) mainly shows two trends of lineations. NW-SE trending lineations disrupted by NE-SW transform faults occur both in the northern edge of the area along

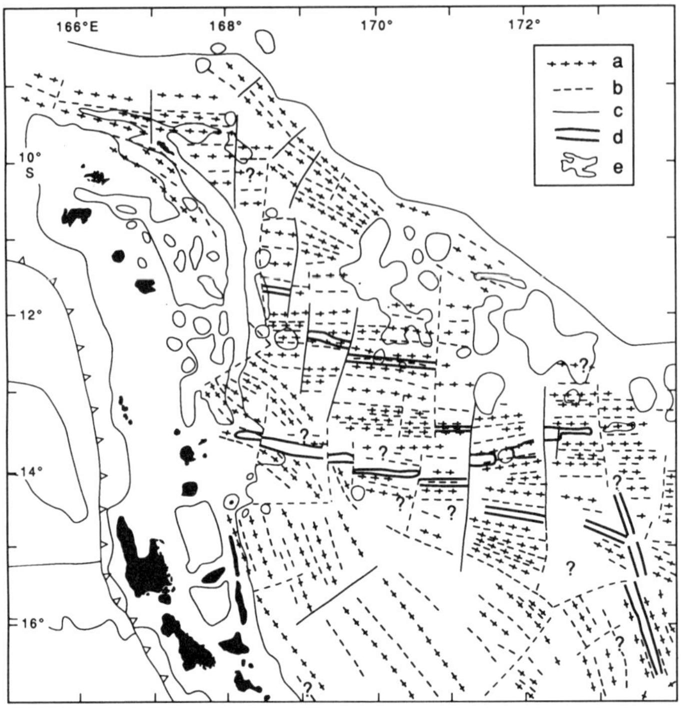

FIGURE 4.12. Map of magnetic anomaly lineations in the northwestern NFB (after Pelletier *et al.*, 1993b). (a) positive magnetic anomaly lineation; (b) negative magnetic anomaly lineation; (c) fault zone; (d) contour of the axial part of the spreading ridges; (e) contour of the main morphostructural features.

the Vitiaz Trench up to 8°30′S and in the southwestern part of the area along the New Hebrides arc, north and south of the western end of the Hazel Holme Ridge. Some NW-SE lineations are also present in the western side of the basin, north of 11°S from the eastern edge of the New Hebrides arc platform to the Duff Islands. The NW-SE lineations identified as anomalies 5A?–5 to 4 (12?–11 to 7 Ma) resulted from the initial opening of the basin in a NE-SW direction. E-W trending magnetic lineations disrupted by N-S transform faults are distributed over the entire central part of the northwestern basin from 9°S to 15°S and are flanked by the NW-SE lineations. The E-W trending lineations identified as anomalies 3A to 2A (7 to 2.5 Ma) resulted from a second stage of opening in a N-S direction and are associated with the South Pandora, Tikopia, and 9°30′S ridges. The latter is interpreted as a spreading center. At 168°30′E, the E-W lineations abut the N-S trending West Tikopia Ridge, which is interpreted as a major transform fault that offsets northward the spreading center from the Tikopia Ridge to the 9°30′ Ridge. The Hazel Holme Ridge is interpreted as an active and very young extensional zone or slow spreading ridge which crosscuts older oceanic crust and connects the N160° trending segment of the central spreading ridge with the southern tip of the northern New Hebrides backarc troughs. E-W trending magnetic lineations seem to be associated with the Hazel Holme Ridge but are not datable.

During the SEAPSO and STARMER French–Japanese joint projects, a detailed magnetic survey (Fig. 4.3) of the central part of the NFB was conducted. It allows us to refine identification of the magnetic anomalies and calculation of the spreading rate on the different segments during the past 3 m.y. (Auzende *et al.*, 1990c; Huchon *et al.*, 1994). The important features are the following.

1. An axial anomaly is identified all along the structural ridge axis. The width of this anomaly varies, giving spreading rates ranging from 50 to 82 mm/yr.

2. The J (Jaramillo) event is only clearly represented on the N-S segment. Its existence along the N15° and the southern part of the N160° segments is debatable. Based on the interpretation of the axial and J (Jaramillo) anomalies, the estimated spreading rate of the N160° segment has varied between 40 and 50 mm/yr, which are rates representative of intermediate rate spreading ridges. Along the entire N160° segment the present-day ridge axis is characterized by a morphology usually encountered at slow spreading ridges (Macdonald *et al.*, 1984). It is made up of a succession of deep grabens reaching more than 4000 m deep and bounded by steep 1000-m-high walls.

3. The anomalies 2 and 2A are clearly identified on both sides of the ridge axis, at least along the N-S and N15° segments. Along the southernmost N-S segment only anomaly 2 has been identified.

In the eastern part of the NFB, west of the Fiji Islands, the identification of magnetic anomalies (Fig. 4.8) also favors the existence of an active spreading center (Auzende *et al.*, 1994b). In the western part of the survey, corresponding to the propagating rift system, a well-defined anomaly is interpreted as anomaly 1. The width of this anomaly decreases from north to south. On both sides of the area, a magnetic lineation could be anomaly J (Jaramillo—1 Ma). The spreading rates calculated from these anomaly identifications are close to 40 mm/yr. The magnetic pattern exactly superimposes the structure. Between both axes, a large transverse E-W trending negative anomaly can be interpreted as the transform zone between the propagating rift and the southern rift (Hey *et al.*, 1986). This anomaly has no morphological expression. From the calculated spreading rate and the angles between the pseudofaults, the southward propagation velocity can be calculated. For an angle of 55° to 40°, as is observed north of 17°35′S, the propagation velocity is respectively 25 or 45 mm/yr. The tip of the propagator is characterized by a 20° angle with the pseudofault, implying an increase of the propagation velocity up to 100 mm/yr in the present phase. The age of this propagation could be related to the emplacement or the reactivation of the north Fiji fracture zone 1 to 1.5 m.y. ago (Lafoy *et al.*, 1990), resulting in the formation of the 16°50′S triple junction.

3. SEISMICITY

Gutenberg and Richter (1954) described isolated seismicity between the Tonga and New Hebrides arc major seismic zones. Sykes (1966) later succeeded in identifying a NE-SW trending seismic zone running from the southern tip of the New Hebrides arc to the center of the NFB. On the basis of seismic evidence, Sykes *et al.* (1969) suggested that the two triangular areas between the Tonga and New Hebrides arcs should be described as two different basins, the Lau and North Fiji basins. Chase (1971) extended this analysis and proposed a model including five microplates for the same area. The deep structure of the Lau and North Fiji basins was studied by Dubois (1971), Aggarwal *et al.* (1972), and Dubois *et al.* (1973). They showed the existence of a low-seismic-velocity zone in the upper mantle. Barazangi *et al.* (1974) demonstrated that this low-velocity zone was related to seismic-wave attenuation due to active spreading.

Crustal deformation in the NFB was addressed by Eguchi (1984), Hamburger and Isacks (1988), and Louat and Pelletier (1989), using a large number of shallow earthquake data and focal mechanism solutions from international catalogues. Despite discrepancies in their interpretations concerning the sense of active crustal motion, they provide a description of the distribution of shallow seismicity in the NFB and the major seismic lineaments (see following). Hamburger and Isacks (1988) claimed that there are no steady-state spreading centers in the NFB but only diffuse extensional zones in a wide strike-slip boundary between the Pacific and Australian major plates. In contrast, Louat and Pelletier (1989) tried to quantify the crustal motions along discrete spreading centers or transform faults, using the seismological and marine geological data.

Figure 4.13 shows the distribution of the shallow seismicity in the NFB (Louat and Pelletier, 1989). Earthquakes concentrate within certain areas delineating linear zones. Some of the seismic lines are associated with known structural features.

1. The north Fiji fracture zone (NFFZ) is clearly marked by a WSW-ENE trending seismic belt between 16° and 17°S from the central triple junction to the north of the Fiji Platform. Numerous focal mechanism solutions attest to E-W left-lateral strike-slip motion

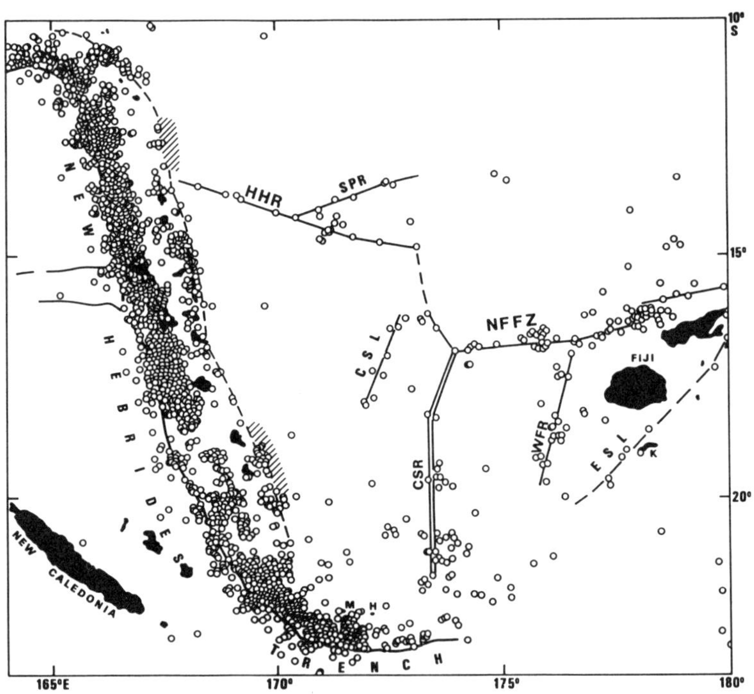

FIGURE 4.13. Shallow seismicity (0–70 km) in the NFB (after Louat and Pelletier, 1989). CSR: central spreading ridge; NFFZ: north Fiji fracture zone; WFR: West Fiji Ridge; HHR: Hazel Holme Ridge; SPR: South Pandora Ridge; CSL: central seismic lineament; ESL: eastern seismic lineament; K: Kandavu Island. See Louat and Pelletier (1989) for the data source.

along the NFFZ. However, events with normal fault solutions have occurred east of the 177°25′E pull-apart basin at 178°E and at the western end of the north Fiji fracture zone.

2. The West Fiji Ridge (WFR) coincides with a NNE-SSW trending seismic line located near 176°E, immediately west of the Fiji Platform between 17° and 20°S. Events with strike-slip or normal fault solutions occur along the west Fiji Ridge.

3. A WNW-ESE seismic belt correlates with the Hazel Holme Ridge between the New Hebrides backarc area and the northern tip of the N160°-trending segment of the central spreading ridge. Focal mechanism solutions indicate strike-slip faulting.

In contrast, the central spreading ridge, which is the major active spreading center of the basin, is almost unidentifiable with shallow seismicity. However, a few events cluster at particular areas: around 21°S at the tip of the southward propagating rift and at the offset of the axis, at 19°40′S where transverse volcanic alignments exist, around 18°10′S at the tip of the northward propagating rift, and between 16° and 17°S along the N160° trending segment. Events with strike-slip faulting solutions occur all along the central spreading ridge, but are especially concentrated at the 21°S ridge axis offset. However, one event with normal fault solution is located on the N160° segment.

Two seismic lineaments are observed in the basin where the bathymetry largely remains unknown (Louat and Pelletier, 1989): a NNE-SSW line (CSL) west of the central spreading ridge and characterized by pure normal faulting solutions, and a NE-SW line (ESL) between the Fiji Platform and Kandavu Island. Moreover, diffuse areas of seismicity are located in the southernmost part of the NFB and in the northeastern basin north of the NFFZ. All these seismic features indicate active crustal deformation which is still not understood.

4. HEAT FLOW DATA

Few heat flow measurements exist for the NFB. They all indicate high heat flow values (Sclater and Menard, 1967; Macdonald *et al.*, 1973; Halunen, 1979) over the entire basin, which is consistent with it being interpreted as a recent oceanic basin. At a smaller scale, two different provinces are distinguished. North of the Hazel Holme Ridge, the average heat flow value is 2.29 HFU; however, Halunen (1979) suggests that the heat flow is very low (0.88 HFU) if the New Hebrides arc volcanism effect is subtracted. These low values can be related to hydrothermal circulation or to the older age of this part of the NFB. South of the Hazel Holme Ridge, the values are higher with an average of 4 HFU. The relatively high heat flow values have been interpreted as indicative of the abnormally shallow depth of the NFB, called for this reason the North Fiji Plateau. In fact, the depths measured in the NFB are "normal" in terms of vertical evolution related to the age of the oceanic crust. Except in some peculiar areas such as the 16°50′S triple junction, the ridge crest depth varies from 2500 to 2800 m, which is the standard for mid-oceanic ridges (Sclater and Francheteau, 1970).

During the STARMER project a few heat flow measurements have been performed within the active rift valley by the submersibles *Nautile* and *Shinkai 6500*, close to the 16°50′S triple junction and in the area of the White Lady active hydrothermal site (see following). They all give low values (less than 1 HFU; Joshima, pers. comm.) that are probably related to water circulation.

5. ROCK GEOCHEMISTRY

The compilation of the geochemical data collected along the North Fiji Basin spreading system between latitude 13°S and 22°S during the STARMER project (Eissen *et al.*, 1994; Lagabrielle *et al.*, 1994; Nohara *et al.*, 1994), in addition to all published data (Sinton *et al.*, 1985; Aggrey *et al.*, 1988; Auzende *et al.*, 1990a; Boespflug, 1990; Eissen *et al.*, 1990, 1991; Johnson and Sinton, 1990; Price *et al.*, 1990; Monjaret *et al.*, 1991; Sinton *et al.*, 1994), shows systematic geochemical variation along the different spreading segments. Crystal fractionation—controlled almost exclusively by olivine for the less fractionated lavas, or by plagioclase + olivine ± clinopyroxene for the most fractionated lavas—can explain part of the observed major element geochemical variations. However, the large ion lithophile elements (LILE), high field strength elements (HFSE), and rare earth elements (REE) contents of the basalts of the NFB as well as their isotopic compositions are strongly heterogeneous. Mixing of mantle sources must play a major role in order to explain the observed LILE, HFSE, REE, and isotopic variations. Thus, three different mantle sources have been identified in the NFB (Eissen *et al.*, 1994):

1. A typical depleted N-MORB source, producing basalts with flat patterns on N-MORB-normalized extended element variation diagrams (Fig. 4.14d), $1 < La/Nb < 2$ (Fig. 5.15; N-MORB field of Gill, 1981), and Sr- and Nd-isotopic ratios close to the Pacific MORB field (Fig. 4.14i).

2. An E-MORB- or OIB-related source giving basalts with a strong enrichment in HFSE, LILE, and LREE, on N-MORB-normalized extended element variation diagrams (Fig. 4.14a), $La/Nb < 1$ (Fig. 4.15; E-MORB field of Gill, 1981), and Sr- and Nd-isotopic ratios trending toward the intra-plate field (e.g. Samoan field in Fig. 4.14f).

3. A subduction-modified (or -related) mantle source, with basalts having generally a higher volatile content (Aggrey *et al.*, 1988), and a clear enrichment in HFSE, LILE, and LREE, on N-MORB-normalized extended element variation diagrams with a negative Nb anomaly indicating some subduction-related contamination (Fig. 4.14d), $La/Nb > 2$ (Fig. 4.14; orogenic field of Gill, 1981), and Sr- and Nd-isotopic ratios ranging from the Pacific MORB field to the New Hebrides arc field (Fig. 4.14h). These basalts derived from this source might also have been previously called BABB (Saunders and Tarney, 1984; Hawkins and Melchior, 1985; Sinton and Fryer, 1987).

The observed variations, reported as a function of their respective spreading segments, are as follows. Along the N-S segment, which also represents morphologically the most regular spreading ridge of the NFB, the magma source produces only basalts with N-MORB characteristics (Figs. 4.14d, 4.15). However, their isotopic compositions extend slightly over a range wider than the Pacific MORB field, with one trend toward the Samoan field and a weak trend toward the New Hebrides field (Fig. 4.14i). Along the N15° segment the three sources coexist, giving a wide range of geochemical signatures for the basalts collected (Fig. 4.15). Their mixing results in a large number of transitional compositions giving patterns of E-MORB types (Fig. 4.14c). Their isotopic compositions are essentially similar to the more Sr radiogenic of the Pacific MORB with two additional groups having, respectively, an OIB- and a subduction-related trend (Fig. 4.14h). The source mixing transitional toward OIB (hot-spot-related) increases northward from 18°20′S to 12°S, as the Rotuma–Samoan hot-spot lineament is approached. Thus, along the N160° segment, the three sources still coexist, but the influence of the OIB source increases, whereas the influence of the subduction-related source decreases (Figs. 4.14b and 4.15). It is particularly

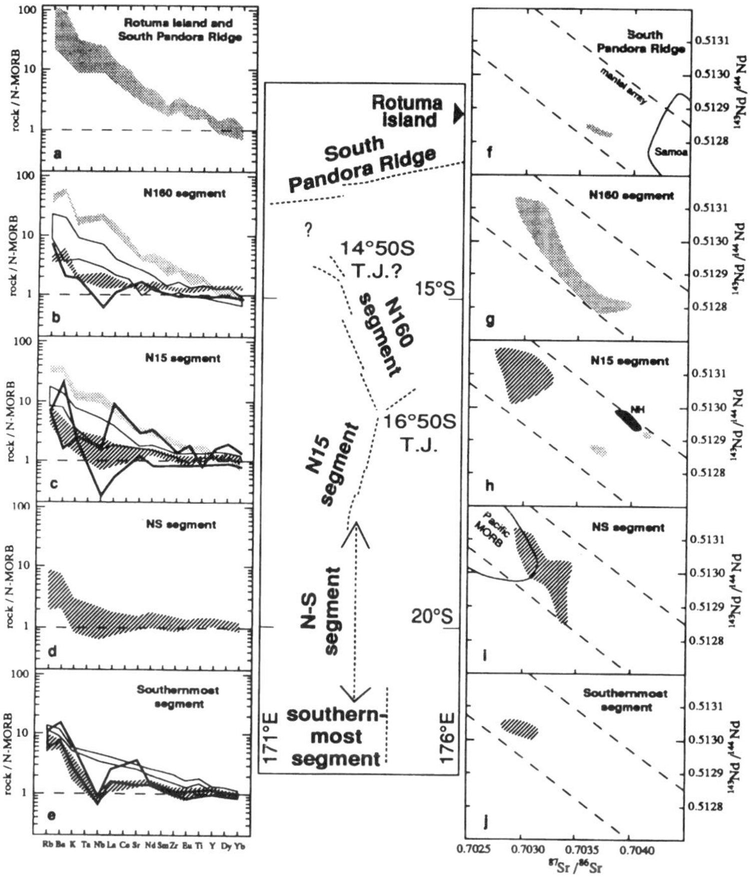

FIGURE 4.14. Incompatible elements and isotopic variations of the North Fiji Basin basalts as a function of a schematic spreading segmentation (central sketch); (a–e) N-MORB-normalized extended element variation diagrams (normalization values after Sun and McDonough, 1989); (f–j) variations of $^{87}Sr/^{86}Sr$ with $^{143}Nd/^{144}Nd$ (Pacific MORB field after Boespflug, 1990; Samoan field after Wright and White, 1987; NH: New Hebrides 1 field after Briqueu *et al.*, 1994). Light gray field: OIB-related source; dark gray field: N-MORB source; thin lines field: transitional toward OIB (or E-MORB) source; heavy lines field (with negative Nb anomaly): subduction related source.

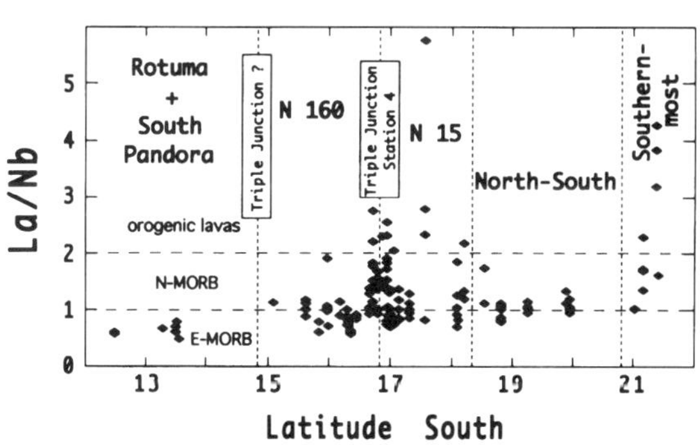

FIGURE 4.15. La/Nb along-strike variations of the North Fiji Basin basalts as a function of the latitude of the sampling site.

well marked on the isotopic composition diagram (Fig. 4.14g) where all the compositions spread between the Pacific MORB and the Samoan fields. Along the Pandora–Rotuma Ridge, only the OIB-derived lava type is present, the eventual contribution from other sources being completely diluted (Figs. 4.14a, 4.14f, 4.15). Mixing calculations based on incompatible elements involving OIB and N-MORB sources show that the OIB source contribution is between 50% and 80% for the Rotuma–Pandora Ridge lavas and is lower than 25% for the N160° and N15° segments (Eissen *et al.*, 1994). On the southernmost segment (N174°E), N-MORB are present, but the dying subduction along the Hunter Ridge still influences several basalt compositions, which show a significant subduction-related contamination with negative Nb anomalies and high volatiles content. A weak E-MORB source contribution is also present in several others and may be related to subducted-OIB seamounts from the South Fiji Basin (Fig. 4.14e, 4.14j). However, no significant isotopic variations are observed along this segment (Fig. 4.15).

Thus, the geochemistry of the NFB magma sources is directly influenced by the surrounding active or dead subduction zones and the regional OIB (or hot-spot)-related sources. The magma genesis is dominantly controlled by magma production along the spreading system, resulting in the dominance of N-MORBs. However, this source mixes to a various extent with the two sources mentioned previously. In the entire northern NFB many basalts result from the mixing of an N-MORB and a OIB source, similar to transitional alkalic lavas from oceanic intraplate magmatism. This source is related either to the Samoan lineament (Wright and White, 1987) or to the Fiji Platform (Gill, 1984) or even to the Wallis lineament (Price *et al.*, 1991). Similarly, if the subduction-related contamination along the 174°E segment is directly linked to the dying subduction at the southern end of the New Hebrides arc, the presence of perceptible subduction-related geochemical characteristics in basalts from the central NFB, 500 km away from the active subduction zone, is more questionable. This influence results in fact from the partial melting of an upper mantle source that was affected by subduction contamination during the clockwise rotation of the New Hebrides arc leading to the opening of the NFB during the past 10 m.y. (Malahoff *et al.*, 1982a; Auzende *et al.*, 1988a) and is no longer directly linked to the presently active subduction.

6. SULFIDE DEPOSITS

Since the beginning of the SEAPSO and STARMER projects, two main hydrothermal sulfide deposit fields have been explored by *Nautile* in 1989 (Auzende *et al.*, 1991b) and *Shinkai 6500* in 1991 (Auzende *et al.*, 1992a) in the vicinity of the 16°50′S triple junction (Lafoy *et al.*, 1987, 1990). The first (named "White Lady") at 16°59′S and the second (named "Père Lachaise") at 16°58′S were both discovered during the *Nautile* dives in 1989.

The main structural features, which characterize the 2-km-wide, 100–150-m-deep rift valley of the NFB in the White Lady area, are steep fault scarps associated with open fissures. Fissures a few meters long and a few centimeters to 1-m wide are commonly observed on the seafloor and on the terraces adjacent to the fault scarps. The fault trend is parallel to the main trend of the central spreading ridge, N10° to N20° (Fig. 4.16). The White Lady chimney (Figs. 4.16 and 4.17) is located on the top of a 6–7-m-high, 50-m-diameter mound composed of sulfides and oxides. The chimney bifurcates and is

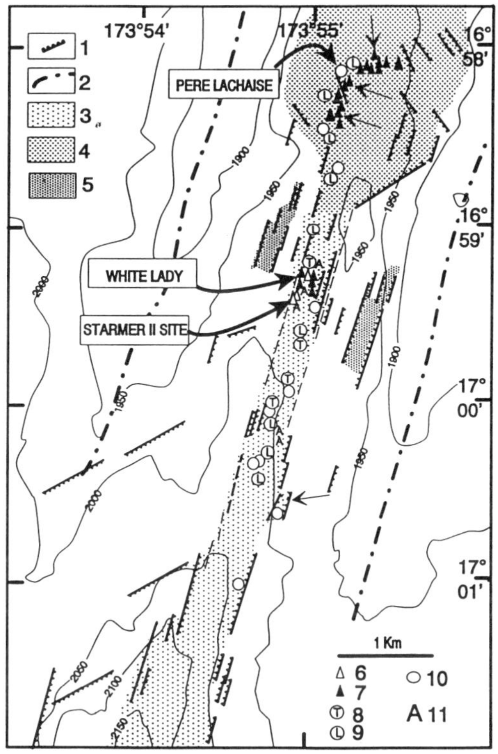

FIGURE 4.16. Detailed structural map of the axial graben close to 17°S. Location of the active hydrothermal sites (White Lady and STARMER) and major sulfides deposit (Père Lachaise) (after Bendel *et al.*, 1993). 1: normal faults; 2: crest; 3: main axial graben; 4: intersection zone; 5: secondary graben; 6: active hydrothermal site; 7: fossil chimneys; 8: temperature anomalies; 9: lava lakes; 10: hydrothermal oxydes staining; 11: mussels and gastropods colonies. The small arrows indicate the sampling areas out of the White Lady site.

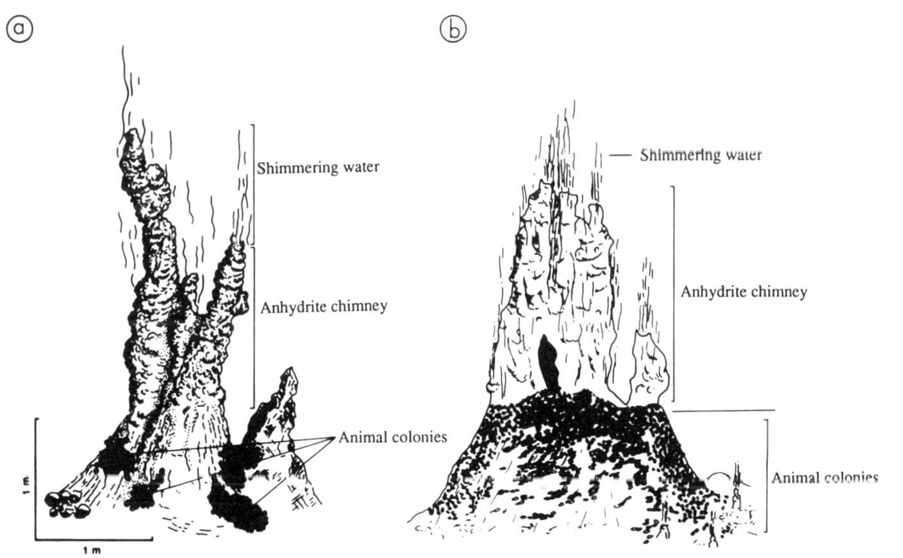

FIGURE 4.17. Artistic view of the White Lady evolution from (a) 1989 and (b) 1991 (after Auzende *et al.*, 1992a).

composed exclusively of anhydrite. It expels shimmering 285°C water, which is poor in particulates. At the foot of the main chimney, six secondary vents expelling the same transparent water are colonized by gastropods, mussels, crabs, and cirripeds. In 1991, six *Shinkai 6500* dives were devoted to the study of the White Lady and of the surrounding active hydrothermal sites. The present-day condition of the White Lady can be summarized as follows: the anhydrite chimney currently comprises a massive 2-m-high, 2-m-diameter main conduit and a 1.5-m-high, 10-cm-diameter secondary chimney. These vents expel the same shimmering water, but the flux is about twice that of two years ago. The measured temperature is 265°C, 20°C less than the measured temperature in 1989. Finally, significant changes (Fig. 4.17) have been observed for the colonies of living fauna. They are in all cases more numerous, and their territories are expanding toward the top of the mound. The sulfates, oxides, and sulfides have been studied in great detail by Bendel (1993) and Bendel *et al.* (1993), who deduced from geological observations and mineralogical study four stages in the evolution of the White Lady site. The sulfide spires are formed in the first stage by mixing of hydrothermal fluid with seawater. The sulfide mound and the maturation of sulfides reflect the second stage. Copper-rich massive sulfides are precipitated at the core of the mound. During the third stage the sulfides are tectonically brecciated, and the resulting fissures are filled with amorphous silica, talc, and barite. The fourth stage corresponds to anhydrite precipitation related to the present-day activity at the top of the sulfide mound.

The Père Lachaise site is characterized by the disappearance of the axial graben and the increasing width of the domain. This area, close to the 16°50′S triple junction, exhibits three sets of normal fault and fissures trending N15°, N140–150°, and N60°. The explored fossil hydrothermal field consists of several tens of individual spires growing directly on basalt scattered over a 2-km × 2-km surface. The deposits sampled from the Père Lachaise site indicate the same compositions as are observed on the White Lady mound. The absence of a massive mound in this area is interpreted as the result of the conjunction of the three directions of faulting, which creates a zone of high porosity where hydrothermal discharge is dispersed over a large area (Bendel *et al.*, 1993).

7. WATER SAMPLING

The first significant water samples on the ridge axis were taken in 1985 during the SEAPSO III cruise (Auzende *et al.*, 1988a) and the Papatua cruise in 1986 (Craig, 1986). Other water samplings have been conducted on the northern part of the NFB ridge during the Moana Wave cruise in 1987. They indicate hydrothermal activity around the South Pandora Ridge (Kroenke *et al.*, 1987).

During the SEAPSO III cruise, seven hydrocasts were taken on the axis between 16°S and 20°S. Four of these indicate significant methane and manganese anomalies close to the seafloor; two hydrocasts are located at the junction of the N-S and N15° segments and the other two are on the N160° segment close to the 16°50′S triple junction. These last hydrocasts show anomalies 10 times higher than background level for the deep Pacific water. During the STARMER project, hydrocasts were taken about every 20 miles all along the ridge axis. They demonstrate the quite constant distribution of the hydrothermal activity along the spreading ridge axis.

In situ sampling was conducted during the STARMER *Nautile* cruise (1989) and the

Shinkai 6500 cruise (1991) on the White Lady active site at 17°S in the triple junction area (Auzende *et al.*, 1991b). The main results can be summarized as follows (Grimaud *et al.*, 1991; Ishibashi *et al.*, 1994): the samples were collected with conventional titanium syringes by *Nautile* or with a pumping system by *Shinkai 6500*. In total, 16 samples were taken on the White Lady site. Their compositions range from a few percent hydrothermal water to nearly pure 285°C temperature fluids. The elemental concentrations show linear trends on element versus Mg diagrams. All the element concentrations of the parent end member are significantly lower than all other hydrothermal systems. Chloride (255 mmol/kg) and sodium (210 mmol/kg) have especially low concentrations compared with fluids collected on the East Pacific Rise (EPR), Juan de Fuca Ridge, Galapagos Ridge (von Damm and Bischoff, 1987) and Mid-Atlantic Ridge (Campbell *et al.*, 1988). The only similar fluids have been sampled in the Ashes vent field on the Juan de Fuca Ridge (Massoth *et al.*, 1989; Butterfield *et al.*, 1990). In spite of these low concentrations the characteristic elemental ratios K/Na and Ca/Sr are close to the values usually found in other hydrothermal systems (von Damm and Bischoff, 1987).

8. DISCUSSION AND CONCLUSION

The evolution of the NFB, located between the opposed New Hebrides and Tonga subduction zones, results from variation in the relative motions of the Australian and Pacific plates. At a large scale the NFB can be interpreted as a deformation zone at the boundary of the two plates. Depending on the geometry of the boundary, the shortening between the plates will be accommodated by compressive stresses such as along New Hebrides and Tonga trenches or by strike-slip motion such as in the northern part of the basin. The geometry of the ridge axis, its segmentation, and its changes of direction are controlled by this relative motion.

8.1. Hectokilometric, Decakilometric, and Kilometric Ridge Segmentation

The precise mapping of the central spreading ridge illustrates different scales of segmentation. At a large scale, five main segments can be considered. Their length varies from about 140 km for the southernmost N-S segment to more than 280 km for the N-S segment. Although the spreading rate is intermediate all along the central spreading ridge, ranging from 50 mm/yr on the N160° and N15° segments to 80 mm/yr on the N-S segment, the segments show very different morphologies. The N-S and N15° segments' morphology is typical of EPR fast-spreading-type morphology (Fig. 4.4) with a central dome 8 km wide cut in its axial part by a graben 50 m to several hundred meters wide and 50 m deep. In contrast, the N160° segment is characterized by an 8-km-wide, 1000-m-deep graben occupied in its axial part by a neovolcanic zone (Fig. 4.4). This morphology is typical of slow-spreading MAR-type morphology. The conclusion drawn from these observations is that there is not a unique correlation between the ridge morphology and the spreading rate. It is also necessary to take into account the magmatic budget of each area and its geodynamic context. The drastic change of morphology north and south of the 16°50′S triple junction suggests that the triple junction area constitutes a magmatic supply boundary between the N15° segment, characterized by significant volcanic construction, and the amagmatic extension of the N160° segment.

Within each major segment, segmentation at the scale of 10 s of kilometers is observed. These subsegments are limited by offsets, OSCs, and propagating rifts as described on the EPR (Macdonald, 1983) and as illustrated in Fig. 4.3. The average lengths of these intermediate subsegments varies from 40 km on the N-S segment to more than 60 km on the N160° segment.

Segmentation at a scale of a kilometer has also been documented on the N-S segment around 19°S by the *in situ* exploration with *Nautile* and *Shinkai 6500* (Ondréas *et al.*, 1993; Gracia-Mont *et al.*, 1994). The active accretion is located in tectonic grabens a kilometer long and tens of meters wide separated by kilometer-wide magmatic saddles between the grabens. The active hydrothermal sites are located in the grabens, whereas very fresh magmatic edifices are observed on the saddles.

8.2. Unusual Tectonic Features

One of the characteristics of spreading in the NFB is the existence all along the ridge axis of unusual tectonic features such as propagating rifts, OSC, offsets, and others, representative of the complex geodynamic environment of the basin. At a global scale the NFB can be considered as a megashear zone at the boundary between the Australian and Pacific plates (Hamburger and Isacks, 1988). The effect of these permanent shear stresses applied on the whole basin is the deformation of the spreading ridge. This deformation is accommodated by changes of morphology and migration of the ridge axis with initiation of triple junctions (examples of 16°50′S and 14°50′S triple junctions) and ridge propagation (northward propagation of the N-S segment at 18°10′S, southward propagation of the N-S segment at 20°30′S) and activation of major transform faults (north Fiji fracture zone, central fracture zone, and south Fiji fracture zone) (Fig. 4.18).

Also, probably as a result of the geodynamic environment, the accretion south of the north Fiji fracture zone is distributed on two parallel active spreading ridges, the central spreading ridge and the West Fiji Ridge, propagating at a large scale, respectively to the north and to the south (Auzende *et al.*, 1993, 1994b). Since at least 1.5 to 1 Ma, this twin-spreading-ridge system isolates an intermediate West Fiji microplate between the central spreading ridge and the West Fiji Ridge (Fig. 4.18). The same phenomenon is observed for the Easter (Francheteau *et al.*, 1987; Larson *et al.*, 1992) and Juan Fernandez microplates. The spreading rates on both ridges vary, increasing from south to north on the West Fiji Ridge and from north to south on the central spreading ridge. The sum of spreading rates is close to 100 mm/yr all along the N-S oceanic crust created during the last 1 to 1.5 m.y.

8.3. Petrology and Geochemistry

The compilation of all the geochemical data available for the basalts of the NFB shows that low-pressure crystal fractionation controls the major element geochemistry, and mantle source heterogeneities and their mixing dominantly control the LILE, HFSE, and REE patterns and isotopic compositions of these basalts. The dominant magma of this backarc basin is an N-MORB type, producing basalts very similar to those emplaced along the most classical intermediate-rate mid-oceanic spreading centers. But other geochemical signatures, representative of the local upper mantle geochemistry, are observed: (1) an OIB source, similar to transitional alkalic lavas from oceanic intraplate magmatism; its influence increases northward from 18°20′S to 12°S, this lava type being the only one present

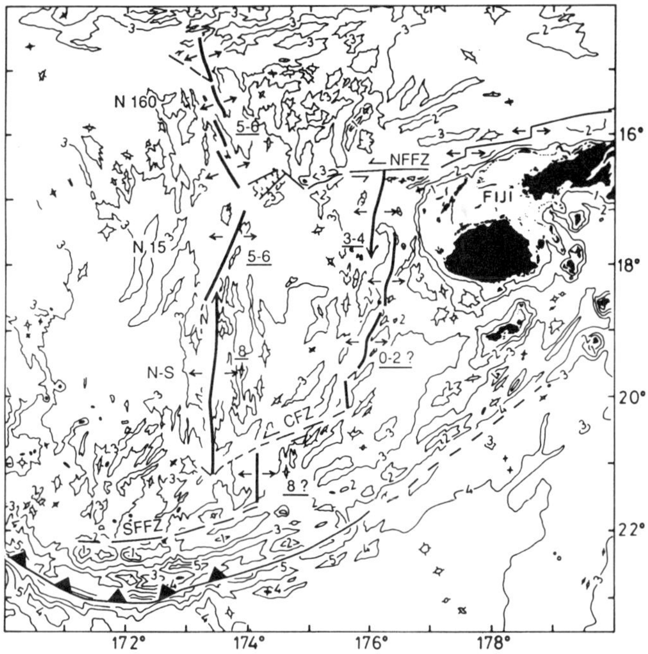

FIGURE 4.18. Kinematic sketch of the twin ridges (central spreading ridge and the West Fiji Ridge) with simplified bathymetry. NFFZ: north Fiji fracture zone; CFZ: central Fiji fracture zone; SFFZ: south Fiji fracture zone; N-S, N15 and N160: the different segments of the main central spreading ridge. *5–6*: spreading rate calculated from magnetic data. *0–2?*: inferred spreading rate in cm/yr. The arrows at the tip of ridge segments indicate the direction of propagation (after Auzende *et al.*, 1994b).

along the Rotuma–Pandora Ridge; and (2) a source that was affected by subduction contamination from the New Hebrides subduction, with a clear enrichment in incompatible elements, a negative Nb anomaly, and a high loss on ignition (LOI). This contamination with variable arc affinity, which affected the entire NFB upper mantle, was left behind after the clockwise rotation of this arc, as a result of the opening of the NFB. This subduction-related material was then recycled during partial melting of the NFB upper mantle beneath the central spreading axis.

8.4. Hydrothermal Activity: Water Chemistry and Sulfides

The explored hydrothermal active sites of the NFB central spreading ridge are probably only representative of a few percent of the hydrothermal activity. The hydrocasts taken all along the ridge axis suggest a widespread activity with the indication in some places of a very strong but unstable activity. For example, the 19°S "megaplume" (Nojiri *et al.*, 1989) discovered during the *Kaiyo* cruise in 1987 was extinct two years later when attempts were made to sample it again.

One of the main characteristics of the NFB ridge axis hydrothermal waters sampled by *Nautile* and *Shinkai 6500* on the White Lady site is their low salinity, which suggests that fluids have undergone phase separation (Grimaud *et al.*, 1991; Ishibashi *et al.*, 1994). Such a phase separation would result in the formation of one vaporlike phase with low salinity and one liquid-like phase with high salinity (brine). Dissolved species will be depleted in the vapor, but phase separation will not significantly modify the elemental ratios as observed in the White Lady waters. To conclude, the NFB hydrothermal waters could result from three-component mixing of normal hydrothermal water, low-salinity fluid from condensed vapor, and brine.

Concerning the sulfides, the analysis of the White Lady site deposits succession implies three different hydrothermal episodes (Bendel, 1993; Bendel *et al.*, 1993): (1) The first episode is characterized by deposition of Cu, Fe, and Zn-rich sulfides and implies that a hot (more than 300°C) water was being expelled. This episode is close to the mid-oceanic ridge type. (2) The second hydrothermal episode produced opal, barite, and talc precipitates. It certainly occurs after a tectonic event which was responsible for the sulfides mound brecciation. This tectonic event caused the fluids to become more oxidized with temperatures ranging from 150°C to 300°C, depending on subsurface mixing. (3) The third hydrothermal episode only produced anhydrite and minor opal and talc. This deposit was associated with low elemental abundances in the expelled water and could be, as suggested by Grimaud *et al.* (1991) and Ishibashi *et al.* (1994), the result of a phase separation after boiling near the surface. The succession of these three hydrothermal phases results in a great instability of the hydrothermal system over a short period of time (less than 1 m.y.). This instability could be related to the tectonic events affecting the area, especially the intense tectonic fracturing associated with the 16°50′S triple junction.

8.5. Evolution of the Basin

The recent data acquired on the northern and the northwestern parts of the NFB (especially those of Pelletier *et al.*, 1988, 1993a,b) allow construction of a previously proposed geodynamical evolution model (Auzende *et al.*, 1988b) to be completed. The evolution of the NFB is illustrated by the six cartoons of Fig. 4.19.

1. At 12 Ma the Vitiaz–New Hebrides–Fiji–Lau–Tonga arc split and NFB rifting started with the change of subduction polarity from the Vitiaz to the New Hebrides system. The main part of the initial arc remained on the southern side of the spreading axis.
2. Spreading along a NW-SE axis was synchronous with the clockwise rotation of the New Hebrides arc and the anticlockwise rotation of the Fiji Platform. The initial axis located north of the Fiji Platform jumped southward between the New Hebrides and Fiji. The crust created during this stage is mainly located in the western basin.
3. At 7 Ma the NW-SE spreading axis stopped and was replaced by an E-W trending spreading center from the northwestern tip of the basin to the north of the Fiji Platform. This center corresponds to the 9°30′S, Tikopia, and proto–south Pandora ridges. This opening induced a new subduction zone along the southern limit of the NFB.
4. A triple junction was active around 3 Ma (anomaly 2A) between the former E-W axis and the newly created N-S-trending spreading center. At this stage there was a major change in the stress direction from N-S to E-W. This period was synchronous with the beginning of spreading in the Lau Basin.
5. Around 1.5 Ma the opening along the E-W axis was reorganized by the development of the north Fiji fracture zone along the Fiji Platform up to the N-S spreading axis, creating the 16°50′S triple junction. The left-lateral strike-slip motion along the north Fiji fracture zone induced the change of the direction of the N-S axis at the 16°50′S triple junction. The West Fiji and Hazel Holme ridges developed to accommodate strike-slip motion along the north Fiji fracture zone.

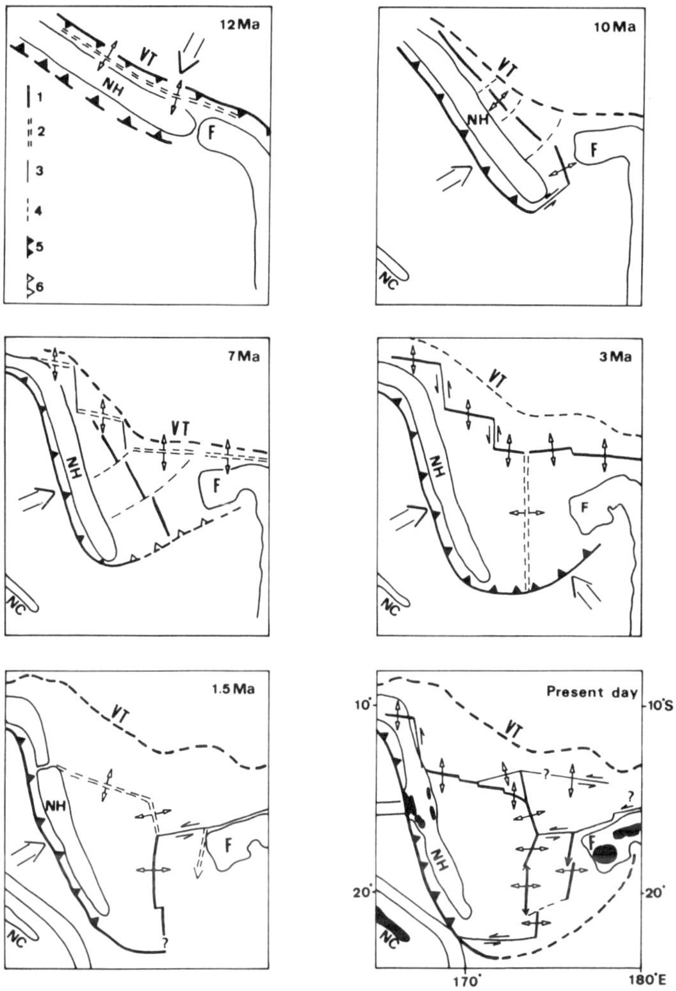

FIGURE 4.19. Geodynamic evolution of the North Fiji Basin. See text for explanation. 1: active ridge axis; 2: incipient ridge axis; 3: transform fault; 4: flowlines; 5: active subduction zone; 6: incipient subduction zone. VT: Vitiaz Trench; NH: New Hebrides arc; F: Fiji Platform; NC: Nouvelle-Calédonie.

6. At present, the spreading south of the north Fiji fracture zone is distributed along the parallel central spreading ridge and West Fiji Ridge isolating an intermediate microplate. To the south, the southernmost spreading axis is connected to the New Hebrides Trench by a large left-lateral strike-slip fault. North of the 16°50′S triple junction, opening occurs along the N160° segment and the Hazel Holme Ridge. The Santa Cruz Trough, which crosscuts the northernmost part of the New Hebrides arc, could be linked to the Hazel Holme spreading system. Crustal motion probably occurs in the north and the northeastern part of the NFB along the South Pandora Ridge and its eastern prolongation, interpreted either as a slow spreading ridge or a strike-slip fault. This lineament is connected with the N160°–Hazel Holme system by the 14°50′S triple junction.

Finally the opening of the NFB can be divided into three major stages: an opening in a NE-SW direction from 12 to 7 Ma, an opening in a N-S direction from 7 to 3 Ma, and an opening in an E-W direction from 3 Ma to the present day. The triangular shape of the basin results from these three successive spreading phases. Since the beginning of the creation of the NFB, the location of the successive spreading centers has migrated southward to accompany the migration of the New Hebrides arc.

Acknowledgments

We thank all the scientific parties and the ships and submersibles crews involved in the studies of NFB during the SEAPSO, EVA, and STARMER projects. We thank Thomas Brocher, Brian Taylor, and an anonymous reviewer for their constructive comments and for their help in improving our English. J. Butscher and J. Perrier are thanked for the preparation of the illustrations.

REFERENCES

Aggarwal, Y. P., Barazangi, B., and Isacks, B. 1972. P and S travel times in the Tonga–Fiji region: A zone of low velocity in the uppermost mantle behind the Tonga Island arc, *J. Geophys. Res.* **77:**6427–6434.

Aggrey, K. E., Muenow, D. W., and Sinton, J. M. 1988. Volatile abundances in submarine glasses from the North Fiji and Lau backarc basins, *Geochim. Cosmochim. Acta* **52:**25401–2506.

Alteris (de), G., Ruellan E., Auzende J. M., Ondréas H., Bendel V., Gracia-Mont E., Lagabrielle Y., Huchon P., and Tanahashi M. 1993. Propagating rifts in the North Fiji Basin (Southwest Pacific), *Geology* **21:**583–586,

Auzende, J. M., Bendel, V., Fujikura, K., Geistdoerfer, P., Gracia-Mont, E., Joshima, M., Kisimoto, K., Mitsuzawa, K., Murai, M., Nojiri, Y., Ondreas, H., Pratt, C., and Ruellan, E. 1992a. Résultats préliminaires des plongées du "Shinkai 6500" sur la dorsale du Bassin Nord-Fidjien (SW Pacifique)–STARMER, *C.R. Acad. Sci. Paris* **314**(II)**:**491–498.

Auzende, J. M., Boespflug, X., Bougault, H., Dosso, L., Foucher, J. P., Joron, J. L., Ruellan, E., and Sibuet, J. C. 1990a. From intraoceanic extension to mature spreading in back arc basins: Examples from the Okinawa, Lau and North Fiji Basin, Actes du Colloque Tour du Monde Jean Charcot, *Oceanol. Acta* **10:**153–163.

Auzende, J. M., Eissen, J. P., Caprais, M. P., Gente, P., Gueneley, S., Harmegnies, F., Lagabrielle, Y., Lapouille, A., Lefèvre, C., Maillet, P., Mazé, J.P., Ondréas, H., Schaaf, A., and Singh, R. 1986a. Accrétion océanique dans la partie méridionale du bassin Nord-Fidjien: Résultats préliminaires de la campagne océanographique SEAPSO III du N.O. Jean Charcot (décembre 1985), *C.R. Acad. Sci. Paris* **303:**93–98.

Auzende, J. M., Eissen, J. P., Lafoy, Y., Gente, P., and Charlou, J. L. 1988a. Seafloor spreading in the North Fiji Basin (Southwest Pacific), *Tectonophysics* **146:**317–351.

Auzende, J. M., Gracia-Mont, E., Bendel, V., Huchon, P., Lafoy, Y., Lagabrielle, Y., de Alteris, G., and Tanahashi, M. 1994a. A possible triple junction at 14°50′S on the North Fiji Basin Ridge (Southwest Pacific), *Mar. Geol.* **116:**25–36.

Auzende, J. M., Gracia-Mont, E., Bendel, V., Lafoy, Y., Lagabrielle, Y., Okuda, Y., and Ruellan, E. in press. Amagmatic extension at intermediate spreading ridge (North Fiji Basin), in *Oceanic Lithosphere*, Special Issue (P. Kapezinskhas, ed.), Wiley, New York.

Auzende, J. M., Hey, R. N., Pelletier, B., Lafoy, Y., and Lagabrielle, Y. 1993. Propagation d'une zone d'accrétion à l'est de la dorsale du bassin Nord Fidjien (SW Pacifique), *C.R. Acad. Sci. Paris* **317**(II)**:**671–678.

Auzende, J. M., Honza, E., Mazé, J. P., and the STARMER Group. 1992b. Comments on the Seabeam map of the North Fiji Basin Ridge between 16°10′S and 21°40′S, *Ofioliti* **17**(1)**:**43–53.

Auzende, J. M., Honza, E., and the STARMER Group. 1990b. Bathymetric map of the North Fiji Basin Ridge between 16°10′S and 21°40′S, published by IFREMER and STA Japan, six colored sheets edited by Beicip, Paris.

Auzende, J. M., Lafoy, Y., and Marsset, B. 1988b. Recent geodynamic evolution of the North Fiji Basin (SW Pacific), *Geology* **16:**925–929.

Auzende, J. M., Lagabrielle, Y., Schaaf, A., Gente, P., and Eissen, J. P. 1986b. Tectonique intra-océanique décrochante à l'ouest des îles Fidji (Bassin Nord Fidjien). Campagne SEAPSO III du N.O. Jean Charcot, *C.R. Acad. Sci. Paris* **303**:241–246.

Auzende, J. M., Okuda, Y., Bendel, V., Ciabrini, J. P., Eissen, J. P., Gracia, E., Hirose, K., Iwabushi, Y., Kisimoto, K., Lafoy, Y., Lagabrielle, Y., Marumo, K., Matsumoto, T., Mitsusawa, K., Momma, H., Mukai, H., Nojiri, Y., Okuda, Y., Ortega-Osorio, A., Ruellan, E., Tanahashi, M., Tupua, E., and Yamaguchi, K. 1991a. Propagation "en échelon" de la dorsale du Bassin Nord Fidjien entre 16°40 et 14°50S (Yokosuka 90–STARMER), *C.R. Acad. Sci. Paris* **312**(II):1531–1538.

Auzende, J. M., Pelletier, B., and Lafoy, Y. 1994b. Twin active spreading ridges in the North Fiji Basin (S.W. Pacific), *Geology* **22**:63–66.

Auzende, J. M., and the STARMER Group. 1990c. Active spreading and hydrothermalism in North Fiji Basin (SW Pacific). Results of Japanese–French cruise *Kaiyo 87*, *Mar. Geophys. Res.* **12**:269–283.

Auzende, J. M., Urabe, T., Bendel, V., Deplus, C., Eissen, J. P., Grimaud, D., Huchon, P., Ishibashi, J., Joshima, M., Lagabrielle, Y., Mevel, C., Naka, J., Ruellan, E., Tanaka, T., and Tanahashi, M. 1991b. In situ geological and geochemical study of an active hydrothermal site on the North Fiji Basin Ridge, *Mar. Geol.* **98**:259–269.

Barazangi, M., Isacks, B. L., Dubois, J., and Pascal, G. 1974. Seismic wave attenuation in the upper mantle beneath the Southwest Pacific, *J. Geophys. Res.* **76**:8493–8516.

Bendel, V. 1993. Cadre gèologique et composition des minéralisations hydrothermales en contexte arrière-arc: exemple de la dorsale du Bassin Nord Fidjien, Thèse Université Brest, France.

Bendel, V., Fouquet, Y., Auzende, J. M., Lagabrielle, Y., Grimaud, D., and Urabe, T. 1993. Metallogenesis at a Triple Junction system: The White Lady hydrothermal field (North Fiji Back-Arc Basin, SW Pacific), *Economical Geol.* **88**:2237–2249.

Boespflug, X. 1990. Evolution géodynamique et géochimique des bassins arrière-arcs. Exemples des bassins d'Okinawa, de Lau et Nord-Fidjien, Thèse, Université Brest, France.

Briqueu, L., Laporte, C., Crawford, A., Hasenaka, T., Baker, P., and Coltorti, M. 1994. Temporal magmatic evolution of the Aoba basin–central New Hebrides Island Arc: Pb, Sr and Nd isotopic evidence for the coexistence of two mantle components beneath the arc, in *Proc. ODP, Sci. Results*, 134 (H. G. Greene, J.-Y. Collet, L. B. Stokking *et al.*, eds.), pp. 393–401, Ocean Drilling Program, College Station, TX.

Brocher, T. M. 1985, On the formation of the Vitiaz Trench lineament and North Fiji Basin, in *Investigations of the Northern Melanesian Borderland* (T. M. Brocher, ed.), pp. 13–34, Circum-Pacific Council for Energy and Mineral Resources, Houston, TX, Earth Science Series, Vol. 3.

Brocher, T. M., and Holmes, R. 1985. The marine geology of sedimentary basins south of Viti Levu, Fiji, in *Investigations of the Northern Melanesian Borderland* (T. M. Brocher, ed.), pp. 123–138, Circum-Pacific Council for Energy and Mineral Resources, Houston, TX, Earth Science Series, Vol. 3.

Butterfield, D. A., Massoth, G. J., McDuff, R. E., Lupton, J. E., and Lilley, D. 1990. Geochemistry of hydrothermal fluids from Axial Seamount Hydrothermal Emissions Study vent field, Juan de Fuca Ridge: Subseafloor boiling and subsequent fluid-rock interaction, *J. Geophys. Res.* **95**:12,895–12,921.

Campbell, A. C., *et al.* 1988. Chemistry of hot springs on the Mid-Atlantic Ridge, *Nature* **335**:514–518.

Cande, S. C., and Kent, D. V. 1992. A new geomagnetic polarity time scale for the late Cretaceous and Cenozoic, *J. Geophys. Res.* **97**(B10):13,917–13,951.

Charvis, P., and Pelletier, B. 1989. The northern New Hebrides backarc troughs: History and relation with the North Fiji Basin, *Tectonophysics* **170**:259–277.

Chase, C. G. 1971. Tectonic history of the Fiji Plateau, *Geol. Soc. Am. Bull.* **82**:3087–3110.

Cherkis, N. Z. 1980. Aeromagnetic investigations and sea floor spreading history in the Lau basin and the northern Fiji Plateau, in *Symposium on Petroleum Potential in Island Arcs, Small Basins, Submerged Margins and Related Areas* (W. J. Clark, ed.), United Nations, ESCAP, CCOP/SOPAC, Technical Bull. **3**:37–45.

Craig, H. 1986. Papatua 86: A grand tour of the Havre Trough and the Lau, North Fiji, Woodlark and Manus Basins, *Eos* **67**:378.

Daniel, J. 1982. Morphologie et structures superficielles de la partie sud de la zone de subduction des Nouvelles-Hébrides, in *Contribution à l'étude géodynamique du Sud-Ouest Pacifique*, Equipe de Géologie- Géophysique du Centre ORSTOM de Nouméa, Travaux et Documents de l'ORSTOM **147**:39–60.

Dubois, J. 1971. Propagation of P waves and Rayleigh waves in Melanesia: structural implications, *J. Geophys. Res.* **76**:7217–7240.

Dubois, J., Launay, J., Récy, J., and Marshall, J. 1977. New Hebrides Trench: Subduction rate from associated lithospheric bulge, *Can. J. Earth Sci.* **14**:250–255.

Dubois, J., Pascal, G., Barazangi, M., Isacks, B. L., and Oliver, J. 1973. Travel times of seismic waves between the

New Hebrides and Fiji Islands: A zone of low velocity beneath the Fiji Plateau, *J. Geophys. Res.* **78:**3431–3436.

Eguchi, T. 1984. Seismotectonics of the Fiji Plateau and Lau Basin, *Tectonophysics* **102:**17–32.

Eissen, J. P., Lefèvre, C., Maillet, P., Morvan, G., and Nohara, M. 1991. Petrology and geochemistry of the central North Fiji Basin spreading center (SW Pacific) between 16°S and 22°S, *Mar. Geol.* **98:**201–239.

Eissen, J. P., Morvan, G., Lefèvre, C., Maillet, P., Urabe, T., Auzende, J. M., and Honza, E. 1990. Pétrologie et géochimie de la zone d'accrétion du centre du bassin Nord Fidjien (SW Pacifique), *C.R. Acad. Sci. Paris* **310**(II)**:**771–778.

Eissen, J. P., Nohara, M., and Cotten, J. 1994. North Fiji Basin basalts and their magma sources. 1: Incompatible element constraints, *Mar. Geol.* **116:**163–178.

Falvey, D. A. 1975. Arc reversals, and a tectonic model for the North Fiji Basin, *Austr. Soc. Explor. Geophys. Bull.* **6:**47–49.

Falvey, D. A. 1978. Analysis of paleomagnetic data from the New Hebrides, *Austr. Soc. Explor. Geophys. Bull.* **9**(3)**:**117–123.

Francheteau, J., Yelles-Chaouche, A., and Craig, H. 1987. The Juan Fernandez microplate north of the Nazca–Pacific–Antarctic plate junction at 35°S, *Earth Planet. Sci. Lett.* **86:**253–286.

Gente, P., Mével, C., Auzende, J. M., Karson, J. A., and Fouquet, Y. 1991. An example of recent accretion on the Mid-Atlantic Ridge: The Snake Pit neovolcanic ridge (MARK area, 23°22′N), *Tectonophysics* **190:**1–29.

Gill, J. B. 1981. *Orogenic Andesites and Plate Tectonics*, Springer-Verlag, Berlin.

Gill, J. B. 1984. Sr-Pb-Nd isotopic evidence that both MORB and OIB sources contribute to oceanic island arc magmas in Fiji, *Earth Planet. Sci. Lett.* **68:**443–458.

Gill, J. B., and Gorton M. 1973. A proposed geological and geochemical history of eastern Melanesia, in *The Western Pacific: Island Arcs, Marginal Seas and Geochemistry* (P. J. Coleman, ed.), pp. 543–566, University of Western Australia Press.

Gracia-Mont, E. 1991. Etude morphostructurale du segment N160° la dorsale du Basin Nord-Fidjien, Analyse des données de la campagne Yokosuka 90, Rapport de D.E.A., Université de Bretagne Occidentale, Brest.

Gracia-Mont, E. 1992. El segment N160 de la Conca Nord-Fijiana (Pacific Sud-Oest): Morfoestructura d'un eix d'acrecio oceanica d'edat quaternaria dins una conca marginal, Tesi de Licenciatura, Universitat de Barcelona, Barcelona.

Gracia-Mont, E., Ondréas, H., Bendel, V., and STARMER Group. 1994. Multiscale morphologic variability of the North Fiji Basin Ridge (Southwest Pacific), *Mar. Geol.* **116:**133–162.

Grimaud, D., Ishibashi, J. I., Lagabrielle, Y., Auzende, J. M., and Urabe, T. 1991. Chemistry of hydrothermal fluids from the 17°S active site on the North Fiji Basin Ridge (SW Pacific), *Chem. Geol.* **93:**209–218.

Gutenberg, B., and Richter C. F. 1954. *Seismicity of the Earth*, 2nd ed., Princeton University Press, Princeton, NJ.

Halunen, A. J., Jr. 1979. Tectonic history of the Fiji Plateau, Ph.D. thesis, University of Hawaii, Honolulu.

Hey, R. N., Kleinrock, M. C., Miller, S. P., Atwater, T. M., and Searle, R. C. 1986. Sea Beam/Deep-Tow investigation of an active oceanic propagating rift system, Galapagos 95.5°W, *J. Geophys. Res.* **91:**3369–3393.

Hamburger, M. W., and Isacks, B. L. 1988. Diffuse backarc deformation in the Southwestern Pacific, *Nature* **332:**599–604.

Hamburger, M. W., and Isacks, B. L. 1994. Shallow seismicity in the North Fiji Basin, in *Basin Formation, Ridge Crest Processes, and Metallogenesis in the North Fiji Basin* (L. W. Kroenke and J. V. Eade, eds.), pp. 21–32, Circum-Pacific Council for Energy and Mineral Resources. Earth Science Series, Vol. 15.

Hawkins, J. W., and Melchior J. T. 1985. Petrology of Mariana Trough and Lau Basin basalts, *J. Geophys. Res.* **90:**11,431–11,468.

Huchon, P., Gracia, E., Ruellan, E., Joshima, M., and Auzende, J. M. 1994. Kinematics of active spreading in the central North Fiji Basin (Southwest Pacific), *Mar. Geol.* **116:**69–88.

Hughes Clarke, J. E., Jarvis, P., Tiffin, D., Price, R., and Kroenke, L. 1993. Tectonic activity and plate boundaries along the northern flank of the Fiji Platform, *Geo-Mar. Lett.* **13:**98–106.

Ishibashi, J.-I., Grimaud, D., Nojiri, Y., Auzende, J. M., and Urabe, T. 1994. Fluctuation of chemical compositions of the phase-separated hydrothermal fluid from the North Fiji Basin Ridge, *Mar. Geol.* **116:**215–226.

James, A., and Falvey, D. A. 1978. Analysis of paleomagnetic data from Viti Levu, Fiji, *Austr. Soc. Explor. Geophys. Bull.* **9:**15–123.

Jarvis, P., Hughes-Clarke, J., Tanahashi, M., Tiffin, D., and Kroenke, L. 1994. The western Fiji Transform Fault and its role in the dismemberment of the Fiji Platform, *Mar. Geol.* **116:**57–68.

Jarvis, P., Kroenke, L., Price, R., and Maillet, P. 1993. GLORIA imagery of sea floor structures in the northern North Fiji Basin, *Geo-Mar. Lett.* **13:**90–97.

Johnson, D. P., Maillet, P., and Price, P. 1993. Regional setting of a complex backarc: New Hebrides arc, northern Vanuatu-eastern Solomon Islands, *Geo-Mar. Lett.* **13:**82–89.

Johnson, K. T. M., and Sinton, J. M. 1990. Petrology, tectonic setting and the formation of backarc basin basalts in the North Fiji Basin, *Geol. Jb. D* **92:**517–545.

Kappel, E. S., and Ryan, W. B. F. 1986. Volcanic episodicity and a non-steady state rift valley along Northeast Pacific spreading centers: Evidence from SeaMarc I, *J. Geophys. Res.* **91**(13)**:**13,925–13,940.

Karson, J. A. 1990. Seafloor spreading of the Mid-Atlantic Ridge: Implications for the structure of ophiolites and oceanic lithosphere produced in slow-spreading environments, in *Proceedings of the Symposium "Troodos 1987"* (J. Malpas, E. M. Moores, A. Panayiotou, and C. Xanophontos, eds.), Geological Survey Department, Nicosia, Cyprus.

Karson, J. A., Thompson, G., Humphries, S. E., Edmond, J. M., Bryan, W. B., Brown, J. R., Winters, A. T., Pockalny, R. A., Casey, J. F., Campbell, A. C., Klinkhammer, G., Palmer, M. R., Kinzler, R. J., and Sulanovska, M. M. 1987. Along axis variations in seafloor spreading in the MARK area, *Nature* **328:** 681–685.

Kleinrock, M. C. in press. The southern boundary of the Juan Fernandez microplate: braking microplate rotation and deformation of the Antarctic plate, *J. Geophys. Res.*

Kroenke, L. W., Eade, J. V., Yan, C. Y., and Smith, R. 1994. Sediment distribution in the north-central North Fiji Basin, in *Basin Formation, Ridge Crest Processes, and Metallogenesis in the North Fiji Basin* (L. W. Kroenke and J. V. Eade, eds.), pp. 63–73, Circum-Pacific Council Energy and Mineral Resources, Springer, Heidelberg. Earth Science Series, Vol. 153.

Kroenke, L. W., Jarvis, P., and Price, R. C. 1987, Morphology of the Fiji Fracture Zone: Recent reorientation of plate boundaries in the vicinity of the North Fiji Basin, *Eos* **68**(44)**:**1445.

Kroenke, L. W., Price R. C., and Jarvis P. A. 1991. North Fiji Basin, 1:250,000, Pacific Seafloor Atlas, Hawaii Institute of Geophysics, Honolulu, HI, sheets 10 to 17.

Lafoy, Y., Auzende, J. M., Gente, P., and Eissen, J. P. 1987. L'extremité occidentale de la zone de fracture Fidjienne et le point triple de 16°40S, Résultats du Leg III de la campagne SEAPSO du N.O. Jean Charcot (Décembre 1985) dans le bassin Nord Fidjien, SW Pacifique, *C.R. Acad. Sci. Paris* **304:**147–152.

Lafoy, Y., Auzende, J. M., Ruellan, E., Huchon, P., and Honza, E. 1990. The 16°40S triple junction in the North Fiji Basin (SW Pacific), *Mar. Geophys. Res.* **12:**285–296.

Lagabrielle, Y., Auzende, J. M., Eissen, J. P., Janin, M. C., and Cotten, J. 1994. Geology and geochemistry of a 800 m section through young upper oceanic crust in the North Fiji Basin (Southwest Pacific), *Mar. Geol.* **116:** 113–132.

Larson, R. L., Searle, R. C., Kleinrock, M. C., Shouten, H., Bird, R. T., Naar, D. F., Rusby, R. I., Hooft, E. E., and Lasthiotakis, H. 1992. Roller-bearing tectonic evolution of the Juan Fernandez microplate, *Nature* **356:** 571–576.

Larue, B. M., Pontoise, B., Malahoff, A., Lapouille, A., and Latham, G. V. 1982. Bassins marginaux actifs du Sud-Ouest Pacifique: Plateau Nord-Fidjien, bassin de Lau, in *Contribution à l'étude géodynamique du Sud-Ouest Pacifique*, Equipe de Géologie-Géophysique du Centre ORSTOM de Nouméa. Travaux et Documents de l'ORSTOM **147:**363–406.

Louat, R. 1982. Séismicité et subduction de la terminaison sud de l'arc insulaire des Nouvelles Hébrides, in *Contribution à l'étude géodynamique du Sud-Ouest Pacifique*, Equipe de Géologie-Géophysique du Centre ORSTOM de Nouméa. Travaux et Documents de l'ORSTOM **147:**179–185.

Louat, R., and Pelletier, B. 1989. Seismotectonic and present-day relative plate motions in the New Hebrides–North Fiji Basin region, *Tectonophysics* **167:**41–55.

Luyendyck, B. P., Bryan, W. B., and Jezek, P. A. 1974. Shallow structure of the New Hebrides island arc, *Geol. Soc. Am. Bull.* **85:**1287–1300.

Macdonald, K. C. 1983. Crustal processes at spreading centers, *Rev. Geophys.* **21:**1441–1453.

Macdonald, K. C., Luyendyck, B. P., and Von Herzen, R. P. 1973. Heat flow and plate boundaries in Melanesia, *J. Geophys. Res.* **78:**2537–2546.

Macdonald, K. C., Sempère, J. C., and Fox, P. J. 1984. The East Pacific Rise from the Siquciros of the Orozco fracture zone: Along strike continuity of the neovolcanic zone and the structure and evolution of overlapping spreading centers, *J. Geophys. Res.* **89:**6049–6069.

Maillet, P., Eissen, J. P., Lapouille, A., Monzier, M., Baleivanualala, V., Butscher, J., Gallois, F., and Lardy, M.

1986. La dorsale active du bassin Nord-Fidjien entre 20°S et 20°53′S: Signature magnétique et morphologique, *C.R. Acad. Sci. Paris* **302**(II):135–140.

Maillet, P., Monzier M., Eissen, J. P., and Louat R. 1989. Geodynamics of an arc ridge junction: The case of the New Hebrides arc–North Fiji Basin, *Tectonophysics* **165**:251–268.

Malahoff, A., Feden, R. H., and Fleming, H. F. 1982a. Magnetic anomalies and tectonic fabric of marginal basins north of New Zealand, *J. Geophys. Res.* **87**:4109–4125.

Malahoff, A., Hammond, S. R., Naughton, J. J., Keeling, D. L., and Richmond, R. N. 1982b. Geophysical evidence for post-Miocene rotation of the island of Viti Levu, Fiji, and its relationship to the tectonic development of the North Fiji Basin, *Earth Planet. Sci. Lett.* **57**:398–414.

Malahoff, A., Kroenke, L. W., Cherkis, N., and Brozena, J. 1994. Magnetic and tectonic fabric of the North Fiji Basin and Lau Basin, in *Basin Formation, Ridge Crest Processes, and Metallogenesis in the North Fiji Basin* (L. W. Kroenke and J. V. Eade, eds.), Earth Science Ser., Vol. 15, pp. 49–61, Circum-Pacific Council for Energy and Mineral Resources.

Mammerickx, J., Chase, T. E., Smith, S. M., and Taylor, I. L. 1971. Bathymetry of the South Pacific, chart n°12. Scripps Institution of Oceanography.

Massoth, G. J., Butterfield, D. A., Lupton, J. E., McDuff, R. E., Lilley, M. D., and Jonasson, I. R. 1989. Submarine venting of phase-separated hydrothermal fluids at axial volcano, Juan de Fuca Ridge, *Nature* **340**:702–705.

Monjaret, M. C., Bellon, H., and Maillet, P. 1991. Magmatism of the troughs behind the New Hebrides island arc (R.V. *Jean Charcot* SEAPSO 2 cruise): K-Ar geochronology and petrology, *J. Volcanol. Geotherm. Res.* **46**:265–280.

Monzier, M., Collot, J. Y., and Daniel, J. 1984. Carte bathymétrique des parties centrale et méridionale de l'arc insulaire des Nouvelles-Hébrides. Office de la Recherche Scientifique et Technique Outre-Mer, Editions de l'ORSTOM, Paris.

Monzier, M., Daniel, J., and Maillet, P. 1990. La collision ride des Loyauté/arc des Nouvelles-Hébrides (Pacifique Sud-Ouest), *Oceanol. Acta* **10**:43–56.

Monzier, M., Maillet, P., and Dupont, J. 1991. Carte bathymétrique des parties méridionales de l'arc insulaire des Nouvelles-Hébrides et du bassin Nord-Fidjien, Institut Français de Recherche Scientifique pour le Développement en Coopération, Editions de l'ORSTOM, Paris.

Nohara, M., Hirose, K., Eissen, J. P., Urabe, T., and Joshima, M. 1994. North Fiji Basin basalts and their magma sources. II: Sr-Nd isotopic and trace element constraints, *Mar. Geol.* **116**:179–196.

Nojiri, Y., Ishibashi, J. I., Kawai, T., Otsuki, A., and Sakai, H. 1989. Hydrothermal plumes along the North Fiji Basin spreading axis, *Nature* **342**:667–670.

Ondréas, H., Ruellan, E., Auzende, J. M., Bendel, V., de Alteriis, G., Gracia-Mont, E., Lagabrielle, Y., and Tanahashi, M. 1993. Variabilité morpho-structurale à l'échelle kilométrique de la dorsale du Bassin Nord Fidjien: Exploration in situ du segment compris enter 18°50′S et 19°S, *C.R. Acad. Sci. Paris* **316**(II):115–122.

Pelletier, B., Charvis, P., Daniel, J., Hello, Y., Jamet, F., Louat, R., Nanau, P., and Rigolot, P. 1988. Structure et linéations magnétiques dans le coin nord-ouest du bassin Nord-Fidjien: Résultats préliminaires de la campagne EVA 14 (août 1987), *C.R. Acad. Sci. Paris,* **306**(II):1247–1254.

Pelletier, B., Lafoy, Y., and Missegue, F. 1993a. Morphostructure and magnetic fabric of the northwestern North Fiji Basin, *Geophys. Res. Lett.* **20**(12):1151–1154.

Pelletier, B., and Louat, R. 1989. Mouvement relatif des plaques dans le Sud-Ouest Pacifique, *C.R. Acad. Sci. Paris* **308**:123–130.

Pelletier, B., Missègue, F., Lafoy, Y., Mollard, L., Decourt, R., Dupont, J., Join, Y., Perrier, J., and Récy, J. 1993b. Extrémités nord du bassin Nord-Fidjien et des fossés arrière-arc des Nouvelles-Hébrides: Morphostructure et signature magnétique, *C.R. Acad. Sci. Paris* **316**(II):637–644.

Price, R. C., Johnson, L. E., and Crawford, A. J. 1990. Basalts of the North Fiji basin: The generation of back arc basin magmas by mixing of depleted and enriched mantle sources, *Contrib. Mineral. Petrol.* **105**:106–121.

Price, R. C., and Kroenke, L. W. 1991. Tectonics and magma genesis in the northern North Fiji Basin, *Mar. Geol.* **98**:241–258.

Price, R. C., Maillet, P., and Johnson, D. A. 1993. Interpretation of GLORIA side-scan sonar imagery for the Coriolis Troughs of the New Hebrides backarc, *Geo-Mar. Lett.* **13**:71–81.

Price, R. C., Maillet, P., McDougall, I., and Dupont, J. 1991. The geochemistry of basalts from the Wallis Islands, Northern Melanesian Borderland: Evidence for a lithospheric origin for Samoan-type basaltic magmas, *J. Volcanol. Geotherm. Res.* **45**:267–288.

Récy, J., Pelletier, B., Charvis, P., Gerard, M., Monjaret, MC., and Maillet, P. 1990. Structure, âge et origine des fossés arrière-arc des Nouvelles-Hébrides (Sud-Ouest Pacifique) *Oceanol. Acta* **10**:165–182.

Ruellan, E., Auzende, J. M., Honza, E., and the STARMER Group. 1989. L'accrétion dans le bassin Nord Fidjien méridional: Premiers résultats de la campagne franco-japonaise STARMER/KAIYO 88, *C.R. Acad. Sci. Paris* **309**(II):1247–1254.

Ruellan, E., Huchon P., Auzende, J. M., and Gracia, E. 1994. Propagating rift and overlapping spreading center in the North Fiji Basin, *Mar. Geol.* **116:**37–56.

Saunders, A. D., and Tarney, J. 1984. Geochemical characteristics of basaltic volcanism within backarc basins, in *Marginal Basins: Geology, Volcanism and Associated Sedimentary and Tectonic Processes in Modern and Ancient Marginal Basins* (B. P. Kokelaar and M. F. Holwells, eds.), pp. 59–76, Blackwell, Oxford.

Sclater, J. G., and Francheteau, J. 1970. The implications of terrestrial heat flow observations on current tectonic and geochemical models of the crust and upper mantle of the earth, *Geophys. J. R. Astron. Soc.* **20:**509–542.

Sclater, J. G., and Menard, H. W. 1967. Topography and heat flow of the Fiji Plateau, *Nature* **216:**991–993.

Sinton, J. M., and Fryer, P. 1987. Mariana lavas from 18°N: Implications for the origin of back arc basin basalts, *J. Geophys. Res.* **92**(B12)**:**782–802.

Sinton, J. M., Johnson K. T. M., and Price R. C. 1985. Petrology and geochemistry of volcanic rocks from the Northern Melanesian Borderland, in *Investigations of the Northern Melanesian Borderland* (T. M. Brocher, ed.), Earth Science Ser., Vol. 3, pp. 35–65, Circum-Pacific Council for Energy and Mineral Resources, Houston, TX.

Sinton J. M., Price, R. C., Johnson, K. T. M., Staudigel, H., and Zindler, A. 1994. Petrology and geochemistry of submarine lavas from the Lau and North Fiji Back-Arc Basins, in *Basin Formation, Ridge Crest Processes, and Metallogenesis in the North Fiji Basin* (L. W. Kroenke and J. V. Eade, eds.), Earth Science Ser., Vol. 15, pp. 119–135, Circum-Pacific Council for Energy and Mineral Resources.

Stackelberg, U. von, and Rad, U. von. 1990. Geological evolution and hydrothermal activity in the Lau and North Fiji basins (Sonne Cruise SO-35)—a synthesis, *Geol. Jb.* **92:**629–660.

Stackelberg, U. von, and Shipboard Scientific Party. 1985. Hydrothermal sufide deposits in backarc spreading centers in the Southwest Pacific, BGR Bundesandstalt für Geowissenschaften und Röstoffe Circular **2:**3–14.

Sun, S. S., and McDonough, W. F. 1989. Chemical and isotopic systematics of oceanic basalts: Implications for mantle composition and processes, in *Magmatism in the Ocean Basins* (A. D. Saunders and M. J. Norry, eds.), Geol. Soc. Spec. Publ. **42:**313–345.

Sykes, L. R. 1966. The seismicity and deep structure of islands arcs, *J. Geophys. Res.* **71:**2981–3006.

Sykes, L. R., Isacks, B. L., and Oliver, J. 1969. Spatial distribution of deep and shallow earthquakes of small magnitudes in the Fiji–Tonga region, *Bull. Seismol. Soc. Am.* **59:**1093–1113.

Urabe, T., Auzende, J. M., *et al.* 1992. Bathymetric map of the central part of the North Fiji Basin, Southwest Pacific, between 14°20S and 21°50S, 2 colored maps, scale: 1/500 000, Published under the Special Coordination for Promoting Science and Technology (Japan).

Von Damm, K. L., and Bischoff, J. L. 1987. Chemistry of hydrothermal solutions from the Juan de Fuca Ridge, *J. Geophys. Res.* **92:**11,334–11,346.

Wright, E., and White, W. M. 1987. The origin of Samoa: New evidence from Sr, Nd, and Pb isotopes, *Earth Planet. Sci. Lett.* **81:**151–162.

5

Tectonics, Magmatism, and Evolution of the New Hebrides Backarc Troughs (Southwest Pacific)

Patrick Maillet, Etienne Ruellan, Martine Gérard, Alain Person, Hervé Bellon, Joseph Cotten, Jean-Louis Joron, Setsuya Nakada, and Richard C. Price

ABSTRACT

In the southwest Pacific, a discontinuous series of narrow and elongated troughs separates the New Hebrides island arc from the adjacent active marginal basin, the North Fiji Basin. This chapter reviews the structural, geophysical, geochronological, and petrological data available for the New Hebrides backarc troughs (NHBAT) and discusses the significance of these structures.

A diffuse horst-and-graben morphology, partly obscured in some places by recent volcanic complexes, characterizes the northern Jean-Charcot troughs (JCT). By contrast, the southern Coriolis troughs (CT) show well-developed flat-bottomed grabens. Moreover, no backarc troughs are observed in the central backarc area, adjacent where the d'Entrecasteaux zone collides with the arc.

Volcanic rocks dredged in the NHBAT show a wide range of SiO_2 contents, with high-Al_2O_3 and low-TiO_2 contents, features typical of their arc/backarc environments. Trace element analyses indicate a much stronger subduction component in the volcanics of the southern CT than in those of the northern JCT. However, large-ionic-radius-lithophile-element (LILE) (Ba, Rb, Sr) enrichments and high-field-strength-elements (HFSE) (Ta, Nb, Zr, Ti, Y, Yb) depletions, relative to N-MORB (mid-ocean ridge basalts), are generally observed in most NHBAT volcanics and are features characteristic of island-arc basic and

Patrick Maillet • ORSTOM Centre de Brest-GDR "GEDO" 910, 29280 Plouzané, France. *Etienne Ruellan* • CNRS Sophia-Antipolis, 06560 Valbonne, France. *Martine Gérard* • ORSTOM, 93143 Bondy, France. *Alain Person* • Laboratoire de Géologie des Bassins Sédimentaires, Université Pierre et Marie Curie, 75252 Paris, France. *Hervé Bellon* and *Joseph Cotten* • CNRS URA 1278-GDR "GEDO" 910, Université de Bretagne Occidentale, 29287 Brest, France. *Jean-Louis Joron* • Groupe des Sciences de la Terre, Laboratoire Pierre-Süe, CEN Saclay, 91191 Gif sur Yvette, France. *Setsuya Nakada* • Kyushu University, Fukuoka 812, Japan. *Richard C. Price* • La Trobe University, Bundoora, Victoria 3083, Australia.

Backarc Basins: Tectonics and Magmatism, edited by Brian Taylor, Plenum Press, New York, 1995.

intermediate volcanics. Backarc basin basalts (BABB) are scarce; the only occurrence was found in the very northern JCT, indicative of an aborted tendency toward oceanic spreading, between 3.9 and 1.1 Ma (K/Ar dating).

Geological long-range inclined asdic (GLORIA) seafloor imagery, manned submersible observations, water chemistry analyses, and sediment heat flow measurements do not provide evidence for widespread hydrothermal activity, or oceanic spreading, in the NHBAT. However, some ferromanganese crusts coating volcanic and volcano-sedimentary formations have a hydrothermal origin (todorokite, birnessite, phillipsite). These hydrothermal crusts are mainly located on the eastern faulted border of the NHBAT.

The volcanic-tectonic evolution of the New Hebrides backarc troughs primarily results from the concomitant effects of nearby subduction (along the New Hebrides subduction zone) and spreading (in the central North Fiji basin) and secondarily from the after-effects of the collision of the d'Entrecasteaux zone with the arc. The NHBAT represent the very first stage of backarc crustal extension, characterized by volcanics with predominantly island-arc tholeiite and some BABB compositional features.

1. INTRODUCTION

Located between an island arc and a marginal basin, backarc troughs may record the concomitant or successive influences of crustal distension, rifting, oceanic spreading, with or without any arc-induced magmatic contamination. A number of recent papers have discussed the structural, petrological, and geochronological characteristics of the New Hebrides backarc troughs (Récy *et al.*, 1986, 1990; Monjaret *et al.*, 1987, 1991; Charvis and Pelletier, 1989; Monjaret, 1989; Sage and Charvis, 1991; Matsumoto *et al.*, 1992; Johnson *et al.*, 1993; Pelletier *et al.*, 1993b; Price *et al.*, 1993; Nakada *et al.*, 1994). Yet the significance of these troughs in the regional tectonic framework of the southwest Pacific remains controversial. Do these troughs represent old and inactive structures or incipient spreading features—that is, a nascent backarc basin with recent active submarine volcanism? Is their formation primarily linked to the evolution of the New Hebrides island arc or to that of the North Fiji Basin?

To answer these questions, we review and discuss available data on the New Hebrides backarc troughs and propose a new interpretation accounting for the occurrence of these unusual submarine structures.

2. GEOLOGICAL AND TECTONIC FRAMEWORK OF THE NEW HEBRIDES ISLAND ARC AND THE NORTH FIJI BASIN

In the area between 10°S and 25°S and 165°E and 170°W, two subduction zones with opposite directions of convergence dip toward each other (Fig. 5.1). Along the New Hebrides subduction zone (i.e., beneath the Santa Cruz, or the eastern outer Solomon, and Vanuatu islands), the consumption of the India–Australia plate is marked by a rapid subduction in an E-NE direction (9–16 cm/yr; N76°E ±11; dip 70°). Along the Tonga subduction zone, the geodynamic scheme is almost symmetrical for the subducting Pacific plate (16.5–18 cm/yr along a W-NW direction, with a dip of 80°) (Isacks *et al.*, 1981; Louat *et al.*, 1988; Louat and Pelletier, 1989; Pelletier and Louat, 1989; Pelletier and Dupont, 1990a).

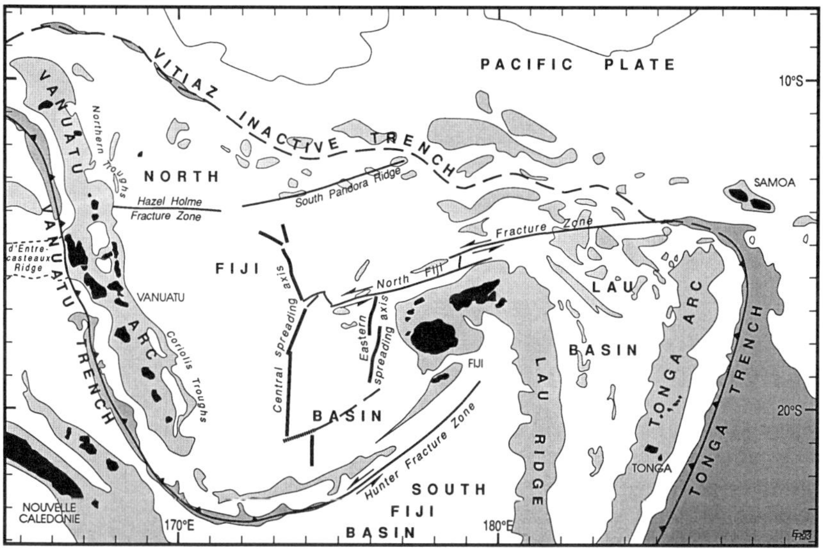

FIGURE 5.1. Tectonic setting of the New Hebrides (Vanuatu) arc in the southwest Pacific. Northern troughs = Jean-Charcot troughs (JCT) in the text.

Plate consumption along these two subduction zones is accompanied by the opening of two active marginal basins: the North Fiji Basin, between Vanuatu and Fiji (Auzende *et al.*, Chapter 4 this volume), and the Lau Basin, between Fiji and Tonga (Hawkins, Chapter 3 this volume).

Interestingly, both subduction zones are affected by the collision of major submarine ridges. The d'Entrecasteaux zone is colliding with the central New Hebrides arc, and the collision zone is moving northward (Maillet *et al.*, 1983; Collot *et al.*, 1985; 1992a,b), whereas the Louisville Ridge is colliding with and sweeping southward along the Tonga arc (Dupont and Herzer, 1985; Pelletier and Dupont, 1990b; Lallemand *et al.*, 1992).

2.1. The New Hebrides Island Arc

The New Hebrides island arc comprises three volcanic chains (Carney *et al.*, 1985): a western belt (late Oligocene to middle Miocene) forming the islands of the Torres Group, Espiritu Santo and Malakula; an eastern belt (late Miocene to early Pliocene), including the islands of Maewo and Pentecost; and a central chain (late Miocene to Holocene), which runs more than 1500 km from Tinakula–Nendo (Santa Cruz Islands) in the north to Matthew and Hunter volcanoes in the south (Fig. 5.2).

The following summary is from Macfarlane *et al.* (1988) and Gorton (1974, 1977).

2.1.1. Western Belt

Volcanism in the islands of the Torres Group, Espiritu Santo and Malakula, commenced in latest Oligocene time but occurred largely during the early to middle Miocene (25–14 Ma), along the former Vitiaz island arc, overlying a southwest dipping Benioff zone. Petrographically, lavas of the western belt range from olivine basalt to hornblende-

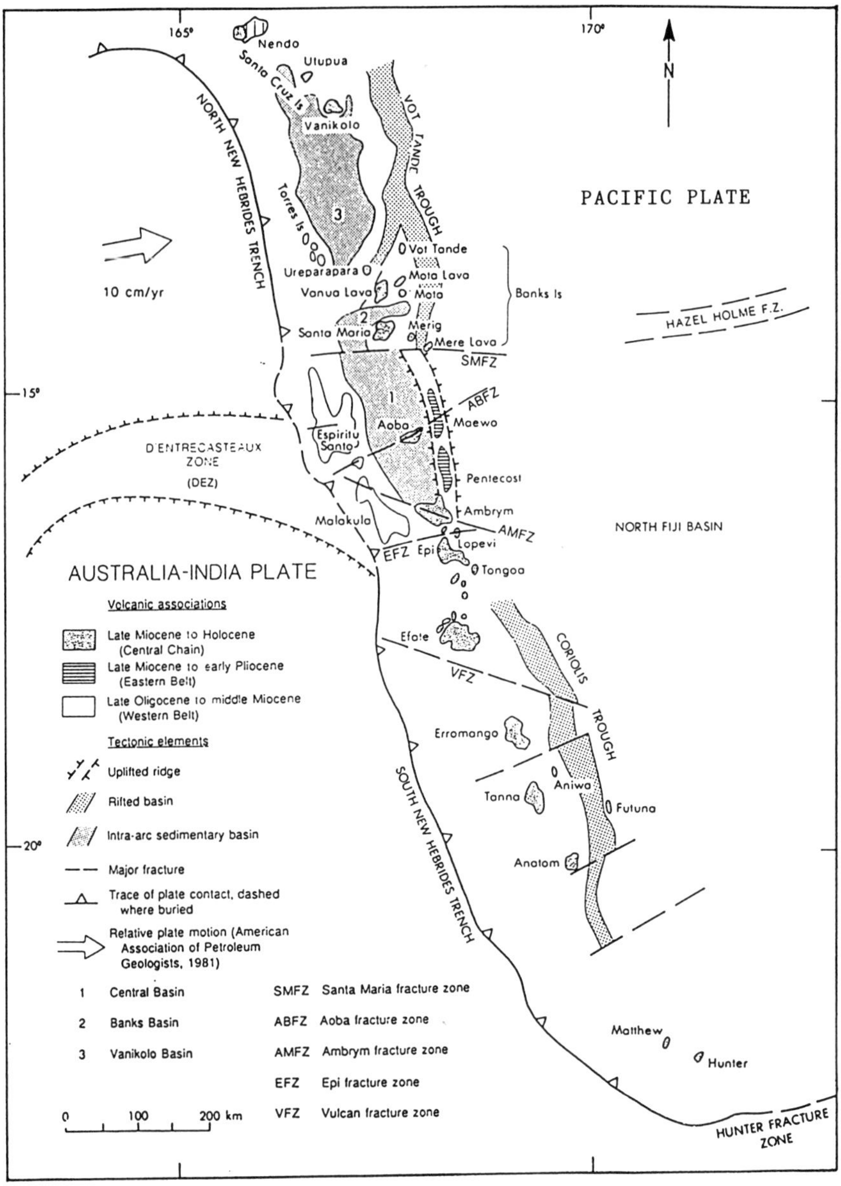

FIGURE 5.2. The principal geological and tectonic features of the New Hebrides island arc (after Greene *et al.*, 1988a). Vot Tande trough = Jean-Charcot troughs (JCT) in the text.

bearing dacite and rhyodacite pumice, but andesite (including both two-pyroxene and hornblende-phyric varieties) is most abundant.

The western belt lavas show transitional calc-alkaline/tholeiitic characteristics and in this respect resemble the Lau Ridge volcanics and lavas from the lower-K_2O suite presently erupting in the New Hebrides arc central chain.

2.1.2. Eastern Belt

The island of Pentecost is the only island of the New Hebrides arc on which a basement complex crops out. This is composed of slices of ultramafic to basaltic rocks interpreted as a fragmented ophiolite, possibly representing sections of the oceanic crust upon which the eastern belt arc volcanics accumulated. Most of these rocks are too altered and metamorphosed for K/Ar dating and have low K_2O contents. Still, two mineral K/Ar ages (35 ± 2 Ma for a hornblende from an amphibolite, and 28 ± 6 Ma for a plagioclase from a gabbro) place younger limits on the age of this basement (Gorton, 1974).

Most of the Maewo and Pentecost volcanics erupted between late Miocene and lower-middle Pliocene time (7–5 Ma for Maewo, 6–3 Ma for Pentecost) from submarine fissures. Pillow lavas, associated intrusions, and pyroclastic rocks range in composition from basal ankaramites, picrites, and mafic-enriched porphyritic basalts to an upper series of feldspar-phyric basalts and basaltic andesites. They range across the medium-K_2O to high-K_2O fields and are differentiated to high-K_2O calc-alkaline andesite.

2.1.3. Central Chain

Volcanism began in the central chain in latest Miocene time (Erromango island: 5.8–5.3 Ma; Bellon *et al.*, 1984) and thus overlapped the eastern belt activity that ceased at about 3.5 Ma. Activity has been virtually continuous, at least from the earliest Pleistocene (1.8 Ma). From Pleistocene to present times, volcanism, which was mainly subaerial, developed extensively along the length of the active arc. Lavas range from picrite to dacite or rhyodacite, with a peak frequency distribution in the basalt range, a trough between 52% and 62% SiO_2, and a smaller peak at 63% SiO_2. This distribution was considered to be representative of all the lavas in the exposed arc, but recent work seems to indicate that the compositional range of the central chain volcanics may be much more continuous (Robin *et al.*, 1991, 1994; Eissen *et al.*, 1992; Picard *et al.*, 1995).

Two broad lava suites can be distinguished, although a compositional spectrum undoubtedly exists between the two. A higher-K_2O suite is represented by picritic to rhyodacitic lavas on the islands of Tongoa, Ambrym, Aoba, and Santa Maria. These volcanics are strongly light-rare-earth-element (LREE) enriched and have $^{87}Sr/^{86}Sr$ ratios ranging from 0.7038 to 0.7043 (Briqueu and Lancelot, 1983). Lavas of the lower-K_2O group dominate the islands of Matthew, Hunter (Maillet *et al.*, 1986), Anatom, Tanna, Erromango, Efate, Lopevi, active submarine volcanoes around Epi, and all the Banks islands except Santa Maria. This lower-K_2O suite characteristically exhibits LREE-enriched REE patterns, but the degree of LREE enrichment is notably less than in the higher-K_2O suite, and $^{87}Sr/^{86}Sr$ ratios are significantly lower (0.7030–0.7033; Briqueu and Lancelot, 1983). The lower-K_2O suite might have been generated by higher degrees of partial melting at shallower levels in the upper mantle than the higher-K_2O suite (Crawford *et al.*, 1988), although the respective mantle sources were clearly different isotopically.

The structural consequences of the collision of the seismically inactive d'Entrecasteaux zone with the central part of the New Hebrides arc (Fig. 5.2) are now well documented (Collot *et al.*, 1985, 1991, 1992a,b; Greene *et al.*, 1988a,b, 1992). This collision, which began 4–3 Ma (Macfarlane *et al.*, 1988), likely affects the central chain as well as the backarc area, tectonically and petrologically. This essential point, though still controversial (Roca, 1978; Monjaret, 1989), will be discussed later.

2.2. The North Fiji Basin

Auzende *et al.* (Chapter 4 this volume) discuss the structure and evolution of the North Fiji basin (NFB) in detail, and the reader is referred to that chapter for a complete description. A summary of the geology and petrology of the NFB is presented here and is taken from Auzende *et al.* (1990), Huchon *et al.* (1994), Nohara *et al.* (1994), and Eissen *et al.* (1994).

The NFB is an active marginal basin bounded by the New Hebrides arc to the west, by the Hunter ridge and fracture zone to the south, by the Fiji Platform to the east, and by the Vitiaz paleosubduction zone to the north (Fig. 5.1). This basin was created by backarc spreading, which commenced behind the New Hebrides arc 10 to 8 Ma ago in response to a clockwise rotation of the arc from an initial position close to the present location of the Vitiaz zone.

The formation of the NFB can be summarized as follows.

10 to 8 Ma (or even in early Miocene time, according to Kroenke, 1984): Collision of the Ontong Java Plateau near the Solomon arc (volcanism ceased on New Hebrides western belt ca. 15–14 Ma) and reversal of the polarity of the subduction from southwestward along the Vitiaz zone to northeastward beneath the New Hebrides arc.

8 to 3 Ma: Beginning of NFB opening, along an axis roughly parallel to the Vitiaz zone; clockwise rotation of the New Hebrides arc (the spreading axis trend rotates progressively from N120° to N150°), and counterclockwise rotation of the Fiji Platform. Active volcanism occurs on the New Hebrides eastern belt until 3.5 Ma, and on the central chain from 5.8 to 5.3 Ma onward.

3 to 0.7 Ma: The spreading axis jumps into the central NFB from a N150° to a N-S direction; synchronously, spreading starts in the Lau Basin. Active volcanism occurs on the New Hebrides central chain.

0.7 Ma to present: Modification of the spreading geometry in the central NFB between 15°S and 18°30′S; migration of a triple junction (ridge-ridge-transform) near 16°40′S, with a southern branch oriented N15°, a northern branch oriented N160°, and the prolongation of the North Fiji fracture zone, oriented N60° in this area. Active volcanism continues on the New Hebrides central chain.

As demonstrated by Eissen *et al.* (1991, 1994), the dominant magma type produced in the NFB is N-MORB, showing depleted large-ion lithophile (LIL), high-field strength (HFS), and light rare-earth (LRE) element patterns. This magma type is the only one present along the most mature N-S spreading segment of the NFB (i.e., between 18°20′S and 21°S).

Several ridges of various origin transect the NFB crust. In particular, the Hazel Holme–South Pandora Ridge is significant, since it may influence the volcano-tectonic evolution of the northern New Hebrides backarc area (Figs. 5.1 and 5.3). As indicated by Pelletier *et al.* (1993a), this seismically active ridge is considered to be an extensional zone in its western part (Hazel Holme extensional zone) and a slow spreading ridge in its easternmost part (South Pandora Ridge; Kroenke *et al.*, 1994; Price and Kroenke, 1991). The western part of the Hazel Holme Ridge trends N85°–N90°E over 100–120 km in width. It is composed of a series of volcanic ridges, scarps, and parallel narrow troughs deeper than 3000 m (Fig. 5.4). To the west of 168°30′E, the width of the whole structure decreases; lateral troughs disappear and a single 2500–3500 m deep E-W trough remains, bounded by symmetrical ridges culminating at a depth of 1700 m (Figs. 5.5, 5.6). This trough abruptly

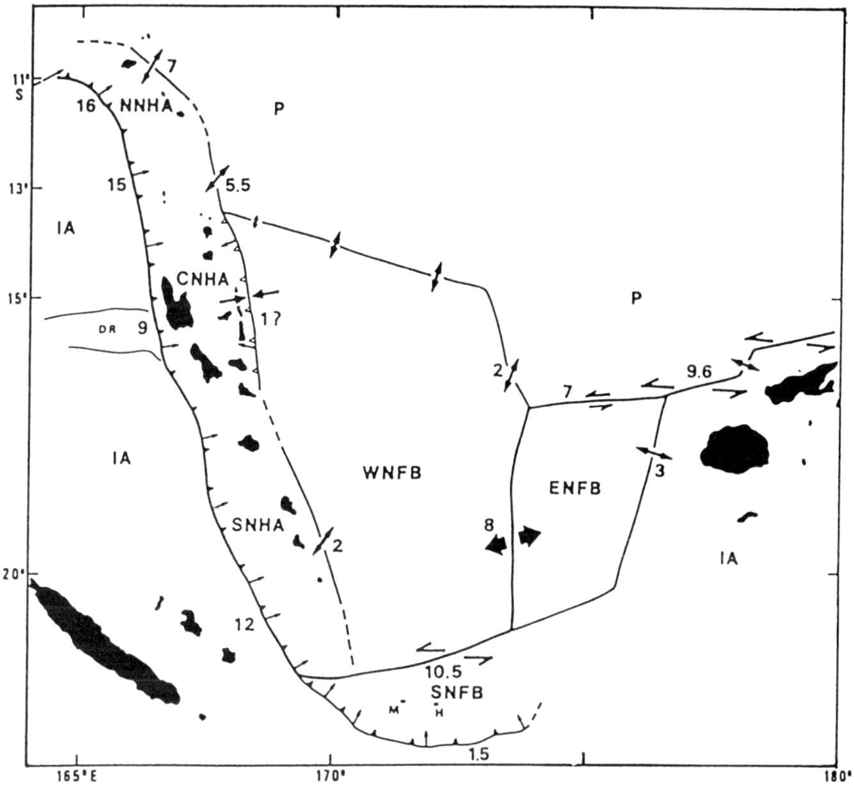

FIGURE 5.3. Present-day relative motions in the New Hebrides–North Fiji Basin region, as proposed by Louat and Pelletier (1989). P: Pacific plate; IA: Indo–Australian plate; DR: d'Entrecasteaux ridge; M and H: Matthew and Hunter volcanoes. WNFB, ENFB, and SNFB: western, eastern and southern North Fiji Basin microplates, respectively; NNHA, CNHA, and SNHA: northern, central and southern segments of the New Hebrides arc microplate, respectively. Numbers and arrows beside plate boundaries indicate the rates (in cm/yr) and trends of the relative motions. Very thick divergent arrows show the motion along the main spreading center of the North Fiji basin. Line with filled barbs corresponds to the New Hebrides Trench. Line with open barbs marks the New Hebrides backarc compressive belt.

terminates at 168°10′E, at the very southernmost tip of the N-S-trending northern New Hebrides backarc area (Pelletier *et al.*, 1993a).

3. THE NEW HEBRIDES BACKARC TROUGHS

3.1. Previous Work and Recent Investigations

A series of narrow and elongated troughs separates the New Hebrides island arc from the adjacent, active, marginal North Fiji Basin (Fig. 5.2). These troughs, however, do not extend along the entire length of the arc, since they are absent in the central backarc area (and, in particular, to the east of Maewo and Pentecost islands). Interestingly, this central area also lies opposite the d'Entrecasteaux zone.

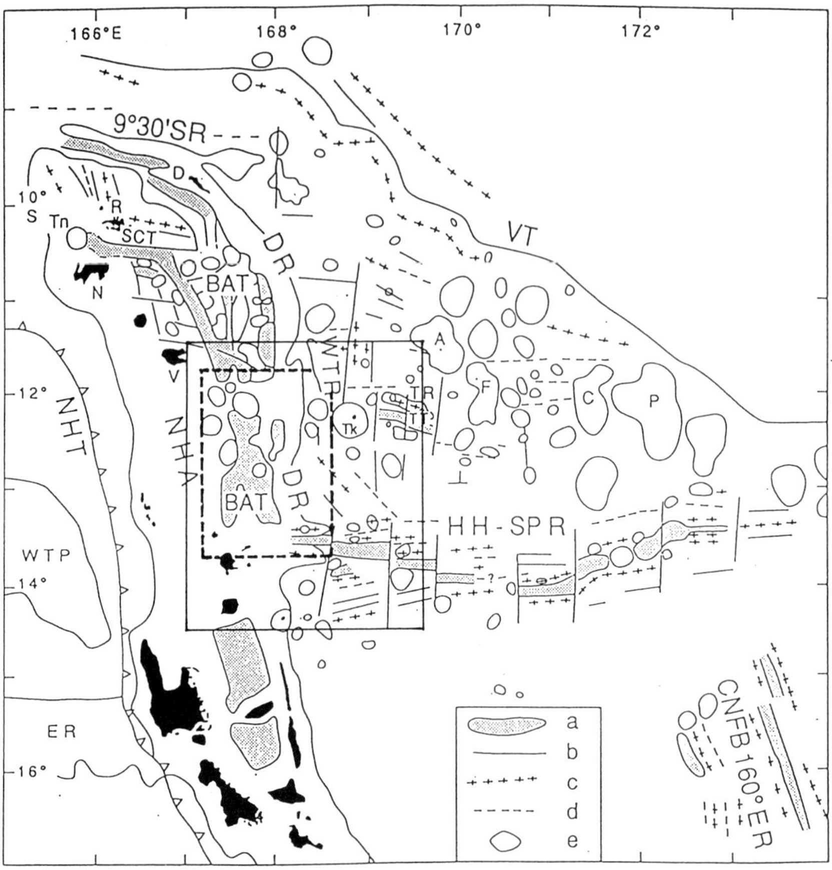

FIGURE 5.4. Structural map of the northern New Hebrides arc/northwestern North Fiji Basin (adapted from Pelletier *et al.*, 1993a). (a) troughs and depressions; (b) structural trends and fractures; (c) structural highs; (d) structural lows; (e) volcanic highs. NHA: New Hebrides arc; VT: Vitiaz Trench; BAT: backarc troughs domain (Jean-Charcot troughs); 9°30′SR: 9°30′S ridge; DR: Duff Ridge; WTR: West Tikopia ridge; TR: Tikopia Ridge; HH-SPR: Hazel Holme–south Pandora ridge; CNFB 160°ER: central North Fiji Basin 160°E ridge; TT: Tikopia Trough; SCT: Santa Cruz Trough; WTP: west Torres plateau; ER: d'Entrecasteaux Ridge. Islands and reefs are in black; D: Duff Islands; R: Reef Islands; Tn: Tinakula Island; N: Nendo Island; V: Vanikoro Island; Tk: Tikopia Island; A: Anuta Island; F: Fatutaka island; C: Charlotte Bank; P: Pandora Bank. Areas detailed in Figs. 5.5–5.7 are outlined.

Thus, between 10°S and 21°S, it is possible to distinguish three main areas in the New Hebrides backarc: (1) the Jean-Charcot troughs (JCT, named after the R/V *Jean Charcot*), or northern troughs, extend between 10°S (Pelletier *et al.*, 1993a,b) and 13°30′S, where they abut the western termination of the Hazel Holme Ridge (Fig. 5.4); a central area, extending from 13°30′S to 17°30′S—that is, roughly from the east of Vanua Lava Island to the east of Tongoa Island, is totally devoid of troughs (Fig. 5.2); (3) the Coriolis troughs (CT; named after the R/V *Coriolis*), or southern troughs, extend between 17°30′S and 21°S, where their southern en echelon termination merges into the structural grain of the island arc (Monzier *et al.*, 1991; Figs. 5.2, 5.9).

Since the first mention of the Coriolis troughs in the literature (Puech and Reichenfeld,

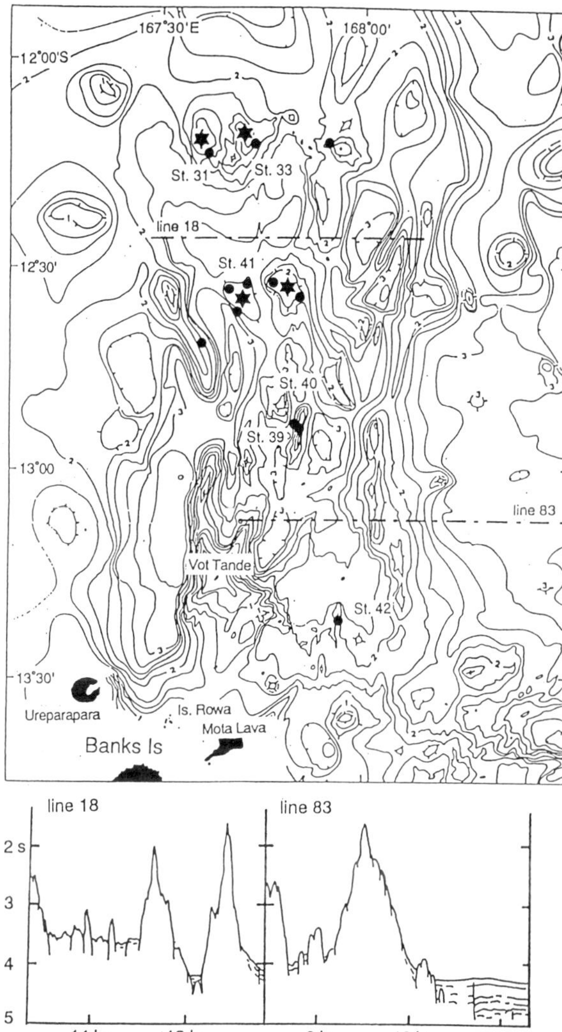

FIGURE 5.5. Bathymetric map of the Jean-Charcot troughs (after Charvis and Pelletier, 1989), showing positions of volcanic cones (stars) and sample localities (dots). Data from SEAPSO 2, KAIYO 89, and SAVANES cruises. Depths in km, with 250 m isobaths. Bottom diagrams: seismic cross sections along survey lines shown above (adapted from Nakada *et al.*, 1994). See Fig. 5.4 for location.

1969), several oceanographic cruises have been partly or fully devoted to the study of the New Hebrides backarc area (Dubois *et al.*, 1978). The most recent, which provide most of the data used in this chapter, include SEAPSO 2 (Nov. 1985, R/V *Jean Charcot*; Récy *et al.*, 1986); EVA 13 (1986, R/V *Coriolis*; Sage and Charvis, 1991); MULTIPSO (May 1987, R/V *Jean Charcot*; Daniel *et al.*, 1989); EVA 14 (Aug. 1987, R/V *Le Noroît*; Pelletier *et al.*, 1988); GLORIA survey (Aug. 1989, HMAS *Cook*; Johnson *et al.*, 1993; Price *et al.*, 1993; Tiffin, 1993); KAIYO 89 (Dec. 1989–Jan. 1990, R/V *Kaiyo*; KAIYO 89 Cruise Report, 1990; Nakada *et al.*, 1994; this chapter); YOKOSUKA 90 (Jan.–Feb. 1991; R/V *Yokosuka*;

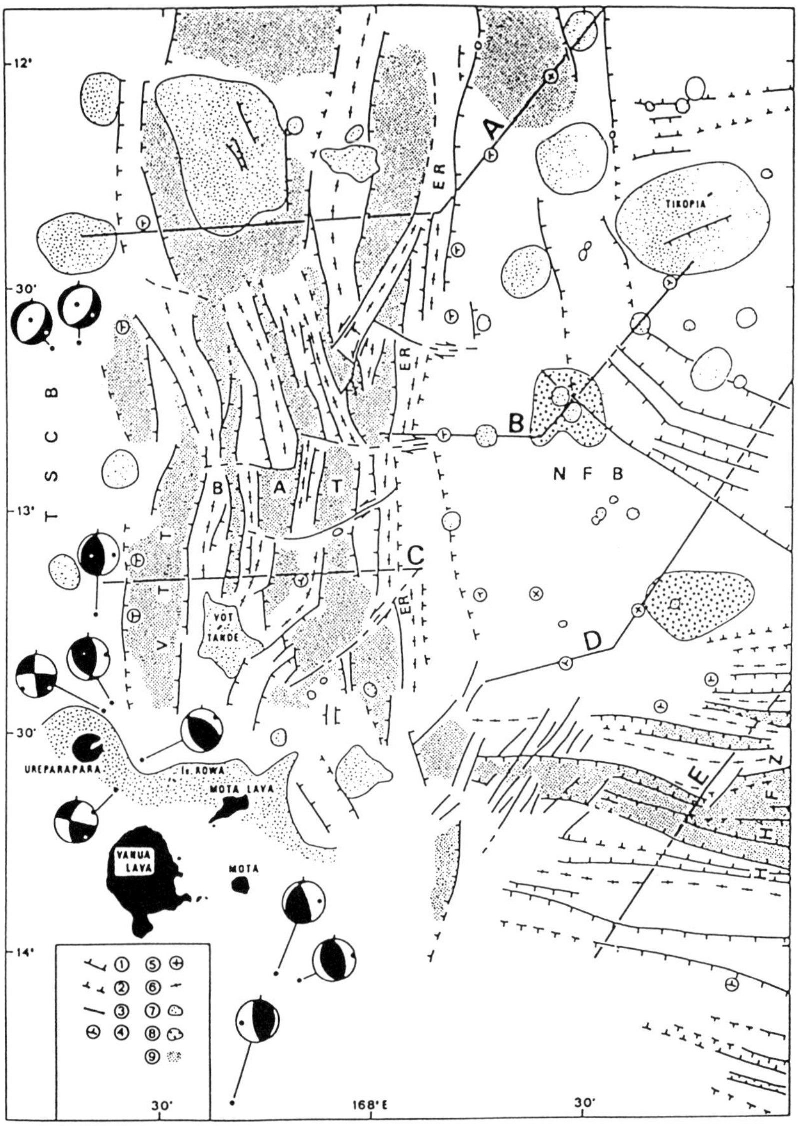

FIGURE 5.6. Structural map of the Jean-Charcot troughs (from Charvis and Pelletier, 1989). 1: normal fault; 2: small and/or inferred normal fault; 3: transverse tectonic lineation (arrows indicate strike-slip fault); 4: inclined bedding; 5: horizontal bedding; 6: structural high; 7: volcano; 8: lava flow; 9: major graben. TSCB: Torres–Santa Cruz sedimentary basin; BAT: backarc troughs (Jean-Charcot troughs); VTT: Vot Tande Trough; ER: Eastern Ridge (Duff Ridge of Fig. 5.4); HHFZ: Hazel Holme fracture zone; NFB: North Fiji Basin. See Fig. 5.4 for location.

Eissen *et al.*, unpublished information, 1994); SANTA CRUZ (Nov.–Dec. 1991; R/V *Le Noroît*; Pelletier *et al.*, 1993b); SAVANES (Dec. 1991–Jan. 1992; R/V *Le Noroît* and *Cyana* submersible; Savanes 91–92 Cruise Report, 1992; Nakada *et al.*, 1994; this chapter).

The KAIYO 89, YOKOSUKA 90, and SAVANES cruises were parts of the STARMER French–Japanese Joint Project (Joint Research Program on Rift System in the Pacific Ocean; Auzende and Urabe, 1994), endorsed by SOPAC (the South Pacific Applied Geoscience Commission, Suva, Fiji).

3.2. Structure and Tectonics

3.2.1. The Jean-Charcot Troughs

The Jean-Charcot troughs (JCT) are a succession of discontinuous and, in some areas, anastomosing, horsts, grabens, and half-grabens, trending slightly oblique to the submeridional northern New Hebrides central chain (Fig. 5.4). Between 12°20′S and 13°20′S, in an area 50–55 km wide (E-W) and 100–120 km long (N-S), at least six individual troughs have been recognized, each 3–9 km wide and 20–35 km long (Figs. 5.5, 5.6; Récy *et al.*, 1986, 1990; Matsumoto *et al.*, 1992; Johnson *et al.*, 1993). The JCT are limited to the west by the submarine extensions of the New Hebrides central chain; to the south by the Hazel Holme ridge; and to the east by the N-S Duff ridge, culminating around 1250 m depth, which separates the JCT from the North Fiji Basin. The northern extension of the JCT, however, is still conjectural. North of a shallow (up to 460 m deep) volcanic complex centered near 12°10′S, the JCT domain widens up to 100 km and abuts the intra-arc E-W-trending Santa Cruz Trough (Fig. 5.4; Pelletier *et al.*, 1993a).

The average depth of the JCT is 3000–3500 m (Fig. 5.5), similar to the mean depth of the North Fiji Basin (Pelletier *et al.*, 1993a; the maximum depth in the whole New Hebrides backarc, 3658 m, was recorded in the JCT, during the KAIYO 89 cruise, at 12°46.9′S–167°41.9′E).

Seismic reflection profiles across the JCT show thick sedimentary layers (more than 1-s twt), which are usually tilted eastward and dip more steeply downsection. Unconformities or tectonic discordances are frequent. The most recent deposits also dip more steeply downsection, indicating coeval tectonics and sedimentation during the entire structural history of the troughs (Charvis and Pelletier, 1989; Récy *et al.*, 1990).

During the period 1977–1987 no shallow earthquake (0–70 km) with magnitude large enough to have a focal mechanism determination has been recorded in the entire JCT area or in the western end of the Hazel Holme Ridge (Charvis and Pelletier, 1989). Yet Louat and Pelletier (1989) suggested a surprisingly high rate of extension in the JCT (5.5 cm/yr along a N45°E direction, at 13°S; 7 cm/yr along a N37°E direction at 11°S; Fig. 5.3), using the RM-2 plate model of Minster and Jordan (1978) combined with a large set of regional data (i.e., shallow seismicity, focal mechanism solutions, and magnetism). Louat and Pelletier (1989) argue that horst-and-graben structures like the JCT can be formed in an aseismic but tectonically active environment. However, GLORIA imagery data show no evidence of any neovolcanism in these troughs (Figs. 5.7, 5.8). The eroded nature of at least some of the seamounts behind the arc, as well as the thick sediment fills in the grabens of the Hazel Holme Ridge, led Johnson *et al.* (1993) to suggest that these structures are not young. However, this interpretation is at odds with the evidence cited for recent faulting producing rotated sedimentary sequences.

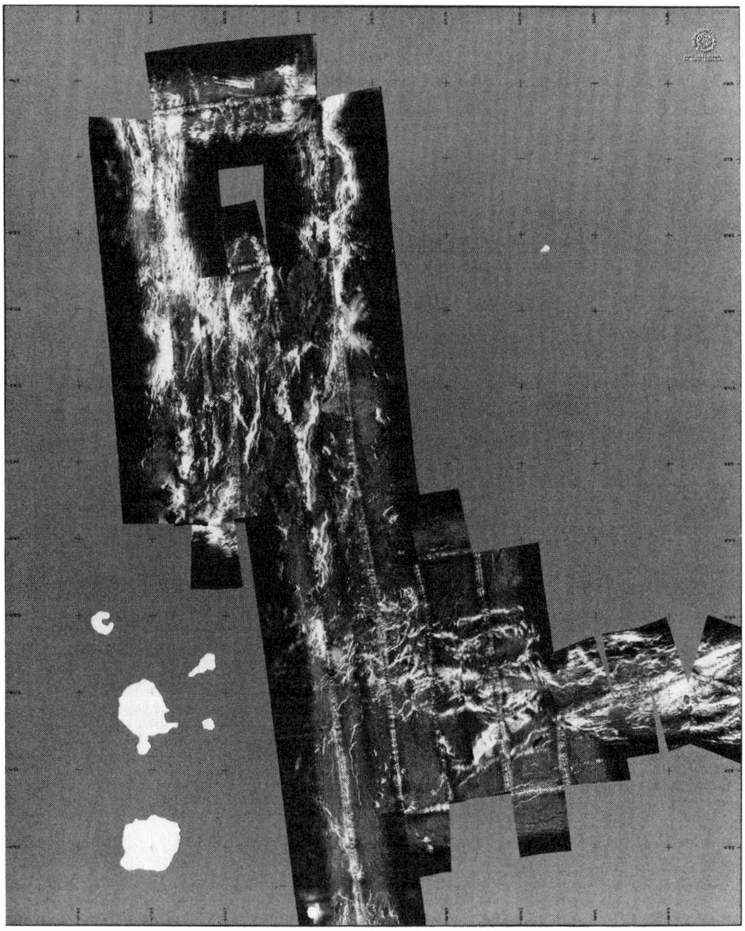

FIGURE 5.7. GLORIA imagery of northern New Hebrides–southeastern Solomon Islands area (from Johnson *et al.*, 1993). Lineations indicated by strong white reflections are associated with the Hazel Holme Ridge in lower right and the Jean-Charcot troughs in upper diagram. See Fig. 5.4 for location.

3.2.2. The Coriolis Troughs

The morphology of the CT looks comparatively simpler (Fig. 5.9). They are made up of well-delineated grabens (Daniel, 1982) discontinuously paralleling the eastern flank of the southern New Hebrides central chain between Efate and Anatom Islands. SeaBeam bathymetry (Récy *et al.*, 1986, 1990) and GLORIA seafloor imagery (Price *et al.*, 1993) allow three main grabens to be distinguished in the CT, namely the Efate, Erromango, and Futuna troughs.

The Efate Trough is a complex double graben, trending NNW-SSE, 10–25 km wide and about 100 km long. A horst at a depth of 1200 m separates a western, small, asymmetrical graben (mean depth: 2150 m) from a larger, deeper, eastern graben (mean depth: 2500–2600 m). The latter is bounded eastward by a very steep normal fault and terminates southward on a structural threshold about 1500 m deep, occupied to the west at 18°32′S–169°35′E by a 1000 m high, less than 500 m deep volcanic seamount (Monzier *et al.*, 1984a;

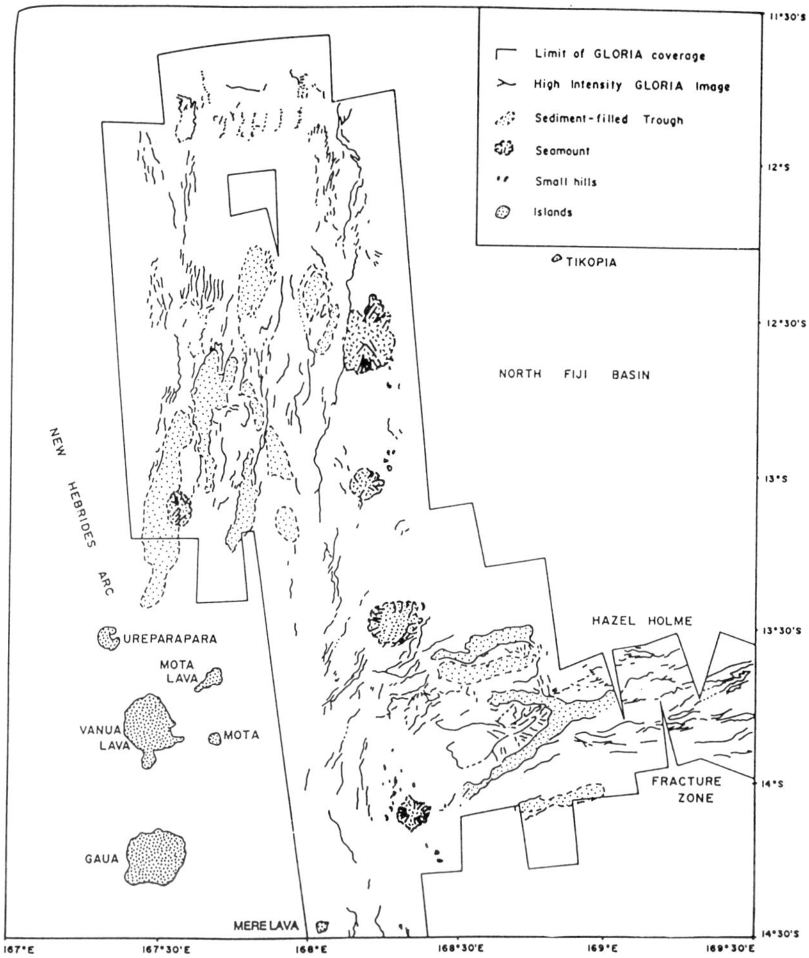

FIGURE 5.8. Geological interpretation of the GLORIA imagery of Fig. 5.7 (from Johnson *et al.*, 1993).

Récy *et al.*, 1990). Downfaulted sediment-draped blocks can be identified in seismic profiles across the Efate Trough, but substantial areas appear on GLORIA imagery to be unsedimented basement outcrop (Fig. 5.10).

Southeastward, the Efate Trough is succeeded by another depression, the Erromango Trough, which is about 75 km long, 30 km wide, and 2500–3100 m deep. Its northeastern and southwestern margins, interpreted as fault scarps, are roughly linear and parallel and rise to about 1500 m depth. An unsampled bathymetric high (2100 m bsl) in the center of the trough (19°05′S–169°52′E) is highly reflective on GLORIA imagery and may correspond to a young, unsedimented volcanic feature (Price *et al.*, 1993).

There is no major structural discontinuity between the Erromango Trough and the Futuna Trough, and the latter can be considered as the southeastward extension of the former (Daniel 1982; Monzier *et al.* 1984a). The Futuna Trough is approximately 75 km long, 20-30 km wide, and 3200 m deep. It is therefore the deepest graben of the CT (the maximum depth recorded during SEAPSO 2 cruise is about 3400 m, at 19°42.5′S–170°09′E, about 20 km south-southwest of Futuna Island). Its asymmetrical morphology is

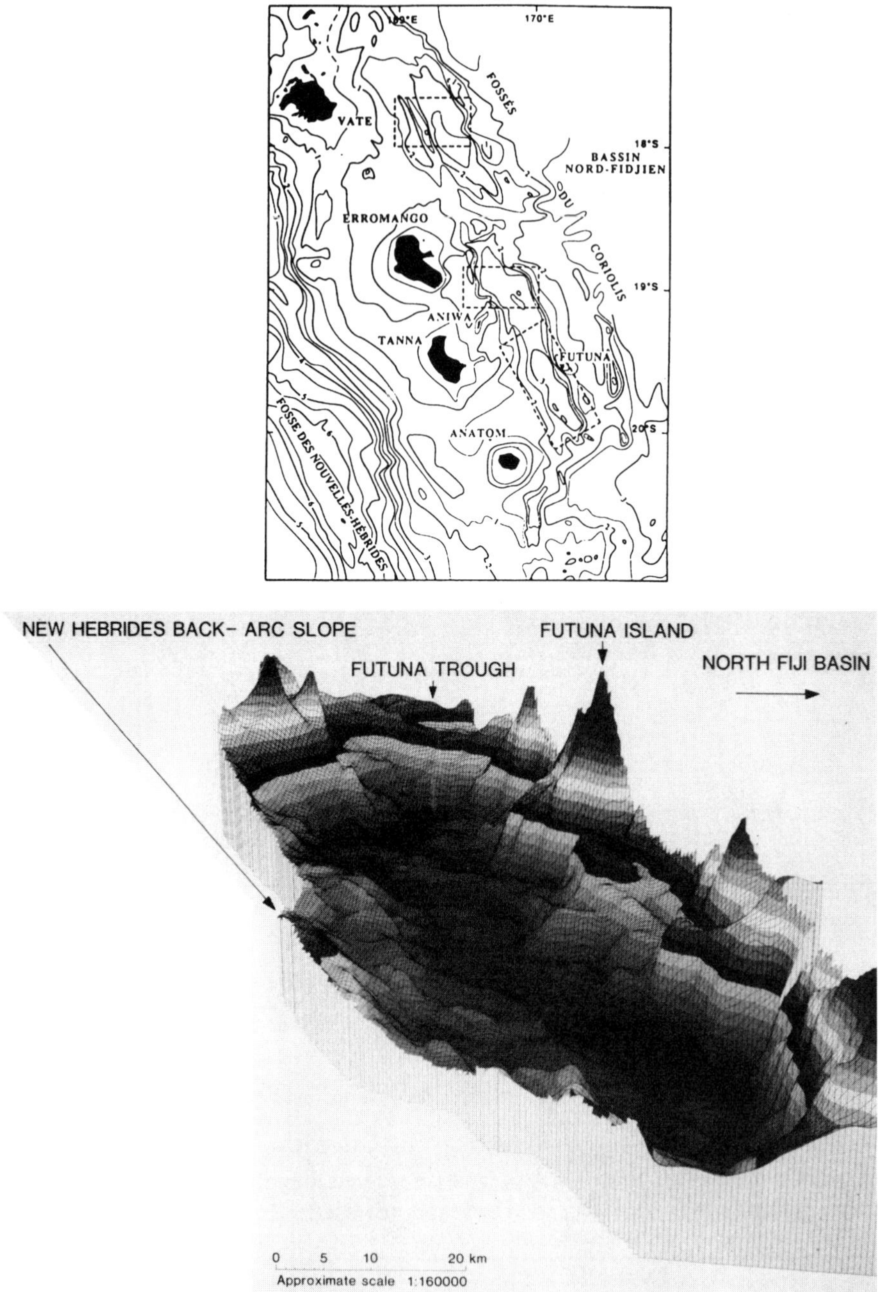

FIGURE 5.9. The Coriolis troughs (from Récy *et al.*, 1990). (Top) Bathymetric map. Depths in km, with 500-m isobaths (from Monzier *et al.*, 1984a). SeaBeam surveyed outlined areas. See Fig. 5.2 for location (Note: Vate = Efate). (Bottom) The Futuna Trough. Block diagram based on SeaBeam bathymetry. Vertical exaggeration: 5/1; mesh size: 300 m; each nuance represents a 150 m depth unit.

obvious, with the eastern wall steeper and shallower (above sea level at Futuna Island) than the western wall (Fig. 5.9). Rough topography, steep fault scarps trending NW-SE, and a thin (less than 80 m) sediment cover characterize most of its floor.

The shallower depth of the Efate Trough may indicate thermal uplift, consistent with extensive recent volcanism apparent on GLORIA imagery (Fig. 5.10). The deeper Erromango and Futuna troughs seem to be relatively immature backarc structures, formed as a consequence of lithospheric extension accompanying uplift or doming (Price *et al.*, 1993).

Numerous shallow earthquakes border the western limit of the CT between 18°50′S

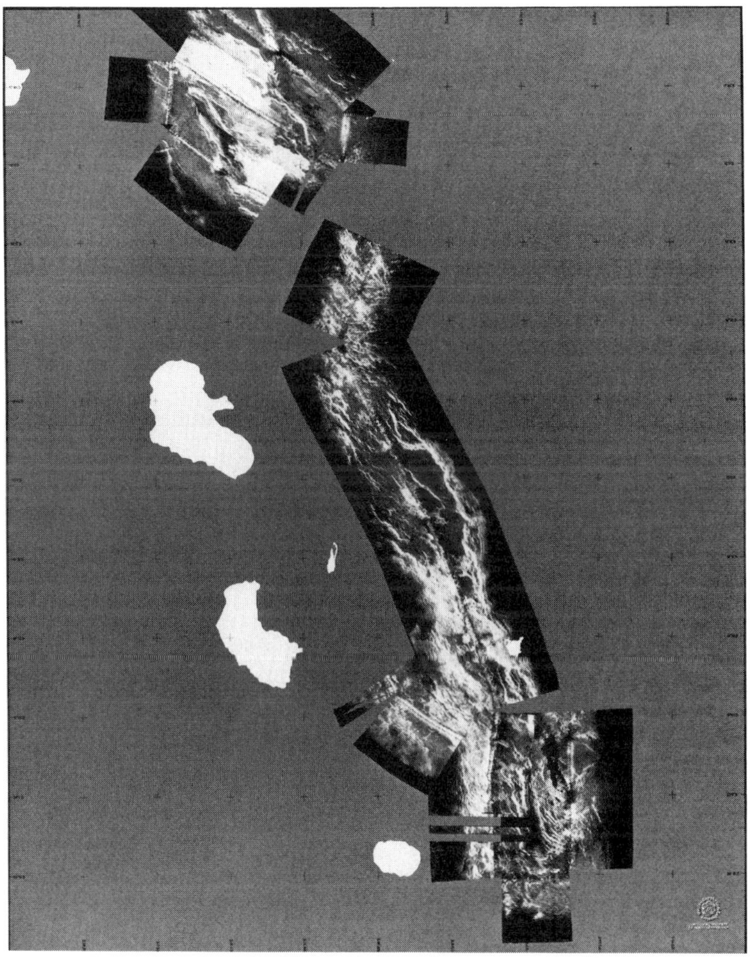

FIGURE 5.10. GLORIA imagery of the Coriolis troughs (from Price *et al.*, 1993). See Fig. 5.2 for location. South of 19°15′S, the line R to W marks the northern wall of the northern Futuna trough. P is an isolated, sediment-filled basin, perched above the Futuna Trough. South of 19°45′S, dark area labeled S is an acoustic shadow on the eastern wall of the southern Futuna Trough. R indicates a series of scarps marking the southwestern wall of the trough. Note: Vate = Efate; Anetium = Anatom.

and 21°S, and normal faulting mechanisms indicate a NE-SW extension. Louat and Pelletier (1989) calculated a rate of 2 cm/yr along a N37°E direction, at 20°S (Fig. 5.3).

Gravity data show a clear positive anomaly paralleling both western and eastern borders of the CT, without any deep-rooted compensation (Collot and Malahoff, 1982), indicative of rift-flank flexural uplift (Weissel and Karner, 1989).

The formation of intra-arc sedimentary basins may have partly influenced the evolution of the New Hebrides backarc area. Among them, the Aoba and Vanikoro–Torres basins deserve special mention.

3.2.3. The Aoba Basins

The central part of the New Hebrides island arc, between Banks Islands and Epi Island, is structurally complex and has recorded the effects of the ongoing collision of the arc with the d'Entrecasteaux zone (Fig. 5.11; Collot *et al.*, 1985, 1992a; Collot and Fisher, 1988; Greene and Johnson, 1988). Deep and thick intra-arc sedimentary basins (North and South Aoba basins) separate the old western belt islands (Espiritu Santo, Malakula) from the younger eastern belt islands (Maewo, Pentecost). Preliminary results of Ocean Drilling Program (ODP) Leg 134 in the North Aoba Basin (Fig. 5.11) indicate that this basin is the product of island-arc volcanism and tectonic deformation—that is, a piggyback subsiding basin pinched and overthrust between two ancient volcanic arcs (Gérard, 1993). At Site 832, an unconformity marks the uplift time of the central part of the western belt in response to the collision of the d'Entrecasteaux zone. The age of this sedimentary hiatus is near the Upper Pliocene–Lower Pleistocene boundary. Recent (Pleistocene) basin-filling recorded at Sites 832 and 833 comes from effusive products of the central chain volcanoes (Aoba, Santa Maria-Gaua, and Mere Lava islands). At Site 833, Pleistocene basaltic sills intruded the Lower Pliocene sedimentary sequence. This recent sill complex indicates that volcanism was active along the eastern belt during the early Pliocene as well as during the Pleistocene (Collot *et al.*, 1992a).

Compression (North Aoba Basin; Greene and Johnson, 1988; Daniel *et al.*, 1989), distension (South Aoba Basin; Greene and Johnson, 1988), and folds and faults with different orientations characterize the Aoba basins area. Moreover, compressive stresses are relayed to the very eastern margins of Maewo and Pentecost, since westward-dipping thrust faults appear on seismic reflection records (Louat and Pelletier, 1989; Récy *et al.*, 1990). Shallow earthquake focal mechanisms confirm this observation (Louat and Pelletier, 1989).

On GLORIA seafloor imagery (Fig. 5.12), the structural limit between the central New Hebrides arc platform and the North Fiji Basin clearly appears as anastomosing fault scarps, between the midpoint of Pentecost Island and the northern end of Maewo Island (Price *et al.*, 1993; Tiffin, 1993).

3.2.4. The Vanikoro–Torres Basin

The poorly known Vanikoro–Torres basin, which extends between Vanikoro and Vanua Lava islands, east of Torres Islands (Fig. 5.13), is a very thick sedimentary basin (up to 6 km thick in its center; Holmes, 1988) lying on the summit platform of the northern New Hebrides arc (Falvey and Greene, 1988). On seismic reflection profiles, three unconformities are distinguished (early-middle Miocene; late Miocene; late Pliocene–early Pleistocene), as well as faulting and folding extending up to the surface (Falvey and

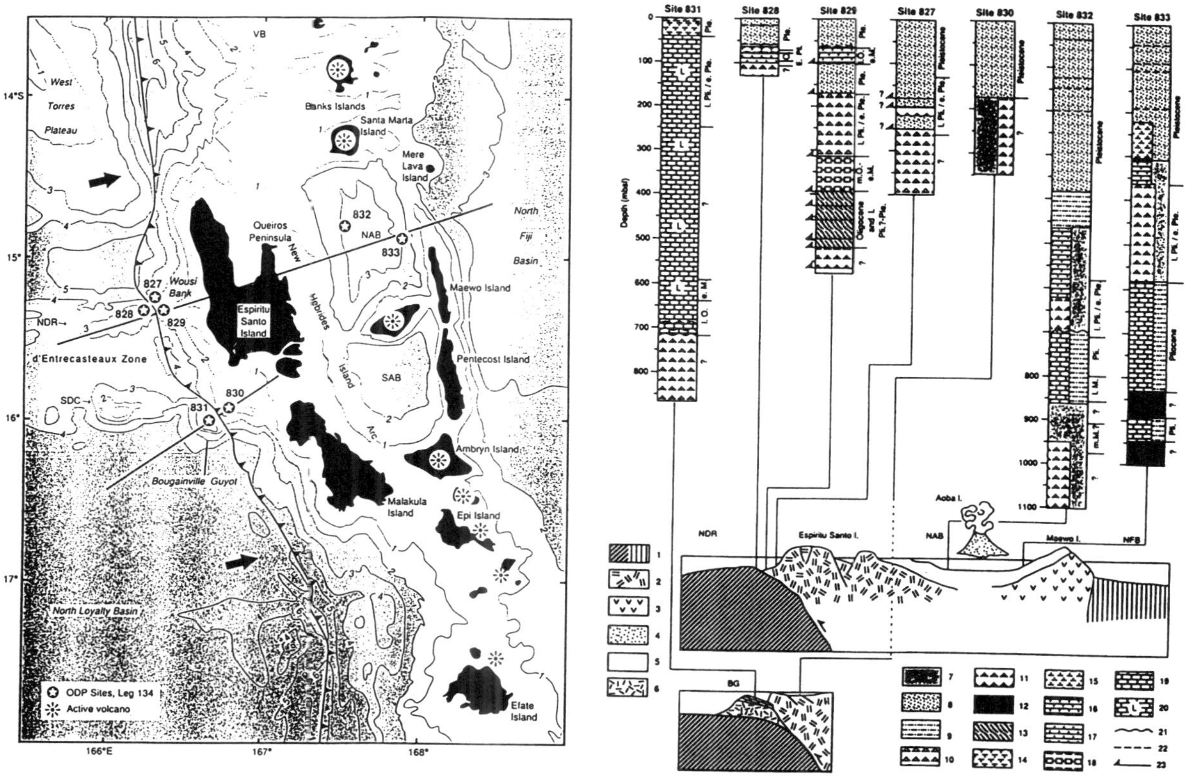

FIGURE 5.11. (Left) Location of sites drilled during ODP Leg 134. Bold lines indicate location of cross sections shown at right. NDR: North d'Entrecasteaux Ridge; SDC: south d'Entrecasteaux chain; NAB: North Aoba Basin; SAB: South Aoba Basin; VB: Vanikoro Basin. Bold line with teeth indicates approximate position of subduction zone; teeth are on upper plate. Arrows indicate direction of plate convergence. Bathymetry in km. (Right) Geologic columns and cross sections, ODP Leg 134. 1: oceanic crust; 2: western belt volcanic rocks; 3: eastern belt volcanic rocks; 4: central chain volcanic rocks; 5: basin fill; 6: guyot volcanic rocks; 7: volcanic sand/sandstone; 8: volcanic silt/siltstone; 9: volcanic sandstone/siltstone/claystone; 10: sed-lithic breccia; 11: volcanic breccia; 12: basalt chalk; 13: multiple slivers of siltstone and chalk; 14: foraminiferal ooze; 15: nanofossil ooze; 16: foraminiferal chalk; 17: nannofossil chalk; 18: calcareous chalk; 19: pelagic limestone; 20: lagoonal limestone; 21: unconformity; 22: ash; 23: thrust fault. NDR: north d'Entrecasteaux Ridge; BG: Bougainville guyot; NAB: North Aoba Basin; NFB: North Fiji Basin (From Collot *et al.*, 1992a).

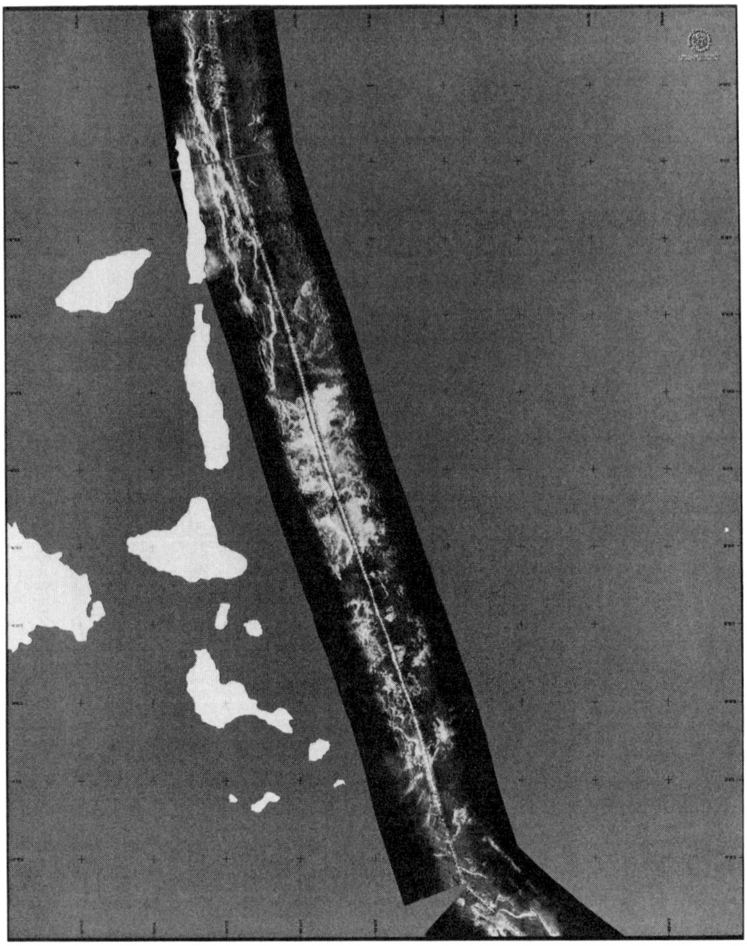

FIGURE 5.12. GLORIA imagery along the New Hebrides backarc between 14°30′S and 17°30′S (from Price *et al.*, 1993). See Fig. 5.2 for location.

Greene, 1988). The easternmost part of the Vanikoro–Torres Basin (i.e., the area closest to the northern NHBAT) is intruded by young (Pleistocene–Holocene) volcanic intrusions from the New Hebrides central chain.

3.3. Volcanic Petrology, Geochronology, and Geochemistry

3.3.1. Volcanic Petrology

Before the SEAPSO 2 cruise (1985), the petrological knowledge of the NHBAT was limited to two studies. Dugas *et al.* (1977) described and analyzed a few volcanic samples from the Futuna Trough (GEORSTOM III CENTRE cruise, 1975), while Vallot (1984) presented the first compilation of dredged samples from along the whole southern New Hebrides arc (i.e., inner trench slope, arc substratum, and Coriolis backarc troughs) during GEORSTOM III CENTRE and EVA V (1977) cruises.

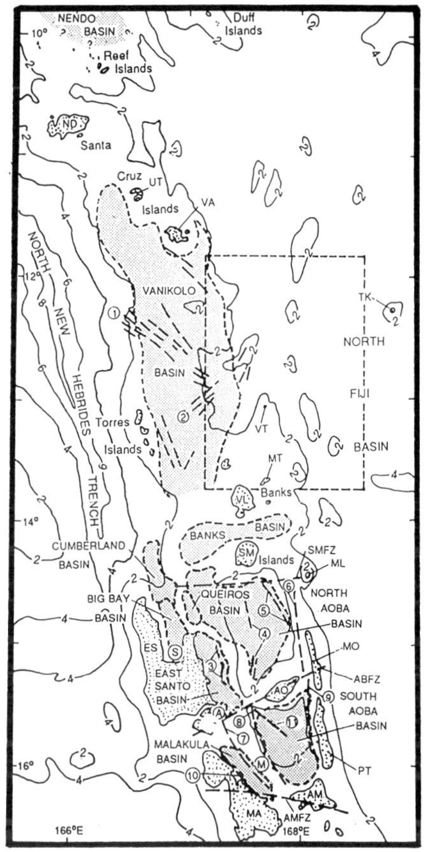

FIGURE 5.13. The central and northern New Hebrides arc, with its intra-arc sedimentary basins (North Aoba, South Aoba and Vanikoro–Torres basins) and fault systems. The Vanikoro-Torres basin appears as Torres–Santa Cruz sedimentary basin in Fig. 5.6, and as Vanikoro Basin in Fig. 5.11. The Jean-Charcot troughs area is outlined (see Fig. 5.5). Vanikolo = Vanikoro. Bathymetry in km. Islands toponymy: VA: Vanikoro; VL: Vanua Lava; VT: Vot Tande; MT: Mota Lava; TK: Tikopia; MO: Maewo; PT: Pentecost; AO: Aoba; ES: Espiritu Santo; MA: Malakula; AM: Ambrym (From Falvey and Greene, 1988).

In order to get a closer view of the structure and volcanology of these troughs, a comprehensive geological, geophysical, and geochronological study was initiated during the 1985 SEAPSO 2 cruise (Récy *et al.*, 1986; Monjaret *et al.*, 1991). Most of the petrological data discussed here come from this study. A few cruises with more specific targets followed more recently (see above), and some data from these are also included in this chapter (Table I).

The dredged samples and samples picked up *in situ* (via submersible) from the NHBAT have been classified geographically from north to south according to neighboring islands or structures as follows: Vanikoro (VAN), Tikopia (TIK), and Vot Tande (VOT), for the Jean-Charcot troughs (JCT); Hazel Holme (HAZ); Efate (EFA), Erromango (ERR), Futuna (FUT), and Anatom (ANA) for the Coriolis troughs (CT). The coordinates of these samples and their structural settings are shown in Table Ia–Ib and Figs. 5.5, 5.14, and 5.15. As summarized by Monjaret *et al.* (1991), basalts (45–53% SiO_2) clearly predominate (60% of lavas dredged during the SEAPSO 2 cruise), whereas basaltic andesites (53–60%) and dacites (SiO_2 > 60%) represent 13% and 27% of dredged samples, respectively. These relative abundances did not drastically change after sampling carried out during KAIYO 89, YOKOSUKA 90 and SAVANES cruises (Table II).

Lavas from the JCT are usually aphyric or weakly porphyritic and show high vesicularity (generally 20% of the rock volume, sometimes up to 40–50%). By contrast, volcanics from the CT are generally porphyritic (15% of phenocrysts: plagioclase, clinopyroxene ± olivine, orthopyroxene, Fe-Ti oxides) or, in some cases, highly porphyritic (up to 50% phenocrysts).

Nine petrological types (Table II: types a to i) can be distinguished in the NHBAT volcanics (Monjaret *et al.*, 1991). Mafic lavas are represented by six types:

- Type a: MORB (Mid-ocean ridge basalts)
- Types b, c, d: IAT (island-arc tholeiites), which may be enriched in TiO_2 (type c), in MgO (type d), or in TiO_2 and MgO (type c+d)
- Type e: CAB (calc-alkaline basalts and andesites), which are also in some cases enriched in TiO_2 (type e+c) or in MgO (type e+d)
- Type f: BABB (backarc basin basalts)—basalts with a composition intermediate between MORB and IAT

Felsic lavas are represented by three types:

- Type g: high-Na/low-K dacites (Nakada *et al.*, 1994)
- Type h: high-K dacites
- Type i: hyper-K dacites

MORB (a), BABB (f), and high-Na/low-K dacites (g) have only been found in the northern troughs (JCT) and in the Hazel Holme area: MORB (a) are restricted to VAN and HAZ areas, BABB (f) to the VAN area, high-Na/low-K dacites (g) to the VAN and TIK areas (Table II); all other petrological types recovered in the NHBAT resemble the ones found on the central chain islands.

3.3.2. Geochronology

Most K-Ar ages presented in Table III originated from SEAPSO 2 cruise samples and were obtained in France (UBO, Université de Bretagne Occidentale, Brest). They have been discussed elsewhere (Monjaret *et al.*, 1987, 1991; Monjaret, 1989; Récy *et al.*, 1990). Complementary data from samples 3152 (Vanikoro area), 3981X and 415X (Tikopia area), recovered during the KAIYO 89 cruise, were analyzed in Japan and have been added to this list.

Sample 7M2 is a MORB dredged on the Duff Ridge (Table I; Fig. 5.14) (i.e., on the eastern border of the Vanikoro area). It gave a K/Ar age of 12.4 ± 0.9 Ma (Table III). This MORB sample represents the oldest known remnant of the North Fiji Basin oceanic crust (Monjaret *et al.*, 1991) and is not considered to be related to oceanic spreading in the NHBAT. K-Ar ages of BABB dredged in the northern JCT (type f in Table III) range between 3.9 and 1.1 Ma. Other geochemical types encountered in the NHBAT are characterized by a rather large spectrum of isotopic ages, most of them, however, being younger than 4 Ma. To Monjaret *et al.* (1991) this succession reveals a polyphased and diachronous trough formation.

3.3.3. Geochemistry

3.3.3.1. Major Elements. As noted by Fryer *et al.* (1990) for the volcanics from the Izu–Bonin backarc rifts, the wide range of SiO_2 contents together with the high-Al_2O_3 contents of the NHBAT volcanics (Table II; Fig. 5.16) clearly emphasize their arc/backarc environment, and this is confirmed by their TiO_2 contents, which are usually lower than 1.2%. However, the SiO_2 contents of the NHBAT volcanics cover a compositional continuum, which is not common in such a backarc environment.

TABLE I

Location (a) and Structural Setting (b) of Volcanic Samples[a]

Location	Start	End	Depth (m)
	(a)		
	Jean Charcot Troughs		
Vanikoro area (VAN)			
1	12°12.9′S–167°34.6′E	12°11.9′S–167°34.9′E	−1470 to −940
2	12°15.3′S–167°38.5′E	12°14.9′S–167°38.8′E	−1400 ± 100
3	12°13.5′S–167°40.6′E	12°10.9′S–167°42.1′E	−1350 to −900
5	12°09.1′S–167°48.5′E	12°08.8′S–167°48.8′E	−1850 to −1650
6	12°14.7′S–167°50.3′E	12°15.8′S–167°51.0′E	−2450 to −2120
7	12°16.3′S–167°51.8′E	12°16.7′S–167°52.7′E	−2165 to −1930
3321	12°10.3′S–167°47.0′E	12°10.3′S–167°46.3′E	−2038 to −1658
3151	12°11.6′S–167°34.9′E	12°11.6′S–167°35.4′E	−915 to −459
3154	12°11.6′S–167°34.9′E	12°11.6′S–167°35.4′E	−915 to −459
3152	12°11.6′S–167°34.9′E	12°11.6′S–167°35.4′E	−915 to −459
3155	12°11.6′S–167°34.9′E	12°11.6′S–167°35.4′E	−915 to −459
Tikopia area (TIK)			
3981	12°51.5′S–167°50.5′E	12°52.2′S–167°51.0′E	−2984 to −2130
3981X	12°51.5′S–167°50.5′E	12°52.2′S–167°51.0′E	−2984 to −2130
3982	12°51.5′S–167°50.5′E	12°52.2′S–167°51.0′E	−2984 to −2130
3983	12°51.5′S–167°50.5′E	12°52.2′S–167°51.0′E	−2984 to −2130
4150	12°33.3′S–167°40.6′E	12°33.2′S–167°40.8′E	−1425 to −1284
4151	12°33.3′S–167°40.6′E	12°33.2′S–167°40.8′E	−1425 to −1284
4153	12°33.3′S–167°40.6′E	12°33.2′S–167°40.8′E	−1425 to −1284
4155	12°33.3′S–167°40.6′E	12°33.2′S–167°40.8′E	−1425 to −1284
4152	12°33.3′S–167°40.6′E	12°33.2′S–167°40.8′E	−1425 to −1284
415X	12°33.3′S–167°40.6′E	12°33.2′S–167°40.8′E	−1425 to −1284
CY11	12°32.0′S–167°40.4′E		−1905
CY31	12°31.7′S–167°45.1′E		−2412
CY34	12°31.5′S–167°45.2′E		−2203
CY36	12°32.4′S–167°47.2′E		−1550
Vot Tande Area (VOT)			
10	13°23.9′S–167°59.7′E	13°24.8′S–167°59.6′E	−2130 to −1880
11	13°20.9′S–167°57.1′E	13°21.7′S–167°55.9′E	−2000 to −1550
12	13°20.9′S–167°49.2′E	13°21.4′S–167°48.1′E	−2200 to −1600
4294	13°19.9′S–167°55.8′E	13°19.9′S–167°56.5′E	−2140 to −1650
4295	13°19.9′S–167°55.8′E	13°19.9′S–167°56.5′E	−2140 to −1650
	Hazel Holme Area (HAZ)		
14	13°40.0′S–168°30.0′E	13°41.6′S–168°28.8′E	−3800 to −2870
15	13°41.0′S–168°29.7′E	13°41.9′S–168°29.6′E	−2900 to −2500
	Coriolis Troughs		
Efate area (EFA)			
26	17°38.6′S–169°24.7′E	17°39.4′S–169°25.6′E	−2080 to −1850
27	17°39.8′S–169°25.5′E	17°39.5′S–169°26.3′E	−1960 to −1200
28	17°38.4′S–169°26.4′E	17°38.2′S–169°25.8′E	−1270 to −700
29	17°38.4′S–169°25.6′E	17°38.2′S–169°26.1′E	−980 to −600
30	17°23.3′S–169°02.5′E	17°23.2′S–169°02.1′E	−1270 to −1200
31	17°23.5′S–169°08.0′E	17°23.6′S–169°09.0′E	−1570 to −1250
Erromango area (ERR)			
22	18°49.6′S–169°39.8′E	18°48.8′S–169°37.9′E	−2800 to −2400
24	18°47.8′S–169°35.1′E	18°47.9′S–169°34.9′E	−1420 to −900

(continued)

TABLE I
(*Continued*)

Location	Start	End	Depth (m)
	(a) Coriolis Troughs (*cont.*)		
Erromango area (ERR) (*cont.*)			
25	18°32.4′S–169°34.3′E	18°31.9′S–169°34.8′E	−910 to −750
Futuna area (FUT)			
16	19°47.8′S–170°16.5′E		−3320 to −2500
17	19°47.6′S–170°17.2′E	19°46.8′S–170°18.3′E	−2900 to −2150
19	19°46.3′S–170°19.1′E	19°46.2′S–170°20.6′E	−1750 to −1550
20	19°25.0′S–169°54.6′E	19°25.6′S–169°55.5′E	−1400 to −1000
21	19°54.1′S–170°17.0′E	19°55.5′S–170°18.6′E	−3280 to −3150
Anatom area (ANA)			
4311	20°31.0′S–170°00.5′E	20°30.9′S–170°00.7′E	−2288 to −2066
4312	20°31.0′S–170°00.5′E	20°30.9′S–170°00.7′E	−2288 to −2066
4313	20°31.0′S–170°00.5′E	20°30.9′S–170°00.7′E	−2288 to −2066
4314	20°31.0′S–170°00.5′E	20°30.9′S–170°00.7′E	−2288 to −2066
4315	20°31.0′S–170°00.5′E	20°30.9′S–170°00.7′E	−2288 to −2066
4316	20°31.0′S–170°00.5′E	20°30.9′S–170°00.7′E	−2288 to −2066
4318	20°31.0′S–170°00.5′E	20°30.9′S–170°00.7′E	−2288 to −2066
43110	20°31.0′S–170°00.5′E	20°30.9′S–170°00.7′E	−2288 to −2066
43112	20°31.0′S–170°00.5′E	20°30.9′S–170°00.7′E	−2288 to −2066
4421	20°52.6′S–170°00.5′E	20°52.3′S–170°00.9′E	−979 to −771
4422	20°52.6′S–170°00.5′E	20°52.3′S–170°00.9′E	−979 to −771
4423	20°52.6′S–170°00.5′E	20°52.3′S–170°00.9′E	−979 to −771
4424	20°52.6′S–170°00.5′E	20°52.3′S–170°00.9′E	−979 to −771
4531	20°34.0′S–170°02.6′E	20°33.9′S–170°02.8′E	−2054 to −1923
4532	20°34.0′S–170°02.6′E	20°33.9′S–170°02.8′E	−2054 to −1923
4533	20°34.0′S–170°02.6′E	20°33.9′S–170°02.8′E	−2054 to −1923
4534	20°34.0′S–170°02.6′E	20°33.9′S–170°02.8′E	−2054 to −1923
4641	20°02.6′S–170°39.8′E	20°02.8′S–170°40.3′E	−1144 to −1055

(b)

Location			Cruise[b]
	Jean Charcot Troughs		
Vanikoro area (VAN)			
1	central volcanic complex		1
2	central volcanic complex		1
3	central volcanic complex		1
5	central volcanic complex		1
6	Duff Ridge (eastern limit of JCT)		1
7	Duff Ridge (eastern limit of JCT)		1
3321	central volcanic complex	(station 33)	2
3151	central volcanic complex	(station 31)	2
3154	central volcanic complex	(station 31)	2
3152	central volcanic complex	(station 31)	2
3155	central volcanic complex	(station 31)	2
Tikopia area (TIK)			
3981	central graben (eastern wall)	(station 39)	2
3981X	central graben (eastern wall)	(station 39)	2
3982	central graben (eastern wall)	(station 39)	2
3983	central graben (eastern wall)	(station 39)	2
4150	western seamount (W slope)	(station 41)	2

(*continued*)

TABLE I
(*Continued*)

Location			Cruise
	(b) Jean Charcot Troughs (*cont.*)		
Tikopia area (TIK) (*cont.*)			
4151	western seamount (W slope)	(station 41)	2
4153	western seamount (W slope)	(station 41)	2
4155	western seamount (W slope)	(station 41)	2
4152	western seamount (W slope)	(station 41)	2
415X	western seamount (W slope)	(station 41)	2
CY11	western seamount (NW slope)	(station 41)	3
CY31	eastern seamount (NW slope)	(station 41)	3
CY34	eastern seamount (NW slope)	(station 41)	3
CY36	eastern seamount (NW slope)	(station 41)	3
Vot Tande area (VOT)			
10	small isolated seamount (east of central horst)		1
11	central horst (E slope)		1
12	Vot Tande island basement (SE flank)		1
4294	central horst (W slope)	(station 42)	2
4295	central horst (W slope)	(station 42)	2
	Hazel Holme Area (HAZ)		
14	southern scarp of western termination (basis)		1
15	southern scarp of western termination (top)		1
	Coriolis Troughs		
Efate area (EFA)			
26	eastern flank (basis) of eastern graben		1
27	eastern flank (basis) of eastern graben		1
28	eastern flank (middle) of eastern graben		1
29	eastern flank (top) of eastern graben		1
30	western volcanic cone to the north of the trough		1
31	eastern volcanic cone to the north of the trough		1
Erromango area (ERR)			
22	western scarp of northwestern termination (basis)		1
24	western scarp of northwestern termination (top)		1
25	isolated volcanic cone to the north of the trough		1
Futuna area (FUT)			
16	southeastern scarp of the trough (basis)		1
17	southeastern scarp of the trough (middle)		1
19	southeastern scarp of the trough (top)		1
20	volcanic cone on the NW flank of the trough		1
21	small southern relief on the trough bottom		1
Anatom area (ANA)			
4311 to 43112	small ridge, within the en echelon termination of the trough	(station 43)	4
4421 to 4424	small seamount outside the trough, on the central chain axis	(station 44)	4
4531 to 4534	eastern flank of the en-echelon termination of the trough	(station 45)	4
4641	volcanic cone (Mt Yokosuka) to the east of the trough	(station 46)	4

[a]The CY samples were picked up in situ by *Cyana* submersible.
[b]1: SEAPSO 2 (1985); 2: KAIYO 89 (1989), with STARMER station number; 3: SAVANES (1991), with STARMER station number; 4: YOKOSUKA 90 (1991), with STARMER station number.

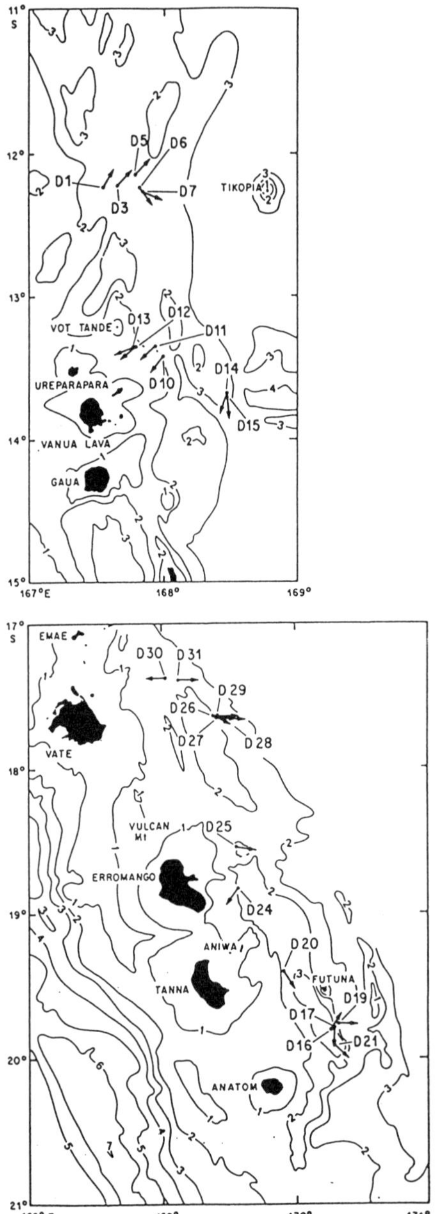

FIGURE 5.14. Location of volcanic samples dredged during the SEAPSO 2 cruise in the Jean-Charcot troughs (top) and Coriolis troughs (bottom). Coordinates and structural setting of these samples are listed in Tables Ia–Ib (sample D1 in Fig. 5.14 = sample 1 in Tables I–III; sample D25 in Fig. 5.14 = sample 25 in Tables I–III, etc.). Bathymetry in km. Vate = Efate, Gaua = Santa Maria in Fig. 5.2 (From Monjaret *et al.*, 1991).

The K_2O versus SiO_2 diagram (Fig. 5.16) emphasizes two trends, similar to those for the New Hebrides central chain volcanics. A lower-K_2O suite encompasses most samples from the northern Jean-Charcot troughs: all the VAN volcanics (except the 7M4 andesite), including the high-Na/low-K dacites studied by Nakada *et al.* (1994); most of the TIK volcanics, except for four samples with higher-K_2O contents (3981, 3981X, 3982, 3983; Table II), which are calc-alkaline andesites; and the island-arc tholeiites of the VOT area. A higher-K_2O suite is represented in the southern Coriolis troughs by the volcanics from the

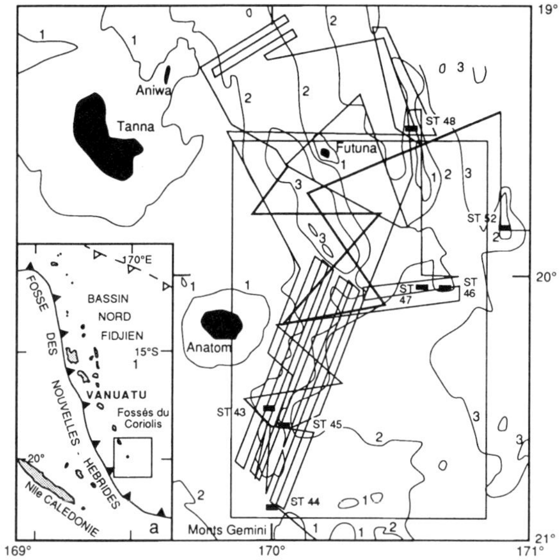

FIGURE 5.15. Location of samples dredged during the YOKOSUKA 90 cruise in the southern Coriolis troughs. Coordinates and structural setting of these samples are listed in Tables Ia–Ib (ST 43 = samples 4311 to 43112, ST 44 = samples 4421 to 4424, ST 45 = samples 4531 to 4534, ST 46 = sample 4641 in Tables I and II). Bathymetry in km (From Eissen *et al.*, unpublished information, 1994).

Efate (EFA) and Anatom (ANA) areas. The ANA backarc volcanics, however, differ from the Anatom Island volcanics, the latter belonging mainly to a lower-K_2O suite; in contrast, hyper-K dacites from Efate area (EFA; i group on Table II) show a close chemical similarity with acidic pumices of Efate island (Coulon *et al.*, 1979).

All other groups of volcanics from the backarc areas display an intermediate K_2O trend with respect to SiO_2 contents.

3.3.3.2. Trace Elements. Various trace elements are plotted against SiO_2 in Fig. 5.17. LILE (e.g., Rb, Ba, and Th) and HFSE (e.g., Nb, Ta, Hf, Sm, Zr and Y) classically show a steady increase with increasing SiO_2. In contrast, Ni behaves compatibly. Rb vs. SiO_2, Ba vs. SiO_2, and Th vs. SiO_2 diagrams (Fig. 5.17) look quite similar, and all emphasize the distinction between the lower-K_2O suites (most of JCT samples) and the higher-K_2O suites (e.g., EFA samples). Contrasting with other LILE, Sr contents decrease with increasing SiO_2 contents. This trend, likely due to plagioclase fractionation, is more pronounced in the felsic lavas (SiO_2 >60%).

The Ba/Zr ratio, often considered as an indicator of the extent of "arc signature" (Fryer *et al.*, 1990), varies greatly in the NHBAT volcanics. VAN and HAZ samples (Table II) have the lowest Ba/Zr ratio, usually between 0.4 and 1. These values are still well above the ratio for N-MORB (0.08; Sun and McDonough, 1989). All other samples have a Ba/Zr ratio ranging between 1 and 5 (or more), typical of volcanics from island-arc environments.

On the Th/Yb vs. Ta/Yb diagram (Pearce, 1983) the distinction between northern and southern troughs is neat (Fig. 5.18). The geochemistry of all Coriolis samples (EFA, ERR, FUT; southern troughs) is influenced by a subduction component, these samples having relatively low Ta/Yb ratios and high Th/Yb ratios. In contrast, this subduction component is less pronounced, or even absent, in the Jean-Charcot (VAN, VOT) and Hazel Holme (HAZ) samples (northern troughs).

The chondrite-normalized REE patterns and N-MORB normalized "spider diagrams" of the different petrological types encountered in the NHBAT are shown in Fig. 5.19. MORB and BABB dredged in the VAN area present similar flat, or slightly LREE-depleted, patterns, as typically observed in N-MORB or BABB (Saunders and Tarney, 1984; Eissen *et al.*, 1991; 1994; Sinton *et al.*, 1994). MORB from the Hazel Holme area are significantly LILE- and LREE-enriched, compared to the VAN N-MORB, and are therefore classified as E-MORB.

The LILE (Ba, Rb, Sr) enrichment and HFSE (Ta, Nb, Zr, Ti, Y, Yb) depletion, relative to N-MORB, characteristic of island-arc basic and intermediate volcanics (Gill, 1981; Thorpe, 1982; Wilson, 1989), is particularly clear in Fig. 5.19 (see, for example, IAT types b, c, d of Table II from the Coriolis troughs). As noted, K-rich felsic lavas—that is, high- and hyper-K dacites (types h and i of Table II)—usually show a strongly-LREE-enriched pattern, often accompanied by a Sr depletion due to plagioclase fractionation.

In summary, even though no oceanic spreading actually occurs in the New Hebrides backarc troughs, some BABB dredged in their very northern part (central volcanic complex of the Vanikoro area, Jean-Charcot troughs) still show evidence of a limited trend toward an aborted marginal basin, with an age between 3.9 and 1.1 Ma (Tables I–III). A neat petrological dichotomy marks these troughs. Volcanics from the southern troughs (CT) present all the classical characteristics of arc-related, orogenic magmas. In contrast, the northern troughs (JCT) are floored with volcanics reflecting more complex influences, with compositions trending between BABB and more typical arc lavas.

The significant extension and thick sedimentary input that characterize the central and northern New Hebrides arc platform have been noted, the Aoba and Vanikoro–Torres basins (Fig. 5.13) having no equivalent in the southern part of the arc. The geographical proximity of the Vanikoro–Torres Basin and JCT, both affected by extensional stresses, raises therefore the question of their possible genetical relationship. No petrological or geophysical argument, however, relates BABB formation in the JCT between 3.9 and 1.1 Ma to extension in the Vanikoro–Torres Basin.

3.4. Backarc Hydrothermal Activity and Ferromanganese Crusts

3.4.1. Backarc Hydrothermal Activity

During the KAIYO 89 cruise, five hydrocasts were performed in the JCT for conductivity-temperature-depth (CTD) profiles and methane (CH_4) analysis. Following onboard analyses, one hydrocast gave a distinct CH_4 anomaly, but this was not accompanied by a thermal anomaly. This water sampling was located in the Vanikoro area, at station 31 (12°12.50′S–167°38.77′E), between 1617 m depth and sea level (Fig. 5.5). The CH_4 anomaly, which is possibly indicative of an hydrothermal plume, was found at 1200 m depth—that is, 400 m above the sea bottom (KAIYO 89 Cruise Report, 1990). It was characterized by an extremely high methane/manganese ratio of 8, similar to that observed in the Okinawa Trough (Nojiri and Ishibashi, 1991).

Five heat flow stations were also run during this cruise, using a gravity corer, in small sedimentary basins within the JCT (three measurements in the Vanikoro area, one in the Tikopia area, one in the Vot Tande area). Heat flow values vary between 23 and 161 mW/m^2,

TABLE II
Chemical Whole-Rock Analyses of Volcanic Samples[a]

Location	VAN		VAN	VAN		VAN		VAN	
Sample	7M2		7M3	2M1		2M5		3M4	
Type	a		a	f		f		f	
SiO_2 (%)	47.00		48.20	49.00		49.00		49.40	
TiO_2	1.36		1.27	1.25		1.15		1.01	
Al_2O_3	17.60		17.39	16.43		16.50		15.78	
Fe_2O_3	1.56		1.43	1.57		1.52		1.46	
FeO	7.96		7.30	7.99		7.76		7.44	
MnO	0.14		0.16	0.15		0.14		0.20	
MgO	8.30		8.87	6.96		8.55		7.15	
CaO	12.22		12.11	12.51		12.40		12.15	
Na_2O	3.20		3.04	3.00		2.30		2.80	
K_2O	0.11		0.09	0.31		0.28		0.15	
P_2O_5	0.05		0.02	0.05		0.05		0.10	
H_2O^+	−0.08		−0.14	0.11		0.12		0.25	
H_2O^-	0.24		0.11	0.10		0.14		0.23	
Total	99.66		99.85	99.43		99.91		98.12	
Source	INA	ICP	AA/ICP	INA	ICP	INA	ICP	INA	ICP
Rb (ppm)	1.1		2	4.0		2.0		1.26	
Sr	297		180	288		314		251	
Ba	7.6		20	54.2		40.6		35.3	
Hf	1.9			1.5		1.2		1.59	
Zr	86	76	75	83	51	60	45	81	51
Ta	0.053			0.12		0.094		0.130	
Th	0.35			0.34		0.32		0.348	
U	0.20			0.17		0.12		0.25	
Cs	0.05			0.06		0.05		0.049	
Sb	0.02			0.02		0.01		0.087	
As								2.84	
La	2.3	2.7	2.5	3.7	4.0	3.4	3.8	3.66	3.8
Ce	6.6	9	9.5	7.6	11	7.0	8	7.6	10
Nd		8	8.0		8		7		8
Sm	2.6			2.3		2.3		2.39	
Eu	1.1	1.10	1.05	0.97	1.00	0.93	0.85	1.13	0.95
Gd									
Tb	0.58			0.53		0.46		0.514	
Dy		4.0	4.2		3.4		3.0		3.8
Er		2.8	2.6		2.3		2.1		2.8
Yb	2.4	2.50	2.55	2.4	2.10	1.95	1.85	1.95	2.25
Lu									
Sc	34.6			39.9		37.3		38.6	
V	167		218	265		216		234	
Cr	255		238	148		274		150	
Co	49.3		50	36.8		41.7		38.6	
Ni	159		159	51		115		53	
Y		26.5	27		22		20		24.5
Nb		1.5	1.5		2.2		1.8		2.3

(*continued*)

TABLE II
(*Continued*)

Location	VAN		VAN	VAN		VAN		VAN	VAN	
Sample	5M5		5M1	6M1		5M4		3321	3M3	
Type	f		f	c		f		b	f	
SiO_2	49.50		50.15	50.80		51.10		49.28	52.20	
TiO_2	0.85		0.79	1.03		0.76		0.55	1.22	
Al_2O_3	16.65		16.53	16.90		15.90		18.01	15.83	
Fe_2O_3	1.52		1.37	1.67		1.39		9.02	1.59	
FeO	7.76		6.99	8.53		7.08			8.10	
MnO	0.16		0.16	0.16		0.16		0.17	0.18	
MgO	7.35		7.51	5.18		7.15		7.11	5.51	
CaO	13.05		12.83	10.80		12.11		13.70	9.86	
Na_2O	2.44		2.33	3.02		2.30		1.74	3.51	
K_2O	0.31		0.32	0.61		0.28		0.28	0.36	
P_2O_5	0.05		0.05	0.10		0.05		0.10	0.15	
H_2O^+	0.03		0.42	0.49		0.13			0.16	
H_2O^-	0.11		0.19	0.27		0.18			0.21	
Total	99.78		99.64	99.56		98.59		99.96	98.88	
Source	INA	ICP	AA/ICP	INA	ICP	INA	ICP	XRF/ICP	INA	ICP
Rb	3.4		4	7.6		3		3	4.3	
Sr	295		227	291		288		298	258	
Ba	42		60	78		45.8		167	72.7	
Hf	1.2			1.8		1.15			2.19	
Zr	40	42	41	66	63	49	43	27	84	74
Ta	0.079			0.14		0.073			0.195	
Th	0.23			0.52		0.237			0.470	
U	0.06			0.26		0.103			0.24	
Cs	0.11			0.17		0.110			0.083	
Sb	0.05			0.07		0.037			0.070	
As				1.1		0.44			1.66	
La	2.5	3.1	3.0	4.9	5.8	2.67	2.7	3.0	4.80	4.9
Ce	7.0	9	10	10.9	13	5.4	10	6.8	9.6	14
Nd		6	6		10		6.5	5.2		10
Sm	1.8			2.8		1.77		2.0	2.95	
Eu	0.72	0.80	0.75	0.95	1.05	0.73	0.75	0.71	1.17	1.25
Gd								2.0		
Tb	0.39			0.51		0.366			0.628	
Dy		2.8	2.8		3.4		2.8	2.2		4.4
Er		1.6	2.0		2.2		1.8	1.4		2.9
Yb	1.9	1.65	1.70	2.3	2.10	1.42	1.75	1.3	2.67	2.70
Lu								0.19		
Sc	38.4			34.0		37.8			32.0	
V	196		243	258		240		240	278	
Cr	141		144	26		153		61	13	
Co	38.1		34	32.2		37.3			32.8	
Ni	52		48	26		48		52	23	
Y		18	18		23		18	11		29
Nb		1.8	2		2.7		1.2	1		3

(*continued*)

TABLE II
(Continued)

Location	VAN		VAN		VAN	VAN		VAN	
Sample	7M4		3M1		3M2	1M1		1M9	
Type	e+c		g		g	g		g	
SiO_2	55.10		57.90		60.50	62.60		64.00	
TiO_2	1.50		1.82		1.63	1.25		0.77	
Al_2O_3	14.52		15.75		14.78	15.93		15.50	
Fe_2O_3	1.74		1.39		1.24	1.02		0.78	
FeO	8.90		7.11		6.33	5.21		4.00	
MnO	0.20		0.19		0.21	0.16		0.18	
MgO	3.24		2.41		2.48	1.73		2.19	
CaO	6.77		5.58		4.71	4.42		4.53	
Na_2O	4.09		5.34		5.07	6.00		5.80	
K_2O	1.25		0.71		0.70	0.75		0.71	
P_2O_5	0.35		0.18		0.40	0.20		0.15	
H_2O^+	0.58		0.79		0.97	0.70		0.78	
H_2O^-	0.27		0.11		0.25	0.07		0.09	
Total	98.51		99.28		99.27	100.04		99.48	
Source	INA	ICP	INA	ICP	AA/ICP	INA	ICP	INA	ICP
Rb	21.0		8.7		7	10.6		11.1	
Sr	268		204		215	207		529	
Ba	202		125		104	158		158	
Hf	3.9		4.1			4.7		4.8	
Zr	157	139	140	149	142	190	168	204	181
Ta	0.54		0.41			0.50		0.51	
Th	1.43		0.83			1.22		1.28	
U	0.65		0.37			0.42		0.50	
Cs	0.45		0.21			0.25		0.25	
Sb	0.12		0.07			0.08		0.07	
As								0.74	
La	12.2	13.0	9.2	9.6	9.2	11.9	12.7	11.7	12.6
Ce	29.7	32	25.9	26	24	31.0	30	30.1	31
Nd		22		21	20		22		22
Sm	5.3		5.6			5.8		5.7	
Eu	1.8	1.70	2.1	1.90	1.90	2.1	1.80	2.0	1.85
Gd									
Tb	0.95		1.16			1.20		1.16	
Dy		6.3		7.9	7.7		8.3		7.9
Er		4.1		5.3	5.0		5.4		5.3
Yb	3.7	3.80	4.7	5.00	4.85	5.3	5.40	5.2	5.30
Lu									
Sc	30.3		21.4			16.1		14.7	
V	337		73		67			70	
Cr	3		5		3	8		41	
Co	26.5		8.9		7.0	5.9		9.6	
Ni	7		2		0	2		16	
Y		40		53	51		54		53
Nb		8.0		5.4	5.0		7.2		7.5

(continued)

TABLE II
(Continued)

Location	VAN		VAN	VAN	VAN	VAN	VAN	VAN
Sample	1M8		1M5	1M3	3151	3154	3152	3155
Type	g		g	g	g	g	g	g
SiO_2	65.20		67.00	67.50	66.43	66.96	66.77	66.09
TiO_2	0.90		0.86	0.78	0.85	0.86	0.95	0.96
Al_2O_3	16.20		14.99	14.60	15.33	15.52	15.38	15.28
Fe_2O_3	0.75		0.68	0.68	1.40	1.20	5.23	5.26
FeO	3.84		3.46	3.48	3.00	3.22		
MnO	0.15		0.15	0.16	0.16	0.16	0.20	0.20
MgO	1.21		1.17	1.04	1.19	1.17	1.16	1.17
CaO	3.56		3.27	2.79	3.18	3.23	3.26	3.29
Na_2O	6.55		6.12	6.12	5.56	6.21	6.76	6.71
K_2O	0.88		0.83	0.93	0.85	0.86	0.92	0.91
P_2O_5	0.15		0.25	0.15	0.24	0.25	0.20	0.22
H_2O^+	0.72		0.73	0.90	0.73	0.60		
H_2O^-	0.05		0.11	0.15	1.28	0.08		
Total	100.16		99.62	99.28	100.20	100.32	100.83	100.09
Source	INA	ICP	AA/ICP	AA/ICP	XRF/ICP	XRF/ICP	XRF/ICP	XRF/ICP
Rb	12.4		11	13	13.5	14.2	13	13
Sr	148		178	149	172	174	164	164
Ba	186		165	163	193	193	211	
Hf	5.4							
Zr	213	205	204	214	220	223	227	229
Ta	0.60							
Th	1.53				1.4			
U	0.50							
Cs	0.30							
Sb	0.08							
As	1.3							
La	13.7	14.5	14.7	15	14.8	15.0	15.4	14.8
Ce	28.0	34	37	37	32	33	36.3	33.9
Nd		24	25	25	24	25	22.8	21.7
Sm	6.1						7.6	7.2
Eu	1.94	2.05	1.95	2.00	2.00	1.95	2.2	2.0
Gd							7.9	7.2
Tb	1.23							
Dy		8.6	8.8	10.0	8.6	8.8	8.7	7.9
Er		5.6	5.9	6.1	5.5	5.8	5.9	5.2
Yb	6.2	5.85	5.95	6.10	5.70	5.75	6.1	5.3
Lu							0.89	0.80
Sc	11.1							
V			38	15	22	20	29	24
Cr			3	0	2	2		
Co	4.9		4	2				
Ni	2.3		0	0	5	5		
Y		59	60	61	52.2	52.6	52	46
Nb		8.7	8.8	9.1	8.5	8.5	10	9

(continued)

TABLE II
(*Continued*)

Location	TIK	TIK	TIK	TIK	TIK	TIK	TIK
Sample	3981	3981X	3982	3983	4150	4151	4153
Type	e	e	e	e	g	g	g
SiO_2	51.92	52.09	52.62	56.97	59.07	59.15	59.12
TiO_2	0.70	0.73	0.76	0.77	0.95	0.92	0.93
Al_2O_3	19.53	20.03	16.01	17.92	17.57	17.59	17.68
Fe_2O_3	3.68	7.76	11.67	7.87	2.55	2.50	2.55
FeO	3.78				3.95	3.93	3.89
MnO	0.14	0.16	0.21	0.18	0.17	0.17	0.17
MgO	2.86	2.65	3.99	2.34	3.07	3.06	3.03
CaO	8.80	8.85	9.19	7.30	6.68	6.70	6.68
Na_2O	2.81	3.59	2.71	3.66	3.60	3.62	3.70
K_2O	2.32	2.43	1.30	1.81	0.94	0.92	0.93
P_2O_5	0.51	0.49	0.22	0.24	0.26	0.26	0.26
H_2O^+	1.29				0.90	0.77	0.81
H_2O^-	1.70				0.21	0.18	0.19
Total	100.04	98.78	98.68	99.06	99.92	99.77	99.94
Source	XRF/ICP	XRF/ICP	XRF/ICP	XRF/ICP	XRF/ICP	XRF/ICP	XRF/ICP
Rb	33.5	38	20	31	12.7	12.1	12.4
Sr	1020	999	357	376	286	288	288
Ba	311	313	515	270	180	180	183
Hf							
Zr	92	109	58	88	154	154	155
Ta							
Th						1.5	
U							
Cs							
Sb							
As							
La	13.8	14.3	6.3	9.0	12.9	12.5	12.5
Ce	29	31.7	12.3	19.4	25	25	25
Nd	19.5	17.8	8.5	10.9	17	17	17
Sm		4.8	2.8	3.4			
Eu	1.35	1.4	0.9	1.1	1.40	1.55	1.45
Gd		3.6	2.8	3.5			
Tb							
Dy	3.2	3.2	3.2	3.6	5.2	4.8	5.0
Er	2.0	1.9	2.1	2.3	3.2	3.2	3.2
Yb	1.75	1.8	2.0	2.3	3.40	3.40	3.50
Lu		0.22	0.34	0.32			
Sc							
V	269				140	133	135
Cr	9				3	2	2
Co							
Ni	12				6	6	6
Y	19.7				31.7	31.5	31.7
Nb	2.5				9.5	9.5	9.0

(*continued*)

TABLE II
(*Continued*)

Location	TIK	TIK	TIK	TIK	TIK	TIK	TIK
Sample	4155	4152	415X	CY11	CY31	CY34	CY36
Type	g	g	g	g	c	c	c
SiO_2	59.10	58.48	58.48	61.55	53.66	53.71	54.79
TiO_2	0.93	1.01	0.99	0.62	1.34	1.37	0.92
Al_2O_3	17.54	17.54	17.60	17.60	16.18	17.76	17.79
Fe_2O_3	2.57	7.29	7.22	5.26	8.91	9.23	9.00
FeO	3.88						
MnO	0.17	0.20	0.20	0.16	0.17	0.17	0.16
MgO	3.07	2.90	2.92	2.19	5.66	5.22	4.28
CaO	6.70	6.77	6.74	5.12	9.07	9.60	9.28
Na_2O	4.41	4.64	4.73	5.27	3.90	3.29	3.40
K_2O	0.94	0.97	0.98	1.06	0.72	0.70	0.65
P_2O_5	0.26	0.23	0.24	0.21	0.22	0.15	0.14
H_2O^+	0.73						
H_2O^-	0.22						
Total	100.52	100.03	100.10	99.04	99.83	101.20	100.41
Source	XRF/ICP	XRF/ICP	XRF/ICP	XRF/ICP	XRF/ICP	XRF/ICP	XRF/ICP
Rb	12.5	13	11	17	11	11	9
Sr	287	273	265	238	178	238	286
Ba	188	217	196				
Hf							
Zr	155	167	163				
Ta							
Th	1.30			3	1	2	1
U				2	1	1	1
Cs							
Sb							
As							
La	12.7	12.6	12.2	13.3	8.4	4.9	5.3
Ce	27	28.7	27.2	27.1	22.4	11.9	11.9
Nd	17	15.8	15.5	14.6	16.5	9.4	8.6
Sm		4.8	4.9	4.0	6.2	3.5	3.0
Eu	1.50	1.5	1.5	1.2	1.8	1.2	1.0
Gd		4.6	4.6	3.6	6.2	4.0	3.0
Tb							
Dy	5.1	5.1	5.1	3.9	6.9	4.3	3.3
Er	3.4	3.4	3.2	2.5	4.2	2.7	2.1
Yb	3.50	3.5	3.3	2.7	4.1	2.6	1.9
Lu		0.51	0.51	0.42	0.60	0.38	0.30
Sc							
V	135	176	189				
Cr	3	5	8				
Co							
Ni	7	9	7				
Y	31.5	29	30	22	38	24	17
Nb	9.5	10	10				

(*continued*)

TABLE II
(Continued)

Location	VOT		VOT		VOT		VOT		VOT	
Sample	11M1		10M1		10M2		12M1		11M2	
Type	e+c		c		c		d		e+c	
SiO_2	47.50		49.40		50.70		50.70		51.00	
TiO_2	1.40		1.03		0.94		0.77		1.36	
Al_2O_3	15.79		18.25		17.90		16.25		16.51	
Fe_2O_3	1.94		1.61		1.51		1.39		1.96	
FeO	9.90		8.22		7.73		7.11		9.99	
MnO	0.17		0.16		0.17		0.15		0.19	
MgO	5.57		5.85		5.79		9.60		4.18	
CaO	10.62		10.82		10.53		10.62		8.80	
Na_2O	2.62		2.97		2.89		2.32		2.60	
K_2O	0.96		0.61		0.64		0.66		1.06	
P_2O_5	0.15		0.10		0.15		0.10		0.20	
H_2O^+	1.32		0.37		0.28		0.33		0.51	
H_2O^-	1.01		0.28		0.26		0.13		0.46	
Total	98.95		99.67		99.49		100.13		98.82	
Source	INA	ICP	INA	ICP	INA	ICP	INA	ICP	INA	ICP
Rb	19.6		7.7		7.5		9.4		20.6	
Sr	344		296		415		457		281	
Ba	127		84		89		115		144	
Hf	1.85		1.5		1.6		1.4		1.8	
Zr	87	69	47	56	67	55	50	50	66	65
Ta	0.31		0.17		0.17		0.26		0.23	
Th	0.43		0.30		0.26		0.56		0.30	
U	0.17		0.11		0.12		0.20		0.18	
Cs	2.0		0.11		0.16		0.22		1.30	
Sb	0.20		0.11		0.11		0.09		0.05	
As			1.3		1.2				0.16	
La	5.7	6.0	4.2	4.9	4.0	4.6	5.0	6.0	4.1	5.4
Ce	15.0	15	8.1	12	8.3	12	12.9	13	9.0	14
Nd		11		9.5		9		8		11
Sm	3.1		2.4		2.3		2.1		3.1	
Eu	1.1	1.10	0.90	0.85	0.89	0.95	0.76	0.80	1.05	1.10
Gd										
Tb	0.64		0.47		0.48		0.38		0.65	
Dy		4.2		3.5		3.3		2.8		4.6
Er		2.7		2.4		2.3		1.8		3.2
Yb	2.5	2.55	2.2	2.2	2.2	2.05	1.5	1.65	2.8	2.75
Lu										
Sc	38.2		31.0		31.1		35.1		35.3	
V	294		238		260		208		370	
Cr	54		31		13		410		6	
Co	41.8		34.3		34.5		38.3		37.4	
Ni	35		39		39		179		17	
Y		28		25		22		17		29
Nb		4.9		3.1		2.6		4		3.3

(continued)

TABLE II
(*Continued*)

Location	VOT	VOT	HAZ		HAZ		HAZ	HAZ	
Sample	4294	4295	14M2		14M1		14M7	14M14	
Type	e	e	a		a		a	a	
SiO_2	54.57	56.87	45.20		46.10			48.20	
TiO_2	0.95	0.80	1.52		1.49			2.06	
Al_2O_3	15.15	15.71	16.05		15.09			14.54	
Fe_2O_3	9.85	9.77	1.63		1.55			1.88	
FeO			8.31		7.93			9.60	
MnO	0.18	0.16	0.14		0.16			0.19	
MgO	4.39	3.85	7.26		7.92			7.25	
CaO	8.54	7.60	10.85		10.59			9.05	
Na_2O	2.86	2.57	2.84		2.88			3.82	
K_2O	1.75	2.31	0.21		0.26			0.40	
P_2O_5	0.87	0.33	0.15		0.10			0.15	
H_2O^+			2.06		2.02			1.94	
H_2O^-			3.00		3.16			0.28	
Total	99.11	99.97	99.22		99.25		0.00	99.36	
Source	XRF/ICP	XRF/ICP	INA	ICP	INA	ICP	INA	INA	ICP
Rb	37	47	2.4		1.7		2.0	4.4	
Sr	395	395	318		304		190	466	
Ba	270	326	36.7		52.4		17.5	118	
Hf			2.4		2.1		2.0	3.0	
Zr	80	76	106	93	76	89	77	139	116
Ta			0.60		0.55		0.29	1.54	
Th			0.68		0.59		0.44	1.74	
U			0.22		0.30		0.20	0.45	
Cs			0.29		0.11		0.02	0.07	
Sb			0.01		0.10		0.07	0.09	
As								1.1	
La	8.5	7.6	6.7	6.9	6.3	7.3	4.1	14.2	15
Ce	18.7	17.1	16.5	17	16.6	17	11.7	24.7	32
Nd	11.1	9.6		12		11			19
Sm	3.4	3.0	2.9		3.0		2.3	4.1	
Eu	1.0	0.84	1.12	1.10	1.23	1.25	0.78	1.45	1.55
Gd	3.4	2.6							
Tb			0.63		0.59		0.48	0.73	
Dy	3.2	2.8		4.1		4.0			5.0
Er	2.1	1.8		3.0		2.5			3.2
Yb	2.1	1.7	2.9	2.55	2.2	2.40	1.8	2.8	2.55
Lu	0.32	0.29							
Sc			37.6		34.2		23.0	40.6	
V			213		195			297	
Cr			356		390		39	115	
Co			48.7		44.8		24.4	41.7	
Ni			199		204		23	48	
Y				27		27			30
Nb				7.2		7.5			20

(*continued*)

TABLE II
(*Continued*)

Location	HAZ		HAZ		HAZ		EFA		EFA	
Sample	15M6		14M5		15M12		27M12		29M6	
Type	b		a		i		d		e+d	
SiO_2	51.20		52.50		62.80		46.50		47.50	
TiO_2	0.72		0.75		0.62		0.85		0.69	
Al_2O_3	17.05		16.30		16.07		15.10		13.65	
Fe_2O_3	1.50		1.44		0.70		1.36		1.21	
FeO	7.65		7.32		3.60		6.95		6.19	
MnO	0.17		0.16		0.11		0.15		0.18	
MgO	5.45		6.08		1.64		9.45		10.30	
CaO	10.74		4.11		0.73		9.55		7.30	
Na_2O	2.42		6.08		5.99		2.47		2.32	
K_2O	0.68		0.10		4.66		0.70		1.87	
P_2O_5	0.12		0.15		0.18		0.10		0.20	
H_2O^+	0.78		3.81		1.59		3.40		2.89	
H_2O^-	0.63		0.69		0.39		2.71		4.65	
Total	99.11		99.49		99.08		99.29		98.95	
Source	INA	ICP	INA	ICP	INA	ICP	INA	ICP	INA	ICP
Rb	8.8		1.4		66.8		7.7		19.1	
Sr	356		158		80		354		757	
Ba	88		6.0		600		54.0		677	
Hf	1.1		1.24		4.6		1.25		0.95	
Zr	48	30	46	47	188	158	46	34	30	33
Ta	0.24		0.15		0.35		0.068		0.052	
Th	0.40		0.27		3.4		0.314		0.368	
U	0.14		0.11		1.48		0.15		0.39	
Cs	0.16		0.01		0.06		0.068		0.46	
Sb	0.13		0.10		0.20		0.061		0.018	
As					2.5				0.68	
La	4.6	5.2	3.9	4.3	17.6	16.0	2.34	2.7	2.92	3.3
Ce	10.2	10.5	10.4	12.5	34.3	33	5.2	9	6.3	9
Nd		7.5		8		20		6		6
Sm	2.0		2.0		4.5		2.02		1.79	
Eu	0.86	0.80	0.75	0.75	0.98	1.05	0.90	0.90	0.73	0.75
Gd										
Tb	0.36		0.39		0.65		0.422		0.328	
Dy		2.5		2.9		4.0		3.0		2.5
Er		1.6		1.8		2.8		2.0		1.8
Yb	1.3	1.35	1.51	1.65	3.2	3.00	1.51	1.75	1.12	1.35
Lu										
Sc	36.1		29.7		7.3		38.6		32.6	
V	249		163				237		240	
Cr	53		51				255		317	
Co	35.2		33.6		6.1		34.6		44.9	
Ni	35		32		5		65		160	
Y		16		18		28		20		14
Nb		3.7		3.0		5.6		1.0		1.6

(*continued*)

TABLE II
(Continued)

Location	EFA		EFA		EFA		EFA		EFA	
Sample	31M2		31M1		30M2		30M1		27M17	
Type	d		d		d		d		i	
SiO_2	51.70		52.00		53.00		54.40		61.80	
TiO_2	0.74		0.72		0.53		0.60		0.80	
Al_2O_3	15.50		15.35		13.22		15.30		15.89	
Fe_2O_3	1.33		1.33		1.42		1.35		0.85	
FeO	6.76		6.76		7.25		6.87		4.34	
MnO	0.15		0.15		0.16		0.15		0.07	
MgO	7.34		7.37		10.28		7.72		2.01	
CaO	11.50		11.55		9.18		8.97		2.82	
Na_2O	2.35		2.27		1.92		2.16		4.50	
K_2O	0.38		0.41		0.62		0.75		5.60	
P_2O_5	0.15		0.15		0.05		0.10		0.30	
H_2O^+	0.30		0.09		0.32		0.27		0.41	
H_2O^-	0.12		0.10		0.12		0.10		0.18	
Total	98.32		98.25		98.07		98.74		99.57	
Source	INA	ICP	INA	ICP	INA	ICP	INA	ICP	INA	ICP
Rb	4.8		5.6		10.2		11.5		96	
Sr	395		402		352		351		257	
Ba	82.4		81.3		178		200		818	
Hf	1.03		1.12		0.98		1.16		4.2	
Zr	38	34	37	35	43	34	47	39	171	90
Ta	0.068		0.075		0.040		0.052		0.21	
Th	0.418		0.438		0.680		0.777		4.1	
U	0.14		0.22		0.23		0.13		2.2	
Cs	0.22		0.22		0.30		0.34		0.68	
Sb	0.037		0.058		0.044		0.050		0.10	
As	1.04		1.41		1.00		1.31		0.8	
La	3.64	3.7	3.43	3.9	3.57	4.0	3.95	4.7	12.6	14.3
Ce	10.3	9	10.7	10	5.8	9.5	7.7	9.5	27.3	31
Nd		7		6.5		5.5		6.5		20
Sm	1.84		1.88		1.47		1.62		3.9	
Eu	0.75	0.70	0.78	0.80	0.52	0.55	0.59	0.60	0.98	1.05
Gd										
Tb	0.339		0.355		0.266		0.298		0.53	
Dy		2.5		2.5		2.2		2.3		3.3
Er		1.6		1.7		1.5		1.7		2.1
Yb	1.36	1.55	1.46	1.50	1.15	1.40	1.24	1.45	2.5	1.90
Lu										
Sc	33.8		35.7		32.3		27.7		8.8	
V	264		252		214		218		140	
Cr	209		236		388		178		42	
Co	33.7		34.9		44.7		34.8		6.7	
Ni	72		74		188		121		17	
Y		16		16		14		15		21
Nb		1.6		1.7		1.0		1.2		3.7

(continued)

TABLE II
(Continued)

Location	EFA		EFA	EFA	EFA	EFA		EFA	ERR
Sample	27M1		27M4	29M3	28M1	26M6		26M7	22M1
Type	i		i	i	i	h		h	d
SiO_2	63.00		63.40	64.00	64.80	65.70		66.30	49.00
TiO_2	0.43		0.50	0.50	0.65	0.55		0.61	0.79
Al_2O_3	14.93		15.09	15.08	14.95	14.36		14.16	14.73
Fe_2O_3	0.65		0.68	0.66	0.60	0.71		0.69	1.54
FeO	3.30		3.46	3.38	3.08	3.63		3.53	7.88
MnO	0.12		0.12	0.12	0.09	0.12		0.12	0.17
MgO	1.00		0.85	0.81	2.47	1.05		1.01	6.51
CaO	2.33		2.32	2.21	0.94	2.58		2.37	12.94
Na_2O	4.64		4.55	4.51	3.47	4.83		4.76	2.08
K_2O	4.96		4.92	4.79	6.25	3.70		3.62	0.81
P_2O_5	0.15		0.15	0.15	0.40	0.15		0.20	0.20
H_2O^+	2.73		2.65	2.26	1.54	2.16		1.91	1.95
H_2O^-	1.59		0.44	0.99	0.61	0.33		0.07	0.68
Total	99.83		99.13	99.46	99.85	99.87		99.35	99.28
Source	INA	ICP	AA/ICP	AA/ICP	AA/ICP	INA	ICP	AA/ICP	AA/ICP
Rb	57.6		97	65	114	65.4		70	13
Sr	297		272	256	153	228		272	405
Ba	912		825	819	1045	1017		950	108
Hf	6.0					6.1			
Zr	235	205	214	218	150	204	210	210	54
Ta	0.34					0.30			
Th	7.4					5.1			
U	2.7					1.85			
Cs	1.78					2.2			
Sb	0.23					0.23			
As	3.7					6.1			
La	23.8	24.5	25	25	15.2	23.6	25	25	4.9
Ce	48.1	50	50	51	34	44.2	51	50	12
Nd		25	24	25	19		28	28	8
Sm	5.0					6.1			
Eu	1.00	1.10	1.10	1.10	0.95	1.3	1.30	1.45	0.90
Gd									
Tb	0.62					0.84			
Dy		4.1	4.1	4.5	3.4		5.3	5.6	2.5
Er		2.8	3.0	3.0	2.3		3.7	4.0	1.7
Yb	3.3	2.85	3.00	3.05	2.20	4.2	3.90	4.05	1.60
Lu									
Sc	8.2					10.3			
V	40		49	50	83	45		52	306
Cr	7		2	3	10	2		2	298
Co	7.7		7	7	7	6.3		5	38
Ni	1.9		1	2	10	2		1	101
Y		27	28	29	22.5		37	38	17
Nb		5.1	5.3	6.2	3.7		4.5	4.8	1.3

(continued)

TABLE II
(*Continued*)

Location	ERR		ERR		ERR	ERR	ERR		FUT	
Sample	25M2		25M4		24M6	24M4	24M3		19M1	
Type	c		c		b	d	h		c+d	
SiO_2	50.30				51.30	52.50	65.00		50.00	
TiO_2	0.94				0.79	0.60	0.62		1.00	
Al_2O_3	16.84				18.13	14.80	14.60		14.90	
Fe_2O_3	1.85				1.50	1.55	0.98		1.21	
FeO	9.45				7.66	7.90	5.00		6.16	
MnO	0.16				0.16	0.14	0.14		0.24	
MgO	5.91				4.31	7.21	1.11		9.62	
CaO	10.55				10.58	10.33	3.60		10.46	
Na_2O	2.47				2.85	2.42	4.19		2.29	
K_2O	0.62				0.85	0.83	2.96		0.83	
P_2O_5	0.15				0.10	0.10	0.25		0.25	
H_2O^+	−0.14				0.72	0.12	0.92		1.12	
H_2O^-	0.13				0.76	0.23	0.27		0.61	
Total	99.23		0.00		99.71	98.73	99.64		98.69	
Source	INA	ICP	INA	ICP	AA/ICP	INA	INA	ICP	INA	ICP
Rb	10.8		7.7		10	12.8	49.4		12.2	
Sr	606		620		410	424	244		789	
Ba	168		152		283	218	626		106	
Hf	1.4		1.3			1.4	4.7		1.8	
Zr	52	42	49	42	43	54	176	153	65	68
Ta	0.13		0.094			0.065	0.25		0.46	
Th	0.71		0.58			0.55	2.0		0.85	
U	0.25		0.26			0.27	1.18		0.32	
Cs	0.35		0.19			0.55	2.2		0.36	
Sb	0.08		0.08			0.12	0.47		0.10	
As	1.5					1.9	6.9		0.9	
La	5.7	5.8	5.4	3.8	6.4	3.7	11.8	12.5	8.6	9.2
Ce	12.8	15	12.9	10	14	6.9	26.1	28	16.2	19.5
Nd		10		7	12			19		12
Sm	2.8		2.8			2.0	4.6		2.8	
Eu	0.95	0.95	0.93	0.70	1.30	0.70	1.14	1.25	0.97	1.00
Gd										
Tb	0.51		0.47			0.40	0.84		0.45	
Dy		3.2		2.5	4.4			5.8		3.1
Er		2.1		1.8	2.7			3.8		2.0
Yb	2.2	1.90	1.9	1.60	2.45	2.1	4.3	3.90	1.7	1.60
Lu										
Sc	37.4		35.4			37.7	14.1		32.9	
V	368				368	255	35		231	
Cr	40		42		58	256			943	
Co	40.2		39.1		24	37.4	9.3		39.2	
Ni	33		33		24	75	0.9		324	
Y		20		17	33			38		18.5
Nb		1.2		1.3	1.3			3.4		6.3

(*continued*)

TABLE II
(*Continued*)

Location	FUT		FUT		FUT		FUT		FUT	
Sample	21M7		20M4		21M1		16M1		17M3	
Type	d		b		e+d		e		e	
SiO_2	50.20		50.50		50.75		53.45		54.10	
TiO_2	0.50		0.86		0.64		0.82		0.68	
Al_2O_3	13.16		18.30		14.17		14.72		13.97	
Fe_2O_3	1.44		1.40		1.45		1.44		1.34	
FeO	7.37		7.14		7.39		7.37		6.85	
MnO	0.16		0.16		0.16		0.18		0.17	
MgO	10.66		4.52		8.73		4.57		5.18	
CaO	11.58		11.48		11.70		6.71		7.67	
Na_2O	1.73		2.57		1.92		3.41		3.19	
K_2O	0.32		0.46		0.96		1.84		1.64	
P_2O_5	0.08		0.10		0.30		0.30		0.30	
H_2O^+	1.13		1.59		0.50		1.99		1.86	
H_2O^-	0.43		0.22		0.53		2.23		1.34	
Total	98.76		99.30		99.20		99.03		98.29	
Source	INA	ICP	INA	ICP	INA	ICP	INA	ICP	INA	ICP
Rb	4.8		5.8		13.7		19.4		28.8	
Sr	321		336		618		547		437	
Ba	52.7		110		157		339		277	
Hf	1.0		1.2		2.1		3.1		2.5	
Zr	28	30	52	37	87	84	135	109	86	87
Ta	0.043		0.021		0.091		0.28		0.22	
Th	0.12		0.29		1.09		2.2		1.76	
U	0.09		0.13		0.80		0.74		0.79	
Cs	0.25		0.17		0.24		0.48		0.84	
Sb	0.06		0.04		0.04		0.14		0.11	
As	0.6		0.6				1.1		1.9	
La	1.4	1.7	2.5	3.3	16.7	17.5	16.5	16	12.7	13.4
Ce	3.9	4	6.1	9	44.3	45	28.0	34	22.8	30
Nd		3		8		32		20		17
Sm	1.1		2.3		6.7		4.4		4.1	
Eu	0.48	0.50	0.92	0.90	1.89	2.00	1.3	1.40	1.12	1.20
Gd										
Tb	0.29		0.44		0.65		0.69		0.61	
Dy		1.9		2.9		3.6		4.2		3.8
Er		1.4		2.0		2.2		2.8		2.6
Yb	1.6	1.35	1.8	1.75	1.58	1.80	3.1	2.7	2.6	2.25
Lu										
Sc	39.7		32.9		41.3		25.5		28.4	
V	216		279		361		225		195	
Cr	574		22		400		59		139	
Co	47.0		30		39.9		28.1		29.3	
Ni	146		32		100		30		46	
Y		14		19		22		28		25
Nb		1.5		1.0		2.3		5.2		3.4

(*continued*)

TABLE II
(Continued)

Location	FUT		FUT	FUT		ANA	ANA	ANA	ANA	ANA
Sample	20M6		20M1	20M3		4311	4312	4313	4314	4315
Type	c		c	c		h	h	h	h	h
SiO_2	54.70		55.10	55.70		60.10	59.40	60.35	59.45	58.70
TiO_2	1.09		1.23	0.95		0.85	0.89	0.87	0.83	0.83
Al_2O_3	19.46		17.60	17.30		14.35	15.16	14.70	13.95	14.09
Fe_2O_3	1.09		1.17	1.13		9.70	9.41	9.17	9.85	9.78
FeO	5.58		5.95	5.77						
MnO	0.13		0.14	0.14		0.20	0.19	0.17	0.21	0.21
MgO	2.98		4.79	4.56		2.04	1.92	1.93	1.99	1.99
CaO	8.68		8.32	8.30		5.11	5.10	5.14	5.10	5.04
Na_2O	3.79		3.87	3.38		4.23	4.73	4.47	4.17	4.34
K_2O	0.80		0.56	0.79		2.28	2.44	2.35	2.37	2.04
P_2O_5	0.15		0.30	0.15		0.42	0.45	0.42	0.41	0.43
H_2O^+	0.10		0.09	0.50		0.87	0.15	0.07	1.47	1.62
H_2O^-	0.08		0.06	0.14						
Total	98.63		99.18	98.81		100.15	99.84	99.64	99.80	99.07
Source	INA	ICP	AA/ICP	INA	ICP	ICP	ICP	ICP	ICP	ICP
Rb	10.1		7	11.6		25	30	42	30	24
Sr	329		249	363		408	420	413	398	400
Ba	147		68	145		270	284	270	273	257
Hf	2.4			2.2						
Zr	92	90	142	86	86	107	112	108	102	105
Ta	0.13			0.13						
Th	0.71			0.55						
U	0.29			0.30						
Cs	0.32			0.38						
Sb	0.06			0.10						
As	1.3			1.7						
La	5.9	6.7	7.5	5.5	6.20	10.85	12.45	11.35	12.10	11.35
Ce	15.0	17	21	16.7	17					
Nd		13	15		12	20	21	20	20	20
Sm	3.6			3.5						
Eu	1.21	1.30	1.30	1.2	1.15	1.55	1.65	1.50	1.70	1.50
Gd										
Tb	0.64			0.62						
Dy		4.2	4.8		3.9	5.5	6.1	5.4	5.7	5.5
Er		2.7	3.4		2.7	3.8	4.0	3.6	3.7	3.7
Yb	3.1	2.60	3.10	2.7	2.55	3.55	3.80	3.30	3.50	3.50
Lu										
Sc	22.0			29.9		19.0	19.0	19.5	19.0	18.5
V	206		240	218		150	150	160	157	150
Cr	15		60	89		2	3	5	3	3
Co	18		23	22.8		20	19	19	20	19
Ni	16		26	50		4	5	5	5	5
Y		28	32		27	36	39	36	36	36
Nb		2.1	2.6		2.3	1.75	1.95	1.95	1.45	0.85

(continued)

TABLE II
(*Continued*)

Location	ANA	ANA	ANA	ANA	ANA	ANA	ANA	ANA	ANA
Sample	4316	4318	43110	43112	4421	4422	4423	4424	4531
Type	h	h	h	h	b	b	b	c	h
SiO_2	58.65	58.60	58.60	58.85	48.60	48.80	48.80	51.00	60.15
TiO_2	0.83	0.82	0.81	0.82	0.70	0.70	0.71	1.12	0.76
Al_2O_3	13.95	13.86	13.90	14.03	17.10	17.05	17.20	14.50	14.20
Fe_2O_3	9.67	9.60	9.65	9.70	11.70	11.71	11.90	14.55	8.79
FeO									
MnO	0.21	0.21	0.21	0.21	0.19	0.19	0.20	0.24	0.20
MgO	1.94	1.95	1.93	1.95	7.13	6.90	6.98	5.22	1.90
CaO	5.03	4.96	5.00	4.98	11.40	11.35	11.52	9.85	4.73
Na_2O	4.30	4.40	4.33	4.27	2.45	2.39	2.43	2.80	4.36
K_2O	2.15	2.10	2.05	2.17	0.23	0.21	0.21	0.32	2.34
P_2O_5	0.40	0.41	0.40	0.40	0.12	0.12	0.12	0.17	0.41
H_2O^+	1.95	3.40	2.73	2.30	−0.27	−0.35	−0.31	−0.20	1.91
H_2O^-									
Total	99.08	100.31	99.61	99.68	99.35	99.07	99.76	99.57	99.75
Source	ICP	ICP	ICP	ICP	ICP	ICP	ICP	ICP	ICP
Rb	30	32	31	45	4	4	4	4	31
Sr	397	390	396	398	390	410	398	402	382
Ba	260	260	260	264	41	42	41	59	273
Hf									
Zr	104	102	102	102	22	23	23	34	116
Ta									
Th									
U									
Cs									
Sb									
As									
La	10.85	10.70	10.70	11.25	2.45	2.40	2.45	3.70	11.95
Ce									
Nd	19	20	19	20	6	6	6	9	20
Sm									
Eu	1.45	1.55	1.45	1.35	0.75	0.75	0.65	1.10	1.45
Gd									
Tb									
Dy	5.5	5.4	5.5	5.5	2.4	2.4	3.0	3.4	5.3
Er	3.7	3.5	3.6	3.5	1.4	1.5	1.5	2.3	3.7
Yb	3.45	3.45	3.38	3.43	1.43	1.45	1.45	2.08	3.45
Lu									
Sc	18.5	18.5	18.0	18.5	38.0	38.0	37.0	45.0	17.0
V	153	150	150	153	371	378	375	520	137
Cr	3	4	3	5	73	63	63	30	9
Co	20	19	18	19	41	40	40	38	18
Ni	5	5	5	6	54	49	50	20	8
Y	36	35	35	37	15	15	15	22	36
Nb	1.45	1.5	1.3	1.7	0.6	0.5	0.5	0.65	1.5

(*continued*)

TABLE II
(*Continued*)

Location	ANA	ANA	ANA	ANA
Sample	4532	4533	4534	4641
Type	h	h	h	e
SiO_2	60.20	60.40	60.10	48.65
TiO_2	0.76	0.76	0.76	0.77
Al_2O_3	14.15	14.15	14.05	14.15
Fe_2O_3	8.65	8.64	8.75	9.62
FeO				
MnO	0.19	0.19	0.21	0.13
MgO	1.87	1.88	1.86	5.72
CaO	4.73	4.64	4.67	12.50
Na_2O	4.54	4.44	4.42	2.11
K_2O	2.32	2.31	2.32	2.23
P_2O_5	0.41	0.40	0.40	0.45
H_2O^+	2.00	1.75	1.89	3.33
H_2O^-				
Total	99.82	99.56	99.43	99.66
Source	ICP	ICP	ICP	ICP
Rb	32	31	32	51
Sr	379	373	375	665
Ba	269	274	272	165
Hf				
Zr	115	117	115	73
Ta				
Th				
U				
Cs				
Sb				
As				
La	12.30	12.45	12.80	10.75
Ce				
Nd	21	21	22	17
Sm				
Eu	1.45	1.50	1.55	1.25
Gd				
Tb				
Dy	5.6	5.5	5.6	3.2
Er	3.8	3.9	3.5	1.8
Yb	3.50	3.50	3.48	1.70
Lu				
Sc	17.0	17.0	16.5	35.0
V	136	136	138	310
Cr	9	11	15	264
Co	18	17	17	36
Ni	8	9	13	76
Y	37	37	37	18
Nb	1.6	1.9	1.25	1.05

TABLE III
K/Ar Ages of Volcanic Samples[a]

Sample	Age (Ma)	Type	Sample	Age (Ma)	Type
Jean-Charcot Troughs			Coriolis Troughs		
Vanikoro area (VAN)			Efate area (EFA)		
7M2	12.4 ± 0.9	a	27M12	3.5 ± 0.3	d
3M4	1.1 ± 0.2	f	29M6	3.2 ± 0.2	e+d
5M1	2.9 ± 0.4	f	31M2	1.5 ± 0.2	d
6M1	2.6 ± 0.5	c	31M1	1.4 ± 0.2	d
5M4	3.9 ± 0.6	f	30M2	1.4 ± 0.2	d
3M3	1.8 ± 0.3	f	30M1	1.1 ± 0.2	d
7M4	2.3 ± 0.2	e+c	27M17	3.4 ± 0.2	i
3M1	1.8 ± 0.1	g	27M1	2.2 ± 0.1	i
3M2	1.5 ± 0.4	g	27M4	2.4 ± 0.1	i
1M9	1.5 ± 0.1	g	29M3	3.0 ± 0.2	i
1M8	1.1 ± 0.2	g	28M1	2.2 ± 0.2	i
1M5	<0.3	g	26M6	0.4 ± 0.05	h
1M3	1.1 ± 0.2	g	26M7	0.5 ± 0.1	h
3152	<0.3	g			
			Erromango area (ERR)		
Tikopia area (TIK)			25M2	4.1 ± 0.3	c
3981X	3.7 ± 0.2	e	25M4	4.0 ± 0.6	c
415X	<0.3	g	24M6	3.6 ± 0.2	b
			24M4	4.1 ± 0.2	d
Vot Tande area (VOT)			24M3	2.7 ± 0.1	h
11M1	4.9 ± 0.2	e+c			
10M1	2.8 ± 0.1	c	Futuna area (FUT)		
10M2	2.7 ± 0.1	c	19M1	2.6 ± 0.2	c+d
12M1	2.8 ± 0.1	d	21M7	6.5 ± 0.5	d
11M2	4.8 ± 0.2	e+c	21M1	6.1 ± 0.3	e+d
			16M1	5.2 ± 0.3	e
Hazel Holme area (HAZ)			17M3	6.1 ± 0.3	e
14M2	5.2 ± 0.8	a	20M6	0.7 ± 0.2	c
14M1	5.5 ± 0.4	a	20M1	0.7 ± 0.3	c
15M6	4.1 ± 0.2	b			
15M12	3.5 ± 0.3	i			

[a]See Monjaret (1989) and Monjaret *et al.* (1991) for analytical procedure.

Footnote to Table II.

[a]Petrological types a to i correspond to mid-ocean ridge basalts (MORB) (a), IAT (island-arc tholeiites) (b, c, d), calc-alkaline basalts and andesites (CAB) (e), backarc basin basalts (BABB) (f), high-Na/low-K dacites (g), high-K dacites (h), and hyper-K dacites (i), respectively (see text for discussion).

Analytical methods: AA = atomic absorption (J. Cotten, UBO). XRF = X-ray fluorescence (R. C. Price, La Trobe University; S. Nakada, Kyushu University). INA = instrumental neutron activation (J. L. Joron, Pierre-Süe; A. Fujinawa, Ibaraki University, for 3321, 3152, 4152, 415X, 3981X, 3982, 3983, 4294, 4295, CY11, CY31, CY34, CY36; see Nakada *et al.*, 1994). ICP = inductively coupled plasma emission spectrometry (ICP-ES, J. Cotten, UBO). Note: For 3151, 3154, 3981, 4150, 4151, 4153, and 4155, Nb and REE by ICP (J. Cotten, UBO); for 3M1, 10M1, 10M2, 27M1, and 24M3, Cr by AA (J. Cotten, UBO); for all analyses, V by AA (J. Cotten, UBO) or XRF (S. Nakada, Kyushu University, for the 13 analyses listed in INA).

Analytical procedure for inductively coupled plasma emission spectrometry (ICP-ES) and atomic absorption spectrometry (AA) analyses (J. Cotten, UBO): Rock powders were digested in closed vessels with 4 ml of a concentrated hydrofluoric/nitric acid mixture. H_3BO_3 was then added to dissolve the precipitated fluorides and to neutralize the excess HF. International standards (JB1, JB2, BEN, Mica-Fe, GSN, ACE) were used for calibration.

Major elements except P_2O_5: AA analyses with a relative standard deviation (RSD) close to 2%. P_2O_5: colorimetry with a standard deviation of 0.05%.

Trace elements: Rb, Sr (AA), limit of detection (LOD): 1 ppm, RSD: 5%; Ba, V (AA), LOD: 25 ppm; RSD: 10%; Cr, Co, Ni (AA): LOD: 2 ppm; RSD: 5%; Ce, Nd, Zr (ICP-ES), LOD: 2 ppm; RSD: 5%; Nb, La, Er (ICP-ES), LOD: 1 ppm; RSD: 5%; Sc, Y, Dy (ICP-ES), LOD: 0.5 ppm, RSD: 5%; Eu, Yb (ICP-ES), LOD: 0.2 ppm; RSD: 5%.

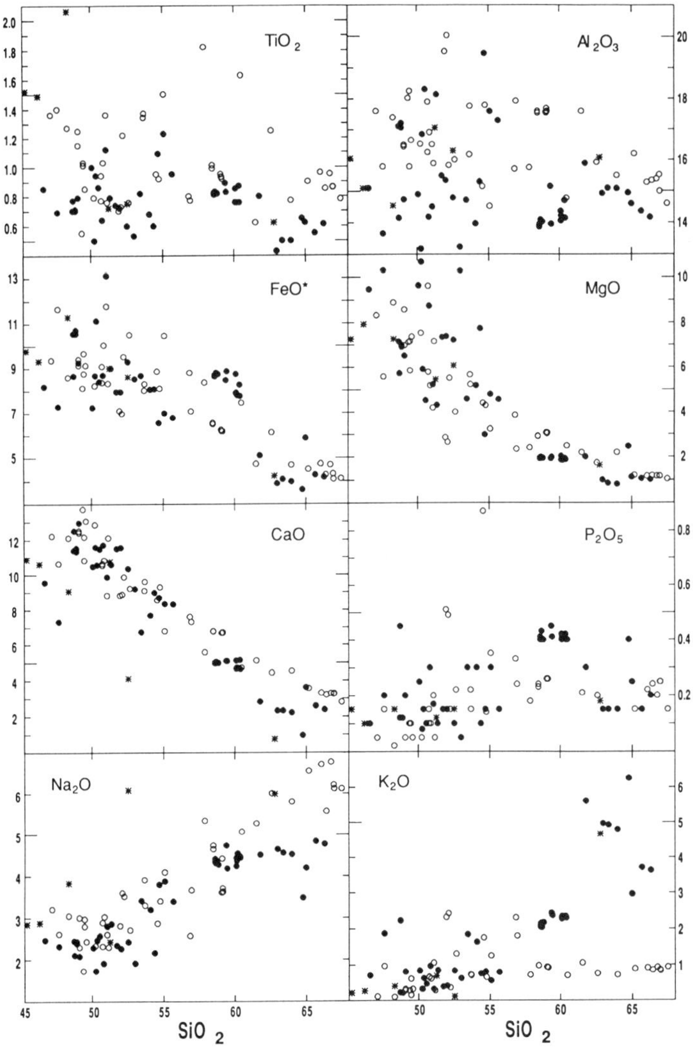

FIGURE 5.16. Major elements (%) versus SiO_2 (%) diagrams for volcanic samples from the NHBAT. Open circles: northern Jean-Charcot troughs (JCT); filled circles: southern Coriolis troughs (CT); asterisks: western termination of the Hazel Holme Ridge. Data from Table II.

with an average of 74 mW/m^2, indicating a rather low heat flow. The highest value (161 mW/m^2) was recorded at station 40 in the Tikopia area (12°46.033′S–167°49.504′E; water depth: 3027 m; Fig. 5.5). This is comparable to the mean of heat flow values on the North Fiji Basin seafloor (150–250 mW/m^2) (KAIYO 89 Cruise Report, 1990).

One of the aims of the SAVANES cruise (1991) was to conduct dives with the *Cyana* submersible on possible active hydrothermal sites in the JCT, previously surveyed during the KAIYO 89 cruise. Diplomatic restrictions imposed by Solomon Islands authorities, relating to their 200 nautical mile boundary, prevented dives at station 31, and consequently the KAIYO 89 observation could not be confirmed. However, a living biological community was discovered southward, at station 41 in the Tikopia area, near sampling site

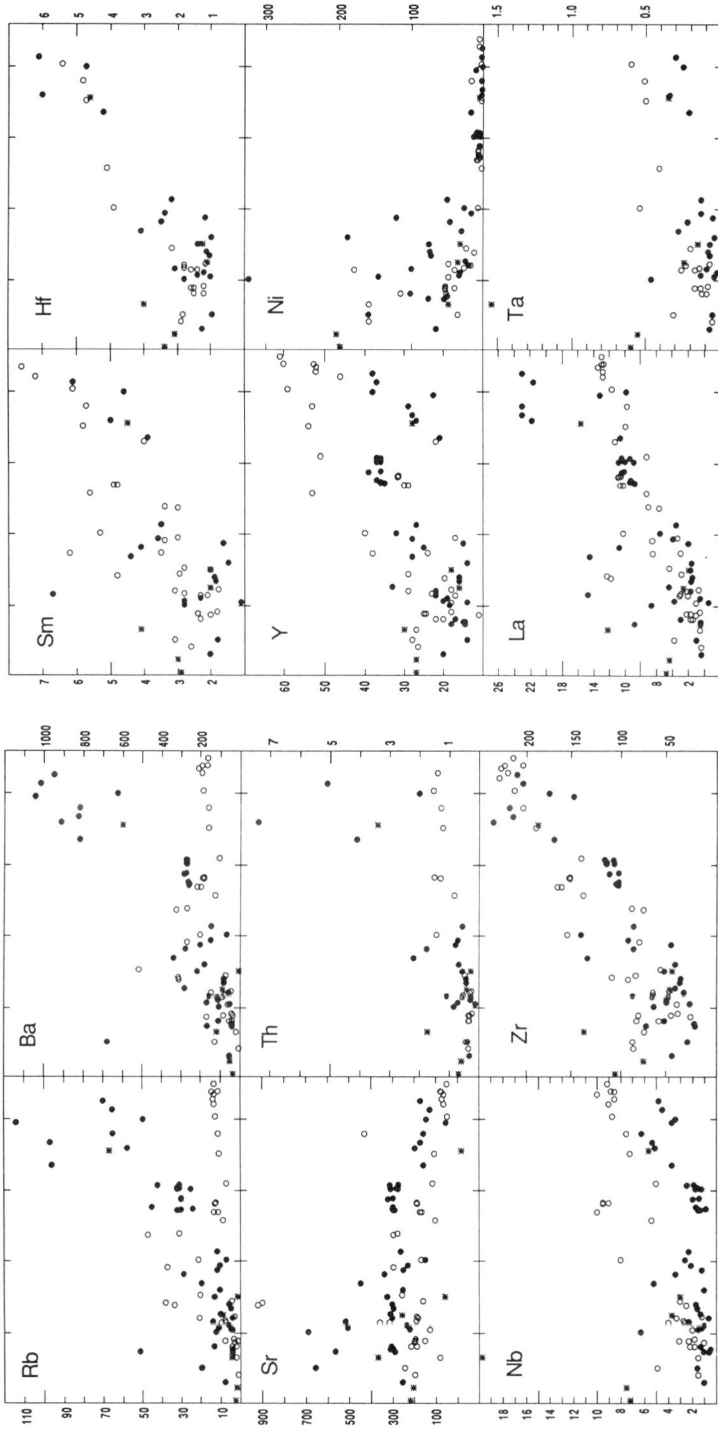

FIGURE 5.17. Trace elements (ppm) versus SiO_2 (%) diagrams for volcanic samples from the NHBAT. Same symbols as in Fig. 5.16. Data from Table II.

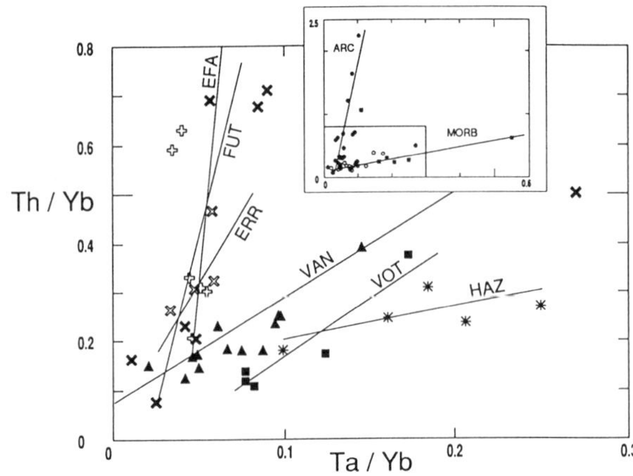

FIGURE 5.18. Th/Yb versus Ta/Yb diagram for volcanic samples from the NHBAT. Data from Table II. JCT samples: filled triangles (Vanikoro area, VAN) and filled squares (Vot Tande area, VOT). Hazel Holme samples: asterisks. CT samples: open crosses (Efate area, EFA), oblique open crosses (Erromango area, ERR) and oblique filled crosses (Futuna area, FUT). An arbitrary correlation line for each area is indicated. The extended diagram is shown in inset. JCT samples: filled circles; Hazel Holme samples: asterisks; CT samples: open circles. ARC and MORB fields are indicated as arbitrary lines.

CY36 (Table I; Fig. 5.5). On the summit of the eastern seamount of station 41 (12°32.6′S-167°47.3′E), at a depth of 1425 m, the *Cyana* submersible team discovered and videotaped a flourishing colony of algae, sponges, corals, galatheas, and shrimp resting on a 50 × 50 m peak of massive to blocky lava. Shimmering water was not observed on the spot during the dive nor on the video records, but a local increase in seawater temperature (3.27°C instead of 2.52°C in the surroundings) might suggest a hydrothermal environment (SAVANES 91-92 Cruise Report, 1992). However, chemical analyses of seawater sampled on this site did not confirm this hypothesis (D. Grimaud, personal communication).

3.4.2. Ferromanganese Crusts

The study of ferromanganese crusts in island arcs gives good environmental criteria to distinguish hydrothermal impact from sedimentary processes.

During the R/V *Jean Charcot* SEAPSO 2 cruise, 18 sites bearing ferromanganese crusts were dredged in the NHBAT, in water depths ranging from 500 to 3000 m, along fault scarps or volcanic cones (Fig. 5.20; Gérard *et al.*, 1987; Gérard, 1993). The crusts were dredged in five Sea Beam-surveyed areas, two in the northern JCT, Vanikoro (VAN) and Vot Tande (VOT), and three in the southern CT, Efate (EFA), Erromango (ERR), and Futuna (FUT). The more complex ferromanganese crusts are essentially located on volcanic cones in the Vanikoro area and along the eastern faulted border of the Futuna Trough. Most of the crusts show coatings 0.1 to 1 cm thick on volcaniclastic rocks. However, the complex crusts from dredges SPS2D4, SPS2D5 (Vanikoro area) and SPS2D19 (Futuna area) cover and impregnate the volcano-sedimentary deposits up to 10 cm thick or show massive figures.

The main oxyhydroxide manganiferous phases present in these crusts are vernadite (a low-crystalline manganese oxide mineral), todorokite, and buserite. Birnessite is less common. The mineralogical structure of todorokite, buserite, and birnessite refers to the evolution of 10-Å and 7-Å manganates (Person, 1980; Usui *et al.*, 1989). Buserite is an instable 10-Å manganate which transforms to a 7-Å manganate upon dehydration; todorokite is a stable 10-Å manganate, and birnessite is a stable 7-Å manganate. These minerals are good criteria to distinguish hydrothermal and hydrogenous processes. Based on microstructural and mineralogical data, two major genetical types of ferromanganese encrustations are found in the NHBAT, a thalassic type and a hydrothermal type (Gérard, 1993).

In the thalassic type, the common crusts consist of $dMnO_2$ (vernadite) and amorphous $FeOOH,xH_2O$. Samples from SPS2D4 and SPS2D5 dredges (Vanikoro area; Fig. 5.20) show such coatings on volcaniclastic rocks which are interpenetrated by buserite with dendritic structures (Fig. 5.21). These crusts of hydrogenous precipitations display botryoidal microstructures associated with bacteriomorph occurrences (Fig 5.21), and are characterized by simultaneous precipitation of Fe and Mn (Table IV).

The hydrothermal type is characterized by stable todorokite (which resists transformation upon dehydration), celadonite (Fe-phyllosilicate), amorphous Fe hydroxide (goethite), or birnessite. Paleohydrothermalism is characterized by a Fe-Mn segregation on a macroscale, with patches of todorokite and celadonite, and amorphous Fe-phase. This todorokite displays a well-crystallized microstructure, namely spheres of todorokite with lamellar shape (Fig. 5.22). Celadonite may be the result of the evolution of hydrothermal, "metastable" nontronite (Alt, 1988; Weaver, 1989); it also may result from the slow evolution of biogenic silica and hydrothermal iron or from direct precipitation associated with low hydrothermal activity (Odin and Desprairies, 1988). Birnessite crusts similar to those of the vents of Teahitia submarine volcano in the central Pacific (Hoffert *et al.*, 1987) are rarer. They are attributed to a more recent hydrothermal phase. A latter stage of hydrothermal activity is also noticeable. Thin cracks of micritic calcite and phillipsite cut the Mn deposits in some todorokite crusts. They do not contain any biogenic fraction, in contrast to the major and older cracks. These crystallizations in the thin cracks are attributed to a last phase of hydrothermal activity.

Chemically, these two types of ferromanganese encrustations can also be distinguished (Table IV): high Mn/Fe ratios and low Co+Ni+Cu contents characterize the hydrothermal encrustations, whereas low Mn/Fe ratios and "higher" Co+Ni+Cu contents characterize the thalassic (hydrogenous) coatings and impregnations. Based on correlation values between major, trace, and rare-earth elements (REE) within the two types of crusts (Gérard, 1993), the principal differentiations are as follows:

1. Si or Fe and REE are positively correlated in the hydrothermal (todorokite) crusts, which is attributed to the adsorption of REE on clays (celadonite), and negatively correlated in the thalassic crusts.

2. P and REE are positively correlated in the thalassic (hydrogenous) group; no significative correlation appears for P in the hydrothermal (todorokite) group. The behavior of REE is mostly attributed to oxidation processes, the todorokite phase being a relatively oxidant phase. The two types of crusts (thalassic and hydrothermal) are therefore essentially controlled by redox conditions. Other determining factors are the low mobility of Al and Ti during the thalassic alteration phase. The high Si/Ti ratio characterizes the hydrothermal influence.

In summary, the ferromanganese crusts dredged along the NHBAT can be classified into two types: thalassic and hydrothermal. The common Fe-Mn crust is a vernadite coating. The more complex crusts are made up of todorokite, buserite, and birnessite, associated with a goethite-celadonite facies.

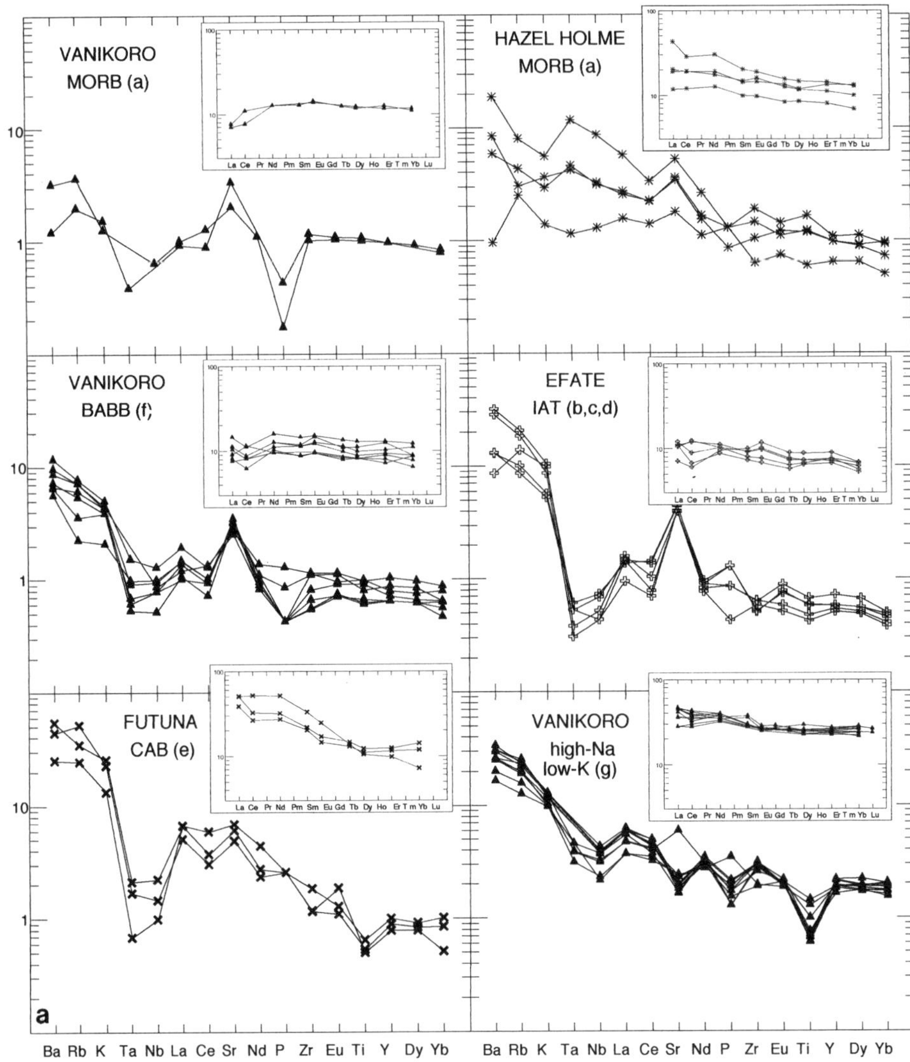

FIGURE 5.19. (a,b). N-MORB normalized "spiderdiagrams" and chondrite-normalized REE patterns (insets) for volcanic samples from the NHBAT. Data from Table II. Normalizing values: Sun and McDonough (1989) for "spiderdiagrams," Nakamura (1974) for REE. Letters in parentheses (a to i) refer to the different petrological types of Table II. Note on Fig. 5.19b the change of vertical scale for the two bottom "spiderdiagrams" (high-K and hyper-K dacites). See text for discussion.

The most evolved thalassic crusts (vernadite and buserite) were dredged in the Vanikoro area (northern Jean-Charcot troughs) on a lateral cone of the 12°10′S central volcanic complex (D5; Fig. 5.20; Table IV).

The hydrothermal crusts were dredged in the Jean-Charcot troughs on a volcanic cone located at the eastern border of the same area (D4; Fig. 5.20; Table IV) and in the southern Coriolis troughs on the eastern faulted border of the Futuna Trough (D19; Fig. 5.20; Table IV). Three successive stages of hydrothermal activity have been identified, each with

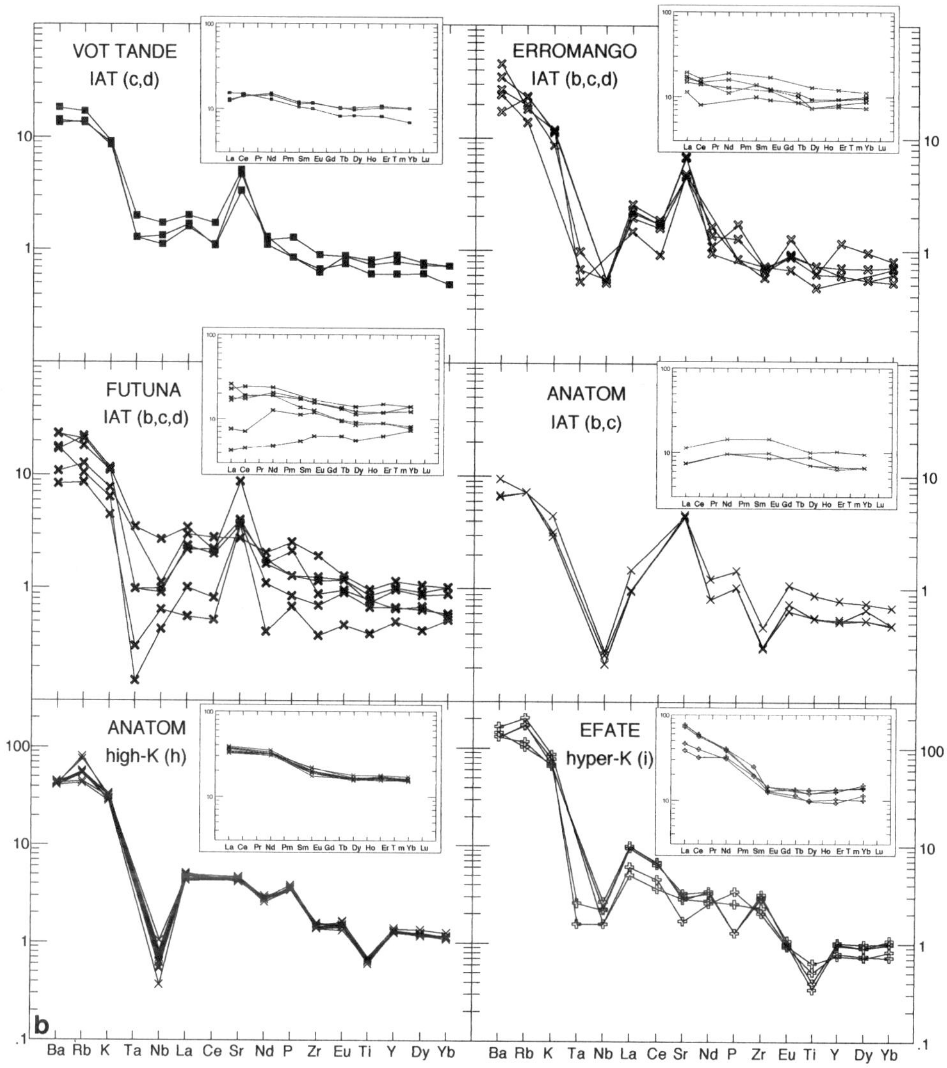

FIGURE 5.19. (*Continued*)

specific mineralizations. An ancient stage is characterized by todorokite and celadonite; a more recent stage is characterized by birnessite; a late hydrothermal stage, marked by micritic calcite and phillipsite in todorokite crusts, is thought to be relatively recent. This low hydrothermal activity seems essentially limited to the eastern border of the New Hebrides backarc troughs.

4. DISCUSSION AND CONCLUSIONS

This review of the NHBAT emphasizes a few regional and thematic points, which are summarized as follows.

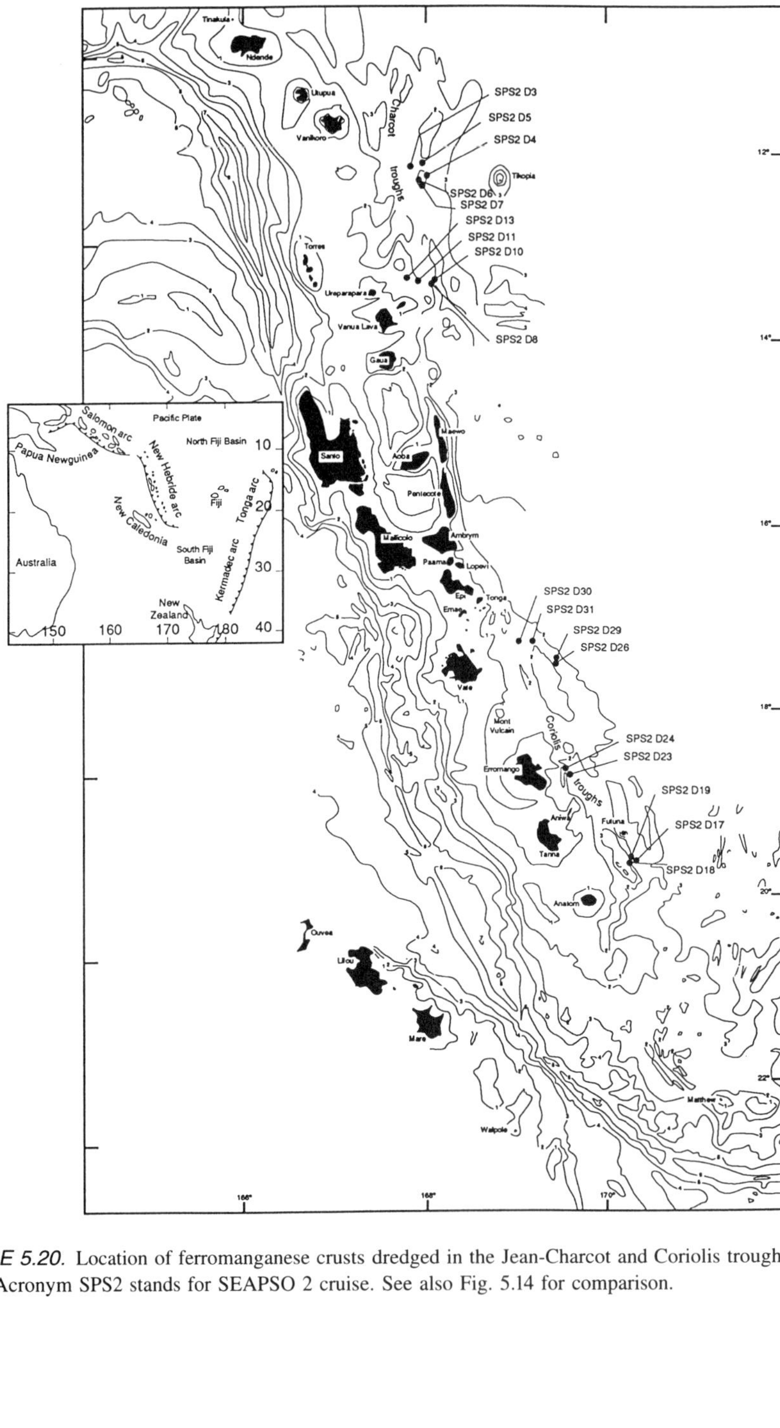

FIGURE 5.20. Location of ferromanganese crusts dredged in the Jean-Charcot and Coriolis troughs (Gérard, 1993). Acronym SPS2 stands for SEAPSO 2 cruise. See also Fig. 5.14 for comparison.

FIGURE 5.21. (Left): Macroscopic view of vernadite crusts with diagenetic buserite dendrites on volcanic sandstone. (Right): SEM view of bacteriomorphs in the vernadite crust.

1. The NHBAT correspond to a discontinuous alignment of tensional structures, bordering the eastern flank of the New Hebrides island arc. A diffuse horst-and-graben morphology, partly obscured in some places by recent volcanic complexes, characterizes the northern JCT. The general orientation of these northern troughs is slightly oblique to the arc central chain. In contrast, the southern CT show a more mature morphology (i.e., well-developed and clearly delineated flat-bottomed grabens), which strictly parallel the arc.

2. Several characteristics of these troughs are paradoxical.

(a) Seismic refraction and gravimetry studies (Collot and Malahoff, 1982; Pontoise *et al.*, 1982; Sage and Charvis, 1991) indicate that the NHBAT partly developed on an ancient oceanic crust similar to that of the NFB. N130°E magnetic lineations recognized in the NHBAT (JCT and CT) are typical of the magnetic pattern of the oldest part of the NFB and corroborate this point (Charvis and Pelletier, 1989; Sage and Charvis, 1991).

(b) The respective positions of the northern and southern troughs in the regional tectonic environment are different (Charvis and Pelletier, 1989). The northern Jean-Charcot troughs abut the western termination of the extensional Hazel Holme Ridge, from where they diffusely extend and widen northward. No subaerial volcanic edifice exists on the arc central chain between Vot Tande and Vanikoro Islands (i.e., opposite the troughs area). On the other hand, the southern Coriolis troughs strictly lie opposite Efate, Erromango, Tanna, and Anatom Islands. They terminate south of the latter island, where they merge into the island arc substratum. Furthermore, they are located well away from the N-S axis of the North Fiji Basin active spreading ridge (Monzier *et al.*, 1984b; Maillet *et al.*, 1989), and, consequently, any direct influence of this expanding ridge on the petrological evolution of these troughs can be excluded.

(c) As noted, two ridges of regional scale frame the backarc area; the d'Entrecasteaux zone on the subducting plate, and the Hazel Holme Ridge on the North Fiji Basin. Their tectonomagmatic influences on the formation of the NHBAT still remain to be deciphered. Two remarks can be made. First, the collision of the d'Entrecasteaux zone with the New Hebrides arc started around 4–3 Ma (Macfarlane *et al.*, 1988); second, the Hazel Holme Ridge acts as an extensional structure in its westernmost termination (Pelletier *et al.*, 1993a), namely, at the southern tip of the Jean-Charcot troughs. It is then possible that the structural and petrological differences observed between the northern and southern troughs may be due, at least in part, to the influences of these two ridges.

TABLE IV
Major Element Contents (wt%), Minor and Rare Earth Element (REE) Contents (ppm) of Some Typical Hydrothermal and Thalassic Ferromanganese Crusts

	Hydrothermal crusts		Thalassic crusts		
Samples	D4G9	D19G3	D5G2X	D5G2Y	D112G6
Fe (wt%)	0.27	2.36	19.17	9.89	18.93
Mn	38.03	38.57	11.33	14.71	10.16
Si	1.31	4.69	6.8	7.8	4.16
Al	0.36	1.7	2.58	3.45	2.75
Mg	3.02	2.27	1.36	1.73	1.08
Ca	5.49	1.89	2.41	4.8	2.87
Na	1.37	2.16	1.73	1.7	1.79
K	0.32	0.68	0.56	0.51	0.5
Ti	tr	0.1	0.89	0.58	1.03
P	0.03	0.04	0.15	0.08	0.16
Co (ppm)	29	51	1238	499	1433
Cr	1294	1227	231	428	160
Cu	22	45	462	199	563
Nb	24	14	44	31	46
Ni	227	157	1829	731	1352
Zn	44	40	695	563	555
Mn/Fe	140.85	16.34	0.59	1.49	0.54
Mn/Ti	3803	385.7	12.73	25.36	9.86
Fe/Ti	27	23.6	21.54	17.05	18.38
Co/Zn	0.66	1.28	1.78	0.89	2.58
Co+Ni+Cu	278	253	3529	1429	3348
Y (ppm)	16.5	7.7	162.1	82.4	185.7
La	2.5	4.1	140.5	64	171.2
Ce	7.1	11.8	214.1	107.3	286
Nd	2	4.5	106.8	51	132.5
Sm	0.8	1.7	24.6	11.9	31.3
Eu	0.2	0.3	6.4	3.1	7.8
Gd	0.9	1	27.9	13.5	33.9
Dy	5.8	9.3	28.5	15.7	35.1
Er	1.1	0.9	16	8.1	19.1
Yb	0.8	0.7	15.7	7.9	18.9
Lu	0.4	0.4	2.2	1.2	2.8
SUM REE	21.48	34.76	582.76	283.7	738.48
La/Yb	3.2	5.9	9	8.1	9.1
Ce/La	2.9	2.9	1.5	1.7	1.7
La/Sm	3.15	2.46	5.72	5.37	5.47

(d) Using seismological arguments, Charvis and Pelletier (1989) proposed to link the formation of the NHBAT to a general NE-SW extensional stress regime affecting the whole western part of the North Fiji Basin as well as its borders (i.e., the Hazel Holme Ridge and the northern and southern NHBAT). This interpretation is contrary to the view of Collot *et al.* (1985), who argued that the aftereffects of the collision-subduction of the d'Entrecasteaux zone is the main factor responsible for the formation of the NHBAT. Thus, there is

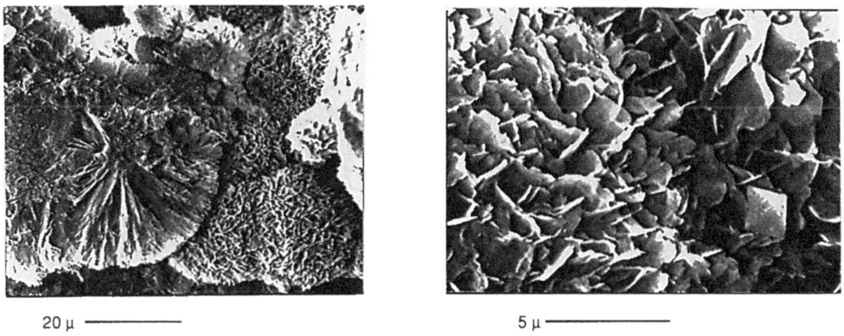

FIGURE 5.22. Todorokite facies. SEM views.

still debate concerning the respective influence of compressional and tensional stresses on the formation of the NHBAT.

3. Petrologically, no pronounced difference exists between the recent volcanic products of the New Hebrides central chain and the volcanics which crop out on the floor or on the faulted edges of the NHBAT. A ubiquitous orogenic, arc-related geochemistry prevails within the whole backarc area. Yet in the very northern part of the northern Jean-Charcot troughs, the presence of some BABB indicates an aborted tendency toward oceanic spreading, effective between 3.9 and 1.1 Ma. These unusual volcanics seem to be spatially restricted to a central volcanic complex in the northern Vanikoro area, where they neighbor coeval low-K, high-Na dacites (Nakada *et al.*, 1994). Manned-submersible surveys, water chemistry analyses, and heat flow measurements do not indicate any widespread modern hydrothermal activity. Yet some ferromanganese crusts coating volcanic and volcano-sedimentary formations result from a recent low hydrothermal activity, which is mainly limited to the eastern faulted border of the NHBAT.

4. The date of initiation of the formation of the NHBAT still remains imprecise, in spite of reliable K/Ar ages (Table III; Monjaret *et al.*, 1991) and micropaleontological determinations (Gérard, 1993). There is a consensus that the age and nature of the volcanic and volcano-sedimentary formations sampled in the troughs or on their flanks are linked to the historical evolution of the New Hebrides central chain. However, Monjaret *et al.* (1991) argued that the formation of the NHBAT progressed from south to north, through successive volcano-tectonic phases. The Coriolis troughs developed first (the Futuna Trough around 6.5–6.1 Ma; the Erromango Trough around 4.1 Ma; the Efate Trough around 3.5 Ma) with the formation of the Jean-Charcot troughs (Vot Tande area, 2.7 Ma; Vanikoro area, 2.3 Ma) being more recent. Collision of the d'Entrecasteaux zone with the arc commenced around 4–3 Ma. This last tectonic phase affected the entire backarc region. In contrast, Récy *et al.* (1990) observed that arc tholeiitic eruptions on the eastern scarps of the NHBAT mostly ceased between 2.8 and 2.3 Ma. Accordingly, they argued that only one major tectonic phase was responsible for the formation of the NHBAT, between those two dates.

5. The NHBAT can be compared with the series of backarc structures and marginal basins from the western Pacific. If these are collectively considered to represent various stages of an evolving process leading to oceanic spreading through crustal extension and rifting, the NHBAT stand at the very least evolved point; that is, they are essentially

characterized by crustal extension. Rifting without spreading occurs in the Okinawa Trough (Sibuet *et al.*, 1987; Chapter 9 this volume) and in the Bonin Trough (Sumisu and Torishima rifts) (Leg 126 Scientific Drilling Party, 1989; Leg 126 Shipboard Scientific Party, 1989; Taylor *et al.*, 1990, 1991, 1992; Taylor, 1992). Oceanic spreading occurs in the Lau Basin (Hawkins and Melchior, 1985; Hawkins *et al.*, 1990; Chapter 3 this volume; Leg 135 Scientific Party, 1992) and the North Fiji Basin (Auzende et al, 1990; Chapter 4 this volume).

In the NHBAT, arc lavas overwhelmingly predominate on BABB, as they do in the Okinawa Trough (Ishizuka *et al.*, 1990). In contrast, rifting in the Bonin Trough is accompanied by a more pronounced BABB signature (Ikeda and Yuasa, 1989; Fryer *et al.*, 1990; Hochstaedter *et al.*, 1990a,b; Tatsumi *et al.*, 1992; Taylor, 1992). Predominating BABB and MORB petrology is found in the Lau and North Fiji basins, where arc influences tend to diminish (Frenzel *et al.*, 1990; Price *et al.*, 1990; von Stackelberg and von Rad, 1990; Johnson and Sinton, 1990; Eissen *et al.*, 1991; 1994; Sigurdsson *et al.* 1993).

In conclusion, the volcanic-tectonic evolution of the New Hebrides backarc troughs results primarily from the concomitant effects of nearby subduction (along the New Hebrides subduction zone) and spreading (in the central North Fiji Basin), and secondarily from the aftereffects of collision (between the d'Entrecasteaux zone and the New Hebrides island arc). The long and complex history of the New Hebrides island arc, the maturity of the North Fiji Basin and its nonrigid plate behavior (Jarvis *et al.*, 1993) are thought to account for the progressive formation of such extensional structures.

Acknowledgments

Thanks are due to captains and crews of research ships and to chief scientists and scientific parties of numerous cruises, who helped, all together, collect data used in this chapter. Thanks also to Jean-Philippe Eissen (ORSTOM Brest), whose competence in computer drafting was appreciated, and to Kevin Speer (CNRS, Brest), who read an early draft of this paper. A. J. Crawford (University of Tasmania, Hobart) and an anonymous reviewer helped to improve this chapter.

REFERENCES

Alt, J. C. 1988. Hydrothermal oxide and nontronite deposits on seamounts in the eastern Pacific, *Mar. Geol.* **81:**227–239.

Auzende, J. M., Honza, H., and Scientific Party. 1990. Active spreading and hydrothermalism in North Fiji Basin (SW Pacific): Results of Japanese French cruise Kaiyo 87, *Mar. Geophys. Res.* **12:**269–283.

Auzende, J. M., and Urabe, T. 1994. The STARMER French–Japanese Joint Project, 1987–1992, *Mar. Geol.* **116:**1–3.

Bellon, H., Marcelot, G., Lefèvre, C., and Maillet, P. 1984. Le volcanisme de l'île d'Erromango (République de Vanuatu): Calendrier de l'activité (données 40K–40Ar), *C. R. Acad. Sci. Paris* **299**(II)**:**257–262.

Briqueu, L., and Lancelot, J. R. 1983. Sr isotopes and K, Rb, Sr balance in sediments and igneous rocks from the subducted plate of the Vanuatu (New Hebrides) active margin, *Geochim. Cosmochim. Acta* **47:**191–200.

Carney, J. N., Macfarlane, A., and Mallick, D. I. J. 1985. The Vanuatu island arc: an outline of the stratigraphy, structure and petrology, in *The Ocean Basins and Margins*, Vol. 7A: *The Pacific Ocean* (A. E. M. Nairn, F. G. Stehli, and S. Uyeda, eds.), pp. 683–718, Plenum Press, New York.

Charvis, P., and Pelletier, B. 1989. The northern New Hebrides backarc troughs: History and relation with the North Fiji basin, *Tectonophysics* **170:**259–277.

Collot, J. Y., Daniel J., and Burne, R. V. 1985. Recent tectonics associated with the subduction/collision of the d'Entrecasteaux zone in the central New Hebrides, *Tectonophysics* **112**:325–356.

Collot, J. Y., and Fisher, M. A. 1988. Crustal structure, from gravity data, of a collision zone in the central New Hebrides island arc, in *Geology and Offshore Resources of Pacific Island Arcs—Vanuatu Region* (H. G. Greene and F. L. Wong, eds.), Earth Science Ser., Vol. 8, pp. 125–139, Circum-Pacific Council for Energy and Mineral Resources, Houston, TX.

Collot, J. Y., Greene, H. G., Stokking, L., et l'équipe du Leg 134. 1991. Résultats préliminaires du Leg 134 de l'Ocean Drilling Program dans la zone de collision entre l'arc insulaire des Nouvelles–Hébrides et la Zone d'Entrecasteaux, *C. R. Acad. Sci. Paris* **313**(II):539–546.

Collot, J. Y., Greene, H. G., Stokking, L. B., *et al.* 1992a. *Proc. ODP, Init. Repts.*, 134, Ocean Drilling Program, College Station, TX.

Collot, J. Y., Lallemand, S., Pelletier, B., Eissen, J. P., Glaçon, G., Fisher, M. A., Greene, H. G., Boulin, J., Daniel, J., and Monzier, M. 1992b. Geology of the d'Entrecasteaux–New Hebrides Arc collision zone: Results from a deep submersible survey, *Tectonophysics* **212**:213–241.

Collot, J. Y., and Malahoff, A. 1982. Anomalies gravimétriques et structure de la zone de subduction des Nouvelles-Hébrides, in *Contribution à l'étude géodynamique du Sud-Ouest Pacifique*, pp. 91–109, Travaux et Documents de l'ORSTOM n°147.

Coulon, C., Maillet, P., and Maury, R. C. 1979. Contribution à l'étude du volcanisme de l'arc des Nouvelles-Hébrides: Données pétrologiques sur les laves de l'île d'Efaté, *Bull. Soc. Géol. France* 7, **21**(5):619–629.

Crawford, A. J., Greene, H. G., and Exon, N. F. 1988. Geology, petrology and geochemistry of submarine volcanoes around Epi Island, New Hebrides island arc, in *Geology and Offshore Resources of Pacific Island Arcs—Vanuatu Region* (H. G. Greene and F. L. Wong, eds.), Earth Science Ser., Vol. 8, pp. 301–327, Circum-Pacific Council for Energy and Mineral Resources, Houston, TX.

Daniel, J. 1982. Morphologie et structures superficielles de la partie sud de la zone de subduction des Nouvelles-Hébrides, in *Contribution à l'étude géodynamique du Sud-Ouest Pacifique*, pp. 39–60, Travaux et Documents de l'ORSTOM n°147.

Daniel, J., Gérard , M., Mauffret, A., Boulanger, D., Cantin, B., Collot, J. Y., Durand, J., Fisher, M. A., Greene, H. G., Michaux, P., Pelletier, B., Pezzimenti, A., Renard, V., Schaming, M., and Tissot, J. D. 1989. Déformation compressive d'un bassin intra-arc dans un contexte de collision ride-arc: Le bassin d'Aoba, arc des Nouvelles-Hébrides, *C. R. Acad. Sci. Paris* **308**(II):239–245.

Dubois, J., Dugas, F., Lapouille, A., and Louat, R. 1978. The troughs at the rear of the New Hebrides island arc: Possible mechanisms of formation, *Can. J. Earth Sci.* **15**:351–360.

Dugas, F., Carney, J. N., Cassignol, C., Jezek, P. A., and Monzier, M. 1977. Dredged rocks along a cross-section in the southern New Hebrides island arc and their bearing on the age of the arc, in *International Symposium on Geodynamics in South-West Pacific, Nouméa (New Caledonia), 27 August–2 September 1976*, pp. 105–116, Editions Technip, Paris.

Dupont, J., and Herzer, R. H. 1985. Effect of subduction of the Louisville Ridge on the structure and morphology of the Tonga arc, in *Geology and Offshore Resources of Pacific Island Arcs–Tonga Region* (D. W. Scholl and T. L. Vallier, eds.), Earth Science Ser., Vol. 2, pp. 323–332, Circum-Pacific Council for Energy and Mineral Resources, Houston, TX.

Eissen, J. P., Lefèvre, C., Maillet, P., Morvan, G., and Nohara, M. 1991. Petrology and geochemistry of the central North Fiji Basin spreading centre (Southwest Pacific) between 16°S and 22°S, *Mar. Geol.* **98**:201–239.

Eissen, J. P., Nohara, M., Cotten, J., and Hirose, K. 1994. North Fiji Basin basalts and their magma sources. I: Incompatible element constraints, *Mar. Geol.* **116**:153–178.

Eissen, J. P., Robin, C., and Monzier, M. 1992. Découverte et interprétation d'ignimbrites basiques à Tanna (Vanuatu, SO Pacifique), *C. R. Acad. Sci. Paris* **315**(II):1253–1260.

Falvey, D. A., and Greene, H. G. 1988. Origin and evolution of the sedimentary basins of the New Hebrides arc, in *Geology and Offshore Resources of Pacific Island Arcs–Vanuatu Region* (H. G. Greene and F. L. Wong, eds.), Earth Science Ser., Vol. 8, pp. 413–442, Circum-Pacific Council for Energy and Mineral Resources, Houston, TX.

Frenzel, G., Mühe, R., and Stoffers, P. 1990. Petrology of the volcanic rocks from the Lau Basin, Southwest Pacific, *Geol. Jb* D **92**:395–479.

Fryer, P., Taylor, B., Langmuir, C. H., and Hochstaedter, A. 1990. Petrology and geochemistry of lavas from the Sumisu and Torishima backarc rifts, *Earth Planet. Sci. Lett.* **100**:161–178.

Gérard, M. 1993. Bassins d'arc et fossés arrière-arc dans un contexte de collision-subduction: l'arc des Nouvelles-Hébrides (Vanuatu). Hydrothermalisme, néogenèses, diagenèse d'une série volcanosédimentaire, Thèse de Doctorat, Université de Paris-Sud, Orsay, France.

Gérard, M., Person, A., Récy, J., and Dubois, J. 1987. Preliminary results of petrological and mineralogical studies of manganesiferous encrustations dredged over the New Hebrides back arc (Vanuatu), E.U.G.–E.G.S. VII, Strasbourg, 13–16 April 1987, *Terra Cognita* **7**:2–3.

Gill, J. B. 1981. *Orogenic Andesites and Plate Tectonics*, Springer-Verlag, Berlin, Heidelberg, New York.

Gorton, M. P. 1974. The geochemistry and geochronology of the New Hebrides, Ph.D. thesis, Australian National University.

Gorton, M. P. 1977. The geochemistry and origin of Quaternary volcanism in the New Hebrides, *Geochim. Cosmochim. Acta* **41**:1257–1270.

Greene, H. G., and Johnson, D. P. 1988. Geology of the central basin region of the New Hebrides arc inferred from single-channel seismic-reflection data, in *Geology and Offshore Resources of Pacific Island Arcs—Vanuatu Region* (H. G. Greene and F. L. Wong, eds.), Earth Science Ser., Vol. 8, pp. 177–199, Circum-Pacific Council for Energy and Mineral Resources, Houston, TX.

Greene, H. G., Macfarlane, A., and Wong, F. L. 1988a. Geology and offshore resources of Vanuatu-Introduction and summary, in *Geology and Offshore Resources of Pacific Island Arcs—Vanuatu Region* (H. G. Greene, and F. L. Wong, eds.), Earth Science Ser., Vol. 8, pp. 1–25, Circum-Pacific Council for Energy and Mineral Resources, Houston, TX.

Greene, H. G., Macfarlane, A., Johnson, D. P., and Crawford, A. J. 1988b. Structure and tectonics of the central New Hebrides arc, in *Geology and Offshore Resources of Pacific Island Arcs—Vanuatu Region* (H. G. Greene, and F. L. Wong, eds.), Earth Science Ser., Vol. 8, pp. 377–412, Circum-Pacific Council for Energy and Mineral Resources, Houston, TX.

Hawkins, J. W., Lonsdale, P. F., Macdougall, J. D., and Volpe, A. M. 1990. Petrology of the axial ridge of the Mariana Trough backarc spreading center, *Earth Planet. Sci. Lett.* **100**:226–250.

Hawkins, J. W., and Melchior, J. T. 1985. Petrology of Mariana Trough and Lau Basin basalts, *J. Geophys. Res.* **90**:11,431–11,468.

Hochstaedter, A. G., Gill, J. B., Kusakabe, M., Newman, S., Pringle, M., Taylor, B., and Fryer, P. 1990a. Volcanism in the Sumisu Rift. I: Major element, volatile, and stable isotope geochemistry, *Earth Planet. Sci. Lett.* **100**:179–194.

Hochstaedter, A. G., Gill, J. B., and Morris, J. D. 1990b. Volcanism in the Sumisu Rift. II: Subduction and non-subduction related components, *Earth Planet. Sci. Lett.* **100**:195–209.

Hoffert, M., Cheminée, J. L., Larque, P., and Person, A. 1987. Dépôt hydrothermal associé au volcanisme sous-marin "intraplaque" océanique. Prélèvements effectués avec Cyana sur le volcan actif de Téahitia, *C. R. Acad. Sci. Paris* **304(II)**:829–833.

Holmes, M. L., 1988, Seismic refraction measurements in the summit basins of the New Hebrides arc, in *Geology and Offshore Resources of Pacific Island Arcs—Vanuatu Region* (H. G. Greene, and F. L. Wong, eds.), Earth Science Ser., Vol. 8, pp. 163–176, Circum-Pacific Council for Energy and Mineral Resources, Houston, TX.

Huchon, Ph., Gràcia, E., Ruellan, E., Joshima, M., and Auzende, J. M. 1994. Kinematics of active spreading in the central North Fiji Basin (Southwest Pacific), *Mar. Geol.* **116**:69–87.

Ikeda, Y., and Yuasa, M. 1989. Volcanism in nascent back-arc basins behind the Shichito Ridge and adjacent areas in the Izu–Ogasawara arc, northwest Pacific: Evidence for mixing between E-type MORB and island arc magmas at the initiation of backarc rifting, *Contrib. Mineral. Petrol.* **101**:377–393.

Isacks, B. L., Cardwell, R. K., Chatelain, J. L., Barazangi, M., Marthelot, J. M., Chinn, D., and Louat, R. 1981. Seismicity and tectonics of the central New Hebrides island arc, in *Earthquake Prediction: An International Review* (D. W. Simpson and P. G. Richards, eds.), Vol. 4, pp. 93–116, American Geophysical Union Maurice Ewing Series.

Ishizuka, H., Kawanobe, Y., and Sakai, H. 1990. Petrology and geochemistry of volcanic rocks dredged from the Okinawa Trough, an active back arc basin, *Geochem. J.* **24**:75–92.

Jarvis, P. A., Kroenke, L. W., Price, R. C., and Maillet, P. 1993. GLORIA imagery of sea floor structures in the northern North Fiji Basin, *Geo-Mar. Lett.* **13**:90–97.

Johnson, D. P., Maillet, P., and Price, R. C. 1993. Regional setting of a complex backarc: New Hebrides Arc, northern Vanuatu-eastern Solomon islands, *Geo-Mar. Lett.* **13**:82–89.

Johnson, K. T. M., and Sinton, J. M. 1990. Petrology, tectonic setting, and the formation of backarc basin basalts in the North Fiji Basin, *Geol. Jb* D **92**:517–545.

KAIYO 89 Cruise Report. 1990. (KAIYO 89 cruise in the North Fiji Basin and Vanuatu backarc troughs, 14 December 1989-13 January 1990), STARMER Cruise Report Vol. V, unpublished.

Kroenke, L. W. 1984. *Cenozoic tectonic development of the Southwest Pacific*, U. N. ESCAP, CCOP/SOPAC Tech. Bull. 6.

Kroenke, L. W., Smith, R., and Nemoto, K. 1994. Morphology and structure of the seafloor in the northern part of the North Fiji Basin, in *Basin Formation, Ridge Crest Processes, and Metallogenesis in the North Fiji Basin* (L. W. Kroenke and J. V. Eade, eds.), Earth Science Ser., Vol. 12, pp. 15–25, Circum-Pacific Council for Energy and Mineral Resources, Springer, Heidelberg.

Lallemand, S. E., Malavieille, J., and Calassou, S. 1992. Effects of oceanic ridge subduction on accretionary wedges: experimental modeling and marine observations, *Tectonics* **11**(6):1301–1313.

Leg 126 Scientific Drilling Party. 1989. ODP Leg 126 drills the Izu–Bonin arc, *Geotimes* **34**(10):36–38.

Leg 126 Shipboard Scientific Party. 1989. Arc volcanism and rifting, *Nature* **342:**18–20.

Leg 135 Scientific Party. 1992. Evolution of backarc basins: ODP Leg 135, Lau Basin, *EOS, Trans. AGU* **73:**22.

Louat, R., Hamburger, M., and Monzier, M. 1988. Shallow and intermediate-depth seismicity in the New Hebrides arc: constraints on the subduction process, in *Geology and Offshore Resources of Pacific Island Arcs—Vanuatu Region* (H. G. Greene and F. L. Wong, eds.), Earth Science Ser., Vol. 8, pp. 329–356, Circum-Pacific Council for Energy and Mineral Resources, Houston, TX.

Louat, R., and Pelletier, B. 1989. Seismotectonics and present-day relative plate motions in the New Hebrides–North Fiji Basin region, *Tectonophysics* **167:**41–55.

Macfarlane, A., Carney, J. N., Crawford, A. J., and Greene, H. G. 1988. Vanuatu—A review of the onshore geology, in *Geology and Offshore Resources of Pacific Island Arcs—Vanuatu Region* (H. G. Greene and F. L. Wong, eds.), Earth Science Ser., Vol. 8, pp. 45–91, Circum-Pacific Council for Energy and Mineral Resources, Houston, TX.

Maillet, P., Monzier, M., Eissen, J. P., and Louat, R. 1989. Geodynamics of an arc-ridge junction: the case of the New Hebrides Arc/North Fiji Basin, *Tectonophysics* **165:**251–268.

Maillet, P., Monzier, M., and Lefèvre, C. 1986. Petrology of Matthew and Hunter volcanoes, south New Hebrides island arc (southwest Pacific), *J. Volcanol. Geotherm. Res.* **30:**1–27.

Maillet, P., Monzier, M., Selo, M., and Storzer, D. 1983. The d'Entrecasteaux zone (southwest Pacific), a petrological and geochronological reappraisal, *Mar. Geol.* **53:**179–197.

Matsumoto, T., Iwabuchi, Y., and Maillet, P. 1992. Tectonics in the Vanuatu backarc basin as derived from precise bottom topography, *International Geological Congress, Kyoto, Japan, August 1992*, Abstracts Volume, p. 36.

Minster, J. B., and Jordan, T. H. 1978. Present day plate motions, *J. Geophys. Res.* **83:**5,331–5,354.

Monjaret, M. C. 1989. Le magmatisme des fossés à l'arrière de l'arc des Nouvelles-Hébrides (Vanuatu) (Campagne SEAPSO 2 du N. O. Jean Charcot). Implications géodynamiques. Chronologie, pétrologie, géochimie, Thèse de Doctorat, Université de Bretagne Occidentale (UBO), Brest, France.

Monjaret, M. C., Bellon, H., and Maillet, P. 1991. Magmatism of the troughs behind the New Hebrides island arc (RV Jean-Charcot SEAPSO 2 cruise): K-Ar geochronology and petrology, *J. Volcanol. Geoth. Res.* **46:** 265–280.

Monjaret, M. C., Bellon, H., Maillet, P., and Récy, J. 1987. Le volcanisme des fossés arrière-arc des Nouvelles-Hébrides (campagne SEAPSO Leg 2 du N/O Jean-Charcot dans le Pacifique Sud-Ouest): Datations K-Ar et données pétrologiques préliminaires, *C. R. Acad. Sci. Paris* **305(II):**605–609.

Monzier, M., Collot, J. Y., and Daniel, J. 1984a. Carte bathymétrique des parties centrale et méridionale de l'arc insulaire des Nouvelles-Hébrides, ORSTOM, Paris.

Monzier, M., Maillet, P., and Dupont, J. 1991. Carte bathymétrique des parties méridionales de l'arc insulaire des Nouvelles-Hébrides et du bassin Nord-Fidjien, Institut Français de Recherche Scientifique pour le Développement en Coopération (ORSTOM), Paris.

Monzier, M., Maillet, P., Foyo Herrera, J., Louat, R., Missègue, F., and Pontoise, B. 1984b. The termination of the southern New Hebrides subduction zone (southwestern Pacific), *Tectonophysics* **101:**177–184.

Nakada, S., Maillet, P., Monjaret, M. C., Fujinawa, A., and Urabe, T. 1994. High-Na dacite from the Jean-Charcot Trough (Vanuatu), Southwest Pacific, *Mar. Geol.* **116:**197–213.

Nakamura, N., 1974, Determination of REE, Ba, Fe, Mg, Na and K in carbonaceous and ordinary chondrites, *Geochim. Cosmochim. Acta* **38:**757–773.

Nohara, M, Hirose, K., Eissen, J. P., Urabe, T., and Joshima, M. 1994. The North Fiji Basin basalts and their magma sources. II: Sr-Nd isotopic and trace element constraints, *Mar. Geol.* **116:**179–195.

Nojiri, Y., and Ishibashi, J. 1991. Hydrothermal plumes observed in the North Fiji Basin, in *STARMER Symposium, Geology and Biology of the Rift System in the North Fiji and Lau Basins, 7–11 Feb. 1991, Noumea, New Caledonia*, Abstracts Volume, p. 41.

Odin, G. S., and Desprairies, A. 1988. Nature and geological significance of celadonite, in *Green Marine Clays* (G. S. Odin, ed.), Developments in Sedimentology 45, Elsevier, Amsterdam.

Pearce, J. A. 1983. The role of sub-continental lithosphere in magma genesis at destructive plate margins, in

Continental Basalts and Mantle Xenoliths (C. J. Hawkesworth and M. J. Norry, eds.), pp. 230–249, Nantwich, Shiva.

Pelletier, B., Charvis, P., Daniel, J., Hello, Y., Jamet, F., Louat, R., Nanau, P., and Rigolot, P. 1988. Structure et linéations magnétiques dans le coin Nord-Ouest du bassin Nord-Fidjien: résultats préliminaires de la campagne Eva 14 (août 1987), *C. R. Acad. Sci. Paris* **306(II)**:1247–1254.

Pelletier, B., and Dupont, J. 1990a. Erosion, accrétion, extension arrière-arc et longueur du plan de subduction le long de la marge active des Kermadec, Pacifique Sud-Ouest, *C. R. Acad. Sci. Paris* **310(II)**:1657–1664.

Pelletier, B., and Dupont, J. 1990b. Effets de la subduction de la ride de Louisville sur l'arc des Tonga-Kermadec, *Oceanol. Acta* **10**:57–76.

Pelletier, B., Lafoy, Y., and Missègue, F. 1993a. Morphostructure and magnetic fabric of the northwestern North Fiji Basin, *Geophys. Res. Lett.* **20**(12):1151–1154.

Pelletier, B., and Louat, R. 1989. Mouvements relatifs des plaques dans le Sud-Ouest Pacifique, *C. R. Acad. Sci. Paris* **308(II)**:123–130.

Pelletier, B., Missègue, F., Lafoy, Y., Mollard, L., Decourt, R., Dupont, J., Join, Y., Perrier, J., and Récy, J. 1993b. Extrêmités nord du bassin Nord-Fidjien et des fossés arrière-arc des Nouvelles-Hébrides: morphostructure et signature magnétique, *C. R. Acad. Sci. Paris* **316(II)**:637–644.

Person, A. 1980. Concrétions polymétalliques des sédiments de l'océan Pacifique équatorial-zone nord est: étude diffractométrique du comportement aux contraintes thermiques de la todorokite, *Bull. Soc. Fr. Mineral. Cristallogr.* **103**(2).

Picard, C., Monzier, M., Eissen, J. P., and Robin, C. 1995. Concomitant evolution of tectonic environment and magma geochemistry, Ambrym volcano (Vanuatu–New Hebrides arc), in *Volcanism Associated with Extension at Consuming Plate Margins* (J. L. Smellie, ed.), Geological Society Special Publication No. 81, pp. 135–154.

Pontoise, B., Latham, G. V., and Ibrahim, A. B. K. 1982. Sismique réfraction: structure de la croûte aux Nouvelles-Hébrides, in *Contribution à l'étude géodynamique du Sud-Ouest Pacifique*, pp. 79–90, Travaux et Documents de l'ORSTOM, n°147.

Price, R. C., and Kroenke, L. W. 1991. Tectonics and magma genesis in the northern North Fiji Basin, *Mar. Geol.* **98**:241–258.

Price, R. C., Johnson, L. E. , and Crawford, A. J. 1990. Basalts of the North Fiji Basin: the generation of back arc basin magmas by mixing of depleted and enriched mantle sources, *Contrib. Mineral. Petrol.* **105**:106–121.

Price, R. C., Maillet, P., and Johnson, D. P. 1993. Interpretation of GLORIA side-scan sonar imagery for the Coriolis troughs of the New Hebrides backarc, *Geo-Mar. Lett.* **13**:71–81.

Puech, J. L., and Reichenfeld, C. 1969. Etudes bathymétriques dans la région des îles Erromango, Tanna et Anatom (Nouvelles-Hébrides), *C. R. Acad. Sci. Paris* **208**:1259–1261.

Récy, J., Charvis, P., Ruellan, E., Monjaret, M. C., Gérard, M., Auclair, G., Baldassari, C., Boirat, J. M., Brown, G. R., Butscher, J., Collot, J. Y., Daniel, J., Louat, R., Monzier, M., and Pontoise, B. 1986. Tectonique et volcanisme sous-marin à l'arrière de l'arc des Nouvelles-Hébrides (Vanuatu): résultats préliminaires de la campagne SEAPSO (leg 2) du N/O Jean Charcot, *C. R. Acad. Sci. Paris* **303(II)**:685–690.

Récy, J., Pelletier, B., Charvis, P., Gérard, M., Monjaret, M. C., and Maillet, P. 1990. Structure, âge et origine des fossés arrière-arc des Nouvelles-Hébrides (Sud-Ouest Pacifique), *Oceanol. Acta* **10**:165–182.

Robin, C., Eissen, J. P., and Monzier, M. 1994. Ignimbrites of basaltic andesite and andesite compositions from Tanna (New Hebrides Arc), *Bull. Volcanol.* **56**:10–22.

Robin, C., Monzier, M., Eissen, J. P., Picard, C., and Camus, G. 1991. Coexistence de lignées HK et MK dans les pyroclastites associées à la caldera d'Ambrym (Vanuatu-Arc des Nouvelles-Hébrides), *C. R. Acad. Sci. Paris* **313(II)**:1425–1432.

Roca, J. L. 1978. Contribution à l'étude pétrologique et structurale des Nouvelles-Hébrides, Thèse de 3ème cycle, Université des Sciences et Techniques du Languedoc, Montpellier, France.

Sage, F., and Charvis, P. 1991. Structure profonde de la transition arc insulaire-bassin marginal dans le nord des Nouvelles-Hébrides (Vanuatu, Pacifique sud-ouest), *C. R. Acad. Sci. Paris* **313(II)**:41–48.

Saunders, A. D., and Tarney, J. 1984. Geochemical characteristics of basaltic volcanism within backarc basins, in *Marginal Basins Geology: Volcanism and Associated Sedimentary and Tectonic Processes in Modern and Ancient Marginal Basins* (B. P. Kokelaar and M. F. Howells, eds.), pp. 59–76, Blackwell Scientific, Cambridge, MA.

SAVANES 91-92 Cruise Report. 1992. (SAVANES 91–92 cruise with *Cyana* submersible in the northern Vanuatu backarc troughs), 19 December 1991–12 January 1992, STARMER Cruise Report Vol. VIII, unpublished.

Sibuet, J. C., Letouzey, J., Barbier, F., Charvet, J., Foucher, J. P., Hilde, T. W. C., Kimura, M., Ling-Yun, C.,

Marsset, B., Muller, C., and Stephan, J. F. 1987. Backarc extension in the Okinawa trough, *J. Geophys. Res.* **92:**14,041–14,063.

Sigurdsson, I. A., Kamenetsky, V. S., Crawford, A. J., Eggins, S. M., and Zlobin, S. K. 1993. Primitive island arc and oceanic lavas from the Hunter Ridge–Hunter fracture zone. Evidence from glass, olivine and spinel compositions, *Mineral. Petrol.* **47:**149–169.

Sinton, J. M., Price, R. C., Johnson, K. T. M., Staudigel, H., and Zindler, A. 1994. Petrology and geochemistry of submarine lavas from the Lau and North Fiji back-arc basins, in *Basin Formation, Ridge Crest Processes, and Metallogenesis in the North Fiji Basin* (L. W. Kroenke and J. V. Eade, eds.), Earth Science Ser., Vol. 12, pp. 155–177, Circum-Pacific Council for Energy and Mineral Resources, Springer, Heidelberg.

Sun, S.-s., and McDonough, W. F. 1989. Chemical and isotopic systematics of oceanic basalts: implications for mantle composition and processes, in *Magmatism in the Ocean Basins* (A. D. Saunders and M. J. Norry, eds.), pp. 313–345, Geological Society Special Publication No. 42.

Tatsumi, Y., Murasaki, M., and Nohda, S. 1992. Across-arc variation of lava chemistry in the Izu–Bonin arc: identification of subduction components, *J. Volcanol. Geotherm. Res.* **49:**179–190.

Taylor, B. 1992. Rifting and the volcanic-tectonic evolution of the Izu-Bonin-Mariana arc, in *Proc. ODP, Sci. Results* Vol. 126 (B. Taylor, K. Fujioka *et al.*, eds.), pp. 627–651, Ocean Drilling Program, College Station, TX.

Taylor, B., Brown, G., Fryer, P., Gill, J. B., Hochstaedter, A. G., Hotta, H., Langmuir, C. H., Leinen, M., Nishimura, A., and Urabe, T. 1990. ALVIN-Sea Beam studies of the Sumisu Rift, Izu–Bonin arc, *Earth Planet. Sci. Lett.* **100:**127–147.

Taylor, B., K. Fujioka, K., *et al.* 1992. *Proc. ODP, Sci. Results,* Vol. 126: College Station, TX (Ocean Drilling Program).

Taylor, B., Klaus, A., Brown, G. R., and Moore, G. F. 1991. Structural development of Sumisu rift, Izu–Bonin arc, *J. Geophys. Res.* **96:**16,113–16,129.

Thorpe, R. S (Editor). 1982. *Andesites, Orogenic Andesites and Related Rocks*, Wiley, New York.

Tiffin, D. L. 1993. Tectonic and structural features of the Pacific/Indo–Australian plate boundary in the North Fiji–Lau Basin regions, southwest Pacific, *Geo-Mar. Lett.* **13:**126–131.

Usui, A., Mellin, T. A., Nohara, M., and Yuasa, M. 1989. Structural stability of marine 10 Å manganates from the Ogasawara (Bonin) arc: implication for low-temperature hydrothermal activity, *Mar. Geol.* **86:**41–56.

Vallot, J. 1984. Volcanites draguées au large de l'arc insulaire des Nouvelles-Hébrides. Implications pétrologiques, Thèse de 3ème cycle, Université de Paris-Sud, Orsay, France.

Von Stackelberg, U., and von Rad, U. 1990. Geological evolution and hydrothermal activity in the Lau and North Fiji basins (SONNE cruise SO-35)—a synthesis, *Geol. Jb* D **92:**629–660.

Weaver, C. E. 1989. *Clays, Muds and Shales*, Developments in Sedimentology, Vol. 44, Elsevier, Amsterdam.

Weissel, J. K., and Karner, G. D. 1989. Flexural uplift of rift flanks due to mechanical unloading of the lithosphere during extension, *J. Geophys. Res.* **94:**13,919–13,950.

Wilson, M. 1989. *Igneous Petrogenesis, A Global Tectonic Approach*, Unwin Hyman, London.

6

Geology of the Mariana Trough

Patricia Fryer

ABSTRACT

The Mariana Trough is an actively spreading, crescent-shaped, backarc basin located in the western Pacific between the active Mariana volcanic island arc and the West Mariana Ridge, a remnant arc. The geologic evolution of the Mariana Trough varies along strike of the basin from the initial opening phase in the north to the mature seafloor spreading phase in the central latitudes. The opening of the basin began with an initial period of stretching and collapse of the preexisting arc followed by development of ridge/ transform structures along an active volcanic and tectonic zone on the eastern side of the basin. Eventually a true spreading center developed within the basin as the principal volcanic and tectonic zone diverged from the active volcanic front. The current along-strike variations in rifting/ spreading history define distinct geographic regions: the northern rifting apex, the central spreading basin, and the southern platform.

The northern apex region of the Mariana Trough typifies the early rifting stage. The zone of principal volcanism and tectonic deformation is close to the active volcanic front, and the western margin is characterized by numerous normal fault blocks downstepping to the east. Magmatic activity has remained primarily on the eastern side of the basin close to its present position, although occasional outbreaks of volcanism occur in the western portion of the basin along boundaries of fault blocks. A block-faulted region approximately 60 km wide exists along the western margin of the Mariana Trough reflecting the early stage of rifting of the basin.

The central spreading region of the Mariana Trough represents the mature spreading section of the basin. The seafloor morphology is typical of slow spreading, with a series of north-south trending valleys and ridges similar to abyssal hills offset by transform fault valleys. Volcanism is predominantly confined to the spreading center. Graben and horst structures at the margin of the abyssal hill fabric in the western part of the basin represent the early phase of opening of the basin. Preexisting structural elements that cut across the arc system persist throughout both the early rifting and the later spreading stages of the basin. Chains of submarine volcanoes (volcanic cross-chains) extend into the backarc basin

Patricia Fryer • Hawaii Institute of Geophysics and Planetology, School of Ocean and Earth Sciences and Technology, University of Hawaii at Manoa, Honolulu, Hawaii 96822.

Backarc Basins: Tectonics and Magmatism, edited by Brian Taylor, Plenum Press, New York, 1995.

along some of these lineaments, permitting the leakage of arc magmas into the backarc basin.

The southern basin platform is a shallow (less than 3 km), relatively low relief region, except where it is separated from the active volcanic front and from the remnant arc by deep, narrow troughs. Perturbations of the typical spreading center morphology occur where the active spreading zone of the platform intersects the volcanic front at about 13°N, 144°E in the southernmost portion of the backarc basin. This position coincides with a major north-south trending fault that crosscuts the forearc, arc, and backarc basin. It has influenced the spreading axis and the eastern margin of the trough.

The distribution of volcanism within the basin varies according to the developmental stages of the spreading center. Volcanism and hydrothermal activity are also influenced by cross-arc and along-arc deformation of the Mariana platelet. Most lavas from the backarc basin spreading ridges show systematic differences from mid-ocean ridge basalt as a consequence of infusion of the source with subduction-related constituents. The compositional variations of lavas within the basin indicate complex local tectonic control over magma genesis and an intricate interplay of mixing from several sources with crystallization at various pressures.

1. INTRODUCTION

The Mariana Trough is an actively opening, crescent-shaped, backarc basin that lies at the eastern edge of the Philippine Sea plate (Fig. 6.1). It is bounded by the West Mariana Ridge (a remnant arc) and the Mariana island arc (an active volcanic arc) (Fig. 6.2). The Mariana Trough is one of several backarc basins that have formed in this position since the Eocene (Karig, 1971a,b, 1974, 1975). Klein and Kobayashi (1980) and Lewis *et al.* (1982) suggested that the Philippine Sea plate, from the Oki–Daito Ridge to the Palau–Kyushu Ridge and the intervening basins, formed as a consequence of southward convergence, arc formation, and episodes of backarc rifting. They suggested that this region moved northward during a translation/rotation episode prior to about 21 Ma. From about 30 to 15 Ma the Parece Vela (Mrozowski and Hayes, 1979) and Shikoku basins (Kobayashi *et al.*, Chapter 10 this volume) were opening and separating the Mariana/Izu–Bonin arcs from the Palau–Kyushu remnant arc.

The Mariana Trough is widest at about 18°N and narrows northward to 24°N where the remnant arc and active arc join at the Volcano Islands (Smoot, 1990). Thick volcaniclastic sediments derived from the active volcanic arc form a wedge-shaped blanket over the eastern side of the trough, and the western side has complex fault-controlled topography with variable sediment cover (Karig *et al.*, 1978; Bibee *et al.*, 1980; Fryer and Hussong, 1981; Baker, 1992). North of 18°N the azimuth of the basin fault scarps lies between 330° and 345° (21°–23° = 330°; 19°–20° = 345°; 19°–18° = 340°) (Baker, 1992; Martinez *et al.*, in press; Smoot, 1990). The zone of rifting intersects the volcanic arc at about 23°N (Fryer, 1986; 1989; Baker *et al.*, 1989; Baker, 1992), marking the northern terminus of the basin. The azimuth of basin fault scarps indicates that the basin is opening in an east-west direction from 18°N to 12°N (Smoot, 1990; Lange, 1992; Fryer, 1993). The axis of spreading intersects the volcanic arc at the southern terminus of the backarc basin at about 13°N and may extend into the forearc (Fryer, 1993).

The first attempts to determine an age for the opening of the Mariana Trough met with

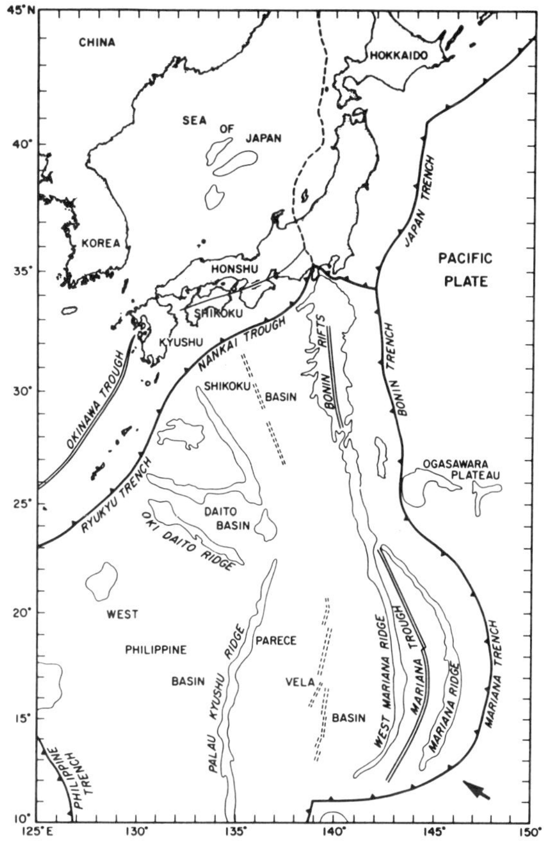

FIGURE 6.1. Bathymetry and geologic features in the Philippines Sea region. Basins and ridges are outlined by the 4-km bathymetric contour, except for the Izu–Bonin arc, West Mariana Ridge and Mariana arc, which are outlined by the 3-km contour. Barbed lines locate axes of trenches; medium double lines locate active spreading centers; dashed lines locate relict spreading centers, single dashed line is a proposed new plate boundary. Arrow shows convergence direction between the Pacific plate and the Philippine Sea plate.

limited success. A search for symmetric spreading anomaly patterns (Karig, 1971a,b; Anderson, 1975; Karig *et al.*, 1978) was unsuccessful because of the proximity of the region to the magnetic equator, the nearly north-south strike of the arc, and the high relief of the seafloor. Later, Bibee *et al.* (1980) correlated observed and synthetic magnetic anomalies near 18°N with a spreading half-rate of 1.5 cm/yr for the last 3 m.y., although the fit was not universally accepted (Eguchi, 1984). Results from the Deep Sea Drilling Project (DSDP) Leg 60 suggest a symmetric spreading rate of 2.15 cm/yr implying initiation of rifting in the late Miocene, about 6.5 Ma (Hussong and Uyeda, 1981). Other workers have suggested initiation of rifting of the Mariana Trough at 3 Ma (Karig, 1975), 8 Ma (Bibee *et al.*, 1980), and less than 10 Ma (Eguchi, 1984). Clearly, the age of initiation of rifting in the basin was not well constrained in these early attempts.

Several models proposed as mechanisms for rifting of the Mariana Trough accounted for the asymmetry of the basin, its crescent shape, and peculiarities of its morphology, compared with typical ocean basins. Karig (1971a,b) invoked nonrigid behavior of the volcanic arc and forearc region (between the trench axis and the volcanic arc) to account for the shape of the basin and apparent variations in spreading rate along strike. Bracey and Ogden (1972) suggested that the basin formed when arc segments rifted apart, beginning in the center of the crescent and migrating northward and southward simultaneously. A

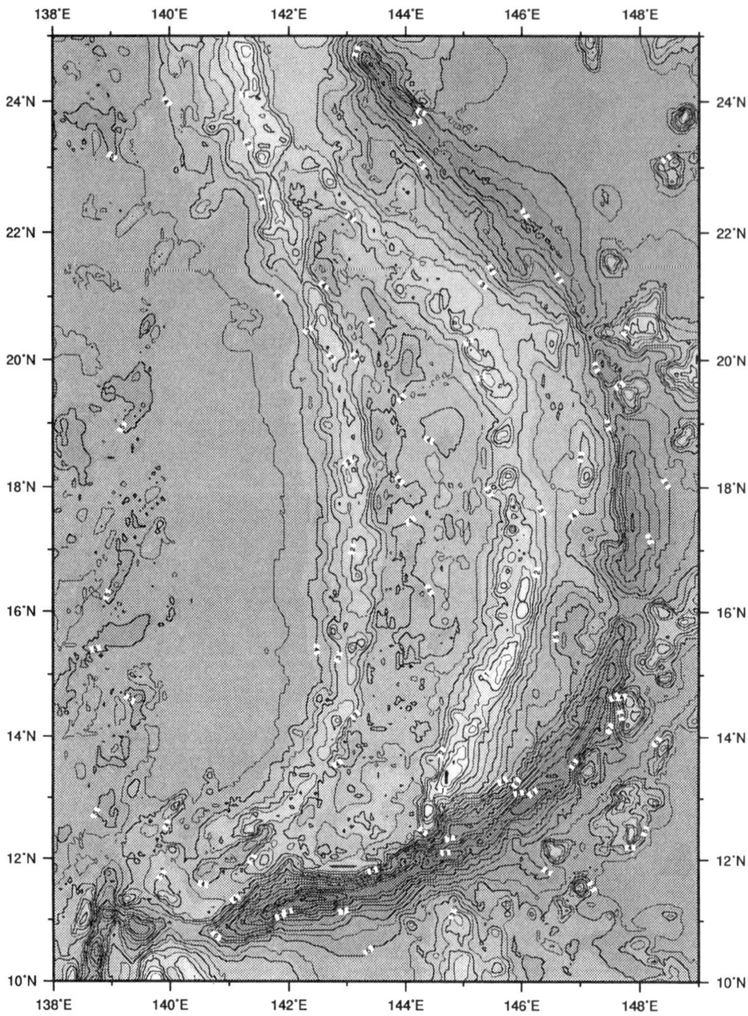

FIGURE 6.2. Bathymetric map of the Mariana system based on the DBDB5 data set of NGDC. Contour interval 500 m. Shading darkens with depth.

rigid-plate model with a pole of rotation about a point at 21°04′N, 138°01′E was suggested by Le Pichon *et al.* (1985). Detailed studies of the match of the east and west bounding margins of the basin (Karig *et al.*, 1978, Fryer and Hussong, 1981; Stern *et al.*, 1984) show that a single pole of rotation cannot exist. Stern *et al.* (1984) suggested that the basin is unzipping northward with the apex of the zipper located at the Volcano Islands, the site of eruption of highly alkalic lavas. These alkalic lavas are suggested to represent volcanism precursory to backarc basin rifting (Stern *et al.*, 1984; Bloomer *et al.*, 1989a). Similar magma types are commonly observed in continental regions undergoing incipient or early stages of rifting. Hsui and Youngquist (1985) suggested that the curvature of the Mariana arc is a consequence of Pacific plate rollback and pinning of the northern and southern apices of the Mariana arc system by the Ogasawara Plateau and the Caroline Ridge respectively. None of these models, however, took into account the details of the basin structures or systematic changes in basin morphology.

A two-stage evolution of the Mariana Trough was first suggested using morphological data from the central part of the Mariana Trough (from about 17°30′ to 18°30′N; Fryer, 1981). Fryer (1981) observed a zone of block-faulted terrain on the western side of the backarc basin, 60 km wide at 18°N. East of the block-faulted zone there is a pronounced change in morphology. The region closest to the spreading axis is characterized by closely spaced ridges and troughs, typical of abyssal hill fabric, and aligned parallel to the strike of the spreading axis (Fryer, 1981). An initial stage of stretching and collapse of the arc in the first few million years of basin extension probably produced the block-faulted terrain. The collapse resulted in normal faulting and down-dropping of fault-controlled terraces within a region at least 60 km wide and, assuming symmetrical rifting, produced a maximum total width of rifted backarc basin of 120 km. A transition to true seafloor spreading resulted in the formation of the ridge segment and transform fault morphology observed in the central Mariana Trough today (Fryer, 1981; Fryer and Hussong, 1981). Early block-faulted terrain of the Mariana Trough is similar in scale and structural style to continental rifts (Cochran and Martinez, 1988; Martinez and Cochran, 1988; Baker, 1992) and to that of incipient rift grabens in the Izu–Bonin arc, located immediately to the north of the Mariana arc (Brown and Taylor, 1988; Brown, 1991; Klaus *et al.*, 1991; Taylor *et al.*, 1991). A two-stage model for opening of the Mariana Trough is reflected in the changes in morphology of the backarc basin along strike. This paper presents the details of the along-strike and across-strike changes in the basin in terms of stages of opening of the Mariana Trough. It also presents a summary of the volcanism and hydrothermal activity in the basin and of the petrology and petrogenesis of basin lavas.

2. STRUCTURE OF THE MARIANA TROUGH

2.1. Initiation of Rifting

Several geophysical studies, side-scan sonar surveys, dredging cruises, and submersible dives indicate that the northern third of the basin is rifting, not spreading. Between 21°N and 23°N the basin is made up of three morphologically distinct regions from north to south (Figs. 6.3 and 6.4), referred to in this section as the northern basin, the central volcanic platform, and the southern basin (southwest rifted basin (SWB) and fault block terrain (FBT)). Each of these three regions is bounded on the east by a zone of active volcanism and tectonism.

SeaMARC II side-scan sonar and bathymetry data suggest that the intersection of the zone of active backarc rifting and the active volcanic front is southeast of the Volcano Islands (Fryer, 1986; Beal, 1987; Baker, 1992). The zone of volcanism and tectonic deformation that represents the center of the rifting region passes through the body of Nikko seamount, located at about 23°N (Fryer, 1986; Fryer *et al.*, 1986; Baker, 1992). Normal faults on the forearc north of Nikko seamount are oriented approximately northwest-southeast, trending along the strike of the eastern boundary fault of the backarc basin. Nikko seamount is a composite of numerous side vents, flow fields, and satellite cones that reflect the interaction between the volcanic front and the backarc rift on which it is built (Fryer, 1986; Baker, 1992).

The northern basin, extending from 23°N to about 22°25′N, is in the early part of the

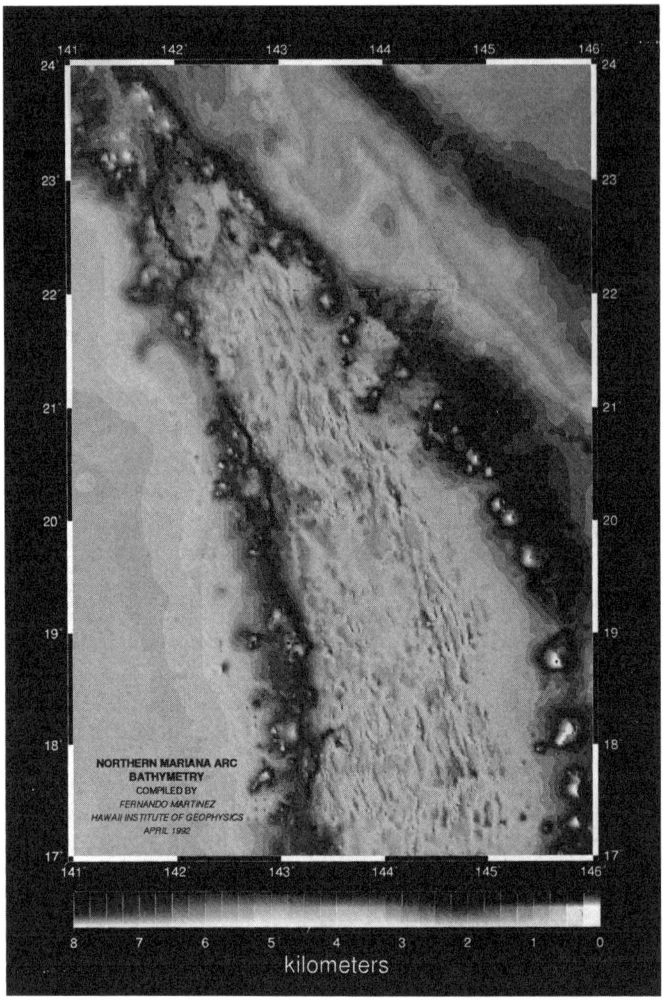

FIGURE 6.3. Bathymetry map of the northern half of the Mariana Trough. The volcanic-tectonic zone, an area of concentration of recent faulting and volcanism within the trough, follows roughly the dark gray region of deeper valleys that lies in the center of the basin at 17°N and is progressively closer to the volcanic front with increasing latitude. A color version of this figure appears opposite page 246.

first stage of development proposed for the Mariana Trough (Fryer, 1981, 1986; Baker, 1992). Side-scan sonar survey of the region shows the diversity of the seafloor terrain in this region (Fig. 6.5). The western margin of the basin is a block-faulted terrain, principally a series of normal faults downdropped to the east (Fig. 6.6), with a variable sediment cover. Similar terrain extends along the entire length of the western margin of the basin. The northern basin lacks a well-developed spreading ridge/transform fault system. However, the eastern side of the basin is characterized by a zone of recent tectonic disturbance and volcanic activity. This zone of recent activity is referred to as the volcanic and tectonic zone (VTZ) and differs from a spreading center in that it does not represent a plate boundary

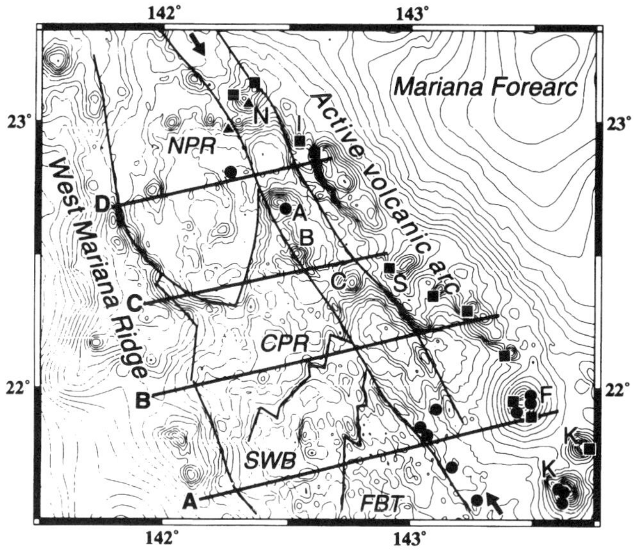

FIGURE 6.4. Location map of structural provinces and features of the northern Mariana Trough, superimposed on bathymetry. Seamounts of the volcanic arc are N-Nikko, I-Ichiyo, S-Syoyo, F-Fukujin, and K-Kasuga cross-chain seamounts. The long narrow region between the two arrows is the volcanic-tectonic zone, an area of concentration of recent faulting and volcanism within the trough. Major backarc volcanic seamounts along the volcanic-tectonic zone are designated A, B, and C. The major structural divisions of the survey area are the NPR (northern province), CPR (central volcanic platform), SWB (southern basin), and FBT (fault block terrain). Lines lettered A-D show the location of GSJ 3.5-kHz profiles in Fig. 6.7. Black symbols illustrate rock dredge locations. (dredge data sources: circles, Jackson, 1989; squares, Stern and Bloomer, 1989; triangles, P. Fryer, unpublished data).

(Baker, 1992). A deeply sedimented graben lies close to the volcanic front of the arc. Recent volcanics have intruded into the sediments or emanate from the walls of faults that bound the graben. A particularly large lava flow has emanated from a fault bounding the western edge of this graben.

The central platform of the northern Mariana Trough (Figs. 6.4 and 6.5) was built up by a locally persistent volcanic center that extends diagonally across the basin between 22°45′N and 22°N (Beal, 1987; Baker, 1992). Volcanism is currently active only on the eastern edge of the platform, and a narrow (5–20 km wide) graben separates the platform from the line of the active volcanic frontal arc. The morphology of the platform indicates an increase in volcanism to the east (Baker, 1992). There are no discrete volcanic edifices along the volcanic front line between Syoyo and Nikko seamounts (Fig. 6.4). The locus of arc volcanism was probably shifted westward, and the volcanoes that lie on the eastern edge of the central volcanic platform probably represent volcanism diverted from the volcanic front. The composition of lavas recovered in dredges from seamount A (22°39′N) and from basin lava flows supports this suggestion (both backarc basin basalt and island-arc basalt sources (Jackson, 1989)). The large lava flow (see Fig. 6.5) from the volcanic ridge in the basin north of this seamount is of island-arc tholeiite composition (Jackson, 1989).

The southern basin from 22°10′N to 21°40′N contains both the early stair-stepped fault blocks on the west (SWB) and the later fault blocks that comprise a basin-and-range-type

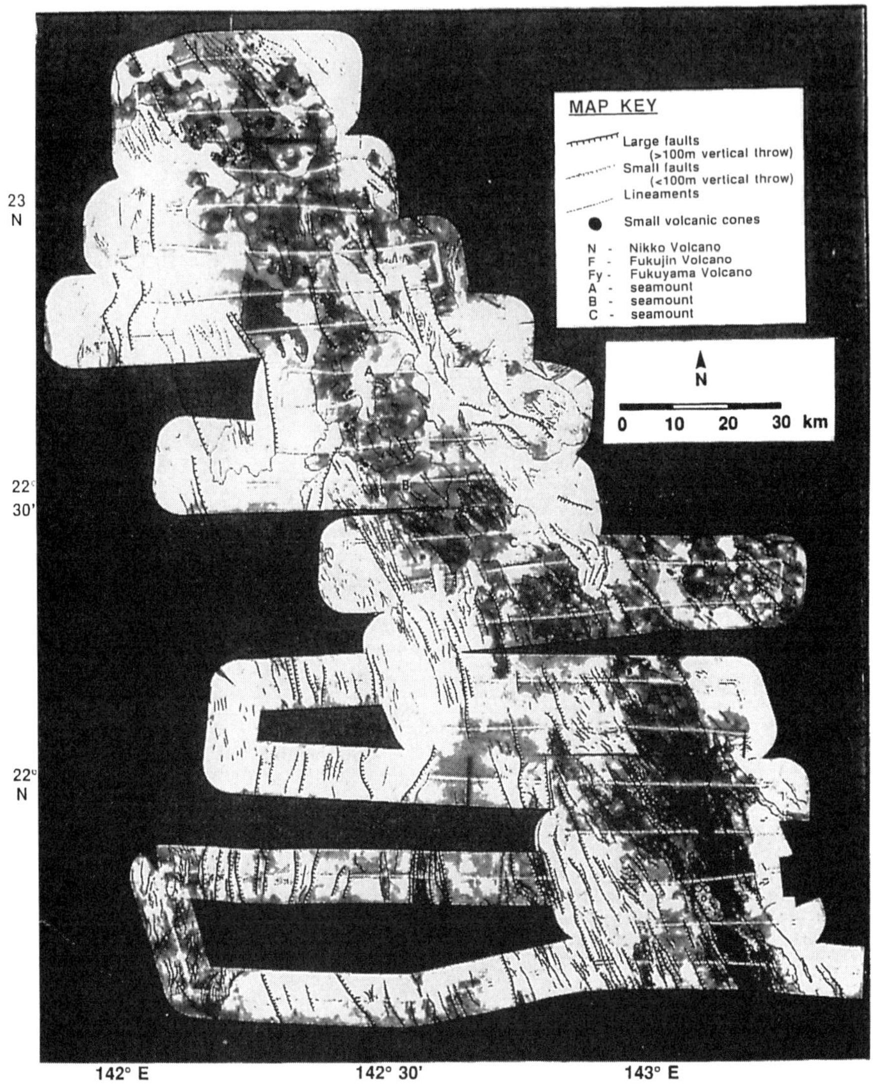

FIGURE 6.5. "Ramped" SeaMARC II sidescan sonar acoustic image of the northern Mariana backarc basin (Sender *et al.*, 1989). Data are plotted so that acoustic signals with strong amplitudes are dark. The ship track is represented by the blank stripe down the center of each 10-km-wide swath. The ramped processing scheme maximizes gray-level contrast to enhance local structural details. The major tectonic and structural elements of the survey region are superimposed on the side-scan imagery.

terrain (FBT) in the center of the backarc basin. The eastern margin of the southern basin contains a series of volcanic ridges on the east (Figs. 6.4 to 6.6). The faulting in the deep western and central parts of the basin was apparently accompanied by volcanism increasing in volume toward the east (Baker, 1992). The stair-step faults of the westernmost part of the southern basin were probably the sites of scattered intrusions or flows in the earliest phases of backarc basin formation (Baker, 1992; Martinez *et al.*, in press). The ridges and troughs of the fault-block terrain located between the western stair-stepped fault blocks and the

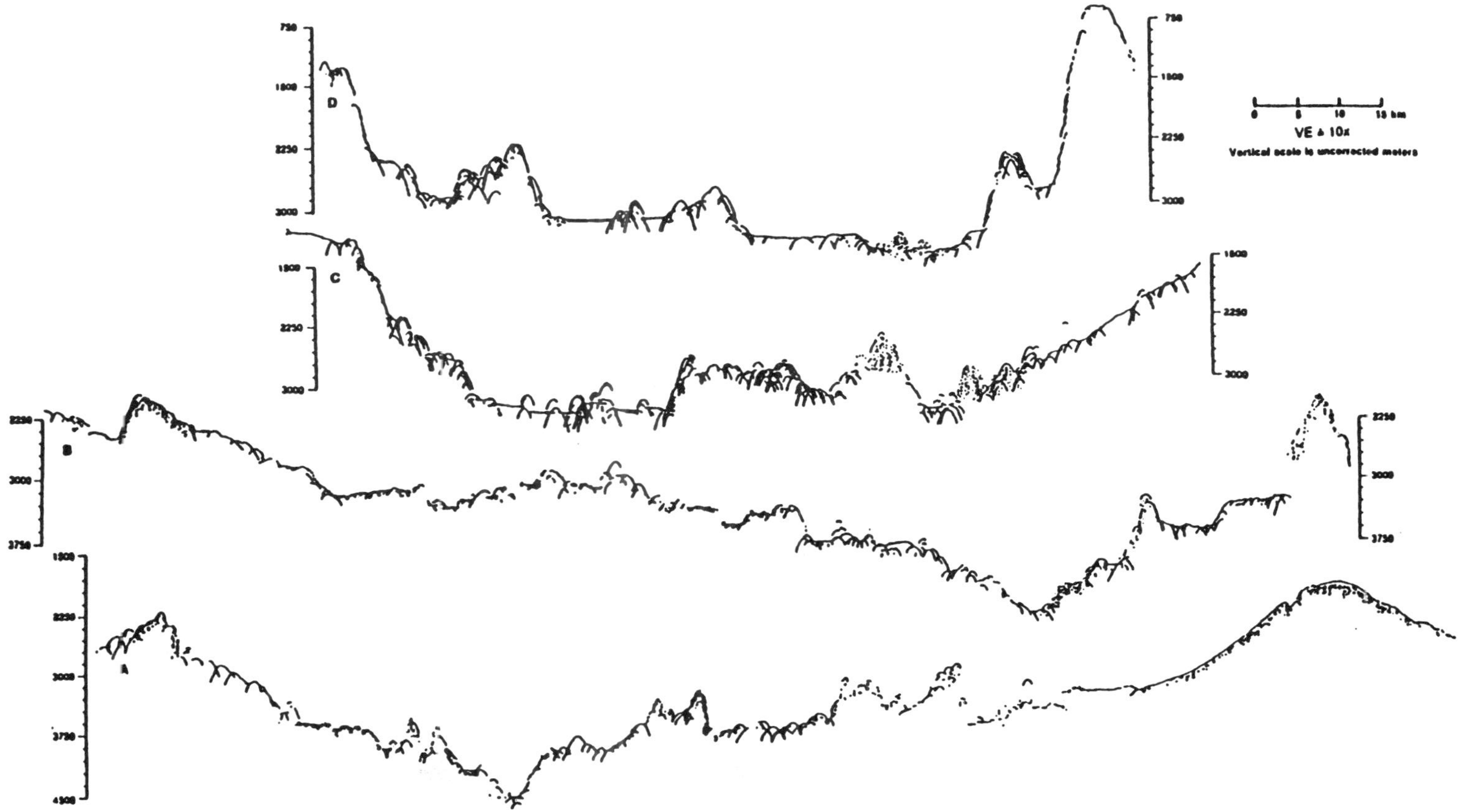

FIGURE 6.6. Profiles A to D across the northern Mariana Trough traced from GSJ 3.5-kHz data. Profile locations are shown in Fig. 6.4. These profiles illustrate the change in width and morphology of the basin from north (profile D) to south (profile A). Large tilted fault blocks characterize the north part of the basin, volcanic seamounts and fault blocks in the central region, and intensely faulted terrain and volcanism in the south, shown by an increasing intensity of hyperbolic echoes. The east end of profile A crosses the south flank of the arc volcano, Fukujin seamount.

VTZ are only partially constructional volcanic ridges (Baker, 1992). Volcanism is superimposed on fault blocks, even in the eastern part of the basin (Baker, 1992,) although recent *Alvin* observations of fissures, numerous vents, and fresh pillow ridges suggest that volcanism dominates the central part of the VTZ (Fryer, 1990; Stern *et al.*, 1990). Based on SeaMARC II interpretations, the VTZ lies in a 6- to 11-km-wide graben from 143°E to 143°10′E between 22°10′N and 21°40′N and is characterized by fault-bounded valleys and ridges, lava flows that show up as high-backscatter regions on sonar imagery, and small volcanic cones (Fig. 6.7). From southeast to northwest the zone widens and deepens to >4500 m at the base of a fault scarp that defines the boundary with the volcanic platform to the north at 22°20′N (see Fig. 6.4). At 22°20′N several small seamounts have developed on the edge of a fault block or accommodation zone (Beal, 1987; Baker, 1992) against which this segment of the VTZ abuts. Several lava flow fields can be seen on the uplifted, south-facing wall of this accommodation zone in the side-scan images (Figs. 6.5 and 6.7). The southward increase in width of the basin and the general relationships of the structures can be seen in seismic reflection profiles across the basin (Fig. 6.6).

2.2. From Rifting to Spreading

The VTZ remains on the eastern (arc) side of the basin during the early rifting stage of the Mariana Trough (Baker, 1992; Martinez *et al.*, in press). It begins to diverge from the volcanic arc once the basin extends to a width of about 120 km (at about 22°N; Fig. 6.3). The VTZ eventually becomes dominated by development of new backarc basin lithosphere at a seafloor spreading center. The transition from rifting to spreading would therefore be marked by a shift from volcanism confined to fault boundaries to development of volcanic ridges and abyssal hill fabric. We can investigate this model for transition either by locating the current point of transition along strike of the basin or by examining the change from rifted, graben-and-horst terrain to abyssal-hill fabric on transects across the basin.

Along strike the basin must undergo a change from rifting to spreading somewhere between 21°40′N, where side-scan data indicate the basin is still rifting, and 19°45′N, where side-scan data show true seafloor spreading. Only bathymetry, magnetics, and gravity data (Martinez *et al.*, in press) are available to trace the continuation of the VTZ between these latitudes. The VTZ progressively diverges from the active volcanic arc and terminates at about 21°20′N, approximately 75 km west of the eastern boundary fault of the basin. Yamazaki *et al.* (1991, 1993) interpreted lineated magnetic anomalies in the basin between 21°N and 22°N as seafloor spreading anomalies. More recently, Martinez *et al.* (in press) argued that the magnetic anomalies in the whole of the region from 21°N to 23°N are the result of structurally controlled intrusions, implying rifting, not seafloor spreading. Bouguer gravity data in the northern Mariana Trough show positive anomalies, implying thinned crust (Martinez *et al.*, in press) and discontinuous lows that suggest low-density material (possibly partial melt zones) underlies the currently active volcanic and tectonic zone. The discontinuous nature of the lows implies that the volcanic segments are not organized in the form of a spreading axis anywhere between 21°N and 23°N (Martinez *et al.*, in press).

Rates of extension for the northern part of the Mariana Trough have generally been estimated to be slow, consistent with the morphology. Slow spreading ridges may be structurally complex; cycles of alternating tectonic and volcanic activity can occur within a given location. The structures produced during tectonic cycles of a slow spreading ridge may not differ significantly from those produced by rifting of preexisting crust. A qualita-

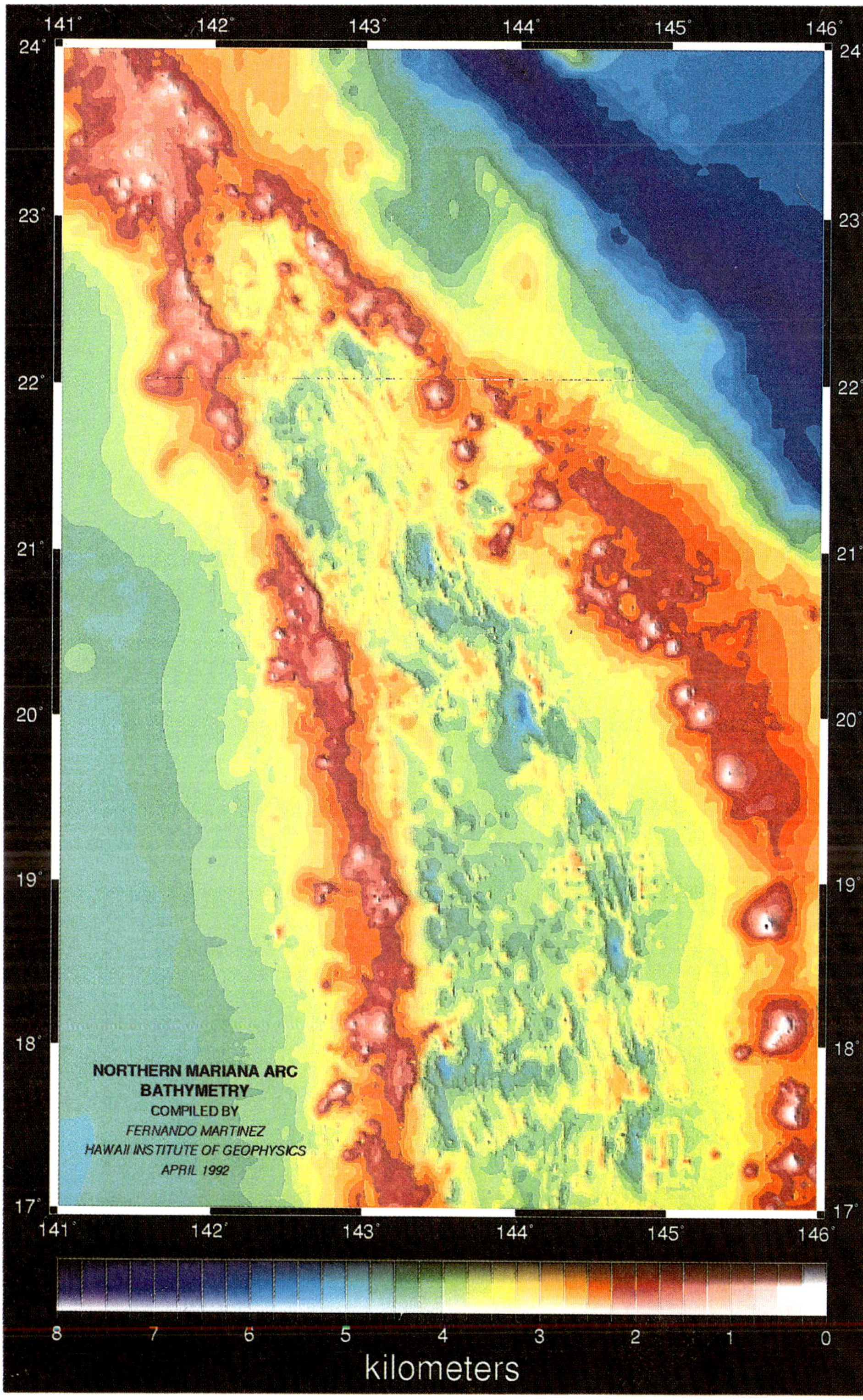

FIGURE 6.3. Color contoured bathymetry map of the northern half of the Mariana Trough. The volcanic-tectonic zone, an area of concentration of recent faulting and volcanism within the trough, follows roughly the dark blue region of deeper valleys that lies in the center of the basin at 17°N and is progressively closer to the volcanic front with increasing latitude.

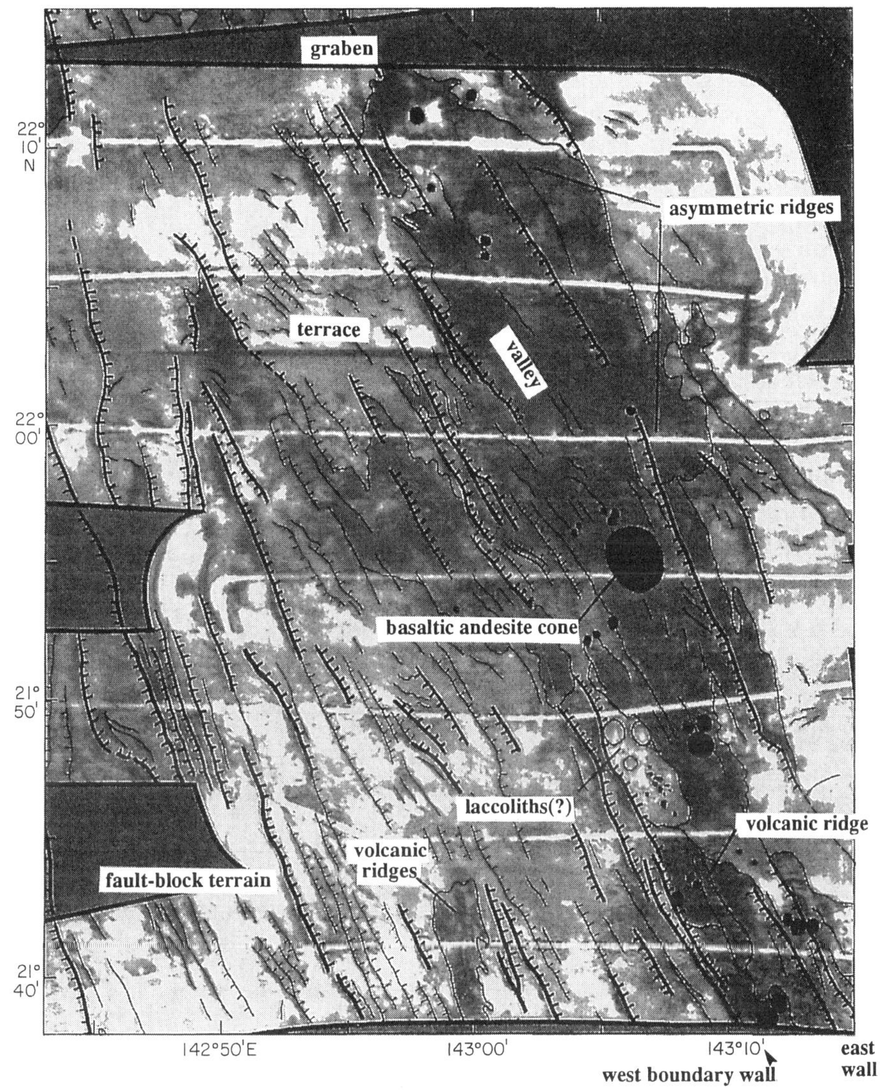

FIGURE 6.7. Geologic interpretation map of the southernmost part of the volcanic and tectonic zone surveyed by SeaMARC II. The interpretation is based on the combination of side-scan sonar, bathymetric, single-channel seismic, and 3.5-kHz data (Baker, 1992). The major tectonic and structural elements of the survey region are superimposed on the side-scan imagery. Faults are shown as solid lines with hachures indicating the downthrown side. Solid lines are for faults with throws of over 100 m. Fine lines are faults with throws less than 100 m. Lineaments visible in side-scan data, but not apparent on seismic profiles as having detectable offsets, are mapped using a solid line lacking hachures. Volcanic cones are designated as solid black regions. Volcanism is concentrated in a fault-bound valley trending 330° which widens into a 4500-m graben at the top of this figure.

tive structural description of the Mariana Trough between 21°40′N and 22°10′N does not adequately distinguish rifting and spreading, and the distinction is important because it bears on the question of general evolution of backarc basins. If the northern portion of the Mariana Trough (north of 21°N) is not spreading, and by spreading it is meant that new lithosphere of backarc basin origin is forming, then there is no new plate boundary in the northern Mariana Trough.

The geology changes markedly between 21°N and 20°N. A series of ~5-km-deep half-grabens oriented roughly N-S occupies the center of the basin in these latitudes (see Fig. 6.3; Smoot, 1990). There is no side-scan imagery of these grabens, and thus the exact distribution and type of volcanic activity associated with them is unknown at present; however, dredging in these basins in 1984 (Shibata and Segawa, 1984; Ishii *et al.*, 1985) recovered fresh basalts compositionally similar to the backarc basin basalts of the ridge segments farther south. Gabbros recently dredged from the grabens are also of backarc basin basalt composition (Stern, pers. comm., 1992). The major bounding faults of the half-grabens are alternately on the east and west sides of the half-grabens, and thus the sense of offset between the half-grabens alternates between right and left lateral. The length of the half-grabens is approximately 20 to 30 km between offsets. This type of half-graben is similar to that observed in the rift grabens of the Izu–Bonin arc to the north (Klaus *et al.*, 1991), and their distribution is similar to the distribution of half-grabens in continental rift areas. The magnetics and gravity data in this part of the basin differ significantly from that of the basin to the north (Martinez *et al.*, in press). There is a break in the continuous magnetic lineaments at about 21°N, which implies that this part of the basin is rifting in a less well-defined, regular manner (Martinez *et al.*, in press). The half-grabens thus represent a different tectonic regime superimposed on the rifting backarc basin. There is insufficient data at present to define the origin of the half-grabens; however, it is possible that cross-arc structural trends and forearc tectonics influence their formation. Hussong and Fryer (1983) suggested that Fukujin and Kasuga seamounts formed within a N-S trending cross-arc rift (represented by a regional bathymetric low and a pronounced reentrant in the inner trench slope). They suggested that this rift cuts across the northern Mariana forearc from the trench axis to the volcanic front and, together with the eastern bounding fault of the basin, controls distribution of the arc volcanoes along its trend. The extension of this rift into the backarc basin would intersect the backarc basin and encompass the half-grabens. Such a cross-arc rift could influence the structural development of the backarc basin to the extent that it would perturb a gradual change from the rift morphology north of 21°30′N to the spreading morphology that characterizes the basin to the south.

A SeaMARC II survey of the central part of the Mariana Trough between 19° and 19°45′N shows a well-developed spreading center with symmetric abyssal hill fabric developed at the margins of a spreading valley that lies in a zone of high backscatter (Figs. 6.8a,b). The high-backscatter region indicates recent volcanism. The fault-bounded spreading valley is less than 10 km wide and is flanked by abyssal hill fabric extending to a total width of about 40 km (widest in the south). Two first-order discontinuities are evident in this image at latitudes of 19°22′N and 19°40′N separating three ridge segments. Bathymetry coverage of the region is not complete; therefore, bathymetric details of secondary discontinuities along the ridge segments are not obvious. However, it is apparent from inspection of the SeaMARC II imagery (Fig. 6.8b) that the volcanic ridges within the three segments surveyed contain several discontinuities along strike (solid strips in Fig. 6.8a). The mean azimuthal trend of the faults bounding the axial valley is about 345°; however, the fault scarps converge on about 19°45′N and are terminated in a V-shaped valley

(Fig. 6.8a), suggesting a propagating rift. The bathymetry of the region suggests that the rift tip may extend to 20°10′N (Fig. 6.3), but it will require side-scan data to verify this suggestion. A change from rifting to true mid-ocean spreading in the Mariana Trough takes place near this point. South of 20°N new backarc basin lithosphere forms along mid-ocean-style volcanic ridges.

If gradual transition from rifting to spreading is the correct model for evolution of the Mariana Trough, this transition in the northern part of the basin is obscured by two abrupt structural changes in the style of opening of the basin; from rifted basin to half-grabens and from half-grabens to propagating ridge tip. It may be possible to identify the transition from rifting to spreading by examining the distinct change in morphology across the basin from rifted block-faulted crust to abyssal hill fabric. The best-studied part of the basin where seafloor spreading is taking place is at about 18°N.

2.3. The Central Mariana Trough

The central latitudes of the Mariana Trough (from 14° to 20°N; see Figs. 6.3, 6.8–6.10) show a series of spreading valley segments offset by a complex series of faults and fault systems comprising transform fault valleys. In general, the center of spreading is about 40–50 km closer to the active Mariana Islands chain than to the West Mariana Ridge at these latitudes (Fig. 6.9a). Fryer and Hussong (1981) estimated that the central portion of the Mariana Trough at 18°N contains the widest section of newly generated backarc basin lithosphere formed by seafloor spreading (approximately 110 km). They also assume that the seafloor spreading is symmetric. Based on a spreading half-rate of 1.5 cm/yr (Seama and Fujiwara, 1993), the newly generated backarc basin lithosphere would have begun forming about 3.67 Ma. The change from block-faulted terrain to abyssal-hill-type topography occurs at a maximum of 80 km east of the West Mariana Ridge. If the total width of the basin at about 18°N is 240 km, assuming that Pagan Island marks the base of the eastern bounding fault of the basin, there must be approximately 50 km of rifted arc lithosphere underlying the volcaniclastic wedge of the arc on the eastern side of the basin (some of which is visible in Fig. 6.9a). The total rifted portion of the basin is about 130 km wide. Estimates of a maximum possible width of 120 km for extension by rifting prior to development of true seafloor spreading in the basin (Fryer, 1981; Fryer and Hussong, 1981) are consistent with this calculation. The only reliable age date on the rifted portion of the basin, 5 Ma, is from DSDP Site 453 (Hussong and Uyeda, 1981), located about 25 km east of the West Mariana Ridge. Thus, it is not possible to date accurately the inception of rifting in the Mariana Trough at this latitude or to determine the rate of extension in the rifting stage.

The locus of spreading between about 17°30′N to 19°45′E is a well-defined valley with symetrically disposed fault blocks (Fig. 6.8b) and containing an elongate volcanic ridge or ridges (Figs. 6.9b,c). The maximum width of the spreading center valley is about 20 km, and it ranges in depth from about 4.0 to 5.3 km (Fig. 6.9c). Broad ridges and troughs also strike ENE across the width of the Mariana Trough (Figs. 6.3 and 6.9b). The ridges are from 20 to 30 km wide (north to south), with intervening troughs about 20 km wide. The troughs are collinear with offsets in spreading rift segments and are interpreted as fracture zone valleys (Fryer and Hussong 1981; Hussong and Sinton, 1983; Kong, 1993). Hussong and Sinton (1983) suggested that there are small offsets in ridge segments within the Pagan fracture zone at 17°30′N. The transform fault valleys are generally 4.2 to 4.4 km deep. The deepest portions of the valley segments (maximum depths are slightly greater than 5 km) occur where they intersect ridge offsets. Two of these intersections contain in excess of

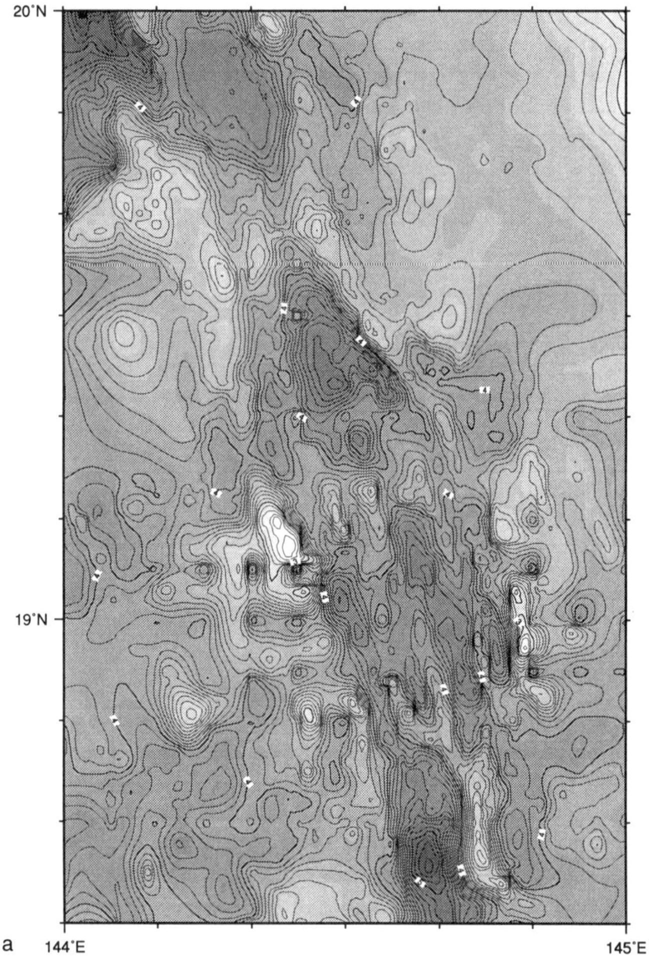

FIGURE 6.8. a. Bathymetry of the region 18°30′ to 20°N. This is not SeaMARC II bathymetry but a compilation of data from several sources (DBDB5, Sea Beam and 3.5-kHz data). (b) SeaMARC II side-scan sonar imagery of a survey of the Mariana backarc basin spreading center from 18°45′N to about 19°45′N. The high-backscatter regions (dark) in the center of the survey denote the active volcanic ridges of the spreading center valley and walls. The small arrows point to several lava flows that have broken out on the flanks of the central valley and to a small cone (second from the top) that has developed on the eastern flank of the valley. Note the symmetric distribution of abyssal hill topography and the sinuosity of the abyssal hill terrain. This is typical of mid-ocean ridge topography. Bathymetry data were not collected on this early survey.

300 m of sediment in the active transform fault zones between major rift segments (Hussong and Sinton, 1983).

The 20- to 30-km-wide ridges that strike east-west across the Mariana Trough contain a series of fault blocks that parallel the strike of the ridge segment (nearly north-south). These fault blocks have symmetrically (toward the rift valley) dipping fault scarps similar to abyssal hill fabric (Figs. 6.9a,b) typical of mid-ocean ridge systems (e.g., Perram and Macdonald, 1990). Although the eastern portion of the basin is covered with a wedge of

b

FIGURE 6.8. (*Continued*)

volcaniclastics from the active arc, basement protrudes through the sediment wedge close to the rift valley (Fig. 6.9a). The distribution of these faults is shown on the SeaMARC II image of the rift segments between 19°N and 19°45′N (Fig. 6.8b). The sinuousity of the faults throughout this survey area is an artifact.

The flanks of the ENE ridges have frequent reentrants (Fig. 6.9b), which give them a jagged, sawtooth appearance. The ridges vary in depth along strike, rising to minima adjacent to intersection highs. Intersection highs have been noted on a number of slow spreading mid-ocean ridges where major fracture zones intersect rift segments (e.g., Atlantis, Vema, and Oceanographer). A study of seismicity at the intersection of the Pagan fracture zone with the spreading valley at 18°N in the Mariana Trough (Hussong and Sinton, 1983) indicates that the formation of the intersection high, and the associated relative uplift of the adjacent ridge edge, may result from high-angle dip-slip faulting which predominates along the scarps bordering the intersection high. The occurrence of these intersection highs and associated steep ridge flanks produces a complex bathymetric grain in the central part of the basin. Between 17° and 19°30′N the ridges have steep, sediment-free flanks, with frequent reentrants. This "abyssal hill" fabric is absent about 80 km from the remnant arc, where the change from the initial rifting terrain to the subsequent

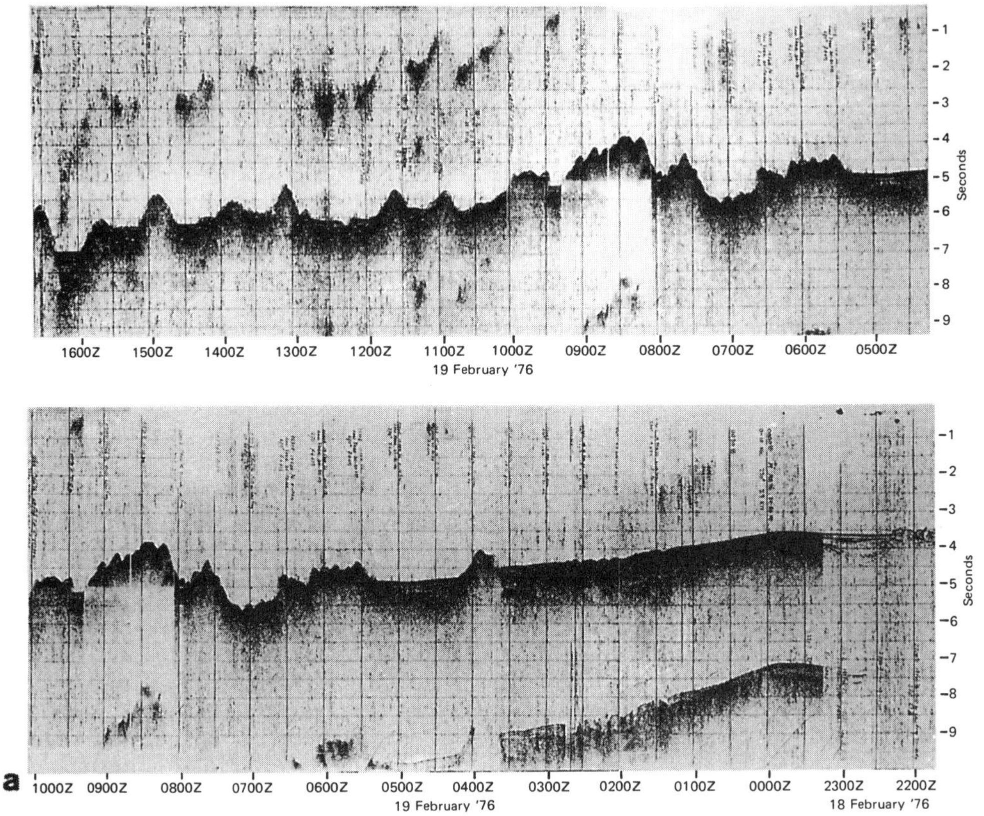

FIGURE 6.9. a. Seismic reflection profile at 17°50′N across the Mariana Trough from the West Mariana Ridge (left of upper profile) to the volcaniclastic apron south of Pagan Island (right of lower profile). The center of the spreading axis of the basin is located at 0700Z. Note angular unconformities in the sediments of the western grabens and normal faulting at 1415Z. (b) Bathymetry of the central Mariana Trough and reflection seismic profiles showing major structural features (Fryer and Hussong, 1981). Profile A (B-B′) shows the transition from the abyssal hills topography to the graben and horst topography beginning with the large block at 1530Z. Profile B (C-C′) shows the morphology along strike of the abyssal hill fabric. (c) Bathymetry of the central Mariana Trough showing the morphology of the central spreading ridge and the locations of dredge samples and dives (Hawkins *et al.*, 1990).

spreading morphology takes place (Fryer and Hussong, 1981). Also, at this distance from the West Mariana Ridge, several of the transform valleys change strike from ENE to NE-SW, suggesting a change in the direction of extension (Fryer and Hussong, 1982).

The locus of spreading in the central Mariana Trough is defined on the basis of SeaMARC II side-scan imagery, bottom photography, detailed bathymetric data, the recovery of fresh, glassy basalt fragments in dredge hauls, seismicity studies, and direct observation using submersibles between about 17°30′ to 19°45′N. A detailed description of the rift valley between 17°30′ and 18°30′N is given by Hawkins *et al.* (1990) and Kong (1993). At these latitudes the complexity of the volcanic axial ridge is apparent (see Fig. 6.9c). There are three major offsets greater than 5 km, which comprise the first-order segmentation of the axial ridge (Kong, 1993), and more than a dozen minor offsets, which

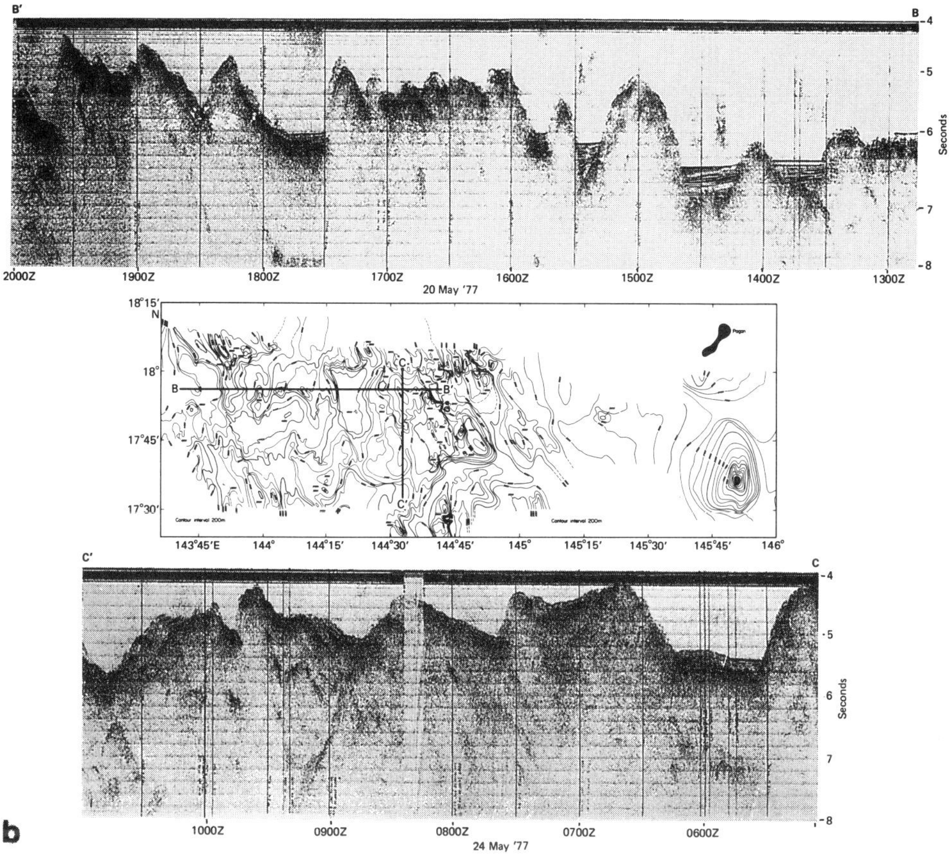

b

FIGURE 6.9. (*Continued*)

comprise the second-order segmentation (Hawkins *et al.*, 1990). These second-order discontinuities are expressed by a series of both right- and left-lateral en echelon offsets of less than 1 km (Hawkins *et al.*, 1990). The ridge segments are broadly shallowest at the centers of the segments and slope steeply toward the deep offset valleys. The shoal regions have maximum relief on the order of 300 m with local highs between the en echelon offsets marking the second-order discontinuities in the ridges.

Interpretation of the location of the spreading center valley from 14° to 17°N is based solely on bathymetry data (Barone *et al.*, 1990; Smoot, 1990; Lange, 1992). The bathymetry coverage of the Mariana Trough south of 15°40′N is the most complete (Fig. 6.10). A series of short spreading valleys is offset westward by a series of right-lateral transform fault valleys (Smoot, 1990; Lange, 1992). The spreading valleys maintain a distance of about 100 km from the active frontal arc between about 14°35′ and 16°N by stepping westward as the curve of the frontal arc swings southwest. The spreading valley segments are about 15 km wide. Only a few identifiable central ridges or volcanic cones are in these valleys. The most prominent of these is from 15°45′N to 16°N. A region of small cones and ridges lies along 144°30′E between about 15°25′ and 15°30′N (Fig. 6.10). From 15°30′ to 15°45′N and from 14°45′ to 15°N there are two deep (>4 km) valleys in which the transform motion between

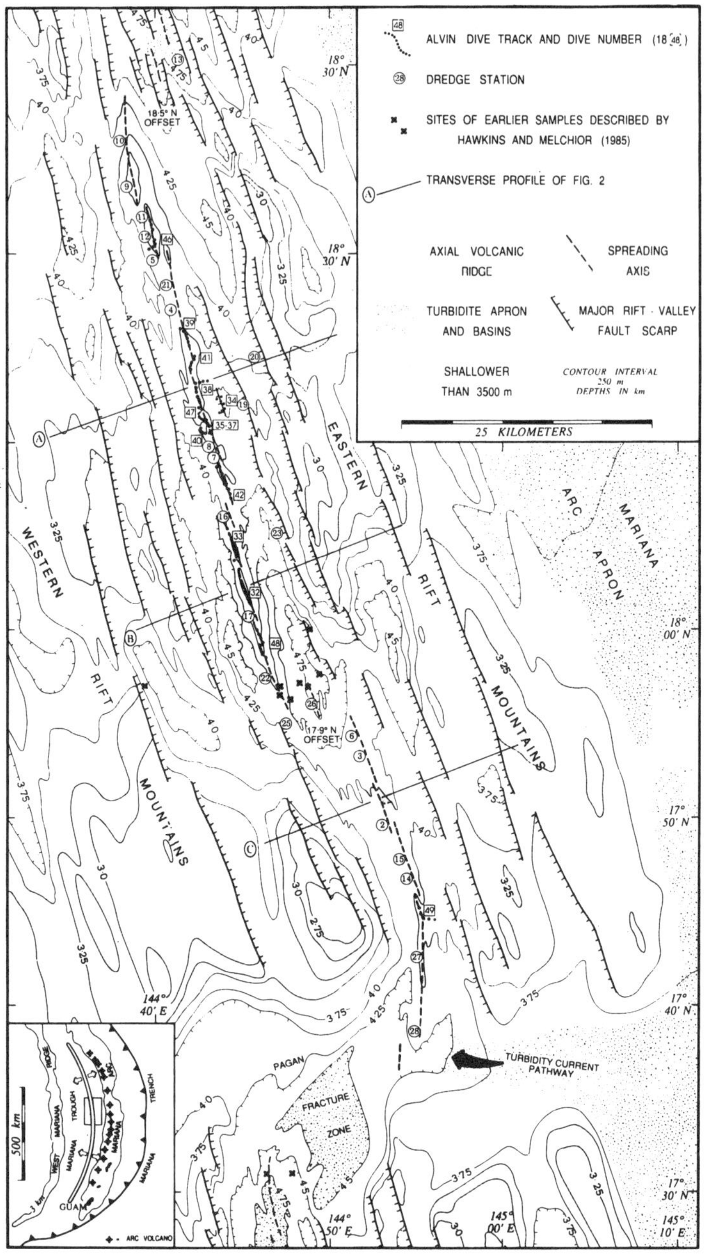

c

FIGURE 6.9. (*Continued*)

the spreading valley segments probably takes place. South of 14°40′N the spreading valley is located at about 144°08′E. There is a gap in published data between 14°49′ and 14°. At 14° the spreading valley is located between ~144°40′E and 144°E, and therefore there must be a transform valley in the data gap. The azimuth of the faults bounding the valleys from 16° to ~14°30′N is about 0°; thus spreading is taking place in almost an east-west direction in this part of the basin.

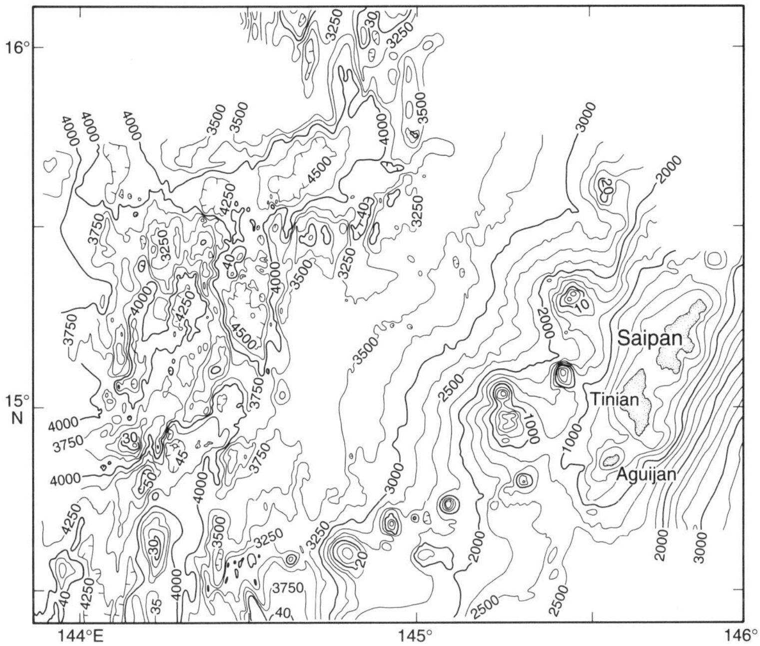

FIGURE 6.10. Bathymetry data from the Mariana Trough contoured every 250 m (data from Lange, 1992 and Smoot, 1990).

The zone of active faulting along a mid-ocean ridge is restricted to a narrow region adjacent to the spreading axis, regardless of spreading rate (e.g., Searle and Laughton, 1977; Macdonald and Atwater, 1978; Kong *et al.*, 1988/89; Edwards *et al.*, 1991). Progressive focusing of extensional deformation to a narrow region during the transition to seafloor spreading (Chorowicz *et al.*, 1987; Cochran and Martinez, 1988; Martinez and Cochran, 1988) is characteristic of continental rifting models. In contrast, the faulting in the Mariana Trough is not restricted to the region immediately adjacent to the volcanic ridges of the volcanic and tectonic zone in the eastern part of the basin. Sediments in the western portions of the valleys show numerous angular unconformities (Fig. 6.9a), implying a history of tectonic disturbance (Fryer and Hussong, 1981). The angle of the unconformities generally indicates westward rotation of fault blocks, and the gradual increase in angle of the sediment layers argues for continuous extension, at least at the margins of the Mariana Trough, throughout its rifting history. Faulting along the base of the West Mariana Ridge has created a deep trough at the western edge of the basin (Fryer and Hussong, 1981; Baker, 1992; Wessel *et al.*, in press). These faults at the basin margins disrupt recent sediments (Fig. 6.9a), indicating that deformation is still active at the outer margins of the basin. The most marked of these faulted basins lies at the eastern base of the West Mariana Ridge. There is a deep trough (average depth 4.5 km) at the western margin of the Mariana Trough along most of its length. This trough lies between the western block-faulted terrain and the West Mariana Ridge (see Fig. 6.2) at the base of the western bounding fault of the backarc basin. Data recently collected in the southern part of the Mariana Trough show structures that are either unsedimented fault blocks or igneous piercement structures in the deep trough at the western margin of the backarc basin (Hagen *et al.*, 1992). The presence of unsedimented, faulted ridges implies recent tectonism in the westernmost Mariana Trough. If, however, the structures are igneous bodies, then either volcanism was rejuvenated near the remnant arc or extension may be taking place in the far western margin of the southern Mariana Trough. As we shall see in the next section, tectonism and volcanism are taking place in a spreading valley in the eastern half of the southern Mariana Trough (Fryer, 1993). If volcanism is also active on the western margin of the southern part of the basin it represents a unique phenomenon for the Mariana Trough.

2.4. Extension in the Southern Mariana Trough

The Mariana Trough undergoes a regional change in morphology at about 14°N (Figs. 6.2 and 11). South of 14°N the central part of the basin is uniformly shallower by about 1 km than the central basin north of that latitude, and the relief of this part of the basin is much less. Bathymetry data from prior to 1991 show a complex distribution of numerous small ridges (a few kilometers wide and tens of kilometers long), many small volcanoes, and

FIGURE 6.11. Bathymetry of the southern Mariana Trough. Contour interval 250 m [data from Bloomer and Hawkins, 1983; Smoot, 1990; Hagen *et al.*, 1992; Fryer, 1993, and DBDB5 data set of National Geophysical Data Center (NGDC)].

several prominent bounding troughs in the southern basin (Smoot, 1988, 1990). The ridges parallel the strike of the arc, for the most part, changing strike southward as the arc curves around to the west (Smoot, 1990). Karig and Ranken (1983) suggested that the active rift zone from 13°N to 14.5°N was located in the approximate center of the basin. Smoot (1990) suggested, as had Hawkins (1977), that the southern part of the Mariana Trough has formed by diffuse spreading and that there is no well-defined spreading axis.

Recent SeaMARC II and Sea Beam data from about 13°45′N indicate a narrow, low-relief spreading axis between 143°50′ and 144°E (Hagen *et al.*, 1992; Fig. 6.12). The transect skirts Tracey seamount, part of the active Mariana arc (Dixon and Stern, 1983). The location of the volcanic rift, suggested by Hagen *et al.* (1992), is a zone of high backscatter with numerous small volcanic cones between about 143°50′E and 144°E at 13°45′N (Fig. 6.12). This rift valley is oriented approximately N-S and is about 10 km wide. SeaMARC II and Sea Beam data at this latitude (Hagen *et al.*, 1992) show this spreading valley is about 400 m deep and flanked by fault-block ridges. The relief (Fig. 6.13) of the faulted boundary of the spreading valley is less than the relief of the faults bounding the spreading valley over most of the rest of the basin (generally over 1000 m and up to ~3000 m; see Figs. 6.3, 6.8a, and 6.9). The high-backscatter area ends abruptly to the east where volcaniclastic sediments from the active arc are ponded against a volcanic ridge. The high backscatter decreases gradually to the west as increasing sediment cover blankets the topography. The side-scan images of the region (Hagen *et al.*, 1992) show sediments particularly in the western half of the basin (Fig. 6.12). Sediments also partially fill the two deep troughs that lie at the margins of the platform, but are lacking from some of the ridges within the western trough. The

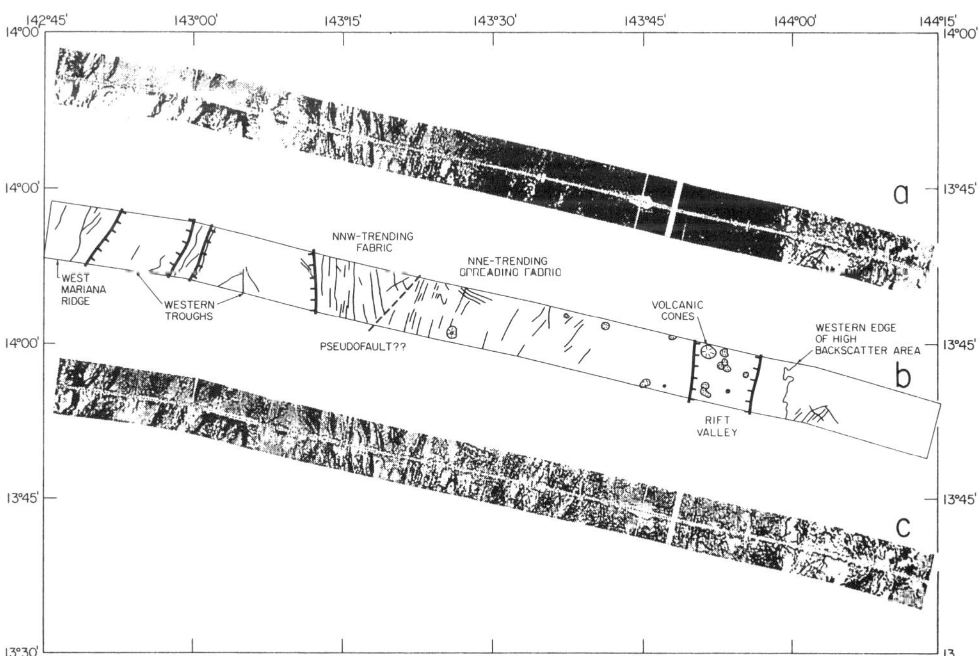

FIGURE 6.12. SeaMARC II side-scan data from the southern Mariana Trough and a sketch of the structures of the swath interpreted by Hagen *et al.* (1992).

deepest basins lie in the eastern bounding trough south of Tracey seamount and in the broader parts of the western bounding trough (Hagen *et al.*, 1992). The trough at the base of the West Mariana Ridge is a continuation of the trough that exists over much of the length of the basin. The trough on the east, however, is restricted to the southern part of the basin. It is related to a major north-south trending fault that cuts across the entire forearc and arc and well into the backarc basin (Fryer, 1993).

Hagen *et al.* (1992) estimated that spreading has been going on for about 3 m.y. in this part of the basin, assuming an extension rate of 2.15 cm/m.y. NNE trending abyssal hill fabric extends from the western boundary of the spreading valley westward to ~143°20′E. The ridges and troughs are spaced 8 to 10 km apart and have relief of 200 to 400 m. A similar set of ridges, but trending NNW, lies at about 143°15′N to 143°20′E. Hagen *et al.* (1992) suggested the change in orientation of fabric may relate to a complex multiple spreading history involving propagation of rifting toward the south in the earliest stage of basin formation. If we accept the model of an initial stage of rifting, not spreading, as the norm for the Mariana Trough, the Hagen *et al.* (1992) suggestion of an early southward propagating ridge does not fit. Rather, the change in orientation of the basin fabric may represent a boundary between the early rifted arc lithosphere and seafloor created over the last 3 m.y. by seafloor spreading.

A 10-km-wide trough separates the eastern slope of the West Mariana Ridge from the southern Mariana Trough platform (Hagen *et al.*, 1992). This trough varies in depth from 4600 to more than 5100 m within the survey area, and the floor of the trough exhibits low backscatter, indicating a sedimented surface. A second, smaller and shallower graben, located just to the west, is separated from the first by a shallow ridge that rises above 3000 m (Hagen *et al.*, 1992). The relief of the fault blocks is greater than that observed in existing data from any other part of the western bounding fault graben along the length of the Mariana Trough. The presence of high-relief, narrow ridges and troughs in the western boundary graben is in sharp contrast to the stair-stepped, fault blocks draped with sediment that are characteristic of the western boundary of the central and northernmost parts of the backarc basin. This difference in morphology possibly reflects a difference in the tectonic style of the basin. It is possible that extension in the Mariana Trough at this latitude is expressed both in the formation of a subdued spreading rift and in the formation of the large grabens that bound the central platform. The abrupt change in depth of the Mariana Trough at about 14°N (shallower to the south) may mark the location of a basinwide structural change, as yet undefined, that facilitates a change in the nature of rifting and volcanism south of that latitude. It is not possible with currently available data to distinguish whether activity in the bounding rift grabens is accompanied by volcanic processes, like that occurring at the periphery of the rift grabens in the northernmost portion of the Mariana Trough, or whether the ridges are being formed solely by continued faulting of the margin of the basin. Some combination of the two is possible.

The pronounced graben at the base of the eastern margin of the southern platform (i.e., south of 13°N) is absent in the rest of the basin. This eastern graben is, on average, about 15 km wide, has a flat-sediment-covered floor, and is generally 3.8–4.0 km deep, with local depths to 4.4 km (Hagen *et al.*, 1992). Piercement structures are apparent within the eastern boundary graben (Hagen *et al.*, 1992). If this graben does project further north, it is apparently filled with volcaniclastics from the active-arc volcanoes. Recent bathymetry data collected during cruise Y9204 of the R/V *Yokosuka* in the southern platform of the Mariana Trough indicate that the eastern graben extends southward to the trench axis and that recent deformation has occurred along its length (Fryer, 1993). Bathymetry data collected from near 144°10′E between 12°10′N and 12°37′N by the R/V *Yokosuka* define the extent of this fault (Fig. 6.13). The position of the fault is consistent with earlier regional structural interpretations (Karig and Ranken, 1983). The bathymetry data show an eastward-facing fault with a throw of nearly 2500 m that strikes approximately north-south at about 144°10′E. Submersible observations of nearly unsedimented talus slopes on the scarp, relatively fresh lava outbreaks on the fault scarp, and unusual fluids in sediments at the base of the scarp suggest recent movement on the fault (Fryer, 1993). The lavas recovered during dives are of island-arc basalt composition; thus it is possible that the fault is an avenue for migration of arc magmas into the forearc region (Fryer, 1993).

Although the data from Hagen *et al.* (1992) suggested that the spreading center in the backarc basin is narrow and of low relief, bathymetry data (Fig. 6.11) collected during the recent R/V *Yokosuka* surveys show that the southern extension of the spreading center broadens southward from 15 km at 13°40′N to over 40 km at 13°10′N (Fryer, 1993). The volcanic edifices located within the broad spreading valley are large and active both tectonically and volcanically (Fryer, 1993). Volcanoes are located along several north-south trending ridges on the eastern margin of the spreading region. A dive on a small dormant volcano on the extreme western edge of the broad rift valley demonstrated that faulting is active throughout the spreading valley (Fryer, 1993). A dive on one of the larger active volcanoes of the eastern side of the rift showed recent volcanism and active hydrothermal activity (Johnson *et al.*, 1993).

If spreading extends farther to the south, the exact location of the spreading center is poorly constrained, particularly in the southwestern part of the basin. Spreading may simply be failing as the spreading axis intersects the arc line. The elevated southern platform may be an expression of diffuse volcanism, as suggested by Hawkins (1977) and Smoot (1990), but this volcanism is not necessarily related to diffuse "spreading," as suggested by these authors. It may relate to intrusion and volcanism of arc magmas within a rifting backarc basin, with true spreading taking place only in the eastermost part of the basin. If spreading does persist to the south, there are at least two possible locations for spreading centers in the southernmost Mariana Trough, and spreading may be taking place in more than one of them.

(1) The spreading center may curve southwestward and lie close to the southwestern extension of the volcanic arc. The general morphology of the southernmost part of the Mariana Trough shows ridges and several localized highs (arc volcanoes) curving to the southwest parallel to the Mariana Trench (Bloomer and Hawkins, 1983; Smoot, 1990). The ridge that bounds the western margin of the spreading valley follows this trend (see Fig. 6.13). Such a phenomenon is consistent with the observations we have made of the tendency of the VTZ to remain close to the arc in the northern part of the Mariana Trough. If the spreading center does follow this trend, the expression of the spreading valley becomes extremely subdued toward the southwest until at least 143°E, where a pair of >4-km-deep troughs lies immediately north of the inner trench slope (at about 12°N, see Figs. 6.2 and 6.11). We have very few data in this part of the basin to help constrain this suggestion.

(2) A far-better-substantiated interpretation is that the spreading valley is offset to the east at 13°10′N and becomes collinear with the major north-south trending fault in the Mariana forearc at 144°10′E. East-west faulting has influenced the development of the volcano at 13°10′N, 143°50′E in the spreading valley. The morphology and marked curvature of the deep graben at 144°E, between 13°N and 13°10′N, toward the southeast are

similar to the morphology of transform fault valleys farther north in the backarc basin. An eastward offset of the spreading center along this transform would result in its lying nearly collinear with the north-south forearc fault. Arc lavas were recovered along the fault well into the forearc (Fryer, 1993). Such a scenario is similar to the extension of backarc spreading from the Three Kings triple junction into the forearc in the northeastern corner of the Lau Basin (Parson and Tiffin, 1993). With currently available data the interpretation that spreading in the southern Mariana Trough extends through the arc line and into the forearc is the best constrained of the two and best fits the regional morphology and the distribution of volcanism in the southernmost part of the Mariana Trough.

3. VOLCANISM OF THE MARIANA TROUGH

3.1. The Volcanic Front and Cross-Chains

Most of the subaerial and submarine volcanoes of the arc have formed immediately west of the eastern boundary fault zone of the Mariana Trough (Fryer and Hussong, 1981, 1982; Hussong and Fryer, 1983; Smoot, 1988). It has been shown recently, however, that arc magmatism has occurred as recently as late Pliocene (1.7 Ma) in the forearc region of the Mariana system at less than 100 km from the Mariana Trench (Marlow *et al.*, 1992; Fryer, 1993). In addition, one of the striking features of the Mariana Trough is a series of chains of volcanoes ("cross-chains"; Fryer and Hussong, 1982) that lie within the backarc basin. These cross-chains of volcanoes are almost all oriented nearly orthogonal to the strike of the volcanic front along structural lineaments that cut across the volcanic front. Cross-chains of volcanoes in the Mariana Trough probably reflect interaction between preexisting cross-arc structures, forearc tectonic activity, magmatic activity at the volcanic front, and backarc basin rifting (Fryer and Hussong, 1982; Fryer, 1993). Fryer and Hussong (1982) suggested that cross-arc structures related to strain between the trench and the volcanic front may be inherited by each successive volcanic arc and backarc basin as an intraplate convergent margin evolves. The locations of volcanoes along island arcs may be influenced by preexisting cross-arc structures. The definition of "frontal volcanoes" in the Mariana arc system must, therefore, be flexible enough to accommodate both forearc and backarc volcanic features.

The subaerial volcanoes of the Mariana arc from 14° to 20°N, called the central island province (CIP) by Bloomer *et al.* (1989b), are active volcanic islands of composite volcanoes (Dixon and Batiza, 1979; Woodhead, 1987, 1989; Stern *et al.*, 1988; Bloomer *et al.*, 1989b). These volcanoes are very similar morphologically. Several of the volcanoes are made up of multiple vents lying along structural lineaments that are oriented approximately orthogonal to the strike of the arc (e.g., Guguan, Pagan, Anatahan). Several of the volcanoes also have structures, grabens or faults, that are oriented parallel to the strike of the arc. For example, Agrigan (Stern, 1979) and Maug (Smoot, 1988) both have north-south striking grabens in their flanks. These arc parallel normal fault grabens indicate that deformation along the eastern boundary of the Mariana Trough continues today.

The largest of the CIP volcanoes are Pagan and Agrigan at about 18°N. The size of these edifices is probably related to regional deformation of the arc at this latitude. These volcanoes lie adjacent to a forearc basin. Studies of the forearc at this latitude indicate forearc-wide tensional deformation at all scales (Mrosowski *et al.*, 1981). There is a distinct break in the inner forearc between 17°30′ and 18°30′N (Fig. 6.3). If this region represents a zone of faulting that cuts across the entire forearc and backarc, the rate of eruption of volcanoes in the gap would be expected to be greater, thus the greater volume of Pagan and Agrigan. Such a cross-arc structure might also explain the more pronounced development of the east-west trending transform fault valleys of the Mariana Trough at this latitude. Cross-arc structures of such magnitude would also explain the presence of arc magmas in the backarc basin even on the active spreading ridge segments (Fryer *et al.*, 1981; Hawkins *et al.*, 1990) by providing avenues for migration of arc magmas into the backarc basin.

The volcanoes north of Uracas, in a region termed the northern seamount province (NSP) by Bloomer *et al.* (1989b), are generally smaller than those of the CIP. The spacing between volcanoes is less regular in the NSP than in the CIP, possibly reflecting the distribution of cross-arc structures. The development of cross-chains of volcanoes extending from the arc line into the backarc basin is more marked in the NSP than elsewhere in the Mariana Trough because the individual cross-chain volcanoes are much larger (Figs. 6.2 and 6.3).

The Kasuga volcanic cross-chain, at about 22°N, has been studied in detail (Figs. 6.4 and 6.13a,b) by side-scan sonar and bathymetry surveys (Hussong and Fryer, 1983) and has been extensively sampled by dredging and submersible (Fryer *et al.*, 1987; Jackson, 1989; McMurtry *et al.*, 1993) It trends nearly north-south and intersects the arc at an angle of about 60° to the strike of the arc (Figs. 6.3, 6.4, and 6.13a,b). A recent analysis of numerous faults and lineaments in the inner forearc at about 21°30′N, adjacent to the Kasuga seamount chain (Mahoney and Fryer, 1988; Wessel, 1991; Wessel *et al.*, in press), indicates that arc-parallel tensional stress is causing rifting in the forearc. This rifting is orthogonal to and tectonically decoupled from the extensional forces involved in backarc basin formation. The forearc rifting may be related to a local tectonic event, the gradual increase in curvature of the forearc with time, or to both (Fryer 1986; Wessel *et al.*, in press). The forearc rifting has apparently crosscut the entire forearc region at 143°30′E and is responsible for development of a bathymetric trough oriented roughly north-south extending from the trench axis at least to the volcanic front. The rifting causes a significant offset in the eastern bounding fault of the backarc basin, and thus does affect the arc front. Activity on this rift may also be responsible for the development of the graben immediately east of the Kasuga volcanoes, for the numerous side vents and lava flow fields on the east flanks of these volcanoes, and ultimately for the growth of the major edifices of the Kasuga cross-chain.

The Kasuga volcanoes are complex edifices reflecting the interplay between cross-arc and arc-parallel structures. There are numerous lava flow fields and regions mantled with volcaniclastic debris on the two southernmost of the three volcanoes imaged (Hussong and Fryer, 1983). The volcanoes were probably built on sundered arc lithosphere, during cross-arc rifting. The magmas from which they were constructed were derived from at least two separate sources, a backarc basin basalt source and an arc source. These volcanoes demonstrate the interrelationships between magma sources and magma mixing processes (Jackson, 1989). Magma mixing processes are evident in the composition of the summit lavas from both volcanoes. Hydrothermal systems were discovered on the summits of the two southernmost volcanoes (McMurtry *et al.*, 1993). A large volcaniclastic debris deposit mantles the southern flank of the second volcano in the chain, and observations of the upper half of the deposit indicate that it may have been emplaced hot in the submarine environment. A comparison of the small grain-size fraction of this deposit with that of Kick-em-Jenny volcano of the Lesser Antilles arc shows remarkable similarity (Ballance *et al.*, 1988).

The presence of flank ridges on the two southern volcanoes of the Kasuga chain

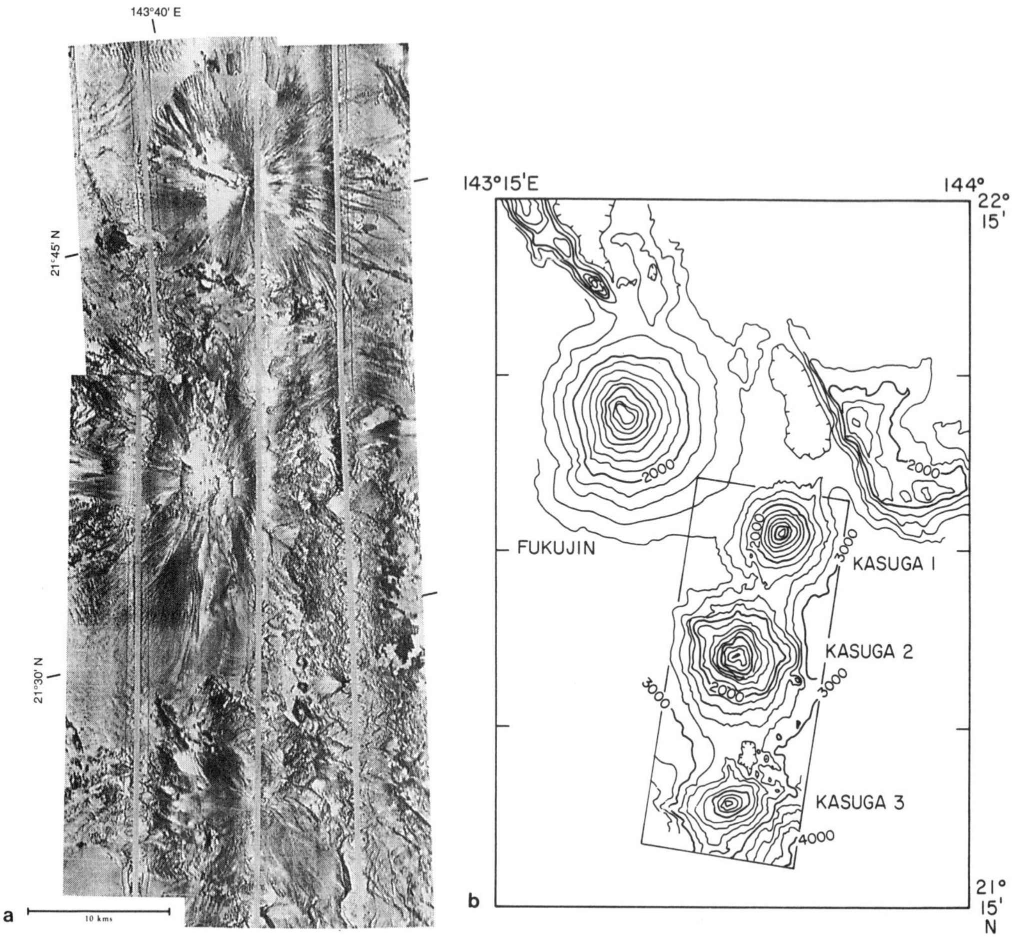

FIGURE 6.13. (a). SeaMARC II side-scan image of the three Kasuga volcanoes of the Mariana backarc basin at 22°N (Hussong and Fryer, 1983). (b) Bathymetry of the three Kasuga volcanoes of the Mariana backarc basin at 22°N (Hussong and Fryer, 1983).

suggests volcanic rift development. Examination of the regional structures, however, indicates instead that these flank ridges represent structural elements of the backarc basin on which the later lavas of the Kasuga volcanoes have been superimposed. It is entirely possible that these structures subsequently also furnished avenues for egress of Kasuga magmas.

The Kasuga volcanoes lie at the top of a fault scarp bounding a graben to the east of the chain (Hussong and Fryer, 1983). Lobate regions of high-backscatter on side-scan images of the eastern flanks of the volcanoes indicate recent volcanism along the face of the fault scarp (Hussong and Fryer, 1983). Satellite cones on the Kasuga volcanoes are most prevalent on the eastern side of the chain, also suggesting a control over distribution of volcanism by the fault underlying the eastern flanks of the volcanoes. The fact that the eastern flanks of the volcanoes are the regions of highest relief on the edifices suggests that the bounding fault of the graben is still active and that deformation of the flanks is ongoing.

The volcanoes south of 16°N comprise the southern seamount province (SSP) of Bloomer *et al.* (1989b). As in the north, these volcanoes are smaller than those of the CIP. Cross-chain development is present, but the individual volcanic edifices are smaller (e.g. see Fig. 6.10). Several of the volcanoes of the SSP have a north-south elongation of vents and satellite cones (Bloomer *et al.*, 1989b). This suggests that the fault system bounding the eastern portion of the basin also plays a significant role in the development of these volcanoes. Until recently, Tracey seamount was considered to be the southernmost active volcano in the frontal arc. The results of recent bathymetric mapping and submersible work in the Mariana backarc basin (Fryer, 1993) show that the arc volcanoes extend farther to the southwest and intersect the spreading ridge of the southernmost backarc basin (Fig. 6.11). Although detailed bathymetry data are not available for the southwest corner of the Mariana Trough, several shoal areas extend to the southwest of the rift intersection and probably represent additional arc volcanoes that lie along a line roughly parallel to the curvature of the Mariana Trench (see Figs. 6.2 and 6.11). The composition of the lavas collected on *Shinkai 6500* dives on two of the volcanoes of this southwestern group at about 144°E have backarc basin basalt signatures (Johnson *et al.*, 1993). Several of the volcanoes have north-south alignment of rift zones. These rift zones are probably related to the intersection of the backarc rifting system with the line of the volcanic arc. To date no samples have been collected from the southwestern volcanoes.

3.2. Intrabasin Faults and the Spreading Ridges

Throughout the basin side-scan sonar images indicate that fault boundaries of major structures are the most likely sites of volcanism. Many of the fault scarps, like that inferred to the east of the Kasuga cross-chain, are draped with lobate high-backscatter regions indicative of recent volcanism. The distribution of volcanism on faults oriented orthogonal to the strike of the arc or backarc basin trends is in discrete edifices, creating cross-chains. Volcanic ridges oriented at high angles to the arc line are nonexistent. Along arc-parallel lineaments within the backarc basin, however, the lava outbreaks tend to occur in fissures. Small volcanic cones or ridges lie parallel to the faults or coalesce to form larger volcanic ridges parallel to local structures. This tendency contributes to the growth of arc parallel ridges that eventually form significant structural elements within the volcanic regions of the backarc basin. Magnetization of the Mariana Trough from the northern apex to about 21°N shows a series of well-defined strips indicating periods of intrusion and volcanism as the basin developed (Martinez *et al.*, in press). The process of concentration of volcanism along arc-parallel fissures eventually results in the development of true mid-ocean ridge spreading segments and the evolution of new backarc basin lithosphere. On the regional scale most outbreaks of lavas occur on the eastern side of the basin. This preference for alignment on the frontal arc side of the basin persists over the length of the Mariana Trough and is probably related to rheological characteristics of the arc lithosphere. It is likely that the combination of high heat flow and more silicic compositions at the frontal arc tend to facilitate rifting on the eastern side of the Mariana Trough (Martinez *et al.*, in press). Although most volcanism occurs in the east, individual volcanic edifices persist in the off-axis regions of the western part of the basin as edifices and volcanic ridges aligned parallel to basin bounding fault activity, as hydrothermal fields, or as scattered edifices (Leinen and

Anderson, 1981; Hobart *et al.*, 1983; Lonsdale and Hawkins, 1985; Martinez *et al.*, in press). This broad distribution implies that volcanic activity can play a significant role along the boundaries between axis-parallel fault blocks at some distance from the currently active ridge segments. The broad distribution of volcanism further supports the suggestion that deformation can continue basinwide throughout the evolution of the Mariana Trough.

The better-formed ridge segments south of 19°45′N show concentration of volcanism along ridges in the center of roughly symmetric valleys. At 19°45′N there appears to be a propagating ridge tip (Figs. 6.8a,b) with a northward- (and possibly a southward-) facing propagator. Farther south along the spreading axis the volcanic ridges are disposed in a series of en echelon segments offset by small displacements (Hawkins *et al.*, 1990) or in segments offset by large transform valleys (Hussong and Sinton, 1983). Volcanism is distributed along these ridges much as it is on normal mid-ocean spreading centers. The ridges are shallowest in the center between offsets (Hawkins *et al.*, 1990). Between 17° and 14°N the spreading center is essentially a series of volcanic fissure ridge segments offset by lateral shifts; however, no side-scan images are available for this region, and thus the details of distribution of volcanism are unknown.

The southern backarc basin has been surveyed recently with SeaMARC II and Sea Beam (Hagen *et al.*, 1992), and the eastern portion has been swath-mapped producing detailed bathymetry (Fryer, 1993). Volcanism there is a combination of spreading (or possibly rifting) fissure activity and eruptions at discrete volcanic edifices. Details of the volcanism of the southwestern extension of the Mariana arc are poorly known at present. Certainly volcanism has occurred in the forearc on an extension southward of the volcanic and tectonic zone in the southernmost part of the backarc basin. Whether this volcanism represents an extension of the spreading axis of the backarc basin, as does the volcanism along the continuation of the zone of volcanism and tectonic deformation north of Nikko seamount, cannot be determined without side-scan sonar mapping of the seafloor.

4. HYDROTHERMAL ACTIVITY

4.1. Cross-Chains and Volcanic Front

Active hydrothermal systems have been discovered on several of the seamounts of the volcanic front and within the backarc basin. Fukujin seamount is located on the volcanic front, immediately north of the Kasuga volcanic cross-chain. A small satellite cone on the northwest flank at a depth of about 1100 m yielded dredge samples of a hydrothermally altered basalt and one of a vuggy silica deposit containing disseminated pyrite (McMurtry *et al.*, 1993).

The summits of the two southern Kasuga volcanoes (Kasuga 2 and Kasuga 3 seamounts) show hydrothermal activity and low-temperature solfatara-type deposits as observed from *Alvin* submersible surveys. The deposits on Kasuga 2 consist of crusty patches of nontronite and sulfur compounds. Very small (less than 10 cm) chimney structures of nontronite are scattered in patches about the fields. The only biological associations with the hydrothermal fields that have been identified are small (10–15 cm long) fish similar to plecostamus. These are abundant in the hydrothermal areas and are probably feeding on microbiological communities associated with the hydrothermal fields. The maximum temperature measured in the hydrothermal solfatara field was 38.8°C. The details of the fluid composition and of the hydrothermal deposits are discussed elsewhere (McMurtry *et al.*, 1993). They indicate a hydrothermal system unique among those sampled in the submarine environment to date. Apparently, addition of significant magmatic CO_2 and SO_2 has contributed to the unusual composition.

Hydrothermal deposits at the summit region of Kasuga 3 consist of a crusty pavement and small (up to 20 cm high) chimneys of amorphous iron oxides and nontronite and linear ridges of birnessite and vernadite about 3 to 5 cm high. Some of these small ramparts are nearly 20 cm high. A few yellowish green chimney structures of nontronite, up to about 30 cm high, were observed along some of the small manganese ramparts. A small vent of shimmering water was discovered at 1140 m on a dacite dome at the summit of the volcano. Two temperature measurements recorded a dT of 4.67°C and 5.53°C. The mineralogy of the hydrothermal deposits is consistent with the low temperatures measured (McMurtry, *et al.*, 1993). The seafloor around the vent is encrusted with patchy reddish deposits of amorphous iron oxide that are very vuggy. White flocculent material, probably bacterial mats, was observed on much of the exposed rock and hydrothermal crusts, but no other biological communities.

Sulfide deposits were recovered recently from dredging the summit of Esmeralda Bank in the southern Mariana seamount province (Stuben *et al.*, 1992).

4.2. Central Spreading Basin

A summary of preliminary results of *Alvin* submersible studies of hydrothermal systems along the Mariana Trough spreading ridge segments of 18°N was presented by Fryer (1990). Recent publications have refined some of the earlier observations. Hydrothermal vent activity was first detected as plumes of methane and high $^3He/^4He$ ratios (Poreda, 1985; Horibe *et al.*, 1986; Craig *et al.*, 1987) 800 m above the seafloor at 18°13′N. Submersible observations show that the axial volcanoes of the spreading axis support two large hydrothermal fields at depths of 3600–3700 m (Craig *et al.*, 1987; Kastner *et al.*, 1990). The vent waters are similar to Loihi vents in $^3He/^4He$ ratio (8.6 × atmospheric (R/R_A = 8.6)) with $CH_4/^3He = (0.5 \pm 0.2) \times 10^6$. These values are very different from the plumes (R/R_A = 3.2 and $CH_4/^3He > 100 \times 10^6$) which had led to their investigation (Craig *et al.*, 1987). Several smaller fields were discovered on a subsequent dive series (Hessler *et al.*, 1987). The measured temperatures of the hydrothermal vents vary from ~6°C to a maximum of 287°C. The high-temperature chimney structures produce "clear smokers" and structures dominantly composed of barite and silica, with lesser amounts of sphalerite, galena, pyrite, and chalcopyrite (Craig *et al.*, 1987; Kastner *et al.*, 1987). The composition of the Mariana vent fluids and chimney structures differs from those of hydrothermal regions of similar temperatures on mid-ocean ridge spreading centers in several respects (Campbell *et al.*, 1987; Kastner *et al.*, 1987; Kusakabe *et al.*, 1990). The pH of the Mariana vent fluids is higher (4.39) than that for mid-ocean ridge vent fluids (3.8), and the alkalinity is higher (0.43 meq/L) compared with mid-ocean ridge vent waters (0.19 meq/L; Campbell *et al.*, 1987). Boron in the vent fluids of the Mariana chimneys is 200% that of ambient seawater and has a $\delta^{11}B$ of +20/mil, whereas mid-ocean ridge vents are usually only 10% to 20% that of seawater and have a $\delta^{11}B$ of +26 to +32/mil (Campbell *et al.*, 1987). It appears from the compositions of the chimney structures and the Mariana vent fluids that precipitation of sulfides must be occurring at depth within the ridge crest hydrothermal system, although no samples of such precipitates were recovered (Campbell *et al.*, 1987; Kastner *et al.*, 1987).

Biological studies of the vents indicated "rich and unusual" communities (Hessler *et al.*, 1987). Abundant widespread anemones and clusters of barnacles around discrete vent openings are characteristic of the low-temperature (~6°C) regions. Galatheid crabs and polynoid polychaetes are associated with the barnacles and anemones in these low-temperature regions. "Hairy" gastropods cluster around the slightly higher temperature fields (from ~15°C to 250°C). Grazing bresilid shrimp and brachyuran crabs, encrustations of barnacles, paralvinellid polychaetes and associated harpactacoid copepods, many limpets, occasional mussels, and dispersed anemones are also found in these locations (Hessler *et al.*, 1987).

A region of off-axis hydrothermal mounds was discovered approximately 50 km west of the Mariana backarc spreading center by heat flow and deep-tow studies (Leinen and Anderson, 1981; Hobart *et al.*, 1983; Lonsdale and Hawkins, 1985). The mounds were examined by *Alvin* submersible dives in 1987 (Leinen *et al.*, 1987; Wheat and McDuff, 1987). They show heat flow values exceeding 10 W/m. Nonlinear, concave thermal gradients indicate an active hydrothermal system, recharge of the system, and potential escape of fluids along fault scarps within the region (Leinen and Anderson, 1981; Hobart *et al.*, 1983; Leinen *et al.*, 1987).

The distribution of the high-heat-flow areas and the associated mounds appears to be controlled by faulting, not by evenly disseminated hydrothermal activity. The mounds are structurally very complex and associated heat flow values indicate variability even within a given mound. The distribution of the mounds and the local topography confirm the structural control over the formation of the mounds (Leinen *et al.*, 1987).

4.3. Southern Platform

Some of the largest volcanoes of the southern arc lie in the Mariana Trough at about 13°N (Fig. 6.11). These volcanoes lie at the intersection of the island arc and the backarc basin spreading center (Fryer, 1993). It may be that within this region both backarc basin lavas and arc lavas are generated and erupt from an interrelated plumbing system, as has been suggested for the northernmost Mariana backarc basin active rifting region (Baker, 1992). The north-south trending volcanic ridges that form at the eastern boundary of the spreading center give clear evidence that magma from the major volcanic centers is leaking along the fault zones, at least at the edge of the spreading center. An intersection of a north-south trending fault mapped in the forearc (Fig. 6.11) with the rift axis would facilitate both magma egress and permit arc or backarc basin magmas to follow the north-south trending fault southward into the forearc region (Fryer, 1993). The volcano that lies at the southern margin of the spreading valley is offset by an east-west striking fault with right lateral displacement (Fryer, 1993). Fryer (1993) has noted that the latitude of the offset of this volcano is nearly identical to that of a zone of east-west striking faults identified by Karig and Ranken (1983) in the forearc east of Guam at 13°N. East-west trending faults extending across the arc into the backarc basin may also facilitate magma egress at the spreading center (Fryer, 1993). Most likely all three phenomena play a role in the development of the volcanoes of the backarc spreading center at 13°N (Fryer, 1993).

Recent submersible dives with the *Shinkai 6500* on volcanoes of the southern platform revealed a region of hydrothermal chimneys at the summit region of the eastern volcano observed using the *Shinkai 6500* submersible (Fryer, 1993). The chimney samples are composed principally of marcasite (Fryer, 1993). The temperature of the vent from which this chimney sample was obtained was 204°C (Johnson *et al.*, 1993). Biological communities observed in association with this vent and others in the vicinity are dominated by the same type of hairy gastropod (*Alvinensis hesslerii*) observed at the 18°N vent fields. Small numbers of shrimps and of galatheid crabs were also observed at the vent sites. Additional work on the composition of the hydrothermal deposits is underway.

5. PETROLOGY AND PETROGENESIS

5.1. Introduction

In the Mariana convergent margin system, arc magmas are generated as a consequence of interaction of the mantle overlying the subducted slab with constituents driven off the descending slab as subduction proceeds (e.g., Fryer, 1981; Fryer *et al.*, 1981; Sinton and Fryer, 1987; Hawkins *et al.*, 1990; Stern and Bloomer, 1992). These arc magmas have erupted within the forearc over the entire history of its formation, although predominantly in the early stages (Marlow *et al.*, 1992; Stern and Bloomer, 1992). Today the arc magmas are predominant in the volcanic front of the arc and produce volcanic seamounts along lineaments that crosscut the arc and extend into the backarc basin (Hussong and Fryer, 1983). Although arc magmas do occur along the rifting and spreading regions of the Mariana Trough (Fryer, 1981; Macdougall *et al.*, 1987; Volpe *et al.*, 1990), magmas similar to mid-ocean ridge basalt (MORB) predominate and constitute the newly forming lithosphere of the spreading portions of the backarc basin.

The similarity between Mariana Trough basalts and MORB is widely recognized (e.g., Hart *et al.*, 1972; Hawkins, 1977; Natland and Tarney, 1981; Hawkins and Melchior, 1985; Stern *et al.*, 1990). Mariana Trough backarc basin basalt (BABB), as defined by Fryer (1981), is indistinguishable from normal MORB on trace element provenance diagrams (Fig. 6.14) used to discriminate between magmas generated in a variety of tectonic settings (Johnson and Fryer, 1990). Glass selvages from basalts of the Mariana Trough at 18N, however, do show consistent differences from MORB (Fig. 6.15). The definition of the term

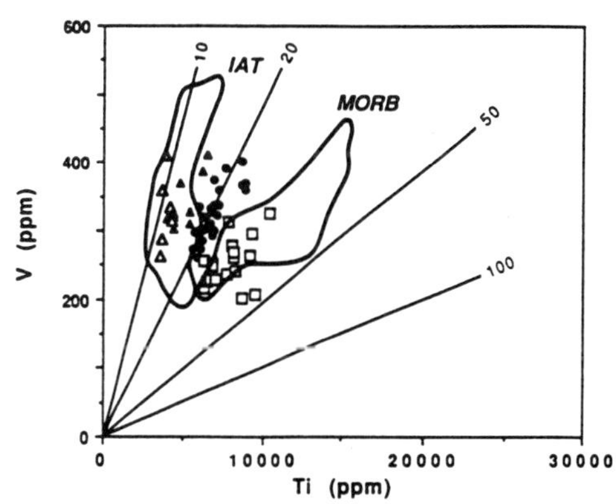

FIGURE 6.14. Ti/V plot of lavas from the Mariana Trough, mid-ocean ridges and Izu–Bonin backarc regions (Fryer *et al.*, 1990). Squares are Mariana Trough glasses from Sinton and Fryer (1987), circles are Izu–Bonin backarc basin samples, and triangles are Izu–Bonin island-arc samples from Fryer *et al.* (1990). The outlined fields are the fields of MORB and IAT from Shervais (1982).

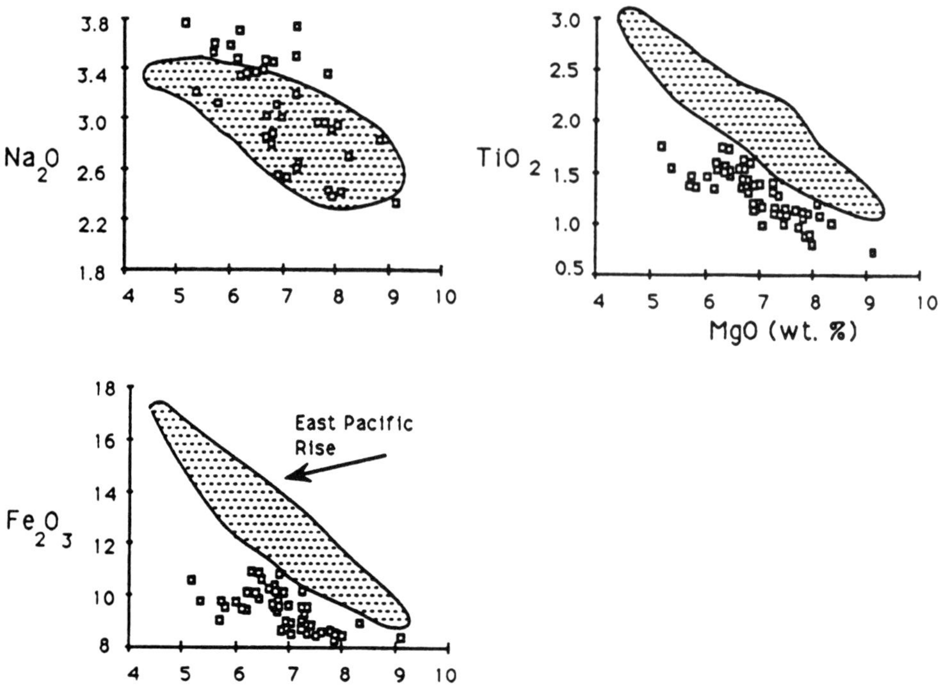

FIGURE 6.15. MgO variation diagrams for Mariana Trough basalts (squares) compared with basalts from the East Pacific Rise (stippled field) (Fryer *et al.*, 1990).

BABB (Fryer, 1981) is based on the fact that, at equivalent MgO contents, glasses from Mariana Trough BABB have higher Al_2O_3, lower FeO, slightly lower TiO_2, and slightly higher Na_2O than MORB. This subtle difference in composition is caused by a slab-related component involved in formation of the Mariana BABB. The nature of this component is the subject of some debate, as we shall see. The problem is in determining whether this component contains sediments subducted with the slab, represents small amounts of melting of the slab itself, or represents purely a volatile component (Meijer, 1976, 1980; Hawkins and Melchior, 1985; Sinton and Fryer, 1987).

The relationship between BABB and island-arc magmas is also controversial. It is a particularly challenging problem to determine the nature of this relationship in the initial stages of formation of the backarc basin, because both types of magmas can be generated in a backarc basin even in the earliest stages of arc rifting (Fryer *et al.*, 1990; Stern *et al.*, 1990). The compositional variability in each can be substantial. The northern Mariana Trough region is an excellent place to examine this problem because it is in the initial stages of rifting, contains a well-developed VTZ on the eastern side, and has a large, active volcanic cross-chain. In addition, it is well sampled and there exist detailed bathymetry and side-scan data with which to place the results of the sample analysis in context.

5.2. Northern Rifting Basin and the Kasuga Cross-Chain

The characteristics of the lavas of the northern Mariana Trough are that they are enriched in large-ion lithophile elements (LILE), are depleted in high-field-strength cations, and have more radiogenic Sr and less radiogenic Nd than MORB (Stern *et al.*, 1990). The general similarity of the northern Mariana Trough lavas to arc lavas suggested to Stern *et al.* (1990) that either an arc component becomes incorporated in a MORB-like source by mixing or that variable metasomatism of that source by subduction-related fluids produces the northern Mariana Trough BABB. Although the lavas recovered in dredging and submersible sampling from the northern Mariana Trough are similar to arc magmas for the most part, the presence of lavas with BABB compositions and lavas with unusual enrichments in subduction-related components (Jackson, 1989; Fryer *et al.*, 1990) argues for caution in developing models for petrogenesis in this region. The necessity for multiple-source compositions for the northern Mariana Trough lavas is clearly indicated by the regional diversity of the magmas. This may explain local variations in the general compositional trends that have been noted for some of the major volcanic edifices of the northern Mariana Trough.

Arc magmas have erupted from fissures on the western boundary of the VTZ in the eastern part of the northern Mariana Trough (Jackson, 1989; Stern *et al.*, 1990). Magmas of both arc and backarc basin composition have erupted from the large volcano (seamount A) at 22°39′N and from a volcanic ridge immediately north of this volcano at the northwestern edge of the VTZ in the northern Mariana Trough (Jackson, 1989; Baker, 1992). Lavas from the summit of seamount A are of two compositions: a BABB and a primitive, low-TiO_2 basalt that lacks plagioclase phenocrysts (Jackson, 1989). The source of this low-TiO_2 basalt was suggested by Jackson (1989) to be similar to that involved in production of forearc boninitic suites. Either a metasomatic fluid rich in subduction-related components or a low-degree melt of some exotic mantle component, or both, would be required for the formation of such lavas. This is the only site in the Mariana Trough where lavas of both a BABB and a "boninitic" source have been erupted in the same volcano.

Both arc basalts and BABB have erupted from the currently active fissures of the eastern portion of the Mariana Trough at about 22°N. The rocks dredged in the northern Mariana Trough (locations indicated in Fig. 6.4), revealed BABB both north and south of 22°N (Jackson, 1989; Stern, 1990). Jackson (1989) suggested that these lavas were formed by a similar degree of melting, but at higher temperatures and pressures than MORB and rocks collected at 18°N.

The proximity of the VTZ to the active volcanism of the frontal arc in the northern Mariana Trough presents a challenge in unraveling the petrogenetic interrelationships of the lavas. This problem is well demonstrated by the lavas of the Kasuga cross-chain. Most lavas from these volcanoes are basalts, although andesites and dacites are present at the summits. The most striking geochemical characteristic of the Kasuga volcanoes is that the along-chain variability of the Kasuga lavas (Jackson, 1989) is as great or greater than the regional variability of lavas from the NSP (Bloomer *et al.*, 1989b; Lin *et al.*, 1989, 1990; Ikeda and Yuasa, 1989) from Uracas to Iwo Jima. For example, the Kasuga cross-chain contains both lavas with the most alkalic composition of any lavas from the NSP and lavas similar to the least alkalic lavas of the CIP (Jackson, 1989; Stern *et al.*, 1990; Gill and Williams, 1990). Lavas from Kasuga 1 at the volcanic front are most similar to the least alkalic of the lavas from the NSP volcanoes and the high-K lavas of Kasuga 2 are most similar to the most alkalic lavas of Iwo Jima.

If Kasuga lavas were to have been generated by the same processes and from the same sources as the lavas erupting along the VTZ immediately north of the chain, they should have similar compositions. Yet all Kasuga rock types differ strikingly from those dredged at the pillow ridges along the volcanic and tectonic zone (Stern *et al.*, 1990). The wide range in compositions of lavas from the Kasuga seamounts of the northern Mariana Trough probably reflects a combination of heterogeneous processes of melt production or source composition, or both (Jackson, 1989). Several source components have been invoked to explain the isotopic diversity of the northern part of the Mariana arc and trough (Bloomer *et al.*, 1989a; Lin *et al.*, 1989, 1990; Hickey-Vargas, 1991; Hickey-Vargas and Mandal, 1992).

The variation in K_2O content of the Kasuga lavas west of the volcanic front partly results from variations in the degree of partial melting and partially from a variety of sources in the northern Mariana Trough region. Higher-K basalts of Kasuga 2 and Kasuga 3 probably preferentially sampled a local source component that shares distinctive geochemical features with some boninites: low ϵNd accompanied by high K, Rb, Zr, P, and LREE relative to Sr, Ba, Pb, HREE, Y, Ti, and Nb (Bloomer *et al.*, 1989a,b; Jackson 1989; Lin *et al.*, 1989, 1990; Gill and Williams, 1990; Stern *et al.*, 1990). A component of this type apparently contributes to the diversity of lavas in the NSP as a whole. The same component appears to be further diluted in lavas within comparably near-front, active volcanic ridges in the Mariana Trough at 22°N (Stern *et al.*, 1990). Seamount A on the northeast tip of the volcanic platform of the Mariana Trough at 22°39′N yielded a lava similar in composition to the boninitic series lavas of the forearc (Jackson, 1989). Thus, this component may be better represented in the northern backarc basin than recognized by Stern *et al.* (1990). The association of K, Zr and LREE enrichment with low Ba/La and eNd ratios is common in subduction zones, especially behind volcanic fronts (e.g., Carr *et al.*, 1990; Tatsumi *et al.*, 1991). Atypical tectonics allows atypically pure melts of this exotic component to erupt. The cross-arc structures such as the faults upon which volcanic cross-chains such as the Kasuga volcanoes develop are likely to provide a better opportunity to explore the heterogeneity of the mantle wedge in the backarc basin settings. The diversity of rock types in close proximity to each other argues for caution in generalizations correlating composition of the rocks with tectonic setting. The distributions of varying magma types in the region probably reflect the interactions between the incipient spreading zones and the rifting arc lithosphere both in the backarc basin and, in the case of the Kasuga seamounts, across the forearc and arc.

5.3. Central Spreading Basin

Most of the detailed petrologic work in the central part of the Mariana Trough has been done in the spreading valley at about 18°N. Dredges and DSDP samples defined the principal characteristics of BABB and pointed out that arc magmas have been erupted in the central spreading region of the Mariana Trough (Fryer *et al.*, 1981; Wood *et al.*, 1981, Sinton and Fryer, 1987). An *Alvin* submersible dive series on the volcanic ridge that lies in the central spreading valley provided structural and petrologic studies of the rift center volcanoes (Hawkins *et al.*, 1990).

The controversy over the origin of BABB arises as a consequence of the fact that compositional differences between BABB and MORB may have several causes. The major element differences are easily explained. Higher H_2O content in the source region for the Mariana Trough backarc basin basalts displaces the olivine-plagioclase-liquid cotectic toward plagioclase in the clinopyroxene-olivine-plagioclase-SiO_2 basalt system, resulting in magmas with higher Al_2O_3 content in olivine-saturated backarc basin liquids relative to (olivine + plagioclase)-saturated basalt liquids from normal mid-ocean ridges (Fryer, 1981; Fryer *et al.*, 1981; Sinton and Fryer, 1987). The fact that the Mariana Trough basalts have low-FeO contents at equivalent MgO by comparison with normal MORB, probably reflects lower temperature and pressures of melting (Sinton and Fryer, 1987). Although Stern *et al.* (1990) have noted that the Mariana BABB lacks enrichment in LIL elements, Sinton and Fryer (1987) observed enrichments in glasses from the central Mariana Trough (18°N) and suggested that several possible phenomena may contribute to the source of the BABB magmas, principal among these is a fluid component from the subducted slab that has been active in metasomatism of an otherwise depleted mantle, as well as a small amount of slab melting (Sinton and Fryer, 1987).

Fryer (1981) noted that some lavas from the spreading ridge segment to the north of the Pagan transform (Fig. 6.9c) have compositions intermediate between the typical BABB of the region and Mariana island-arc tholeiite; however, the spatial distribution of samples from the dredge sites was too poorly constrained spatially to permit interpretations of this observation. Alvin submersible investigations of the volcanic ridge in the Mariana Trough spreading center valley at 18°N (Macdougall *et al.*, 1987; Volpe *et al.*, 1990) recovered basalts that span a range between BABB and island-arc tholeiite. Isotopic analyses and the Rb and Sr concentrations of the basaltic glasses from some of the samples show an intimate mixing of the MORB source with an arclike source component (Macdougal *et al.*, 1987). The lavas recovered during the dives confirm the general compositional trends described by Fryer *et al.* (1981), but provide far greater detail in terms of distribution of various rock types along axis. Lavas with greatest similarity to MORB occupy the ridge segment immediately north of the Pagan transform, those of more typical BABB composition are located on the ridge segment further north from the transform, and the arclike lavas occupy one position on the northern segment of the spreading ridge (Macdougall *et al.*, 1987; Volpe *et al.*, 1990). Volpe *et al.* (1990) interpret the diversity in the rock types on such small scale as indicating that the source region for the lavas is equally diverse. They suggest that the heterogeneity of the source may indicate the presence of fragments of previously subducted lithosphere in the source region. This would imply variability on a very small scale for the sources of backarc basin lavas, and, as we have shown, the studies of the northern portion of the Mariana Trough would support this interpretation. Furthermore, the studies of the Kasuga lavas support the contention that an unusual local source has contributed to the magmas of the backarc basin. It may be that in the incipient phases of opening of the basin and before new plate boundaries are established the heterogeneities of the backarc basin sources are most prevalent.

5.4. Southern Platform

The seamounts of the frontal arc south of 16°30′N lie at the base of the fault scarp that forms the eastern boundary of the Mariana Trough, but are better separated from that scarp than are the islands of the CIP or the seamounts of the NSP. The lavas of the SSP edifices show a compositional diversity similar to that in the NSP and greater than that of the CIP (Dixon and Stern, 1983; Stern and Bibee, 1984; Stern *et al.*, 1989).

Bloomer and Hawkins (1983) show that arc-related rocks ranging from ultramafics to dacites are exposed in the inner trench wall region south of 12°N. The petrologic and geochemical results of their study indicate that no true mid-ocean ridge material is exposed in the inner trench wall of the Mariana convergence zone within their sample area. This prompted them to suggest that most of the crust of the southern backarc basin is composed of pre-late Eocene arc complex. These findings are consistent with an evolutionary model for the backarc platform in the southern part of the Mariana Trough that requires it to be composed of fractured, subsided, and possibly rotated blocks of the arc massif. Very little information exists on the nature of the transition from the composition of rocks from this deformed arc complex to the lavas of the active spreading zone to the north.

Additional rock samples were collected from the forearc fault scarp south of Guam during recent *Shinkai 6500* dives (Fryer, 1993). The lavas recovered are primarily massive mafic rocks: basaltic and gabbroic. Details of the compositions are given by Fryer (1993), but all are of island-arc composition.

6. SUMMARY AND CONCLUSIONS

The Mariana Trough is an active backarc basin that represents the latest in a series of arc rifting and spreading events that have taken place at the eastern edge of the Philippine Sea plate. During its opening, tectonic and, to some extent, volcanic activity sometimes occurred basinwide, although the locus of greatest activity remained close to the volcanic front.

An initial period of stretching and collapse of the preexisting arc is indicated by a broad normal-faulted terrain, particularly at the western margin of the basin. During this period of extension, deformation and volcanism were focused on a VTZ close to the volcanic front. Subsequently, ridge transform structures developed as seafloor spreading began. Arc construction continued on the eastern rifted crust. The basin thus has the appearance of having *spread* asymmetrically, even in the more mature portion of the basin (between 14° and 20°N), but in reality the basin *rifted* asymmetrically. Compositional variations of lavas within the basin suggest both local tectonic control over magma genesis and an intricate interplay of mixing from several magmatic sources. The fundamental aspects of basin development can be traced in each of three major geographic provinces of the Mariana Trough, the northern rifting apex, the central spreading basin, and the southern platform.

North of 20°N, graben-and-horst structures predominate, and maximum depths are in excess of 3500 m. Either the rifting of the arc in this basin is propagating northward and the northern part of the Mariana Trough has only recently begun to separate, or rifting has been going on much more slowly in this part of the basin, possibly pivoting about a point at the northern apex of the basin. The volcanic-tectonic zone remains within 15 km of the frontal arc between 234° and 22°N; southward of that latitude the zone diverges from the arc, and by 21°30′N it is 70 km west of the arc line. At this point the volcanic-tectonic zone undergoes a transition to a series of half-graben structures and at about 20°N there is a change to true seafloor spreading farther south. The composition of lavas within the northern region is more variable than in any other portion of the basin studied to date. Both island-arc tholeiite basalts and backarc basin basalts erupt from fissures within the currently active VTZ, and one dredge recovered lavas that are similar to those from boninitic source magmas from forearc regions. Some fault scarps within the basin are mantled with recent lava flows. A few are also the sites of construction of major volcanic edifices in the backarc basin and thus give rise to volcanic cross-chains. Some of these major edifices appear to have supplanted the adjacent volcanic front as the primary locus of along-strike eruption of arc magmas.

The spreading center is best developed between about 14° and 20°N, in the widest part of the basin. The bathymetric grain reflects the structures of the spreading axis and transform fault valleys. Abyssal hill fabric, parallel to the ridge segments, is offset by broad valleys that cut across the basin and are continuous with lateral offsets between the spreading ridge segments. Structural lineaments exist even in the early block-faulted terrain at the western margin of the basin. The tectonic control effected by these preexisting structural elements apparently persisted throughout both the early rifting and the later mid-ocean ridge/transform spreading stages of opening in the basin. Cross-chains of submarine volcanoes extend westward from volcanic centers at the volcanic front into the backarc basin along some of these lineaments, suggesting that they are avenues for migration of arc magmas into the backarc basin. Magmas of the spreading ridge segments are composed of basalts that are very similar to mid-ocean ridge basalts. They differ from normal mid-ocean ridge basalts in having major element compositions affected by the higher volatile content of the source and in having trace element signatures indicative of a component of slab-related fluids. Arc lavas and basalts transitional to arc lavas have also been collected at a few localities from the central spreading ridge segments. This diversity of lava types suggests a complex interplay of sources even in the more mature central portion of the basin.

Hydrothermal systems of the Mariana Trough are well developed along the volcanic fissures that mark the spreading ridge segments. Well-developed hydrothermal systems are also evident on the large volcanic centers of the southernmost intersection of the spreading axis with the volcanic front. These systems produce hydrothermal deposits which reflect the cooler temperatures (by comparison with the mid-ocean ridge systems) observed in the basin hydrothermal fields. The presence of off-axis hydrothermal systems suggests that magmatic activity is not confined strictly to the ridge crests even in the most mature portion of the central spreading basin. Hydrothermal systems of the volcanoes along the frontal arc and the cross-chains are similar to one another and differ from those of the backarc basin spreading center only in subtle geochemical characteristics related to the individual edifices.

South of 14°N the basin is marked by a shallow (less than 3 km) platform. The southern platform is separated from the active arc and the remnant arc by deep fault-controlled grabens. The southern platform is furrowed by a series of narrow ridges and troughs. The strike of these features changes from roughly north-south to roughly east-west with decreasing latitude. In the western margin of the southern platform there is a series of ridges and troughs with a trend that suggests a possible earlier strain regime that is different from the rest of the southern platform. The position of the spreading center south of 14°N has recently been defined by side-scan surveys, swath mapping, and submersible observations at 13°40′N to be close to the volcanic front. At that latitude the rift valley is a narrow feature only approximately 15 km wide. Spreading at 13°10′N, 144°10′E in the southernmost portion of the backarc basin is taking place in a broader zone and may represent the intersection of the spreading axis with the frontal arc line. Large volcanic centers lie atop volcanic ridges that are only 30 km or less from the eastern bounding fault of the backarc basin. Preliminary studies of lavas collected recently from these features indicate BABB from two sources are being erupted from one of the edifices on the eastern portion of the spreading ridge (Johnson *et al.*, 1993). There is evidence of tectonic disturbance in sediments near the western boundary of the spreading valley, as observed from submersibles, and thus the entire valley may be tectonically active. Southward of this position the spreading axis is not well defined but may be influenced by forearc faulting. A major N-S striking fault in the Mariana forearc lies south of the spreading axis and is offset about 15 km to the east of the easternmost volcanic ridge. Exposures of fresh basalts in the forearc along this fault scarp suggest that the magmas generated in the southern portion of the arc may be involved to some extent in forearc volcanism.

Understanding the history of suprasubduction zone tectonics and magma genesis in the Mariana Trough is important for the ultimate understanding of geochemical mass balance in convergent margin systems and of evolution of continents. Profound changes in the models of arc evolution have resulted from recent studies of the Mariana interoceanic arc system. Vast portions of the suprasubduction zone have been subjected to metamorphism driven primarily by fluid flux through that region in response to dehydration of the subducted oceanic slab. Clearly, variations in source composition as a consequence of previous subduction episodes, possible incorporation of fragments of older lithosphere, migrations of mantle sources, and influx from variations in convection play a part in the generation of magmas in this region. The wide variety of rock types recovered from the Mariana Trough and the discovery of a new model for evolution of the backarc basin requires us to reevaluate our concepts of the petrogenetic evolution of the backarc crust and mantle. The study of the lavas of the northern Mariana arc and of the Mariana backarc basin requires us to evaluate the interaction of magmas generated by shallow partial melting of an exotic source with those of the deeper-seated arc magma sources and to try to trace the influence on both of the subduction derived constituents. The apparently shallow generated backarc basin basalts are emplaced in close proximity to true arc-generated lavas. Mixing and magma differentiation processes on cross-chain volcanoes in particular tend to obscure the end-member source components in the lavas. The importance of understanding these regions is becoming increasingly evident as more and more ophiolite terrains are identified as having formed in an arc or suprasubduction zone environment. Thus, the comparisons of subaerial terrains to the recent studies of the interoceanic convergent margins will continue to add immeasurably to our understanding of plate interactions and of the evolution of the volcanic arcs and backarc basins generated within such environments.

Acknowledgments

The author is indebted to F. Martinez for his invaluable assistance in producing several of the bathymetry maps. Thoughtful reviews of this manuscript by J. Natland and A. Klaus and comments by B. Taylor and R. Stern have improved it significantly. This manuscript is SOEST contribution no. 3581 and HIGP contribution no. 767.

REFERENCES

Anderson, R. N. 1975. Heat flow in the Mariana marginal basin, *J. Geophys. Res.*, **80:**4043–4048.

Baker, N. 1992. Rifting and volcanism in the northern Mariana Trough: a SeaMARC II and seismic reflection study, Master's thesis, Univ. Hawaii.

Baker, N., Fryer, P., and Brown, G. 1989. Rifting in the northern Mariana backarc basin (abstract), *EOS Trans. AGU*, **70:**1313.

Ballance, P. F., Carey, S., Sigurdsson, H., and Fryer, P. 1988. Production of basaltic sediment on Kick'em Jenny submarine volcano, Lesser Antilles arc, and central seamount, Mariana arc (abstract), Abstracts with program, Geol. Soc. N.Z., Ann. Mtg., Geol. Soc. N.Z. Misc Publ. 41a:30.

Barone, A. M., Lonsdale, P. F., and Puteanus, D. 1990. Formation of axial volcanic ridges of the Mariana slow spreading center from spatial magma sources (abstract), *EOS Trans. AGU* **71:**1637–1638.

Beal, K. L. 1987. The opening of a backarc basin: northern Mariana Trough, Master's thesis, Univ. Hawaii.

Bloomer, S. H., and Hawkins, J. W. 1983a. Gabbroic and ultramafic rocks from the Mariana trench: an island arc ophiolite, in *The Tectonic and Geologic Evolution of Southeast Asian Seas and Islands*: Part II. (D. E. Hayes, ed.) Geophys. Monogr. Ser., Vol. 27, pp. 294–317, American Geophysical Union, Washington, DC.

Bloomer, S. H., and Hawkins, J. W. 1983b. Petrology and geochemistry of boninite series volcanic rocks from the Mariana trench, *Contrib. Mineral. Petrol.* **97:**361–377.

Bloomer, S. H., Stern, R. J., Fisk, E., and Geshwind, C. H. 1989a. Shoshonitic volcanism in the northern Mariana Arc. 1: mineralogic and major and trace element characteristics, *J. Geophys. Res.* **94** (B4):4469–4496.

Bloomer, S. H., Stern, J. R., and Smoot, N. C. 1989b. Physical volcanology of the submarine Mariana and volcano arcs, *Bull. Volcanol.* **51:**210–224.

Bracey, D. R., and Ogden, T. A. 1972. Southern Mariana arc geophysical observations and hypothesis of evolution, *Bull. Geol. Soc. Am.*, **83:**1509–1522.

Brown, G. 1991. Rifting of the Bonin Island arc, Doctoral dissertation, Univ. Hawaii.

Brown, G., and Taylor, B. 1988. Seafloor mapping of the Sumisu Rift, Izu–Ogasawara (Bonin) island arc, *Bull. Geol. Surv. Jpn.* **39:**23–3.

Bibee, L. D., Shor, G. G., Jr., and Lu, R.S. 1980. Inter-arc spreading in the Mariana Trough, *Mar. Geol.* **35:** 183–197.

Campbell, A. C., Edmond, J. M., Colodner, D., Palmer, M. R, and Falkner, K. K. 1987. Chemistry of hydrothermal fluids from the Mariana Trough backarc basin in comparison to mid-ocean ridge fluids (abstract), *EOS Trans. AGU* **68**(44):1531.

Carr, M. J., Feigenson, M. D., and Bennett, E. A. 1990. Incompatible element and isotopic evidence for tectonic control of source mixing and melt exteraction along the Central American Arc, *Contrib. Mineral. Petrol.* **105:**369–380.

Chorowicz, J., Le Fournier, J., and Vidal, G. 1987. A model for rift development in eastern Africa, *Geol. J.* **22:**495–513.

Cochran, J. F., and Martinez, F. 1988. Evidence from the northern Red Sea on the transition from continental to oceanic rifting, *Tectonophysics* **153:**25–53.

Craig, H., Horibe, Y., and Farley, K. A. 1987. Hydrothermal vents in the Mariana Trough: results of the first Alvin dives (abstract), *EOS Trans. AGU* **68**(44):1531.

Dixon, T. H., and Batiza, R. 1979. Petrology and chemistry of recent volcanics in the northern Marianas, *Contrib. Mineral. Petrol.* **70:**167–181.

Dixon, T. H., and Stern, R. J. 1983. Petrology, chemistry and isotopic composition of submarine volcanoes in the southern Mariana arc, *Geol. Soc. Am. Bull.* **94:**1159–1172.

Edwards, M. H., Fornari, D. J., Malinverno, A., and Ryan, W. B. F. 1991. The regional tectonic fabric of the East Pacific Rise from 12°50′N to 15°10′N, *J. Geophys. Res.* **96:**7995–8017.

Eguchi, T. 1984. Seismotectonics around the Mariana Trough, *Tectonophysics* **102:**33–5.

Fryer, P. 1981. Petrogenesis of basaltic rocks from the Mariana Trough, Doctoral dissertation, Univ. Hawaii.

Fryer, P. 1986. Tectonics of the Bonin/Mariana arc intersection and effects on alkaline magmatism of the volcano islands (abstract), *EOS, Trans. AGU* **67**(44)**:**1277.

Fryer, P. 1989. Tectonic evolution of the Mariana backarc basin, Abstracts Vol. 1, 28th Int. Geol. Cong., Washington, DC, p. 51.

Fryer, P. 1990. Recent marine geological research in the Mariana and Izu–Bonin island arcs, *Pac. Sci.* **44**(2)**:** 59–114.

Fryer, P. 1993. The relationship between tectonic deformation, volcanism, and fluid venting in the southeastern Mariana convergent margin, Proc. JAMSTEC Symp. Deep Sea Res., Vol. 9, pp. 161–179.

Fryer, P., Beal, K. L., and Jackson, M. C. 1986. Volcanism of the northern Mariana backarc rift (abstract), *EOS Trans. AGU* **67**(44):1239.

Fryer, P., Gill, J., Jackson, M., Fiske, R., Hochstaedter, A., McMurtry, G., Sedwick, P., Malahoff, A., Mouginis-Mark, P., and Horikoshi, E. 1987. Results of recent Alvin studies of the Kasuga volcanoes, northern Mariana arc (abstract), *EOS Trans. AGU* **68**(44)**:**153.

Fryer, P., Beal, K. L., and Jackson, M. C., 1986. Volcanism of the northern Mariana backarc rift (abstract), *EOS Trans AGU* **67**(44)**:**1239.

Fryer, P., and Hussong, D. M. 1981. Seafloor spreading in the Mariana Trough: results of Leg 60 drill site selection surveys, in *Init. Repts. DSDP*, 60 (D. M. Hussong and S. Uyeda *et al.*, eds.), pp. 45–55, U.S. Govt. Printing Office, Washington, DC.

Fryer, P., and Hussong, D. M. 1982. Arc volcanism in the Mariana Trough (abstract), *EOS Trans. AGU* **63:**1135.

Fryer, P., Saboda, K. L., Johnson, L. E., MacKay, M. E., Moore, G. F., and Stoffers, P. 1990. Conical seamount: SeaMARC II, Alvin submersible, and seismic reflection studies, in *Proc. ODP, Init. Repts.* 125. (P. Fryer, J. A. Pearce, L. B. Stokking, *et al.*, eds.), pp. 69–80, Ocean Drilling Program, College Station, TX.

Fryer, P., Sinton, J. M., and Philpotts, J. A. 1981. Basaltic glasses from the Mariana Trough, in *Init. Repts. DSDP* (D. M. Hussong and S. Uyeda *et al.*, eds.), pp. 601–607, U.S. Govt. Printing Office, Washington DC.

Gill, J. B., and Williams, R. 1990. Th isotopes and U-series disequilibrium in subduction-related volcanic rocks, *Geochim. Cosmochim. Acta* **54**(5)**:**1427–1442.

Hagen, R., Shor, A., and Fryer, P. 1992. SeaMARC II evidence for the locus of seafloor spreading in the southern Mariana Trough, *Mar. Geol.* **103:**311–322.

Hart, S. R., Glassey, W. E., and Karig, D. E. 1972. Basalts and seafloor spreading behind the Mariana island arc, *Earth Planet. Sci. Lett.* **15:**12–18.

Hawkins, J. W. 1977. Petrology and geochemical characteristics of marginal basin basalt, in *Island Arcs, Deep Sea Trenches, and Back-Arc Basins,* (D. E. Hayes, ed.), Geophys. Monogr. Ser., Vol. 23, pp. 355–365, American Geophysical Union, Washington, DC.

Hawkins, J. W., Lonsdale, P. F., Macdougall J. D., and A. M. 1990. Petrology of the axial ridge of the Mariana Trough backarc spreading center, *Earth Planet. Sci. Lett.* **100**(1/3)**:**226–250.

Hawkins, J. W., and Melchior, J. T. 1985. Petrology of Mariana trough and Lau basin basalts, *J. Geophys. Res.* **90:**11,431–11,468.

Hessler, R. R., France, S. C., and Boudrias, M. A. 1987. Hydrothermal vent communities of the Mariana backarc basin (abstract), *EOS Trans. AGU* **68**(44)**:**1531.

Hickey-Vargas, R. 1991. Isotope characteristics of submarine submarine lavas from the Philippine Sea: implications for the origin of arc and basin magmas of the Philippine plate, *Earth Planet. Sci. Lett.* **107:**290–304.

Hickey-Vargas, R., and Mandal, G. K. 1992. The changing contribution of mantle and slab derived components in arc and backarc basin magmas: a model based on the Eocene to Recent, Palau-Kyiushu–West Mariana–Mariana arc system (abstract), *EOS Trans. AGU* **73:**342.

Hobart, M. A., Anderson, R. N., Fujii, N., and Uyeda, S. 1983. Heat flow from hydrothermal mounds in two million year old crust of the Mariana Trough which exceeds two watts per meter (abstract), *EOS Trans. AGU* **64:**315.

Horibe, Y., Kim, K. Craig, H. 1986. Hydrothermal methane plumes in the Mariana backarc spreading center, *Nature* **324:**131–133.

Hsui, A. T., and Youngquist, S. 1985. A dynamic model of the curvature of the Mariana Trench, *Nature* **318**(6045)**:**455–457.

Hussong, D. M., and Fryer, P. 1983. Back-arc seamounts and the SeaMARC II seafloor mapping system (abstract), *EOS Trans. AGU* **64**(45)**:**627–632.

Hussong, D. M., and Sinton, J. B. 1983. Seismicity associated with back-arc crustal spreading in the central Mariana Trough, in *The Tectonic and Geologic Evolution of Southeast Asian Seas and Islands*: Part 2, (D. E. Hayes, ed.), Geophys. Monog. Ser., Vol. 27, pp. 217–235, American Geophysical Union, Washington, DC.

Hussong, D. M., and Uyeda, S. 1981. Tectonic processes and the history of the Mariana arc, a synthesis of the results of Deep Sea Drilling leg 60, in *Init. Repts, DSDP*, 60 (D. M. Hussong and S. Uyeda *et al.*, eds.) pp. 909–929, U.S. Govt. Printing Office, Washington, DC.

Ikeda, Y., and Yuasa, M. 1989. Volcanism in nascent back-arc basins behind the Shichito Ridge and adjacent areas in the Izu–Ogasawara arc, northwest Pacific: evidence for mixing between E-type MORB and island arc magmas at the initiation of backarc rifting, *Contrib. Mineral. Petrol.* **101**(4)**:**377–393.

Ishii, T., Kobayashi, K., Shibata, T., Naka, J., Johnson, K., Ikehara, K., Iguchi, M., Konishi, K.,Wakita, H., Zhang, F., Nakamura, Y., and Kayane, H. 1985. Description of samples from Ogasawara forearc, Ogasawara Plateau and Mariana Trough, during KH84-1 cruise, 1985, in *Preliminary Report of the Hakuho Maru Cruise KH84-1; Geological and Geophysical Investigation of Seafloor East of Ogasawara (Bonin) Islands, Northern Mariana Trough, Southern Mariana, Yap, and Palau Trench and Arcs* (K. Kobayashi, ed.), pp. 89–168, Ocean Research Insititute, University of Tokyo.

Jackson, M. C. 1989. Petrology and petrogenesis of recent submarine volcanoes from the northern Mariana arc and back-arc basin, Doctoral dissertation, Univ. Hawaii.

Johnson, L. E., Fryer, P., Masuda, H., Ishii, T., and Gamo, T. 1993. Hydrothermal vent deposits and two magma sources for volcanoes near 13°20′N in the Mariana backarc basin: a view from Shinkai 6500 (abstract), *EOS Trans. AGU Fall Meeting Suppl.* **74**(43)**:**681.

Karig, D. E. 1971a. Structural history of the Mariana island arc system, *Geol. Soc. Am. Bull.* **83:**323–344.

Karig, D. E. 1971b. Origin and development of marginal basins in the western Pacific, *J. Geophys. Res.* **76**:2452–2561.

Karig, D. E. 1974. Evolution of arc systems in the western Pacific, *Ann. Rev. Earth Planet. Sci.* **2**:51–75.

Karig, D. E. 1975. Basin genesis in the Philippine Sea, in *Init. Repts. DSDP*, 31 (D. E. Karig *et al.*, ed.), pp. 857–879, U.S. Govt. Printing Office, Washington DC.

Karig, D. E., Anderson, R. N., and Bibee, L. D., 1978, Characteristics of back arc spreading in the Mariana Trough, *J. Geophys. Res.* **83**:1213–1226.

Karig, D. E., and Ranken, B. 1983. Marine geology of the forearc region, southern Mariana island arc, in *The Tectonic and Geologic Evolution of Southeast Asian Seas and Islands*, Part 2 (D. E. Hayes, ed.), Geophys. Monogr. Ser., Vol. 27, pp. 266–280, American Geophysical Union, Washington, DC.

Kastner, M., Craig, H., and Sturz, A. in press. Hydrothermal deposits in the Mariana Trough: preliminary mineralogical investigation (abstracted), *EOS, Trans. Am. Geophys. Un.* **68**:1531.

Klein, G. deV., and Kobayashi, K. 1980. Geologic summary of the north Philippine Sea, based on Deep Sea Drilling Project Leg 58 results, in *Init. Repts. DSDP* (G. deV. Klein and K. Kobayashi *et al.*, eds.) Vol. 58, pp. 951–962, U.S. Govt. Printing Office, Washington DC.

Klaus, A., Taylor, B., Moore, G. F., Murakami, F., and Okamura, Y. 1991. Structural evolution of Hachijo and Aoga Shima rifts: backarc rifting in the Izu–Bónin island arc, *EOS Trans. AGU, Fall Meeting Suppl.*, **72**(44):243–244.

Kong, L. S. 1993. Seafloor spreading in the Mariana Trough, in *Preliminary Report of the Hakuho-Maru Cruise KH92-1* (J. Segawa, ed.), pp. 5–16, Ocean Research Institute, Univ. Tokyo.

Kong, L. S., Detrick, R. S., Fox, P. J., Mayer, L. A., and Ryan, W. B. F. 1988/89. The morphology and tectonics of the MARK area from Sea Beam and SeaMARC I: Observations (mid-Atlantic Ridge 23°N), *Mar. Geophys. Res.* **10**:59–90.

Kusakabe, M., Mayeda, S., and Nakamura, E. 1990. S, O, and Sr isotope systematics of active vent materials from the Mariana backarc basin spreading axis at 18°N, *Earth Planet. Sci. Lett.* **100**(1/3):275–282.

Lange, K. 1992. Auswertung eines weitwinkelreflexion- und refraktionsseismischen Profils im sudlichen Marianen Backarc Becken, Diplomarbeit, Institut fur Geophysik, Univ. Hamburg.

Le Pichon, X., Huchon, P., and Barrier, E. 1985. Geoid and the evolution of the western margin of the Pacific Ocean, in *Formation of Active Ocean Margins* (N. Nasu, ed.), pp. 3–42, Terrapub, Tokyo.

Lewis, S. D., Hayes, D. E., and Mrozowski, C. L. 1982. The origin of the West Philippine Basin by interarc spreading (thesis), in *Geology and Tectonics of the Luzon–Marianas Region*, in Philippine Seatar Committee Special Publication 1 (G. R. Balce and A. S. Zanoria, eds.), pp. 31–51.

Lin, P.-N., Stern, R. J., Bloomer, S. H. 1989. Shoshonitic volcanism in the northern Mariana arc. 2: Large-ion lithophile and rare earth element abundances evidence for the source of incompatible element enrichments in intraoceanic arcs, *J. Geophys. Res.* **94**(B4):4497–4514.

Lin, P.-N., Stern, R. J., Morris, J., and Bloomer, S. H. 1990. Nd- and Sr-isotopic composition of lavas from the northern Mariana and southern volcano arcs: implications for the origin of island arc melts, *Contrib. Mineral. Petrol.* **105**:381–392.

Leinen, M., and Anderson, R. N. 1981. Hydrothermal sediment from the Mariana Trough (abstract), *EOS Trans. AGU* **62**:914.

Leinen, M., MacDuff, R. E., Delaney, J., Becker, K., and Schultheiss, P. 1987. Off-axis hydrothermal activity in the mariana mounds field (abstract), *EOS Trans. AGU* **68**(44):1531.

Lonsdale, P., and Hawkins, J. W. 1985. Silicic volcanism at an off-axis geothermal field in the Mariana Trough backarc basin, *Geol. Soc. Am. Bull.* **96**:940–951.

Macdonald, K. C., and Atwater, T. M. 1978. Evolution of rifted ocean ridges, *Earth Planet. Sci. Lett.* **39**:319–327.

Macdougall, J. D., Volpe, A., and Hawkins, J. W. 1987. An arc-like component in Mariana Trough lavas (abstract), *EOS Trans. AGU* **68**(44):1531.

Mahoney J., and Fryer, P. 1988. Stress field re-orientation of the inner Mariana forearc at 22°N (abstract), *EOS Trans. AGU* **69**(44):1443.

Marlow, M S., Johnson, L. E., Pearce, J. A., Fryer, P., Pickthorn, L. G. and Murton, B. J. 1992. Upper Cenozoic volcanic rocks in the Mariana forearc recovered from drilling at Ocean Drilling Program Site 781: implications for forearc magmatism, *J. Geophys. Res.* **97**:15,085–15,098.

Martinez, F., and Cochran, J. R. 1988. Structure and tectonics of the northern Red Sea: catching a continental margin between rifting and drifting, *Tectonophysics* **150**:1–32.

Martinez, F., Fryer, P., Baker, N., and Yamazaki, T. in press. Evolution of backarc rifting Mariana Trough 20–24°N, *J. Geophys. Res.*

McMurtry, G., Sedwick, P., Fryer, P., Vonderhaar, D. L., and Yeh, H.-W. 1993. Unusual geochemistry of hydrothermal vents on submarine arc volcanoes: Kasuga seamounts, northern Mariana arc, *Earth Planet. Sci. Lett.* **114**:517–528.

Meijer, A. 1976. Pb and Sr isotope data bearing on the origin of volcanic rocks from the Mariana island arc system, *Geol. Soc. Am. Bull.* **87**(9):1359–1369.

Meijer, A. 1980. Primitive arc volcanism and a boninite series; examples from western Pacific island arcs, in *The Tectonic and Geologic Evolution of Southeast Asian Seas and Islands* (D. E. Hayes, ed.) Geophys, Monogr. Ser., Vol. 23, pp. 271–282, American Geophysical Union, Washington, DC.

Mrozowski, C. L., Hayes, D. E. 1979. The evolution of the Parece Vela Basin, eastern Philippine Sea, *Earth Planet. Sci. Lett.* **46**:49–67.

Mrozowski, C. L., Hayes, D. E., and Taylor, B. 1981. Multichannel seismic reflection surveys of Leg 60 Sites, Deep Sea Drilling Project, in *Init. Repts. DSDP*, 60 (D. M. Hussong and S. Uyeda *et al.*, eds.), pp. 57–71, U.S. Govt. Printing Office, Washington, DC.

Natland, J. H., and Tarney, J. 1981. Petrologic evolution of the Mariana arc and backarc basin system—a synthesis of drilling results in the south Philippine Sea, in *Init. Repts. DSDP* (D. M. Hussong and S. Uyeda, eds.), Vol. 60, pp. 877–907, U.S. Govt. Printing Office, Washington DC.

Parson, L. M., and Tiffin, D. L. 1993. Northern Lau Basin: Backarc extension at the leading edge of the Indo–Australian plate, *Geo-Mar. Lett.* **13**(2):107–115.

Perram, L. M., and Macdonald, K. C. 1990. A one-million-year history of the 11°45′N East Pacific Rise discontinuity, *J. Geophys. Res.* **95**:21,363–21,381.

Poreda, R. 1985. Helium-3 and deuterium in backarc basalts: Lau and Mariana troughs, *Earth Planet. Sci. Lett.* **73**:244–254.

Seama, N., and Fujiwarra, T. 1993. Geomagnetic anomalies in the Mariana Trough 18°N, in *Preliminary Report of the Hakuho–Maru Cruise KH92-1* (J. Segawa, ed.), pp. 70–71, Ocean Research Institute, Univ. Tokyo.

Searle, R. C., and Laughton, A. S. 1977. Sonar studies of the Mid-Atlantic Ridge and Kurchatov fracture zone, *J. Geophys. Res.* **82**:5313–5328.

Sender, K. L., Shor, A. N., and Hagen, R. A. 1989. SeaMARC II sidescan processing techniques (abstract), *EOS Trans. AGU* **70**(43):1304.

Shervais, J. W. 1982. Ti-V plots and the petrogenesis of modern and ophiolitic lavas, *Earth Planet. Sci. Lett.* **59**:101–118.

Shibata, T. and Segawa, K. 1984. Basaltic glasses from the northern Mariana Trough, in *Preliminary Report of the Hakuhu Maru cruise KH84-1 April, 16–May 30 1984* (K. Kobayashi, ed.), pp. 215–224, Ocean Inst., Univ. Tokyo.

Sinton, J. M., and Fryer, P. 1987. Mariana Trough lavas from 18°N: Implications for the origin of backarc basin basalts, *J. Geophys. Res.* **92**(12):12,782–12,802.

Smoot, N. C. 1988. The growth rate of submarine volcanoes on the South Honshu and East Mariana ridges, *J. Volc. Geotherm. Res.* **35**:1–15.

Smoot, N. C. 1990. Mariana Trough by multi-beam sonar, *Geo-Mar. Lett.* **10**:137–144.

Stern, R. J. 1979. On the origin of andesite in the northern Mariana islands: implications from Agrigan, *Contrib. Mineral. Petrol.* **68**:207–219.

Stern, R. J., and Bibee, L. D. 1984. Esmeralda Bank: geochemistry of an active submarine volcano in the Southern Mariana island arc, *Contrib. Mineral. Petrol.* **86**:159–169.

Stern, R. J., and Bloomer, S. H. 1992. Subduction zone infancy, *Bull. Geol. Soc. Am.* **104**(12):1621.

Stern, R. J., Bloomer, S. H., Lin, P.-N., Ito, E., and Morris, J. 1988. Shoshonitic magmas in nascent arcs: new evidence from submarine volcanoes in the northern Marianas, *Geology* **16**:426–430.

Stern, R. J., Bloomer, S. H., Lin, P.-N., Ito, E., and Smoot, N. C. 1989. Submarine arc volcanism in the southern Mariana arc as an ophiolite analogue, *Tectonophysics* **168**:151–170.

Stern, R. J., Morris, J. D., Jackson, M. C., Fryer, P., Bloomer, S. H., and Ito, E. 1990. Enriched backarc basin basalts from the northern Mariana Trough: implications for evolution of backarc basins, *Earth Planet. Sci. Lett.*, **100**(1/3):210–225.

Stern, R. J., Smoot, N. C., and Rubin, M. 1984. Unzipping of the volcano arc, Japan, *Tectonophysics* **102**:153–174.

Stuben, D., Bloomer, S. H., Taibi, N. E., Neumann, T. Bendel, V., Puschel, U., Barone, A., Lange, A. Shiying, W., Cuizhong, L., and Deyu, Z. 1992. First results of study of sulphur-rich hydrothermal activity from an island-arc environment: Esmeralda Bank in the Mariana arc, *Mar. Geol.* **103**:521–528.

Tatsumi, Y., Murasaki, M., Arsade, E. M. and Nohda, S. 1991. Geochemistry of Quaternary lavas from NE Sulawesi: transfer of subduction components into the mantle wedge, *Contrib. Mineral. Petrol.* **107**:137–149.

Taylor, B., Klaus, A., Brown G. R., and Moore, G. F. 1991. Structural development of Sumisu rift, Izu–Bonin arc, *J. Geophys. Res.* **96**(B10):16,113–16,129.

Volpe, A. M., Macdougall, J. D., Lugmair, G. W., Hawkins J. W., and Lonsdale, P. 1990. Fine-scale isotopic variation in Mariana Trough basalts: evidence for heterogeneity and a recycled component in backarc basin mantle, *Earth Planet. Sci. Lett.* **100**(1/3):251–264.

Wessel, J. K. 1991. A structural analysis of the Mariana inner forearc at 22°N, Master's thesis, Univ. Hawaii.

Wessel, J., Fryer, P., Wessel, P., and Taylor, B. 1994. Extension in the northern Mariana inner forearc, *J. Geophys. Res.*, **99**(B8):15,181–15,203.

Wheat, C. G., and McDuff, R. E. 1987. Advection of pore waters in the Marianas mounds hydrothermal region as determined from nutrient profiles (abstract), *EOS Trans. AGU* **68**(44):1531.

Wood D. A., Marsh, N. G., Tarney, J., Joron, J.-L., Fryer, P., and Treuil, M. 1981. Geochemistry of igneous rocks recovered from a transect across the Mariana Trough, arc, forearc, and trench, Sites 453 through 461, Deep Sea Drilling Project Leg 60, in *Init. Repts. DSDP* (D. M. Hussong and S. Uyeda *et al.*, eds.), Vol. 60, pp. 611–645, U. S. Govt. Printing Office, Washington DC.

Woodhead, J. D. 1987. Geochemistry of volcanic rocks from the northern Mariana islands, west Pacific, Doctoral dissertation, Oxford Univ., UK.

Woodhead, J. D. 1989. Geochemistry of the Mariana arc (western Pacific): source composition and process, *Chem. Geol.* **76**:1–24.

Yamazaki, T., Ishihara, T., Murakami, F. 1991. Magnetic anomalies over the Izu–Ogasawara (Bonin) arc, Mariana arc and Mariana Trough, *Bull. Geol. Surv. Jpn.* **42**(12):655–686.

Yamazaki, T., Murakami, F., and Saito, E. 1993. Mode of seafloor spreading in the northern Mariana Trough, *Tectonophysics* **221**:207–222.

7

Tectonic Framework of the East Scotia Sea

Peter F. Barker

ABSTRACT

This chapter reviews the tectonic evolution of the East Scotia Sea, testing and extending previously published conclusions in light of the additional and expanded data sets now available. The East Scotia Sea floor was generated behind the east-migrating South Sandwich Trench, at a spreading center now lying along 30°W. On its western flank, lineated magnetic anomalies are identified out to at least anomaly 5 (10–11 Ma) and probably out to anomaly 5B (ca. 15 Ma). Spreading was essentially symmetric at about 27 mm/year from 15 Ma to about 5–7 Ma, then slowly accelerated. From 4 Ma to 1.7 Ma, spreading was at 50 mm/year and slightly asymmetric. Since 1.7 Ma, spreading has been up to 15% asymmetric, favoring accretion to the arc flank, within an overall rate of 65 mm/year. Asymmetry is confined within segments bounded by fracture zones that in some cases were created only at 1.7 Ma. A relation between asymmetric spreading, segmentation, and ridge migration seems likely. The median valley is between 6 and 20 km wide and exceptionally is up to 1200 m deep, but usually is smaller and the ridge flanks smooth, as is typical of faster spreading. The ridge crest depth is 500 m or more deeper than the global MOR average. Before 3–4 Ma the ridge was rougher and probably the ridge crest shallower.

In the south, the extensional zone is narrower, and the present spreading probably started only about 3 Ma, after an eastward ridge jump associated with ridge crest–trench collision in the South Sandwich forearc. The ridge jump caused fragments of the previous South Sandwich arc and forearc to be transferred to the Scotia and Antarctic plates, as part of the inevitable adjustment of plate boundaries following ridge crest collision. The detailed history of collision along the South Scotia Ridge is poorly known, but previous collisions involved similar transfers of arc and forearc fragments and may have influenced previous episodes of backarc extension.

On the eastern flank, volcanoes of the South Sandwich island arc lie on ocean floor aged from about 10 Ma to as young as 3 Ma, formed during the present spreading episode. Both island-arc and backarc extensional volcanic geochemistry seems to reflect varying

---

Peter F. Barker • British Antarctic Survey, Natural Environment Research Council, Cambridge CB3 0ET, England.

Backarc Basins: Tectonics and Magmatism, edited by Brian Taylor, Plenum Press, New York, 1995.

degrees of contamination of the mantle, by fluids from the subducting South American lithosphere, and of partial melting. The rocks appear similar to but simpler than those from other intraoceanic backarc environments, showing only minor prior source depletion. However, the arc chemistry is geographically heterogeneous and does not reflect systematic north-south variation in the age and sediment cover of the subducted slab, as might have been expected.

The distribution of older magnetic anomalies on the western flank suggests that congruent ocean floor of the eastern flank should occupy most of the present forearc. If so, there has been significant tectonic erosion of the northeast corner of the forearc, where also serpentinized ultrabasic rocks have been dredged. This corner lies directly above the locus of tearing of the subducting slab at its northern end. In the southern forearc off Montagu Island, dredge hauls have identified an abnormally elevated block as a fragment of a 31 Ma calc-alkaline arc volcano, presumably associated with the early stages of Scotia Sea evolution.

There is no correlation between the level of development of the accretionary prism in the lower forearc and the sediment cover of the subducting slab. Variations along strike in the elevation of the forearc mid-slope high have controlled the transport of arc volcaniclastic sediment to the trench, and hence influenced accretionary prism development to some extent, but the main control appears to have been tectonic.

1. INTRODUCTION

The East Scotia Sea backarc basin is in many respects an end member of the wide range of backarc extensional environments, and its overall setting and behavior are sufficiently unusual to merit close examination. However it is less well known than many backarc basins because of its geographic isolation and the rumored unpleasantness of its climate. A principal aim of this chapter is to expose its inherent interest to a wider audience so as to encourage further work.

Knowledge of the East Scotia Sea and its surroundings has accumulated only slowly. Historically, the two most significant data sets have been bathymetric and magnetic. These were last reviewed and interpreted in detail by Barker and Hill (1981). Those findings were incorporated with minor revision into a tectonic map of the entire Scotia Sea region and review of tectonic evolution (Tectonic Map, 1985; Barker *et al.*, 1991). Meanwhile, these data sets have grown by collection on passage and by specific survey of a young ridge crest collision at the southern end of the backarc basin (Hamilton, 1989). Of other, lesser data sets, earthquake distribution (Isacks and Molnar, 1971; Forsyth, 1975; Brett, 1977; Pelayo and Wiens, 1989) and dredge hauls (Saunders and Tarney, 1979; Hamilton, 1989) have been described and discussed at intervals. Seismic reflection profiles remain sparse. More recently, a small geologic long-range inclined asdic (GLORIA) survey was accomplished in 1989 in the area considered by Hamilton (1989), and release in 1992 of Geosat GM altimetry of the region south of 30°S has provided a major new free-air gravity data set.

Since this region was last reviewed, significant additional data have accumulated, only published in part, and it is not sufficient here merely to review published work. In Section 2 therefore I set in context the main conclusions drawn from published work. In Sections 3 and 4 I evaluate the new and enlarged data sets now available, then use them to reassess and extend those conclusions.

2. REGIONAL CONTEXT OF EAST SCOTIA SEA DEVELOPMENT

2.1. Early Exploration

The South Sandwich volcanic island arc and trench lie at the eastern end of a long eastward loop of islands and submarine ridges, loosely termed the Scotia arc, that connects southernmost South America and the Antarctic Peninsula via the North and South Scotia ridges (Fig. 7.1). The southern islands were discovered by Cook (1777), the northern islands by von Bellingshausen in 1819 (Debenham, 1945), and the trench by the *Meteor* expedition of 1926 (Maurer and Stocks, 1933). The similar geology of the Antarctic Peninsula and South America had long been recognized (e.g., Barrow, 1830; Suess, 1909) and had led many to anticipate a continental connection between them, around the Scotia arc. Further marine survey by RRS *Discovery* (Kemp and Nelson, 1931; Herdman, 1948) and USNS *Eltanin* (Heezen and Johnson, 1965) improved the regional bathymetry and supported this view. Rocks from the South Sandwich Islands, though showing none of the continental affinities of subaerial components of the North and South Scotia Ridge such as South Georgia and the South Orkney Islands (Matthews, 1959; Hawkes, 1962), were found to be arc tholeiites akin to Lesser Antilles rocks (Tyrrell, 1945; Baker, 1968); the apparent parallels between the Scotia and Caribbean arcs dominated regional tectonic models well into the 1960s.

2.2. Marine Geophysics

Marine geophysical exploration of the Scotia Sea began in 1958 (Griffiths, 1963; Griffiths *et al.*, 1964; Heezen and Johnson, 1965; Allen, 1966; Ewing *et al.*, 1971), also

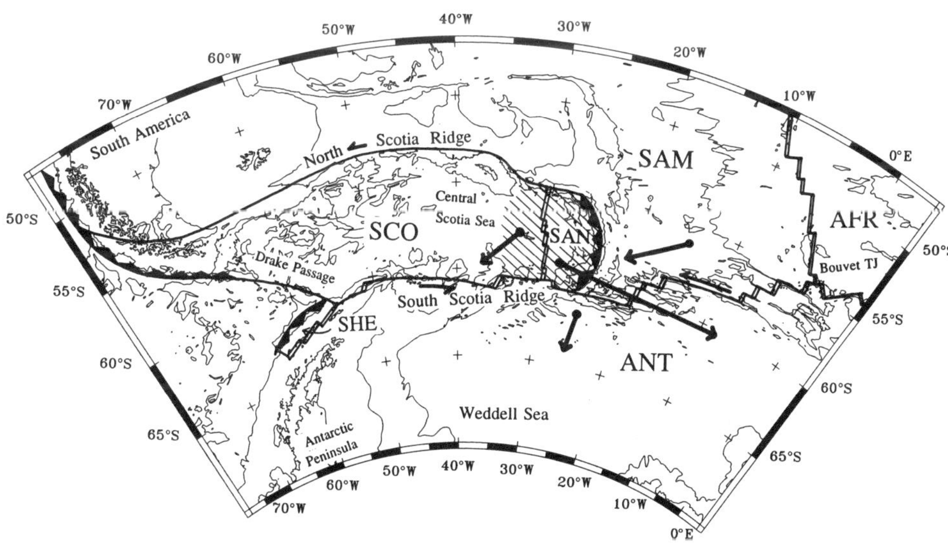

FIGURE 7.1. Present plate boundaries in the Scotia Arc region. "Absolute" plate motion vectors shown by thick arrows. Half-arrows show relative motion at plate boundaries. Major plates are South American (SAM), African (AFR), and Antarctic (ANT). Smaller plates are Scotia (SCO), South Shetland (SHE), and Sandwich (SAN). East Scotia Sea backarc region is shaded.

inspired partly by early successes in the Caribbean. The result, however, was to emphasize the differences between those two regions and the similarities between the Scotia Sea and western Pacific backarc regions. Active backarc extension in the East Scotia Sea was proposed by Barker (1970) and confirmed by a more precisely positioned marine magnetic survey shortly afterward (Barker, 1972). These were the first well-formed magnetic anomalies to be identified in an active backarc, and they showed the essential identity of ocean-floor and backarc spreading processes. Brett and Griffiths (1975) demonstrated attenuation and low velocities of P-wave energy crossing the backarc basin, compatible with the presence of high temperatures at shallow mantle depths and thus with active extension.

Work elsewhere in and around the Scotia Sea served to date the opening of Drake Passage and the Central Scotia Sea (Barker and Burrell, 1977; Hill and Barker, 1980) and map the structure of the North and South Scotia Ridge (e.g., Harrington *et al.*, 1972; Ludwig and Rabinowitz, 1982; Simpson and Griffiths, 1982). It became possible to set the development of the East Scotia Sea into context, as the latest in a series of overlapping extensional episodes that had built the Scotia Sea and extended the North and South Scotia ridges eastward over the past 30–40 m.y. (Barker and Hill, 1981; Barker *et al.*, 1984; Fig. 7.2). That this extension *was* essentially backarc became clear because it had created more ocean floor, at faster spreading rates, than was required by major plate (South American–Antarctic: SAM-ANT) motion alone. Collisions between the advancing trench and ridge crest sections of the SAM-ANT plate boundary were likely triggers for rearrangement of the mode of backarc extension, including the onset of East Scotia Sea opening (Barker *et al.*, 1982, 1984).

2.3. Neotectonics: Regional Context

Present-day plate boundaries and motions in the southwest Atlantic, shown in Fig. 7.1, demonstrate one of the main features of East Scotia Sea backarc extension: its virtual independence of major plate motion. In general terms, the Scotia Sea region appears as a complication of the SAM-ANT plate boundary. The eastern SAM-ANT boundary is a simple, long-offset spreading center from the Bouvet triple junction to the southern end of the South Sandwich Trench at 61°S, 26°W, and becomes simple again in the far west at the slowly subducting Chile Trench between 52° and 46°S. In between lie the moderately large Scotia (SCO) plate and the Sandwich (SAN) and Shetland (SHE) microplates. The East Scotia Sea (shaded in Fig. 7.1) is developing along the divergent SCO-SAN boundary, close to 30°W.

SAM-ANT relative plate motion is west-east and slow—about 18 mm/year, whether estimated locally (Lawver and Dick, 1983; Barker and Lawver, 1988) or globally (NUVEL 1: De Mets *et al.*, 1990)—and has changed little over the past 20 m.y. (Barker and Lawver, 1988). SAM-ANT motion is shared between SAM-SCO and SCO-ANT boundaries, the North and South Scotia ridges respectively. Earthquake focal mechanisms (Forsyth, 1975; Pelayo and Wiens, 1989) indicate sinistral motion along each, so each is slow. Pelayo and Wiens calculated motion along the South Scotia Ridge to be about twice as fast as along the North Scotia Ridge.

"Absolute" motions of the South American and Antarctic plates (with respect to the mean Pacific hot-spot reference frame HS-2: Minster and Jordan, 1978; Gripp and Gordon, 1990) are also slow. In the vicinity of the Sandwich plate, Antarctic plate motion in this reference frame is about 12 mm/year toward 188° and South American plate motion 22 mm/year toward 237° (Fig. 7.1). Because of the large uncertainties in hot-spot motion, the

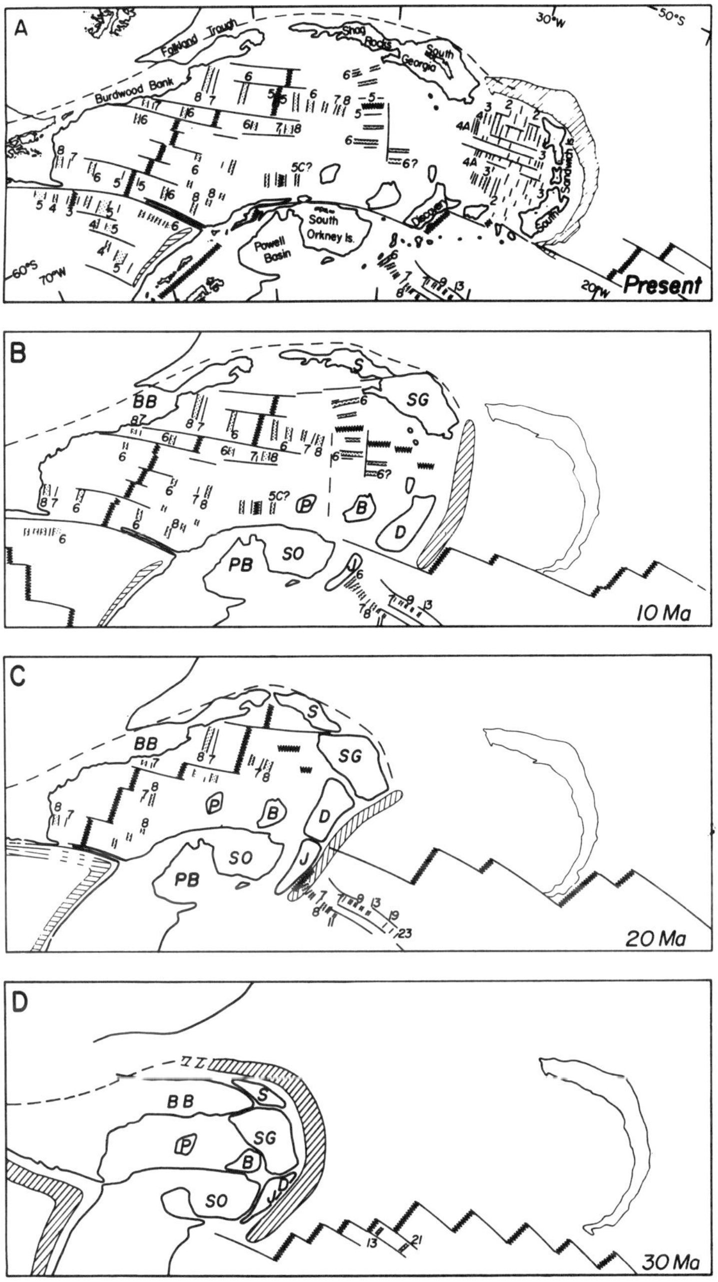

FIGURE 7.2. Tectonic evolution of the Scotia Sea region over the past 30 m.y. (Barker *et al.*, 1984, 1991). Small blocks are BB, Burdwood Bank; S, Shag Rocks; SG, South Georgia; H, Herdman Bank; D, Discovery Bank; J, Jane Bank; SO, South Orkney; B, Bruce Bank; P, Pirie Bank. These are one or more of microcontinental fragment, dissected intraoceanic island arc, accretionary prism. PB is Powell Basin. (A) Present: showing identified magnetic anomalies in the Scotia Sea and northern Weddell Sea. South Sandwich Trench is shaded. Reconstructions (B) To 10 Ma: showing trench shaded, and "ghost" of present S. Sandwich Trench for scale. East Scotia Sea opening here assumed all younger than 10 Ma (but see Section 4 and Fig. 7.12). (C) To 20 Ma: showing much smaller Scotia Sea, and Jane Bank ridge crest-trench collision current. (D) To 30 Ma: showing very much smaller Scotia Sea and Powell Basin. Note persistence of South Georgia in the northern forearc.

Antarctic plate motion is barely significant. A small southwesterly vector of absolute motion of the Scotia plate may also be deduced, but is equally uncertain.

Essentially, therefore, both relative and "absolute" motions of the major (SAM and ANT) plates and of the Scotia plate are slow. Within this slow-motion environment, the rapid eastward migration of the Sandwich plate stands out. Present East Scotia Sea extension was estimated by Barker and Hill (1981) from magnetic anomalies at about 70 mm/year, in a direction close to east-west. Assuming Pelayo and Wiens's (1989) estimate of SCO motion is correct, then SAN-SAM convergence at the South Sandwich Trench may be estimated at about 75 mm/year along 085°. "Absolute" motion of the Sandwich plate is about 57 mm/year toward 095°. If the forearc is regarded as being in dynamic equilibrium, then this last is the rate of eastward roll-back of the trench hinge. Revised estimates of these motion vectors, using East Scotia Sea spreading rates determined in Section 4.3, are only slightly different.

Estimates of plate motion are pertinent to a discussion of the reasons for backarc extension in the East Scotia Sea. Chase (1978) first noted the correlation between backarc extension and retreat (in the hot-spot reference frame) of the major plate behind the arc. Following Elsasser (1971), he postulated that the subducting slab would not merely glide along its own length, forming an anchor in the sublithospheric mantle, but would also tend to sink vertically because of its weight, leading to rollback of the trench hinge. If the overriding major plate was not advancing, then backarc extension might be induced to fill the gap. In the case of the East Scotia Sea, rollback is remarkably fast (only the New Hebrides is faster: Jarrard, 1986), and in the place of the major plate is the Scotia plate, not particularly large but apparently retreating slowly southwestward. As noted elsewhere (Section 2.5), it is considered that almost all of the Scotia plate has itself formed by backarc extension during earlier episodes, and the present East Scotia Sea opening represents merely the latest phase of a long-lived backarc extensional regime that presumably was also reacting to rollback of the trench hinge. It has been suggested by Alvarez (1982) that rollback here is facilitated by shallow eastward sublithospheric mantle return flow that is balancing the net asthenospheric loss (to the lithosphere) in the growing Atlantic and Indian Oceans and the net gain in the shrinking Pacific.

2.4. Neotectonics: The Subduction Zone

Oceanic lithosphere of the South American plate is being subducted at the South Sandwich Trench, the eastern boundary of the Sandwich plate (Fig. 7.1). Ocean floor at the trench ranges in age from 27 Ma (C8r) in the south to perhaps 80 Ma (C33r) in the north (Barker and Lawver, 1988). It is likely that the oceanic lithosphere now at moderate depth within the subduction zone (say less than 200 km) was all produced at slow spreading rates at the South American–Antarctic Ridge and therefore has a similar age range (probably younger in the far south) to that in the trench.

Isacks and Molnar (1971), Forsyth (1975), and Brett (1977) have analyzed the seismicity of the dipping slab. Along most of the trench, the slab dips at 45–55° down to 180 km (possibly to 254 km near 58°S; Brett, 1977). A zone of more intense activity with a steeper apparent dip occurs at the northern end (between 55.8° and 56.3°S). Since the period of activity examined in these studies, four deeper earthquakes have been reported in NOAA PDE lists, at 196-, 217- and 288-km depth at the northern end, within a continuation of the zone of steep apparent dip.

The steeply dipping northern zone was associated by Forsyth (1975) and Brett (1977) with tearing of the downwarped South American plate, as the hinge of subduction has advanced eastward (Fig. 7.3). At the southern end of the South Sandwich Trench, the long eastward line of the South Sandwich fracture zone forms the southern boundary of the South American plate and facilitates future subduction and rollback, but at the northern end the subducting slab must tear. Earthquake focal mechanisms within the steeply dipping zone were interpreted by Forsyth as hinge-faulting or bending-stress release, compatible with tearing of a slab downwarped to 30–120 km, along a west-east line close to 56°S.

Elsewhere within the subduction zone, earthquake focal mechanisms are distributed similarly to those in other such regions in most respects, but there are notable differences. While intermediate-depth earthquakes in the northern half are downdip extensional, those in the southern half are downdip compressional. Forsyth (1975) and Brett (1977) attributed this to the greater buoyancy of the younger, southern part of the sinking slab, which may be being towed down by the older, denser northern part. Similarly, the greater age of the northern part might be expected to result in deeper earthquake foci there, given that the younger, thinner southern part should reheat sooner to temperatures too high to permit failure by brittle fracture (e.g., Vlaar and Wortel, 1976; Jarrard, 1986). At the time, no such effect was seen, but the distribution of the more recent, deeper events supports this notion. Again, the age-related density difference could explain the greater slab dips seen by Brett (1977) north of 58°S.

2.5. Scotia Sea Reconstructions

The continuity and vigor of backarc extension, driven by eastward rollback of the trench hinge, are the key features of Scotia Sea evolution. Several spreading regimes have been identified within the Scotia Sea, overlapping in time. For example, Drake Passage and Central Scotia Sea opening (Barker and Burrell, 1977; Hill and Barker, 1980), though now stopped, are thought to have continued for a while after East Scotia Sea opening started. Other, short-lived spreading systems are inferred, in basins too small to produce unambiguously identifiable magnetic anomaly sequences. In total, far too much oceanic lithosphere was being produced in the Scotia Sea, for at least the past 21 m.y. (Barker and Hill, 1981) and most probably over its entire 30–40 Ma lifetime, to be explained without recourse to the decoupling provided by subduction: even without a complete knowledge of

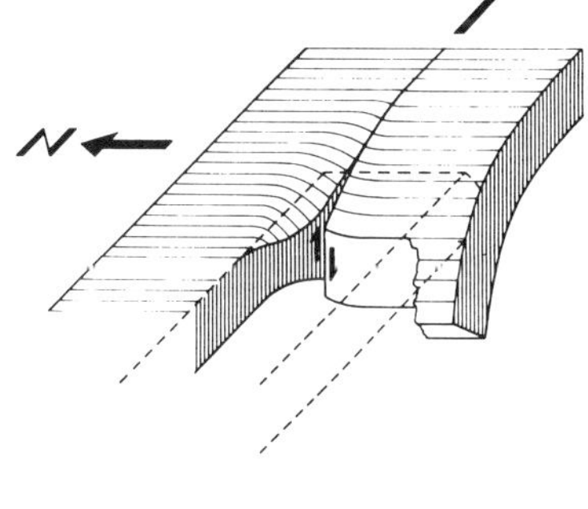

FIGURE 7.3. Sketch of tearing of 80-m.y.-old South American oceanic lithosphere at the northern end of the South Sandwich subduction zone, based on earthquake fault plane solutions (from Forsyth, 1975). The young Sandwich plate (shown dashed) overlies the downwarped edge of the South American plate to the north of the tear, preventing its complete recovery.

the spreading history, or of contemporaneous arc volcanism, the essential backarc setting of Scotia Sea extension was apparent.

Figure 7.2 shows the regions of the Scotia Sea that are dated by marine magnetic anomalies, and reconstructions of Scotia Sea evolution through 10 Ma, 20 Ma, and 30 Ma (Barker *et al.*, 1984, 1991). Several assumptions are necessary in order to make these reconstructions, principally concerning the age of several undated small ocean basins and the nature of coupling between contemporaneous spreading provinces. It is not appropriate to discuss those further here (see Barker *et al.*, 1991), but it is useful to consider Scotia Sea evolution in order to understand the processes acting.

No single spreading regime persisted for more than a few million years, and it has been suggested that collisions between ancestors of the present South Sandwich Trench and more westerly spreading sections of the South American–Antarctic plate boundary triggered changes of regime. The South Scotia Ridge is the locus of these past collisions, younging eastward. To illustrate this process, it is clear from Fig. 7.1 that, if present-day spreading rates and directions persist, the southern part of the South Sandwich Trench will overtake the slower spreading ridge crest sections now at 19–20°W, in about 5 m.y. The presumption is that spreading and subduction at the collision zone would then both stop, but subduction would continue at the northern half of the trench, possibly with a different rate, direction, and site of complementary backarc extension. Somehow, the Sandwich plate would fracture, its southern part accreting to the Antarctic plate to produce a longer, more complicated Scotia–Antarctic plate boundary. The key feature of past collisions seems to have been that all of the southern oceanic part of the South American plate was subducted as the arc and trench migrated eastward along the plate boundary, but none of the Antarctic plate.

Attempts have been made to define and date previous ridge crest–trench collisions along the South Scotia Ridge. Rocks dredged from Discovery Bank (Barker *et al.*, 1982) and Jane Bank (Barker *et al.*, 1984)—D and J in Fig. 7.2—show strong chemical similarities to South Sandwich Island lavas: Discovery Bank rocks are of Miocene age (12–20 Ma). A magnetic anomaly sequence in the northern Weddell Sea that younged toward Jane Bank was interpreted by Barker *et al.* (1984) to imply a collision there ca. 20 Ma (Fig. 7.2C). The magnetic anomaly identifications have been queried (Hamilton, 1989; Livermore and Woollett, 1993) in light of rate constraints placed on SAM-ANT motion by Barker and Lawver (1988). However, it is possible that, for a while before ridge crest collision, the subducting slab became decoupled from the South American plate (as happened repeatedly off the Pacific margin of North America in similar circumstances; Menard, 1978; Atwater, 1989), thereby removing such constraints. Age uncertainties do not affect the reality of the collisions or their backarc implications, merely their timing.

Further definition of the Jane Bank collision zone was hampered by sediment cover. Hamilton (1989) surveyed and sampled the youngest collision zone, in the area directly south of the present backarc spreading center, that promised to be more accessible. He found the collision zone to be narrow, and probably oblique to the convergence direction, because of complexities of the SAM-ANT boundary. Also, SAM-ANT spreading was slow over that period, producing barely recognizable marine magnetic anomalies. The history of collision was therefore difficult to extract. However, the additional data do allow an improved description of the history of extension in the southernmost part of the backarc (Section 4.6) and better definition of present plate boundaries. The GLORIA data from the collision zone, which will not be considered in detail here, appear largely to confirm Hamilton's (1989) conclusions.

2.6. East Scotia Sea Development and Structure

Several features of East Scotia Sea development and structure are significant in global terms. Its location remote from other backarc basins, its well-formed magnetic anomalies and simple chemistry, and the rapid rollback of the trench hinge make it a useful end member among backarc environments. Other, lesser features gain significance from this uniqueness. They include aspects of the spreading history; the bathymetric expression of spreading, the forearc, arc, and remnant arc; and the anomalous southern margin. Because of the primitive nature of knowledge of the East Scotia Sea at the time they were described (Barker, 1972; Saunders and Tarney, 1979; Barker and Hill, 1981), compared with present knowledge of this and other backarc regions, it is important to reexamine these features and their significance in the light of the expanded data sets now available. In preparation for this, I summarize them briefly below.

Most of East Scotia Sea spreading was asymmetric, with accretion favoring the eastern, island-arc flank. Spreading accelerated from about 50 to 70 mm/year, between 3 and 1.5 Ma. The oldest identified anomaly on the western flank was 4A. Little difference was seen between the amount of ocean floor formed at 56°S and at 59°S, over the past 5–6 m.y., suggesting that spreading did not contribute to the curvature of the arc.

Ridge crest depth of the East Scotia Sea spreading center is about 3000 m, about 500 m deeper than the global average for mid-ocean ridges (e.g., Parsons and Sclater, 1977). The eastern arc flank is buried beneath a thick arc-derived volcaniclastic apron, but the western flank is covered sparsely and unevenly by pelagic and hemipelagic sediment deposited under the influence of the Antarctic Circumpolar Current (ACC). Although the western flank shows rough, elevated topography, no simple grouping of elevations with east-facing scarps could be identified with a remnant arc that might mark the beginning of East Scotia Sea extension. East Scotia Sea spreading overlapped in time extent with the adjacent Central Scotia Sea spreading, so a complicated early history of spreading might be expected.

Many of the volcanic islands of the arc are built on ocean floor produced during the present spreading episode, no more than a few million years old. The oldest radiometric ages from the arc (3.1 Ma; Baker, 1990) are compatible with this. Although no marine magnetic lineations were seen in the forearc (Barker and Hill, 1981), the upper forearc, too, at least in the north, may comprise ocean floor formed during the present spreading episode. The upper forearc in the south is significantly more elevated than in the north, which some have linked to the greater elevation of the younger SAM ocean floor being subducted in the southern half of the trench. The lower forearc is narrow and poorly developed.

In the south, the distance between the arc and the elevated topography of the South Scotia Ridge (specifically, Herdman Bank) is less than the east-west extent of magnetic anomalies farther north. Either spreading started more recently in the south or (since the orientation of the magnetic anomalies appeared to change) an additional Herdman Bank microplate was involved, so spreading here has been slower.

All these published conclusions (Barker, 1972; Saunders and Tarney, 1979; Barker and Hill, 1981) are reassessed in Section 4.

3. NEW OR ENLARGED DATA SETS

3.1. Bathymetry

The bathymetry of the Tectonic Map (1985) was based on all available soundings to 1980, including, with discretion, some early data that may not have been accurately located. The amount of data now available for the backarc in general has increased by perhaps 10%, which has not justified revision of the contour map to date. In the southern area around the youngest ridge crest collision zone, the increase in soundings has been greater, and Hamilton's (1989) recontoured chart of this area is incorporated in Fig. 7.4. Contours are at 500-m intervals. Ship tracks are not shown, to avoid confusion, but the magnetic tracks (Section 3.3) show the locations of most of the post-1969 soundings.

3.2. Shipboard and Geosat GM Free-Air Gravity

The Geosat "Geodetic Mission" (GM) radar altimetric data for the ocean areas south of 30°S were declassified by the U.S. Navy in late 1992. The long repeat period of the GM orbit produced a ground track separation in these latitudes of 2–3 km, which is short compared with the 2–9 km altimeter footprint (McAdoo and Marks, 1992) and implies essentially complete coverage. Free-air gravity data at a grid spacing of 0.05° longitude by

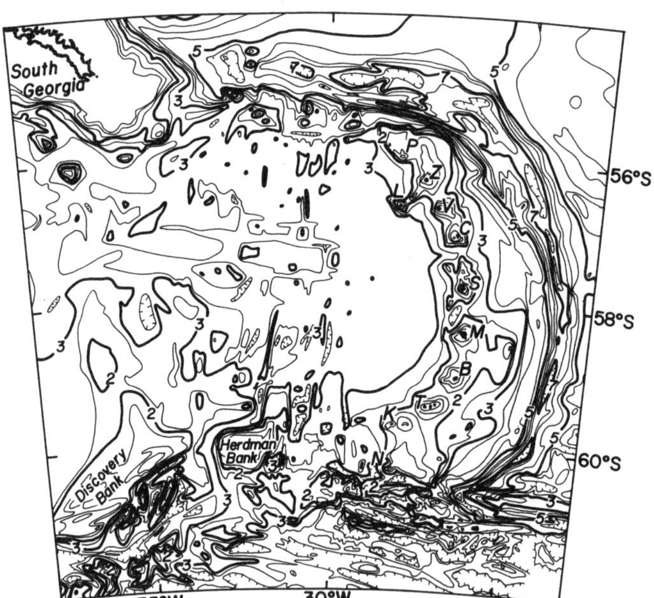

FIGURE 7.4. Bathymetry of the East Scotia Sea region, replotted on Lambert Conical Orthomorphic projection, from Tectonic Map (1985) with revisions by Hamilton (1989). Contours are at 500-m intervals, and the 2-, 3-, 5-, and 7-km contours are thickened to aid definition of tectonic provinces. Deepest contour is 8000 m in northeast South Sandwich Trench. South Sandwich Islands are Z, Zavodovski; L, Leskov; V, Visokoi; C, Candlemas (with Vindication); S, Saunders; M, Montagu; B, Bristol; T, Thule (with Cook and Bellingshausen). Other, submarine arc volcanoes are P, Protector Shoal; K, Kemp; and N, Nelson seamounts.

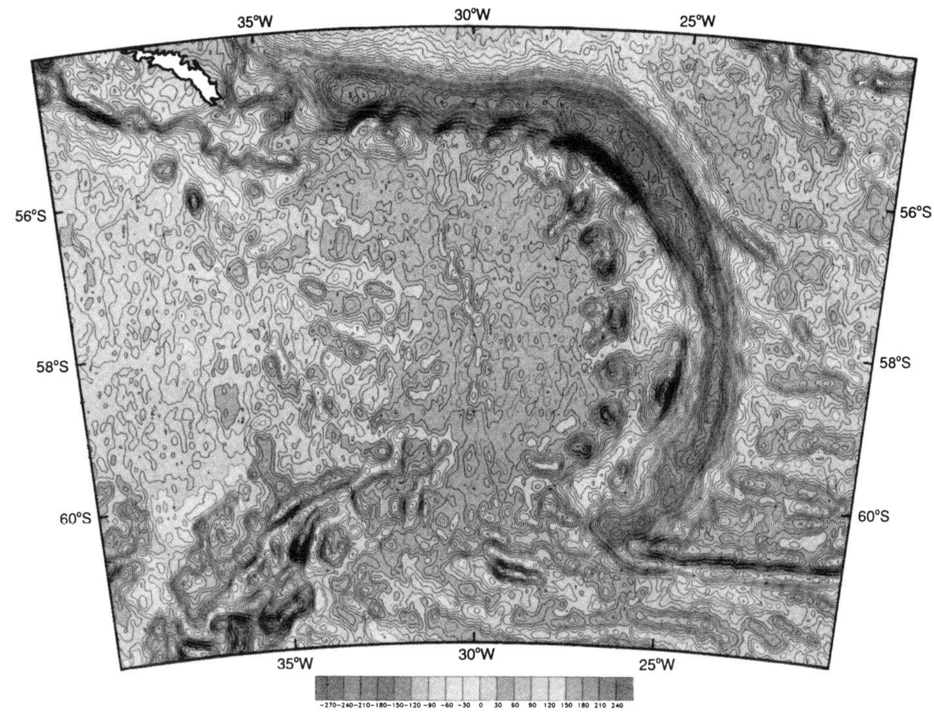

FIGURE 7.5. Free-air gravity map of the East Scotia Sea region derived from Geosat (GM and ERM) altimetric data (NGDC-NOAA, 1993), using Lambert Conical Orthomorphic projection. Contours at 10-mgal intervals. A color version of this figure appears opposite page 310.

0.04° latitude are now available on CD-ROM (NGDC-NOAA, 1993). The limit of short-wavelength response of the gridded data is generally considered to be about 12–20 km (McAdoo and Marks, 1992), comparable with surface ship measurements in oceanic areas. To produce Fig. 7.5, I projected the grid points onto a Lambert Conical Orthomorphic projection and regridded at approximately the same spacing before contouring, using proprietary minimum-tension software.

Gravity measurements were made along about one third of the ship tracks shown in Fig. 7.6 (Section 3.3). Compared with the Geosat gravity data, therefore, the shipboard data are few and their distribution uneven. Comparison of typical ship tracks with satellite data shows that over almost all of the area of Fig. 7.5, except areas of shallow water, the correspondence is very close. Discrepancies in shallow water stem from a combination of the short-wavelength gravity anomalies from shallow sources, better reproduced by shipboard measurements, and distortion introduced by gridding and contouring algorithms close to land where there are no altimetry data.

The bathymetric chart and gravity contour map (Figs. 7.4 and 7.5) are best considered together. In a purely oceanic region, the overwhelming contribution to the gravity anomaly map at short to intermediate wavelengths comes from the topographic relief on oceanic basement, either at the seafloor or, reduced in amplitude, buried beneath sediments (McAdoo and Marks, 1992). Complications arise when the ocean floor is loaded locally (as with volcanic islands) after gaining rigidity so that the load is regionally rather than locally

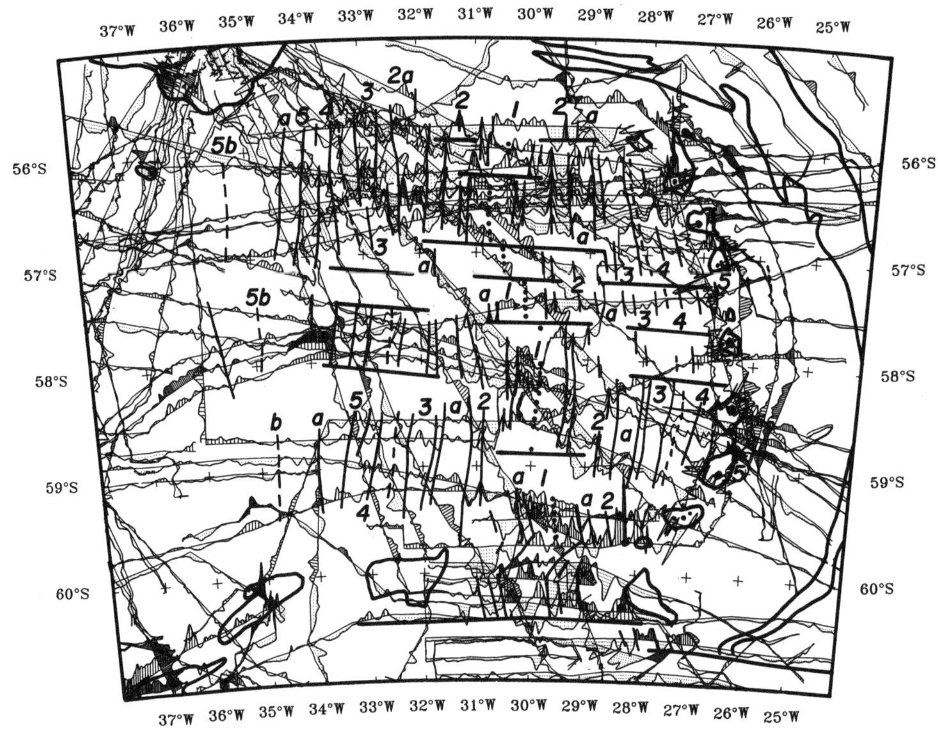

FIGURE 7.6. Magnetic anomalies in the East Scotia Sea region, projected along ship tracks (solid line ornament for positive values). Identified anomalies are marked and major anomalies numbered. For ease of identification, the ridge crest is dotted, where seen on bathymetric profiles or in Fig. 7.5, and anomalies 3A and 5B are dashed. Comparisons with synthetic anomalies are shown in Figs. 7.10 and 7.11. Thicker, straight lines are transform faults and fault traces. The 1000-m and 6000-m contours from Fig. 7.4 are used to identify the South Sandwich Islands, other submarine elevations, and the South Sandwich Trench. Lambert Conical Orthomorphic projection.

compensated and the crustal structure possibly modified. Other complications concern continental fragments that may have an entirely different, and typically more heterogeneous, crustal structure. Nevertheless there are strong similarities between gravity and bathymetric maps of largely oceanic areas, and the former can be used to infer bathymetric features in areas where shipboard bathymetric data are sparse.

Almost all of the area shown in Figs. 7.4 and 7.5 is floored by oceanic lithosphere of the South American, Antarctic, Scotia, or Sandwich plates. The South Georgia block is a continental fragment, and the South Sandwich volcanic arc may lie on ocean floor modified to a greater or lesser extent by previous subduction-related intrusion. Parts of the South Scotia Ridge, such as Herdman Bank, may be continental fragments or parts of older intraoceanic arcs, rendered inactive by ridge crest collision and left behind by subsequent backarc extension. Both figures show the main tectonic features well. There are differences (notably in the definition of ocean-floor fabric *outside* the Scotia Sea), but few are major, a reflection perhaps of the high density of existing bathymetric data over much of the region. Note, however, the gravity lows indenting the northern and southern margins of the South Georgia block, interpreted in Section 4.8.

3.3. Magnetic Anomalies

Magnetic anomalies in the East Scotia Sea have been attributed to east-west backarc extension over the past 8 m.y. (Barker, 1970, 1972; Barker and Hill, 1981). Since these studies, the magnetic data set has increased about 40% by acquisition along passage tracks (not always in a useful direction) and also by detailed study of ridge crest–trench collision in the far south (Hamilton, 1989). Among the principal findings of previous work were the fast rates of modern seafloor spreading in the backarc, the acceleration in spreading at 1.5 to 3 Ma, and a pronounced asymmetry of spreading. The spreading rates have been combined with other similar data in computing unknown plate motions, both relative and "absolute" (see, for example, Section 2.3), the acceleration bears on studies of driving forces and causation, and the asymmetry may reflect conditions at the base of the lithosphere. Because of their significance, it is useful to reassess these original findings in light of the enlarged data set.

Also, the spreading rates and particularly the acceleration depend upon the accuracy of the magnetic reversal time scale (MRTS) used in their determination, which has been revised meanwhile. Barker and Hill (1981) used the MRTS of LaBrecque *et al.* (1977); De Mets *et al.* (1990), in determining NUVEL-1, used that of Harland *et al.* (1982) to fix the time of the center of anomaly 2A (C2An.2n). For this chapter I have used the MRTS of Cande and Kent (1992), which is a significant improvement in accuracy over the previous generation. However, further revision of the MRTS may follow detailed comparisons of magnetic and orbital (Milankovich) time scales (e.g., Wilson, 1993).

Magnetic anomalies are displayed along track in Fig. 7.6 after removal of the International Geomagnetic Reference Field (IGRF) (Langel, 1992). No attempt has been made to eliminate the diurnal variation, but data were checked for magnetic storm influence before acceptance. Ship tracks for all profiles are satellite-positioned (transit or GPS), mostly to better than 1.5 km.

3.4. Seismic Reflection Profiles

Few seismic reflection profiles have been acquired in the South Sandwich arc and backarc region. Those known to me are the tracks shown in Fig. 7.7, which are almost all single-channel profiles and largely unpublished, and the older, astronavigated RV *Zapiola* Cruise II profile shown by Heezen and Johnson (1965). Data from two of the profiles crossing the backarc are incorporated into Fig. 7.8 (from Barker and Hill, 1981), and others have been described by Hamilton (1989). Most of those in the far south and the far northeast were acquired in conjunction with the GLORIA survey during RRS *Charles Darwin* Cruise 37 in 1989 and will not be considered specifically here.

3.5. Onshore Geology and Dredged Rocks

The South Sandwich Islands have been sampled several times during opportunistic visits, but the most significant mapping and sampling exercise by far was that undertaken in the period 1961–1964 from RRS *Shackleton* and HMS *Protector* and described by Baker (1978), Holdgate and Baker (1979), and Tomblin (1979). Geochemical analyses of these rocks by Luff (1982) have been supplemented or superseded by more extensive recent work (Pearce *et al.*, in press).

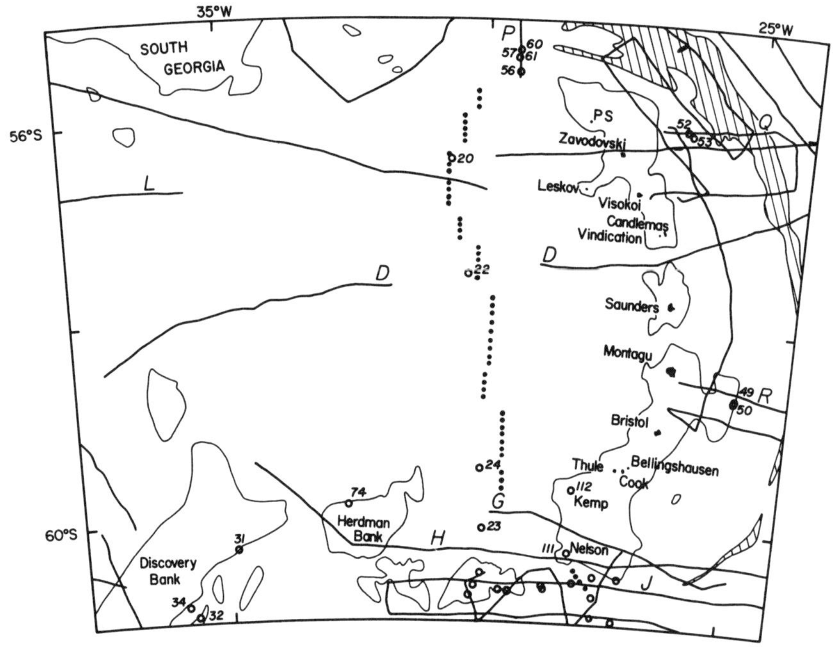

FIGURE 7.7. Locations of seismic reflection profiles and dredge hauls in the East Scotia Sea region. Profiles referred to in the text or shown in other figures are lettered, and dredge hauls discussed are numbered. No dredge sites are shown where an *in situ* component could not be identified. PS is Protector Shoal. The East Scotia Sea spreading center, where known from median valley or magnetic anomaly, is shown dotted. The South Sandwich Trench (7-km contour) is shaded.

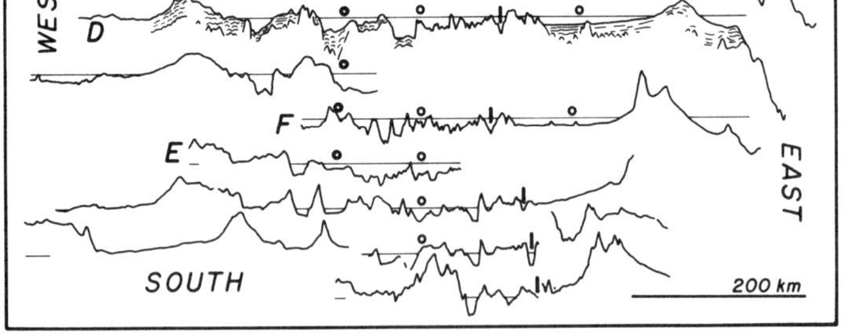

FIGURE 7.8. Bathymetric profiles crossing the East Scotia Sea (from Barker and Hill, 1981). Profiles have been lettered where referred to in the text or shown in other figures. Vertical scale is km.

Dredging in the Scotia Sea region is made more difficult by the need to establish that a dredged rock was *in situ*, against a background of abundant ice-rafted detritus. Fig. 7.7 shows the locations of successful dredges, in which an *in situ* component has been established, that have been obtained during Birmingham University and British Antarctic Survey (BAS) cruises over the past 20 years. These dredges were undertaken principally for "mapping" purposes, to determine an environment from its chemistry rather than to examine the chemistry of a known environment. For example, dredge hauls from Discovery Bank showed it was an intraoceanic island arc ancestral to the South Sandwich arc (Barker *et al.*, 1982). The main exception was a suite of four dredges from the backarc spreading center, described by Saunders and Tarney (1979). Unpublished dredge hauls from the northern, eastern, and southern margins of the Sandwich plate have been described, and x-ray fluorescence (XRF) analyses were carried out at Birmingham University by P. L. Barber (pers. comm.) and by Hamilton (1989). The unnumbered dredge hauls in the south are from the youngest ridge crest collision zone (Hamilton, 1989) and will not be considered here because there is no independent evidence of the tectonic environment that they sampled.

4. DATA INTERPRETATION

4.1. Context: The South Sandwich Island Arc and Trench

4.1.1. South Sandwich Island Arc

All of the volcanic islands, except Leskov Island, lie along a smooth arc (Baker, 1990; Fig. 7.7). That arc continues in the north to include Protector Shoal, at 27 m the shoalest part of a large bank that the gravity map suggests extends to Zavodovski Island, the northernmost of the chain. Protector Shoal was considered by Gass *et al.* (1963) and Holdgate and Baker (1979) the likely source of a 1962 pumice eruption. The arc extends southward also to include Kemp Seamount, a submarine volcano near 59.5°S, 28°W with a least measured depth of 89 m, dredge-sampled from RRS *Discovery* in 1985 (DR 112 in Fig. 7.7). Hamilton (1989) considered this volcano to have been recently active because of its shallow depth: many small, isolated elevations in this area that are not actively being enlarged are flat-topped at 500- to 800-m depths, the likely result of planation by icebergs from the southern Weddell Sea.

Several submarine volcanoes lie, like Leskov Island, off the main line of the arc. The gravity and bathymetric maps (Figs. 7.4 and 7.5) indicate subsidiary submarine elevations west of Saunders, Montagu, and Bristol Islands that are almost certainly volcanic. East of the main arc, in the forearc, there are also isolated elevations, but the two largest of these, near 58.5°S, 25°W and 60.3°S, 27.5°W, are old (see Section 4.7). Others, notably that at 59.6°S, 26°W, have an uncertain origin. Only Nelson Seamount, the small peak at 60.3°S, 28.2°W (Fig. 7.4) with a least measured depth of 280 m, is known to be volcanic: Hamilton (1989) notes similarities to South Sandwich Islands lava chemistry in rocks dredged from this peak from RRS *Discovery* in 1985 (DR 111 in Fig. 7.7). Nelson Seamount was also considered by Hamilton to be young (rather than an older volcano dissected by extension following ridge crest collision), because it is shallow yet not flat-topped. This is borne out by the GLORIA survey, which shows its apron overlapping young faults and ocean floor.

However, it is very close to the southern limit of the Sandwich plate and closer to the trench than other arc volcanoes.

The dominant rock type of the islands is a low-K tholeiitic basalt, with successively smaller amounts of more acidic equivalents. In these rocks the alkali metals are enriched with respect to mid-ocean ridge basalts (MORB). The rare-earth elements (REE), particularly the light-rare-earth elements (LREE), and high-field-strength (HFS) elements are depleted. They are thought to have been produced by high degrees of partial melting of depleted mantle in the region above the subducting slab, fluxed by hydrous fluids or melts from the slab (Luff, 1982; Pearce, 1982; Baker, 1990; Pearce *et al.*, in press). They are similar to rocks from the Mariana, Izu, and Tonga arcs.

The systematic tectonic differences that one might expect to see reflected in arc chemistry are the north-south age gradient in the subducted slab, that will have been consistent for several million years, and a consequent thicker sediment cover in the north. A secondary feature reflecting the difference in basement age and latitude of the subducting slab will be an almost total dominance of biosiliceous sediments in the south, but a more even calcareous/siliceous mix in the north. In addition, to the extent that the arc *does not* lie on ocean floor of the present spreading episode, but on that produced during the earlier Central Scotia Sea opening (Barker and Hill, 1981; Fig. 7.2B), the mantle beneath the northern and southern South Sandwich arcs could be different. All of these factors might be expected to produce systematic chemical differences between northern and southern parts of the arc, but no such differences have been detected.

Inter-island chemical variation is significant but not obviously linked to regional tectonic variation. About the clearest distinction is a low-K, LREE-depleted group of Zavodovski, Candlemas, Vindication, Montagu, and Bristol Islands, most of which also have low incompatible-element ratios and abundances. These islands do not form a geographically distinct group. One possible exception identified by Pearce *et al.* (in press) concerns the more calc-alkaline character of rocks from Leskov Island (which lies west of the main arc) and Freezland Rock (5 km west of Bristol Island and possibly the oldest exposed rocks, at 3.1 Ma; Baker, 1990).

Hamilton (1989) reported somewhat limited geochemical data from the Kemp and Nelson seamount dredge hauls (DR 112 and 111 respectively; Fig. 7.7). Kemp Seamount, which lies on a southern extension of the main arc, yielded a basalt most closely resembling rocks from Zavodovski and Visokoi Islands. This result accords with the apparently random geographic distribution of chemical diversity around the arc. A fresh dacite dredged from Nelson Seamount most closely resembles dacites from Leskov Island and Freezland Rock. This correspondence also reflects geographic heterogeneity since, unlike these sites, Nelson Seamount is (a) young and (b) closer to the trench than the main line of the arc.

4.1.2. South Sandwich Trench

The trench extends from 56°S, 33°W in the northwest to 60.7°S, 26°W in the southeast (Figs. 7.4 and 7.5). The outer trench wall is uneven, the effect of a linear ocean-floor topography oriented between east-west and southeast-northwest, created at the SAM-ANT plate boundary between 25 and 80 Ma (Barker and Lawver, 1988) and clearly seen in Fig. 7.5. Oceanic basement of the outer wall is covered by up to 1.5 km of pelagic sediments, distributed unevenly because of the rough basement topography (profile Q of Fig. 7.9) and

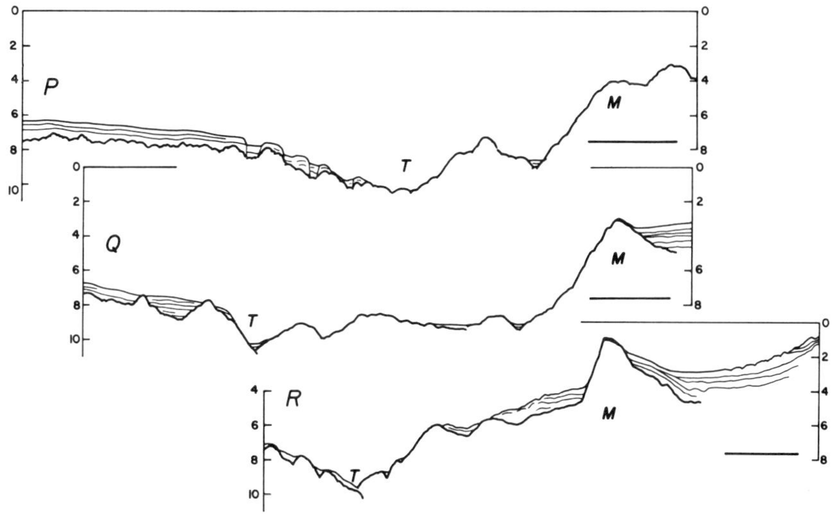

FIGURE 7.9. Single-channel seismic reflection profiles P, Q, and R (located in Fig. 7.7) crossing the northern and eastern (forearc) margins of the Sandwich plate. M is the mid-slope high in profiles Q and R, and the edge of normal backarc in profile P. T is the trench axis. Vertical scale is seconds of two-way time. Horizontal bars are 20 km. Dredge hauls from the outer (trench) slope of M on each profile are reported in the text.

influence of the ACC. Sediments are thicker in the north because of the greater basement age.

A mid-slope high (labeled M in Fig. 7.9, profiles Q and R) with a steep outer slope and gravity gradient separates the upper and lower forearcs of the inner trench wall. The trench is deepest, below 8000 m, in the northeast corner where also the trench is broadest, the upper forearc narrowest, the slope to the arc steepest, and the lower forearc least well developed (Figs. 7.4, 7.5, and 7.9). Southward, the trench shoals and narrows and both lower and upper forearcs become broader, more continuous, and more elevated. Although the outer slope of the mid-slope high forms a smooth curve along the forearc, its elevation is highly variable along strike.

On reflection profiles (for example, profiles Q and R of Fig. 7.9) the lower forearc is mainly acoustically opaque, as is common within accretionary prisms. Although the gravity effect of the prism has not been modeled, it is clear that, paradoxically, the prism is least well developed in the north, where the subducting ocean floor brings the most sediment to the trench.

The other contributor to trench (thence prism) sediments is the arc. Some reflection profiles (e.g., profile Q of Fig. 7.9) show arc-derived sediment ponded behind the mid-slope basement high, but on others (e.g., profile D of Fig. 7.8) there appears to be no barrier to direct downslope transport to the trench floor. The variable depth to the mid-slope basement high therefore ensures that arc sediment is channeled to certain sections of the trench. Seismic reflection profiles and the gravity map show a coincidence between the presence or absence of a prominent, shallow mid-slope basement high and the local accumulation of sediments on the trench floor. Fig. 7.5 shows that, where the mid-slope high is least well developed, as between 56.8 and 57.8°S, there is also a gentler slope to the gravity profile over the lower forearc and trench, indicating a better-developed accretionary prism. Similarly, where the mid-slope high is most prominent, as between 55.3 and 56.0°S (Fig. 7.9, profile Q), the accretionary prism is less well developed. These are second-order effects, however, superimposed on the general north-south contrast in accretionary prism development. The question of the origin of this contrast remains, and the possibility must be considered of tectonic erosion, concentrated in the north because of the ready subsidence of the older oceanic lithosphere (but also, possibly, in the far south associated with ridge crest subduction).

Tectonic erosion of a forearc is generally inferred when transport of loose material down the subduction zone regularly exceeds what is being supplied at the seabed in the trench. It has been associated with ridge crest subduction off Chile (Cande *et al.*, 1987) and the Antarctic Peninsula (Barker, 1982), in the Woodlark Basin (Taylor, 1987), and along the South Scotia Ridge (Barker *et al.*, 1984). The term is vague, however, in the sense that the processes involved may be active in all trenches, to some degree. Tectonic erosion can lead, as off Peru and northern Chile (Von Huene *et al.*, 1985), to a trench with a long subduction history having only a poorly developed accretionary prism and exposing relatively old rocks on the inner trench slope. The question arises, is erosion confined to the loose debris of the accretionary prism, or is material from the edge of the rigid overriding plate also being removed, and subducted? I return to this question in Section 4.7, after considering also the backarc spreading history, which bears on the nature and extent of forearc basement.

In the north along the east-west arm of the trench, the southern, inner trench slope is the northern margin of the East Scotia Sea backarc basin. Its profile (for example, Fig. 7.9, profile P) corresponds closely to a normal trench profile, despite its having undergone only eastward transcurrent motion with respect to the South American plate. An elongated ridge and trough occupy the lower part of the inner trench slope and resemble a poorly developed lower forearc accretionary prism in their topography and gravity signature. The main upper slope is very steep. Dredge hauls from this steep wall (described in Section 4.4) are mostly fresh lavas, showing that it is original, the product of backarc extensional volcanism rather than of any subsequent tectonic erosion.

4.2. Character of the Backarc

The East Scotia Sea backarc region is bounded in the north by the east-west prolongation of the South Sandwich Trench, and in the south by a complex boundary with the Antarctic plate, lying approximately along 60.5°S, and involving the elevated Herdman and Discovery banks (both of which in places lie above 750 m). In the east, the formal boundary is the island arc but, as has often been noted in this region and elsewhere, both arc and backarc extension are secondary features of subduction, so the boundary of "backarc" extension is not necessarily the arc. The nature of subarc and upper forearc basement is therefore at issue.

In the west, the boundary between the East Scotia Sea and Central Scotia Sea, current and previous backarc extensional provinces (e.g., Fig. 7.2A), is uncertain. There is a wide gap in published magnetic lineations of each province. Topographically (seen also in the gravity map, Fig. 7.5) there is a change at about 36°W from rough, elevated to smoother, deeper seafloor, more pronounced in the south than in the north. As a boundary to the modern backarc, however, this elevated area is more diffuse and irregular than the remnant arc province of most backarc regions.

The dominant feature of the backarc is its spreading system, centered approximately along 30°W, and here defined very precisely by its lineated oceanic magnetic anomalies. Fig. 7.6 shows the available magnetic profiles and identified anomalies. Almost all of the identifications previously made along older profiles by Barker and Hill (1981) over most of the backarc region are confirmed. The newer data have filled gaps and in places have allowed additional identifications to be made on existing profiles. In particular, some older anomalies can be identified, and the history of the complicated southern end of the spreading center is now perhaps clearer (Section 4.6). The very northern end, where the spreading center meets the trench at a presumed RFF triple junction, remains poorly known.

4.3. Magnetic Anomaly Identifications, Spreading Rates, Asymmetry

In the northern and central sections of the backarc (north of 59.5°S; Fig. 7.6), well-formed magnetic anomalies are seen out to at least anomaly 5 (ca. 11 Ma) on the western flank, but in the east the magnetic signature of the island arc overprints most profiles. Farther south, anomalies can be identified only out to 2 or 2A, within a narrower band of rough topography flanked in the west by the flat-topped Herdman Bank and in the east by the volcanic arc. Spreading today is essentially east-west, in a simple two-plate system, but the slight discordance between the older anomalies on each flank suggests the possibility of significant earlier complexity. The positions of transform faults and fracture zones in Fig. 7.6 have been determined mainly from bathymetric data (Barker and Hill, 1981; Fig. 7.4) and from the gravity map (Fig. 7.5, which shows the median valley well) and partly by the need to accommodate offsets in the magnetic pattern. Farther from the ridge crest, particularly on the eastern flank, fracture zone locations and orientations are not well constrained. Many transforms appear to have been short-lived.

Figure 7.10 shows a selection of long magnetic and bathymetric profiles across the backarc, arc, and forearc together with distance plots reduced to 25 mm/year (Barker, 1979; Barker and Hill, 1981) and a synthetic profile, generated using the MRTS of Cande and Kent (1992). The reduced distance method is more sensitive to small variations in spreading rate and was used by Barker and Hill (1981) (with the MRTS of LaBrecque *et al.*, 1977 and on profile A only) to identify anomalies out to 4A and to propose acceleration and a persistent asymmetry. The better coverage of long profiles now available prompts a revised and extended interpretation.

Spreading since 1.7 Ma or so has been fast, averaging about 65 mm/year over the full length of the backarc. Between about 4 and 1.7 Ma it was slower, approximately 50 mm/year, and before 5–7 Ma it was slower still, less than 30 mm/year. On most bathymetric profiles crossing the spreading center, ocean floor produced during the past 3–4 m.y. is smoother than that produced earlier at lower spreading rates. Figure 7.8 (from Barker and Hill, 1981) shows several east-west bathymetric profiles across the backarc, arc, and forearc that illustrate this feature. The profiles are aligned along western anomaly 2A, and the positions of other anomalies and the median valley are marked.

There is a general asymmetry of spreading, favoring accretion to the arc flank, but not as uniform or as pronounced as previously thought (Barker and Hill, 1980, 1981). It is not seen in ocean floor produced before 4 Ma and is barely detectable until 1.7 Ma. Thereafter, at the fastest spreading rates, it is more obvious, but even so is not ubiquitous. It appears to occur within distinct spreading segments bounded by transform faults (Fig. 7.6), but the

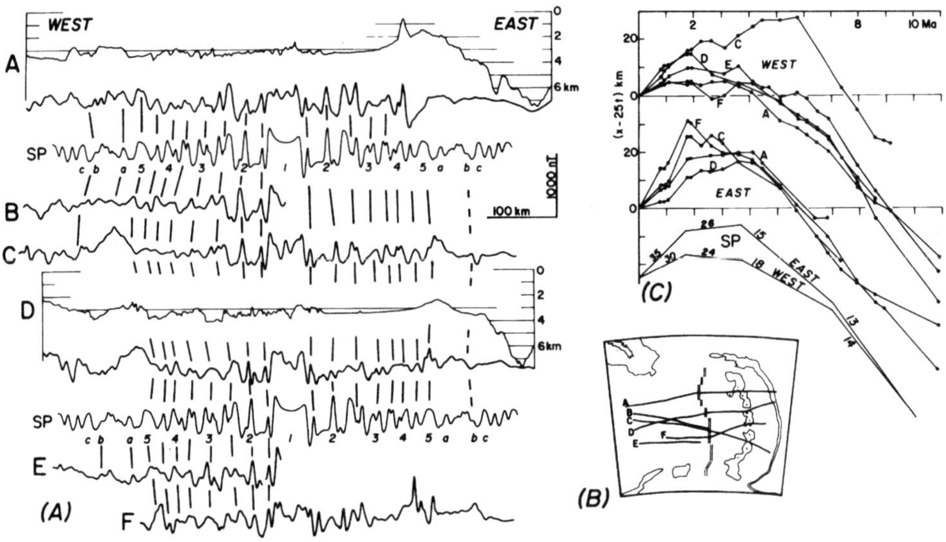

FIGURE 7.10. (A) Magnetic profiles A to F (with bathymetry for A and D) crossing the East Scotia Sea, compared with a synthetic anomaly profile (SP) generated using the MRTS of Cande and Kent (1992). Located in (B). Also shown (C) is a reduced distance plot (Barker, 1979; Barker and Hill, 1981) of identified anomalies from the lettered profiles, and the set of spreading rates (SP) used to generate the synthetic anomaly profile, all reduced to 25 mm/year.

asymmetry in one segment is in the opposite sense. In the slower-spreading flank areas in such cases, there is no initial anomaly offset; the transforms appear to have been created at the time of the acceleration in order to accommodate asymmetric accretion.

Barker and Hill (1980) pointed to the correlation between sense of asymmetry and "absolute" migration of the ridge crest in some areas of the main ocean basins and in backarc regions as supporting a thermal origin for asymmetric spreading. In this explanation, isotherms at the base of the lithosphere become asymmetrically dipping as a result of the migration, and accretion favors the cooler, leading flank. This correlation is largely supported by the new results, since the East Scotia Sea ridge crest will only migrate now when spreading is faster than about 30 mm/year (from Section 2.3). However, it is clear that ridge crest migration is not the whole story. An additional enabling mechanism for asymmetric accretion might be involved. Although more detailed survey and direct sampling would be required to examine this fully, there may be a causal relationship between spreading rate, asymmetry, and ridge segmentation.

4.4. The Ridge Crest and Spreading Center Chemistry

The topographic expression of the ridge crest varies along strike. In Fig. 7.5, gravity lows associated with a median valley can be recognized along most of the central 70% of ridge crest length. Similarly, the median valley is seen on most bathymetric crossings (for example, Fig. 7.8). The lengths of ridge crest where either bathymetric or gravity expression of the median valley can be detected are dotted in Fig. 7.6. Where seen, the median valley reaches 1200 m in depth and 20 km in width but more typically is 500–800 m deep and 6 to 10 km wide. Gravity anomalies associated with these typical dimensions are 20–30 mgal. In the northernmost ridge segment, there are no magnetic data to constrain the ridge crest position, and more than one elongated gravity low. In the far south, crossings show neither gravity nor topographic expression (except that a central ridge may occur at the southern end; Section 4.6), and the average elevation at the ridge crest is shoaler than farther north.

Four dredge hauls from close to the ridge crest (DR 20, 22, 23, 24 in Fig. 7.7, all younger than 0.7 Ma in age from their positions) were reported by Saunders and Tarney (1979). The lavas are all basalts and basaltic andesites. Major and trace element chemistry, Nd and Sr, Pb and C isotopes (Hawkesworth *et al.*, 1977; Cohen and O'Nions, 1982; Mattey *et al.*, 1984 respectively) and volatiles (Muenow *et al.*, 1980) all show features intermediate between MORB and arc tholeiites and point to an N-type MORB source variably contaminated by a slab-derived component and to varying degrees of partial melting (Saunders and Tarney, 1991). The slab-derived component is similar to that responsible for island-arc chemical compositions (generally considered to be an enriched aqueous fluid produced by low-temperature dehydration of the subducting slab; Saunders and Tarney, 1991) but is less abundant. These Scotia Sea rocks show no signs of additional complexity beyond a light carbon-isotope ratio that might reflect a sedimentary component (Mattey *et al.*, 1984).

The apparent simplicity of East Scotia Sea magma genesis is underlined by a comparative study of high field strength and transition elements in arc-backarc systems by Woodhead *et al.* (1993). The analyzed rocks of the East Scotia Sea show only minor signs of depletion of the MORB source by melt extraction before the production of the backarc melt: for the South Sandwich island arc, prior depletion is estimated at 2% (Pearce *et al.*, in press).

There is a wide range of compositional variation between the four dredge sites, with DR 24 being the most clearly subduction related, DR 22 rather less so, and the other two only slightly. Sites of DR 24 and DR 22 are the central locations, and DR 23 lies within the more recently rifted southern province (Section 4.6), but it would be unwise to read any geometric significance into data from only four sites. The transect of sites dredged up the northern wall of the backarc (DR 56, 57, 60, 61 in Figures 7.7 and 7.9) underlines this point. The exact ages of these rocks are unknown, for lack of magnetic data from the most northerly spreading segment, but they are probably less than 2 Ma. The rocks are fresh basalts and basaltic andesites, and preliminary XRF major and trace element geochemical data (P. L. Barber, pers. comm., 1992) point to a strong similarity to DR 24 with, in particular, low Zr, Ti, P, and Nb and high (with respect to N-MORB) K, Rb, and Ba. The scale of heterogeneity within the backarc cannot be established without a much more detailed sampling grid.

4.5. Western Flank: Older Anomalies

The age of the older part of the western flank of the East Scotia Sea spreading center bears on several other matters—the nature of the forearc, possible tectonic erosion in the forearc, interaction with Central Scotia Sea spreading, reconstructions—that are discussed in subsequent sections. There are no published magnetic anomaly identifications in the area between 34° and 38°W, and East Scotia Sea opening was only documented back to 8 Ma.

Previous anomaly identifications out to 4A on the western flank (Barker and Hill, 1981) still hold, and it has been possible to identify western anomaly 5 on many profiles to the north of about 59.5°S. On some of the profiles in Fig. 7.10 (profiles A and E in particular), magnetic anomalies can be seen that are a reasonable fit to a synthetic magnetic profile produced by extrapolating back in time using the same half spreading rate (14 mm/year) out to anomaly 5B (about 15 Ma). One strong lineation is seen even farther west in Fig. 7.6. As noted, the bathymetry in this area shows no clear sign of a remnant arc ridge with a synchronous or consistent onset of spreading. There are topographic elevations (Figs. 7.4 and 7.8) generally shoaler in the south, but they appear disordered, and all lie eastward of the 36°W boundary already noted (Section 4.2). On the gravity map (Fig. 7.5), this topographic roughness shows at first sight a strong linearity, trending about 300°. The linearity could be associated with fracture zones, since magnetic anomaly sequences separated by one or more lineations are offset, which would imply an early direction of opening along 120°–300°. However, the magnetic anomaly orientations, though variable, retain a mean north-south orientation and, on closer inspection, the gross lineation of the gravity anomalies is not continuous. There may of course be additional complications that could generate the kind of bathymetric complexity observed. According to Hill and Barker (1980), the Central Scotia Sea did not stop opening until about 6 Ma (7 Ma by the MRTS of Cande and Kent, 1992), after East Scotia Sea opening had started: The boundary between these two synchronous, orthogonal spreading systems could itself have been chaotic. Also, the synthetic magnetic profile shows that short sections of the time scale are not distinctive at slow spreading rates. However, these older anomaly identifications out to C5b may well be correct, and the rough, elevated topography (Fig. 7.8) may reflect the slower spreading of that time coupled with a more "normal" ridge crest elevation than the present, anomalously deep average value.

4.6. Southern Province: Ridge Crest Collision and Rifting

South of 59.5°S lies the narrower area of rough, ocean-depth topography between Herdman Bank and the arc. South again lies the zone of the youngest collision between an ancestral South Sandwich Trench and a segment of the South American–Antarctic Ridge. In the area between Herdman Bank and the arc, magnetic anomalies are similar in amplitude and wavelength to those to the north, and the evidence of dredge DR 23 (Saunders and Tarney, 1979; Fig. 7.7) and earthquake epicenters (Forsyth, 1975; Pelayo and Wiens, 1989) confirms it as an area of active backarc extension. However, Barker and Hill (1981) failed to identify any magnetic lineations there, a measure of its greater complexity and the sparse data base of the time. Hamilton (1989) has since investigated the youngest collision zone in detail, using magnetic, bathymetric and gravity data, a short length of seismic profile and several dredge hauls intended to identify the tectonic origin of the major features from their chemical composition. Hamilton also tentatively identified magnetic anomalies in the narrow, southern backarc area. Subsequently, a small GLORIA survey of the collision zone was undertaken during *Charles Darwin* Cruise 37 in 1989 that also included the southern margin of the backarc.

The collision zone is extremely complicated and not yet fully understood. It is well beyond the scope of this chapter to present in full the evidence that supports our current, imperfect understanding of its structure and evolution. However, it cannot be neglected: ridge crest collision has interacted strongly with the developing backarc of the East Scotia Sea and (as argued in Section 2.5) will do so again in the future. I therefore attempt in this section to present the essentials: (a) a model of spreading history in the southernmost backarc; (b) the relationship between backarc extension and ridge crest collision.

4.6.1. Backarc Spreading History

Figure 7.1 shows the likely disposition of active plate boundaries at the southern end of the backarc basin. The backarc spreading center ends at an RFF triple junction: East from there the dextral transcurrent SAN-ANT boundary is offset southward at a short ridge crest section, on its way to another, TFF triple junction at the southern end of the South Sandwich Trench. To the west, the sinistral transcurrent ANT-SCO boundary is also offset south at a small, slowly spreading ridge section in the basin south of Discovery Bank. These features appear also in the inset location diagram for profiles G, H, and J in Fig. 7.11. Also shown in Fig. 7.11 (inset) are Herdman and Discovery banks, the deep trough (dashed line) striking about 070° that divides them, and Hamilton's (1989) "Area B", an elevated (at 1000–2500 m), largely non-magnetic area with a likely collision suture along its southeast margin. South of the suture lie long ridges striking slightly south of east that are old short-offset fracture zones on the Antarctic plate, produced by SAM-ANT separation shortly before collision (Tectonic Map, 1985; Barker and Lawver, 1988; Hamilton, 1989; Livermore and Woollett, 1993; Figs. 7.4 and 7.5). These are the several elements essential to an understanding of tectonic evolution.

Profiles G and H in Fig. 7.11 show a central 160–200-km-wide zone of strongly magnetized rough topography at 2–3-km depth, typical of young ocean floor, flanked by Herdman Bank in the west and part of the volcanic arc and forearc in the east; elevations at 1-km depth or less. These elevations are nonmagnetic, except for the young volcano of

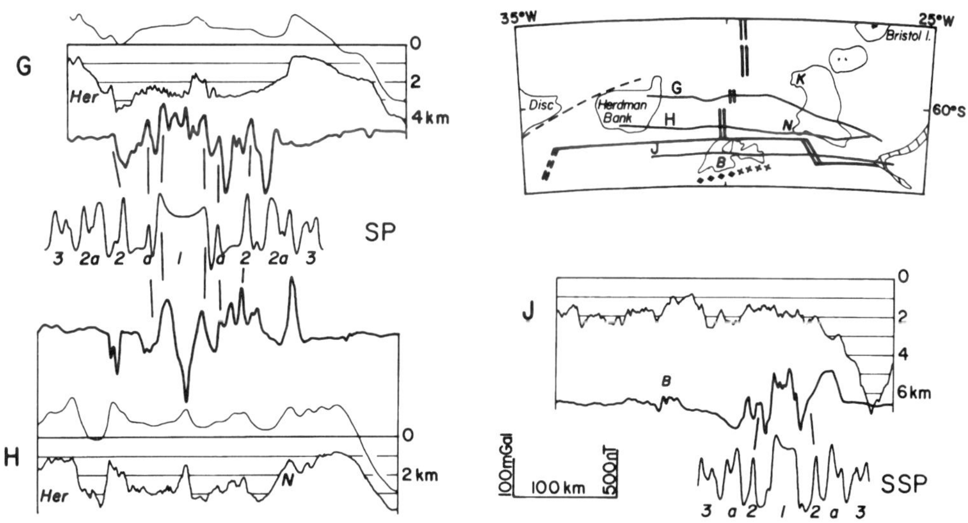

FIGURE 7.11. Magnetic, bathymetric, and shipboard gravity profiles along three tracks crossing the southern part of the East Scotia Sea. The synthetic magnetic anomaly profile (SP) compared with profiles G and H is that used in Fig. 7.10, but that (SSP) compared with profile J has a uniform half spreading rate of 16 mm/year. Her is Herdman Bank, Disc is Discovery Bank, N and K are Nelson and Kemp seamounts; B is Hamilton's (1989) "Area B," an elevated forearc remnant. The gravity anomaly is the thin line in profiles G and H; zeroes of bathymetry and gravity are common, and 2-km water depth is -100 mgal. The inset location map also shows the ridge crest and active transforms. The dashed line off Discovery and Herdman banks is a sediment-free trough imaged in Fig. 7.5 and considered to have been the line of relative eastward motion of Herdman Bank before 3.2 Ma. Diamonds and crosses south of Area B mark the sutures of two recent ridge crest collisions.

Nelson Seamount (N on profile H). At the edges of the oceanic province are deeps (buried by sediment at the east end of profile G but revealed by the gravity low). The magnetic anomalies on profile G match the central part of the synthetic magnetic profile (SP) well, with spreading rates over the past 1–2 m.y. identical to those farther north. This shows that Herdman Bank was part of the Scotia plate over that time. If that was the case over the full period of opening of the basin, then opening would have started 3–3.5 m.y. ago. Kemp Seamount appears to lie on ocean floor no more than 3 m.y. old. The outer parts of profile G are not as good a match to the synthetic anomalies as the central part: Hamilton (1989) explained the offset of the central, Brunhes-age anomaly from the basin center by a small ridge jump before 2 Ma. Profile H is a much poorer match to SP and is included to show the possibility of greater complexity. Hamilton's (1989) correlations are shown in Fig. 7.11 for profile H: the central anomaly includes an unusually large axial magnetic low and a central topographic high, which are not unknown at ridge crests but are unusual here. Alternatively, one or more ridge jumps may also have affected this section.

Profile J of Fig. 7.11 is from a different, more southerly spreading province, offset eastward and nominally representing SAN-ANT motion. The rough topography is more elevated, and there are no flanking topographic highs. To the east lies the South Sandwich trench and to the west Hamilton's "Area B," which he concludes is a fragment of precollision forearc. Both the magnetic province and the well-formed central anomaly are narrower than on profile G. They match a synthetic anomaly (SSP) generated using a full spreading rate of 32 mm/year. At that rate, the magnetic region would also be generated in 3–3.5 m.y. However, the rate is only 60% of estimated SAN-ANT motion, about 20 mm/year too slow.

There are many possible explanations for this discrepancy. Some call into question the correctness of the path of the present Scotia–ANT and Sandwich–ANT plate boundaries in Figs. 7.1 and 7.11, and the present state of knowledge leaves open a range of alternatives. However, the simplest and perhaps most plausible is active stretching of Area B and its western flank. The six dredges from Area B considered by Hamilton (1989) yielded a wide range of lithologies which, if dredged *in situ*, would signify a continental origin. Additionally, however, there were indications of young igneous activity, mostly acidic. Moreover, Area B is nonmagnetic except for a small area (marked "B" on profile J) directly south of the present ridge crest (the central ridge on profile H). It is possible that Area B and its western flank are being pervasively stretched, with thinning and remelting of continental crust, and are thus accommodating part of the SAN-ANT motion.

4.6.2. Ridge Crest–Trench Collision

The relationship between rifting in the southern backarc and ridge crest-trench collision directly to the south is important. That rifting was caused by collision is a part of the general hypothesis concerning the effects of successive collisions on backarc extension in the Scotia Sea over the past 30 Ma or more. It would have been useful to examine the time relationships of collision and rifting here, to test that hypothesis. Unfortunately the evidence is incomplete, so I offer what is essentially a restatement of the hypothesis as it applies in this particular case.

Hamilton (1989) identified two narrow ridge crest–trench collision zones in the south of this area. Their ages could not be determined independently of data from the backarc, so they were assumed to be virtually contemporaneous and to coincide with the time of rifting of Herdman Bank and Area B from the forearc. The magnetic data discussed in Section 4.6 suggest this happened about 3.2 Ma. The extent of these two collision zones is marked (with respect to Area B) by crosses and diamonds in Fig. 7.11. Strictly, Hamilton considers that the ridge crest would have arrived at the trench earlier and that a short period of accelerated tectonic erosion of the lower forearc accretionary prism would have preceded collision with the rigid Sandwich plate: Area B and Herdman Bank are considered to have been parts of the upper forearc.

It is assumed that prior to 3.2 Ma, as part of the Sandwich plate, Herdman Bank and Area B were moving eastward away from Discovery Bank along the line of the deep trough shown dashed in Fig. 7.11 (inset). A reconstruction to about 3.2 Ma (within the older part of anomaly 2A) of the southern province of the backarc is shown in Fig. 7.12A. It incorporates a restored Herdman Bank and Area B, west of the present forearc. The reconstruction is referred to a presumed rigid Scotia plate.

The hypothesis that rifting followed ridge crest collision is attractive. Ridge crest subduction could have provided the heat source needed for pervasive extension of Area B and might have instigated the rifting to adjust the line of the Scotia–ANT and Sandwich–ANT plate boundaries. Prior to collision, the southern backarc lay west of Herdman Bank and south of Discovery Bank. That region is now being dissected further by Scotia–ANT motion, as previously inferred (Tectonic Map, 1985; Pelayo and Wiens, 1989, Barker *et al.*, 1991). Discovery Bank has yielded Miocene arc volcanic rocks (12–20 Ma; Barker *et al.*, 1982). A forearc comprising the largely nonmagnetic and possibly continental Area B and Herdman Bank to the southeast would have been matched by a backarc volcaniclastic apron in the gently sloping sedimented area northwest of Discovery Bank (Fig. 7.4). Parts of both Herdman Bank and the congruent eastern forearc flank of the backarc basin are flat-topped,

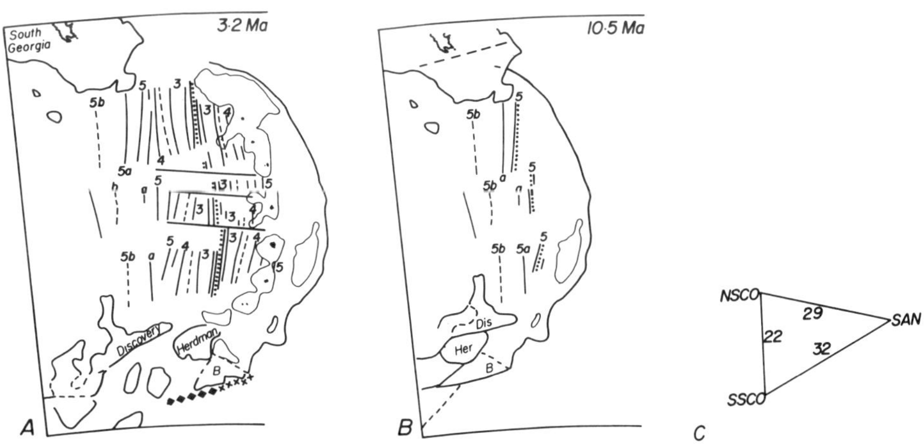

FIGURE 7.12. Revised reconstructions of East Scotia Sea evolution (A) to ca. 3.2 Ma (B) to 10.5 Ma. Anomalies 3A and 5B are dashed as in Fig. 7.6, and the ridge crest position is dotted where constrained by magnetic anomaly. (C) Vectors of SAN motion with respect to Central Scotia Sea opening (SSCO and NSCO) before 7 Ma, compatible with East Scotia Sea fabric and Herdman/Discovery Bank motion. In (A) the −60-mgal contour from the forearc in Fig. 7.5 defines the eastern edge of the Sandwich plate, and the +80-mgal contour defines the elevations within the arc and forearc. The 2-km contour from Fig. 7.4 defines the South Georgia block. Diamonds and crosses mark the sutures of recent ridge crest–trench collisions south of Area B.

which Hamilton (1989) attributed to planation by Weddell Sea icebergs. It seems likely that both bodies would have been elevated by the thermal effects of rifting.

Thus, in summary, the effects of the youngest collision(s) were to erode the lower forearc, to cause subduction to cease at the collision zone, to transfer older forearc blocks from the Sandwich to the Antarctic and Scotia plates, and to modify the locus of backarc extension. This is compatible with the general hypothesis of the effects of collision.

For the next collision zone to the west, Hamilton (1989) correlated magnetic anomalies with two alternative sections of the MRTS, implying collision ages of 3 Ma or 10 Ma, both using Barker and Lawver's (1988) rotations to constrain spreading rates. If this constraint is relaxed (by "pivoting subduction"; Section 2.5), other identifications are possible. The older age makes more sense of the backarc basin history: a 10 Ma collision opposite Discovery Bank could have triggered the dissection of its forearc, with eastward migration of Herdman Bank as proposed above continuing until the 3.2 Ma rifting event. This older, western collision zone lies west of the zone marked by diamonds in Fig. 7.12A.

Backarc extension has accelerated over the past few million years (Fig. 7.10). Is this too related to the collisions? Certainly, to the extent that rollback drives backarc extension, the cessation of extension in Drake Passage and the Central Scotia Sea (Barker and Burrell, 1977; Hill and Barker, 1980), about 6 m.y. ago (7 Ma using the Cande and Kent, 1992 MRTS), would have required faster extension within (most probably) the only backarc spreading regime that remained. The acceleration apparent in Fig. 7.10 began at about 7 Ma. In addition, however, several factors associated with the collisions may have contributed to the more recent (ca. 2 Ma) acceleration:

(a) The very young, buoyant ocean floor in the far south, being subducted with difficulty before collision, would remelt and become detached from the sinking SAM plate, allowing the main slab to sink more quickly.
(b) In the backarc, the long transform offset along the line S of Discovery Bank was eliminated, reducing drag.
(c) The rifting process required by collision was completed, and new ridge segments, providing ridge push, were put in place in the backarc.

4.7. Eastern Flank: Volcanic Arc, Forearc and Trench

On the eastern flank of the backarc basin, many members of the volcanic island arc lie on ocean floor produced during the past 10 m.y. The main islands of the central section of the arc may lie on ocean floor possibly as old as 11 Ma, but the northern and southern extremities lie on younger ocean floor. Kemp and Nelson seamounts, already mentioned, lie on or close to ocean floor no older than about 3 Ma, and the more westerly of the seamounts inferred from the gravity map (Fig. 7.5) lie on ocean floor as young as 6 to 8 Ma. The magnetic signature of the arc rocks overprints any seafloor spreading record, but the older anomalies may perhaps be seen along some passage tracks through the arc. Profiles C and D in Fig. 7.10 show a possible anomaly 5 on the eastern flank, and the sequence to anomaly 5B may even be preserved in the forearc (though this identification is highly speculative).

Previous studies (e.g., Barker and Hill, 1981) had suggested that before the start of east-west extension in the East Scotia Sea, a north-south backarc extensional province in the Central Scotia Sea, active between 21 and 7 Ma, may have reached as far east as the trench (Fig. 7.2B,C). It was thought that the northern half of the forearc province may have been produced by this extensional regime, whereas the southern half was already in existence and may even have been continental. This, rather than the lesser age of the subducting South American ocean floor, was thought to account for the greater elevation of the forearc in the south.

Although there are now more magnetic data from the forearc than were considered by Barker and Hill (1981) it remains impossible to map magnetic lineations there with any confidence. The identifications in Fig. 7.10 are tentative, although if the older anomalies on the western flank are correctly identified then the presence of corresponding lineations in the forearc is to be expected.

As important to an understanding of the forearc are the other data (single-channel seismic profiles, dredge hauls, and Geosat gravity) that are now available. The gravity signature of the forearc is dominated by the steep gradient, common in forearcs, that separates the upper forearc, underlain by rigid lithosphere of the Sandwich plate, from the deforming accretionary prism of the lower forearc resting directly on the subducting South American lithosphere. In the upper forearc, gravity varies considerably along strike. The simple concept of a north-south contrast needs revision: There is high gravity both in the far northeast and in the center south, but there are also low areas in the south as well as the center north. In discussing the trench (Section 4.1.2), I pointed to variations in the height of the mid-slope basement high as the cause of variation in the amount of sediment that could be retained on the upper forearc. These variations in basement elevation and sediment cover probably account for most of the gravity variation.

The outer slopes of both sections of the mid-slope high that show the major gravity highs have been dredged: the northeast corner east of Zavodovski Island and the large gravity high east of Montagu Island (Fig. 7.7, and profiles Q and R of Fig. 7.9). Two dredges from the northeast corner of the upper forearc (DR 52 and DR 53, 56.1°S, 26.2°W: 4100-3100 m depth, and profile Q) recovered *in situ* serpentinized peridotite (P. L. Barber, pers. comm., 1992). Two others from east of Montagu Island (DR 49 and DR 50, 58.6°S, 24.8°W: 2100–1100 m depth, and profile R) yielded calc-alkaline basalts, basaltic andesites and andesites, both fresh and altered. Three samples from DR 49 and DR 50 yielded K-Ar ages in the range 28.5 to 32.8 Ma (weighted average 31.0 Ma; Table I).

TABLE I
K/Ar Ages of Rocks Dredged from the South Sandwich Forearc (DR 49 and DR 50, Fig. 7.7), Determined by R. J. Pankhurst and I. Millar[a]

Sample	K (%)	^{40}Ar (nl/g)	Percent (atm)	Age (Ma)
DR 49/8	0.950	1.0810	32.1	29.1 ± 0.8
		1.0608	37.2	28.5 ± 0.9
DR 50/1	1.097	1.3604	55.8	31.7 ± 0.8
		1.3530	45.4	31.5 ± 1.0
DR 50/3	0.668	0.8586	64.8	32.8 ± 3.2
		0.8511	63.3	32.4 ± 1.5

[a]Decay constants as in Steiger and Jaeger (1977).

The origin of the peridotites from dredges DR 52 and 53 is uncertain. Among the possibilities are

(a) a rafted fragment of a late Jurassic backarc basin at the Pacific margin (seen also on South Georgia; Macdonald *et al.*, 1987); this is less likely than either
(b) serpentinite diapirs, seen also in the Marianas upper forearc and considered there to result from diapiric uplift of buoyant serpentinized fault blocks, and serpentinite mud volcanoes (e.g., Fryer, 1992), within an extensional forearc environment; or
(c) part of the mantle of the Sandwich plate, tectonically exposed within the trench inner wall. This would imply an additional tectonic unroofing mechanism, such as tectonic erosion at the plate margin (Section 4.1) or uplift and subaerial erosion of the overlying oceanic crust of the Sandwich plate at its northeast corner. It may be highly significant that the scarp from which these rocks were dredged lies almost exactly on the line of the steeply dipping zone of deep earthquake activity along 56°S, associated by Forsyth (1975) with tearing of the downwarped South American plate at depth, at the northern end of the subducting slab (Section 2.4 and Fig. 7.3).

The mid-Oligocene calc-alkaline basalts and andesites dredged from the elevation east of Montagu Island may be explained in terms of Scotia Sea evolution. The Scotia Sea has grown by backarc extension over the past 30–40 Ma, and has carried eastward fragments of an original continuous continental connection of South America and the Antarctic Peninsula (see, for example, Barker *et al.*, 1991). The elevation sampled by these dredges may have lain at that margin originally, having formed as a result of subduction of Pacific ocean floor. More probably, it was produced at an early stage of Scotia Arc development, before the arc products of east-directed subduction had acquired their present intraoceanic low-K tholeiitic character (Discovery Bank rocks, dated at 12–20 Ma, *are* low-K tholeiites: Barker *et al.*, 1982). It is plausible that successive episodes of backarc extension subsequently could have allowed a fragment of that original arc to survive within the east-migrating forearc.

Identification of the two main elevated areas within the upper forearc as (a) tectonically exposed mantle and (b) a fragment of a much older arc is compatible with, but does not prove, a simple ocean-floor origin for the remaining, less-elevated parts. Strong support for this interpretation comes from the older magnetic anomalies identified on the *western* flank of the backarc basin and from the coincidence within the forearc of sub-sediment basement depths and free-air gravity anomalies similar to those of "normal" ocean floor of the same likely age.

4.8. Reconstructions

Figure 7.12 shows reconstructions of East Scotia Sea evolution at (a) 3.2 Ma and (b) 10.5 Ma. These times were chosen (a) to demonstrate the effects of the two most recent ridge crest collisions and (b) to compare the results of this reassessment with the reconstructions of Fig. 7.2B. The eastern boundary of the Sandwich plate in Fig. 7.12 is taken from the free-air gravity map (Fig. 7.5) at the −60-mgal contour, within the steep gravity gradient that separates the upper and lower forearc. Choice of a point on that gradient is unimportant in most areas, provided it is unlikely to reflect "normal" oceanic crustal structure. The island-arc and anomalous forearc elevations are defined by the +80-mgal contour for the same reason. For other bodies, not within the Sandwich plate, the level chosen varies to take into account their assumed structure and present surroundings. For the boundary of the South Georgia microcontinent, the 2000-m bathymetric contour is used.

The 3.2 Ma reconstruction (Fig. 7.12A) closes the young southern basin of the backarc, under assumptions discussed in Section 4.6. Anomalies 2A on either flank fit precisely under a rigid two-plate assumption, and Hamilton's (1989) elevated Area B and Herdman Bank become more southwesterly components of the forearc. Although this reconstruction is not placed in major plate coordinates (the Scotia plate remains fixed), it is assumed that two ridge crest collisions have taken place in the past 3.2 Ma, so that the SAM–ANT boundary met the forearc then farther to the southwest than at present, and that Scotia–ANT motion had been about 12 mm/year (from Pelayo and Wiens, 1989) throughout. The crosses and diamonds mark the extent of these young collisions.

Little is known of the northern limit of the backarc, except that dredges DR 56–61 suggest that most of the present northern margin is original. It seems most likely (by comparison with the corresponding western ridge flank in Fig. 7.12A) that the presently elevated northeastern corner of the Sandwich plate was formed by simple backarc extension and has been subsequently elevated, eroded, and intruded by subduction-related magma.

A reconstruction to 10.5 Ma is not so straightforward to achieve: anomalies older than 3A on either flank do not quite coincide under a similar simple two-plate assumption. The reconstruction in Fig. 7.12b is therefore speculative and involves additional complexity and less constraint. Hill and Barker (1980) concluded that north-south spreading in the Central Scotia Sea did not cease until about 6 Ma (7 Ma using the Cande and Kent, 1992 time scale). Those Central Scotia Sea age determinations should be re-assessed in light of additional data now available, but if the original ages are accepted then the western flank of the East Scotia Sea backarc spreading center could have been two separating plates until 7 Ma. As mentioned, the ocean floor fabric on the western flank of the East Scotia Sea spreading center is confused: The apparent 120°–300° gross orientation visible in the Geosat gravity map (Fig. 7.5) is discontinuous in detail, and it is difficult to see where a triple junction, implied by the coexistence of two orthogonal spreading centers before 7 Ma, could have lain. The main fracture zones in the East Scotia Sea (Fig. 7.6) are oblique to the prominent lineation south of Discovery Bank, which is presumed to show the direction of Sandwich plate motion relative to a "southern Scotia plate" (SSCO) before the 3.2 Ma rifting. Fig. 7.12C shows this obliquity to be roughly compatible with a three-plate system with the observed spreading rates. One possible scheme for the period before 7 Ma would therefore locate the triple junction in the south, somewhere close to Discovery Bank.

In the north, a 10.5 Ma reconstruction brings the forearc to the eastern end of South Georgia. Although the South Georgia block seems likely to have occupied a similar forearc or arc position through the previous 20–30 m.y. of Scotia Sea evolution (see Fig. 7.2), there is no sign of Cenozoic arc magmatism on the island or continental shelf (Simpson and Griffiths, 1982; Macdonald *et al.*, 1987). A likely sinistral strike-slip fault with about 70 km offset, crossing the continental shelf of the South Georgia block, is picked out by the gravity anomaly (Fig. 7.5) and shown dashed in Fig. 7.12B. It is not restored in the 10.5 Ma reconstruction, since there are no constraints upon its age.

This reconstruction is essentially compatible with the 10 Ma reconstruction shown in Fig. 7.2B. The backarc nature of the extensional environment is unchanged, and the history of opening of Drake Passage farther west is common to both. The balance of ocean-floor creation by the central and eastern Scotia Sea spreading regimes is different, because the latter started earlier than previously realized. In combination, however, they had the same effect. No attempt is made in Fig. 7.12B to place the reconstruction in a context of slow east-west major plate motion, but it seems plausible that the north-south opening of the Central Scotia Sea would have involved *both* a northward component of South Georgia motion with respect to South America (until collision with the Northeast Georgia Rise, perhaps) *and* a southward component of motion of Discovery Bank (leading to collision with sections of the SAM-ANT plate boundary). These are also assumptions of earlier models but cannot be tested until the detailed history of ridge crest collision along the South Scotia Ridge is known.

In the 10.5 Ma reconstruction, islands of the present arc are not shown. None appears likely to have been erupted on ocean floor older than 11 Ma. The Discovery arc to the south (now exposed in Discovery Bank) would have been active before 10 Ma, and one unsampled elevation in the present southern forearc may also have been an older arc volcano. Other possible components of a pre–10 Ma island arc are two seamounts lying to the south of South Georgia (Figs. 7.4 and 7.5, 7.12). Less speculatively, if the identification of anomaly 5B south of South Georgia is correct in the reconstruction in Fig. 7.12B, then some subsequent tectonic erosion of the northeast corner of the Sandwich plate is implied, since the congruent ocean floor of anomaly 5B age on the Sandwich plate is missing.

The combination of implied tectonic erosion (above) with abnormal elevation and exposure of serpentinized peridotite at shallow depth, at the present northeast corner of the Sandwich plate above the northern end of the subducting slab, is quite compelling. Moreover, it raises an additional question. The exact line of the tear of the subducting South American oceanic plate (Fig. 7.3) lies *beneath* the newly formed Sandwich plate, so the downwarped South American oceanic lithosphere to the north of the tear cannot return to its original surface position; it is held down by the backarc lithosphere. The bathymetric profile across the northern, east-west arm of the South Sandwich trench (profile P, Fig. 7.9) closely resembles a subducting trench profile, although relative motion across it has been sinistral strike slip. How close are these circumstances to those of the initiation of subduction? If ridge crest collisions have at several times in the past caused subduction at a southern part of the subduction zone to stop, has a similar asymmetry at the northern end of the trench ever developed into subduction so that subduction and complementary backarc extension might continue?

5. CONCLUSIONS

This has been an opportunity to reexamine and extend a suite of conclusions concerning the evolution of the East Scotia Sea, a site of active backarc extension, in the light of a larger and more diverse data set than was previously available.

(a) The East Scotia Sea floor was generated at a spreading center now lying approximately along 30°W longitude. Well-formed magnetic anomalies may be identified back to anomaly 5, and probably to anomaly 5B (about 15 Ma) on the western flank. To the east, the island arc lies on ocean floor aged from about 10 Ma to as young as 3 Ma, all produced during the present spreading episode. The time extent of East Scotia Sea spreading and the area of ocean floor produced are greater than previously thought.

(b) Spreading was largely symmetric at about 27 mm/year full rate, from 15 Ma to about 5–7 Ma, when it started to accelerate. From about 4 Ma to 1.7 Ma spreading was at 50

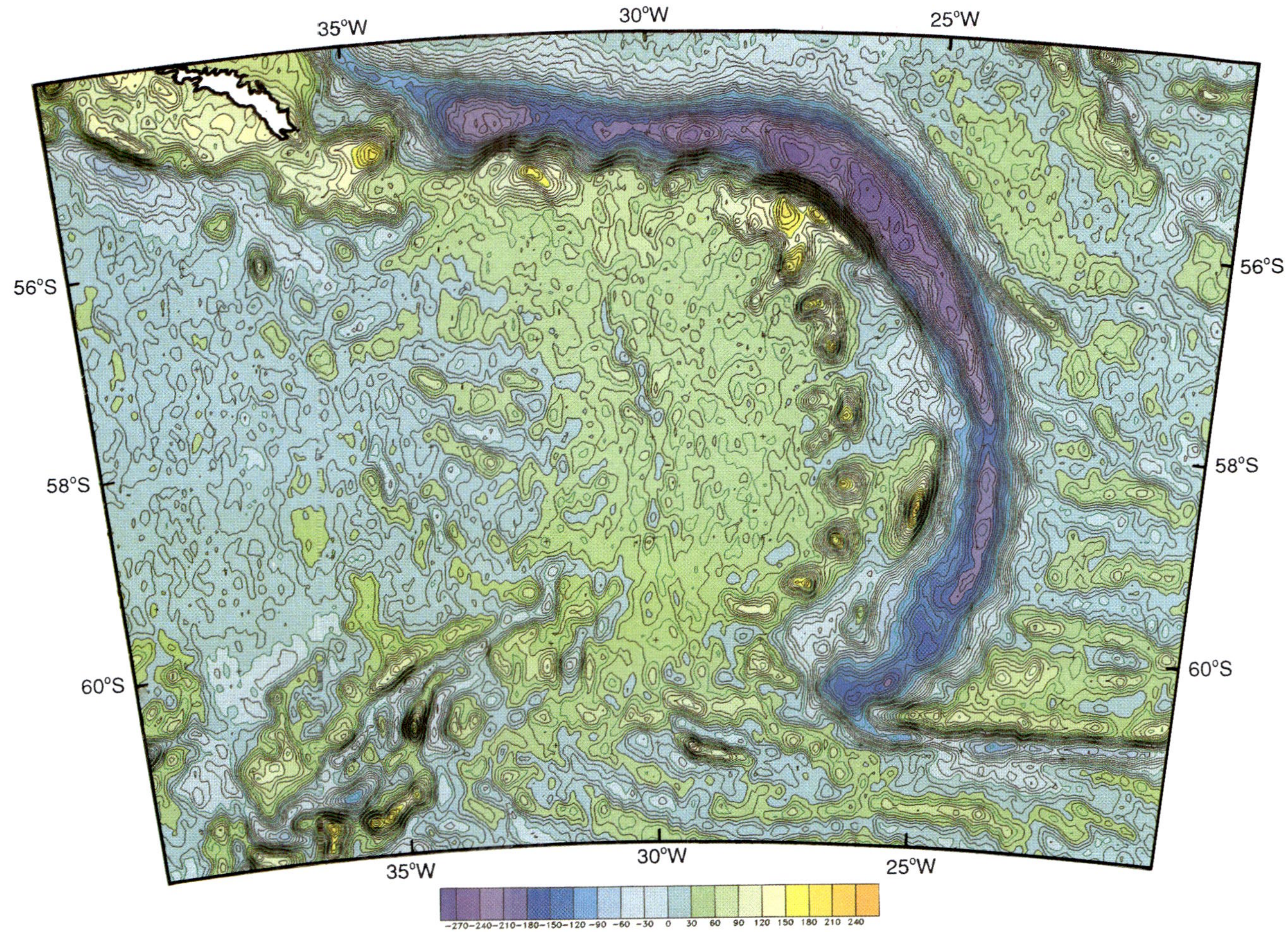

FIGURE 7.5. See p. 291. for full caption.

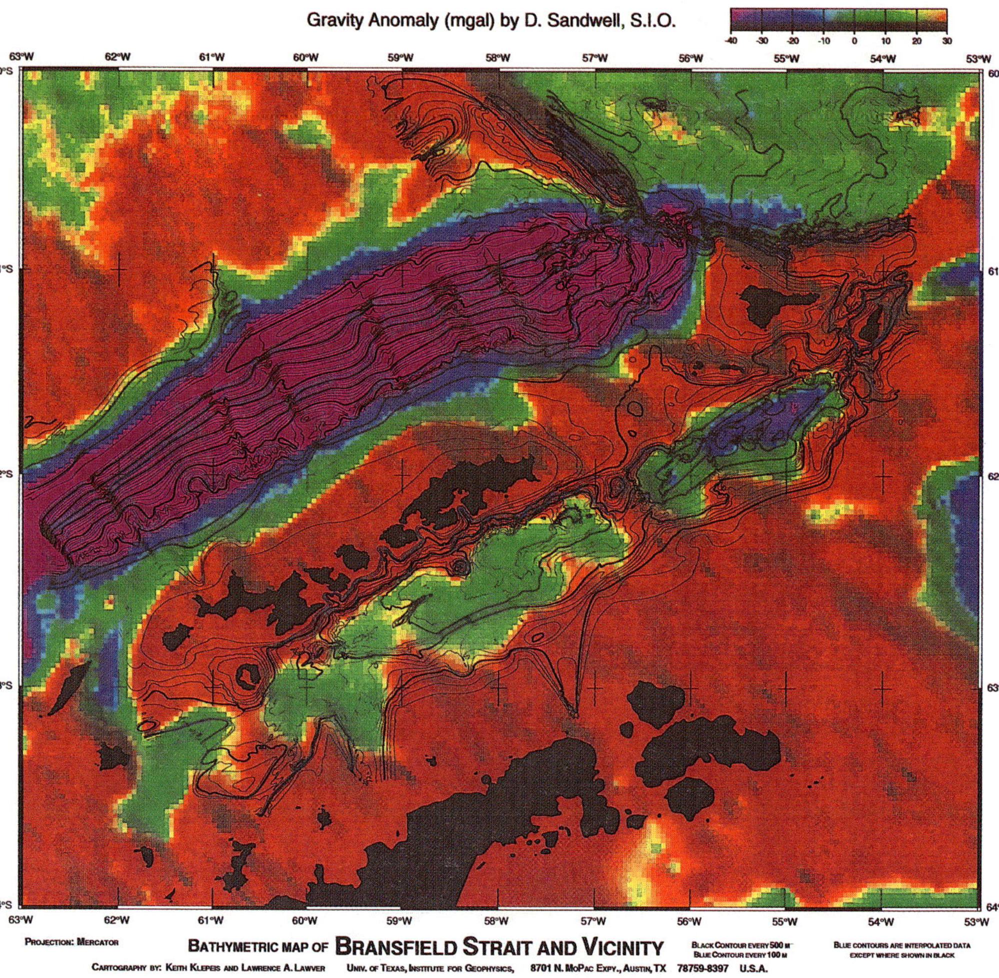

FIGURE 8.5. Geostat/ERS-1 gravity data from Sandwell and Smith (1992) with digital bathymetric data of Klepeis and Lawver (1994) superimposed. The gravity data color bar indicates gravity values from <-40 mgal to $>+30$ mgal. Bold black bathymetric contours are shown every 500 m, while thin black contours are observed bathymetry contoured every 100 m. Thin blue contours are interpolated bathymetric data contoured every 100 m.

mm/year and slightly asymmetric, favoring the eastern, arc flank. At 1.7 Ma a further acceleration to 65 mm/year occurred, with up to 15% asymmetry in many places, favoring the arc flank in most cases. In places, transform faults appeared about 1.7 Ma, separating segments having different degrees of asymmetric spreading. There may be a link between spreading rate, ridge crest migration, asymmetry, and ridge segmentation.

(c) The ridge crest depth exceeds 3000 m, more than 500 m deeper than the global average for mid-ocean ridges. A median valley is present along about 70% of the ridge crest length; typical dimensions are 6–10 km wide and 500-800 m deep, with a free-air gravity anomaly of 20–30 mgal. The ridge crest depth may have been closer to the global average before the accelerations in spreading.

(d) Sediment cover on the east flank is a thick arc-derived volcaniclastic apron. On the west flank, sediment is pelagic and hemipelagic, deposited under the influence of the Antarctic Circumpolar Current. The west flank becomes rougher westward, reflecting a slower spreading rate and perhaps interaction with contemporaneous north-south extension in the Central Scotia Sea. This rough zone ends at about 36°W.

(e) In the south, the backarc basin is much narrower. This zone was formed after Herdman Bank and another forearc fragment rifted from the Sandwich plate ca. 3.2 Ma, probably in response to collision of the trench with two sections of the South American–Antarctic Ridge to the west of any surviving part of the SAM-ANT plate boundary. Herdman Bank became a part of the Scotia plate following rifting, and the other fragment a part of the Antarctic plate. Before then, and following an earlier collision farther west (possibly 10 Ma), Herdman Bank was moving eastward from its original position as forearc to the Discovery arc, as part of the Sandwich plate. The precise timing of these collisions remains uncertain.

(f) The present SCO-ANT boundary runs south of Herdman Bank. To the east lies the RFF SAN-SCO-ANT triple junction, then a slow spreading (32 mm/year) SAN-ANT ridge segment, and a TFF SAN-SAM-ANT boundary. Some crustal thinning and intrusion of "Area B" may be required, to complete SAN-ANT motion.

(g) Ten previously unpublished dredge hauls are reported briefly: four from the forearc, four from the backarc, and two from previously unknown seamounts related to the island arc. The chemistry of all arc and backarc rocks appears to reflect production from N-MORB, variably contaminated by fluids enriched in alkali metals from the subducted slab. They are similar to rocks from other intraoceanic arc-backarc systems, but possibly simpler, having suffered only minor prior source depletion. However, chemical variation in the arc rocks is geographically heterogeneous and bears no visible relationship to known systematic north-south variation in age and sediment cover of the subducting slab. The sparse distribution of backarc samples permits no conclusion on the scale of spatial or temporal heterogeneity.

(h) The upper forearc is probably ocean floor formed on the east flank during the early stages of the present spreading episode, for the most part. An exception is a more elevated area east of Montagu Island, shown from dredge hauls to be a fragment of a ca. 31 Ma calc-alkaline volcanic arc, active during the early stages of Scotia Sea evolution.

(i) The other major forearc elevation, at the northern end, has yielded serpentinized ultrabasic rocks from its steep northeastern scarp. The magnetic anomaly pattern and the ultrabasic outcrop suggest an extensional forearc tectonic regime and likely tectonic erosion. This scarp lies directly beneath the east-west line of tearing of the subducted slab, defined by earthquake fault-plane solutions. Since the northern edge of the torn South

American lithosphere underlies newly created backarc, it cannot return to its original position: the asymmetry may create the conditions necessary for incipient subduction.

(j) The lower forearc accretionary prism is poorly developed. Variations in the elevation of the lip to the upper forearc seem to control the transport of arc volcaniclastics to the trench; basement elevation and sediment thickness dictate free-air gravity levels.

Acknowledgments

I am grateful to Paul Barber and Ian Hamilton for permission to summarize unpublished petrographic and geochemical data from 10 dredge hauls from the East Scotia Sea and South Sandwich forearc. Bob Pankhurst and Ian Millar produced K-Ar ages for rocks from two of those dredge hauls. Julian Pearce and Peter Baker kindly provided an early draft of their paper describing South Sandwich arc volcanic rocks.

REFERENCES

Allen, A. 1966. Seismic refraction investigations in the Scotia Sea, *Br. Antarct. Surv. Sci. Rep.* **55:** 44 pp.

Alvarez, W. 1982. Geological evidence for the geographical pattern of mantle return flow and the driving mechanism of plate tectonics, *J. Geophys. Res.* **87:**6697–6710.

Atwater, T. 1989. Plate tectonic history of the northeast Pacific and western North America, in *The Eastern Pacific Ocean and Hawaii* (E. L. Winterer, D. M. Hussong, and R. W. Decker, eds.), pp. 21–72, Geol. Soc. America, Boulder.

Baker, P. E. 1968. Comparative volcanology and petrology of the Atlantic island arcs, *Bull. Volcanol.* **32:**189–206.

Baker, P. E. 1978. The South Sandwich Islands: iii. petrology of the volcanic rocks, *Br. Antarct. Surv. Sci. Rept.* **93:** 34 pp.

Baker, P. E. 1990. E. South Sandwich Islands, in *Volcanoes of the Antarctic Plate and Southern Oceans* (W. E. LeMasurier and J. W. Thomson, eds.), Vol. 48, pp. 361–395, Antarctic Res. Ser., American Geophysical Union, Washington, DC.

Barker, P. F. 1970. Plate tectonics of the Scotia Sea region, *Nature* **228:**1293–1296.

Barker, P. F. 1972. A spreading centre in the East Scotia Sea, *Earth Planet. Sci. Lett.* **15:**123–132.

Barker, P. F. 1979. The history of ridge-crest offset at the Falkland–Agulhas fracture zone from a small-circle geophysical profile, *Geophys. J. R. Astron. Soc.* **59:**131–145.

Barker, P. F. 1982. The Cenozoic subduction history of the Pacific margin of the Antarctic Peninsula: ridge crest-trench interactions, *J. Geol. Soc. London* **139:**787–801.

Barker, P. F., Barber, P. L., and King, E. C. 1984. An early Miocene ridge crest-trench collision on the South Scotia Ridge near 36W, *Tectonophysics* **102:**315–332.

Barker, P. F., and Burrell, J. 1977. The opening of Drake Passage, *Mar. Geol.* **25:**15–34.

Barker, P. F., Dalziel, I. W. D., and Storey, B. C. 1991. Tectonic development of the Scotia Arc region, in *Geology of Antarctica* (R. J. Tingey, ed.), pp. 215–248, Oxford University Press, Oxford.

Barker, P. F., and Hill, I. A. 1980. Asymmetric spreading in back-arc basins, *Nature* **285:**562–564.

Barker, P. F., and Hill, I. A. 1981. Back-arc extension in the Scotia Sea, *Phil. Trans. R. Soc. Lond. A* **300:**249–262.

Barker, P. F., Hill, I. A., Weaver, S. D., and Pankhurst, R. J. 1982. The origin of the eastern South Scotia Ridge as an intra-oceanic island arc, in *Antarctic Geoscience* (C. Craddock, ed.), pp. 203–211, Univ. Wisconsin Press, Madison.

Barker, P. F., and Lawver, L. A. 1988. South American–Antarctic plate motion over the past 50 Myr, and the evolution of the South American–Antarctic Ridge, *Geophys. J. R. Astron. Soc.* **94:**377–386.

Barrow, J. Sir. 1830. Account of the island of Deception, one of the New Shetland Isles, *J. R. Geogr. Soc.* **1:**62–66.

Brett, C. P. 1977. Seismicity of the South Sandwich Islands region, *Geophys. J. R. Astron. Soc.* **51:**453–464.

Brett, C. P., and Griffiths, D. H. 1975. Seismic wave attenuation and velocity anomalies in the eastern Scotia Sea, *Nature* **253:**613–614.

Cande, S. C., and Kent, D. V. 1992. A new geomagnetic polarity time scale for the Late Cretaceous and Cenozoic, *J. Geophys. Res.* **97:**13,917–13,951.

Cande, S. C., Leslie, R. B., Parra, J. C., and Hobart, M. 1987. Interaction between the Chile Ridge and Chile Trench: geophysical and geothermal evidence, *J. Geophys. Res.* **92**:495–520.

Chase, C. G. 1978. Extension behind island arcs and motions relative to hotspots, *J. Geophys. Res.* **83**:5385–5388.

Cohen, R. S., and O'Nions, R. K. 1982. Identification of recycled continental material in the mantle from Sr, Nd and Pb isotope investigation, *Earth Planet. Sci. Lett.* **61**:73–84.

Cook, J. 1777. *A voyage towards the South Pole and around the World Performed in His Majesty's Ships the "Resolution" and "Adventure" in the Years 1772–75*, Shanan and Cadell, London.

Debenham, F. 1945. *The Voyage of Captain Bellingshausen to the Antarctic Seas, 1819–1821* (translated from the Russian), Vols. 1, 2, Hakluyt Society, London.

DeMets, C., Gordon, R. G., Argus, D. F., and Stein, S. 1990. Current plate motions, *Geophys. J. Int.* **101**:425–478.

Elsasser, W. M. 1971. Sea floor spreading as thermal convection, *J. Geophys. Res.* **76**:1101–1112.

Ewing, J. I., Ludwig, W. J., Ewing, M., and Eittreim, S. L. 1971. Structure of the Scotia Sea and Falkland Plateau, *J. Geophys. Res.* **76**:7118–7137.

Forsyth, D. W. 1975. Fault plane solutions and tectonics of the South Atlantic and Scotia Sea, *J. Geophys. Res.* **80**:1429–1443.

Fryer, P. 1992. A synthesis of Leg 125 drilling of serpentine seamounts on the Mariana and Izu–Bonin fore-arcs, in *Proc. ODP, Sci Results*, 125 (P. Fryer, J. A. Pearce, *et al.*, eds.), pp. 593–614, Ocean Drilling Program, College Station, TX.

Gass, I. G., Harris, P. G., and Holdgate, M. W. 1963. Pumice eruption in the area of the South Sandwich Islands, *Geol. Mag.* **100**:321–330.

Griffiths, D. H. 1963. Geophysical investigations in the Scotia Arc and Graham Land, *Br. Antarct. Surv. Bull.* **1**: 27–32.

Griffiths, D. H., Riddihough, R. P., Cameron, H. A. D., and Kennet, P. 1964. Geophysical investigations of the Scotia Arc, *Br. Antarct. Surv. Sci. Rep.* **46**: 43 pp.

Gripp, A. E., and Gordon, R. J. 1990. Current plate velocities relative to the hotspots incorporating the NUVEL-1 global plate motion model, *Geophys. Res. Lett.* **17**:1109–1112.

Hamilton, I. W. 1989. Geophysical investigations of subduction-related processes in the Scotia Sea, unpublished Ph.D. thesis, Birmingham University, U.K.

Harland, W. B., Cox, A. V., Llewellyn, P. G., Pickton, C. A. G., Smith, A. G., and Walters, R. 1982. *A Geologic Time Scale*, Cambridge University Press, Cambridge.

Harrington, P. K., Barker, P. F., and Griffiths, D. H. 1972. Crustal structure of the South Orkney Islands area from seismic refraction and magnetic measurements, in *Antarctic Geology and Geophysics* (R. J. Adie, ed.), pp. 27–32, Universitetsforlaget, Oslo.

Hawkes, D. D. 1962. The structure of the Scotia Arc, *Geol. Mag.* **99**:85–91.

Hawkesworth, C. J., O'Nions, R. K., Pankhurst, R. J., and Evensen, N. M. 1977. A geochemical study of island-arc and back-arc tholeiites from the Scotia Sea, *Earth Planet. Sci. Lett.* **36**:253–262.

Heezen, B. C., and Johnson, G. L. 1965. The South Sandwich Trench, *Deep-Sea Res.* **12**:185–197.

Herdman, H. F. P. 1948. Soundings taken during *Discovery* investigations 1932–1939, *Discovery Reports* **25**: 39–106.

Hill, I. A., and Barker, P. F. 1980. Evidence for Miocene back-arc spreading in the central Scotia Sea, *Geophys. J. R. Astron. Soc.* **63**:427–440.

Holdgate, M. W., and Baker, P. E. 1979. The South Sandwich Islands. I: General description, *Br. Antarct. Surv. Sci. Rept.* **91**: 76 pp.

Isacks, B., and Molnar, P. 1971. Distribution of stresses in the descending lithosphere from a global survey of focal mechanism solutions of mantle earthquakes, *Rev. Geophys. Space Phys.* **9**:103–174.

Jarrard, R. D. 1986. Relations among subduction parameters, *Rev. Geophys.* **24**:217–284.

Kemp, S., and Nelson, A. L. 1931. The South Sandwich Islands, *Discovery Reports* **3**:133–198.

LaBrecque, J. L., Kent, D. V., and Cande, S. C. 1977. Revised magnetic polarity time scale for late Cretaceous and Cenozoic time, *Geology* **5**:330–335.

Langel, R. A. 1992. International geomagnetic reference field: The sixth generation, *J. Geomagn. Geoelectr.* **44**:679–707.

Lawver, L. A., and Dick, H. J. B. 1983. The American–Antarctic Ridge. *J. Geophys. Res.* **88**:8193–8202.

Livermore, R. A., and Woollett, R. W. 1993. Seafloor spreading in the Weddell Sea and southwest Atlantic since the late Cretaceous, *Earth Planet. Sci. Lett.* **117**:475–495.

Ludwig, W. J., and Rabinowitz, P. D. 1982. The collision complex of the North Scotia Ridge, *J. Geophys. Res.* **87**:3731–40.

Luff, I. W., 1982. Petrogenesis of the island arc tholeiite series of the South Sandwich Islands, unpublished Ph.D. thesis, University of Leeds, U.K.

Macdonald, D. I. M., Storey, B. C., and Thomson, J. W. 1987. *South Georgia*, p. 63, BAS GEOMAP Series, Sheet 1, 1:250,000, Geological map and supplementary text, British Antarctic Survey, Cambridge.

Mattey, D. P., Carr, R. H., Wright, I. P., and Pillinger, C. P. 1984. Carbon isotopes in submarine basalts, *Earth Planet. Sci. Lett.* **70:**196–206.

Matthews, D. H. 1959. Aspects of the geology of the Scotia Arc, *Geol. Mag.* **95:**425–441.

Maurer, H., and Stocks, T. 1933. Die echolotungen des *Meteor. Wiss. Ergebn. Deut. Atlantic Expedition Meteor, 1925–27*, pp. 1–309.

McAdoo, D. C., and Marks, K. M. 1992. Gravity fields of the Southern Ocean from Geosat data, *J. Geophys. Res.* **97:**3247–3260.

Menard, H. W. 1978. Fragmentation of the Farallon plate by pivoting subduction, *J. Geol.* **86:**99–110.

Minster, J. B., and Jordan, T. H. 1978. Present-day plate motions, *J. Geophys Res.* **83:**5331–5354.

Muenow, D. W., Liu, N. W. K., Garcia, M. O., and Saunders, A. D. 1980. Volatiles in submarine volcanic rocks from the spreading axis of the East Scotia Sea back-arc basin, *Earth Planet. Sci. Lett.* **47:**272–278.

NGDC-NOAA. 1993. *Global Relief CD-ROM Data Set*, National Geophysical Data Center, Boulder.

Parsons, B. L., and Sclater, J. G. 1977. An analysis of the variation of ocean floor bathymetry and heat flow with age, *J. Geophys. Res.* **82:**803–827.

Pearce, J. A. 1982. Trace element characteristics of lavas from destructive plate boundaries, in *Andesites: Orogenic Andesites and Related Rocks* (R. S. Thorpe, ed.), pp. 525–548, Wiley, Chichester.

Pearce, J. A., Baker, P. E., Harvey, P. E., and Luff, I. W. in press. Geochemical evidence for subduction fluxes, mantle melting and fractional crystallization beneath the South Sandwich island arc, *J. Petrol.*

Pelayo, A. M., and Wiens D. A. 1989. Seismotectonics and relative plate motions in the Scotia Sea region, *J. Geophys. Res.* **94:**7293–7320.

Saunders, A. D., and Tarney, J. 1979. The geochemistry of basalt from a back-arc spreading centre in the East Scotia Sea, *Geochim. Cosmochim. Acta* **43:**555–572.

Saunders, A. D., and Tarney, J. 1991. Back-arc basins, in *Oceanic Basalts* (P. A. Floyd, ed.), pp. 219–263, Blackie, Van Nostrand, Reinhold, Glasgow.

Simpson, P., and Griffiths, D. H. 1982. The structure of the South Georgia continental block, in *Antarctic Geoscience* (C. Craddock, ed.), pp. 185–192, University of Wisconsin Press, Madison.

Steiger, R. H., and Jaeger, E. 1977. Subcommission on geochronology: Convention on the use of decay constants in geo- and cosmochronology, *Earth Planet. Sci. Lett.* **36:**359–362.

Suess, E. 1909. *Das Antlitz der Erde*, III Band, Freytag, Leipzig.

Taylor, B. 1987. A geophysical survey of the Woodlark-Solomons region, in *Marine Geology, Geophysics, and Geochemistry of the Woodlark Basin–Solomon Islands* (B. Taylor and N. F. Exon, eds.), Earth Science Ser., Vol. 7, pp. 25–48, Circum-Pacific Council for Energy and Mineral Resources, Earth Science Ser., Houston, TX.

Tectonic Map of the Scotia Arc. 1985. 1:3000000. BAS (Misc) 3, British Antarctic Survey, Cambridge.

Tomblin, J. F. 1979. The South Sandwich Islands. II: The geology of Candlemas Island, *Br. Antarct. Surv. Sci. Rept.* **92:** 33 pp.

Tyrrell, G. W. 1945. Report on rocks from West Antarctica and the Scotia Arc, *Disc. Rep.* **23:**37–102.

Vlaar, N. J., and Wortel, M. J. R. 1976. Lithospheric ageing, instability and subduction, *Tectonophysics* **32:** 331–351.

Von Huene, R., Kulm, L. D., and Miller, J. 1985. Structure of the frontal part of the Andean convergent margin, *J. Geophys. Res.* **90:**5429–5442.

Wilson, D. S. 1993. Confirmation of the astronomical calibration of the magnetic polarity timescale from sea-floor spreading rates, *Nature* **364:**788–790.

Woodhead, J., Eggins, S., and Gamble, J. 1993. High field strength and transition element systematics in island arc and back-arc basin basalts: evidence for multi-phase melt extraction and a depleted mantle wedge, *Earth Planet. Sci. Lett.* **114:**491–504.

8

Bransfield Strait, Antarctic Peninsula Active Extension behind a Dead Arc

Lawrence A. Lawver, Randall A. Keller, Martin R. Fisk, and Jorge A. Strelin

ABSTRACT

Bransfield Strait is a marginal basin landward of the South Shetland Trench. It lies between the South Shetland Islands and the tip of the Antarctic Peninsula and is an example of an extensional basin formed by rifting within a continental volcanic arc. The Antarctic Peninsula is the product of at least 200 m.y. of subduction with the majority of the exposed rocks related to continental arc volcanism older than 20 Ma. Volcanism in Bransfield Strait started by 0.3 Ma and continues today. This new volcanism maintains some of the chemical signatures of the old arc volcanism but also has signatures transitional between arc rocks and backarc basin rocks. On the basis of high heat flow, active volcanism, extensional faulting, and earthquake fault plane mechanisms, Bransfield Strait is an active extensional basin forming within the Antarctic Peninsula. Seismic refraction work in Bransfield Strait indicates some thinning of continental crust, but the basin itself is underlain by as much as 30 km of anomalous crustal material. The observed extension seems to be confined to Bransfield Strait, which is bounded by the landward projections of the Hero and Shackleton fracture zones. The present extension in Bransfield Rift started less than 4 m.y. ago, and possibly less than 1.5 m.y. ago, following the demise of the Antarctic–Phoenix spreading center (ANT–PHO ridge), which ceased spreading about 4 Ma. Apparent, continued subduction at the South Shetland Trench after the ANT–PHO ridge stopped spreading may occur as trench rollback. The amount of trench rollback should be comparable to the amount of extension in Bransfield Strait.

Lawrence A. Lawver • Institute for Geophysics, University of Texas at Austin, Austin, Texas 78759-8397. *Randall A. Keller and Martin R. Fisk* • College of Oceanic and Atmospheric Sciences, Oregon State University, Corvallis, Oregon 97331-5503. *Jorge A. Strelin* • Departmento de Ciencias de la Tierra, Instituto Antartico Argentino, Buenos Aires, Argentina.

Backarc Basins: Tectonics and Magmatism, edited by Brian Taylor, Plenum Press, New York, 1995.

1. INTRODUCTION

Bransfield Strait (Fig. 8.1) is an example of a basin that formed by rifting within a continental volcanic arc. For most of the past 200 m.y., the Antarctic Peninsula (Fig. 8.1) has been the site of subduction of oceanic lithosphere that formed at spreading centers in the South Pacific (Tanner *et al.*, 1982; Barker and Dalziel, 1983; Pankhurst, 1983; Hole *et al.*, 1991). Major plate motions in the vicinity of Bransfield Strait and the Antarctic Peninsula are well constrained by magnetic anomaly patterns preserved on the Antarctic, Pacific, Scotia, and former Phoenix plates (Barker, 1982). At about 4 Ma, the Antarctic–Phoenix (ANT–PHO) spreading center ceased spreading and the remnant Phoenix plate was incorporated into the Antarctic plate. Previous authors have referred to the remnant plate as the Aluk plate (Herron and Tucholke, 1976; Barker, 1982; Mayes *et al.*, 1990; GRAPE Team, 1990), the Drake plate (Barker, 1982; Mayes *et al.*, 1990), and/or the Phoenix plate (Barker, 1982; Larter and Barker, 1991). Larson and Chase (1972) used the terms *Phoenix lineations*, *Phoenix model*, and *Phoenix plate*, and related the terms to the Phoenix Islands found in the central Pacific Ocean. Although the Phoenix lineations and the Phoenix Islands both lie on the Pacific plate, they assigned the term *Phoenix plate* to the "southern" half of the Pacific–Phoenix spreading center and traced a possible Pacific–Phoenix fracture zone from the Phoenix lineations to the Eltanin fracture zone system's intersection with the present-day Pacific–Antarctic Ridge. They correctly pointed out that "It is unlikely that the Phoenix plate corresponded to the Antarctic plate until about 85 m.y.B.P. ...," because "Antarctica and New Zealand did not begin to separate until just prior to anomaly 32." The designation

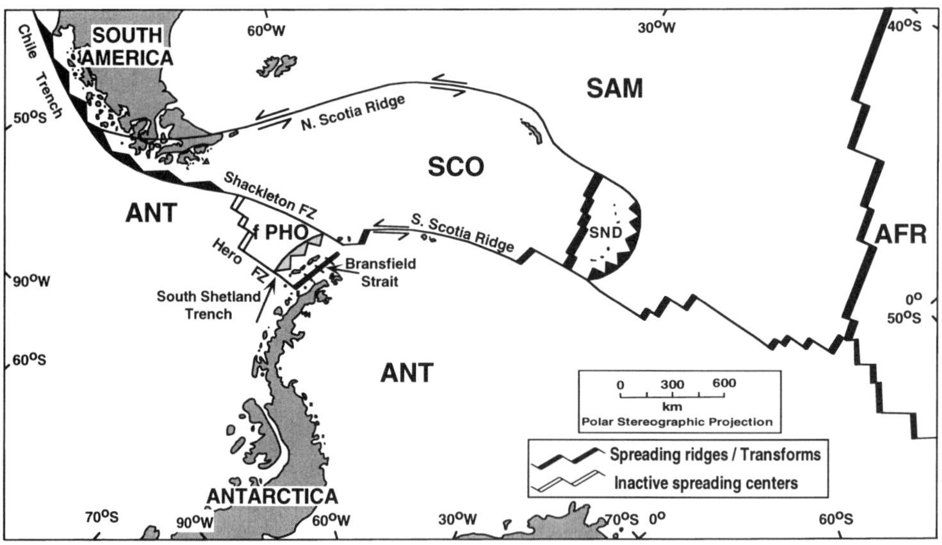

FIGURE 8.1. Present-day, polar stereographic map of the regional tectonic setting of the Bransfield Strait area modified from the Tectonic Map of the Scotia Arc [1985; British Antarctic Survey Sheet (Miscellaneous) 3]. Abbreviations for the major plates are AFR = Africa, ANT = Antarctic, f PHO = former Phoenix, SCO = Scotia, and SND = Sandwich plate. The South Shetland plate lies between the South Shetland Trench and the active rift shown in Bransfield Strait. Double arrows indicate direction of strike-slip motion along the North and South Scotia ridges. Active trenches are shown as black. The South Shetland Trench is shown shaded.

Phoenix plate for the remnant plate off Bransfield Strait is therefore tectonically most accurate and should take precedence.

Seismic reflection and refraction work, gravity, magnetics, and heat flow data, and the history and composition of volcanism on the peninsula and in Bransfield Strait place constraints on the processes associated with the formation of Bransfield Strait as a marginal basin. Recent dredges from new sites in Bransfield Strait (Keller *et al.*, 1994) will soon be analyzed and will give better control on active basin formation.

2. REGIONAL TECTONIC SETTING

Present-day plate boundaries are indicated in Fig. 8.1. Plate reconstructions (Barker, 1982; Mayes *et al.*, 1990) suggest that at least a thousand kilometers of the Phoenix plate have been subducted beneath Bransfield Strait since 50 Ma. Between 50 and 43 Ma, a segment of the ANT-PHO spreading center between the Tharp and Heezen fracture zones (Fig. 8.2) was subducted (Barker, 1982). At 20 Ma, the active South Shetland Trench was

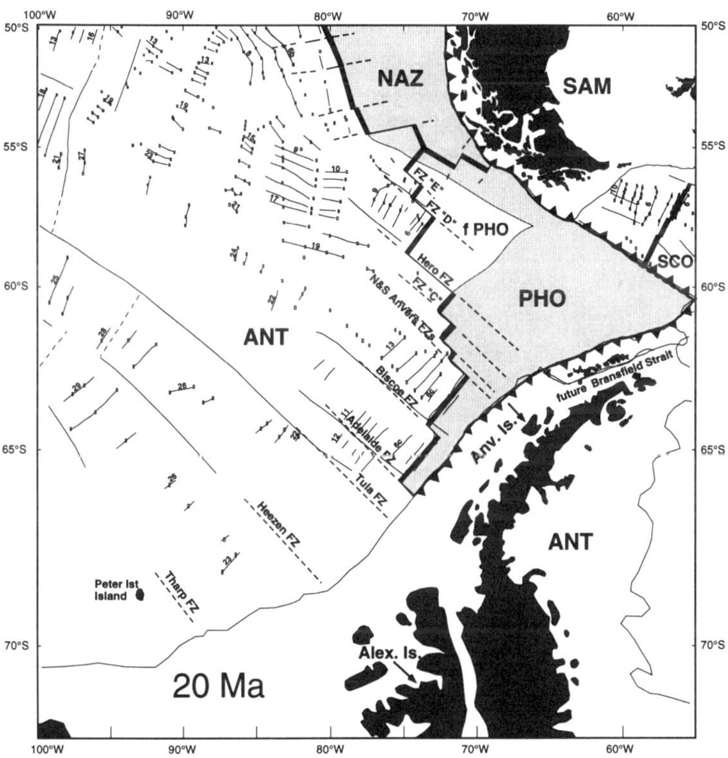

FIGURE 8.2. Reconstruction of Southeast Pacific and Antarctic Peninsula regions for 20 Ma. The reconstruction is based on the PLATES database. Dashed fracture zones and spreading centers are taken from Larter and Barker (1991). Plate abbreviations are as in Fig. 8.1: NAZ = Nazca plate, PHO = Phoenix, Alex. = Alexander Island, Anv. = Anvers Island. Magnetic anomaly picks are shown as small squares. Magnetic lineations connect the small squares. Rifting in Bransfield Strait has been closed up so that the South Shetland Islands are closer to the Antarctic Peninsula than they are at present. The shaded region is the seafloor that was subducted since 20 Ma.

nearly twice the length it was after about 14 Ma, when the sections between the Adelaide and Anvers fracture zones were subducted. Through time, additional segments of the ridge were subducted to the northeast, until about 4 Ma, when the spreading center immediately southwest of the Hero fracture zone reached the trench (Barker, 1982). The last segment of the ANT-PHO spreading center (between the Hero and Shackleton fracture zones) was then abandoned offshore of the trench and the last remnant of the Phoenix plate (f PHO, Fig. 8.2) became part of the Antarctic plate.

Until 4 Ma, the subducting plate was decoupled from the overriding plate as the lower Phoenix plate slid beneath the upper Antarctic plate. As the Phoenix–Antarctic spreading centers were subducted, the subducted slab continued to sink, leaving in its wake a slab window (Hole and Larter, 1993). After 4 Ma, when the last Phoenix–Antarctic spreading center southwest of the Hero fracture zone was subducted, the remnant, unsubducted Phoenix plate (former Phoenix plate = f PHO) was frozen in place between the Hero and Shackleton fracture zones. Whether this resulted in the subducted Phoenix slab, to the southwest of the Hero fracture zone, to also be frozen in place with respect to the underlying mantle or if it continued to sink independently is a matter of speculation. If the subducted slab to the southwest of the Hero fracture zone continued to sink but the slab beneath Bransfield Strait was held in place by the f PHO, then there would have to be tearing or differential motion along the subducted section of the Hero fracture zone, and there would be no slab window beneath Bransfield Strait. On the other hand, if the slab beneath Bransfield Strait is detached from the f PHO and sinking, then there should be deep seismicity and evidence of a slab window in the geochemistry.

Other marginal basins along the outer edge of the Antarctic Peninsula may have formed in response to the subduction of ridge segments to the southwest of the Hero fracture zone (Gambôa and Maldonado, 1990), although none of these basins seem to have reached the advanced state of Bransfield Strait. GEOSAT gravity data from 72°S to the northeast along the western margin of the Antarctic Peninsula (Lawver *et al.*, 1993) show a marginal gravity low that extends from George VI Sound (70°S, 70°W) between Alexander Island and the peninsula, to the southern edge of Anvers Island at 65°W. This gravity signature is similar to the one seen in Bransfield Strait. Single-channel seismic reflection lines from along this margin (Anderson *et al.*, 1990) show features that resemble inner shelf troughs and old forearc basins that might be interpreted as precursors to Bransfield Strait, although these basins do not show any evidence for extension. Larter and Barker (1989) suggested that the forearc to the south of Bransfield Strait was simply uplifted and eroded but not extended after ridge-trench collision.

The age of the subducted Phoenix plate at the South Shetland Trench (Fig. 8.1) increases from 14 Ma in the southwest near the Hero fracture zone to 23 Ma in the northeast near the Shackleton fracture zone (Barker, 1982). Between the Hero and Shackleton fracture zones, there are at least two additional fracture zones, D and E (Fig. 8.2). Their landward extensions would cross the neovolcanic zone of Bransfield Strait at about 59°W (FZ D) and 57°50′W (FZ E). At the trench, the age difference across FZ D is 3 m.y.; across FZ E, 3.5 m.y. (Larter and Barker, 1991).

Presently, the subducted slab can be detected beneath the South Shetland Islands (Grad *et al.*, 1993) dipping at an angle of 25° to the southeast (Fig. 8.3), although there is virtually no seismic activity associated with the lower plate slab (Pelayo and Wiens, 1989). This may be explained by either slow, aseismic descent of the slab, or the slab beneath the South Shetland Islands and Bransfield Strait is frozen in place such that there is no detectable motion of the slab relative to the asthenosphere below it. If the lower or

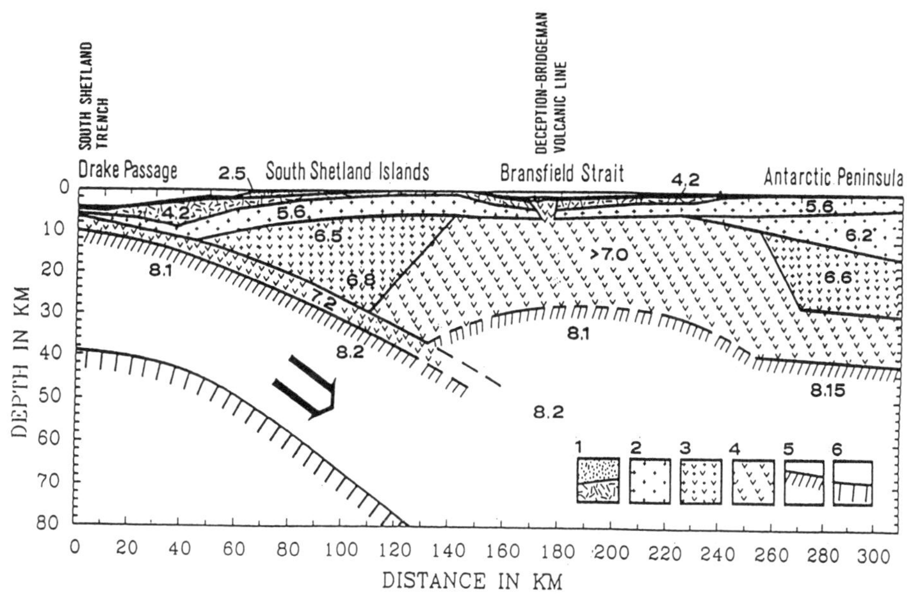

FIGURE 8.3. Seismic model of the lithosphere from the deep-ocean side of the South Shetland Islands through Bransfield Strait and onto the Antarctic Peninsula taken from Grad *et al.* (1993). The symbols used are (1) sediments, $v_p = 2.5–4.2$ km s^{-1}, (2) upper crust, $v_p = 5.4–6.3$ km s^{-1}, (3) middle crust, $v_p = 6.4–6.8$ km s^{-1}, (4) lower crust and high-velocity body in Bransfield Strait, $v_p > 7.0$ km s^{-1}, (5) Moho boundary, $v_p > 8.0$ km s^{-1}, (6) reflection boundary in the lower lithosphere. Notice that the volcanic line is not centered in the middle of Bransfield Strait.

subducted plate remains attached to the f PHO, then any extension in the upper plate must be accompanied by trench rollback. While some may argue that trench rollback is evidence for subduction, we make the distinction that in this case, if the lower plate is still attached to the Antarctic plate, it cannot be subducting with respect to itself, and the independent South Shetland plate must therefore be moving with respect to the lower plate. This in turn produces the observed rollback or seaward motion of the trench. If the slab has broken from the former Phoenix plate or is sinking along a hinge line at the trench, then there should be deep seismicity and there would not be evidence for deformation of the accretionary wedge.

Prior to the demise of the ANT–PHO spreading center, the Shackleton FZ to the north of Bransfield Strait was the boundary between the Antarctic and Phoenix plates on one side and the Scotia plate on the other (Fig. 8.1). Seafloor spreading stopped in the western Scotia Sea immediately to the east of the Shackleton FZ at 6 Ma (Barker and Burrell, 1977). At about 4 Ma, when the remaining Phoenix plate became a part of the Antarctic plate, the Scotia–Phoenix plate boundary became part of the Scotia–Antarctic plate boundary. From 4 Ma to recently, the Scotia–Antarctic plate boundary to the north of Bransfield Strait has trended NW-SE (Shackleton fracture zone), while to the east of Bransfield Strait it has trended E-W (South Scotia Ridge). The consequence of this 110° bend in the plate boundary would be compression on the inside of the bend and extension on the outside of the bend. This should produce extension in the North Bransfield Basin, the part of Bransfield Strait between Bridgeman Island (Fig. 8.4) and Clarence Island. Single-channel seismic reflection profiles across the North Bransfield Basin (Lawver and Villinger, 1989) show continentward tilted blocks, reminiscent of the continental extension models of Wernicke

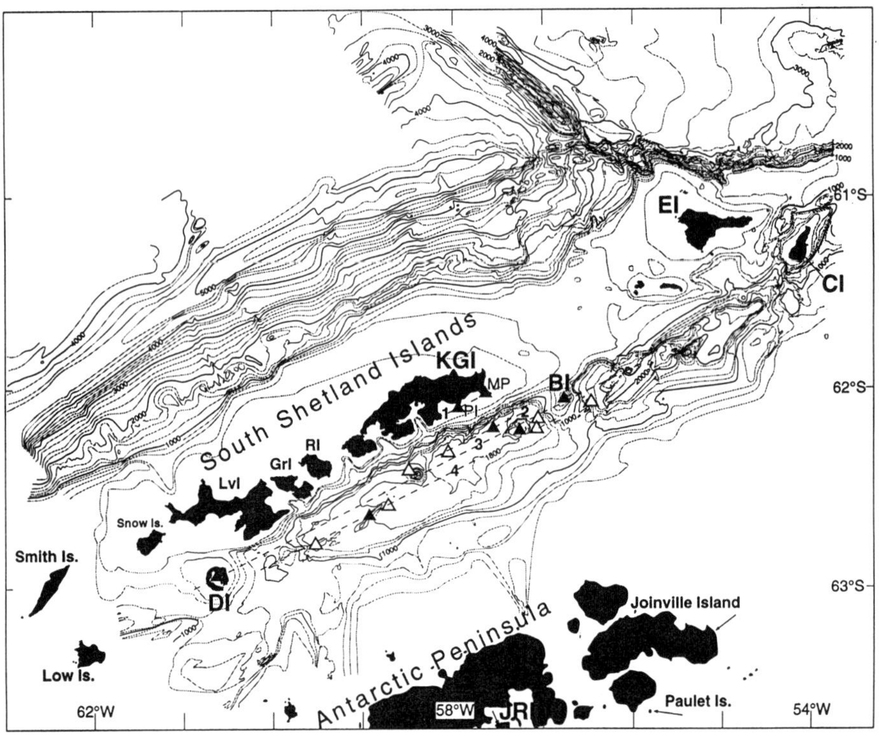

FIGURE 8.4. Bathymetric map of Bransfield Strait region (modified from Klepeis and Lawver, 1994). Bransfield Strait lies between the South Shetland Islands and the Antarctic Peninsula. Contour interval is 200 m, 1000-m contour lines are shown as bold lines and some are labeled. The 1800-m contour southwest of King George Island (KGI) outlines the King George Basin. Solid triangles are locations of submarine and subaerial Quaternary rift-related volcanism that have been sampled and analyzed (Keller *et al.*, 1992). Hollow triangles are submarine locations where fresh, glassy basalts have been recovered (Keller *et al.*, 1994) but not yet analyzed. BI = Bridgman Island, CI = Clarence Island, DI = Deception Island, EI = Elephant Island, GrI = Greenwich Island, JRI = James Ross Island, LvI = Livingston Island, MP = Melville Peak, PI = Penguin Island, and RI = Robert Island. Dashed lines numbered 1 through 4 are parallel or subparallel volcanic lineations, ranging in size from tens of meters in height to 2 km, or more for Deception Island and the submarine volcano south of KGI.

(1981). Evidence for compression includes the extraordinarily steep SW face of Clarence Island which varies from −1200 m to +2300 m in a distance of 10 km or less. The low temperature, high-pressure metamorphics found on Clarence Island and on the eastern half of Elephant Island (Dalziel, 1984) also support the idea of compression and uplift within the bend.

3. TECTONIC SETTING OF BRANSFIELD STRAIT

The strait itself is approximately 100 km wide by 400 km long and is bounded on the northwest by steep normal faults with as much as 4 km of downthrow to the southeast between the South Shetland Islands and the Bransfield basins (Ashcroft, 1972). Birkenmajer (1992) distinguishes between the 100-km-wide Bransfield Strait and a narrowly defined (15–20 km wide) Bransfield rift. From field mapping on King George Island,

Birkenmajer (1992) found fossil evidence that a marine basin existed during the early Eocene (fossiliferous glaciomarine strata from the Kraków Glaciation) and during the Eocene part of the succeeding interglacial time. He reports that during early Oligocene the region was covered by an ice sheet and then later flooded by a shallow sea. Regional uplift caused the area to be above sea level until the late Oligocene when incipient rifting started at the end of the Oligocene (26 to 22 Ma).

The basins of Bransfield Strait are thought to be mainly Quaternary features, although the age of initiation of rifting is poorly constrained. Much of the early faulting is associated with the arc-building period of the peninsula. Evidence for early extension consists of a system of antithetic faults that cut the upper Oligocene and older rocks along the margin of the rift. Arc tension continued through early Miocene when several stages of basaltic to andesitic dyke intrusions occurred between 22 and 20 Ma and at 14 Ma (Birkenmajer, 1992). From the late Miocene to Pliocene there is no evidence of faulting or volcanism within Bransfield Strait. The present Bransfield rift is the site of Pleistocene to Recent, mildly alkaline to calcalkaline volcanic activity.

The present plate configuration in the Bransfield Strait region places the South Shetland Islands on a separate (South Shetland) microplate, bounded on the northwest by the South Shetland Trench and on the southeast by the Bransfield rift. The southeast boundary of Bransfield Strait rises more gently toward the Antarctic Peninsula through a series of widely spaced normal faults (Ashcroft, 1972). The active rift appears to extend 400 km (Gonzalez-Ferran, 1985) from Low Island (Fig. 8.4) above the landward extension of the Hero fracture zone to Clarence Island above the landward extension of the Shackleton fracture zone. Based on GEOSAT/ERS-1 satellite-derived gravity data (Fig. 8.5), Bransfield Strait can be subdivided into three basins: the south, central, and north Bransfield basins. The south Bransfield Basin shows up as a ~100-km-long gravity anomaly trending southwest of Deception Island. The south Bransfield Basin approximately follows the 1000-m contour shown in Larter and Barker (1991, Fig. 8.4). The central Bransfield Basin is a graben up to 2 km deep and roughly 40 km wide by 200 km long that is easily seen on the bathymetric map (Fig. 8.4). It extends from Deception Island to Bridgeman Island and contains the 20- by 50-km, flat-floored King George Basin southeast of King George Island. The north Bransfield Basin is actually the deepest part of Bransfield Strait with a maximum depth over 2740 m and the lowest gravity anomaly, <-30 mgal. The lowest gravity anomaly is not coincident with the deepest part of the basin but is to the northeast toward Clarence Island. If the positive gravity anomalies produced by Deception and Bridgeman Islands (which are relatively young structures) are removed from the GEOSAT data shown in Fig. 8.5, the subdivisions of Bransfield Strait lose their definition and a single major gravity low extends from just south of Low Island to Clarence Island.

The digital bathymetric map (Fig. 8.4) of Bransfield Strait and the nearby region (Klepeis and Lawver, 1994) shows numerous small bathymetric features that appear to be mounds and are lineated, parallel to the basin axis defined by a line through Deception and Bridgeman Islands. Some of these mounds can be seen in Fig. 8.4. There are at least four parallel lines of active and incipient volcanism. The first is discussed by Birkenmajer (1992) and includes Penguin Island, an active cinder cone just off King George Island. This lineation is within the shelf area of the South Shetland Islands. The second line is defined in King George Basin by at least six to eight closed contours indicating circular structures from a few tens of meters in height to a couple of hundred meters. The third line is the main rift axis and includes Deception and Bridgeman Islands and at least two major submarine volcanoes that are up to 1.4 km in height above the seafloor and as much as 8 km in

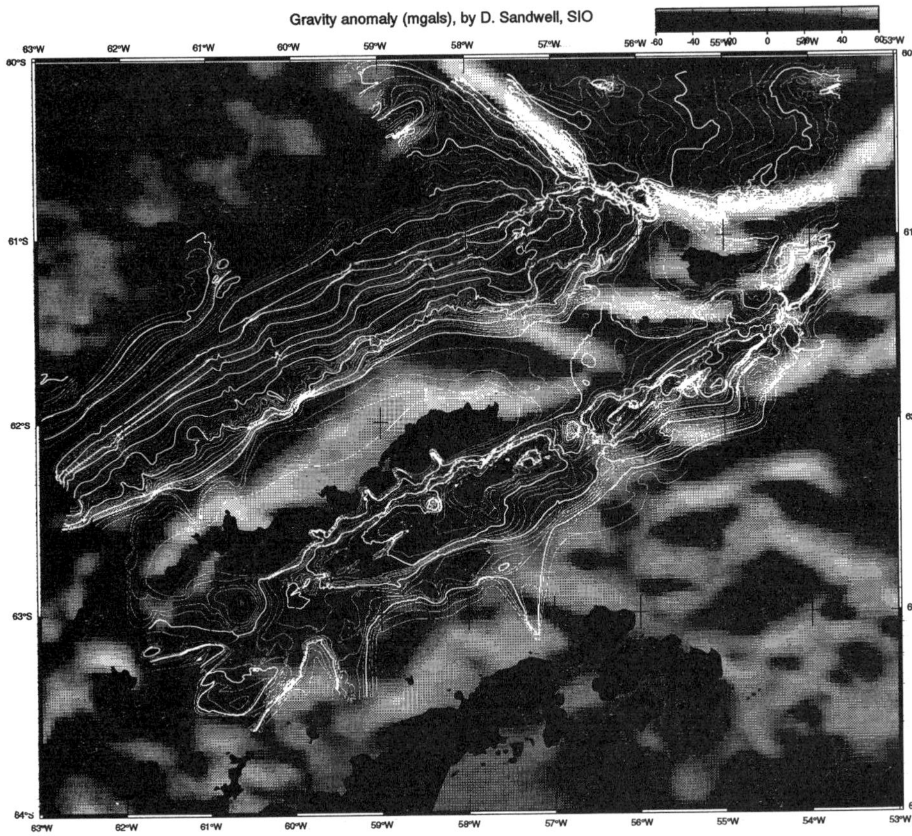

FIGURE 8.5. Geosat/ERS-1 gravity data from Sandwell and Smith (1992) with digital bathymetric data of Klepeis and Lawver (1994) superimposed. The gravity data bar indicates gravity values from <-40 mgal to $>+30$ mgal. Bold white bathymetric contours are shown every 500 m, while thin white contours are observed bathymetry contoured every 100 m. Thin gray contours are interpolated bathymetric data contoured every 100 m. A color version of this figure appears opposite page 311.

diameter. The final volcanic lineation is defined by a ridge (Fig. 8.4), the highest heat flow in King George Basin, and a very recent extrusion at 57°W.

The observed recent faulting (Fig. 8.6) seen in the multichannel seismic data (Barker and Austin, 1994) may define yet a fifth zone of extension and deformation, and although diapirism is suggested, no actual intrusion is observed. Barker and Austin (1994) show an alignment of intracrustal diapirs that trend subparallel to the other neovolcanic zones but in fact diverge from them by about 5°. They suggest that the divergence shows a temporal/spatial evolution of stress in Bransfield Strait, possibly related to the complex plate interactions to the northeast.

4. EARTHQUAKES

Most of the earthquakes in Bransfield Strait are shallow events related to extension and volcanic activity (Fig. 8.7). Deep events in the subducted plate are rare. The vast

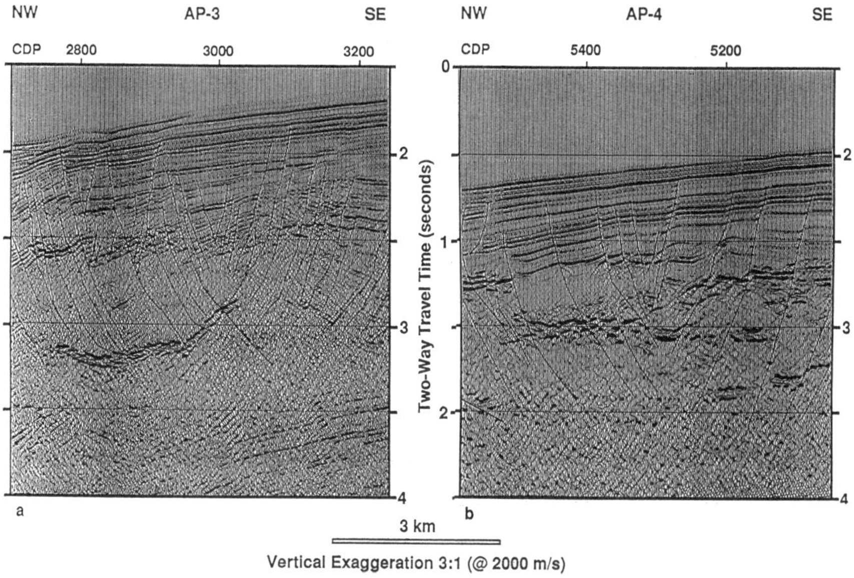

FIGURE 8.6. Seismic cross sections taken from Barker and Austin (1994). Locations shown in Fig. 8.8. Profile shows fan-shaped, normal faulting pattern where the faults reverse their dip about a midpoint (at ~CDP 3000 for 8.6a, ~CDP 5300 for 8.6b). The fault blocks rotate away from the center of the structure, and central faults are confined to the upper part of the section. The inferred diapiric body is unreflective.

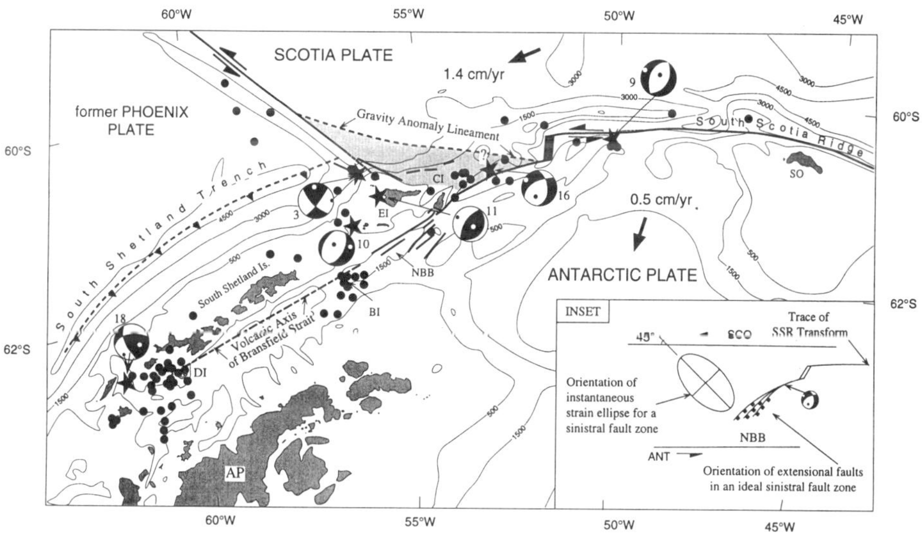

FIGURE 8.7. Tectonic map of the area north of the Antarctic Peninsula (AP) showing interpreted geometry of the plate boundaries between the Shackleton fracture zone and the South Scotia Ridge. EI is Elephant Island, CI is Clarence Island, BI is Bridgeman Island, and NBB is North Bransfield Basin, SSR (inset) is South Scotia Ridge, dashed line labeled BS axis is the volcanic rift axis of Bransfield Strait. Bold lines located north EI/CI represent faults interpreted to parallel the steep scarps of the EI platform's northern margin, bold lines south of EI/CI are the major fault zones taken from Klepeis and Lawver (1994). GEOSAT gravity anomaly is after Sandwell (1992). Circles are earthquake epicenters from the Tectonic Map of the Scotia Arc (1985). Focal mechanisms are for shallow earthquakes from Pelayo and Wiens (1989). Shaded area between GEOSAT gravity anomaly lineament and EI/CI represents a broad zone of tectonic disruption where the Shackleton fracture zone transform apparently is stepping eastward to join with a transform along the South Scotia Ridge.

majority of the teleseismic events between 1964 and 1992 are swarms of moderate-magnitude events which lasted up to six months (International Seismological Centre Earthquake Catalogs (ISC), 1964–1993).

The December 1967 eruption of Deception Island produced three teleseismic events, although two of them were mislocated almost 30 km from Deception Island. One event occurred in connection with the 1969 eruption. The 1970 eruption produced four events located at depths of 45 to 49 km, one event at 9 km with no magnitude determination, and one magnitude 4.6 event with no depth determination (ISC catalogue, 1970). Three swarms of five to eight events not associated with volcanic eruptions occurred near Deception Island in 1971, 1974, and 1982. In 1989 a single event at 10 km occurred near Deception Island (ISC catalogue, 1989). The events in 1971 and 1982, at 15 and 12 km, respectively, produced normal fault plane solutions indicative of extension (Pelayo and Wiens, 1989).

A large swarm of 24 separate events (magnitude 4.5 to 5.3) occurred near Bridgeman Island in the first six months of 1975. Nearly all the events were assigned depths of 33 km or less, except for one at 167 km. [This event was undoubtedly mislocated depthwise because of its low magnitude, 4.5, and the few stations which recorded it. Pelayo and Wiens (1989) do not show any events at such a depth.] One event near Bridgeman Island was recorded in 1990 at a depth of 10 km. These earthquakes may be related to ongoing extension rather than volcanic activity, because there is no historical evidence for volcanic activity on Bridgeman Island.

Elsewhere in Bransfield Strait two events relocated by Pelayo and Wiens (1989) lie at depths of 35 and 55 km (their events 10 and 18). Event 10 (61.4°S, 56.4°W) was a normal faulting event that may be in the upper plate if the seismic refraction work of Guterch *et al.* (1991) is correct. Event 18 (63°S, 61.9°W) at 55 km must be in the subducted slab although the Harvard catalogue, for that event lists its depth as 42 km. It is located almost directly on strike with the onshore projection of the Hero fracture zone, and may in fact be evidence of tearing along the subducted fracture zone.

In addition to the teleseismic events mentioned, a local network on Deception Island has operated for two months each austral summer since 1986 (Vila *et al.*, 1992). It has consistently recorded 1000 events per month with a total released seismic energy of about 3.0×10^{13} ergs/day. These events seem to be aligned parallel to the main fault system which crosses the island (E-W), and they follow the regional tectonic trend of the Bransfield rift (Vila *et al.*, 1992). The 1986–1992 events were grouped into two depth ranges, those from 0 to 2.5 km and those from 2.5 to 7.0 km. During the 1992–1993 austral summer this same network recorded five deep (55–85 km) events of magnitude 1.6–2.4 (SEAN, 1993), but it is unclear how such a closely spaced network could resolve deep events of such low magnitude.

5. SEISMIC REFRACTION

Early seismic refraction work found an anomalously thick layer beneath Bransfield Strait with a seismic velocity lower than that of normal mantle material but higher than typical continental crust (Ashcroft, 1972). Later seismic refraction work in Bransfield Strait (Guterch *et al.*, 1985, 1991, 1992; Grad *et al.*, 1993) identified returns from the Moho that dips from 10 km in the South Shetland Trench to 40 km below the South Shetland Islands, and also at 40 km below the Antarctic Peninsula. Grad *et al.* (1993) recorded a seismic boundary in the lower lithosphere that ranged in depth from 35 to 80 km as shown in Fig. 8.3. Both the Moho and the lower seismic boundary dip at an angle of 25°, which was interpreted to be the dip of the remnant Phoenix plate under the Antarctic plate. Although published gravity coverage is not extensive, Bouguer gravity anomalies (Renner *et al.*, 1985) of up to +100 mgal correlate with the anomalous crustal structure found by Grad *et al.* (1993) under the South Shetland Islands and Bransfield Strait.

The sequence stratigraphy of Bransfield Basin is discussed by Jeffers and Anderson (1990) and was used by Grad *et al.* (1993) to constrain their seismic velocity model. They found an anomalous layer under Bransfield Basin with seismic wave velocities of $v_p >$ 7.0 km s^{-1}. Unfortunately this high-velocity layer obscures the normal Moho transition as seen under the South Shetland Islands and under the Antarctic Peninsula. This anomalous layer is overlain by layers with velocities of 2.5, 4.2, and 5.6 km s^{-1}. The upper layer consists of 0.2 to 1.0 km of unconsolidated to poorly consolidated young sediments with substantial amounts of lava and tuff. The second layer consists of 1.2 to 2.5 km of older and better consolidated sediments, lava, and tuff. In the 2- to 7-km depth range, Grad *et al.* (1993) found an approximately 10-km-wide body with velocity v_p = 6.8 km s^{-1}. They attribute this body to be the active Bransfield rift. The active rift is manifested in the basin itself as submarine volcanoes, intrusive dykes, large magnetic anomalies, and high heat flow.

6. SEISMIC REFLECTION

There are numerous multichannel seismic and single-channel seismic lines known to have been collected in Bransfield Strait, including ones done by the British Antarctic Survey (Larter, 1991; Barker *et al.*, 1992), German scientists as part of the GRAPE (*G*eophysical *R*esearch Group for the *A*ntarctic *PE*ninsula) Team (1990), Polish scientists (Guterch *et al.*, 1985), Brazilian scientists (Gambôa and Maldonado, 1990), Spanish scientists (Acosta *et al.*, 1992a,b), U.S. scientists (Lawver and Villinger, 1989; Nagihara and Lawver, 1989; Jeffers and Anderson, 1990; Jeffers *et al.*, 1991; Barker and Austin, 1994), and other presently unpublished work from Italian, Chinese, Japanese, and Korean cruises. Barker (1976) published the first seismic profile for Bransfield Strait, while Guterch *et al.* (1985) presented the first seismic reflection lines to demonstrate the "complex crustal structure in the trough of Bransfield Strait." Their profile crosses the southern end of the central Bransfield Basin and clearly shows a neovolcanic intrusion zone. Jeffers and Anderson (1990) published a number of line drawings of seismic reflection data from Bransfield Strait, including one line that nearly duplicated the earlier Polish line. They determined that the sedimentation is dominated by glacial-marine processes with associated lithofacies. The central basin has received terrigenous sediment from the Trinity Peninsula and from the South Shetland Islands (Jeffers and Anderson, 1990). They discuss basinward transport of sediment through deeply incised troughs that form a prograding complex of coalescing trough-sourced wedges, although when the troughs are not active, principally during interglacial periods, the supply of terrigenous sediment to the basin is relatively low and sedimentation is predominantly biogenic. Although they state that the volcanic ridge acts as a barrier to sediment transport across the basin, later bathymetric mapping of the basin revealed that the volcanic ridge is not continuous, so it would not act as a complete barrier (Klepeis and Lawver, 1994). In places, the neovolcanic ridge is easily defined as a linear wall less than half a kilometer across with nearly sheer sides that are at least a few hundred meters high. It was recently dredged (Keller *et al.*, 1994) and produced fresh glassy basalts.

Gambôa and Maldonado (1990) presented multichannel seismic reflection data collected in 1987 and 1988 by the Brazilian Antarctic Program. In each of their crossings they show a "spreading center" and ponded turbidites. On all multichannel seismic lines across Bransfield Strait, there is never more than 1 s of recent ponded turbiditic sediment. The GRAPE team (1990) published preliminary results from 1400 km of reflection seismic profiles from Bransfield Strait and the adjoining peninsula region. Their profile 9 crossed the major submarine volcano just off King George Island (also shown in Barker, 1976), and they identified an additional "magmatically influenced" zone just to the southeast of the volcanic edifice toward the Antarctic Peninsula. Larter (1991) questioned the designation of certain crust in Bransfield Strait as oceanic crust based solely on the reflection character of the top of acoustic basement, as the GRAPE team (1990) did. The unusual thickness of the crust found from the seismic refraction data would also suggest a lack of true oceanic crust.

Acosta *et al.* (1992b) present line drawings of single channel seismic reflection profiles that show the southeastern flank of the South Shetland Islands to be broken by faults into horsts and grabens which have been intruded by dikes and plugs. Barker and Austin (1994) display initial results of the R/V *Ewing* multichannel seismic reflection cruise, where they found complex fan-faulting patterns (Fig. 8.6) along the Antarctic Peninsula side of

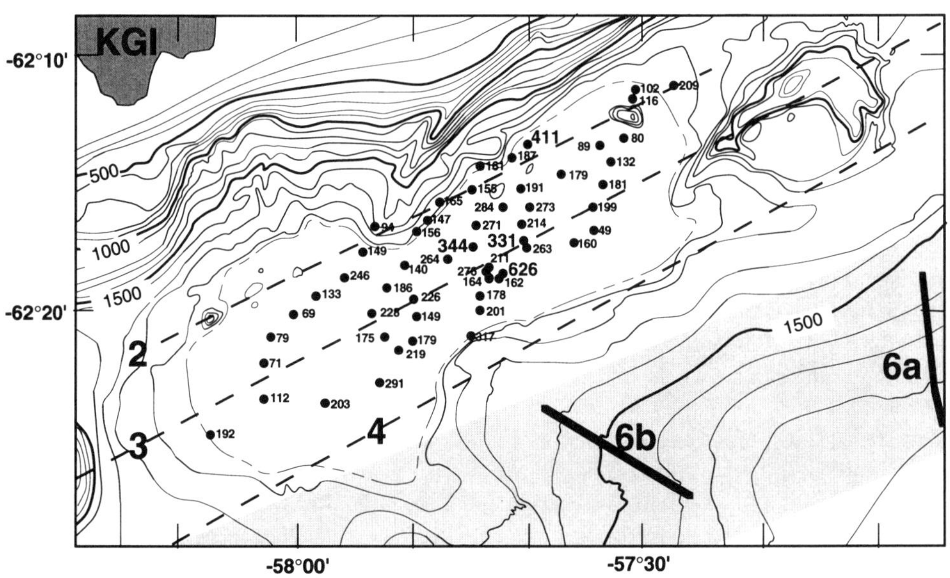

FIGURE 8.8. Heat flow map of the King George Basin. Heat flow in mW m^{-2}. Four highest values are shown with larger type size. Bathymetric contours are shown every 100 m with 500-m contours shown as bold lines. The inner, dashed, closed contour is the 1980-m contour showing the flat floor of the King George Basin that is uniformly between 1980 m and 2000 m with the exception of the linear volcanic mounds found along dashed line 2. Dashed lines labeled 2, 3, and 4 correspond to parallel volcanic lineations shown in Fig. 8.4. Thick lines labeled 6a and 6b are the locations of the seismic sections shown in Fig. 8.6. Shaded area is diapir zone of recent? activity taken from Barker and Austin (1994).

Bransfield Strait. They interpret the fanfaulting to be intracrustal diapirism that indicates a new zone of diffuse extension. They found a new zone of active extension on the margin of King George Basin toward the Peninsula from the four zones discussed above.

7. MAGNETICS

Aeromagnetic surveys of the Bransfield Strait region show a central positive anomaly along the axis of the deepest part of the strait (Garrett, 1990; Gonzalez-Ferran, 1991; Maslanyj *et al.*, 1991). This central high has been modeled as a large, positively magnetized igneous body associated with the inferred axis of rifting (Parra *et al.*, 1984; Gonzalez-Ferran, 1991). Gonzalez-Ferran (1991) concluded that reversely magnetized crust at the flanks of the central body best fit the data, and that a total of approximately 15 km of new crust has been created by 2 m.y. of spreading in Bransfield Strait.

Roach (1978) modeled detailed marine magnetic data from Bransfield Strait and found that only normally magnetized crust existed beneath the axial trough. Roach later reported (in Barker and Dalziel, 1983) that narrow strips of reversely magnetized crust were required at the margins of the axial trough, and a total of about 30 km of spreading had occurred since 1.3 Ma. Unfortunately his data and the details of his modeling were not published, although his results are now widely cited in the literature as evidence for a 1–2 Ma age for Bransfield Strait. Such a short magnetic anomaly profile is ambiguous, and models of it are poorly constrained. It may also be inaccurate to assume for the purpose of modeling that there is true seafloor spreading in Bransfield Strait, since seismic refraction data (see Section 5) suggest that the crust in Bransfield Strait is too thick to be normal oceanic crust. No spreading center has been identified, and the recent volcanism is dispersed along several lineaments in Bransfield Strait (Klepeis and Lawver, 1994). The thick sediment cover contains interbedded basalt flows and hides basement topography that affect the analyses of marine magnetic anomalies in the basin and complicate any modeling, as discussed by Lawver and Hawkins (1978).

8. HEAT FLOW

Heat flow measurements have been made in the King George Basin (Fig. 8.8) immediately south of King George Island. In the King George Basin the seafloor is flat (1.96 to 1.99 km), well sedimented, and covers an area of roughly 10 to 18 km by 45 km. Thermal gradients were measured at 63 locations, and *in situ* thermal conductivities were measured at 25 sites. The *in situ* conductivity matches thermal conductivity measurements made on piston core samples. The heat flow values range from 49 to 626 mW m^{-2} and are generally high, with about 25% of the values greater than 220 mW m^{-2}. There is significant local variation in values, with the highest values in the central part of the basin and along the southeast and northeast edges of the basin. In the western part of the basin values are generally less than 100 mW m^{-2}.

The highest value (626 mW m^{-2}) is near the center of the basin, although five additional measurements within a few hundred meters of this high value vary from 150 to 250 mW m^{-2}, typical of regions with active hydrothermal circulation. While this region of very high heat flow is not in line with the volcanic islands and submarine volcanoes of Bransfield rift, it is close to a region of recent faulting seen in the multichannel seismic reflection data (Fig. 8.8). The second highest recorded heat flow value (411 mW m^{-2}) was found on the northeast edge of the basin in an area of mounds a few meters in height and up to a kilometer in diameter.

9. HYDROTHERMAL ACTIVITY

Evidence for hydrothermal activity in Bransfield Strait has been found in the water column and in the sediments. In the waters of Bransfield Strait, δ^3He increases from less than zero near the surface to >7 at depth. This has been interpreted as evidence for injection of ^{3}He into the water by backarc rifting (Schlosser *et al.*, 1988). Mn concentrations in the water also increase with depth to a maximum of almost 7 nM. Temperature changes are less pronounced, but appear to mimic the Mn profiles (Suess *et al.*, 1988). Maturation of biogenic components in the sediments (Whiticar *et al.*, 1985; Brault and Simoneit, 1990) is similar to the thermogenic hydrocarbons found to be associated with hydrothermal activity in the Gulf of California (Simoneit, 1983). Significant downcore decreases in amino acid abundances despite relatively constant δ^{13}C, δ^{15}N, C/N, and percent carbonate C in the sediments suggest hydrothermally accelerated complexation of amino acids (Silfer *et al.*, 1990).

Within the area of high and variable heat flow in King George Basin, several hundred grams of sediment were recovered in the heat flow probe. Qualitative analysis of this sediment by an electron microprobe indicated the presence of grains of Fe sulfide, Fe-Zn sulfide, Fe-Zn-Cu sulfide, Zn chloride, and metals or oxides of Zn and Fe, all presumed to be products of hydrothermal activity. This indicates hydrothermal activity occurred in the region of high heat flow (62°14′S, 57°44′W) and water-column measurements indicate that hydrothermal circulation is still active.

10. PETROLOGY AND GEOCHEMISTRY

10.1. Regional

Volcanism near the northern end of the Antarctic Peninsula can be divided into the main arc phase and a later phase associated with rifting of the peninsula (Bransfield Strait) and volcanism on the Weddell Sea side of the peninsula (James Ross Island Volcanic Group). Volcanism is presently active on both sides of the peninsula; however, the composition of the rocks on the two sides are chemically distinct (Fig. 8.9) and are also chemically distinct from the earlier arc volcanism.

10.1.1. Antarctic Peninsula Arc

Although the crust beneath the South Shetland Islands is assumed to be continental, actual basement rocks are unknown. According to Birkenmajer (1992), the basement of Bransfield Strait consists of continental-type crust modified by basic intrusions. The oldest exposed rocks are metasediments on Livingston Island (Fig. 8.4) and are thought to be between late Carboniferous and Triassic in age (Thomson, 1992). Similar rocks on the Antarctic Peninsula are between Permian and Cretaceous in age (Pankhurst, 1983). The

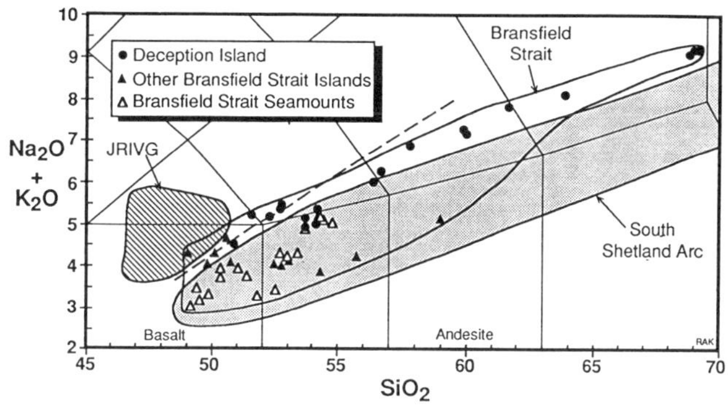

FIGURE 8.9. Weight percent SiO_2 versus weight percent Na_2O+K_2O for samples from the South Shetland island arc, Bransfield Strait, and the James Ross Island Volcanic Group (JRIVG). South Shetland arc data are from Saunders *et al.* (1980), Smellie (1988), Smellie *et al.* (1984, 1988), and Keller *et al.* (1992). Bransfield Strait data are from Keller and Fisk (1992) and Keller *et al.* (1992). Additional data from Bransfield Strait consistent with those shown here can be found in Weaver *et al.* (1979). JRIVG field is from data in Table III. Classification fields are from Le Bas *et al.* (1986). Dashed line is silica saturation line from Irvine and Baragar (1971).

oldest igneous rocks in the area are 155 to 185 Ma plutonic rocks on the northern Antarctic Peninsula near James Ross Island (Saunders *et al.*, 1982). K-Ar dates of 351 to 384 Ma of micas from granites from the same area have been reported (Rex, 1976) but are not confirmed by Rb-Sr dating. Plutonic rocks along the Bransfield Strait side of the peninsula are mostly 70 to 125 Ma, but some are as old as 130 to 145 Ma (Saunders *et al.*, 1982). The oldest extrusive rocks on the South Shetland Islands are late Jurassic–early Cretaceous volcaniclastic rocks on islands at the western end of the chain (Smellie *et al.*, 1980). A granodiorite intruding the volcaniclastic beds yielded a K-Ar age of 121 Ma (Smellie *et al.*, 1984). Additional isotopic ages suggest relatively continuous arc volcanism in the South Shetland Islands from late Jurassic to late Tertiary (Pankhurst and Smellie, 1983), although dates younger than about 20 Ma are extremely rare in the subduction-related volcanic rocks (Rex and Baker, 1973; Smellie *et al.*, 1984; Birkenmajer *et al.*, 1986).

Arc volcanism on the South Shetland Islands consisted mainly of calc-alkaline, high-alumina basalts, basaltic andesites, and low-silica andesites. Dacites and rhyolites are rare (Smellie, 1983; Smellie *et al.*, 1984). Exposures of intrusive rocks are less common and are mainly gabbros, tonalites, and granodiorites interpreted to be cogenetic with the extrusive rocks (Smellie, 1983). The chemistry and mineralogy of the pre-Quaternary South Shetland Islands have been described in detail by Smellie *et al.* (1984). Islands in the southern part of Bransfield Strait, close to the Antarctic Peninsula, are Eocene–Oligocene arc tholeiites similar to rocks on the South Shetland Islands (Rex, 1976; Baker *et al.*, 1977).

Pre-Quaternary volcanic arc rocks from the South Shetland Islands and northern Antarctic Peninsula are calc-alkaline with mild tholeiitic tendencies (Smellie *et al.*, 1984). Chemical variations within the arc magmas can be accounted for by fractional crystallization and slight differences in extent of partial melting of the source (Smellie, 1983). Their alkalinity (Fig. 8.9) and trace element abundances (Saunders *et al.*, 1980; Smellie *et al.*, 1984) are typical of island-arc rocks; that is, the large-ion lithophile elements (e.g., Rb, Ba, K, and Ce) are enriched relative to the high-field-strength elements (e.g., Zr, Ti, and Nb). Normalized multielement plots show an obvious Nb depletion (Fig. 8.10).

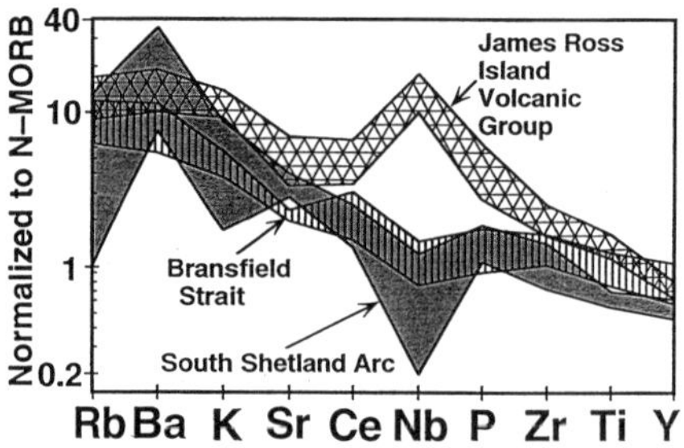

FIGURE 8.10. N-MORB-normalized multielement plot for basalts from the South Shetland arc (Tertiary basalts only), Bransfield Strait (seamounts only), and James Ross Island Volcanic Group. Data sources as in Fig. 8.9. N-MORB normalization values are from Saunders and Tarney (1984). Increasing distance from the South Shetland Trench corresponds to increasing Nb and decreasing Ba/Nb.

Strontium isotopic analyses are limited for the South Shetland Arc, but mainly range from 0.7033 to 0.7043 (Smellie *et al.*, 1984; Barbieri *et al.*, 1989; Jin Qingmin *et al.*, 1991; Keller *et al.*, 1992). The only Nd and Pb isotopic data available (excepting a single analysis in Keller *et al.*, 1992) are from the Fildes Peninsula on western King George Island and are published as ranges only: 0.51288 to 0.51301 for $^{143}Nd/^{144}Nd$, 18.503 to 18.641 for $^{206}Pb/^{204}Pb$, and 15.588 to 15.625 for $^{207}Pb/^{204}Pb$ (Jin Qingmin *et al.*, 1991). No $^{208}Pb/^{204}Pb$ data were included in Jin Qingmin *et al.* (1991). A single South Shetland arc basalt from the Low Head area of King George Island had $^{87}Sr/^{86}Sr$, $^{143}Nd/^{144}Nd$, and $^{207}Pb/^{204}Pb$ within the ranges reported by Jin Qingmin *et al.* (1991) but had higher $^{206}Pb/^{204}Pb$ (Keller *et al.*, 1992).

10.1.2. Bransfield Strait

Quaternary volcanism in Bransfield Strait occurred on Deception, Penguin, and Bridgeman Islands (Weaver *et al.*, 1979), and at a few isolated locations on King George, Livingston, and Greenwich Islands (Smellie, 1990), and on seamounts and ridges between Deception and Bridgeman Islands (Fig. 8.4; Fisk, 1990; Keller *et al.*, 1994). All reported K-Ar dates from the Bransfield Strait islands and seamounts are 300 ka or less (Table I). Deception Island is historically active (e.g., Smellie, 1990), and the most recent volcanism on Penguin Island was probably within the last 300 years (Birkenmajer, 1980). All of the sampled seamounts yielded fresh, glassy basalts (Fisk, 1990; Keller *et al.*, 1994).

The Quaternary volcanic centers in Bransfield Strait are mainly olivine basalts and basaltic andesites, although Deception Island erupted a complete basalt-to-trachydacite evolutionary suite (Keller *et al.*, 1992). Bridgeman Island is composed of basaltic andesites, while Melville Peak and Penguin Island are olivine basalts and scoria (Weaver *et al.*, 1979; Keller *et al.*, 1992). The seamounts are aphyric to olivine-plagioclase-phyric, vesicular, tholeiitic basalts, and basaltic andesites (Fig. 8.9; Fisk, 1990; Keller and Fisk, 1992).

The Bransfield Strait samples, with the exceptions of the sodic samples from Penguin and Deception Islands, can be classified as subalkaline basalts and basaltic andesites (Table II). They all have higher concentrations of the alkali and alkali earth elements (e.g., Na, K, Rb, Sr, and Ba), and less of an Fe enrichment trend than typical mid-ocean ridge tholeiites (Fisk, 1990; Keller *et al.*, 1992). Many of the Quaternary basalts are similar to the older-arc rocks in silica versus total alkalies (Fig. 8.9). Trace element concentrations are slightly to moderately enriched relative to N-MORB, and show a subtle negative Nb anomaly when normalized to N-MORB (Fig. 8.10). Low-pressure, fractional crystallization of olivine, spinel, and plagioclase can account for the chemical variation between samples from an individual volcano (Keller and Fisk, 1992; Keller *et al.*, 1992). Sr-isotopic ratios

TABLE I
K-Ar Data for Basalts from Bransfield Strait and the James Ross Island Volcanic Group

Location	Sample	K (wt%)	$^{40}Ar_{rad}$ (10^{-9} cc/g)	$^{40}Ar_{rad}$ (%)	Age ± 1 σ (10^3 yr)	Ref.[a]
Bransfield Strait						
King George Island						
Melville Peak	MP372	0.694	6.229	6.7	231 ± 19	3
Melville Peak	MP375	0.767	2.137	2.2	72 ± 15	3
Melville Peak	MP406	0.802	9.240	5.3	296 ± 27	3
Livingston Island						
Gleaner Heights	P.51.1	0.433	0.001	0.2	100 ± 400	4
Greenwich Island						
Mt. Plymouth	P.54.1	0.305	0.002	0.4	200 ± 300	4
Mt. Plymouth	P.55.1	0.300	0.002	0.5	200 ± 400	4
Deception Island	D1048	0.390	2.266	0.928	105 ± 46	1
Bridgeman Island	B2	0.581	1.414	0.932	63 ± 25	1
Eastern Seamount	300.13	0.435	0.8779	0.2744	53 ± 36	2
Western Seamount	310.14	0.418	1.631	0.6808	103 ± 35	2
James Ross Island Volcanic Group						
James Ross Island, Prince Gustav Channel islands, Vega Island, Humps Island, and Paulet Island				300 ± 100 to 7130 ± 490		5,6
James Ross Island						
Dreadnought	4M18	0.838	216.6	26.3	6643 ± 102	1
Villar Fabre	5M117	1.063	162.9	13.7	3940 ± 85	1
Cockburn Island	CK3	0.992	129.7	31.3	2781 ± 32	1

[a]Reference codes: 1—This paper; 2—Fisk (1990); 3—Birkenmajer and Keller (1990); 4—Smellie et al. (1984); 5—Smellie *et al.* (1988); 6—Sykes (1988). Samples with reference codes 1, 2, and 3 were analyzed at Oregon State University using techniques described in Fisk (1990).

range from 0.70304 to 0.70386, and Nd-isotopic ratios from 0.51287 to 0.51301. Pb isotopes range from 18.691 to 18.757 ($^{206}Pb/^{204}Pb$), 15.594 to 15.624 ($^{207}Pb/^{204}Pb$), and 38.441 to 38.556 ($^{208}Pb/^{204}Pb$) (Keller *et al.*, 1992).

10.1.3. James Ross Island Volcanic Group

The James Ross Island Volcanic Group (JRIVG) crops out on the northern tip of the Antarctic Peninsula and on numerous islands in the northwestern Weddell Sea in a backarc setting that may be analogous to the Patagonia Plateau basalts. Many outcrops of the JRIVG are sheer cliffs of palagonitized basaltic hyaloclastite and pillow breccia capped by coeval basalt flows (Nelson, 1966). Basaltic dikes intrude the underlying Cretaceous sediments and Tertiary glacial deposits. The basalts are mainly alkalic (Fig. 8.9), but hypersthene- and quartz-normative tholeiites have also been reported (Smellie, 1987). The oldest rocks in the JRIVG are 7 Ma basalt clasts in glacial deposits beneath the main volcanic pile on James Ross Island (Table I). Volcanic activity continued through the Pliocene and Quaternary, with the youngest K-Ar date being less than 0.3 Ma (Table I). The most recent activity on James Ross Island occurred at three volcanic cones near the margin of the ice cap. These cones are undisrupted by glacial activity and are undoubtedly of a very young age (Strelin *et al.*, 1993).

The alkalic basalts of the JRIVG are chemically distinct from both the South Shetland

TABLE II

Representative Whole Rock Major and Trace Element Concentration for Quaternary Volcanic Rocks from Bransfield Strait[a]

	Western seamount		Eastern seamount		Axial ridge	Deception Island						Bridgeman Island	Penguin Island	Melville Peak
Sample	292.18	310.07	297.01	309.01	DF86.32	D1048	D1082	D1051	D1055	D1075	D1060	B1	P174	MP405
SiO_2	49.5	50.4	52.5	54.3	54.3	50.9	53.8	51.6	68.9	63.9	57.8	55.8	50.6	52.8
TiO_2	1.13	1.54	1.05	1.71	1.82	1.58	1.82	2.44	0.60	1.12	1.86	0.78	1.30	0.92
Al_2O_3	14.1	16.3	16.9	15.7	15.8	17.3	16.2	15.2	14.5	15.6	15.6	18.7	17.2	16.7
FeO*	8.6	9.0	7.1	10.1	9.6	8.4	9.0	11.0	4.3	6.7	8.6	6.6	8.7	7.1
MnO	0.16	0.16	0.14	0.17	0.19	0.15	0.17	0.19	0.15	0.17	0.17	0.12	0.16	0.14
MgO	12.3	6.7	7.0	3.7	4.4	6.1	4.6	4.9	0.3	1.2	2.6	5.0	7.9	7.1
CaO	10.4	10.9	11.3	7.4	8.1	10.3	8.9	9.1	1.8	3.6	6.0	9.4	10.2	11.3
Na_2O	2.73	3.40	3.04	4.23	4.72	4.05	4.40	4.71	7.17	6.67	5.90	3.52	3.99	3.09
K_20	0.46	0.55	0.39	0.97	0.45	0.47	0.51	0.50	1.89	1.43	0.96	0.70	0.66	0.91
P_2O_5	0.16	0.19	0.12	0.23	0.30	0.25	0.29	0.36	0.12	0.42	0.37	0.08	0.29	0.14
Total	99.56	99.08	99.56	98.49	99.65	99.55	99.66	100.06	99.71	100.76	99.77	100.64	100.96	100.16
Sc	35	32	31	30	27	32	35	37	11	17	20	29	29	42
V	256	328	220	351	235	245	265	336	13	50	194	210	282	249
Cr	618	192	228	8	23	106	42	44	0	2	9	40	218	157
Ni	283	80	84	3	8	28	14	14	9	7	1	27	95	41
Cu	68	76	54	70	29	31	30	47	8	17	54	92	154	79
Zn	63	69	56	92	85	66	77	81	87	90	85	56	71	55
Ga	16	19	16	17	21	17	19	21	24	23	20	17	22	18
Rb	10	10	7	17	4	7	9	5	30	21	14	17	7	17
Sr	255	266	280	330	189	465	374	350	130	254	325	357	636	598
Y	23	26	21	37	52	26	33	36	62	56	43	14	16	21
Zr	93	104	94	154	197	143	167	171	444	355	249	72	90	112
Nb	2	4	2	4	7	7	7	8	16	14	10	1	2	2
Ba	91	98	65	144	6	40	62	29	256	167	130	76	146	173
Zr/Y	4.0	4.0	4.5	4.2	3.8	5.5	5.1	4.8	7.2	6.3	5.8	5.1	5.6	5.3

[a]Concentrations were determined by x-ray fluorescence at Washington State University, and are given as weight percent oxides for the major elements and parts per million for the trace elements. FeO* is total Fe as FeO. Data for Western and Eastern seamounts are from Keller and Fisk (1992). Other data are from Keller *et al.* (1992). Data for Deception, Bridgeman, and Penguin Islands and Melville Peak have also been published by other workers (e.g., Weaver *et al.*, 1982; Smellie, 1990) and are comparable to data given here.

arc and from the <1 Ma Bransfield Strait basalts. The JRIVG tends to have lower SiO_2 (Fig. 8.9), and higher K_2O and TiO_2 (Table III). Paulet Island, especially, is high in total alkalies, Al_2O_3, and many incompatible trace elements. High concentrations of alkalies and incompatible trace elements in the JRIVG are reminiscent of many continental-rift basalts and alkalic ocean island basalts. Their prominent Nb hump on the MORB-normalized multi-element plot (Fig. 8.10) is a common characteristic of intraplate basalts not associated with a subduction zone and is in marked contrast to the Nb depletions of the South Shetland arc and Bransfield Strait backarc basalts. The $^{87}Sr/^{86}Sr$ values range from 0.70306 to 0.70348, and $^{143}Nd/^{144}Nd$ from 0.51282 to 0.51293 (Keller *et al.*, 1993), which places them in the lower left quadrant in $^{87}Sr/^{86}Sr$ versus $^{143}Nd/^{144}Nd$ space, displaced toward St. Helena–type values.

10.1.4. Temporal Comparison

The hiatus between the end of arc volcanism on the South Shetland Islands and the beginning of volcanism in Bransfield Strait was approximately 20 m.y., but the major and trace element compositions are similar in some cases (Figs. 8.9 and 8.10). Bridgeman Island, in particular, could be mistaken for an island-arc tholeiite (Table II). The submarine samples, on the other hand, are more similar to backarc basin basalt (i.e., a hybrid of arc and mid-ocean ridge basalt compositions; Fisk, 1990). The isotopic ratios of the arc and backarc rocks are also similar, which implies that they all were derived from similar sources. However, the lack of trace element and isotopic data from the South Shetland Islands and the lack of samples from Bransfield Strait (partially rectified in 1993) preclude thorough comparisons.

10.1.5. Across-Arc Comparison

The importance of the chemical contribution from a subducting slab has been shown to decrease with distance from the trench in southern South America (Stern *et al.*, 1990). On the northern Antarctic Peninsula, the occurrence of volcanism at a variety of distances from the South Shetland Trench (50–125 km for the South Shetland arc, 125–150 km for the Bransfield Strait, and 300–400 km for the James Ross Island Volcanic Group) is an excellent chance to constrain the spatial extent of contamination of the mantle by a subducting slab. On the northern Antarctic Peninsula we see an obvious decrease in contribution from the subducted slab with increasing distance from the trench. The high Ba and low Nb of the South Shetland arc samples are typical of the subduction-zone influence found in many volcanic arcs. The Bransfield Strait basalts are farther from the trench, have a slight negative Nb anomaly, and are moderately enriched in the alkalies relative to MORB. This is typical of backarc basin basalts, and suggests less of a contribution from the subducted slab. The JRIVG basalts are farthest from the South Shetland Trench and do not show any evidence for the influence of the slab subducting at the trench. Their high Nb is typical of intraplate basalts and is the opposite of what would be expected if their source had been contaminated by subduction (Keller *et al.*, 1993). Apparently, between 150 and 300 km is the greatest distance that the subducting slab can influence the chemistry of the volcanism above it.

Some published comparisons of arc and backarc basalts suggest that the upwelling mantle in backarc regions contains an ocean island basalt (OIB) component (Stern *et al.*, 1990; Hickey-Vargas, 1992). This OIB component can be seen in the JRIVG data but not

TABLE III

James Ross Island Volcanic Group Whole Rock Major and Trace Element Concentrations by X-ray Fluorescence[a]

	James Ross Island																
	Hidden Lake		Dreadnought		Massey				Conico	Ventisca	Kotick		Villar Fabre				
Lat (S)	64°02′		64°00′		63°57′				63°56′	63°57′	64°01′		64°06′				
Lon (W)	58°18′		57°50′		57°58′				58°08′	58°04′	58°22′		58°22′				
SiO_2	48.1	47.9	48.5	48.2	50.2	47.1	47.4	48.5	50.4	47.1	46.8	48.4	48.3	48.4	47.9	49.8	48.2
TiO_2	1.78	2.01	2.09	1.89	1.84	1.71	1.95	2.04	1.87	1.71	1.75	1.55	1.69	1.59	1.64	1.77	1.74
Al_2O_3	15.1	16.1	16.0	15.7	16.4	15.4	15.7	16.1	16.3	15.1	15.1	14.9	15.3	15.5	15.0	15.8	15.1
FeO*	10.5	10.0	9.8	10.4	9.5	10.9	10.2	10.6	9.7	10.7	10.5	10.9	10.6	10.8	11.1	10.4	12.0
MnO	0.16	0.17	0.16	0.16	0.15	0.17	0.16	0.17	0.16	0.16	0.16	0.17	0.15	0.16	0.17	0.15	0.18
MgO	8.9	7.2	7.6	7.9	6.7	8.1	8.6	8.4	6.8	10.3	10.0	9.2	9.1	8.9	8.6	7.8	9.1
CaO	8.2	9.9	8.9	9.2	8.3	8.4	8.4	9.5	9.2	8.0	8.5	8.6	8.0	7.9	7.8	7.9	8.0
Na_2O	3.34	3.56	3.39	3.37	3.83	3.34	3.64	3.57	3.58	2.58	2.96	3.11	3.27	3.16	3.43	3.71	3.27
K_2O	1.20	0.93	1.13	1.01	1.59	0.83	1.44	1.19	1.37	1.13	0.89	0.72	1.01	0.92	1.05	1.28	0.95
P_2O_5	0.38	0.38	0.47	0.49	0.49	0.35	0.63	0.50	0.36	0.42	0.33	0.28	0.38	0.36	0.37	0.54	0.21
Total	97.57	98.16	98.10	98.38	98.96	96.17	97.99	100.55	99.71	97.15	96.94	97.73	97.64	97.66	97.02	99.17	98.72
Sc	19	27	28	27	23	24	22	29	30	20	26	20	22	21	19	13	25
V	176	207	191	180	169	159	169	197	195	149	167	166	146	154	151	150	148
Cr	276	178	207	256	188	223	228	239	184	313	304	272	233	231	226	199	225
Ni	166	75	94	115	96	146	157	116	86	200	180	180	177	162	168	145	170
Cu	41	39	48	50	37	58	56	49	44	48	42	52	44	51	41	47	35
Zn	92	83	85	82	90	93	87	81	87	88	90	90	101	99	99	97	90
Ga	16	17	19	15	17	16	18	15	19	21	16	18	18	18	15	17	20
Rb	17	11	13	12	19	10	17	13	25	10	13	11	10	9	13	12	10
Sr	506	510	607	611	701	453	742	618	481	686	510	359	482	446	475	653	395
Y	24	25	24	26	24	24	25	26	28	22	22	25	24	22	24	21	19
Zr	154	168	191	178	203	162	226	180	164	174	134	129	142	141	142	197	139
Nb	27	29	36	40	41	38	41	39	28	28	24	17	24	25	24	38	23

	James Ross Island																	
	Sta Marta	Dobson			Ekelof	Coley	Marina			Elba			Paulet Island				Cockburn Island	
Lat (S)	63°54′	64°02′			64°12′	64°11′	64°10′			64°15′			63°35′				64°12′	
Lon (W)	57°54′	57°56′			57°14′	57°10′	57°22′			57°30′			55°46′				56°54′	
SiO_2	47.2	48.0	47.8	48.3	46.1	48.3	48.9	48.5	48.8	48.7	48.8	48.5	48.2	47.4	46.9	47.8	48.0	48.8
TiO_2	1.73	1.56	1.65	1.64	2.12	2.11	1.99	2.00	1.95	2.44	2.45	2.46	2.26	2.26	2.26	2.30	1.82	1.86
Al_2O_3	15.4	15.5	15.2	15.4	14.6	15.7	15.4	15.3	15.3	18.2	18.4	18.3	16.9	16.9	16.7	17.0	16.0	16.4
FeO*	10.3	10.9	10.6	11.0	10.9	10.7	10.4	10.4	10.2	10.4	10.2	10.8	9.7	9.6	9.5	9.3	10.5	10.3
MnO	0.17	0.17	0.17	0.17	0.16	0.16	0.16	0.16	0.15	0.16	0.16	0.16	0.16	0.16	0.16	0.16	0.16	0.16
MgO	8.5	9.5	9.3	8.7	9.4	7.6	8.9	9.0	8.9	4.7	4.8	4.6	7.7	6.6	6.9	6.7	7.0	6.7
CaO	7.8	8.7	8.7	8.6	8.2	9.5	8.0	7.9	7.8	7.6	7.4	7.3	9.3	8.9	9.1	9.2	8.6	8.7
Na_2O	3.74	3.11	3.27	3.37	4.03	3.50	4.10	3.99	4.35	4.89	5.32	4.93	4.21	4.16	4.30	4.33	3.69	3.59
K_2O	1.38	0.82	0.91	0.88	1.17	1.05	1.57	1.56	1.65	1.55	1.97	1.82	1.40	1.43	1.37	1.44	1.16	1.22
P_2O_5	0.52	0.32	0.37	0.36	0.96	0.38	0.63	0.62	0.64	0.75	0.84	0.84	0.81	0.75	0.75	0.86	0.40	0.43
Total	96.79	98.54	98.02	98.56	97.63	98.97	99.96	99.48	99.66	99.29	100.24	99.72	100.68	98.19	97.83	99.03	97.27	97.99
Sc	17	30	27	23	18	26	23	22	21	16	17	18	28	23	23	22	24	25
V	151	185	195	189	137	211	165	177	185	126	151	121	185	187	204	191	173	177
Cr	237	295	294	260	288	224	288	288	282	10	10	10	180	102	114	115	180	150
Ni	191	189	188	164	206	86	132	138	135	12	17	17	99	69	78	74	96	77
Cu	47	53	59	61	47	21	29	34	29	29	27	21	29	31	41	43	19	29
Zn	92	85	87	88	98	90	93	87	91	78	82	86	74	75	81	75	85	81
Ga	17	17	17	18	15	21	21	20	19	18	18	22	19	19	16	20	18	20
Rb	11	13	13	15	11	11	17	17	17	13	17	16	17	17	16	16	16	16
Sr	593	435	431	433	1383	536	728	720	732	945	911	917	951	940	943	937	536	547
Y	22	22	25	25	21	25	23	23	23	29	28	29	28	30	29	28	23	25
Zr	179	120	131	135	296	149	206	205	208	254	250	254	225	229	216	228	163	170
Nb	34	19	21	24	91	26	43	43	43	55	55	55	45	47	46	48	26	27

[a]Concentrations were determined by x-ray fluorescence at Washington State University using techniques described in Keller *et al.* (1992), and are given as weight percent oxides for the major elements and parts per million for the trace elements. FeO* is total Fe as FeO. Data for Paulet Island and some of these locations on James Ross Island have also been published by other workers (Nelson, 1966; Baker *et al.*, 1977; Smellie, 1987) and are comparable to the data given here.

in the Bransfield Strait data (Keller *et al.*, 1992). Apparently the OIB component is not available or is not melted beneath Bransfield Strait, just as the subducted component does not appear in the JRIVG lavas. The presence of a subducted slab beneath James Ross Island should rule out the possibility that a rising plume or hot spot is responsible for the OIB-like compositions of the JRIVG. Also, the "slab-window" hypothesis that has been suggested for the origin of similar alkalic basalts found farther south on the Antarctic Peninsula at Seal Nunataks and Alexander Island (Hole *et al.*, 1991) cannot apply to James Ross Island.

10.2. Intra-Bransfield Strait Comparison

All of the recent (≤300 ka) episode of volcanic activity in Bransfield Strait can be classified as basalts to basaltic andesites, and on Deception Island as basalts to dacites. These rocks appear to be the product of <5% to 15% melting of mantle that contained 0.5% to 2% of a subducted component (Keller *et al.*, 1992). Within these rocks, however, we identify three contrasts in chemistry that are correlated with differences in mode or location of eruption: (1) on-axis versus off-axis, (2) submarine versus subaerial, and (3) northeast of 57°50′W versus southwest of 57°50′W. These differences are probably the result of different mantle sources and processes.

10.2.1. On Axis versus off Axis

The two off-axis volcanoes, Penguin Island and Melville Peak (Fig. 8.4), have high Sr, Ba, and K, and low Y relative to the on-axis basalts. Off-axis volcanic remnants on Livingston and Greenwich islands, possibly contemporary with Penguin Island and Melville Peak, also have high Sr and Ba and low Y, similar to Penguin Island and Melville Peak (Smellie *et al.*, 1984). The high-Sr abundances of Penguin Island and Melville Peak are not the result of crustal assimilation because other chemical changes that would be associated with this assimilation are not observed (Keller *et al.*, 1992). High Sr is also typical of the older (>20 Ma) volcanic arc rocks of the South Shetland Islands, so the off-axis Bransfield Strait volcanoes may have tapped the older arc mantle source. The lower $^{206}Pb/^{204}Pb$ of the off-axis samples also shows that the mantle that produced the off-axis lavas was chemically different from that which produced the on-axis basalts. The high Sr of Penguin Island and Melville Peak excludes the possibility of residual plagioclase in the mantle source (Keller *et al.*, 1992). Small amounts of melting of a deeper, mineralogically different (garnet-bearing) source is also suggested by the higher Ce/Sm (Keller *et al.*, 1992) and other trace elements (Weaver *et al.*, 1979) of the off-axis basalts. Thus, off-axis volcanoes appear to be the result of a smaller degree of melting of a deeper and chemically distinct mantle than on-axis volcanoes (Weaver *et al.*, 1979; Keller *et al.*, 1992).

10.2.2. Submarine versus Subaerial

The seamounts have lower Zr/Y ratios (~4; Table II) than the subaerial volcanoes (mostly >5). This difference is not a result of alteration, since both Zr and Y are relatively immobile during weathering and most of the rocks are fresh. The degree of melting of the mantle can affect the Zr/Y ratio; however, this effect is only significant for small (<5%) amounts of melt, which would also be expressed in much larger amounts of incompatible elements in the subaerial volcanoes than in the submarine ones. Some trace elements

(Ba, Rb, K), however, are lower in the subaerial volcanoes than in the submarine volcanoes, indicating that chemical differences between these two groups are not related to the degree of melting.

Most of the subaerial volcanoes are larger than the submarine volcanoes, so their higher Zr/Y may represent a characteristic of a volcano as it grows; that is, continuous partial melting of the mantle beneath the larger volcanoes could produce high Zr/Y, and the smaller submarine volcanoes have not yet evolved to this chemistry. Alternatively, the lower Zr/Y of the submarine basalts (which are closer to MORB values of 2.5 to 3.5) may indicate that as the rift develops and deepens, the volcanoes become MORB-like. The rift sample (DF86.32 in Table II) has the lowest Zr/Y (3.8) of any Bransfield Strait rock and appears to be the most similar to MORB. It also has the lowest Rb, Sr, and Ba. Finally, the larger volcanoes, which are mainly subaerial, may represent melting of deeper mantle where garnet is present.

10.2.3. Along Axis

The chemistry of basalts from backarc rifts can vary significantly along strike of the rift, as has been found in the Mariana Trough, Lau Basin, and Gulf of California (Saunders and Tarney, 1984; Hawkins and Melchior, 1985; and Hawkins *et al.*, 1990). This is also true of Bransfield Strait, where Na_2O increases and Rb/Sr decreases abruptly from NE to SW at approximately 57°50′W. The change occurs between Melville Peak and Penguin Island and is evident in all of the analyzed submarine and subaerial samples. Basalts that are northeast of this boundary also have lower total alkalies and lower $^{87}Sr/^{86}Sr$ at the same $^{143}Nd/^{144}Nd$ compared with basalts southwest of this boundary. The chemical boundary is found in the same general region as the highest heat flow.

The abrupt change in isotopes along axis indicates there is a change in mantle sources across the boundary, perhaps due to a change in age or composition of the subducted oceanic crust. The boundary approximately coincides with the onshore extrapolation of fracture zone E (Fig. 8.4), which separates younger crust to the southwest and older crust to the northeast (Larter and Barker, 1991). Alternatively, the chemical boundary may reflect the contrast between subducted oceanic crust created at the Antarctic–Phoenix Ridge versus oceanic crust created at the Nazca–Phoenix Ridge. Different inputs to the mantle wedge from the subducted slab could account for the isotopic differences as well as induce changes in melting or mineralogy in the mantle. Elsewhere, along-arc variations in basalt chemistry have been attributed to variations in the composition of the subducted material (e.g., Lin *et al.*, 1990).

11. CONCLUSIONS

Regional tectonics of the Antarctic Peninsula support the hypothesis that little or no active subduction is occurring beneath Bransfield Strait and the South Shetland Islands. What little subduction that may be occurring would be equal to the amount of extension in Bransfield Strait. Spreading ceased on the Antarctic–Phoenix spreading center at approximately 4 Ma. The remnant of the Phoenix plate was pinned under the South Shetland Islands, where it has been imaged to a depth of 40 km by seismic refraction results (Grad *et al.*, 1993). Teleseismic earthquakes do not indicate any motion of a subducted slab beneath Bransfield Strait, although one event (event 18; Pelayo and Wiens, 1989) may have been produced along the downdip extension of the Hero fracture zone. Plate reorganization since ~4 Ma resulted in a 110° bend in the Antarctic–Scotia plate boundary that seems to have imposed extensional stress on the Bransfield Strait region.

Bransfield Strait is a region of recent extension based on active and recent volcanism, high heat flow, earthquake fault plane solutions, and seismic reflection data. This extension may also be associated with trench rollback at the South Shetland Trench and defines the SE boundary of the South Shetland Islands microplate. Seismic refraction results (Grad *et al.*, 1993) indicate that the crust under Bransfield Strait is anomalously thick, up to 30 km, with an unusual layer of crustal material with $v_p > 7.0$ km s^{-1}. They also found a 10-km-wide body with seismic velocity $v_p = 6.8$ km s^{-1}, which they interpreted to be the active Bransfield rift. Fault plane solutions (Pelayo and Wiens, 1989) from the Bransfield Strait region support active extension, and some earthquake swarms can be directly correlated to episodes of active volcanism. Barker and Austin (1994) and Lawver and Villinger (1989) show evidence of active crustal extension from seismic reflection results, supportive of continental crustal extension rather than normal seafloor spreading.

Active arc volcanism seems to have ceased at about 20 Ma. Recent Bransfield Strait basalts are chemically transitional between island-arc and ocean ridge basalts, a feature they share with many marginal basin basalts. Such chemistry agrees with the transitional nature of the tectonics of the strait (i.e., transitional between arc and spreading regimes). Considerable data supporting the existence of hydrothermal activity in Bransfield Strait have already been published. The new discovery of sulfide minerals in the sediments at a high heat flow site in Bransfield Strait is additional support for recent hydrothermal activity.

Acknowledgments

This work was supported by a number of National Science Foundation grants to the authors and to those that have provided us with their results. Grants include, but are not limited to, DPP-8916436 and DPP-9019247 to L. Lawver and DPP-8817126 to M. Fisk. Field work on James Ross Island was supported by the Argentine Antarctic Institute and Oregon State University. We are grateful to S. Porebski for donating the Cockburn Island samples, and to E. Godoy for donating the Paulet Island samples. GPS-navigated multi-beam bathymetry were collected on cruise 91-01 of R/V *Maurice Ewing* and two cruises of NOAA ship *Surveyor*. Heat flow data was collected on R/V *Polar Duke* cruise PD89-IV. We wish to thank the captains and crews from all those ships, without whose assistance this work would not have been possible. We would particularly like to thank Rob Larter, Dallas Abbott, and Daniel Barker for reviewing an unacceptably rough draft of this manuscript. This is UTIG contribution number 1062.

REFERENCES

Acosta, J., Herranz, P., and Sanz, J. L. 1992a. Perfiles sismicos en el rift de Bransfield, Campaña Exantarte 90/91, in *Geologica de la Antártida Occidental* (J. López-Martinez, ed.), pp. 195–202, Simposios T 3, III Congreso Geológico de España y VIII Congreso Latinoamericano de Geologia, Salamanca.

Acosta, J., Herranz, P., Sanz, J. L., and Uchupi, E. 1992b. Antarctic continental margin: Geological image of the

Bransfield Trough, an incipient oceanic basin, in *Geological Evolution of Atlantic Continental Rises* (C. W. Poag and P. C. Graciansky, eds.), pp. 49–61, Van Nostrand Reinhold, New York.
Anderson, J. A., Pope, P. G., and Thomas, M. A. 1990. Evolution and hydrocarbon potential of the northern Antarctica Peninsula continental shelf, in *Antarctica as an Exploration Frontier: Hydrocarbon Potential, Geology and Hazards* (B. St. John, ed.), pp. 1–12, Am. Assoc. Petrol. Geol., Studies in Geology 31.
Ashcroft, W. A. 1972. Crustal structure of the South Shetland Islands and Bransfield Strait, *Br. Antarc. Surv. Sci. Rep.* **66**.
Baker, P. E., Buckley, F., and Rex, D. C. 1977. Cenozoic volcanism in the Antarctic, *Phil. Trans. R. Soc. London* **279**:131–142.
Barbieri, M., Birkenmajer, K., Delitala, M. C., Francalanci, L., Narebski, W., Nicoletti, M., Peccerillo, A., Petrucciani, C., Todaro, M. L., Tolomeo, L., and Trudu, C. 1989. Preliminary petrological, geochemical and Sr isotopic investigation on Mesozoic to Cainozoic magmatism of King George Island, South Shetland Islands (West Antarctica), *Mineral. Petrogr. Acta (Bologna)* **32**:37–49.
Barker, D. H. N., and Austin, J.A., Jr. 1994. Crustal diapirism in Bransfield Strait, West Antarctica—evidence for distributed extension in marginal basin formation, *Geology* **22**:657–660.
Barker, P. F. 1976. The tectonic framework of Cenozoic volcanism in the Scotia Sea region, a review, in *Andean and Antarctic Volcanology Problems* (O. González-Ferrán, ed.), pp. 330–346, International Association of Volcanology and Chemistry of the Earth's Interior, Rome.
Barker, P. F. 1982. The Cenozoic subduction history of the pacific margin of the Antarctic Peninsula: Ridge crest-trench interactions, *J. Geol. Soc. London* **139**:787–801.
Barker, P. F., and Burrell, J. 1977. The opening of Drake Passage, *Mar. Geol.* **25**:15–34.
Barker, P. F., and Dalziel, I. W. D. 1983. Progress in geodynamics of the Scotia Arc region, in *Geodynamics of the Eastern Pacific Region, Caribbean and Scotia Arcs* (R. Cabre, ed.), Geodyn. Ser., Vol. 9, pp. 137–170, American Geophysical Union, Washington DC.
Barker, P. F., Dalziel, I. W. D., and Storey, B.C. 1992. Tectonic development of the Scotia arc region, in *Antarctic Geology* (R.J. Tingey, ed.), pp. 215–248, Oxford University Press, Oxford.
Birkenmajer, K. 1980. Age of the Penguin Island volcano, South Shetland Islands (West Antarctica) by the lichenometric method, *Bull. Acad. Pol. Sci. Ser. Sci. Terre* **27**:69–76.
Birkenmajer, K. 1992. Evolution of the Bransfield Basin and rift, West Antarctica, in *Recent Progress in Antarctic Earth Science* (Y. Yoshida, K. Kaminuma, and K. Shiraishi, eds.), pp. 405–410, Terra, Tokyo.
Birkenmajer, K., Delitala, M. C., Narebski, W., Nicoletti, M., and Petrucciani, C. 1986. Geochronology of Tertiary island-arc volcanics and glacigenic deposits, King George Island, South Shetland Islands (West Antarctica), *Bull. Acad. Pol. Sci. Ser. Sci. Terre* **34**:257–273.
Birkenmajer, K., and Keller, R. A. 1990. Pleistocene age of the Melville Peak volcano, King George Island, West Antarctica, by K-Ar dating, *Bull. Polish Acad. Sci., Earth Sci.* **38**:17–24.
Brault, M., and Simoneit, B. R. T. 1990. Mild hydrothermal alteration of immature organic matter in sediments from the Bransfield Strait, Antarctica, *Appl. Geochem.* **5**:149–158.
Dalziel, I. W. D. 1984. *Tectonic Evolution of a Forearc Terrane, Southern Scotia Ridge, Antarctica*, Geological Society of America Special Publication 200.
Fisk, M. R. 1990. Back arc volcanism in the Bransfield Strait, Antarctica, *J. South Am. Earth Sci.* **3**:91–101.
Gambôa, L. A. P., and Maldonado, P. R. 1990. Geophysical investigations in the Bransfield Strait and in the Bellingshausen Sea, Antarctica, in *Antarctica as an Exploration Frontier: Hydrocarbon Potential, Geology and Hazards* (B. St. John, ed.), pp. 127–141, Am. Assoc. Petrol. Geol., Studies in Geology 31.
Garrett, S. W. 1990. Interpretation of reconnaissance gravity and aeromagnetic surveys of the Antarctic Peninsula, *J. Geophys. Res.* **95**:6759–6777.
Gonzalez-Ferran, O. 1985. Volcanic and tectonic evolution of the northern Antarctic Peninsula—late Cenozoic to recent, *Tectonophysics* **114**:389–409.
Gonzalez-Ferran, O. 1991. The Bransfield rift and its active volcanism, in *Geological Evolution of Antarctica* (M. R. A. Thomson, J. A. Crame, and J. W. Thomson, eds.), pp. 505–509, Cambridge Univ. Press, Cambridge.
Grad, M., Guterch, A., and Janik, T. 1993. Seismic structure of the lithosphere across the zone of subducted Drake plate under the Antarctic plate, West Antarctica, *Geophys. J. Int.* **115**:586–600.
GRAPE Team. 1990. Preliminary results of seismic reflection investigations and associated geophysical studies in the area of the Antarctic Peninsula, *Antarctic Sci.* **2**:223–234.
Guterch, A., Grad, M., Janik, T., and Perchuc, E. 1992. Tectonophysical models of the crust between the Antarctic Peninsula and the South Shetland trench, in *Geological Evolution of Antarctica* (M. R. A. Thomson, J. A. Crame, and J. W. Thomson, eds.), pp. 499–504, Cambridge University Press, Cambridge.

Guterch, A., Grad, M., Janik, T., Perchuc, E., and Pajchel, J. 1985. Seismic studies of the crustal structure in West Antarctica 1979–1980—Preliminary results, *Tectonophysics* **114:**411–429.
Guterch, A., Shimamura, H., and Polish–Japan–Argentina Research Group. 1991. An OBS-land refraction seismological experiment in the Bransfield trough, West Antarctica, 1990/1991: Abstracts, Sixth International Symposium on Antarctic Earth Sciences, 9–13 September, 1991, Japan, pp. 201–202.
Hawkins, J. W., Lonsdale, P. F., Macdougall, J. D., and Volpe, A. M. 1990. Petrology of the axial ridge of the Mariana Trough backarc spreading center, *Earth Planet. Sci. Lett.* **100:**226–250.
Hawkins, J. W., and Melchior, J. T. 1985. Petrology of Mariana Trough and Lau Basin basalts, *J. Geophys. Res.* **90:**11431–11468.
Herron, E. M., and Tucholke, B. E. 1976. Sea-floor magnetic patterns and basement structure in the southeastern Pacific, in *Init. Repts. DSDP*, 35 (C. D. Hollister and C. Craddock, *et al.*, eds.), pp. 263–278, U.S. Government Printing Office, Washington, DC.
Hickey-Vargas, R. 1992. A refractory HIMU component in the sources of island-arc magma, *Nature* **360:**57–59.
Hole, M. J., and Larter, R. D. 1993. Trench-proximal volcanism following ridge crest—trench collision along the Antarctic Peninsula, *Tectonics* **12:**897–910.
Hole, M. J., Rogers, G., Saunders, A. D., and Storey, M. 1991. Relation between alkalic volcanism and slab-window formation, *Geology* **19:**657–660.
International Seismological Centre Earthquake Catalogs. 1964–1993.
Irvine, T. N., and Baragar, W. R. A. 1971. A guide to the chemical classification of the common volcanic rocks, *Can. J. Earth Sci.* **8:**523–548.
Jeffers, J. D., and Anderson, J. B. 1990. Sequence stratigraphy of the Bransfield Basin, Antarctica: Implications for tectonic history and hydrocarbon potential, in *Antarctica as an Exploration Frontier: Hydrocarbon Potential, Geology and Hazards* (B. St. John, ed.), pp. 13–30, Am. Assoc. Petrol. Geol., Studies in Geology 31.
Jeffers, J. D., Anderson, J. B., and Lawver, L. A. 1991. Evolution of Bransfield basin, Antarctic Peninsula, in *Geological Evolution of Antarctica* (M. R. A. Thomson, J. A. Crame, and J. W. Thomson, eds.), pp. 481–485, Cambridge University Press, Cambridge.
Jin, Q., Kuang, F., Ruan, H., and Xing, G. 1991. Island arc volcanism and magmatic evolution in Fildes Peninsula, King George Island, Antarctica, Abstracts, Sixth International Symp. Antarctic Earth Sci., Ranzan, Japan, pp. 250–255.
Keller, R. A., and Fisk, M. R. 1992. Quaternary marginal basin volcanism in the Bransfield Strait as a modern analogue of the southern Chilean ophiolites, in *Ophiolites and Their Modern Analogues* (L. M. Parson, B. J. Murton, and P. Browning, eds.), pp. 155–170, Geol. Soc. Lond. Spec. Pub. No. 60.
Keller, R. A., Fisk, M. R., and Strelin J. A. 1993. Correlating distance from a trench with subducted component in recent basalts from the northern Antarctic Peninsula [abs.], *EOS* supplement, October 26, 1993, p. 663.
Keller, R. A., Fisk, M. R., White, W. M., and Birkenmajer, K. 1992. Isotopic and trace element constraints on mixing and melting models of marginal basin volcanism, Bransfield Strait, Antarctica, *Earth Planet. Sci. Lett.* **111:**287–303.
Keller, R. A., Strelin, J. A., Lawver, L. A., and Fisk, M. R. 1994. Dredging young volcanic rocks in Bransfield Strait, *Antarctic J. U.S.* **XXVIII**(1993 review issue)**:**98–100.
Klepeis, K. A., and Lawver, L. A. 1994. Bathymetry of the Bransfield Strait, southeastern Shackleton fracture zone and South Shetland Trench, *Antarctic J. U.S.*, **XXVIII**(1993 review issue)**:** 103–104.
Larter, R. D. 1991. Debate: Preliminary results of seismic reflection investigations and associated geophysical studies in the area of the Antarctic Peninsula, *Antarctic Sci.* **3:**217–222.
Larter, R. D., and Barker, P. F. 1989. Seismic stratigraphy of the Antarctic Peninsula Pacific margin: A record of Pliocene–Pleistocene ice volume and paleoclimate, *Geology* **17:**731–734.
Larter, R. D., and Barker, P. F. 1991. Effects of ridge crest-trench interaction on Antarctic–Phoenix spreading:forces on a young subducting plate, *J. Geophys. Res.* **96:**19,583–19,607.
Larson, R. L., and Chase, C. G. 1972. Late Mesozoic evolution of the western Pacific Ocean, *Geol. Soc. Am. Bull.* **83:**3627–3644.
Lawver, L. A., Dalziel, I. W. D., and Sandwell, D. T. 1993. Antarctic plate: tectonics from a gravity anomaly and infrared satellite image, *GSA Today* **3:**117–122.
Lawver, L. A., and Hawkins, J. W. 1978. Diffuse magnetic anomalies in marginal basins: Their possible tectonic and petrologic significance, *Tectonophysics* **45:**323–338.
Lawver, L. A., and Villinger, H. 1989. North Bransfield Basin: R/V POLAR DUKE cruise PD VI-88, *Antarctic J. Sci.* **23:**117–120.
Le Bas, M. J., La Maitre, R. W., Streckeisen, A., and Zanettin, B. 1986. A chemical classification of volcanic rocks based on the total alkali-silica diagram, *J. Petrol.* **27:**45–750.

Lin, P. N., Stern, R. J., Morris, J., and Bloomer, S. H. 1990. Nd- and Sr-isotopic compositions of lavas from the northern Mariana and southern volcano arcs: Implications for the origin of island arc melts, *Contrib. Mineral. Petrol.* **105:**381–392.

Maslanyj, M. P., Garret, S. W., Johnson, A. C., Renner, R. G. B., and Smith, A. M. 1991. Aeromagnetic anomaly of West Antarctica (Weddell Sea sector): BAS GEOMAP Series, Sheet 2, 1:2,500,000, with supplementary text, Cambridge, British Antarctic Survey.

Mayes, C. L., Lawver, L. A., and Sandwell, D. T. 1990. Tectonic history and new isochron chart of the South Pacific, *J. Geophys. Res.* **95:**8543–8568.

Nagihara, S., and Lawver, L. A. 1989. Heat flow measurements in the King George Basin, Bransfield Strait, *Antarctic J. Sci.* **23:**123–125.

Nelson, P. H. H. 1966. The James Ross Island Volcanic Group of north-east Graham Land, *Br. Antarct. Surv. Sci. Reports*, No. 54.

Pankhurst, R. J. 1983. Rb-Sr constraints on the ages of basement rocks of the Antarctic Peninsula, in *Antarctic Earth Science* (R. L. Oliver, P. L. James, and J. B. Jago, eds.), pp. 367–371, Austral. Acad. Sci., Canberra, and Cambridge Univ. Press, Cambridge.

Pankhurst, R. J., and Smellie, J. L. 1983. K-Ar geochronology of the South Shetland Islands, Lesser Antarctica: Apparent lateral migration of Jurassic to Quaternary island arc volcanism, *Earth Planet. Sci. Lett.* **66:** 214–222.

Parra, J. C., Gonzalez-Ferran, O., and Bannister, J. 1984. Aeromagnetic survey over the South Shetland Islands, Bransfield Strait and part of the Antarctic Peninsula, *Rev. Geol. Chile* **23:**3–20.

Pelayo, A. M., and Wiens, D. A. 1989. Seismotectonics and relative plate motions in the Scotia Sea region, *J. Geophys. Res.* **94:**7293–7320.

Renner, R. G. B., Sturgeon, L. J. S., and Garrett, S. W. 1985. Reconnaissance gravity and aeromagnetic surveys of the Antarctic Peninsula, *Br. Antarct. Surv. Sci. Rep.* No. 110.

Rex, D. C. 1976. Geochronology in relation to the stratigraphy of the Antarctic Peninsula, *Bull. Br. Antarct. Surv.* **43:**49–58.

Rex, D. C., and Baker, P. E. 1973. Age and petrology of Cornwallis Island granodiorite, *Br. Antarct. Surv. Bull.* **32:**55–61.

Roach, P. J. 1978. The nature of backarc extension in Bransfield Strait [abstract], *Geophys. J. R. Astron. Soc.* **53:**165.

Sandwell, T. T. 1992. Antarctic marine gravity field from high-density satellite altimetry, *Geophys. J. Int.* **109:**437–448.

Sandwell, D. T., and Smith, W. H. F. 1992. Global marine gravity from ERS-1 Geosat and Seasat reveals new tectonic fabric, *EOS Trans. AGU* **73:**133.

Saunders, A. D., and Tarney, J. 1984. Geochemical characteristics of basaltic volcanism within backarc basins, in *Marginal Basin Geology: Volcanic and Associated Sedimentary and Tectonic Processes in Modern and Ancient Marginal Basins* (B. P. Kokelaar and M. F. Howells, eds.), pp. 59–76, Geol. Soc. Sp. Pub. No. 16.

Saunders, A. D., Tarney, J., and Weaver, S. D. 1980. Tranverse geochemical variations across the Antarctic Peninsula: Implications for the genesis of calc-alkaline magmas, *Earth Planet. Sci. Lett.* **46:**344–360.

Saunders, A. D., Weaver, S. D., and Tarney, J. 1982. The pattern of Antarctic Peninsula plutonism, in *Antarctic Geosciences* (C. Craddock, ed.), pp. 305–314, University of Wisconsin Press, Madison.

Schlosser, P., Suess, E., Bayer, R., and Rhein, M. 1988. ^{3}He in the Bransfield Strait waters: Indication for local injection from backarc rifting, *Deep-Sea Res.* **35:**1919–1935.

SEAN. 1993. Scientific Event Alert Network Bulletin for 31 March 1993, Smithsonian Institute, Washington, DC, p. 8.

Silfer, J. A., Engel, M. H., and Macko, S. A. 1990. The effect of hydrothermal processes on the distribution and stereochemistry of amino acids in recent Antarctic sediments, *Appl. Geochem.* **5:**159–167.

Simoneit, B. R. T. 1983. Effects of hydrothermal activity on sedimentary organic matter: Guaymas Basin, Gulf of California—Petroleum genesis of protokerogen degradation, in *Hydrothermal Processes at Seafloor Spreading Centers* (P. A. Rona, K. Bostrom, L. Laubier, and K. L. Smith, eds.), pp. 451–471, Plenum, New York.

Smellie, J. L. 1983. A geochemical overview of subduction-related igneous activity in the South Shetland Islands, Lesser Antarctica, in: *Antarctic Earth Science* (R. L. Oliver, P. L. James, and J. B. Jago, eds.), pp. 352–356, Austral. Acad. Sci., Canberra, and Cambridge Univ. Press, Cambridge.

Smellie, J. L. 1987. Geochemistry and tectonic setting of alkaline volcanic rocks in the Antarctic Peninsula: A review, *J. Volcanol. Geotherm. Res.* **32:**269–285.

Smellie, J. L. 1988. Recent observations on the volcanic history of Deception Island, South Shetland Islands, *Br. Antarct. Surv. Bull.* **81:**83–85.

Smellie, J. L. 1990. D. Graham Land and South Shetland Islands, in *Volcanoes of the Antarctic Plate and Southern Oceans* (W. E. LeMasurier and J. W. Thomson, eds.), pp. 302–359, Antarct. Res. Ser., American Geophysical Union, Washington, DC, 48.

Smellie, J. L., Davies, R. E. S., and Thomson, M. R. A. 1980. Geology of a Mesozoic intra-arc sequence on Byers Peninsula, Livingston Island, South Shetland Islands, *Br. Antarct. Surv. Bull.* **50:**55–76.

Smellie, J. L., Pankhurst, R. J., Hole, M. J., and Thomson, J. W. 1988. Age, distribution and eruptive conditions of late Cenozoic alkaline volcanism in the Antarctic Peninsula and eastern Ellsworthland: Review, *Br. Antarct. Surv. Bull.* **80:**21–49.

Smellie, J. L., Pankhurst, R. J., Thomson, M. R. A., and Davies, R. E. S. 1984. The Geology of the South Shetland Islands. VI: Stratigraphy, Geochemistry and Evolution, *Br. Antarct. Surv. Sci. Rep.* No. 87.

Stern, C. R., Frey, F. A., Futa, K., Zartman, R. E., Peng, Z., and Kyser, T. K. 1990. Trace-element and Sr, Nd, Pb, and O isotopic composition of Pliocene and Quaternary alkali basalts of the Patagonian Plateau lavas of southernmost South America, *Contrib. Min. Petrol.* **104:**294–308.

Strelin, J. A., Carrizo, H., Lopez, A., and Torielli, C. 1993. Actividad volcanica holocena en la Isla James Ross. *Segundas Jornadas de Comunicaciones sobre investigaciones Antarticas, Actes* pp. 335–340, Buenos Aires.

Suess, E., Fisk, M., and Kadko, D. 1988. Thermal interaction between backarc volcanism and basin sediments in the Bransfield Strait, Antarctica, *Antarctic J. U. S.* **22**(5)**:**46–49.

Sykes, M. A. 1988. New K-Ar age determinations on the James Ross Island Volcanic Group, north-east Graham Land, Antarctica, *Brit. Antarct. Surv. Bull.* **80:**51–56.

Tanner, P. W. G., Pankhurst, R. J., and Hyden, G. 1982. Radiometric evidence for the age of the subduction complex of the South Orkney and South Shetland Islands, West Antarctica, *J. Geol. Soc. Lond.* **139:**683–690.

Thomson, M. R. A. 1992. Stratigraphy and age of pre-Cenozoic stratified rocks of the South Shetland Islands: review, in *Geologia de la Antartida occidental* (J. Lopez-Martinez, ed.), pp. 75–92, Simposios T3, III Congreso Geologico de Espana y VIII Congreso Latinamericano de Geologica, Salamanca.

Vila, J., Ortiz, R., Correig, A. M., and Garcia, A. 1992. Seismic activity on Deception Island, in *Geological evolution of Antarctica* (M. R. A. Thomson, J. A. Crame, and J. W. Thomson, eds.), pp. 449–456, Cambridge University Press, Cambridge.

Weaver, S. D., Saunders, A. D., Pankhurst, R. J., and Tarney, J. 1979. A geochemical study of magmatism associated with the initial stages of backarc spreading: The Quaternary volcanics of the Bransfield Strait from South Shetland Islands, *Contrib. Mineral. Petrol.* **68:**151–169.

Weaver, S. D., Saunders, A. D., and Tarney, J. 1982. Mesozoic-Cenozoic volcanism in the South Shetland Islands and the Antarctic Peninsula: Geochemical nature and plate tectonic significance, in *Antarctic Geoscience* (C. Craddock, ed.), pp. 263–273, Univ. Wisconsin Press, Madison.

Wernicke, B. 1981. Low-angle normal faults in the Basin and Range province: Nappe tectonics in an extending orogen, *Nature* **291:**645–648.

Whiticar, M. J., Suess, E., and Wehner, H. 1985. Thermogenic hydrocarbons in surface sediments of the Bransfield Strait, Antarctic Peninsula, *Nature* **314:**87–90.

9

Structural and Kinematic Evolutions of the Okinawa Trough Backarc Basin

Jean-Claude Sibuet, Shu-Kun Hsu, Chuen-Tien Shyu, and Char-Shine Liu

ABSTRACT

Refraction data acquired in the Okinawa Trough show that the crust is of continental origin and that its thickness increases from 10 km in the southern Okinawa Trough to 30 km in the northern Okinawa Trough. The Okinawa Trough is still in a rifting stage. In the southern Okinawa Trough, magnetic anomalies are linked either to limited arc volcanic intrusions located on the northern side of the Ryukyu arc, to the axis of the main depressions where basaltic intrusions with volcanic arc affinities are present, or to volcanic products lying at different depths within the thinned continental crust. A link among partial melting occurring just above the Benioff zone, ascent of volcanic products through extended continental crust, and the amount of crustal extension are proposed. We also suggest that both arc and backarc basin volcanism are derived from the same source located above the Benioff zone. Melt preferentially rises through the fissures and faults created by extension in the continental crust. The total amount of extension across the Okinawa Trough has been estimated from refraction and gravity data. It decreases slightly from 80 km in the southern Okinawa Trough to 74 km in the northern Okinawa Trough. Parameters of rotation have been established for the entire opening of the Okinawa Trough and for the three phases of extension already identified: (1) The late Pleistocene to Recent phase of extension is characterized by normal faults with vertical offsets of a few meters changing progressively in direction along the Okinawa Trough. The amount of extension is about 5 km in the middle Okinawa Trough. (2) The early Pleistocene phase of extension is characterized by large tilted blocks which affect late Pliocene–early Pleistocene sediments. Directions of these normal faults change progressively along the Okinawa Trough. Extension in the northern Okinawa Trough is estimated as 25 km. The difference in azimuth of the normal faults for these two tectonic phases increases toward the northern Okinawa Trough. (3) The early Miocene phase of extension is poorly characterized from geological data but has been

Jean-Claude Sibuet and *Shu-Kun Hsu* • IFREMER, Centre de Brest, 29280 Plouzané, France. *Chuen-Tien Shyu* and *Char-Shine Liu* • Institute of Oceanography, National Taiwan University, Taipei, Taiwan, Republic of China.

Backarc Basins: Tectonics and Magmatism, edited by Brian Taylor, Plenum Press, New York, 1995.

calculated from parameters of other rotations. The corresponding pole of rotation is located 1500 km northeast of Kyushu; 50 km of extension in the northern Okinawa Trough and 75 km in the southern Okinawa Trough are expected. The proposed kinematic evolution of the Okinawa Trough differs considerably from the previous reconstructions and raises numerous questions, as for example, the interpretation of paleomagnetic measurements obtained in the southeastern Ryukyu Islands. The bending of the Okinawa Trough developed since the onset of the backarc basin extension, except for the southwestern portion of the Okinawa platelet whose pronounced bending is due to the collision in Taiwan.

1. INTRODUCTION

The island of Taiwan is located at the junction between the East China Sea continental margin and the Luzon arc (Fig. 9.1). East of Taiwan, the Okinawa Trough (OT) is linked to the subduction of the Philippine Sea plate beneath the Eurasian plate. South of Taiwan, the South China Sea on the Eurasia plate is subducted beneath the Luzon arc, which belongs to

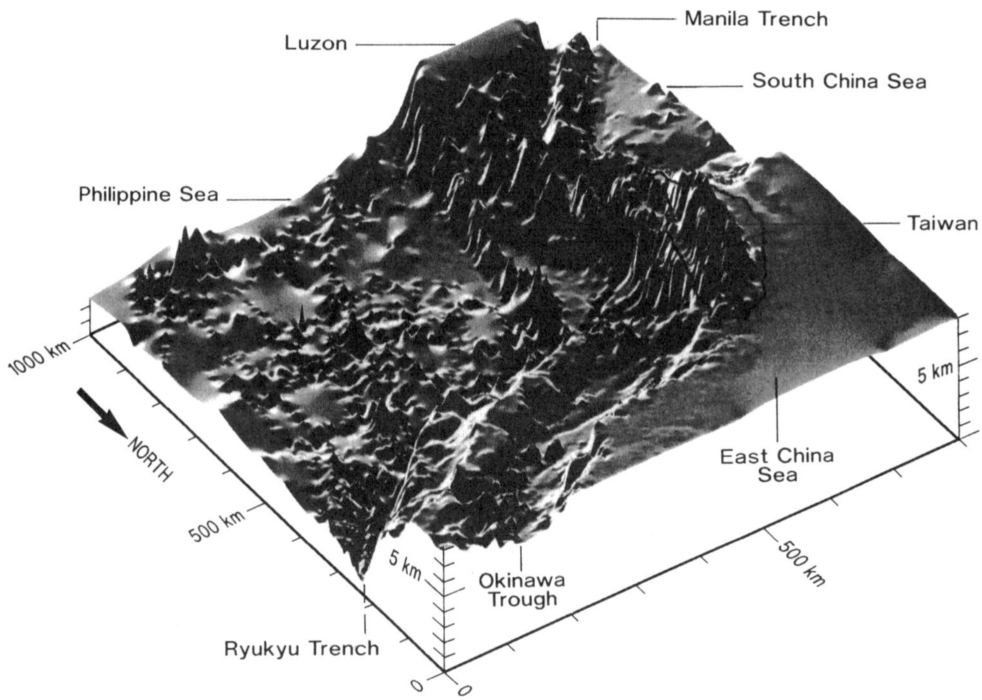

FIGURE 9.1. Block diagram of the deep ocean around Taiwan. Bathymetric data have been collected by numerous institutions and are available through the National Geographic Data Center (NGDC). Complementary data acquired east and south of Taiwan have been made available by the Institute of Oceanography of National Taiwan University. After cleaning and crossover adjustments (Hsu, in press), data have been gridded and represented with the GMT software package (Wessel and Smith, 1991). The viewing angle is chosen in order to image the southern Okinawa Trough and its relationship with the Ryukyu subduction zone but also the transfer of the convergent motion from the Manila Trench to the collision in Taiwan. South of Taiwan, the Philippine Sea plate overrides the Eurasian plate, whereas northeast of Taiwan, the Eurasian plate overrides the Philippine Sea plate. Shading from the south (azimuth 190°).

the Philippine Sea plate. The main convergent motion is thus transferred from the Manila Trench to the collision in Taiwan. In other words, south of Taiwan, the Philippine Sea plate overrides the Eurasian plate, whereas northeast of Taiwan the Eurasian plate overrides the Philippine Sea plate.

The Okinawa Trough is a curved backarc basin convex to the southeast and located behind the Ryukyu Trench and Ryukyu Islands (Fig. 9.2). It extends from the Ilan Plain in northern Taiwan to Kyushu Island in southern Japan. Its width increases from 100 km in the south to more than 200 km in the north, and its curvature is mainly restricted to the south. It abuts the narrow parallel curved belts of Taiwan, which are convex to the northwest over approximately the same distance. Several authors have suggested that the bending of the OT was due to the collision and the following indentation of the Philippine Sea plate within the Eurasian plate (Jarrard and Sasajima, 1980; Letouzey and Kimura, 1986), the indenter being the undeformed Luzon volcanic arc, located on the northwestern border of the Philippine Sea plate, against the continental part of the Eurasian plate.

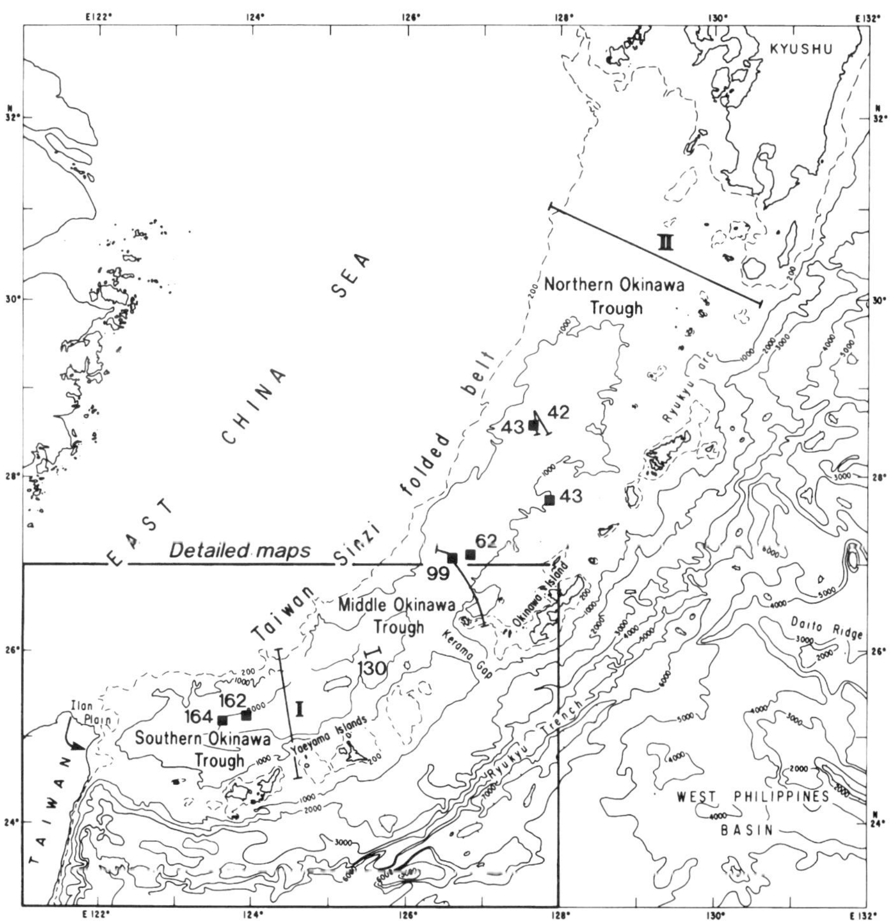

FIGURE 9.2. Map of the Okinawa Trough, Ryukyu arc, and surrounding areas with location of seismic, 3.5-kHz, and Sea Beam profiles shown in this study. Black squares represent Sea Beam detailed maps shown in this study with the corresponding profile numbers.

The purpose of this chapter is to determine the direction of normal faults that accommodate the different phases of extension along the OT, to determine if tensional features are inherited from previous tensional phases, to evaluate the amount of extension associated with each tensional phase, and to propose a coherent kinematic evolution of the OT. We will try to establish if the pronounced bending of the southern OT occurs during the formation of the backarc basin or if it is an inherited curvature since the initiation of the OT. Basic data used in this chapter are detailed swath bathymetyric data, single-channel seismic and 3.5-kHz data, and new compilations of all available bathymetric and magnetic data acquired east of Taiwan.

2. TECTONIC SETTING OF THE OKINAWA TROUGH

The OT formed by extension within the continental lithosphere, as first suggested by Uyeda (1977). The first rifting phase is dated middle to late Miocene and is associated with a broad uplift of the continental Ryukyu nonvolcanic arc and Taiwan–Sinzi folded belt, followed by normal faulting and subsidence (Sun, 1981; Letouzey and Kimura, 1986). This continental rifting phase (Table I) occurred after a major change in the direction of the Philippine Sea plate moving relative to the Eurasia plate in early Miocene (Le Pichon *et al.*, 1985). After 5 m.y. cessation of tectonic activity (Kimura, 1985), the second rifting phase (Table I), which is thought to be much more important than the previous one and resulted in the main formation of the OT, started about 2 m.y. ago at the Plio–Pleistocene boundary. The initiation of this phase is associated with the recent uplift of the Ryukyu arc at the Plio–Pleistocene boundary (Ujiie, 1980). Based on seismic correlations with drilling stratigraphy (Tsuburaya and Sato, 1985), an upper Pleistocene phase of extension, called the third phase of extension but much less developed than the lower Pleistocene phase, has been identified (Furukawa *et al.*, 1991b). This third phase of extension very probably extends to Recent times, as it is clearly expressed within the sediments deposited since late Pleistocene as normal faults characterized by small vertical offsets (Sibuet *et al.*, 1987). Consequently, the Okinawa backarc basin was formed through at least three tectonic phases (Table I) that occurred within a relatively short period of time. To explain the occurrence of closely consecutive tectonic phases in backarc basins, Sibuet *et al.* (1987) suggested that very slight changes in the subduction parameters (such as convergence rate, angle of convergence, angle of the Benioff zone) would induce major tectonic change in the backarc basin evolution.

2.1. Northern Okinawa Trough

The northern OT is considered to be in a rifting stage as revealed by refraction experiments of Ludwig *et al.* (1973), Kasahara *et al.* (1985), and Iwasaki *et al.* (1990). From a 190-km-long profile shot from Kyushu Island along the axis of the trough, Iwasaki *et al.* (1990) demonstrated that the velocity structure is characteristic of continental crust and that the Moho shallows toward the southwest from 27 to 30 km below the continental Kyushu Island to about 23–24 km at the SW end of the profile, where the trough is 1,000 m deep. Independently, Sibuet *et al.* (1987) showed that the thick sedimentary cover of the northern OT results from a massive supply of sediments from the Yangtze and Yellow rivers and was affected by tensional processes. Normal faults 120 to 150 km long and trending N040° to N060° limit tilted fault blocks to about 10 to 30 km wide. These faults related to the second tectonic phase and were active in late Pliocene–early Pleistocene

TABLE I
Main Phases of Extension in the Okinawa Trough

Age (m.y.)	Period	OT main tensional phases	Normal faults in the OT: Southern OT (25°N; 124.5°E) Mean direction[a]	Southern OT (25°N; 124.5°E) Amount of extension[b]	Northern OT (29.5°N; 128.5°E) Mean direction[a]	Northern OT (29.5°N; 128.5°E) Amount of extension[b]
0.01	Holocene (Recent)	3rd phase	90° ± 10°	5 km	74° ± 15°	6 km
0.7	Upper Pleistocene	2nd phase	80° ± 10°	8 km	55° ± 15°	25 km
1.7	Lower Pleistocene					
5	Pliocene	Cessation of tectonic activity				
12	Upper Miocene	1st phase	54° (calculated)	75 km (calculated)	58° (calculated)	57 km (calculated)
17	Middle Miocene					
	Lower Miocene					

[a]The mean directions of normal faults are extracted from Figs. 9.18a and b.
[b]The amounts of extension is computed from parameters of rotation (Table II).

(Fig. 9.3). Onlapping Pleistocene sediments partially filled half-grabens but are, in turn, normally faulted by the present-day third phase of extension. This phase of extension is characterized by faults of a different strike, which is about 25° clockwise from the previous trends. However, most of the deformation occurs during the early Pleistocene. All the early Pleistocene tilted fault blocks and half-grabens are arranged in a right-lateral en echelon pattern without evidence of transform motion between blocks, which suggests that the amount of extension varies along the blocks. However, the amount of total extension is approximately the same when local extensions are summed up perpendicularly to the OT over several en echelon tilted blocks (Sibuet *et al.*, 1987).

2.2. Southern Okinawa Trough

The southern OT located SW of Okinawa Island, consists of a broad flat basin with over 2 km of sediments which were deposited mostly at a very high deposition rate in the

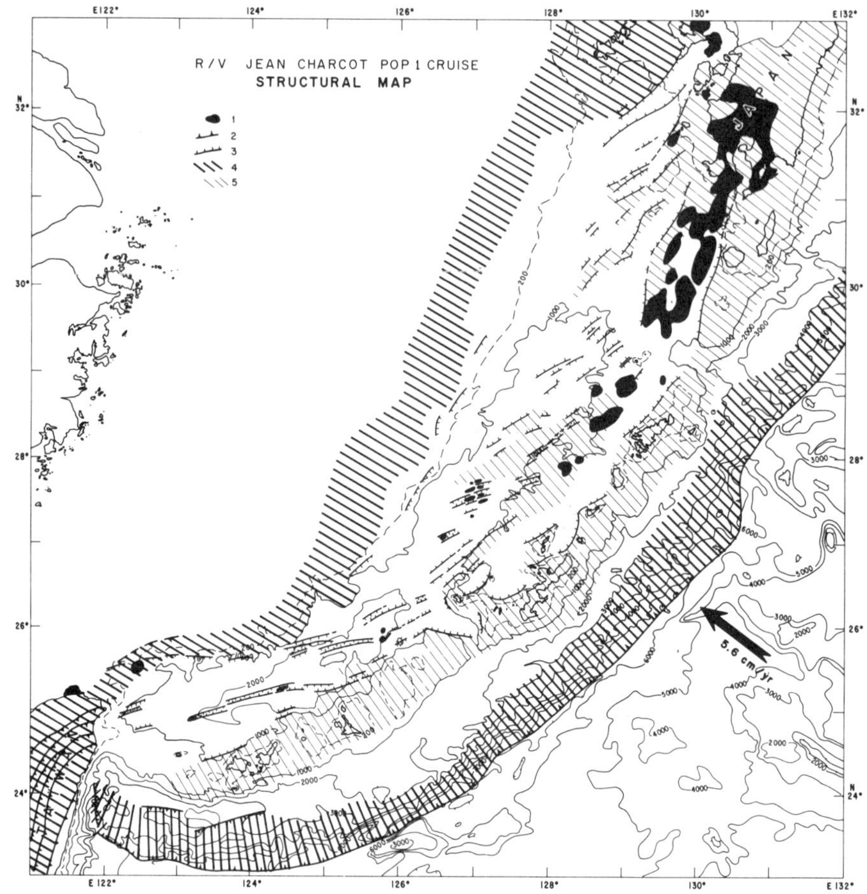

FIGURE 9.3. Geological and structural map of the Okinawa Trough from Sibuet *et al.* (1987): 1, present-day arc volcanism; 2, trench axis; 3, normal faults; 4, trench slopes and accretionary complex; 5, structural highs.

last 0.5 m.y. (Hilde *et al.*, 1984; Kimura, 1985). The trough ends at the eastern termination of the Ilan Plain right above the abrupt end of the Ryukyu Benioff zone defined by the associated cloud of deep earthquakes (Tsai, 1986). Based on the interpretation of multichannel seismic (MCS) lines, Kimura (1985) suggested that most of the Okinawa depression (at least a band 80 km wide) was of oceanic origin. This was first questioned by Sibuet *et al.* (1987), who demonstrated that the backarc spreading phase only started very recently, as shown by the presence of en echelon and, in some cases, overlapping depressions oriented N070°–N080° with the presence of a few intruded volcanics (Fig. 9.3). During this period, extension and subsidence were recorded both within the whole sedimentary column by the downward curvature of seismic reflectors toward the center of elongated bathymetric depressions and within the depressions by normal faulting and intrusion of volcanics (Sibuet *et al.*, 1987). Magnetic anomalies up to 300 nT (Furukawa *et al.*, 1991a) are associated with these volcanic ridges, where vesicular basalts have been photographed and dredged. Sibuet *et al.* (1987) restricted the width of the oceanic part to a maximum of 20 km on each side of the axial depressions, and even suggest that the oceanic domain could have been limited only to the volcanic ridges themselves. In 1988, a refraction seismic experiment with a deployment of 18 ocean bottom seismometers was conducted across the southern OT from the East China shelf to the Ryukyu nonvolcanic arc at about 124.5°E longitude (Hirata *et al.*, 1991; profile I, Fig. 9.2). In the central part of the trough, the Moho is at least 18 km below the sea surface. Even if the Moho cannot be followed outside the trough axis, this is the first complete and comprehensive refraction data set available across the OT. As underlined by Hirata *et al.* (1991), the major contribution of this refraction profile is that the southern OT is still in a continental rifting stage and that the oceanic domain, if any, is limited to the volcanic ridges in the axial portion of the trough. This interpretation is confirmed by trace element analyses and isotopic ratios of basaltic samples collected during the R/V *Jean Charcot* POP1 cruise (October 1984) on the southern OT volcanic ridges, which show that these lavas are characteristic of arc volcanism (Sibuet *et al.*, 1986; Boespflug, 1990). Analyses results are similar to those performed on samples of the Ryukyu arc (e.g., the Iizuma arc volcano in Japan; Ishizaka *et al.*, 1977) which show either that the magmatic output comes from the Benioff zone at an optimal depth for arc magmatism generation or that the arc magmatism ascends through the crustal tensional features of the backarc basin. Since the major part of the southern OT is underlain by at least 10 km of continental crust, the southern OT is still in the stage of incipient rifting of a backarc basin with an incomplete thinning of the continental crust.

2.3. Middle Okinawa Trough

This system of backarc volcanic ridges of the southern OT ends in the middle OT, close to Okinawa Island where a series of parallel basaltic ridges oriented N075° (Fig. 9.3) associated with linear magnetic anomalies have been identified (Sibuet *et al.*, 1987). These magnetic anomalies could be interpreted either in terms of dike intrusions or emplacement of early oceanic crust (Davagnier *et al.*, 1987). This specific area represents the transition between the thinned continental crust and the first signs of possible backarc basin oceanic crust but also the transition from present-day arc volcanism, which decreases in intensity from Japan to north of Okinawa Island, to volcanism occurring within the backarc basin itself. It has been named the volcanic arc rift migration phenomenon (VAMP) area (Sibuet *et al.*, 1987). Refraction data collected by Nagumo *et al.* (1986) show that a 10-km-thick upper crust of continental origin overlies a lower crust (6.8 km/s) of unknown thickness. Consequently, the crust is still of continental origin and more than 10 km thick. Lavas and pumices were sampled during the POP1 (R/V *Jean Charcot* cruise, 1984), DELP-84 (*Wakashio* cruise, 1984), and Shinkai 2000 dives (Uyeda *et al.*, 1985a). Dredged olivine basalts are of island-arc type (Uyeda *et al.*, 1985a,b; Kimura *et al.*, 1986; Sibuet *et al.*, 1986; Boespflug, 1990) and are younger than 1 m.y. (K-Ar and fission track ages; Kimura *et al.*, 1986). Some dredged rhyolite, dacite, and dacitic andesite indicate a bimodal volcanism. Thus, as for the basalts dredged in the southern OT, petrologic and geochemical evidences do not support the existence of a typical oceanic crust in the VAMP area and the middle OT.

In summary, the whole OT is still in a rifting stage. The minimal crustal thickness is 10 km in the deepest southern OT but increases in the northeast direction, along the axis of the trough, to attain about 30 km in southern Japan as beneath the continental shelves of the East China Sea and Ryukyu arc. On tensional continental margins, the thickness of thinned crusts could reach a few kilometers, and synrift volcanism has been currently observed (e.g., northeast Atlantic continental margins; Sibuet, 1992). Complementary continental rifting, in order to reduce the thickness of the thinned continental crust within the OT, is still anticipated before true oceanic crust would be emplaced. Consequently, the present-day observed backarc volcanism is considered there as products of arc magmatism emplaced through the system of normal faults where thinning is maximum rather than true backarc volcanism emplaced through a fully thinned continental crust. However, in the middle OT, as basalts have been emplaced over a width of 30 km (Sibuet *et al.*, 1987), a possibility exists there that no more continental thinning would occur. In the two schematic lithospheric cross sections proposed in Fig. 9.4, we postulate that the magmatic supply has the same origin for the arc and backarc basin volcanism and is issued from approximately the same area above the Benioff zone.

3. EXTENSIONAL TECTONICS IN THE OKINAWA TROUGH

Sibuet *et al.* (1987) demonstrated that normal faults that bound the tilted fault blocks (second phase) vary in direction from N040° to N060° in the northern OT and from N075° to N090° in the southern OT, following the direction of the northward concavity of the trough, but with a systematic offset with respect to the mean direction of the OT (Fig. 9.3). This is obvious on the OT structural map of Sibuet *et al.* (1987). These authors also noticed that small normal faults affecting the Upper Pleistocene and Quaternary sediments (third phase) were oriented slightly different from the trend of main normal faults related to the previous Lower Pleistocene tectonic phase (second phase). We performed detailed analyses of all Sea Beam data acquired during the POP1 cruise in order to discriminate the two families of normal faults.

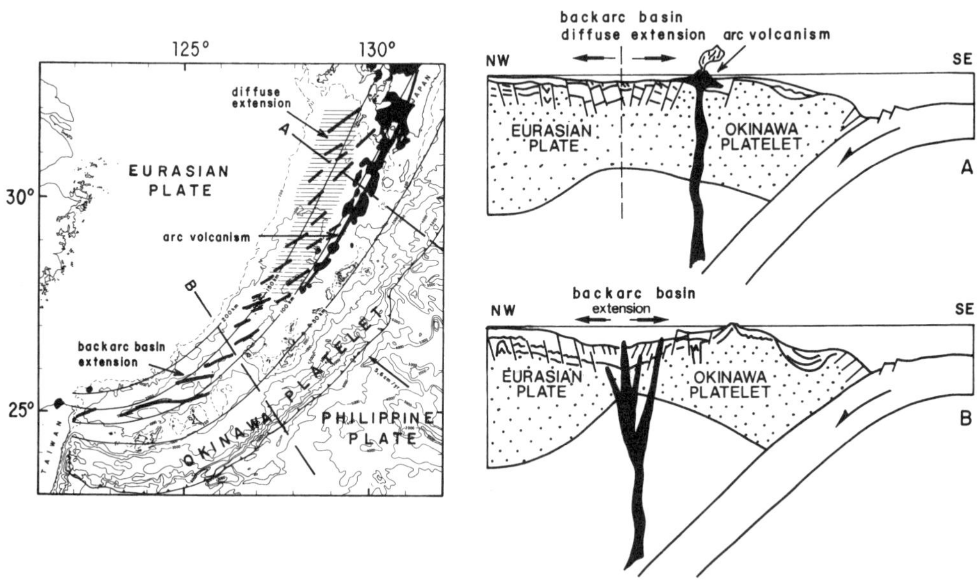

FIGURE 9.4. Map of the Okinawa Trough region showing the position of the trench, the present-day active volcanic arc, and the axes of depressions and elongated ridges of the Okinawa Trough. Depths of the Benioff zone are from Eguchi and Uyeda (1983). Cross sections of the backarc basin and forearc across the northern and southern OT are adapted from Sibuet *et al.* (1987). The magma supply has the same origin for the arc and backarc basin volcanisms and is generated from approximately the same area above the Benioff zone.

3.1. Identification of Two Families of Faults

Figures 9.5 and 9.6 show seismic profiles shot in the northern and middle OT. These profiles complement those already published (Lee *et al.*, 1980; Kato *et al.*, 1982; Sibuet *et al.*, 1987) in illustrating the shallow crustal structures. Large tilted fault blocks generally 10 to 30 km wide are clearly evident on all the seismic sections.

On profiles 42 and 43 located in the northern OT (Figs. 9.5a, 9.5b), the major tilted block, which is the same on both sections, is about 10 km wide and presents a vertical offset of at least 1 km. The formation of this block is late Pliocene–early Pleistocene in age and is related to the second phase of extension. Depressions created during the second tectonic phase are almost completely filled with post–early Pleistocene sediments. A slight tectonic phase of early late Pleistocene could exist just after the deposition of the oldest infilled sediments as shown by the onlapping reflectors located over the slightly deformed deep sedimentary layer (located at 2.5 s.d.t.t. between 00H15 and 01H00 on profile 42). The late Pliocene to Recent third phase of extension is evidenced by normal faults that affect the upper part of the sedimentary section and by the small changes in the topography of the seabed (thin arrows in Fig. 9.5a). Thus, seismic data tell us that the second and third phases of extension are two different successive tectonic phases. However, no clear sign of remobilization of the large offset normal faults belonging to the second phase exists during the third phase.

In the middle OT, tilted blocks are especially well imaged from the arc platform to a distance of a few kilometers to the axis of the backarc basin (Fig. 9.6, 07H45 to 12H15 on profile 99). The 20-km large tilted block located on this profile between 08H10 and 09H30 is a textbook example. It displays all the expected conventional features for such a block formed in a tensional environment: tilting of the block over two thirds of its length without internal deformations, fast rotation of the block attested by the presence of a poorly developed V-shaped body of synrift sediments, and internal deformation of the deeper portion of the block between 08H10 and 08H40 by drag along the normal fault of the adjacent eastern block. Another large tilted block (between 10H30 and 11H30 on profile 99) displays a significant brittle deformation by drag during its tilting and subsidence. Three normal faults are observed within the southeastern half portion of the block, which allows us to reconstruct the initial shape of the block before extension. From dredged samples collected during the POP1 cruise (30°04′N; 128°31′E) and the Toka 1 oil well results (Nash, 1979), tilting occurred in late Pliocene–early Pleistocene (second phase) and has affected the whole sedimentary cover. Onlapping Upper Pleistocene to Recent sediments partly fill the depressions of the OT axial portion where subsidence is maximum (Fig. 9.6). These sediments are in turn normally faulted during the more recent phase of extension. Vertical offsets of the present-day third generation of faults are considerably smaller than those of the second generation. The recent normal faults observed on the seismic sections correspond to minor deformations of the sedimentary column between 12H20 and 14H30 and to a hummocky sea bottom between 12H00 and 12H40 on profile 99. A basaltic intrusion appears in the deepest portion of the OT where the amount of extension is maximum. Large normal faults developed during the second tectonic phase could have been reactivated, but, once again, there is no evidence for that from the available seismic data.

The 3.5-kHz records corresponding to the same portions of seismic profiles 42 and 43 shown in Fig. 9.5 much more clearly reveal normal faults that developed during the last two tectonic phases (Fig. 9.7). The vertical exaggeration of the 3.5-kHz profiles is 40. The large normal faults at 00H10 and 02H15 limit the two tilted blocks of the second generation

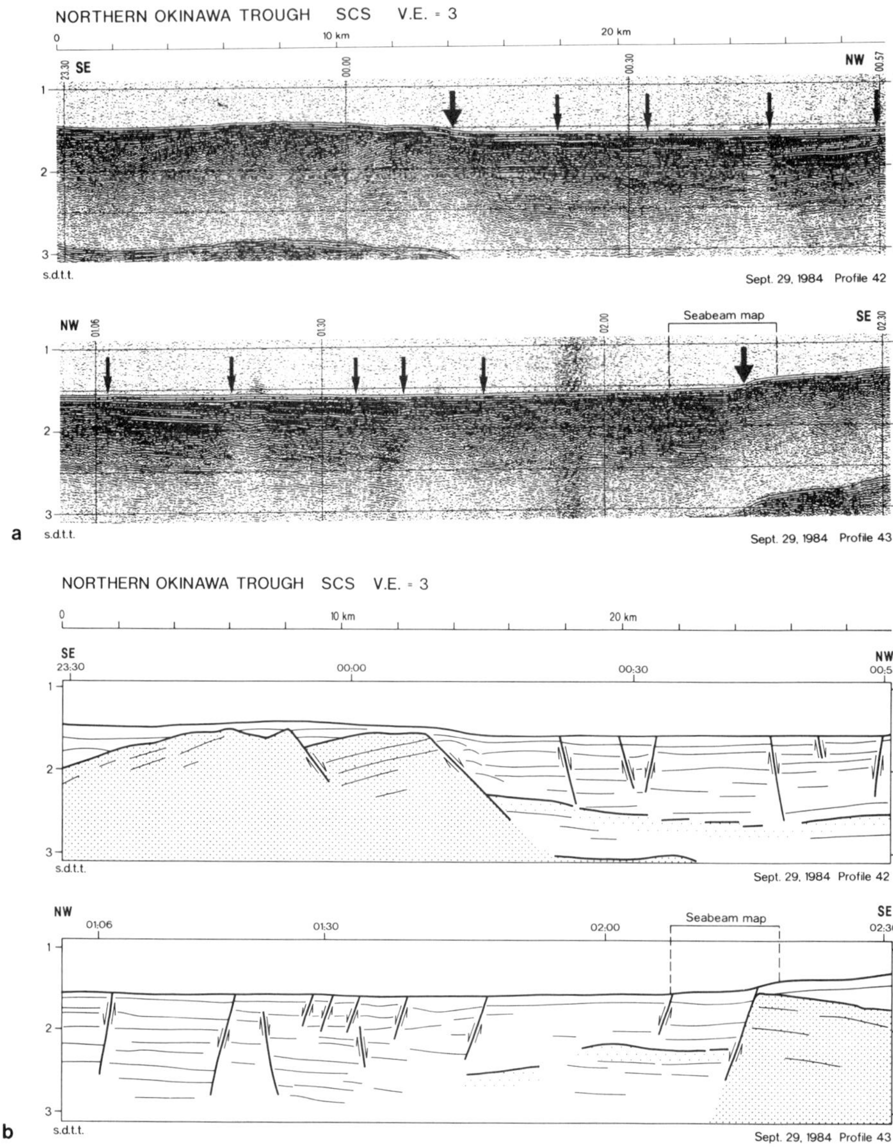

FIGURE 9.5. (a) Examples of the Plio–Pleistocene tilted fault block geometry (large arrows) and Recent normal faulting (small arrows) in the northern Okinawa Trough: IFREMER high-speed (10 knots), single-channel seismic profiles 42 and 43. Locations of profiles are shown in Fig. 9.2. (b) Structural interpretation of seismic profiles shown in (a). Tilted blocks formed during the second tensional phase (Table I) appear in gray. In light gray, deep sediments slightly deformed during early Pleistocene or early late Pleistocene by tensional movements assigned to the second phase of extension.

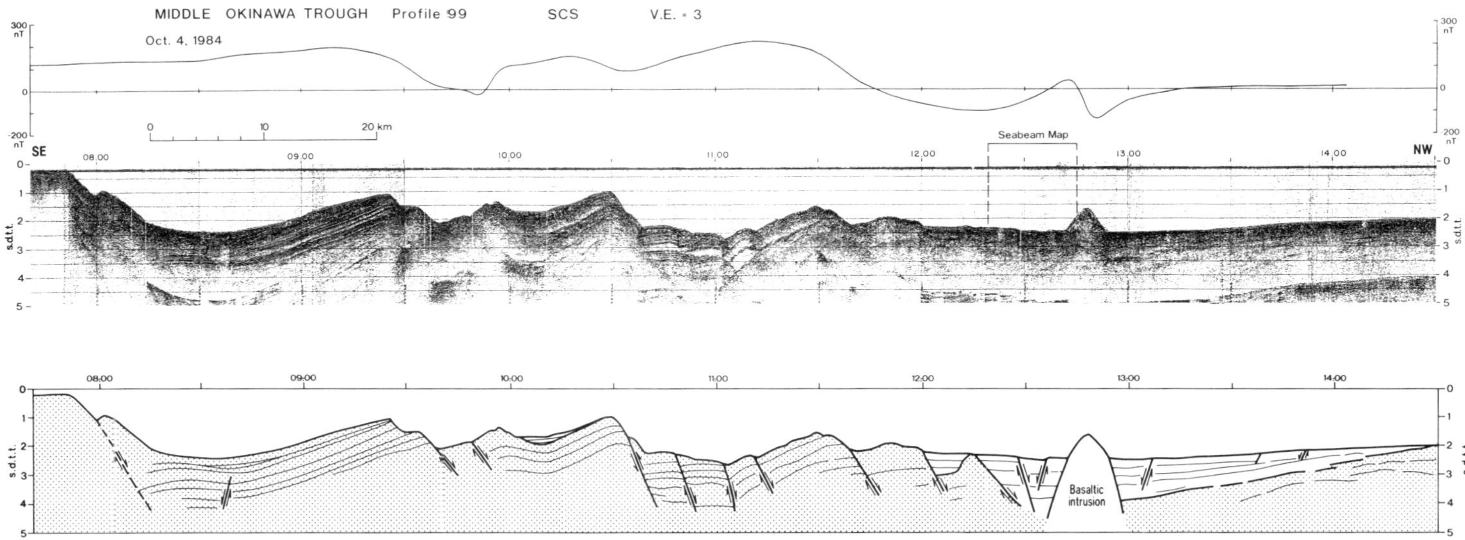

FIGURE 9.6. IFREMER high-speed (10 knots), single-channel seismic profile 99 shot in the middle Okinawa Trough from the arc platform to the axis of the trough. Location shown in Fig. 9.2. Plio–Pleistocene tilted fault blocks are especially well imaged, except in the axial part of the trough where present-day normal faults are only marked by minor deformations of sediments and seabed. A volcanic ridge associated with a 200-nT magnetic anomaly is emplaced in the axis of the depression, where subsidence is maximum. Tilted blocks formed during the second tensional phase (Table I) appear in gray in the structural interpretation of the seismic profile.

already identified in Fig. 9.5. However, between 12H20 and 14H30, numerous normal faults of the third generation, which are not imaged on the seismic sections because their topographic expression does not exceed a few meters, appear on the 3.5-kHz records as nearly vertical features. Profiles 42 and 43 differ only about 20° in azimuth with a course change at 01H03. The same normal faults appear between 01H11 and 01H20 on profile 43 and between 00H45 to 01H00 on profile 42. As profiles 42 and 43 are 2 to 3 km apart, this suggests that normal faults of the third generation present a lateral continuity of at least a few kilometers.

A section of the 3.5-kHz profile recorded in the axial part of the southern OT (Fig. 9.8, profile 130) completes the information given above. Two sedimentary units are identified: The lower folded unit corresponds to the upper part of the Pliocene sedimentary section slightly deformed by the Lower Pleistocene extension; the upper unit corresponds to onlapping Upper Pleistocene to Recent sediments and is still not much affected there by the last tectonic phase. The only evidence of recent deformation is the small bulge of Upper Pleistocene to Recent sediments which appears between 00H55 and 01H10. The recent

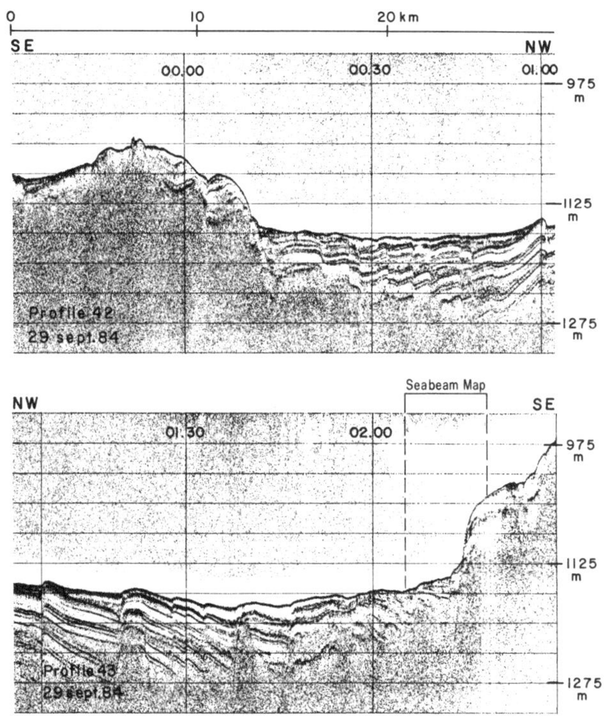

FIGURE 9.7. Portions of the 3.5-kHz profiles 42 and 43 (in northern Okinawa Trough) corresponding to the seismic sections shown in Fig. 9.5a. Location shown in Fig. 9.2. Vertical exaggeration is about 40. The Plio–Pleistocene tilted fault block is bounded by a 30-m-offset normal fault (00H10 on profile 42 and 02H15 on profile 43). Recent normal faults with a few-meter offsets affect Quaternary sediments.

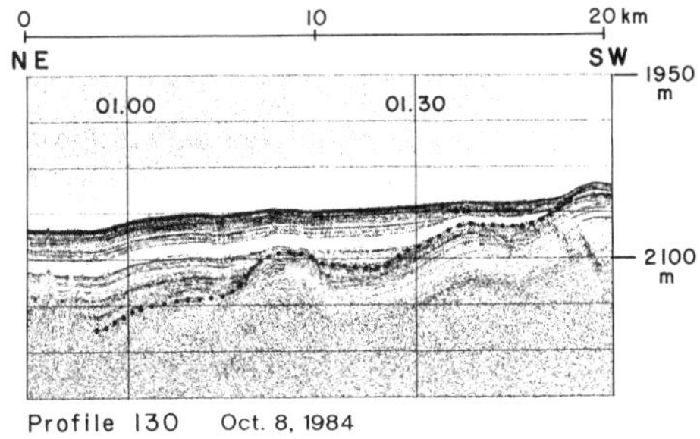

FIGURE 9.8. Portion of the 3.5-kHz profile 130 (in southern Okinawa Trough) located in Fig. 9.2. Vertical exaggeration is about 40. The gently folded Plio–Pleistocene sedimentary unit is overlain by onlapping Upper Pleistocene Quaternary sediments slightly deformed between 00H55 and 01H10. The dotted line separates the two sedimentary units.

deformation seems to occur independently from the previous generation of faults and does not seem to systematically mobilize preexisting normal faults or zones of weakness.

The interpretation of seismic profiles shows that the second and third phases of extension are distinct and that normal faults belonging to each of these tensional phases could be easily differentiated.

3.2. Spatial Distribution of the Two Families of Faults

Normal faults which belong to the two fault families described previously have been identified from the POP1 3.5-kHz records, where they display a topographic offset larger than 4 m, and from the seismic profiles, where the offset is greater than 20 m. Most of the Sea Beam data acquired during this cruise have been reprocessed with a contour spacing of 2 m and displayed with an appropriate scale in order to examine the lateral extension and direction of the normal faults. The working scale is generally between 1/10,000 and 1/50,000, depending on the slope of the normal faults and on the water depth, as the width of the recorded band is about two thirds of the water depth. Examples are given in Figs. 9.9 to 9.14.

In the northern OT, swath bathymetric contours every 2 m are shown for a small section of the profile 43 (Fig. 9.9) along both the seismic and 3.5-kHz sections of Figs. 9.5 and 9.7. The lateral extension of the normal faults is at least as wide as the width of the Sea Beam band, which is 650 m. A normal fault of the second generation is oriented N052° and presents a vertical offset of 38 m. Two normal faults of the third generation are oriented N075° and present vertical offsets of 10 and 4 m. The difference in the orientation of these two different families of faults is 20° to 25° in the northern OT. The detailed Sea Beam map shown in Fig. 9.10 reveals seafloor topography along a small portion of the seismic profile 99 as shown in Fig. 9.6, located in the axial portion of the middle OT where present-day sediments are deformed by the third generation of faults. The basaltic intrusion appears in the northwest corner of the map (Fig. 9.10). Though the hummmocky character of the sea bottom is confirmed by the irregularity of bathymetric contours, the linearity of the normal faults is still expressed over a minimum of 1 km across the profile. The three normal faults and the southeastern limit of the basaltic intrusion belong to the last generation of faults. Two faults with offsets of 20 and 22 m are oriented N065°, the other two, with larger offsets, are oriented N045° (Fig. 9.10). The difference of 20° in the orientation of faults could be explained by the reactivation of faults oriented N045° which belong to the previous

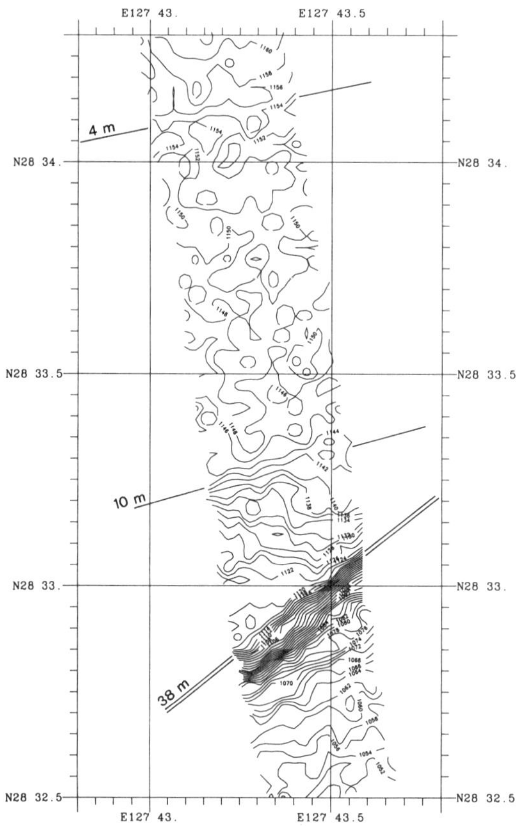

FIGURE 9.9. Sea Beam bathymetric map of a portion of profile 43 located in Figs. 9.2, 9.5, and 9.7, plotted in Mercator projection and contoured every 2 m. Directions and vertical offset values of the normal faults are also shown. The azimuths of early Pleistocene and Recent normal faults differ by 20° to 25° in the northern Okinawa Trough.

generation. Due to the massive sediment supply from the China continental shelf, deposition rates up to 4 km/m.y. have been noticed for the last 0.5 m.y. (Hilde *et al.*, 1984; Kimura, 1985). Consequently, the subsidence at the axial portion of OT imaged on seismic profile 99 (Fig. 9.6) as well as on the swath bathymetric section (Fig. 9.10) is the expression of a significant amount of extension. A large part of the present-day extension seems to be absorbed by older faults.

Thus, the two families of faults may differ about 20° in azimuth locally. Small-offset normal faults are generally associated with the present-day phase of extension, and large-offset normal faults with the previous tectonic phase. Figures 9.11 and 9.12 (profiles 43 and 62 located in the middle OT) show that exceptions exist concerning the magnitude of offsets. On profile 62 (Fig. 9.12), a fault with 8-m offset belongs to the second generation of faults, while on profile 43 (Fig. 9.11), a fault with 30-m offset belongs to the third generation of faults.

However, in the southern OT, the two families of faults may not be differentiated (as shown in profiles 162 and 164, Figs. 9.13 and 9.14 respectively). All the hfaults, whatever is the magnitude of their vertical offsets, present strike directions which differ only by 10° to 15°. Consequently, the difference in strike azimuth between the two families of faults decreases from about 25° in the northern OT to about 20° in the middle OT and is not discernible in the southern OT.

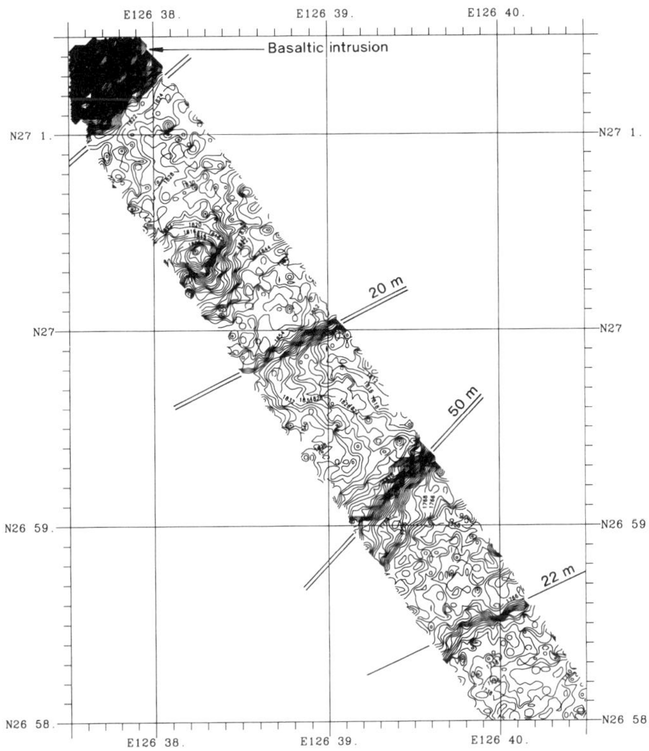

FIGURE 9.10. Sea Beam bathymetric map of a portion of profile 99 located in Figs. 9.2 and 9.6, plotted in Mercator projection and contoured every 2 m. Directions and vertical offset values of the normal faults are also shown. The azimuths of late Pleistocene and Recent normal faults differ by 20° in the middle Okinawa Trough. The subsidence of the axial part of the trough is due to the present-day normal faulting.

These observations have several implications. As extensional faults are well expressed in the sedimentary sections as a result of the high detritic input from mainland China, the present-day tectonic phase of extension activates again the directions of the previous phase of extension, as already known, but a new generation of faults also appears.

The new generation of faults progressively changes in azimuth along the OT, which means that these faults probably record the changes in stress regime.

In the Izu–Bonin arc, extension occurs in an intraoceanic island arc. Two families of normal faults, oriented 337° and 355°, are simultaneously active and have been identified in the 50-km-wide Sumisu rift (Taylor *et al.*, 1991). This rift exhibits a zigzag pattern in plan view, and directions of faults are the same than those of the rift borders. Taylor *et al.* (1991) have shown that faults with both trends form a geometry with an orthorhombic symmetry which is consistent with theoretical models of fault formation in three-dimensional strain fields (Reches, 1983). This model predicts that the maximum strain direction is perpendicular to the bisector of the acute angle between the two fault trends. For the Sumisu rift, the extension direction is N076°. Though the difference in the directions of the two families of faults for the OT and Sumisu rift are similar (about 20°), the tectonic contexts are different: the two families are active simultaneously in an oceanic environment for the Sumisu rift

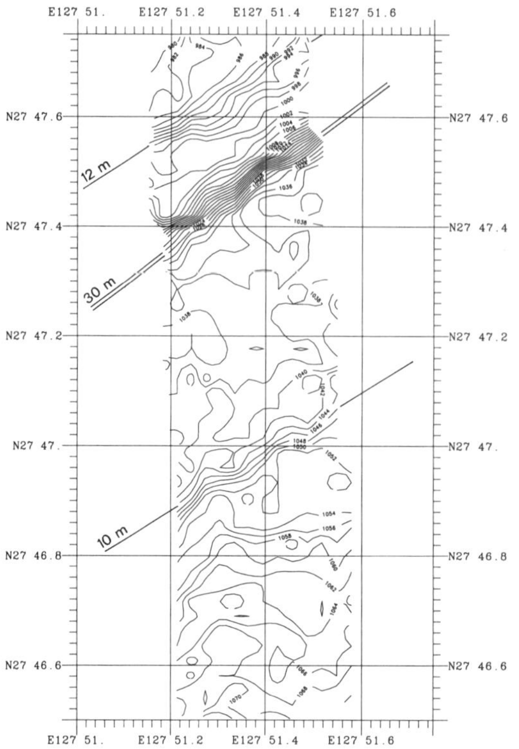

FIGURE 9.11. Sea Beam bathymetric map of a portion of profile 43 (middle Okinawa Trough) located in Fig. 9.2, plotted in Mercator projection and contoured every 2 m. Directions and vertical offset values of the normal faults are given. Small-offset normal faults are generally associated with the present-day phase of extension, but exceptions exist, as the 30-m-offset normal faults shows.

and belong to two consecutive tectonic phases in a thinned continental crust environment for the OT. Following the observations of Sibuet *et al.* (1991) in the Bay of Biscay, another explanation concerning the simultaneous presence of two families of faults in the Sumisu rift would be also possible. There, the plate boundary, which was active during the Pyrenean phase, follows previous lines of weakness as the former Bay of Biscay oceanic rift system. When the motion along the plate boundary is tensional, rift directions of the new oceanic crust are parallel to the plate boundaries and are not coincident with the stress directions given by the accurately defined position of the related pole of rotation. In the Sumisu rift, the location of the zigzag rift pattern could be also related to the presence of a previous zone of weakness such as the intraoceanic island arc. Rift directions that develop within the oceanic rift system would be parallel to the rift borders and independent of the stress directions.

3.3. Quantification of the Spatial Distribution of the Two Families of Faults

Based on the previous assumptions, a systematic study of the directions of these two families of faults has been performed. For each fault that presents a linear trend across the surveyed bathymetric swath, the following parameters are collected: latitude, longitude, vertical offset, azimuth, and dipping direction. As many as 510 measurements have been performed from southern Japan to Taiwan (Fig. 9.15). Directions of faults change from N090° east of Taiwan to N040° southwest of Kyushu (Fig. 9.15). The distribution of faults

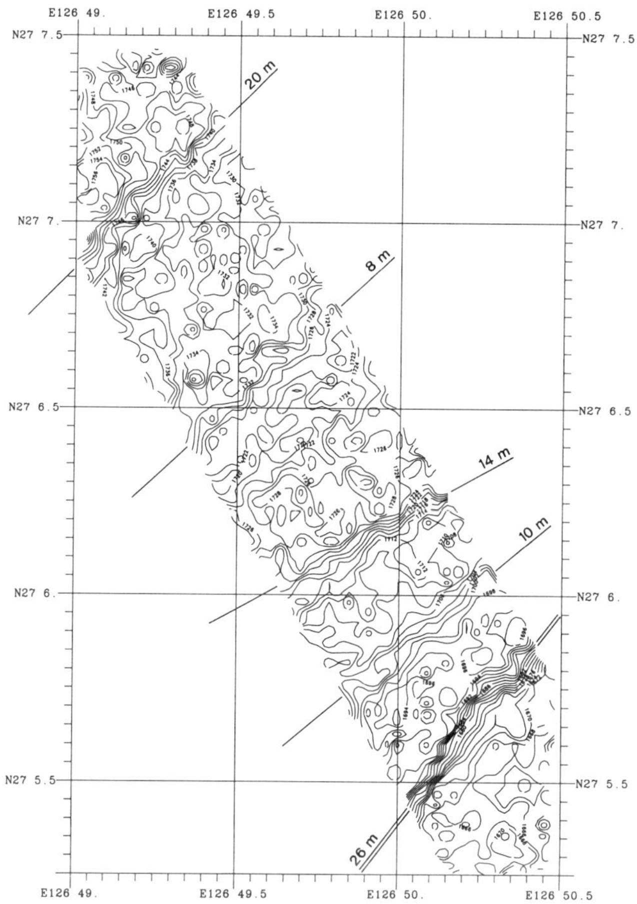

FIGURE 9.12. Sea Beam bathymetric map of a portion of profile 62 (middle Okinawa Trough) located in Fig. 9.2, plotted in Mercator projection and contoured every 2 m. Directions and vertical offset values of the normal faults are given. Large-offset normal faults are generally associated with the early Pleistocene phase of extension, but exceptions exist, as the 8-m-offset normal faults shows.

summed in 5° intervals of azimuths (Fig. 9.16) has been calculated for each of the four geographical boxes displayed in Fig. 9.15. These diagrams show that peaks of dominant fault directions change from 80–95° in box A, to 65–80° in box B, to 50–80° in box C, and to 40–80° in box D.

Figure 9.17 displays the direction of the faults as a function of the longitude. The continuous decrease of the azimuth as a function of the longitude is confirmed. The dispersion of azimuths around the mean value is ±20° at longitude 123°E (close to Taiwan) and ±30° at longitude 129°E (close to Japan). This observation confirms the existence of two families of faults that diverge in the northward direction as previously suggested.

In Fig. 9.17, the directions of the OT axis and of the trench axis are also displayed as a function of the longitude. All the directions of normal faults are located on the right-hand side of the direction of the trough axis and also preferentially on the right-hand side of the

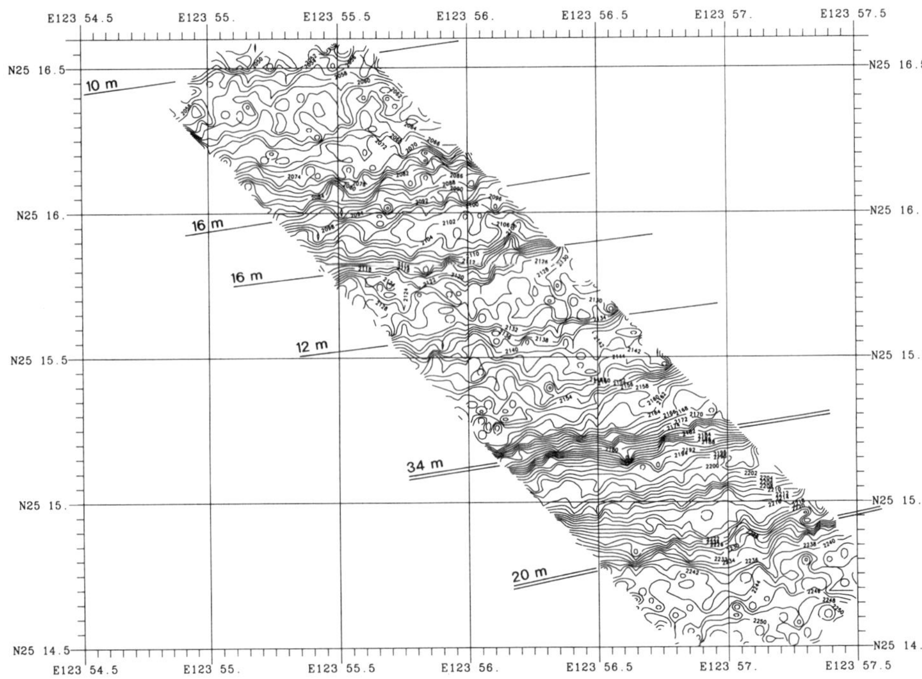

FIGURE 9.13. Sea Beam bathymetric map of a portion of profile 162 (southern Okinawa Trough) located in Fig. 9.2, plotted in Mercator projection and contoured every 2 m. Directions and vertical offset values of the normal faults are given. Normal faults belonging to the early Pleistocene and Recent phases of extension cannot be differentiated in the southern Okinawa Trough.

direction of the trench axis. This means that both the large normal faults related to Lower Pleistocene (as already noticed by Sibuet *et al.* 1987) and the late Pleistocene to Recent small offset normal faults are systematically offset with respect to the directions of the OT axis and continental slopes. The geodynamic context is different from the one of the Izu–Bonin arc and of the western Bay of Biscay in the sense that several tensional phases occur in the OT instead of a single tensional phase where rift features are parallel to the rift borders. Because the two families of normal faults belonging to the second and third phases of extension only differ by a maximum of 20° to 25° in direction, we have selected a few normal faults that clearly belong to one of the two families by using the following criteria.

Seismic data are used to identify large tilted blocks formed during the second phase of extension, bound by normal faults which present a significant vertical offset larger than a few hundred meters, and which display a significant topographic effect larger than 30 m (Fig. 9.18a).

Seismic data also unambiguously indicate Recent faults that affect the late Pleistocene to Recent sedimentary layers. Among them, we have selected normal faults characterized by small topographic vertical offsets and not related to possible deep normal faults bounding large tilted blocks formed during the second phase of extension (Fig. 9.18b).

With these specific criteria, the two families of faults are clearly discriminated. The mean direction of normal faults is estimated as a function of the longitude with an error of ±15° for faults of the second generation and ±10° for faults of the third generation (see mean direction of faults in the southern and northern OT; Table I, and in the four geographical

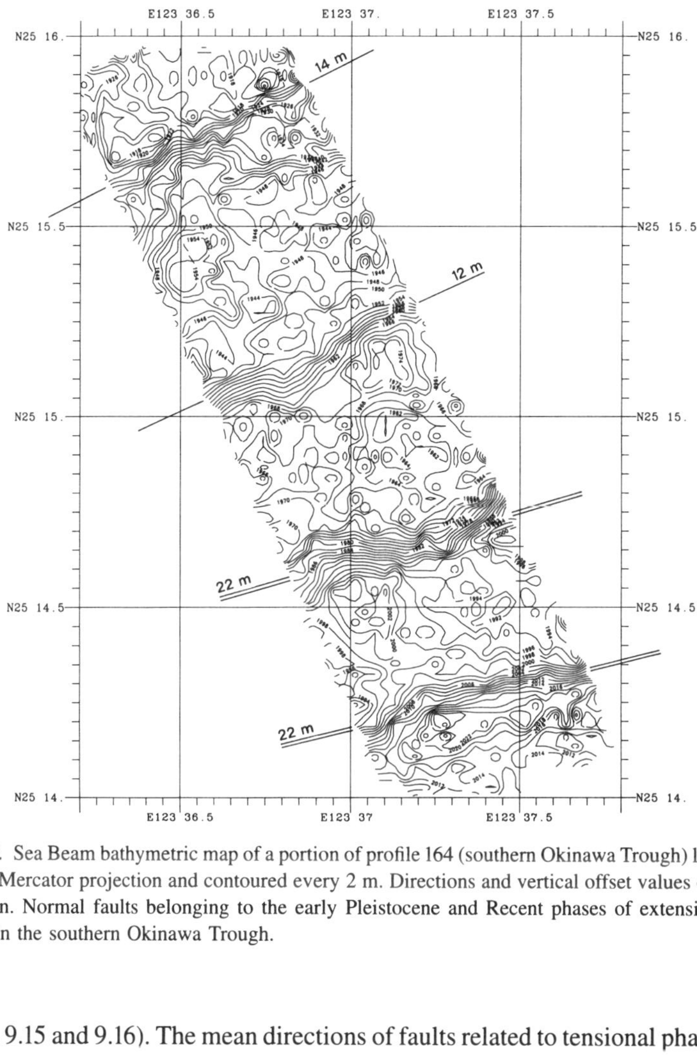

FIGURE 9.14. Sea Beam bathymetric map of a portion of profile 164 (southern Okinawa Trough) located in Fig. 9.2, plotted in Mercator projection and contoured every 2 m. Directions and vertical offset values of the normal faults are given. Normal faults belonging to the early Pleistocene and Recent phases of extension cannot be differentiated in the southern Okinawa Trough.

boxes; Figs. 9.15 and 9.16). The mean directions of faults related to tensional phases 2 and 3 and calculated from the selection of data shown in Fig. 9.18 correspond to the whole spectrum of directions displayed for each of the four geographical boxes (Figs. 9.15 and 9.16), except for the secondary peak, which appears in Fig. 9.18a between 60° and 75°. In fact, this peak corresponds to data collected west of 123°E (Fig. 9.15) on the northern OT continental slope, which changes trend in its southwestern termination.

Consequently, as the change of direction of faults seems to be progressive, it confirms, as previously seen, that the strain directions could correspond to the stress directions. In this hypothesis the extensional stress regime in the backarc basin would have the minimum principal stress $\sigma 3$ perpendicular to the mean directions of faults belonging to each of the two families. However, for both tectonic phases, $\sigma 3$ is not parallel to the future breakup line between the Okinawa platelet (defined by the Ryukyu arc and forearc) and the Eurasia plate (i.e., the trends of continental margins of the OT). This means that $\sigma 3$ has changed through time and that the initial tensional Miocene stress regime was different from the Lower Pleistocene and Recent regimes. We propose to examine if the two families of normal faults could be explained by single rotations.

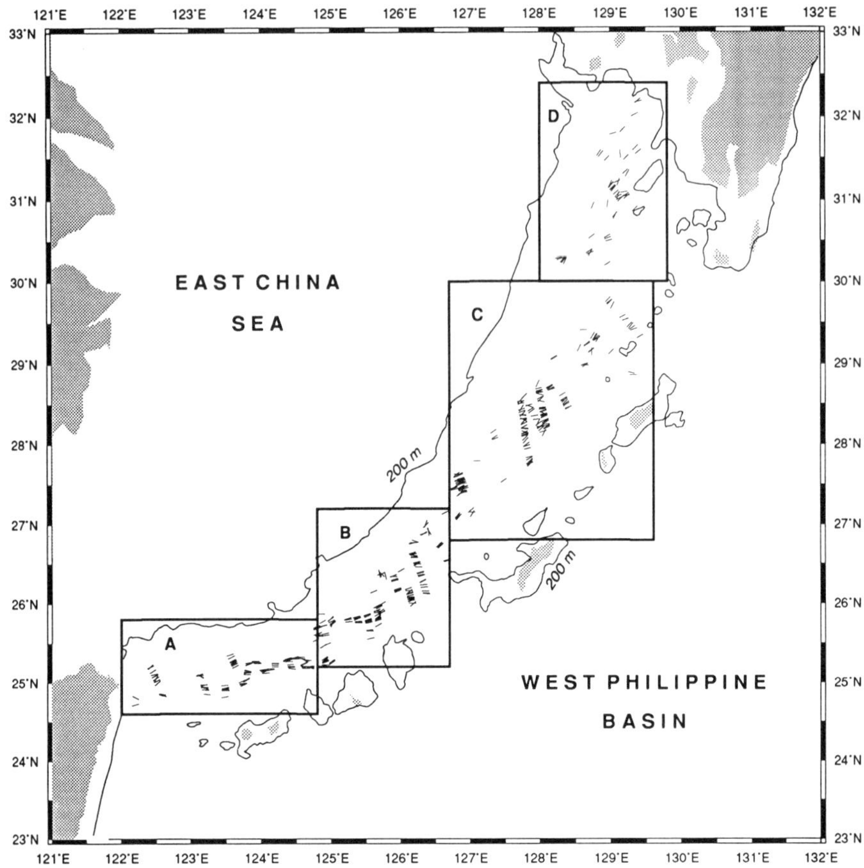

FIGURE 9.15. Directions of normal faults identified on large-scale Sea Beam maps contoured every 2 m. The POP1 data yielded 510 measurements (R/V *Jean Charcot* cruise, September–October 1984) and show a progressive change in direction from N090° east of Taiwan to N040° south of Japan.

4. KINEMATICS OF THE OKINAWA TROUGH OPENING

4.1. Determination of the Poles of Rotation Corresponding to the Early Pleistocene and Recent Tectonic Phases

From previous discussion, the orientation of the normal faults probably represents the stress regime for both the second and third tensional phases. Position of the poles of rotation can thus be computed for these two tensional phases. For each family of faults, the position of the pole of rotation has been computed by minimizing the error on a grid. The error σ at each point of the grid is equal to

$$\sigma = \sqrt{\frac{\sum_{n=1}^{N} (\theta_1 - \theta_2)^2}{N - 1}}$$

where θ_1 is the observed azimuth of a measurement, θ_2 is the azimuth of the segment connecting the computed point to the location of the measured point, and N is the total

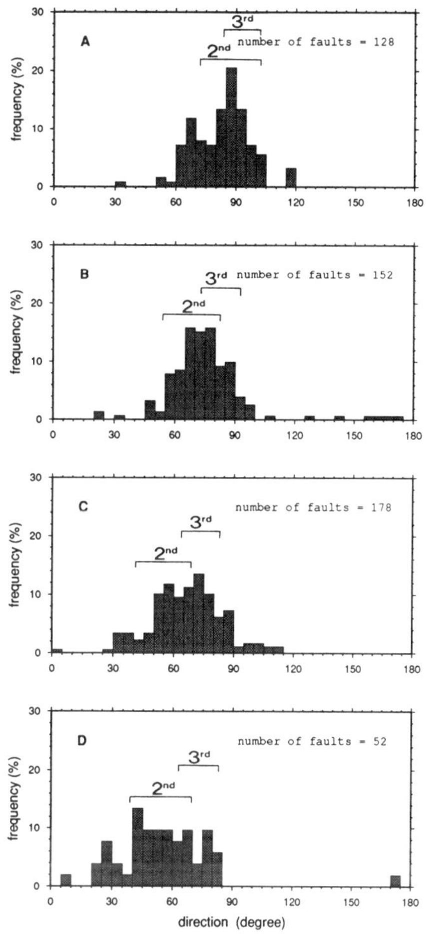

FIGURE 9.16. Distribution of fault directions by summing the faults within 5° intervals of azimuth for each of the four geographical boxes displayed in Fig. 9.15. Azimuths of normal faults belonging to the second and third phases of extension are from the mean curves of Figs. 9.18a and b.

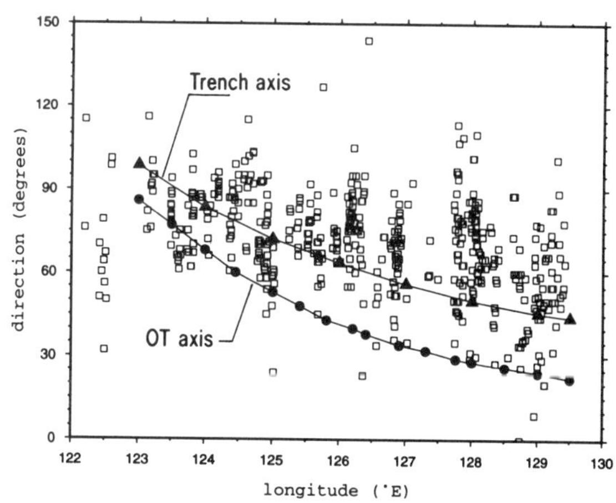

FIGURE 9.17. Directions of the normal faults as a function of the longitude (small open squares). Directions of the axis of the Okinawa Trough (black dots) and of the trench axis (black triangles) are also displayed as functions of the longitude. Directions of the normal faults are systematically deviated from the direction of the Okinawa Trough axis.

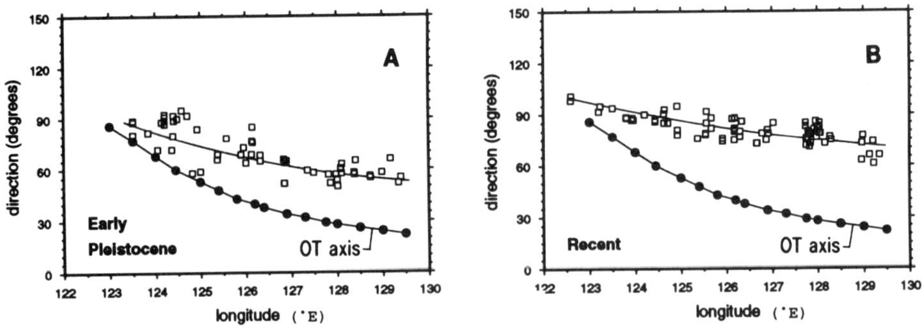

FIGURE 9.18. (a) Selected directions of normal faults belonging to the second phase of extension (early Pleistocene, Table I) as a function of the longitude (small open squares) and mean value (continuous line). Black dots represent the direction of the Okinawa Trough axis. (b) Selected directions of normal faults belonging to the third phase of extension (late Pleistocene to Recent, Table I) as a function of the longitude (small open squares) and mean value (continuous line). Black dots represent the direction of the Okinawa Trough axis.

number of measurements. The error has been computed every 0.2°, and the error values have been contoured and plotted with the GMT software package (Wessel and Smith, 1991). Poles of rotation corresponding to the second and the third generations of faults have been derived from 56 and 63 data points, respectively. The location of the pole of the second generation of faults (24.8°N, 121.6°E) lies in the eastern portion of the Ilan Plain (Fig. 9.19a). Consequently, extension in the southern OT and Ilan Plain would be weaker compared to the northern OT. This explains why large tilted fault blocks with significant offsets are observed in the northern OT and not in the southern OT. The location of the pole for the third generation of faults (24.6°N, 113.6°E) is about 800 km west of the previous one (Fig. 9.19b). A kinematic consequence is that the extension in the Ilan Plain was weak during the early Pleistocene but more active during the Recent phase. In fact, normal faults observed in the Ilan Plain affect the Plio–Pleistocene erosional surface. However, the Tatun and Kilung volcano groups, which are located in northern Taiwan behind the backarc basin but are related to the Ryukyu subduction system, are mostly of Quaternary age (Ho, 1986). High-heat-flow values (Yamano *et al.*, 1989) and hydrothermal manifestations (Yamano *et al.*, 1986) show that N075°-oriented volcanic ridges of the VAMP area located in the middle OT (Davagnier *et al.*, 1987; Sibuet *et al.*, 1987) are present-day to Recent features emplaced along trends that are in the direction of the pole of rotation of the third phase. This observation confirms the validity of the computed pole position and that present-day volcanism is emplaced along the rift direction of the third phase.

4.2. Determination of the Total Amount of Extension in the Okinawa Trough

It has been shown that extension occurs in the continental domain along the whole OT, from Taiwan to Japan. The bathymetric map (Fig. 9.2) shows that the OT is wider on the Japanese side, but the maximum water depth decreases toward the northeast; thus we do not know if the amount of total extension increases toward Japan. From available refraction and gravity data, we have established two crustal sections for the northern and southern OT, respectively, in order to quantify the amount of extension along these two transects.

Two techniques could be used to define the geometry of the Moho if part of the Moho depth is known: (1) By assuming that the crust is in local isostatic equilibrium, the geometry

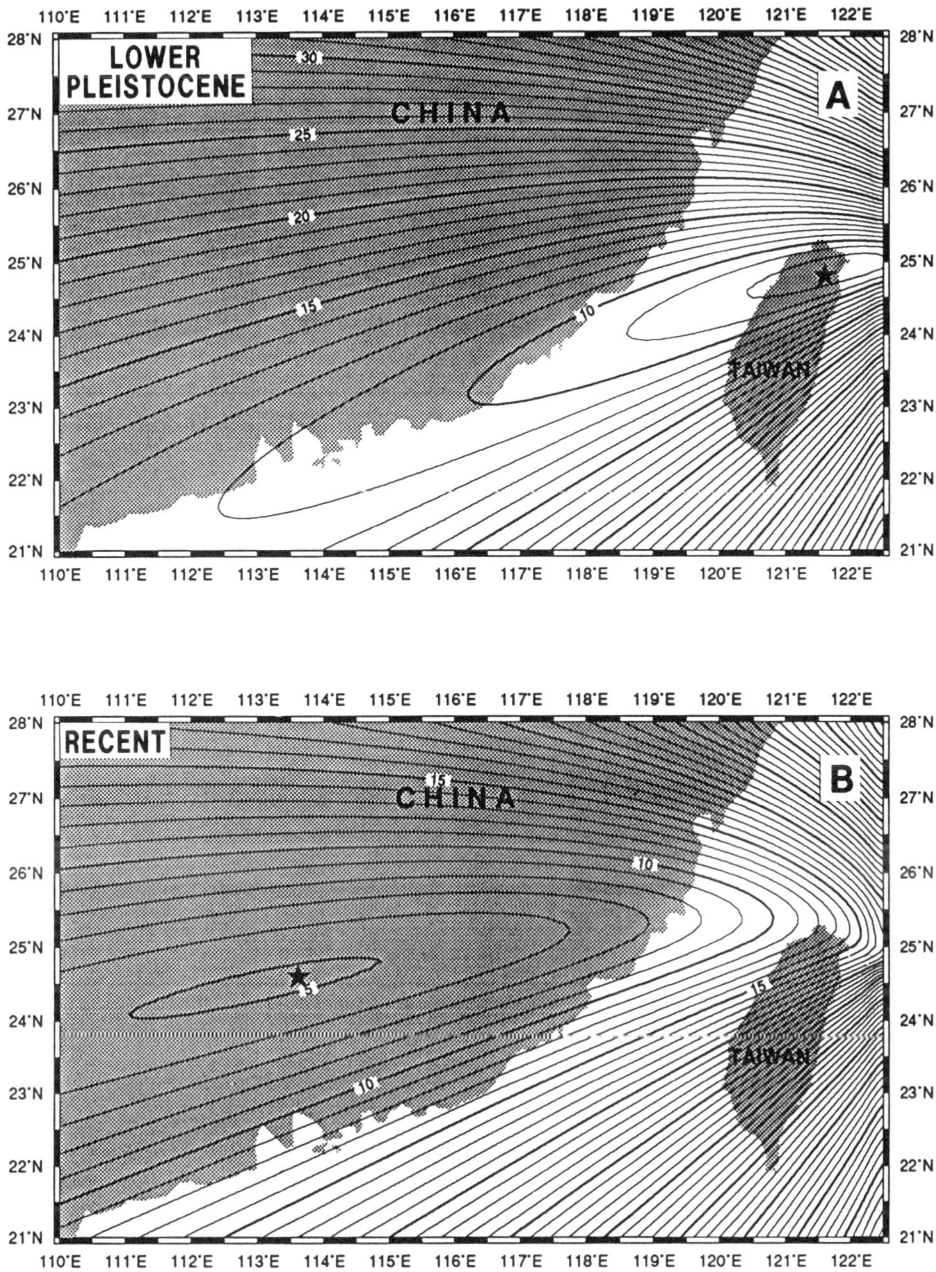

FIGURE 9.19. (A) Location of the pole of rotation (star) computed by minimization of the error on a 0.2° grid (24.8°N, 121.6°E). Data from 56 normal faults that unambiguously belong to the early Pleistocene phase of extension have been used in this calculation. Error values have been contoured and plotted by using the GMT software package (Wessel and Smith, 1991). (B) Location of the pole of rotation (star) computed by minimization of the error on a 0.2° grid (24.6°N; 113.6°E). Data from 63 normal faults which unambiguously belong to the late Pleistocene to Recent phase of extension have been used in this calculation. Error values have been contoured and plotted by using the GMT software package (Wessel and Smith, 1991).

of the Moho can be deduced by using an appropriate relationship between refraction velocities and densities. (2) If the gravity along the profile is known, the geometry of the Moho can be laterally extended by fitting the results of a two-dimensional gravity model with observed seismic refraction data.

We have used the first technique to obtain a model in local isostatic equilibrium, which was in agreement with the limited available refraction constraints. Then the calculated gravity values from this geometrical model have been compared with the observed gravity in order to provide geometrical models across the entire OT reasonably in agreement with both refraction and gravity data.

The seismic refraction experiment of Hirata *et al.* (1991) was conducted in the southern OT along profile I (Fig. 9.2). The Moho has been identified in the axial part of the trough at a depth of 18 km, with no evidence of existing oceanic crust beneath the central axis of the trough. However, in their ray-tracing modeling, Hirata *et al.* (1991) assumed that both the Moho and the boundary between the upper and lower crusts were flat outside of the trough axis. This assumption is not valid in the context of a narrow basin because it strongly violates the isostatic equilibrium principle. Consequently, we can only retain with caution a depth of 18 km for the Moho at the axis of the trough. Lee *et al.* (1980) reported refraction results along a profile parallel to the strike of the trough intersecting profile I. The P-wave velocity distributions in the crust are similar for both studies, except Lee *et al.* (1980) found that the Moho is at a depth of 15 km. Figure 9.20 presents models of profile I in local isostatic equilibrium with Moho depths beneath the axis of the southern OT at 15 and 18 km. Densities are obtained by using the theoretical relationship of Warner (1987) for sediments

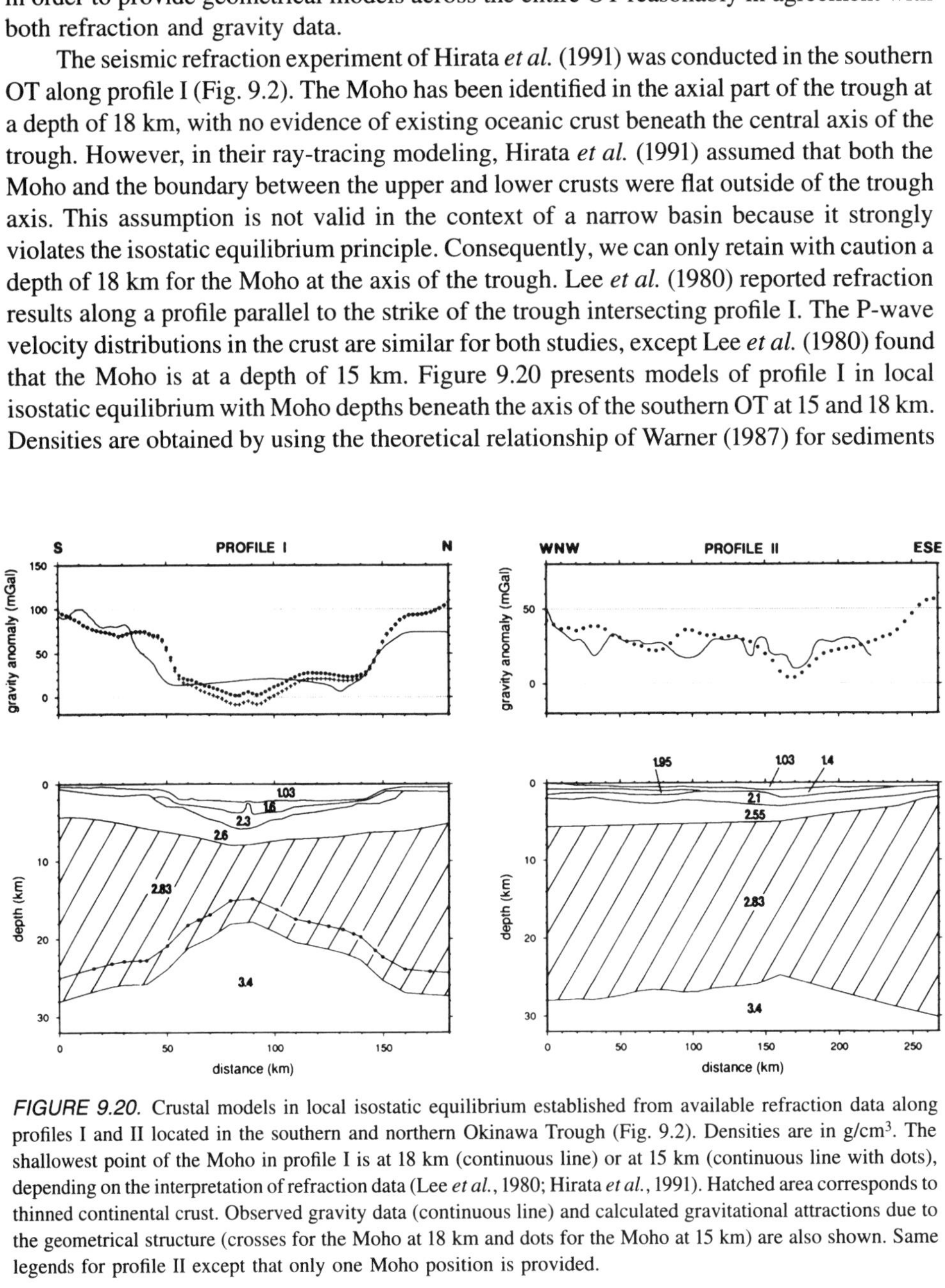

FIGURE 9.20. Crustal models in local isostatic equilibrium established from available refraction data along profiles I and II located in the southern and northern Okinawa Trough (Fig. 9.2). Densities are in g/cm^3. The shallowest point of the Moho in profile I is at 18 km (continuous line) or at 15 km (continuous line with dots), depending on the interpretation of refraction data (Lee *et al.*, 1980; Hirata *et al.*, 1991). Hatched area corresponds to thinned continental crust. Observed gravity data (continuous line) and calculated gravitational attractions due to the geometrical structure (crosses for the Moho at 18 km and dots for the Moho at 15 km) are also shown. Same legends for profile II except that only one Moho position is provided.

and the velocity density relationship for crustal materials of Sibuet *et al.* (1990). The high value of the mantle density (3.4 g/cm^3) has been chosen because continental rifting is still active and the whole lithosphere has not yet been heated. The geometrical model was established by using a reference hydrostatic column located in the axial portion of the trough. The deduced initial crustal thicknesses are 30.8 and 27.8 km, respectively, for axial Moho depths at 18 and 15 km. The observed gravity data are extracted from a free-air gravity anomaly chart of the Okinawa Trough and vicinity (Oshima *et al.*, 1988). The fit between the calculated and the observed gravity data is reasonably good except on both edges of the OT. Shallow Moho gives a better fit.

In the northern OT the refraction experiments of Iwasaki *et al.* (1990) in the western and central parts and Hayes *et al.* (1978) in the eastern part of the trough have been projected onto profile II (Fig. 9.2). The Moho has been identified in the western and central parts of the trough but not in the eastern part, where refraction data only provide information on the thickness of sediments and on the depth of the continental crust. Gravity data and density values are derived by using the same approaches as for profile I. The geometrical model is established by using a reference hydrostatic column located at 160 km where a Moho depth estimation is available. The deduced initial crustal thickness is 31 km, a value very close to the one obtained for profile I. The calculated gravitational attraction of this model in local isostatic equilibrium fits fairly well the general trend of the observed gravity data (Fig. 9.20).

If we assume that the extension is accommodated by stretching (pure shear) in the whole lithosphere, then knowing the initial thickness of the crust, the total amount of extension which affected the nonthinned continental lithosphere can be determined. It is the only way to get a rough estimate of the amount of extension in the OT. Note that this estimate is not significantly affected by the presence of dike intrusions in the OT. If the shallowest Moho is 15 km and the initial crustal thickness is 27.8 km, the amount of extension along profile I is 99 km—that is, 97 km perpendicularly to the axial direction of the trough. With a Moho at 18 km and an initial crustal thickness of 30.8 km, the amount of extension is 82 km—that is, 80 km perpendicularly to the direction of the trough at the location of profile I. For profile II, the total amount of extension is 74 km for an initial crustal thickness of 31 km.

For coherence of Moho depths along the entire OT, an amount of 80 km of extension has been assumed for the southern OT. This value is large compared to the width of the southern OT (100 km between the two continental slopes). However, the cross section of profile I (Fig. 9.20) shows that the Moho topography extends laterally, outside of the topographic expression of the OT located between 50 and 150 km. The existence of a thick pile of sediments above the continental crust is the reason for the slow decay of Moho depth to 31 km below the Eurasian platform. Consequently, part of the 80 km of extension is due to the thinning of the continental crust outside of the trough *sensu stricto*. The amount of extension is quite constant along the OT (Fig. 9.20), but the initial width of continental crust extended during the process of creation of the backarc basin is 100 km in the southeastern OT (along profile I) and 200 km in the northern OT (along profile II). This explains why the Moho depth increases from southern to northern OT.

Though a crude hypothesis has been made on the tensional mechanism to estimate the total amount of extension, an important indication is that the amount of extension across the OT does not increase from Taiwan toward Japan, as has been commonly implied in most of the geodynamic sketches except that of Vander Zouwen (1984), who proposed a pole of rotation located in southern Japan (Kyushu Island). The depth to Moho map of Jin *et al.*

(1983) grossly shows the northeastward increasing of the Moho depth and the increasing width of the trough. The Moho is at about 30-km depth beneath the East China Sea continental shelf. If we assume that the crustal thickness before rifting was 31 km, the total amount of extension along the OT is roughly constant or, to be more precise, slightly decreasing from 80 km at the location of profile I to 74 km at the location of profile II (Fig. 9.2). This means that the pole of rotation is located at about 90° from the OT location. From a kinematic point of view, if the poles of rotation of the second and the third phases of extension are both located west of Taiwan, the pole of rotation of the first phase of extension, which occurred in middle Miocene, was located northeast of the OT but far from the position given by Vander Zouwen (1984).

4.3. Determination of Parameters of the Total Rotation

Knowing the amount of extension perpendicular to the OT in two different places, we can obtain parameters of rotation only if a third constraint is available. This constraint comes from the shape of the continental slopes that bound the trough. At the base of slopes, between 1000 and 1500 m, a large bathymetric gradient could be followed on both edges of the OT (bathymetric chart of the middle and northern OT (Oshima *et al.*, 1988) and bathymetric chart of the southern OT presented in this study). These two boundaries could be matched remarkably well because of their arcuate shapes and the presence of conjugate irregularities as, for example, at 26.2°N, 124.9°E and at 25.5°N, 125.3°E. The amount of extension has been maintained perpendicularly to the trough at 74 and 80 km in the northern and southern OT, respectively. In such conditions, the pole of rotation has been determined at 35°S, 50°N with a rotation angle of 0.77° (Table II). Figure 9.21 shows such a reconstruction with present-day contours and the rotated position of the Okinawa platelet. The rotated 1000-m isobath located north of the Ryukyu Islands fits very well with the conjugate isobath along the northern edge of the trough, except for the southwestern edge of the OT (west of 123.5°E), which has no bathymetric counterpart on the northwestern edge, and possibly for the southern segment located between 123.5°E and 124.5°E, which includes the two main islands of Iriomote Sima (24.3°N, 123.8°E) and Isigaki Sima (24.4°N, 124.2°E).

Consequently, east of 123.5°E or possibly east of 124.5°E, the curvature of the entire trough was not acquired during the formation of the backarc basin but was already in existence at the beginning of the formation of the trough and persisted through time. The main curvature of the trough is already given by the northeastern OT slope, which has not been significantly affected during the collision of Taiwan. The relative uniformity of the amount of extension along the trough (Fig. 9.21) also suggests a decoupling between the OT opening and the compressive tectonics in Taiwan, at least between profiles I and II located

TABLE II
Okinawa Platelet Motions with Respect to Eurasia

OT main tensional phases		Latitude (°N)	Longitude (°E)	Angle (°)
1st phase	Middle to late Miocene	37.6	144.0	−1.93
2nd phase	Early Pleistocene	24.8	121.6	1.58
3rd phase	Late Pleistocene to Recent	24.6	113.6	0.19
	Total extension	−35.0	50.0	0.77

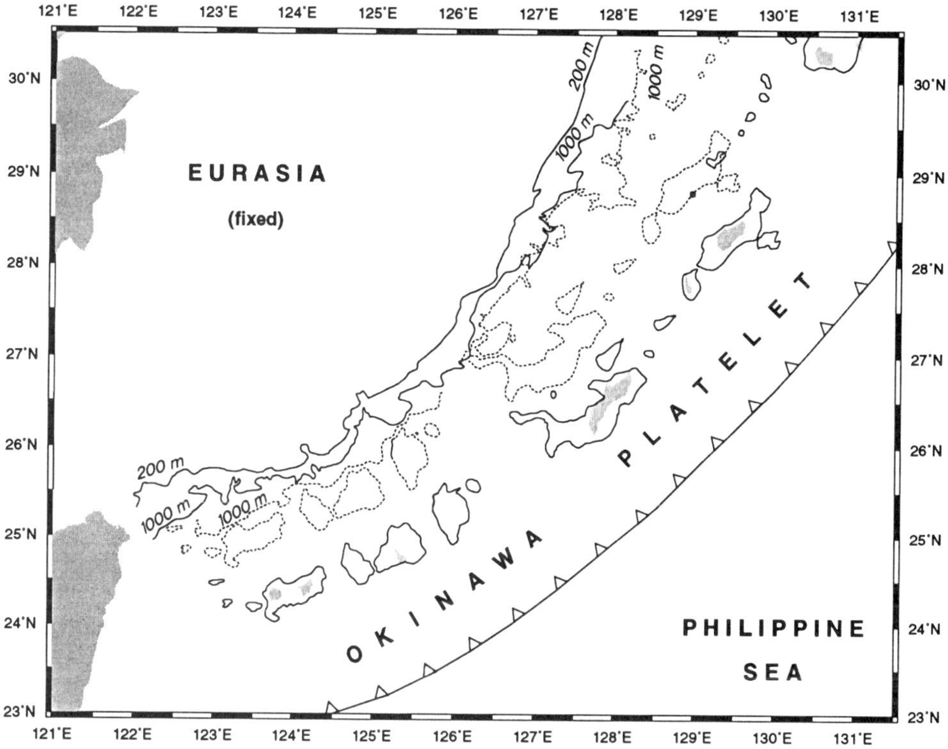

FIGURE 9.21. Reconstructions of the Okinawa Trough opening. Present-day positions are in continuous lines. The Ryukyu Islands and their 200-m isobaths represent the Okinawa platelet, which extends as far as the trench axis (Fig. 9.4). The continuous line with open triangles shows the position of the trench. Eurasia is fixed. The dotted lines represent the present-day position of the Okinawa platelet at the closure of the Okinawa backarc basin before any tensional motion in the area occurred, i.e., in middle Miocene time. The parallelism of the 1000-m isobath of the Eurasian continental platform with the rotated 1000-m of the northwestern border of the Okinawa platelet shows the quality of the fit.

in Fig. 9.2. Another consequence of this reconstruction is that the mean azimuth of the relative motion of the Okinawa platelet with respect to Eurasia is N143°, which is quite similar to the mean direction of motion (N126°) of the Philippine Sea plate with respect to Eurasia (Seno and Maruyama, 1984). The decoupling of the tectonic evolutions of the backarc basin and of the collision in Taiwan is maintained through time, except for the southwestern portion of the trough, west of 123.5°E or 124.5°E, where the Okinawa platelet and the Ryukyu Trench change significantly in direction with respect to the rest of the OT. Numerous earthquakes occur over a maximum distance of 140 km from Taiwan in the EW forearc Nanao basin (Tsai, 1986). Focal mechanisms indicate a compressive motion which is related to the collisional processes in Taiwan (Kao and Chen, 1991; Cheng *et al.*, 1992) and could explain the ESE-WNW orientation of the southwestern extremity of the Ryukyu forearc and trench and the narrowing of the backarc basin itself, though abnormally high tensional motions persist in the southern OT east of the Ilan Plain over a distance of 170 km as confirmed by the high earthquake density, focal mechanisms (Kao and Chen, 1991; Cheng *et al.*, 1992), and seismic reflection profiles which display present-day normal faults parallel to the OT axis (Sibuet, 1991).

4.4. Trial to Quantify the Amount of Extension Corresponding to the Two Last Tensional Phases

To better define the last two phases of extension, we attempt to evaluate the horizontal offsets of faults related to each of the two phases across two transects of the OT. The amount of extension has been determined by assuming that the brittle surface deformation was representative of the crustal deformation following the method of Le Pichon and Sibuet (1981). Because pure shear is the assumed mechanism of extension, the surface extension is identical to crustal or lithospheric extensions. Though this method has been criticized in the past (Chenet *et al.*, 1983), in our opinion it is the only way to try to quantify the amount of extension. Horizontal offsets have been determined from the dip and the vertical offset of each fault and summed for all faults across the entire trough. Numerous seismic profiles of the POP1 cruise have been used for that purpose. Final estimates of the amount of extension perpendicular to the trough direction are given for two complete transects of the trough.

The Recent phase of extension is characterized by numerous small offset normal faults and also by the reactivation of normal faults that belong to the preceding phase of extension. Because the amount of extension cannot be estimated from faults that present a large topographic expression, but only from those faults which offset the sedimentary cover, this bias is eliminated by choosing only seismic sections where normal faults of the previous phase are buried, which is mainly in the southern and middle OT. The estimates of the cumulative extension amount on several OT cross sections are about 3 km in the southern OT and 5 km in the middle OT. Though there may be large errors, which could be a factor of 2 on these estimates, the final accepted estimate is 5 km in the middle OT (at 29°N, 128.5°E) which gives a 0.2° rotation angle for the last tensional phase (Table II).

Concerning the amount of extension that occurred in the second phase of extension (early Pleistocene), the only locations where such estimations can be done are in the middle or northern OT (example on profile 99, Fig. 9.6). Because of the high sedimentation rate, the subsequent thick sediments mask the tilted blocks relative to this phase in the southern OT. For badly defined blocks, the amount of extension has been estimated by using the downward prolongation of both the Plio–Pleistocene surface of the blocks and the normal faults that bound the blocks. Estimates performed in the middle and northern OT are similar and equal to 25 ± 10 km. Though the amount of error cannot be accurately defined, we retain 25 km of extension at 29.7°N, 129°E, which gives a 1.4° rotation angle for the second tensional phase (Table II).

4.5. Parameters of Rotations for the Three Tensional Phases

Parameters of the first phase of extension can be deduced from the total rotation and the two last rotations (Table II). Figure 9.22 shows the relative position of the Okinawa platelet (represented by the Ryukyu Islands and their 200-m isobaths) before extension (at the time of closure) and at the end of each of the first, second, and third (present-day position) phases of extension. As anticipated, the largest amount of motion corresponds to the first phase, which in fact is the least-documented phase from geological data. The corresponding pole of rotation is located in the northeastern prolongation of the OT, at about 1500 km beyond the northern extremity of the OT, which gives about 50 km of extension in the northern OT and 75 km of extension in the southern OT. From paleo-

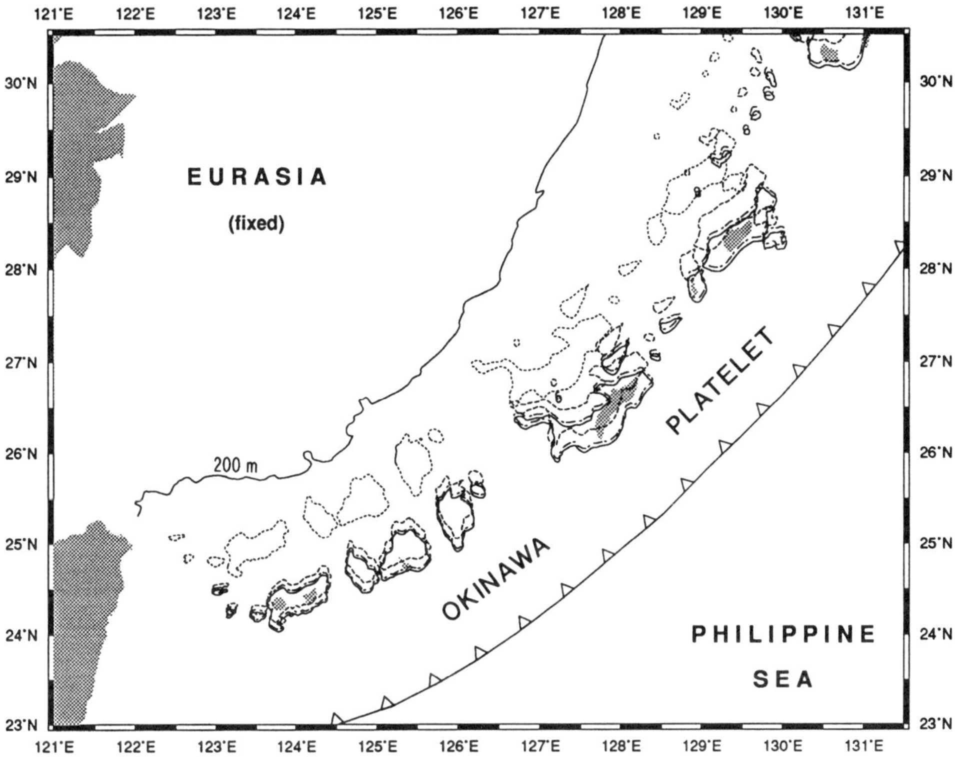

FIGURE 9.22. Reconstructions of the three extensional phases of the Okinawa Trough. Present-day positions are in continuous lines. The Ryukyu Islands and their 200-m isobaths represent the Okinawa platelet. The continuous line with open triangles shows the present-day position of the trench. Eurasia is fixed. The noncontinuous lines represent the position of the Okinawa platelet at the closure of the Okinawa backarc basin (about 15 Ma, dotted lines), at the end of the first tensional phase (about 7 Ma, dashed lines) and at the end of the second tensional phase (about 0.5 Ma, dotted and dashed lines).

magnetic constraints coming from the southeastern Ryukyu Islands, Miki *et al.* (1990) explain the motion of these islands (at least south of the Miyako depression) with respect to Eurasia by a clockwise rotation of 19° around a pole located in northern Taiwan (24.7°N, 121.7°E) for the portion of the Okinawa platelet located south of the Miyako depression. The amount of extension in the OT north of Isigaki Sima Island (Fig. 9.2, 24.4°N, 124.2°E) is 70 km, a value close to the one proposed in this study. However, using these parameters of rotation, the amount of extension would be 150 km near the Miyako depression and 340 km in the northern OT just south of Kyushu. Both values are unacceptable, which questions the validity of the interpretation of these paleomagnetic measurements. The most coherent interpretation of the 19° clockwise rotation of the southwestern portion of the Okinawa platelet would be to link this motion to the collision of the Luzon arc in Taiwan. Paleomagnetic measurements were performed on samples older than 10 Ma (Miki *et al.*, 1990), which do not give any information on the age of the rotation. We suggest that this rotation of the southwestern portion of the Okinawa platelet occurred during the collision in Taiwan (i.e., during the last 4 m.y.) and that the pole of rotation was not located in northern Taiwan but close to or east of Iriomote Sima and Isigaki Sima Islands. This interpretation links the curvature of the southwestern Okinawa platelet west of 123.5°E or 124.5°E to the collision in Taiwan. The pole position of the OT total closure (Table II) also differs considerably from that proposed by Vander Zouwen (1984), which is located in Kyushu. That pole of rotation allows only a few kilometers of extension in the northern OT instead of several tens of kilometers as established in this study. To test the proposed kinematic evolution of the OT backarc basin, we examine newly compiled bathymetric and magnetic anomaly maps of the southern OT (Fig. 9.2).

5. NEWLY COMPILED BATHYMETRIC AND MAGNETIC DATA IN THE SOUTHERN OKINAWA TROUGH

Bathymetric and magnetic data from the National Geographic Data Center (NGDC) and new data acquired east and south of Taiwan by the Institute of Oceanography, National Taiwan University, have been compiled in the area east of Taiwan, in the OT, and in the Philippine Sea Basin.

5.1. New Bathymetric Map

Bathymetric data have been acquired in 95 different cruises (Fig. 9.23). The root-mean-squares (r.m.s.) error at the intersections of all tracklines is 119.3 m and the mean absolute value is 55.1 m. To avoid the appearance of pseudostructures, several cruises

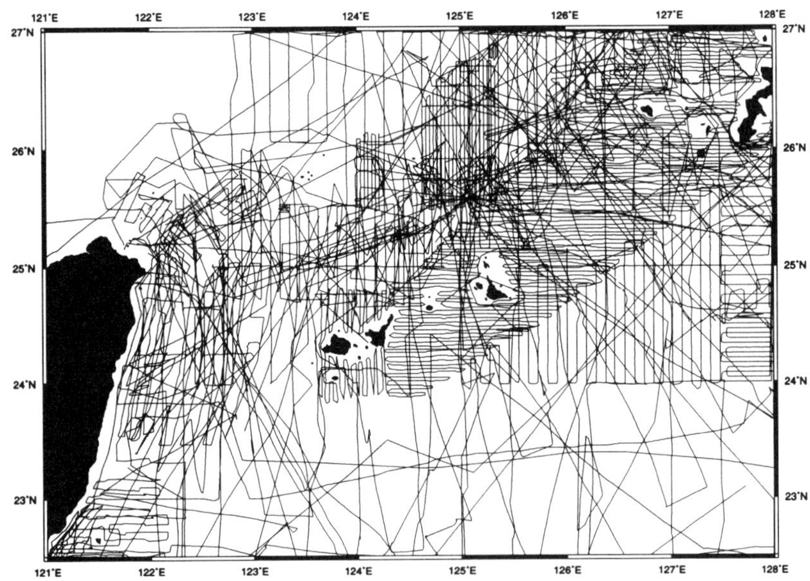

FIGURE 9.23. Tracklines show where bathymetric data have been acquired (95 cruises). Data have been collected by numerous institutions and are available through the National Geographic Data Center. Complementary data acquired east and south of Taiwan were made available by the Institute of Oceanography, National Taiwan University.

which give crossover errors larger than 200 m were removed first. The rest of the bathymetric data were adjusted by linear interpolation (Hsu, in press) to eliminate the crossover errors. Then data were gridded and contoured using the GMT software (Wessel and Smith, 1991). Figure 9.24 shows a portion of the bathymetric map over the southern OT, the direction of normal faults identified from Sea Beam data, together with offset and dip indications. Two major depressions located at about 25.2°N, 124°E and 25.8°N, 125.5°E in the southern OT are underlined by several segments of faults.

5.2. New Magnetic Map

Magnetic data have been acquired in 57 different cruises (Fig. 9.25). Magnetic anomalies are computed using the International Geomagnetic Reference Field (IGRF) 1990. The r.m.s. error at the intersections of all tracklines is 55.9 nT, and the mean absolute value is 38.8 nT. To avoid the appearance of pseudomagnetic features, we removed several cruises that give crossover errors larger than 100 nT and applied a regional correction to magnetic data of those cruises that do not have diurnal informations, using the general method of Regan and Rodriguez (1981). The remaining data have been adjusted by linear interpolation (Hsu, in press) to eliminate crossover errors. Then data were gridded and contoured with the GMT software (Wessel and Smith, 1991). A regional magnetic anomaly was extracted from the magnetic anomaly grid by second-order filtering and then was substracted from the initial grid to give the residual magnetic anomaly map (Fig. 9.26).

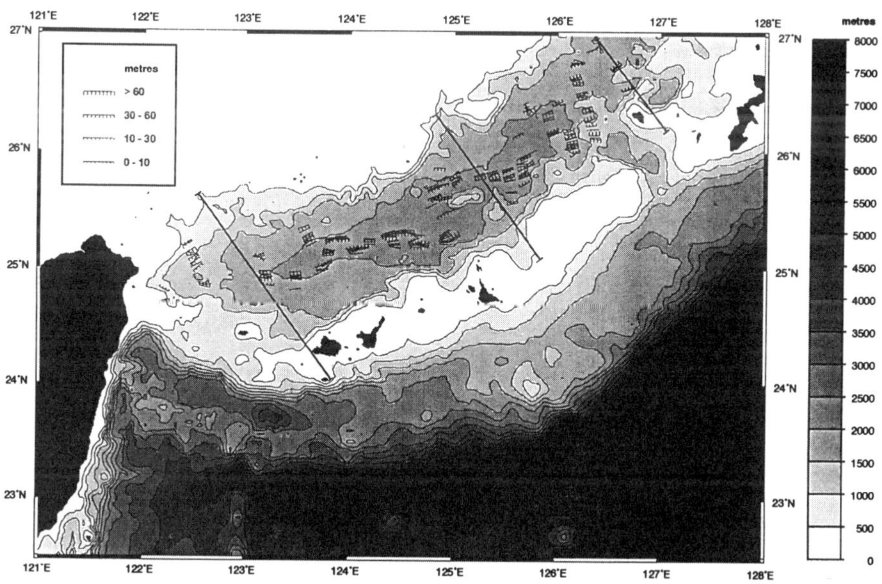

FIGURE 9.24. Bathymetric map of the southern Okinawa Trough in Mercator projection. Contours every 500 m. Gray scale on the right. Measurements of normal fault directions identified on Sea Beam data appear with indications of dips. Scale of vertical offsets in the upper left corner of the map. The N142° mean direction for the opening of the Okinawa Trough (Table II) shows the good correlation between topographic features located on either side of the trough, except for the southeastern portion of the Okinawa platelet, west of 123.5°E or 124.5°E, which has no morphologic counterpart on the northern continental slope of the Okinawa Trough.

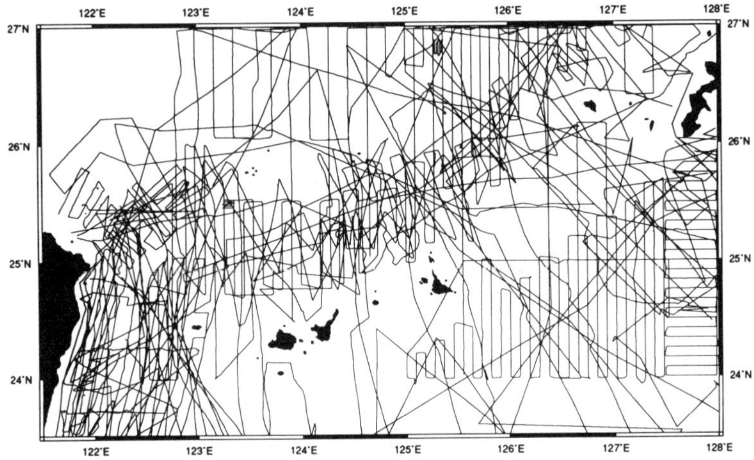

FIGURE 9.25. Tracklines show where magnetic data have been acquired (57 cruises). Data have been collected by numerous institutions and are available through the National Geographic Data Center. Complementary data acquired east and south of Taiwan were made available by the Institute of Oceanography, National Taiwan University.

5.3. DISCUSSION

A major feature of the residual magnetic anomaly map is the strong magnetic contrast between the OT backarc basin and the Okinawa platelet. The southern OT, defined by the 200-m isobath, is characterized by magnetic anomalies with magnitude of a few hundred nanoteslas and wavelengths of a few tens of kilometers. In contrast, the Okinawa platelet, defined by the 200-m isobath on the OT side and the trench axis, is characterized by weak magnetic anomalies, which confirm that the southern Ryukyu arc is a nonvolcanic arc

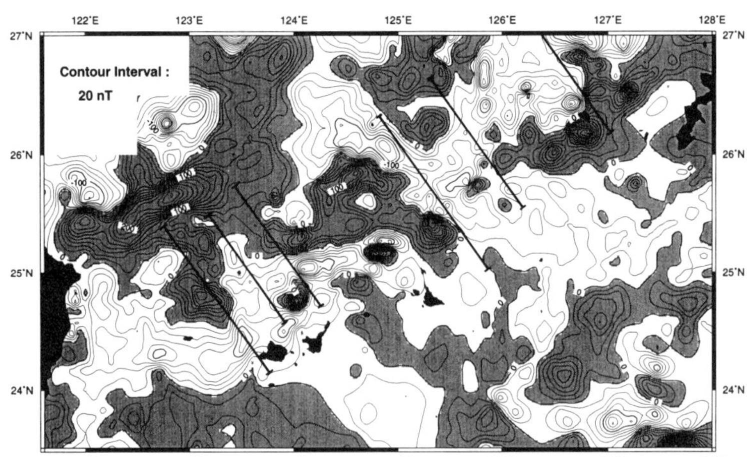

FIGURE 9.26. Residual magnetic anomaly map of the southern Okinawa Trough in Mercator projection obtained by subtraction of a regional field from the magnetic anomaly data. Contours every 20 nT. The N142° direction shows boundaries between areas of similar magnetic character.

(Letouzey and Kimura, 1986). However, except for the high-amplitude anomalies located northeast of Taiwan, which are linked to volcanics of the Taiwan Sinzi folded belt, the East China Sea continental margin is characterized by magnetic anomalies of similar wavelength but lower amplitude than in the OT. Two lines of interpretations can be proposed for the origin of the OT magnetic anomalies.

The initial continental domain, which was lately thinned and extended 80 km, belongs to the domain of the Taiwan Sinzi folded belt, where volcanics have been identified. As the initial continental domain has been extended by a factor of about 2 during rifting, an increase of the wavelength of magnetic anomalies and a reduction of the amplitude due to the subsidence of magnetic sources were anticipated. However, these were not observed.

As a result of tensional processes, the whole OT continental domain has been thinned and crustal and lithospheric thicknesses significantly reduced (Fig. 9.20). Products of partial melting initiated in the lower part of the lithosphere, just above the Benioff zone, probably come up through the lithosphere, lying as arc volcanic intrusions at different depths within the crust, and giving rise to significant magnetic anomalies. We tend to favor this hypothesis.

In an earlier magnetic compilation, Oshida *et al.* (1992) described relationships between magnetic anomalies and geologic features. They demonstrated that the numerous dipole-type anomalies observed along the southern side of the trough are linked to volcanic knolls that are considered as the southwestern extension of the Ryukyu volcanic arc (Ueda, 1986; Oshida *et al.*, 1988). Geochemical analyses performed on basaltic samples collected on these seamounts confirm the arc volcanic affinity of these seamounts (Sibuet *et al.*, 1986; Boespflug, 1990).

Because of the reduced amount of partial melting, only a small percentage of the crust would be composed of volcanics. It seems reasonable to assume that refraction data cannot resolve the presence of such crustal material. A major question concerns the mode of ascent of magmatism through the crust. Does the basaltic magma go indifferently or randomly through the crust? Or does it go preferentially along zones of weakness parallel to rift directions or transform directions? In addition to localized dipole anomalies linked to arc volcanism (Oshida *et al.*, 1992) along the southern side of the trough, a close examination of the shape of magnetic anomalies in the OT shows a good correlation between the location of the elongated magnetic anomalies and the axes of the two depressions already evidenced on the bathymetric map and where basaltic elongated ridges have been mapped and sampled (Sibuet *et al.*, 1987; Fig. 9.26). Oshida *et al.* (1992) modeled these magnetic anomalies with a normally magnetized crust (2.5 A/m), supporting the view that seafloor spreading has not started in the southern OT. In the southern OT only a few basaltic intrusions have been identified in these depressions (Fig. 9.3) where surface extension and consequent crustal extension are maximum. Thus, the residual magnetic anomaly map clearly shows a correspondence between magnetic anomalies and rift directions related to the last two phases of the OT opening. No correspondence seems to exist with rift directions of the first phase, oriented N052° from the kinematic analysis. Previous existing weak magnetic anomalies could have been obliterated by large anomalies related to the second and third phases of extension and, in any case, better developed because extension and partial melting increase with time.

The mean direction of the entire OT opening has been plotted on the bathymetric (Fig. 9.24) and residual magnetic anomaly (Fig. 9.26) maps. Though transform faults are not observed in the OT (Sibuet *et al.*, 1987), the N142° segments limit areas of similar magnetic character. A plausible interpretation of this observation is to relate magnetic anomalies and associated crustal magmatic intrusions with different sections of the continental crust experiencing different tensional factors. Extension could be different in adjacent sections. For example, OT continental slopes are very steep on both sides of the section located between 123.8°E and 125.4°E (Fig. 9.26). Most of the extension occurs in the deepest part of the OT. In the adjacent southwestern section and extending to Taiwan, continental slopes are wider and the amount of extension in the deep part of the trough seems to be reduced compared with other sections. We consequently suggest that a link exists among partial melting, ascent of volcanic products within the crust, and amount of crustal extension.

6. CONCLUSIONS

The major conclusions concerning the structural and kinematic evolution of the OT are the following.

(1) The whole OT is still in a rifting stage as shown by refraction and magnetic data. The crustal thickness increases from 10 km in the deepest part of the southern OT to 30 km in the northern OT, close to Japan.

(2) The amount of extension across the OT has been estimated from refraction and gravity data. It slightly decreases from 80 km in the southern OT to 74 km in the northern OT.

(3) Three phases of extension are identified and their parameters of rotation are established: (a) The late Pleistocene to Recent phase of extension is characterized by normal faults with vertical offsets of a few meters and changing directions progressively along the OT. The amount of extension that occurred during this phase is about 5 km in the middle OT. (b) The early Pleistocene phase of extension is characterized by tilted blocks that affect late Pliocene–early Pleistocene sediments and change direction progressively along the OT. The difference in azimuths of these two last phases increases toward the northern OT. The amount of extension for this phase is estimated to be 25 km in the northern OT. (c) The middle to late Miocene phase of extension, poorly characterized from geological data, has been calculated from parameters of the total rotation and of the two last rotations. This first phase of extension is a major tectonic phase with about 50 km of extension in the northern OT and 75 km in the southern OT. The corresponding pole of rotation is located 1500 km northeast of Kyushu Island.

(4) Directions of normal faults of the last two phases correspond to the stress directions because of their progressive change along the OT axis resulting from a partial relaxation of the motion along zones of weakness active during previous tectonic phases.

(5) The proposed kinematic evolution of the OT differs considerably from previous reconstructions and raises questions concerning, for example, the interpretation of paleomagnetic measurements obtained in the southeastern Ryukyu Islands.

(6) The curved shape of the OT was acquired since the onset of the backarc basin extension except for the southwestern portion of the Okinawa platelet, west of 123.5°E or 124.5°E longitude, which has rotated 19° clockwise (Miki *et al.*, 1990) during the collision of Taiwan around a pole located east or close to the southern Ryukyu Islands.

(7) New bathymetric and residual magnetic anomaly maps are in agreement with the proposed kinematic evolution. Magnetic anomalies in the southern OT and geochemical analyses performed on collected basaltic samples suggest that part of the anomalies are linked to crustal arc volcanic intrusions emplaced close to the southern continental slope

of the OT and within topographic depressions located in the central part of the southern OT. For the remaining magnetic anomalies, we suggest a link among partial melting, ascent of volcanic products within the crust, and amount of crustal extension.

Acknowledgments

We thank Benoît Loubrieu, Alain Normand, and Michel Voisset for their help in processing Sea Beam data with the IFREMER TRISMUS software package, and Daniel Carré for drafting some figures. We acknowledge discussions with H. D. Needham about the kinematics of the early stage of ocean opening. The GMT software package was used to display some of the figures (Wessel and Smith, 1991). Brian Taylor gave us a lot of support and facilities for the submission and revision of this chapter and also numerous scientific comments on the first version of the manuscript.

REFERENCES

Boespflug, X. 1990. Evolution géodynamique et géochimique des bassins arrière-arcs. Exemples des bassins d'Okinawa, de Lau et Nord-Fidjien, Université de Bretagne Occidentale, Brest, France.

Chenet, P.-Y., Montadert, L., Gairaud, H., and Roberts, D. G. 1983. Extension ratio measurements on the Galicia, Portugal and northern Biscay continental margins: Implications for evolutionary models of passive continental margins, in *Studies in Continental Margin Geology* (J. S. Watkins and C. L. Drake, eds.), Am. Assoc. Petrol. Geol. Mem. **34:**703–715.

Cheng, S.-N., Lee, C.-T., and Yeh, Y. T. 1992. Seismotectonics of the Ryukyu arc, in *Proc. 4th Taiwan Symp. Geophys.*, Taipei, pp. 507–516.

Davagnier, M., Marsset, B., Sibuet, J.-C., Letouzey, J., and Foucher, J.-P. 1987. Mécanismes actuels d'extension dans le bassin d'Okinawa, *Bull. Soc. Géol. France* **8:**525–531.

Eguchi, T., and Uyeda, S. 1983. Seismotectonics of the Okinawa Trough and Ryukyu arc, *Mem. Geol. Soc. China* **5:**189–210.

Furukawa, M., Kondo, S., Miki, M., and Isezaki, N. 1991a. Report on DELP 1988 cruises in the Okinawa Trough. Part V: Measurement of the three components and total intensity of the geomagnetic field in the Okinawa Trough, *Bull. Earthquake Res. Inst., U. Tokyo* **66:**91–150.

Furukawa, M., Tokuyama, H., Abe, S., Nishizawa, A., and Kinoshita, H. 1991b. Report on DELP 1988 cruises in the Okinawa Trough. Part II: Seismic reflection studies in the southwestern part of the Okinawa Trough, *Bull. Earthquake Res. Inst., U. Tokyo* **66:**17–36.

Hayes, D. E., Houtz, R. E., Jarrard, R. D., Mrozowski, C. L., and Watanabe, T. 1978. Crustal structure, 1 sheet, scale 1:6,442,194, in *A Geophysical Atlas of the East and Southeast Asian Seas, Map Chart, MC25* (D.E. Hayes, ed.), Geological Society of America, Boulder, CO.

Hilde, T. W. C., Lee, C.-S., and Vander Zouwen, D. E. 1984. Tectonic and sedimentation history of Okinawa Trough: Implications for development of the East China and Yellow seas. Korea–U.S. Conference on the Yellow Sea.

Hirata, N., Kinoshita, H., Katao, H., Baba, H., Kaiho, Y., Koresawa, S., Ono, Y., and Hayashi, K. 1991. Report on DELP 1988 cruises in the Okinawa Trough. Part III: Crustal structure of the southern Okinawa Trough, *Bull. Earthquake Res. Inst., U. Tokyo* **66:**37–70.

Ho, C. S. 1986. A synthesis of the geologic evolution of Taiwan, *Tectonophysics* **125:**1–16.

Hsu, S.-K. in press. XCORR: A cross-over technique to adjust track data, *Comput. Geosci.*

Ishizaka, K., Yanagi, K., and Hayatsu, K. 1977. A strontium study of the volcanic rocks of the Myoko volcano group, central Japan, *Contrib. Mineral. Petrol.* **63:**295–307.

Iwasaki, T., Hirata, N., Kanazawa, T., Melles, J., Suyehiro, K., Urabe, T., Möller, L., Makris, J., and Shimamura, H. 1990. Crustal and upper mantle structure in the Ryukyu island arc deduced from deep seismic sounding, *Geophys. J. Int.* **102:**631–651.

Jarrard, R. D., and Sasajima, S. 1980. Paleomagnetic synthesis for Southeast Asia: constraints on plate motions, in *The Tectonic and Geologic Evolution of Southeast Asian Seas and Islands* (D. E. Hayes, ed.), Geophys. Monogr. Ser., Vol. 23, pp. 293–317, American Geophysical Union, Washington, DC.

Jin, X., Yu, P., Lin, M., Li, C., and Wang, H. 1983. Preliminary study on the characteristics of crustal structure in the Okinawa Trough (in Chinese with English abstract), *Oceanol. Limnol. Sin.* **14:**105–116.

Kao, H., and Chen, W.-P. 1991. Earthquakes along the Ryukyu–Kyushu arc: Strain segmentation, lateral compression, and thermomechanical state of the plate interface, *J. Geophys. Res.* **96:**21,443–21,485.

Kasahara, J., Nagumo, S., Koresawa, S., Ouchi, T., and Kinosita, H. 1985. Seismic features in the central Okinawa Trough—an active incipient rifting, Abstract for the 23rd General Assembly of IASPEI, Tokyo **1:**279.

Kato, S., Katsura, T., and Hirano, K. 1982. Submarine geology off Okinawa Island, *Rep. Hydrogr. Res.* **17:**31–70.

Kimura, M. 1985. Back-arc rifting in the Okinawa Trough, *Mar. Petrol. Geol.* **2:**222–240.

Kimura, M., Kaneoka, I., Kato, Y., Yamamoto, S., Kushiro, I., Tokuyama, H., Kinoshita, H., Isezaki, N., Masaki, H., Oshida, A., Uyeda, S., and Hilde, T. W. C. 1986. Report on DELP 1984 cruises in the middle Okinawa Trough. Part V: Topography and geology of the central grabens and their vicinity, *Bull. Earthquake Res. Inst., U. Tokyo* **61:**269–310.

Le Pichon, X., Huchon, P., and Barrier, E. 1985. Geoid and the evolution of the western margin of the Pacific Ocean, in *Formation of Active Ocean Margins* (N. Nasu, ed.), pp. 3–42, Tokyo, Terrapub.

Le Pichon, X., and Sibuet, J.-C. 1981. Passive margins: A model of formation, *J. Geophys. Res.* **86:**3708–3710.

Lee, C. S., Shor, G. G., Jr., Bibee, L. D., Lu, R. S., and Hilde, T. W. C. 1980. Okinawa Trough: Origin of a back-arc basin, *Mar. Geol.* **35:**219–241.

Letouzey, J., and Kimura, M. 1986, The Okinawa Trough: Genesis of a back-arc basin developing along a continental margin, *Tectonophysics* **125:**209–230.

Ludwig, W., Murauchi, S., Den, N., Bull, P., Hotta, H., Ewing, M., Asanuma, T., Yoshii, T., and Sakajiri, N. 1973. Structure of the East China Sea–West Philippine Sea margin of southern Kyushu, *J. Geophys. Res.* **78:**2526–2536.

Miki, M., Matsuda, T., and Otofuji, Y. 1990. Opening mode of the Okinawa Trough: Paleomagnetic evidence from the South Ryukyu arc, *Tectonophysics* **175:**335–347.

Nagumo, S., Hinoshita, H., Kasahara, J., Ouchi, T., Tokuyama, H., Asamuma, T., Koresawa, S., and Akiyoshi, H. 1986. Report on DELP 1984 cruises in the Middle Okinawa Trough, Part II: Seismic structural studies, *Bull. Earthquake Res. Inst., U. Tokyo* **61:**167–202.

Nash, D. F. 1979. The geological development of the north Okinawa Trough area from Neogene times to Recent, *J. Jpn. Assoc. Petrol. Technol.* **44:**121–133.

Oshida, A., Midorikawa, Y., Kawabata, K., Kanazawa, J., Kimura, M., and Kato, Y. 1988. Submarine acoustic and geomagnetic surveys in the north of Yaeyama–Gunto, Ryukyu arc (RN86 and RN87 cruises), *Bull. Coll. Sci., U. Ryukyus* **46:**123–138.

Oshida, A., Tamaki, K., and Kimura, M. 1992. Origin of the magnetic anomalies in the southern Okinawa Trough, *J. Geomagn. Geoelectr.* **44:**345–359.

Oshima, S., Takanashi, M., Kato, S., Uchida, M., Okazaki, I., Kasuga, S., Kawashiri, C., Kaneko, Y., Ogawa, M., Kawai, K., Seta, H., and Kato, Y. 1988. Geological and geophysical survey in the Okinawa Trough and the adjoining seas of Nansei Syoto, *Rep. Hydrogr. Res.* **24.**

Reches, Z. 1983. Faulting of rocks in three-dimensional strain fields. II: theoretical analysis, *Tectonophysics* **95:**133–156.

Regan, R. D., and Rodriguez, P. 1981. An overview of the external field with regard to magnetic surveys, *Geophys. Surv.* **4:**255–296.

Seno, T., and Maruyama, S. 1984. Paleogeographic reconstruction and origin of the Philippine Sea, *Tectonophysics* **102:**53–54.

Sibuet, J.-C. 1991. The southern Okinawa Trough, in *Taicrust Workshop Proc.*, June 10–12, 1991, National Taiwan University, Taipei, Taiwan, R.O.C., pp. 117–126.

Sibuet, J.-C. 1992. New constraints on the formation of non-volcanic continental Galicia–Flemish Cap conjugate margins, *J. Geol. Soc. London* **149:**829–840.

Sibuet, J.-C., Dyment, J., Bois, C., Pinet, B., and Ondréas, H. 1990. Crustal structure of the Celtic Sea and western approaches from gravity data and deep seismic profiles: Constraints on the formation of continental basins, *J. Geophys. Res.* **95:**10,999–11,020.

Sibuet, J.-C., Letouzey, J., Barbier, F., Charvet, J., Foucher, J.-P., Hilde, T. W. C., Kimura, M., Ling-Yun, C., Marsset, B., Müller, C., and Stéphan, J.-F. 1987. Backarc extension in the Okinawa Trough, *J. Geophys. Res.* **92:**14,041–14,063.

Sibuet, J.-C., Letouzey, J., Marsset, B., Davagnier, M., Foucher, J.-P., Bougault, H., Dosso, L., Maury, R., and Joron, J.-L. 1986. Tectonic evolution and volcanism of Okinawa Trough (abstract), *Am. Assoc. Petrol. Geol. Bull.* **70:**934.

Sibuet, J.-C., Monti, S., Réhault, J.-P., Durand, C., Gueguen, E., and Louvel, V. 1991. Quantification de l'extension liée à la phase pyréneenne et géométrie de la frontière de plaques dans la partie ouest du golfe de Gascogne, *C.R. Acad. Sci. Paris* **317:**1207–1214.

Sun, S. C. 1981. The Tertiary basins of off-shore Taiwan, ASCOPE, Manila.

Taylor, B., Klaus, A., Brown, G. R., and Moore, G. F. 1991. Structural development of Sumisu rift, Izu-Bonin arc, *J. Geophys. Res.* **96:**16,113–16,129.

Tsai, Y.-B. 1986. Seismotectonics of Taïwan, *Tectonophysics* **125:**17–37.

Tsuburaya, H., and Sato, T. 1985. Petroleum exploration well Miyakojima–Oki, *J. Jpn. Assoc. Petrol. Technol.* **50:**25–53.

Ueda, Y. 1986. Geomagnetic anomalies around the Nansei Soto (Ryukyu Islands) and their tectonic implications, *Bull. Volcanol. Soc. Jpn.* **31:**177–192.

Ujiie, H. 1980. Significance of "500 m deep island shelf" surrounding the southern Ryukyu Island arc for its Quaternary geological history, *Quat. Res.* **18:**209–219.

Uyeda, S. 1977. Some basic problems in trench-arc-back-arc-system, in *Island Arcs, Deep Sea Trenches and Back-Arc Basins* (M. Talwani and W. C. Pitman III, eds.), Maurice Ewing Ser., Vol. 1, pp. 1–14, American Geophysical Union, Washington, DC.

Uyeda, S., Kimura, M., Tanaka, T., Kaneoka, J., Kato, Y., and Kushiro, I. 1985a. Spreading center of the Okinawa Trough, *Tech. Rep. JAMSTEC*, pp. 123–142.

Uyeda, S., Nagumo, S., and Hilde, T. W. C. 1985b, Okinawa Trough—An early stage of continental margin rifting, in *1985 Geodynamics Symposium on Intraplate Deformation: Characteristics, Processes, and Causes*, Texas A&M University, College Station, TX.

Vander Zouwen, D. E. 1984. Structure and evolution of Southern Okinawa Trough, Master's thesis, Texas A&M University, College Station.

Warner, M. R. 1987. Seismic reflections from the Moho: The effect of isostasy, *Geophys. J. R. Astron. Soc. London* **88:**425–435.

Wessel, P., and Smith, W. M. F. 1991. Free software helps map and display data, *EOS, Trans. AGU* **72:**441–446.

Yamano, M., Uyeda, S., Foucher, J.-P., and Sibuet, J.-C. 1989. Heat flow anomaly in the middle Okinawa Trough, *Tectonophysics* **159:**307–318.

Yamano, M., Uyeda, A., Kinoshita, H., and Hilde, T. W. C. 1986. Report on DELP 1984 cruises in the Middle Okinawa Trough. Part IV: Heat flow measurements, *Bull. Earthquake Res. Inst., U. Tokyo* **61:**251–267.

10

Shikoku Basin and Its Margins

Kazuo Kobayashi, Shigeru Kasuga, and Kyoko Okino

ABSTRACT

The Shikoku Basin is an inactive backarc basin located south of the southwest Japan arc in the northwestern Pacific margin. Its characteristic features are summarized on the basis of updated geophysical and geological data, including swath bathymetry, gravity, magnetics, and seismic reflection profiling records as well as results from the DSDP/ODP drilling holes. It has been proposed that the Shikoku Basin was born as a rift separating the N-S trending paleo-Kyushu–Palau Ridge at its northern end. The rifting rapidly propagated southward. It was succeeded by seafloor spreading that formed a narrow triangular trough bounded by steep scarps on both east and west margins. Magnetic and bathymetric data indicate that the spreading center has changed its trend at least twice, first at 23 Ma and then at 19 Ma. Widespread off-ridge volcanism occurred after extinction of spreading at 15 Ma. The Kinan seamount chain was formed at this stage. Most of the rocks constituting the basin are tholeiite similar to MORB, whereas some of the off-ridge magmas are alkali basalt. Igneous basement is overlain mostly by hemipelagic sediments containing dispersed detrital clays and interbedded tephra layers. The Shikoku Basin lithosphere is subducting at the Nankai Trough beneath southwest Japan in a NNW direction. Deep-focus earthquakes are not observed at depths greater than 80 km, implying that the subducted young lithosphere loses its rigidity below such depths. It seems plausible to presume that the Shikoku Basin was subducting at the Nankai Trough at 15–12 Ma while the basin floor underwent extensive volcanic intrusions. Emplacement of intermediate to felsic rocks at the outer zone and Setouchi Province of southwest Japan probably originated from this unusual circumstance.

1. INTRODUCTION

The Shikoku Basin is a N-S elongated fan-shaped basin surrounded by the Kyushu–Palau Ridge on its west and by the Nishi–Shichito Ridge (westernmost ridge of the Izu–Bonin arc) on its east. Its northern margin is bounded by the Nankai Trough at which the

---

Kazuo Kobayashi • Japan Marine Science and Technology Center, Yokosuka 237, Japan. *Shigeru Kasuga and Kyoko Okino* • Hydrographic Department, Maritime Safety Agency, Chuo-ku, Tokyo 104, Japan.

Backarc Basins: Tectonics and Magmatism, edited by Brian Taylor, Plenum Press, New York, 1995.

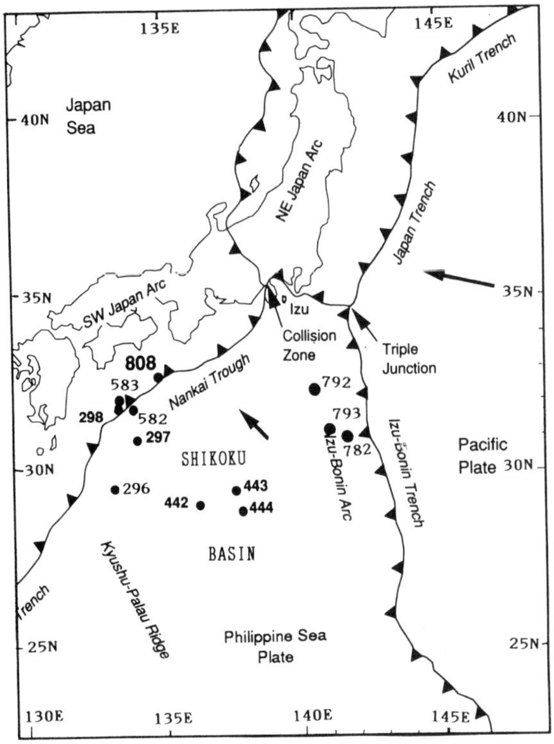

FIGURE 10.1. Index map of the Shikoku Basin and its margins. Positions of relevant DSDP and ODP sites are shown by numerical figures beside solid circles. Arrows denote directions of relative plate motions.

basin is subducted under the southwest Japan arc (Figs. 10.1 and 10.2). On the south it merges with the Parece Vela Basin (or West Mariana Basin in Japanese maps), with which it shares a similar evolution.

The Shikoku Basin is a backarc basin that formed to the west of the Izu–Bonin arc-trench system, which is active at present. Opening of the basin is, however, no longer occurring. The fundamental nature of the Shikoku Basin as an inactive backarc basin was first pointed out by Karig (1971).

Bathymetry, magnetic and gravity anomalies, and subbottom structure of the Shikoku Basin have been surveyed comprehensively by S/V *Takuyo* of the Hydrographic Department, Maritime Safety Agency (HD-MSA) of Japan (Kasuga *et al.*, 1987, 1992; HD-MSA, 1990). The majority of survey tracklines are along an E-W direction with spacing of 5 nautical miles (approx. 9 km). Bathymetric coverage of the area by this swath survey is roughly 40%. Denser tracks were chosen for several selected areas having complex topography. The north-central margin of the basin close to the Nankai Trough was surveyed in detail with nearly 100% swath coverage for bathymetric mapping by the French ship *Jean Charcot* as part of the French–Japanese cooperative KAIKO Project (KAIKO I Research Group, 1986; Le Pichon *et al.*, 1987).

The crustal structure of this area was investigated by two-ship refraction studies as early as the 1960s (Murauchi *et al.*, 1968), and typical oceanic characteristics of the crust under the basin were clearly demonstrated. The distribution of sediment cover was surveyed by single-channel seismic reflection profiling along the HD-MSA's 5-mile spaced EW tracks cited above, as well as by several multichannel reflection lines carried out by

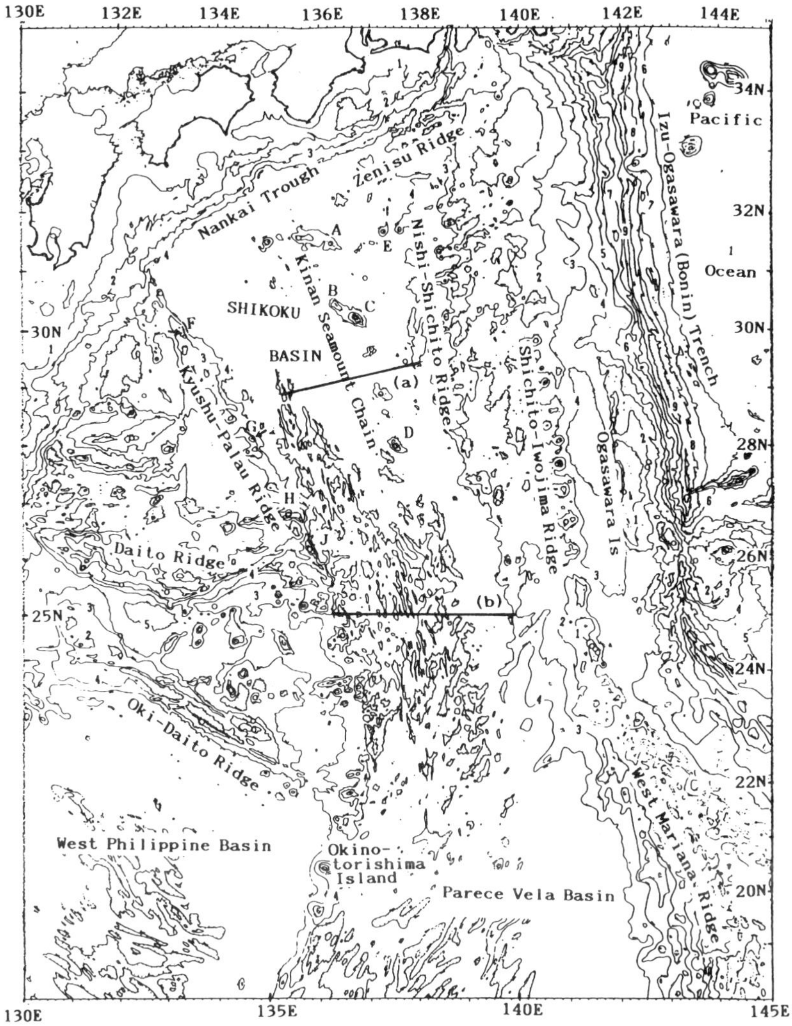

FIGURE 10.2. Major topographic trends in the area surrounding the Shikoku Basin. The contour interval is 1 km. Seamount names are shown by letters as follows: A, Koshu; B, Daiichi–Kinan; C, Daini–Kinan; D, Hakuho; E, Komahashi–Daisan; F, Komahashi–Daini; G, Komahashi; H, Kita–Koho, J, Minami–Koho. Lines (a) and (b) locate seismic profiles shown in Fig. 10.4.

various investigators (e.g., IPOD-Japan, 1977; HD-MSA, 1989, 1990). Most topographic features near the margins of the Shikoku Basin are masked by thick clastic wedges overlying local topographic lows filled by sedimentation in an early stage of basin formation (White *et al.*, 1980; Klein and Kobayashi, 1981).

A Deep Sea Drilling Project (DSDP) hole was first drilled in the northern margin of the basin (at Site 297) during Leg 31, although its penetration was insufficient to reveal the structure and age of basement. On the other hand, hole 296 on the Kyushu–Palau Ridge penetrated the igneous crust and provided crucial information on the origin and subsidence history of the remnant ridge (Ingle *et al.*, 1975). A set of holes was drilled in the inner and axial zones of the Shikoku Basin during Leg 58 (Klein *et al.*, 1980; Klein and Kobayashi,

1981). Holes at Site 442 west of the axis of the basin penetrated the pillow lava layer overlain by sediments intercalated with massive intrusive volcanic flows. Sites 443 and 444 in the eastern zone of the basin provided only the minimum age and sedimentary history of the area, because sediments immediately above the basement pillow were not recovered. Drilling of the east and west margins of the basin has not been attempted.

Hole 298 of Leg 31 and holes 582 and 583 of Leg 87 (Kagami *et al.*, 1987) were drilled in the landward slope of the Nankai Trough and provided information only on the basin sediment. Holes at Site 808 in Ocean Drilling Program (ODP) Leg 131 penetrated the entire accretionary wedge of the western Nankai Trough south of Cape Muroto on Shikoku Island into the underlying oceanic basement rocks, which were subducted from the Shikoku Basin (Taira *et al.*, 1991; Hill *et al.*, 1993) so that the recovered cores yielded information on the northern margin of the Shikoku Basin, which is now overlain by the landward wedge. The present article is based upon these results combined with other survey data.

2. TOPOGRAPHY OF THE BASIN AND ITS MARGINS

2.1. General Bathymetric Features and Basement Topography

Fig. 10.3 shows a bathymetric map of this area with 200-m contour intervals. Water depths gradually increase southwestward, and the maximum depth of over 5000 m is observed in the western margin close to the Kyushu–Palau Ridge. A sharp contrast in topography is seen between the axial zone and both wings of the basin floor. The axial zone is composed of small ridges and troughs trending N40°W with several seamounts (the Kinan seamount chain, Fig. 10.2), whereas the wings show linearity nearly parallel to the trend of the whole basin, implying an episodic change in the opening pattern. A triangle bordered by longitude 135°E, northern Kyushu–Palau Ridge and western Nankai Trough has flat topography caused by a thick cover of sediments. The east wing is relatively flat under the wedge of clastic deposits supplied from the Izu–Bonin arc. In contrast, the west wing south of 30°N is particularly rugged. It consists of small ridges and troughs with lengths of 30 to 50 km, widths of 5 to 10 km, and relative heights of 500 to 1000 m, trending along NNW-SSE.

Figure 10.4 represents a composite bathymetric profile of the Shikoku Basin in which 13 parallel profiles trending N67.5°E from 25°N to 33°N are stacked relative to the Kinan seamount chain (Park *et al.*, 1990). It shows that water depths of the Shikoku Basin are quite asymmetric with a general tendency of westward deepening. Figure 10.5 shows multichannel seismic reflection profiles traversing the Shikoku Basin at about 29°N and at 25°N. Sediment cover in the axial and inner zones is about 300 m except for the north margin close to the Nankai Trough, whereas both west and east wings north of 27°N are covered by horizontally layered sediments thicker than 1000 m. Sediments in the east wing are particularly thick, as seen in the right-hand side of the profiles. General westward-deepening topography of the eastern limb may partly be explained by the thick sediment covers.

The western margin of the basin is distinguished by steep and stepwise scarps along the Kyushu–Palau Ridge as indicated in the left-hand corner of the seismic profile in Fig. 10.5(b). The boundary in the eastern margin along the en echelon Nishi–Shichito Ridge is not so distinct as in the western scarps. Instead a distinct steep westward-dipping scarp trending in a direction of N10°W is recognized in the east wing of the basin floor around the

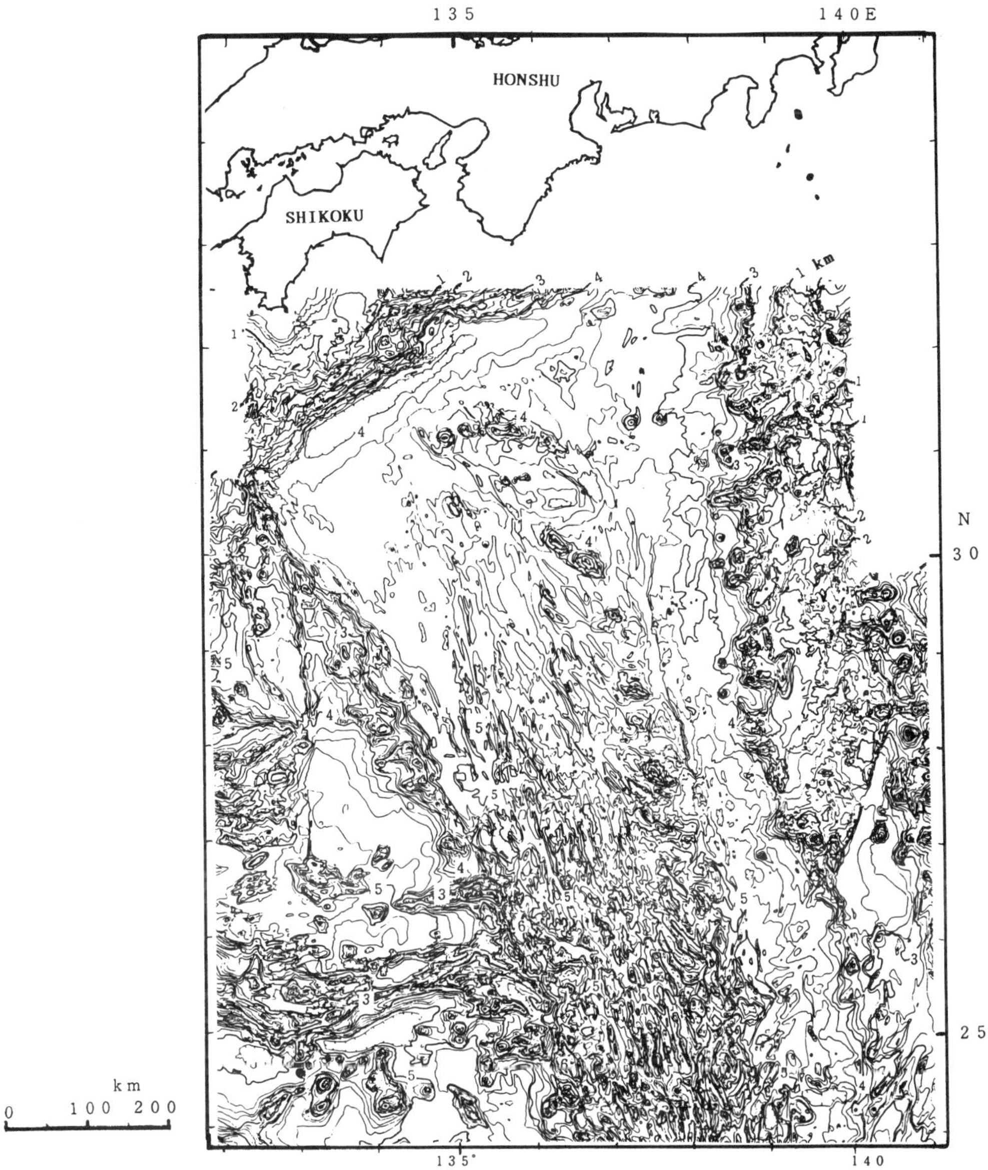

FIGURE 10.3. Bathymetric map of the Shikoku Basin and its margins (from Kasuga *et al.*, 1992), 200-m contours. Numerical figures beside thick contour lines denote depths in km. Lambert Projection.

longitude of 137°30′E north of 27°N. The linear Kinan escarpment has a maximum elevation of 800 m and a length of 500 km at 29°30′N, 137°30′E. Detailed analysis of swath bathymetric maps shows that this scarp is composed of segments of en echelon cliffs each trending along NNW-SSE with lengths of about 30 km in average. Dives of the submersible Shinkai 6500 (dive nos. 176 and 177 in 1993) indicated that the surface of the cliffs and taluses is completely covered by Mn oxides (Okino, 1993). This observation implies that the faulting that forms the scarps is relatively old, perhaps as old as the age of the basin

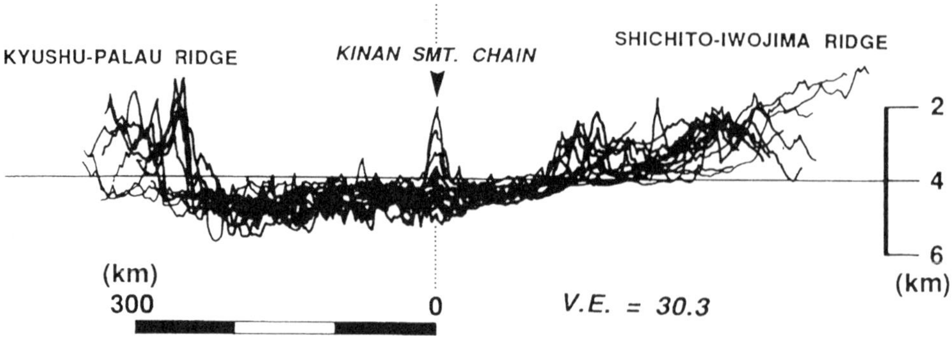

FIGURE 10.4. Stacked profiles of water depths of the Shikoku Basin between 25°–33°N along N67.5°E trending lines (from Park *et al.*, 1990).

generation (19 Ma). East of the escarpment, arrays of small cliffs cutting the upper layer of sediment are recognized. They appear to have been formed by recent tectonic activity.

2.2. The Southern Border of the Basin

The southern margin of the Shikoku Basin merges with the N-S-elongated Parece Vela Basin. Although no distinctly lineated ridge or trough separating the Shikoku Basin and Parece Vela Basin has been found, the Shikoku Basin can be distinguished from the southern basin by a line trending in a direction of N40°E at latitude of about 23 to 25°N. The following viewpoints support this subdivision:

1. Width of the basin is narrowest along this border: the maximum width of the fan-shaped Shikoku Basin is roughly 900 km along its north margin (on the Nankai Trough), and that of the Parece Vela Basin exceeds 1200 km along a latitude of 17°N, whereas the width along this presumed border is only 550 km.
2. The predominant topographic trend changes north and south of this border. In the north it is generally parallel to the Kyushu–Palau Ridge, whereas it trends N40°E in the south.
3. A chain of seamounts exist in the north. In contrast, there are deep troughs along the axis of the south basin.
4. The east side of the Shikoku Basin is fringed by the Izu–Bonin arc with embryonic rifting, whereas the south adjoins the West Mariana Ridge and the well-developed Mariana Trough to its east.

2.3. Axial Seamount Chain

Along the axis of the northern part of the basin, a chain of seamounts has long been known and named the Kinan seamount chain, which is composed of Koshu, Daiichi–Kinan, Daini–Kinan, and Hakuho seamount from north to south (see Fig. 10.2). The northernmost seamount consists of a linear 80-km-long chain with multiple crests trending N70°W (Shino *et al.*, 1991). Water depth of the shallowest peak is 2060 m. Two independent seamounts with crestal depths of 3130 m and 1990 m are situated just west of the northern end of Koshu seamount. One conical seamount exists in the east wing of the basin and is named Komahashi Daisan seamount. Its 1770-m-deep crest is nearly 150 km ENE of the peak of Koshu seamount.

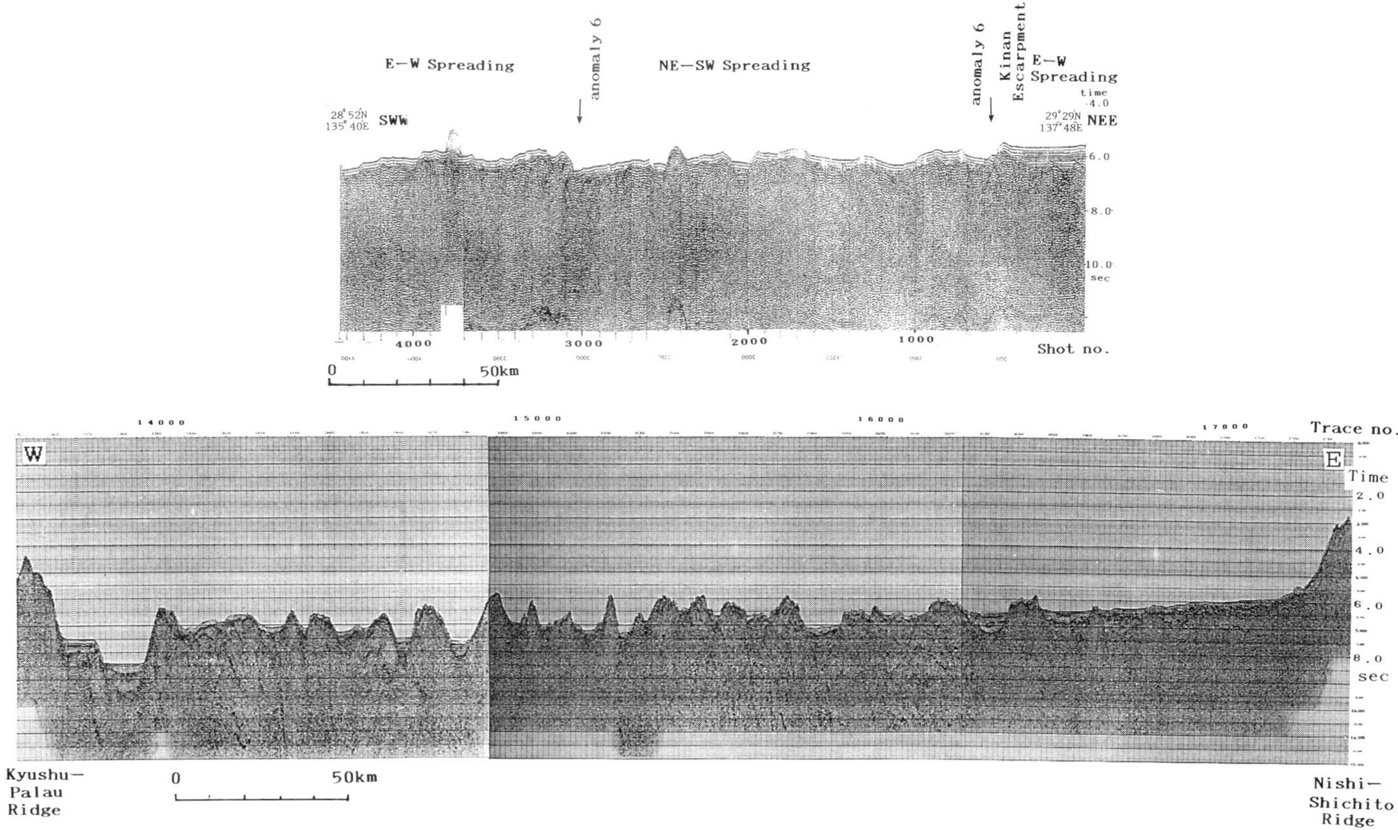

FIGURE 10.5. Multichannel seismic reflection profiles traversing the Shikoku Basin (migrated time sections). For positions see Fig. 10.2. (a) IPOD-Japan (1977) section (220 km long) in the central portion of the basin. Shot-point spacing = 50 m, 24-fold stack, deconvolution and time variant filtering. (b) HD-MSA section in a EW direction at 25°N (HD-MSA, 1989). Shot-point spacing = 50 m, 48-fold stack, deconvolution and time variant filtering. Numbers on the profile denote trace numbers corresponding to one every 25 m.

Two relatively large features, called Daiichi– and Daini–Kinan seamounts, are located on the axial zone roughly 120 km south of Koshu seamount. The twin peaks of each seamount, with water depths as shallow as 700 m, are aligned N40°W, parallel to linearity of the basin in the axial zone. Each seamount body has an elongated shape that also parallels the general trend. Hakuho seamount located at latitude of 28°00′N is also elongated in a direction parallel to the trend of the axial zone (Furuta *et al.*, 1980; Kobayashi, 1984; Kobayashi and Fujiwara, 1993).

2.4. Nishi–Shichito Ridge—The Eastern Border of the Basin

The Izu–Bonin arc, the east-bordering arc-trench system of the Shikoku Basin, appears to be composed of at least three linear N-S trending ridges: the Nishi–Shichito Ridge, Shichito–Iwojima Ridge, and Ogasawara (Bonin) Islands. The Nishi–Shichito Ridge consists of an en echelon alignment of ridges trending N60°E (Fig. 10.2). Seafloor ruggedness is particularly distinct in the region north of 30°N. Separated crests of seamounts are recognizable in each ridge. The size of seamounts generally increases southwestward (toward the Shikoku Basin). The northernmost ridge, called Zenisu Ridge, is the most prominent in morphology and is the longest. Its northeastern end extends toward the line of islands Kozushima, Toshima, and Shikineshima.

In the region between 30°N and 26°N the trend of the en echelon alignment is N45°E. Seamounts are nonexistent in the Nishi–Shichito Ridge south of 26°N, although NE-SW trending troughs or canyons such as Iwo Canyon are still distinguishable there. To the east, a series of rifts, including the Hachijo, Aogashima, Sumisu, Torishima, Sofu Gan, and Nishino–shima rifts (from 33°N to 28°N) segregates the Nishi–Shichito Ridge from the Shichito–Iwojima Ridge, which comprises a chain of active volcanic islands and submarine volcanoes in the Izu–Bonin arc (Brown and Taylor, 1988; Taylor, 1992).

2.5. Kyushu–Palau Ridge—The Western Border of the Basin

The Kyushu–Palau Ridge is a 3000-km-long submarine ridge that connects Kyushu, Japan, at 32°N and the Palau Islands located at 8°N. It divides the floor of the whole Philippine Sea into east and west. North of 24°N it trends NNW-SSE, whereas it runs along a NNE-SSW direction further south. The northern segment of the Kyushu–Palau Ridge is composed of an elongated ridge morphology overlain by seamounts. Saddles between adjacent seamounts are still nearly 2000 m higher than adjacent basins. Particularly large seamounts are named Komahashi-Daini [30°N, CD (crestal depth) = 289 m], Komahashi (28°N, CD = 440 m), Kita–Koho (26°45′N, CD = 329 m) and Minami–Koho (26°10′N, CD = 367 m).

On each seamount more than two crests are aligned NE-SW, oblique to the ridge axis. The seamount bodies on the NE side are usually larger than those to the SW. Okinotorishima at 20°25′N, 136°03′E is the only emerged island (uplifted atoll) in the Kyushu–Palau Ridge and is situated close to its junction with the E-W trending Oki–Daito Ridge, which borders the north of the West Philippine Basin (Fig. 10.2). The southern portion of the Kyushu–Palau Ridge beyond this emergent island changes its morphology to an elongated ridge with fewer seamounts.

The boundary of the Kyushu–Palau Ridge with the Shikoku Basin is marked by sharp, steep scarps. The escarpment is supposed to have been formed during the initial rifting of the Shikoku Basin and subsequent fast subsidence of the basin. In some places the east-dipping scarps are stepwise with sedimented depressions between them as shown in Fig. 10.5b. The west margin of the Kyushu–Palau Ridge, on the other hand, gradually increases its water depth to reach relatively flat basins covered by thick sediments such as the Kikai, north Daito, and south Daito basins.

3. GRAVITY ANOMALIES AND ISOSTATIC COMPENSATION OF THE BASIN AND ITS MARGINS

The gravity field of the northwestern Pacific, including the Shikoku Basin, has long been measured by shipboard gravity meters (e.g., Tomoda *et al.*, 1968), and several maps of free-air anomalies have been published (Tomoda, 1974; Watts, 1976; Kasuga *et al.*, 1987, 1992). Free-air anomalies in the Shikoku Basin, as well as those of the West Philippine Basin, are near 0 mgal except for seamounts and knolls, indicating isostatic compensation of the basin crust. Bouguer gravity anomalies in the Shikoku Basin gradually increase southwestward in harmony with thinning of the sediment cover. In the eastern margin of the basin close to the Nishi–Shichito Ridge the Bouguer anomalies decrease, implying thicker crust there.

Along the Kyushu–Palau Ridge free-air anomalies amounting to 50–100 mgal are observed on each individual topographic high (Fig. 10.6). No continuous anomaly belt exists along the ridge, whereas distinct Bouguer anomalies are seen in the portion north of 26°N. The results seem to imply that the crust of the ridge is thicker than that of the basin floor but that it is isostatically compensated.

Free-air anomalies of +100 to +120 mgal are aligned en echelon along the Nishi–Shichito Ridge in a manner similar to topography. In contrast, such an en echelon alignment is not seen with Bouguer anomalies, suggesting that topographic features are caused only by rugged shallow crust but are not related to the deeper structure and that they are regionally supported.

4. STRATIGRAPHY OF THE SHIKOKU BASIN

The structure and geological history of bottom sediments were revealed by seismic reflection profiling and by ocean drilling. The accretionary wedge sediments recovered by drilling at DSDP/ODP Sites 298, 582, 583, and 808 consist of a succession of turbidites and hemipelagic clays (Ingle *et al.*, 1975; Kagami *et al.*, 1987; Taira *et al.*, 1991; Pickering *et al.*, 1993). Most turbidites recovered from these sites appear to have been accreted from the Nankai Trough to which they were supplied from river and land shelves northeast of the basin via channels along its axis, given that the thickness of turbidite layers revealed by seismic reflection profiling decreases westward (Taira and Niitsuma, 1986; Le Pichon *et al.*, 1987).

Most of the cores from Site 297, located at the northwestern corner of the Shikoku Basin south of the axis of Nankai Trough, are composed of hemipelagic sediments. Only the cores in a limited range between 3 and 5 Ma contain many turbidites. Karig (1975) postulated that the trough topography, which now traps turbidites transported from the northern islands, disappeared in that period. An alternative explanation is fill-up and

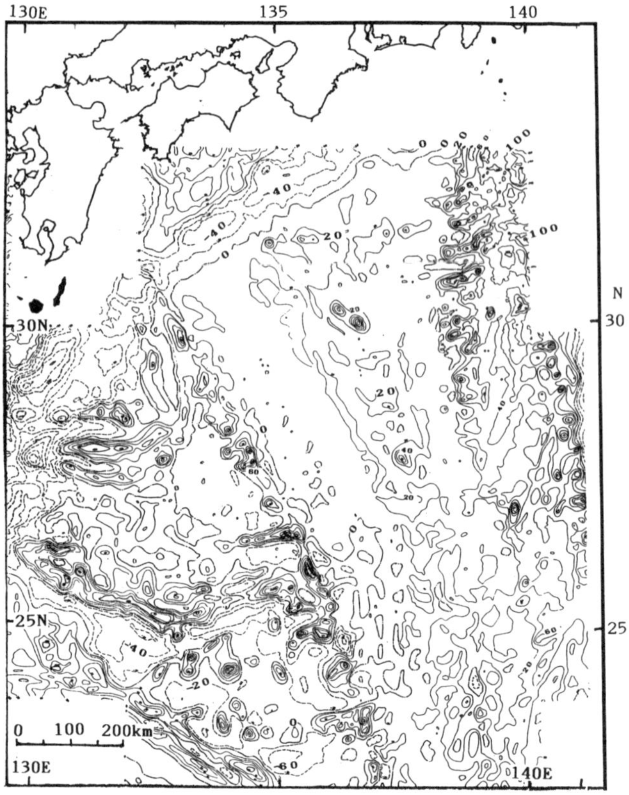

FIGURE 10.6. Free-air gravity anomalies of the Shikoku Basin area (from Kasuga *et al.*, 1992). The contour interval is 20 mgal. Solid lines denote positive and broken lines negative anomalies.

overflow of turbidites at the trough axis occurring when a large amount of turbidites was transported from the highly elevated central mountains in Honshu. The Shikoku Basin facies of sediments in the lower 620-m column (between 620 m and 1240 m below seafloor) at Site 808 consists of hemipelagic sediments (Underwood *et al.*, 1993a).

Drilled cores from the inner wings of the basin (442, 443, and 444) contain hemipelagic sediments interbedded with a number of volcanic tephra layers and dispersed detrital clays (Fig. 10.7). Ages of each horizon in the cores were determined by micropaleontological correlation using nannofossils (Okada, 1980) and, in less frequent cases, foraminifera and radiolarians. Ages of the sediment interbed recovered from a level between massive intrusives and underlying pillow basalt at Site 442 are correlated to be 18 to 21 Ma (*Discoaster drugii* subzone of nannofossils), whereas sediment immediately overlying the upper intrusive sill basalt at the same site is 15 to 17 Ma (*Helicosphaera ampliaptera* zone of nannofossils). The oldest sediment age here is consistent with the magnetic isochron 6 (see the next section), suggesting a seafloor spreading origin for the underlying pillow layer. The age of the red-colored mudstone interbedded with basaltic intrusives at Site 808 was determined to be 15 Ma by using nannofossils (Olafsson, 1993). This age is consistent with that of the final stage of opening of the Shikoku Basin, as Site 808 is situated immediately west of the presumed spreading axis of the subducted basin.

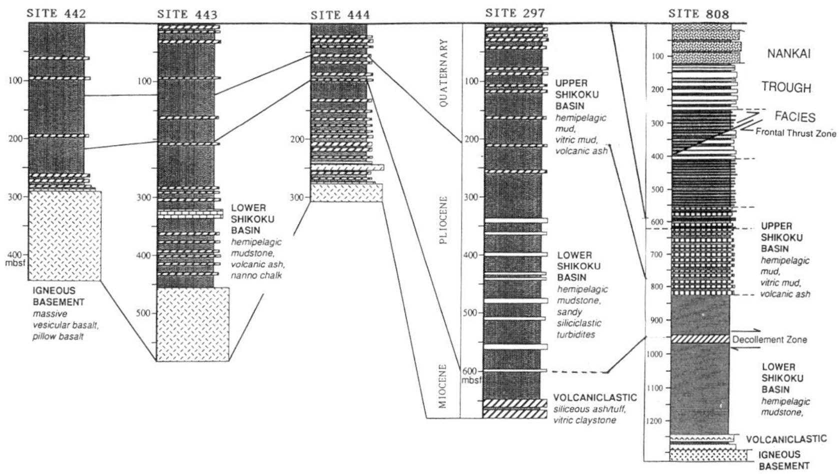

FIGURE 10.7. Stratigraphic correlation of DSDP Sites 442, 443, 444, and 297 in the Shikoku Basin (Klein and Kobayashi, 1981) and ODP Site 808 in the Nankai Trough wedge (Pickering *et al.*, 1993).

The age of the oldest sediment overlying massive basalts at Sites 443 and 444 situated east of the basin axis is 14 to 15 Ma (*Spheonolithus heteromorphus* zone). There is a discrepancy between this age and the magnetic isochron ages (6A according to Watts and Weissel, 1975, and Kobayashi and Nakada, 1978, or 5D-5C identified by Shih, 1980). The lower strata were not recovered at these sites owing to technological difficulty and time limitation. The occurrence of postspreading off-ridge basaltic intrusions widespread in the basin (Klein *et al.*, 1978) obscured the paleontological age relations of Sites 443 and 444.

More than 100 layers of tephras have been recognized in the cores from Sites 442, 443, and 444. Volcanic glass shards are well preserved, since they are overlain by fine-grain sediments with low water permeability. Refractive index and chemical composition of glass shards revealed a sharp contrast between Site 442 in the west wing and the other two in the east wing. Site 442 contains rhyolitic and dacitic tephras throughout the cores exclusively, whereas distinct layers of basaltic tephras occur in middle to early Miocene sediment at Sites 443 and 444 situated east of the axis (Furuta and Arai, 1980). This result seem to be explained by the prevailing northwesterly wind in this region. In Pliocene and Pleistocene cores collected at these sites, tephras display felsic to intermediate compositions, indicating predominance of such volcanism in southwestern Japan. Plentiful rhyolitic tephras and volcaniclastics have been found at the base of a Site 808 sedimentary section immediately above the basaltic basement. Source of the rhyolitic volcanics is presumed to be the episodic forearc volcanism in southwestern Japan taking place in middle Miocene (13–15 Ma). They were most probably transported to Site 808 by wind and/or by turbidity currents. As Site 808 is now situated about 10 km landward from the axis of the trough, it was located roughly 200 km oceanward side of the trough at 13 Ma, if the rates of subduction at the Nankai Trough are assumed to be constant (~2 cm/yr) throughout the geological period. Its distance from the source region (forearc volcanoes) seems to be still close for rhyolitic volcanics to reach the site.

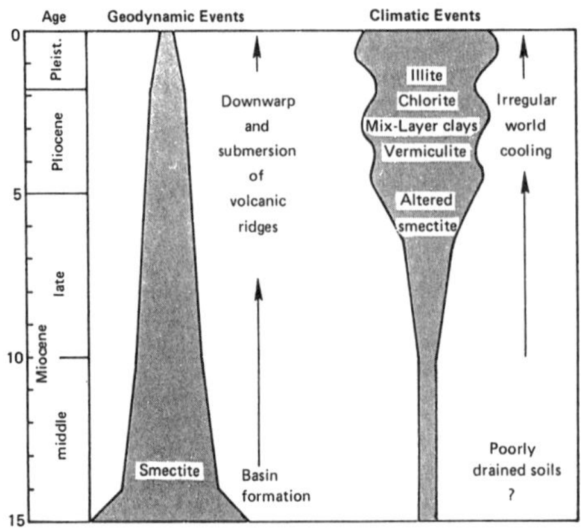

FIGURE 10.8. Chronological variations in clay sedimentation in the Shikoku Basin since its opening based on DSDP 297, 442, 443, and 444 cores (Chamley, 1980).

Chamley (1980) reported changes in content of clay minerals in the cores drilled from the Shikoku Basin. As shown in Fig. 10.8, sedimentation of brown pelagic clay dominated by *in situ* smectite formation began immediately after the basin formed in the early-middle Miocene (15–14 Ma). Latest Miocene to Pleistocene clay sedimentation was progressively influenced by continental climatic factors. Smectite came from volcanic islands such as the Izu–Bonin arc. Illite and chlorite were supplied from southwest Japanese Islands, while kaolinite was transported from the southwest by the Kuroshio current. The amount of detritus has changed with time, depending on vertical tectonic movements of Izu–Bonin and Kyushu–Palau ridges as barriers against ocean circulation. Vermiculite, attapulgite, and most of the irregular mixed-layer clays were blown by wind from the Asiatic continent (Fig. 10.8). A different pattern of detrital clay content was reported at Site 808 (Underwood *et al.*, 1993b). One plausible interpretation of these differences may be proximity of the site to the extinct spreading axis at which hydrothermal circulation to the seafloor was active for some time after spreading ceased.

The present bottom of the Shikoku Basin is below the calcium carbonate compensation depth (CCD, nearly 4500 m at present) and the carbonate lysocline (generally 1000 m shallower than CCD), since $CaCO_3$ content in bottom surface sediments is very low. The DSDP cores from Sites 442–443 indicated that the bottom of this basin in Miocene time was also below the CCD, as evidenced by badly dissolved foraminiferal tests of that age (Echols, 1980). On the other hand, occurrence of a 40-cm-thick limestone layer immediately overlying basalt at hole 442A and chalk at the bottom of hole 443 cores implies that the depth of the Shikoku Basin was shallower than the CCD and lysocline when the basin was just born. This result seems to imply two possibilities: that the young Shikoku Basin was much shallower than now, or the CCD and lysocline at about 20 Ma were much deeper than at present. Van Andel *et al.* (1975) showed that the CCD in the Miocene was nearly 500 m shallower than at present in the Pacific Ocean. If this is valid for the Shikoku Basin at 20 Ma, the occurrence of carbonates indicates that the basin probably was much shallower than at present. Pillow basalts recovered from Site 442 are extremely vesicular, whereas sills contain fewer vesicules. Dick (1980) interpreted the origin of high vesicularity to be caused by high volatile contents of backarc basin magma, whereas Kobayashi (1984) postulated shallow water depths at the time of eruption of these rocks.

5. PETROLOGY AND AGES OF THE BASIN AND MARGINS

Igneous rocks of the Shikoku Basin "basement" were collected by the DSDP and ODP. Pillow basalts recovered from Site 442 are supposed to have erupted on the seafloor at the spreading center. Their petrology and texture are similar to those of the mid-ocean ridge basalt (MORB; Dick, 1982; Dick *et al.*, 1980). Major and trace elements analyses indicate that they are tholeiite with Mg number between 0.6 and 0.7, similar to MORB. Many of the Site 442 massive units are diabases with chemical composition of subalkaline tholeiite resembling N-type MORB (Marsh *et al.*, 1980; Wood *et al.*, 1980). Most other massive flows and sills collected at Sites 443 and 444 have similar MORB composition, except for one sill of alkali olivine basalt at Site 444.

Igneous rocks were reached at a depth of 1289.9 m below the seafloor under an accretionary wedge, and 37.1 m in total of basalts were penetrated (Taira *et al.*, 1991). The upper unit overlain by middle Miocene sediment is composed of sill, and the lower 10 m is pillow basalt, both analogous to the normal MORB similar to Site 442 (Siena *et al.*, 1993).

Isotope ages of some basaltic rocks were determined. The $^{40}Ar/^{39}Ar$ ages of two basalts and one dolerite from Site 443 (total fusion ages) are 15.6 ± 1.9, 8.3 ± 7.8, and 10.9 ± 3.5 Ma, respectively (Ozima *et al.*, 1977; 1980). The K/Ar ages of a pillow basalt from Site 443 and an intrusive sill from site 444 are 17.2 ± 3.2 and 14.7 ± 2.1 Ma (McKee and Klock, 1980).

A boulder of basalt was collected by dredge haul from a slope of the Hakuho seamount in the Kinan seamount chain (28°01′N, 137°27′E, D = 2700–2200 m) (Furuta *et al.*, 1980). Petrography of the sample appears to be intermediate between MORB and BABB (backarc basin basalt), although exact judgment is difficult because of alteration. Several boulders of igneous rocks coated by Mn oxides were collected from other seamounts in the Kinan seamount chain. Among them a piece of basalt from Koshu seamount (31°32′N, 135°36′E, D = 2400 m) was dated by the K/Ar method to be 7.91 ± 0.19 Ma (Katsura *et al.*, 1994).

In contrast to basaltic properties of rocks from the Shikoku Basin floor and axial seamounts, igneous rocks recovered from marginal ridges have petrographic and chemical affinities to island-arc magmatism. Volcaniclastic materials with composition of two-pyroxene andesite were recovered at DSDP Site 296 on a saddle point between seamounts along the Kyushu–Palau Ridge (29°20′N, 133°32′E, D = 2920 m). Their $^{40}Ar/^{39}Ar$ age is 47.5 Ma (Ozima *et al.*, 1977). A number of granodioritic fragments coated by thick Mn oxides have been collected since 1973 by dredge hauls from the slopes of Komahashi–Daini (29°56′N, 133°19′E, D = 2250 m; Kobayashi, 1984), Kita–Koho (26°46′N, 135°27′E, D = 1830 m; Koyama *et al.*, 1986), and Minami-Koho seamounts (26°06′N, 135°49′E, D = 503 m; Mizuno *et al.*, 1977). The K/Ar ages of these samples are plotted in Fig. 10.9, indicating that the Kyushu-Palau Ridge is much older than the Shikoku Basin.

Rocks from Nishi–Shichito Ridge are generally andesitic. One K/Ar age (6 Ma) reported with a sample from a crest at 29°17′N, 138°38′E (D = 903 m) (Hayashida *et al.*, 1989) is much younger than those of either the Shikoku Basin and Kyushu–Palau Ridge, indicating that this ridge was active at periods after the basin ceased to open. A rock sample from Komahashi–Daisan seamount situated in the northeastern portion of the basin (31°38′N, 137°16′E, D = 1900 m) has andesitic composition and age of 11.3 ± 0.4 Ma (Shino *et al.*, 1991), suggesting that this seamount has affinities to the Nishi–Shichito Ridge activities regardless of its position in the east wing of the basin.

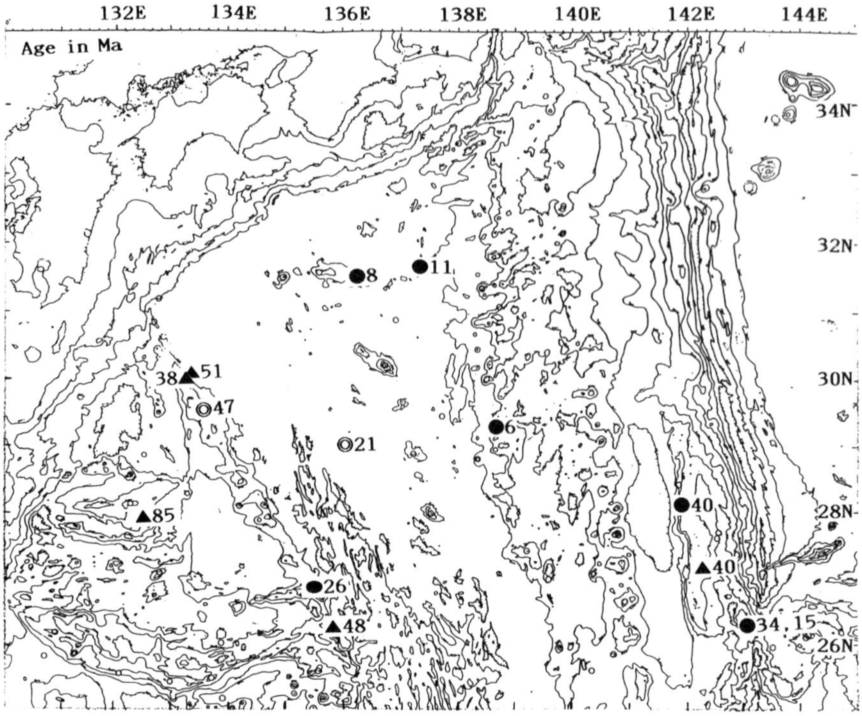

FIGURE 10.9. Isotope ages (in Ma) of rocks collected from the Shikoku Basin and its margins (from various sources mentioned in the text). Double circles denote ages of basement rocks from DSDP holes. For solid triangles see Kobayashi (1983) and for solid circles see Shino *et al.* (1991).

6. MAGNETIC ANOMALIES AND SPREADING HISTORY OF THE SHIKOKU BASIN

The existence of linear magnetic anomalies in the Shikoku Basin was first reported by Tomoda *et al.* (1968). Watts and Weissel (1975) published magnetic anomaly profiles and their correlation with the standard polarity reversal chronology. They proposed correlatable linear anomaly patterns in the western limb of the basin, while they suggested the possibility of either symmetric or single-limb spreading for the eastern portion. Tomoda *et al.* (1975) and Kobayashi and Isezaki (1976) postulated a symmetric spreading model for the Shikoku Basin. Kobayashi and Nakada (1978) and Shih (1980) added survey data and extended interpretation.

Figure 10.10 represents a magnetic anomaly map based upon the HD-MSA's survey data. In this figure the axial zone can be clearly distinguished from the west and east wings of the basin in latitudes north of 26°N. The western wing appears to be divided into the marginal and inner zones, although the eastern margin is overprinted by recent activity of the Nishi–Shichito Ridge.

Good control of magnetic anomaly ages was provided by DSDP site 442, which penetrated the pillow basalt basement underlying a sequence of intrusive sill and sediment (see Section 5). The micropaleontological age of the oldest sediment-pillow basalt contact is 18–21 Ma, which is consistent with the magnetic isochron age of anomaly 6 first

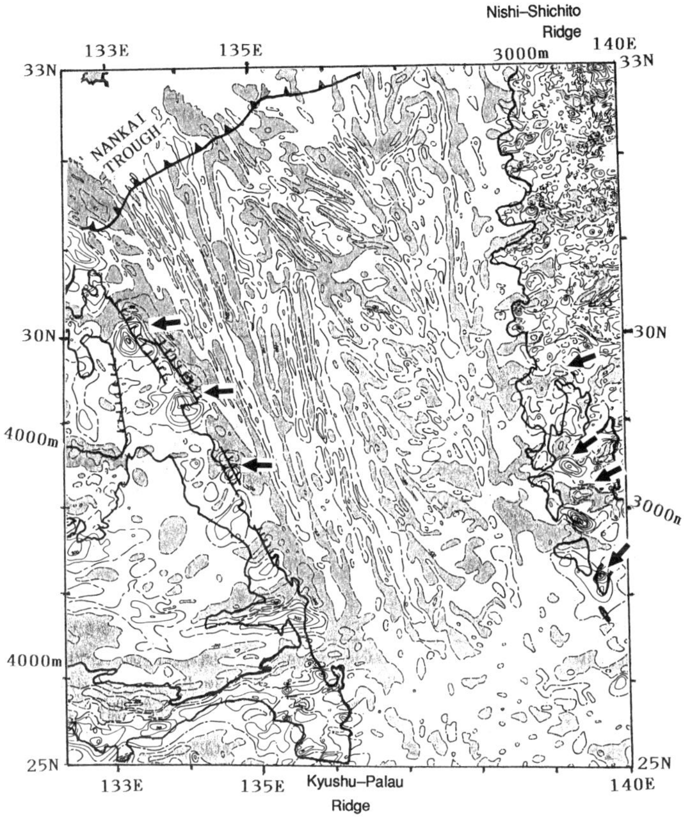

FIGURE 10.10. Magnetic anomalies of the Shikoku Basin and its margins (from Kasuga *et al.*, 1992). The contour interval is 100 nT. Areas with negative anomalies exceeding −100 nT are shaded. Dipole anomalies without seamount topography are indicated by solid arrows. Steep scarps in the eastern margin of Kyushu–Palau Ridge and topographic boundary of Nishi–Shichito Ridge are shown by 4000-m and 3000-m depth contours, respectively.

proposed by Watts and Weissel (1975). Sites 443 and 444 drilled in the eastern part of the basin led to ambiguous crustal ages, because these holes failed to date pillow basalt basement. Nevertheless, the single-limb model seems to be untenable as the hole bottom ages are 17 and 15 Ma for Sites 443 and 444, respectively, which are far older than predicted by such a model.

Okino *et al.* (1994) postulated an updated identification of magnetic isochrons (Fig. 10.11). They defined three distinct fracture zones in both west and east wings and attempted to correlate both wings by the fracture zones assuming symmetric spreading of the basin. Five episodic stages were thus recognized by their analysis: rifting, N70°E opening, E-W opening, and NE-SW opening, followed by off-ridge volcanism and tectonic deformation.

Taylor (1992) postulated a revised identification of Shikoku Basin lineations, suggesting existence of a failed rift at subchron 5D in the west wing of the basin. He proposed the initial age of spreading to be 22 Ma. However, we prefer an earlier age of the initiation of spreading on the basis of evidence and rationale mentioned below.

Chamot-Rooke *et al.* (1987) interpreted isochrons in the northern margin of the basin

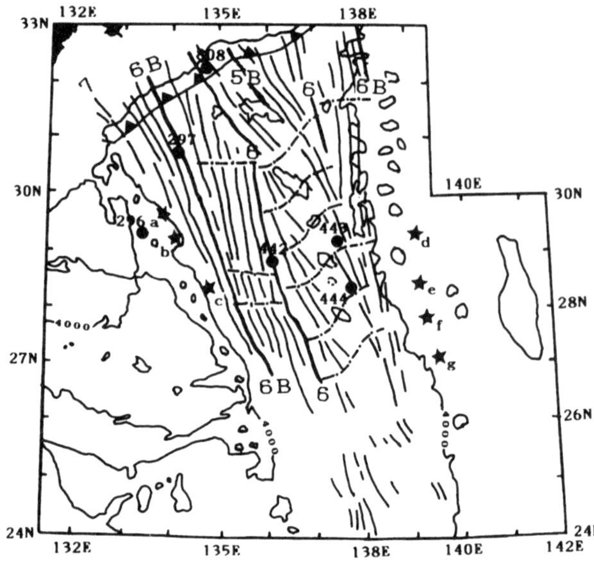

FIGURE 10.11. Magnetic isochrons and fracture zones identified by Okino *et al.* (1994). In the axial zone fracture zones have S-shape. Positions of magnetic dipole anomalies are shown by asterisks. Solid circles with numerical figures indicate positions of DSDP/ODP Sites.

based upon finding three fracture zones during the KAIKO survey. They proposed N10°W trending offsets associated with N-S opening in the axial zone close to the Nankai Trough. Okino *et al.* (1994) questioned such offsets and suggested existence of NNW-SSE trending grabens associated with intense deformation of the basement. They proposed that the rotation of spreading axis resulted from tectonic deformation of the northern margin of the basin caused by the southward drift of Japanese Islands during opening of the Japan Sea at about 15–20 Ma (Tamaki *et al.*, 1992).

Okino *et al.* (1994) found pairs of magnetic dipole anomalies unaccompanied by basement highs both in the flat terrace in the eastern slope of the Kyushu–Palau Ridge and along the trough east of the Nishi–Shichito Ridge. As chains of these anomalies appear to be symmetric relative to the basin axis, it seems plausible to assume that they formed a single line at a time prior to the opening of the Shikoku Basin and separated afterwards to the marginal ridges, although no age constraints have been obtained for these anomalies.

In the following (see Fig. 10.12), the history of the Shikoku Basin is interpreted based on Okino *et al.* (1994) and other earlier considerations.

6.1. Rifting of the Shikoku Basin

The oldest magnetic lineation identified in the northwestern corner of the basin is inferred to be the isochron 7 (26 Ma using the magnetic time scale of Cande and Kent, 1992). Prior to that the Kyushu–Palau Ridge and Izu–Bonin arc were joined to form a single active island arc. The rifting was initiated at the northern edge of the arc and propagated southward. It was then succeeded by seafloor spreading of the floor to form a fan-shaped basin. The exact age of the beginning of rifting is unknown, because the oldest

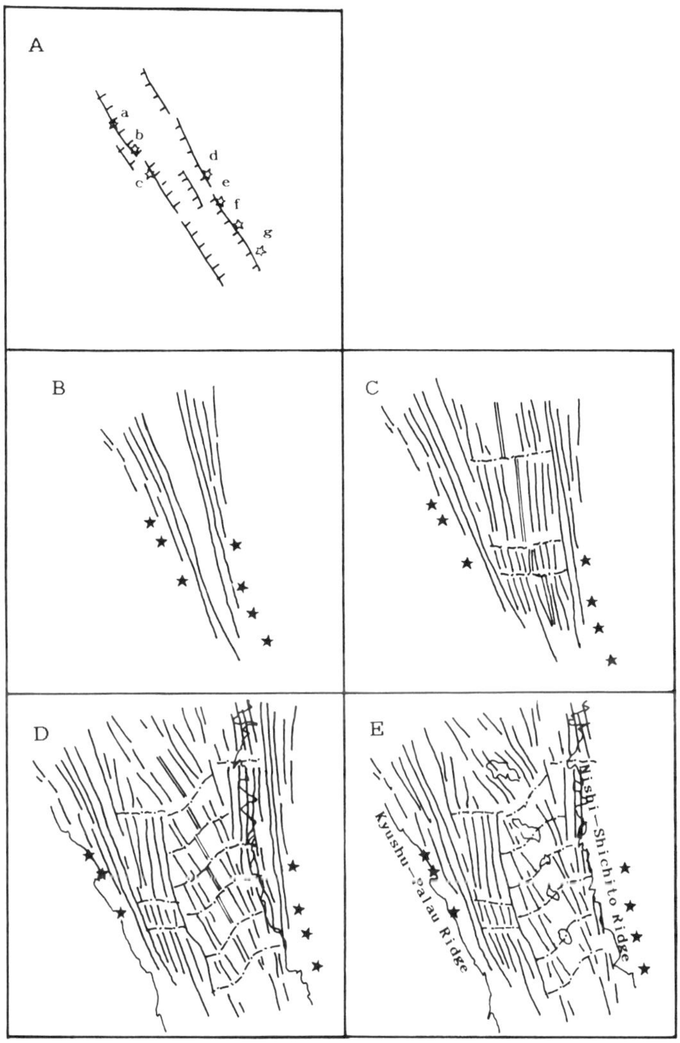

FIGURE 10.12. Five stages of the Shikoku Basin evolution based upon magnetic and related results (Okino *et al.*, 1994). (A) Magmatic injection and rifting, (B) southward propagation of rift and fan-shaped spreading, (C) E-W spreading with transform offsets, (D) NE-SW spreading with en echelon centers, (E) postspreading volcanism. Scale of figure is the same in all stages except for (A) showing schematic illustration of events only.

part of the basin has been subducted along the Nankai Trough. Considering timing of early stages of subsidence of the Kyushu–Palau Ridge described in the next section, the rifting at the latitude of site 296 (29.3°N) started most probably by 30 Ma. Onset of the Shikoku Basin rifting revealed by ODP Sites 787, 792, and 793 was 30 Ma, in harmony with the western margin data (Taylor, 1992).

6.2. Early Opening of the Shikoku Basin

Opening of the basin proceeded at a rate of 2.3–4.6 cm/year in the early stage. Chamot-Rooke *et al.* (1987) pointed out that isochron 6C (triplet anomalies) and 6B abut successively with the Kyushu–Palau margin so that the rate of southward propagation of rifting can be estimated to be about 10–30 cm/year, which is one order of magnitude greater than the spreading rate. At about 23 Ma the rifting and initial stage of opening reached the southern boundary zone between the Shikoku and Parece Vela basins. Mrozowski and Hayes (1979) postulated that the Parece Vela Basin began opening at anomaly 10 (29 Ma) from the central latitudes and propagated north and south. Complicated but deep topography and diffuse magnetic anomalies formed at the narrow transition between the two basins.

6.3. Spreading Phase from Anomaly 6B to Anomaly 6

After 23 Ma the spreading center changed its trend from NNW-SSE to N10°W. The center was cut by at least three transform faults and several offsets, but spreading proceeded nearly constantly to form parallel lineations, although rates of spreading decreased from 4.7 to 2.2 cm/year at anomaly 6A (21 Ma). Although magnetic lineations are obscure in the eastern wing, three fracture zones apparently symmetric to those in the west wing were identified there.

6.4. The Last Stage of Spreading and Eruption of the Kinan Seamounts

At anomaly 6 or slightly later (20 to 19 Ma) the spreading center reoriented to NW-SE. The adjustment appears to have been gradual, since analysis by Okino *et al.* (1994) revealed S-shaped fracture zones as shown in Fig. 10.11. The linear anomalies and topography are aligned en echelon along the NNW-SSE trending axis of the basin. This type of readjustment of the spreading centers seems to have been caused by balancing of areas in the rigid plate surrounded by remnant and active arcs. Similar en echelon alignment in the extinct axis has been found at the central basin fault in the axis of the West Philippine Basin (Andrews, 1978).

Opening of the Shikoku Basin finally ceased at a time of approximately 15 Ma, but eruption of the Kinan seamounts continued until about 12 Ma. Magnetic dipole anomalies associated with the Kinan seamount chain are relatively small compared to their appreciable sizes and somehow masked by the linear anomalies of the surrounding basin, as is seen with many seamounts existing in the areas close to the spreading ridge. This may be caused by mixed normal and reversed polarities of igneous rocks composing each seamount body.

6.5. Postspreading Off-Ridge Volcanism in the Shikoku Basin

DSDP and ODP holes have revealed evidence of widespread off-ridge volcanism throughout the Shikoku Basin (Klein *et al.*, 1978). Some of the volcanism occurred 4–7

Ma at a flank of the spreading ridge after cessation of spreading (most frequently at 12–15 Ma). Such postspreading activity was discovered in many other areas (e.g., Reykjanes Ridge; Luyendyk *et al.*, 1979). In any case it seems likely that the amplitudes of linear magnetic anomalies formed at the spreading center were reduced by overprinting of remanant magnetization of off-ridge intrusions, since magnetic polarity opposite that of the observed anomaly was found with these intrusive rocks (Faller *et al.*, 1979; Klein *et al.*, 1980).

7. UPLIFT AND SUBSIDENCE OF THE BASIN AND ITS MARGINS

Cores from DSDP Hole 296 display a history of vertical movements of the Kyushu–Palau Ridge in a period prior to and during the rifting and initial stage of opening of the Shikoku Basin. Shoal fossils of roughly 30 Ma were obtained from the volcaniclastic sediment overlying the 47.5 Ma rocks contained in the bottom 650 m of the hole. The hiatus between 30 and 47.5 Ma and occurrence of shoal fossils indicate that the ridge crest was upheaved and emerged until 30 Ma.

The Kyushu–Palau Ridge was uplifted by at least 2 km above the present elevation before 30 Ma. Crests of seamounts in the Kyushu–Palau Ridge were emerged above sea level as evidenced by a late Oligocene reefal limestone sample dredged from the upper slope (D = 1350 m) of Komahashi Seamount (28°08′N, 134°40′E, crestal depth = 944 m). The upheaval of the ridge was caused by upwelling and, in some spots, by eruption and intrusion of magma from the wedge mantle overlying the inclined subducting slab. Magnetic dipoles found along both Kyushu–Palau and Nishi–Shichito ridges seem to be correlatable to this magmatism. Petrography of the volcaniclastics at Site 296 (andesitic) and seamount rocks (granodiorite) in the Kyushu–Palau Ridge show that the magma has the chemical composition of arc volcanism.

It may be reasonable to presume that the paleo-Kyushu–Palau Ridge began rifting at 30 Ma along a chain of the magma reservoirs, because the crust of the ridge became thin and soft by heating from below in a similar manner to the process of the initial rifting of the Atlantic Ocean preceded by a chain of hot-spot activities (Burke, 1976). Since no holes have been drilled at the margins of the basin, we cannot determine the exact age and depth of the rift valley in the initial stage of rifting. Nevertheless, it seems likely that the rifting was associated with the normal faults which caused the linear and steep eastward-dipping scarps in the eastern margin of the Kyushu–Palau Ridge amounting to 2000 m in total offset. Extensional force in a direction roughly perpendicular to the ridge axis may have originated from eastward retreat of the Izu–Bonin–Mariana trenches together with the westward drift of the West Philippine Basin plate associated with initiation of subduction at the Ryukyu and Philippine trenches.

The data from Site 296 also revealed subsidence of the western half of the paleo-ridge as rapidly as 325 m/m.y. to form the present Kyushu–Palau Ridge until the hole bottom reached a depth of 650 m at 28 Ma (Fig. 10.13). Such a fast rate of subsidence was associated with stretching and normal faulting of the crust.

Lithology of hole 442 implies that the active center of the Shikoku Basin was still as shallow as 3000 m (the lysocline) in the rifting and initial spreading stages. The crests of the Kinan seamounts situated at the axis of the basin emerged above sea level and emitted anhydrous mafic tephras found in the DSDP cores in the Shikoku Basin (Furuta and Arai, 1980; Kobayashi, 1984). The bottom of the basin subsided synchronously with or even

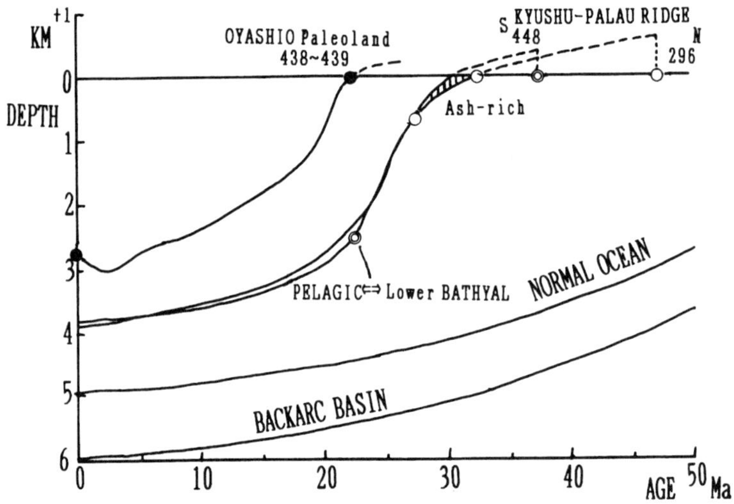

FIGURE 10.13. Postulated subsidence curves of two sites in the Kyushu–Palau Ridge (north and south) with those for sites 438–439, the normal oceans and backarc basins after Kobayashi (1984). Subsidence of remnant arc and backarc basin is at first very fast following a curve defined by DSDP Site 296 and later approaches the $\sqrt{t}$ formula of the "normal" backarc basins (Park *et al.*, 1990).

faster than subsidence of the Kyushu–Palau Ridge after the entire activity ended. The Shikoku Basin has finally subsided following the $\sqrt{t}$-law of the ocean basin subsidence by lithosphere cooling (Kobayashi, 1984; Park *et al.*, 1990). The difference in the present elevations between the Kyushu–Palau Ridge and the Shikoku Basin is isostatically controlled by crustal thickness which is greater beneath the ridge than the basin.

8. INTERACTION OF THE SHIKOKU BASIN WITH SOUTHWEST JAPAN AND NANKAI TROUGH

The Shikoku Basin has been subducted beneath southwest Japan at the Nankai Trough. The present convergence rates in the Nankai Trough are 2 to 4 cm/year in a NNW direction according to the NUVEL-1 (DeMets *et al.*, 1990). The landward slope is now very compressional, and much of the sedimentary section is deformed and accreted to the upper wedge.

The length of the Wadati–Benioff zone recognized by deep-focus earthquakes along the Nankai Trough is so short that the time duration since its initiation appears to be roughly 3 Ma (Kanamori and Tsumura, 1971), in harmony with Karig's postulation of a cessation period of 3–5 Ma (Karig, 1975). However, Sugi and Uyeda (1984) and Shiono and Sugi (1985) pointed out that oceanic lithosphere younger than 15 Ma easily loses the rigidity required to generate deep-focus earthquakes. The shape of the Wadati–Benioff zone beneath southwest Japan (Fig. 10.14) seems to be consistent with their conclusion, as the zone is shortest at the axis of the Shikoku Basin and longer in both wings. Assuming that subduction along the Nankai Trough has continued at the present rate since about 15 Ma, at least 300 km of the northern margin of the Shikoku Basin has been lost under the trough.

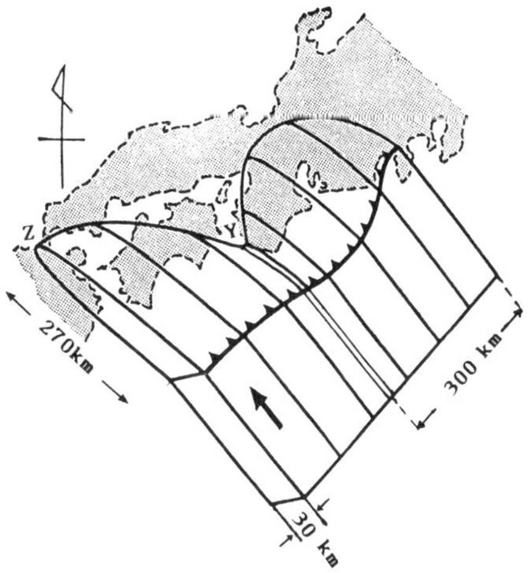

FIGURE 10.14. Schematic illustration of deep-focus earthquakes plane (Wadati–Benioff zone) showing the Shikoku Basin plate subducting at the Nankai Trough (modified from Shiono and Sugi, 1985). Double line denotes inactive spreading center of the Shikoku Basin. Depths of epicenters at points *Y* and *Z* are 70 km and 90 km, respectively. Length of slab from trough axis to point *Y* is roughly 150 km. Aware of the fan-shaped edge at which the subducted lithosphere loses rigidity for earthquake generation at shallower depths at its spreading center than older margins.

It seems very plausible that rates and direction of subduction at the Nankai Trough before 15 Ma were substantially different from those at present, because southwest Japan was rotated clockwise in this period in association with the opening of the Japan Sea.

Whatever the history of the Nankai Trough subduction was, interaction between the Shikoku Basin and southwest Japan has influenced both sides of the trough. Widespread emplacement of felsic to intermediate magmas at 12–15 Ma in the outer zone and Setouchi province of southwest Japan (e.g., Nakada and Takahashi, 1979; Kobayashi, 1983) can be explained by subduction of a hot lithosphere in which widespread off-ridge magmatism occurred immediately before and even after subduction at the Nankai Trough (Klein *et al.*, 1978). Abnormally high heat flow in the Nankai Trough region (Yamano *et al.*, 1984) may be caused by this origin, although fluid circulation in the accretionary wedge may also contribute to heat transport (Yamano *et al.*, 1992). North-south compression of the northern portion of the Shikoku Basin may have formed the late-stage faulted structures as revealed by KAIKO and later surveys, either by jumps of spreading centers (Chamot-Rooke *et al.*, 1987) or by tectonic deformation after the spreading (Okino *et al.*, 1994).

9. SUMMARY, DISCUSSION, AND FURTHER PROBLEMS

Analyses of geophysical and geological data, including those from the DSDP/ ODP cores, have revealed the processes of tectonic evolution of the Shikoku Basin, a backarc basin surrounded by island arcs. In the period prior to 30 Ma, the present Kyushu–Palau Ridge and the Izu–Bonin arc were joined to constitute a large NNW-SSE trending arc-trench system with the crests of local highs above sea level. The arc was then rifted, beginning in the north and propagating southward at a speed of 10 cm/year. The rift valley was bounded by steep scarps on both sides. The western ridge quickly subsided after the rift was formed. Rifting was succeeded by spreading of the basin. The whole Shikoku Basin was rifted by 27 Ma and formed a triangular trough during embryonic spreading which proceeded until 23 Ma.

At 23 Ma the spreading center changed its trend to N10°W and continued to open the basin nearly symmetrically until 20–19 Ma. Major portions of the flat basin were formed during this stage. Rocks constituting the basement are tholeiite with chemical composition similar to MORB. The center was offset by at least three transform faults, which are recognized in the present basin as fracture zones. The western limb of the spreading center remains in the west wing of the basin, whereas the eastern one underwent volcanic activities of the Nishi–Shichito Ridge and lost its original morphology and linear magnetic anomalies.

Widespread intrusion of magma occurred roughly 5 m.y. after the end of opening of the basin. Some of the magma was much differentiated or contaminated with surrounding rocks to generate alkali-basalt, as recovered from a sill at Site 444. Such off-ridge volcanism has been known at a number of locations in the world oceans. It may modify the water depth and complicate magnetic patterns in the area and heat flow values, although we have not yet discovered any remnant volcanic vents supplying the intrusive lavas.

It is recognized that the Shikoku Basin is one of the type localities for study of ancient backarc processes. Comparative investigation of this region with active backarc basins and rifts such as the Mariana Trough, Sumisu Rift, and the Okinawa Trough will provide much crucial information on backarc processes and their driving mechanisms.

We still need more data in the Shikoku Basin itself, particularly on the basement morphology and ages of the marginal wings which are overlain by thick sediment. Deep drilling through the sediment layers to the pillow basement in the east and west margins of the Shikoku basin will be required. Drilling of the magnetic dipoles in Kyushu–Palau and Nishi–Shichito ridges for the purpose of determining their ages and petrography will be extraordinarily important for elucidation of the role of arc volcanism in the initiation of the rifting. Further comprehensive surveys of the slopes of the Nankai Trough will improve understanding of the physical and chemical processes associated with young basin subduction.

Acknowledgments

The authors would like to thank our colleagues in the Hydrographic Department, Ocean Research Institute, University of Tokyo, and JAMSTEC for their help and encouragement to us to complete this chapter.

REFERENCES

Andrews, J. E. 1978. Morphological evidence for the evolution of the central basin fault spreading center in the west Philippine basin, GSA Annual Meeting, Toronto, abstract.

Brown, G., and Taylor, B. 1988. Sea-floor mapping of the Sumisu Rift, Izu–Ogasawara (Bonin) island arc, *Bull. Geol. Surv. Jpn.* **39:**23–38.

Burke, K. 1976. Development of graben associated with the initial rupture of the Atlantic Ocean, *Tectonophysics* **36:**93–112.

Cande, S. C., and Kent, D. V. 1992. A new geomagnetic time scale for the Late Cretaceous and Cenozoic, *J. Geophys. Res.* **97**(B10)**:**13,917–13,952.

Chamley, H. 1980. Clay sedimentation and paleoenvironment in the Shikoku Basin since the middle Miocene (Deep Sea Drilling Project Leg 58, north Philippine Sea), in *Init. Repts. DSDP*, 58 (G. deV. Klein and K. Kobayashi, eds.), pp. 669–682, U.S. Govt. Printing Office, Washington, DC.

Chamot-Rooke, N., Renard, V., and Le Pichon, X. 1987. Magnetic anomalies in the Shikoku Basin: a new interpretation, *Earth Planet. Sci. Lett.* **83**:214–218.
DeMets, C., Gordon, R. G., Argus, D. F., and Stein, S. 1990. Current plate motions, *Geophys. J. Int.* **101**:425–478.
Dick, H. J. B. 1980. Vesicularity of Shikoku Basin basalt: a possible correlation with the anomalous depth of backarc basins, in *Init. Repts. DSDP*, 58 (G. deV. Klein and K. Kobayashi, eds.), pp. 895–904, U.S. Govt. Printing Office, Washington, DC.
Dick, H. J. B. 1982. The petrology of basaltic rocks of the northern Philippine Sea, *Am. J. Sci.* **282**:644–700.
Dick, H. J. B., Marsh, N. G., and Bullen, T. D. 1980. Deep Sea Drilling Project Leg 58 abyssal basalts from the Shikoku Basin: Their petrology and major-element geochemistry, in *Init. Repts. DSDP*, 58 (G. deV. Klein and K. Kobayashi, eds.), pp. 843–872, U.S. Govt. Printing Office, Washington, DC.
Echols, D. J. 1980. Foraminifer biostratigraphy, north Philippine Sea, Deep Sea Drilling Project Leg 58, in *Init. Repts. DSDP*, 58 (G. deV. Klein and K. Kobayashi, eds.), pp. 567–586, U.S. Govt. Printing Office, Washington, DC.
Faller, A. M., Steiner, M., and Kobayashi, K. 1979. Paleomagnetism of basalts and interlayered sediments drilled during DSDP Leg 49 (N-S transect of the northern mid-Atlantic Ridge), in *Init. Repts. DSDP*, 49 (B. P. Luyendyk, J. R. Cann, eds.), pp. 769–780, U.S. Govt. Printing Office, Washington, DC.
Furuta, T. and Arai, F. 1980. Petrographic and geochemical properties of tephras in Deep Sea Drilling Project cores from the North Philippine Sea, in *Init. Repts. DSDP*, 58 (G. deV. Klein and K. Kobayashi, eds.), pp. 617–627, U.S. Govt. Printing Office, Washington, DC.
Furuta, T., Tonouchi S., and Nakada, M. 1980. Magnetic properties of pillow basalt from the Kinan seamount chain, the Shikoku Basin, *J. Geomagn. Geoelectr.* **32**:567–573.
Hayashida, M., Takanashi, M., Ikeda, K., Kaneko, Y., Kato, Y., Ogawa, M., and Kasuga, S. 1989. Preliminary report of continental shelf surveys of "Middle part of Shikoku Basin" quadrangle (in Japanese), *Tech. Bull. Hydrogr.* **8**:85–91.
Hill, I. A., Taira, A., Firth, J. V., *et al.* 1993. *Proc. ODP, Sci. Results*, 131, College Station TX.
Hydrographic Department, Maritime Safety Agency of Japan. 1989. Data Report of Hydrographic Observation, Series of Continental Shelf Survey, HD-MSA, Tokyo 6.
Hydrographic Department, Maritime Safety Agency of Japan. 1990. Data Report of Hydrographic Observation, Series of Continental Shelf Survey, HD-MSA, Tokyo 7.
Hydrographic Department, Maritime Safety Agency of Japan. 1991. Data Report of Hydrographic Observation, Series of Continental Shelf Survey, HD-MSA, Tokyo 8.
Ingle, J. C., Karig, D. E., *et al.* 1975. *Init. Repts DSDP*, 31, U.S. Govt. Printing Office, Washington, DC, 31.
IPOD-Japan. 1977. Multichannel seismic reflection records of the Shikoku Basin and the Daito Ridges, 1976, Part 1, IPOD-Japan Data Series no. 1, Ocean Res. Inst., Univ. of Tokyo.
Kagami, H., Karig, D. E., Coulbourn, W. T. *et al.* 1987. *Init. Repts. DSDP*, 31, U.S. Govt. Printing Office, Washington, DC.
KAIKO I Research Group. 1986. *Topography and Structure of Trenches around Japan*, Univ. of Tokyo Press.
Kanamori, H., and Tsumura, K. 1971. Spatial distribution of earthquakes in the Kii peninsula, Japan, south of the median tectonic line, *Tectonophysics* **12**:327–342.
Karig, D. E. 1971. Origin and development of marginal basins in the western Pacific, *J. Geophys. Res.* **76**:2542–2561.
Karig, D. E. 1975. Basin genesis in the Philippine Sea, in *Init. Repts. DSDP*, 31 (J. C. Ingle, D. E. Karig *et al.*, eds.), pp. 847–879, U.S. Govt. Printing Office, Washington, DC.
Kasuga, S., Iwabuchi, H., and Kato, S. 1987. Results of ocean bottom surveys in junction of the Shikoku Basin and West Mariana Basin (in Japanese with English Abstr.), *Rept. Hydrogr. Res.* **22**:13–134.
Kasuga, S., Kato, Y., Kimura, S., and Okino, K. 1992. Present and former members of the Continental Shelf Surveys Office, characteristics of arc-trench systems and backarc basins in the southern waters of Japan—Outline of the geophysical survey by the Hydrographic Department of Japan (in Japanese with English Abstr), *Rept. Hydrogr. Res.* **28**:9–45.
Katsura, T., Shimamura, K., and Collaborators in Continental Shelf Surveys Office. 1994. Geological, geochemical research on bottom samples, from continental shelf surveys, H.D. Japan (part 1)—Preliminary study for ocean floor on the Japanese continental shelves, *Rept. Hydrogr. Res.* **30**:345–381.
Klein, G. deV., and Kobayashi, K. 1981. Geological summary of the Shikoku Basin and northwestern Philippine Sea, Leg 58, DSDP/IPOD drilling results, *Oceanol. Acta*, Spec. No. 181-192.
Klein, G. deV., Kobayashi, K., Chamley, H., Curtis, D. M., Dick, H. J. B., Echols, D. J., Fountain, D. M.,

Kinoshita, H., Marsh, N. G., Mizuno, A., Nisterenko, G. V., Okada, H., Sloan, J. R., Waples, D. M., and White, S. M. 1978. Off-ridge volcanism and seafloor spreading in the Shikoku Basin, *Nature* **273:**746–748.

Klein, G. deV., Kobayashi, K. *et al.* 1980. *Init. Repts. DSDP*, 58, U.S. Govt. Printing Office, Washington, DC, pp. 567–586.

Kobayashi, K. 1983. Cycles of subduction and Cenozoic arc activity in the northwestern Pacific margin, in *Geodynamics of the Western Pacific–Indonesian Region* (T. W. C. Hilde and S. Uyeda, eds.), Geodyn. Ser., Vol. 11, pp. 287–302, American Geophysical Union/Geological Society of America, Washington, DC.

Kobayashi, K. 1984. Subsidence of the Shikoku backarc basin, *Tectonophysics* **102:**105–117.

Kobayashi, K., and Fujiwara, T. 1993. Survey of HAKUHO Seamount, in *The Shikoku Basin* (J. Segawa, ed.), pp. 243–246, Preliminary Rept. Hakuho Maru Cruise KH92-1, ORI Univ. of Tokyo.

Kobayashi, K., and Isezaki, N. 1976. Magnetic anomalies in the Sea of Japan and the Shikoku Basin: Possible tectonic implications, in *The Geophysics of the Pacific Ocean Basin and Its Margin* (G. H. Sutton, M. H. Manghnani, and R. Moberly, eds.), Geophys. Monogr. Ser., Vol. 19, pp. 239–251, American Geophysical Union, Washington, DC.

Kobayashi, K., and Nakada, M. 1978. Magnetic anomalies and tectonic evolution of the Shikoku inter-arc basin, *J. Phys. Earth* **26:**S391–S402.

Koyama, K., Katsura, T., Ikeda, K., Uchida, M., Kasuga, S., Nagano, M., and Hayashida, M. 1986. Preliminary report of continental shelf survey of Minami–Koho Seamount and adjacent areas (in Japanese), *Tech. Bull. Hydrogr.* **4:**39–46.

Le Pichon, X., Iiyama, T., Chamley, H., Charvet, J., Faure, M., Fujimoto, H., Furuta, T., Ida, Y., Kagami, H., Lallemant, S., Leggett, J., Murata, A., Okada, H., Rangin, C., Renard, V., Taira, A., and Tokuyama, H. 1987. Nankai Trough and the fossil Shikoku Ridge: Results of Box 6 Kaiko survey, *Earth Planet. Sci. Lett.* **83:** 186–198.

Luyendyk, B. P., Cann, J. R. *et al.* 1979. *Init. Repts, DSDP*, 49, U.S. Govt. Printing Office, Washington, DC.

Marsh, N. G., Sanders, A. D., Tarney, J. and Dick, J. B. 1980. Geochemistry of basalts from the Shikoku and Daito Basins, Deep Sea Drilling Leg 58, in *Init. Repts, DSDP*, 58 (G. DeV. Klein, K. Kobayashi *et al.*, eds.), pp. 805–842, U.S. Govt. Printing Office, Washington, DC.

McKee, E. H., and Klock, P. 1980. K-Ar ages of basalt sills from Deep Sea Drilling Project Sites 444 and 446, Shikoku Basin and Daito Basin, Philippine Sea, in *Init. Repts, DSDP*, 58 (G. DeV. Klein, K. Kobayashi *et al.*, eds.), pp. 921–922, U.S. Govt. Printing Office, Washington, DC.

Mizuno, A., Shibata, K., Uchiumi, S., Yuasa, M., Okuda, Y., Nohara, M., and Kinoshita, Y. 1977. Granodiorite from the Minami–Koho Seamount on the Kyushu-Palau Ridge and its K/Ar age, *Bull. Geol. Surv. Jpn.* **28:**507–511.

Mrozowski, C. L., and Hayes, D. E. 1979. The evolution of the Parece Vela Basin, eastern Philippine Sea, *Earth Planet. Sci. Lett.* **46:**49–67.

Murauchi, S., Den, N., Asano, S., Hotta, H., Yoshii, T., Asanuma, T., Hagiwara, K., Ichikawa, K., Sato, T., Ludwig, W. J., Ewing, J., Edgar, N. T., and Houtz, R. E. 1968. Crustal structure of the Philippine Sea, *J. Geophys. Res.* **73:**3143–3171.

Nakada, S., and Takahashi, M. 1979. Regional variation in chemistry of the Miocene intermediate to felsic magmas in the outer zone and the Setouchi province of southwest Japan, *J. Geol. Soc. Jpn.* **85:**571–582.

Okada, H. 1980. Calcareous nannofossils from Deep Sea Drilling Project Sites 442 through 446, Philippine Sea, in *Init. Repts, DSDP*, 58 (G. DeV. Klein, K. Kobayashi *et al.*, eds.), pp. 549–565, U.S. Govt. Printing Office, Washington, DC.

Okino, K. 1993. Formation processes of a backarc basin—Shikoku Basin (in Japanese), Abstr. Papers presented at the 10th Shinkai Symp., JAMSTEC, pp. 23–25.

Okino, K., Shimakawa, Y., and Nagaoka, S. 1994. Evolution of the Shikoku Basin, *J. Geomagn. Geoelectr.* **46,** Spec. Issue: 463–479.

Olafsson, G. 1993. Calcareous nannofossil biostratigraphy of the Nankai Trough, in *Proc. ODP, Sci. Results*, 131 (I. A. Hill, A. Taira, J. V. Firth, eds.), pp. 15–34, Ocean Drilling Program, College Station, TX.

Ozima, M., Kaneoka, I., and Ujiie, H. 1977. ^{40}Ar-^{39}Ar age of rocks and development mode of the Philippine Sea, *Nature* **267:**816–818.

Ozima, M., Takigami, Y., and Kaneoka, I. 1980. ^{40}Ar-^{39}Ar geochronological studies on rocks of Deep Sea Drilling Project Sites 443, 445, and 446, in *Init. Repts, DSDP*, 58 (G. DeV. Klein, K. Kobayashi, eds.), pp. 917–920, U.S. Govt. Printing Office, Washington, DC.

Park, C.-H., Tamaki, K., and Kobayashi, K. 1990. Age-depth correlation of the Philippine Sea backarc basins and other marginal basins in the world, *Tectonophysics* **181:**351–371.

Pickering, K. T., Underwood, M. B, and Taira, A. 1993. Stratigraphic synthesis of the DSDP-ODP sites in the Shikoku Basin, Nankai Trough, and accretionary prism, in *Proc. ODP, Sci. Results*, 131 (I. A. Hill, A. Taira, and J. V. Firth, eds.), pp. 313–330, Ocean Drilling Program, College Station, TX.

Shih, T. C. 1980. Magnetic lineations in the Shikoku Basin, in *Init. Repts, DSDP*, 58 (G. DeV. Klein and K. Kobayashi, eds.), pp. 783–788, U.S. Govt. Printing Office, Washington, DC.

Shino, M., Ikeda, K., Tozaki, T., Nagaoka, S., Kato, Y., Shimakawa, Y., and Seta, H. 1991. Preliminary report of continental shelf surveys of "Kosyu Seamount" quadrangle (in Japanese), *Tech. Bull. Hydrogr.* **9:** 26–31.

Shiono, S., and Sugi, N. 1985. Life of an oceanic plate: Cooling time and assimilation time, *Tectonophysics* **112:**35–50.

Siena, F., Coltorti, M., Saccani, E., and Vaccaro, C. 1993. Petrology of the basaltic rocks of the Nankai Trough basement, in *Proc. ODP, Sci. Results*, 131 (I. A. Hill, A. Taira, and J. V. Firth, eds.), pp. 197–210, Ocean Drilling Program, College Station, TX.

Sugi, N., and Uyeda, S. 1984. Subduction of young oceanic plates without deep focus earthquakes, *Bull. Soc. Geol. France* **26:**245–254.

Taira, A., Hill, I. A., Firth, J. V., *et al.* 1991. *Proc. ODP, Init. Repts.*, 131, Ocean Drilling Program, College Station, TX.

Taira, A., and Niitsuma, N. 1986. Turbidite sedimentation in the Nankai Trough as interpreted from magnetic fabric, grain size, and detrital modal analyses, in *Init. Repts. DSDP*, 87 (H. Kagami, D. E. Karig, and W. T. Coulbourn, eds.), pp. 611–632, U.S. Govt. Printing Office, Washington, DC.

Tamaki, K., Suyehiro, K., Allan, J., McWilliams, M., *et al.* 1992. *Proc. ODP, Sci. Results*, 127/128, Pt. 2, Ocean Drilling Program, College Station, TX, pp. 779–1478.

Taylor, B. 1992. Rifting and the volcanic-tectonic evolution of the Izu–Bonin–Mariana arc, in *Proc. ODP, Sci. Results*, 128 (B. Taylor and K. Fujioka, eds.), pp. 627–651, College Station, TX.

Tomoda, Y. 1974. *Reference Book for Gravity, Magnetic and Bathymetric Data of the Pacific Ocean and Adjacent Seas*, 1963–71, Univ. of Tokyo Press.

Tomoda, Y., Kobayashi, K., Segawa, J., Nomura, M., Kimura, K., and Saki, T. 1975. Linear magnetic anomalies in the Shikoku Basin, northeastern Philippine Sea, *J. Geomagn. Geoelectr.* **28:**47–56.

Tomoda, Y., Ozawa, K., and Segawa, J. 1968. Measurement of gravity and magnetic field on board a cruising vessel, *Bull. Ocean Res. Inst. Univ. of Tokyo* **3:**1–170.

Underwood, M. B., Orr, R., Pickering, K. T., and Taira, A. 1993a. Province and dispersal patterns of sediments in the turbidite wedge of Nankai Trough, in *Proc. ODP, Sci. Results*, 131 (I. A. Hill, A. Taira, and J. V. Firth, eds.), pp. 15–34, Ocean Drilling Program, College Station, TX.

Underwood, M. B., Pickering, K., Gieskes, J. M., Kastner, M., and Orr, R. 1993b. Sediment geochemistry, clay mineralogy, and diagenesis: A synthesis of data from leg 131, Nankai Trough, in *Proc. ODP, Sci. Results*, 131 (I. A. Hill, A. Taira, and J. V. Firth, eds.), pp. 343–364, Ocean Drilling Program, College Station, TX.

Van Andel, T. H., Heath, G. R., and Moore, T. C., Jr. 1975. Cenozoic history and paleooceanography of the central equatorial Pacific Ocean, *Geol. Soc. Am., Mem.* **143**.

Watts, A. B. 1976. Gravity field of the northwest Pacific Ocean basin and its margin: Philippine Sea, *Geol. Soc. Am.* **MC-12**.

Watts, A. B., and Weissel, J. 1975. Tectonic history of the Shikoku marginal basin: Philippine Sea, *Earth Planet. Sci. Lett.* **25:**315–328.

White, S. M., Chamley, H., Curtis, D., Klein, G. DeV., and Mizuno, A., with a special contribution on physical properties by D. M. Fountain. 1980. Sediment synthesis: Deep Sea Drilling Project Leg 58, Philippine Sea, in *Init. Repts. DSDP*, 58 (G. DeV. Klein, and K. Kobayashi, eds.), pp. 963–1014, U.S. Govt. Printing Office, Washington, DC.

Wood, D. A., Joron, J. L., Marsh, N. G., Tarney, J., and Treuil, M. M. 1980. Major- and trace-element variations in basalts from the north Philippine Sea drilled during Deep Sea Drilling Project Leg 58: A comparative study of backarc basin basalts with lava series from Japan and mid-oceanic ridges, in *Init. Repts. DSDP*, 58 (G. DeV. Klein, and K. Kobayashi, eds.), pp. 873–894, U.S. Govt. Printing Office, Washington, DC.

Yamano, M., Honda, S., and Uyeda, S. 1984. Nankai Trough: A hot trench? *Mar. Geophys. Res.* **6:**187–203.

Yamano, M., Foucher, J.-P., Kinoshita, M., Fisher, A., Hyndman, R. D., and ODP Leg 131 Shipboard Scientific Party. 1992. Heat flow and fluid flow regime in the western Nankai accretionary prism, *Earth Planet. Sci. Lett.* **109:**451–462.

11

Opening Tectonics of the Japan Sea

Kensaku Tamaki

ABSTRACT

The examination of the crustal structure and ODP deep-sea drilling results introduced the following opening model of the Japan Sea. The opening of the Japan Sea was initiated by the extension and thinning of the proto-Japan island arc, which was situated on the margin of the Eurasian continent at ca. 30 Ma. During the extension process a large strike-slip fault was generated at the eastern margin of the proto-Japan Sea. The strike-slip movement cut through the entire lithosphere and caused it to split, triggering development of a seafloor spreading system at ca. 28 Ma. The seafloor spreading system propagated to the WSW direction into the thinned crustal zone and formed the eastern part of the Japan Basin that is presently underlain by oceanic crust. In the meantime, the southwestern part of the Japan Sea suffered continuing extension and thinning of the arc crust and formed basins and rises. The basins are composed of thinned lower arc crust overlain by volcanic layers, and the rises are composed of fragmented upper arc crusts. The propagation ceased at ca. 18 Ma, leaving the contrasting topography of the present northern and the southern Japan Sea. The extension and thinning of arc or continental crusts and subsequent development of a propagating spreading system, which is initiated at a basin margin strike-slip zone, are common and fundamental processes of most backarc basins.

1. INTRODUCTION

The backarc basins are divided into two types in terms of their occurrence. One group is the oceanic type that develops at the backarc side of oceanic island arcs such as the Marina Trough of the Mariana arc and the Lau Basin of the Tonga arc. The other group is the continental type that develops at the backarc side of continental arc. The Japan Sea is one of the most typical backarc basins of the continental type; another may be the South China Sea. A 200-m contour line surrounding the Japan Sea shows a closed depressional feature (Fig. 11.1). The Japan Sea is characterized by a complicated seafloor topography with many ridges, rises, banks, and seamounts. The Japan Sea is the most topographically complicatcd marginal basin of thc world.

Kensaku Tamaki • Ocean Research Institute, University of Tokyo, Nakano-ku, Tokyo 164, Japan.

Backarc Basins: Tectonics and Magmatism, edited by Brian Taylor, Plenum Press, New York, 1995.

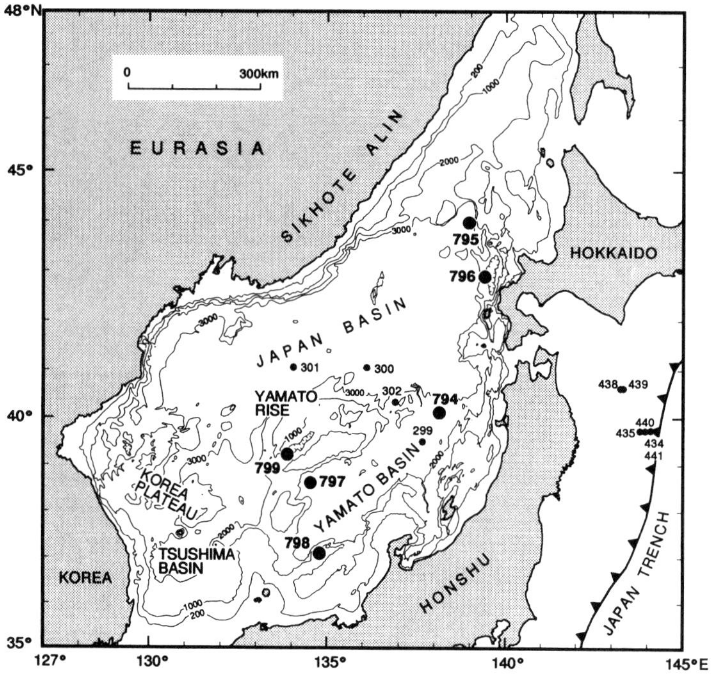

FIGURE 11.1. Topography of the Japan Sea. Solid circles are DSDP and ODP drilling sites with site number.

The cause of the formation of backarc basins has not been answered, in spite of advances in our understanding of seafloor spreading theory and plate tectonics. General observations of backarc basin geology and geophysics suggest that backarc basins are formed under an extensional tectonic condition at the continental margin. However, one major enigma on the formation of backarc basins is why the extensional basins are formed at the plate convergent margin. Another major enigma is why the backarc basins are mostly distributed in the western Pacific and why they are not associated with all island arcs. The origin and the opening processes of the backarc basins are better understood because of the recent data accumulation by Ocean Drilling Program (ODP) deep-sea drilling in several western Pacific backarc basins. In this chapter I summarize principal aspects of geology and geophysics of the Japan Sea and discuss the opening process of the Japan Sea by introducing a model for the opening of the Japan Sea. I further discuss the fundamental aspects of the introduced model by applying it to other backarc basins.

2. TOPOGRAPHY

The physiographic contrast is striking between the southern and northern parts of the Japan Sea (Figs. 11.1 and 11.2). The southern part is characterized by a complicated topography with several rises and intervening basins. The northern part shows a rather simple morphology with a wide, deep-sea basin. This striking contrast is closely related to the opening tectonics of the Japan Sea.

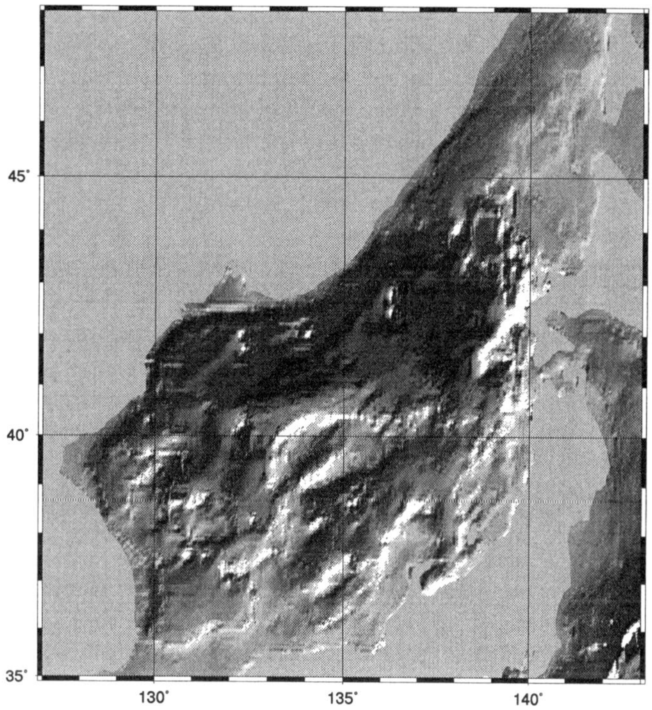

FIGURE 11.2. Shaded relief topography of the Japan Sea. Minor topographic features are discriminated.

The wide, deep basin in the northern part of the Japan Sea is called the Japan Basin. The eastern part of the Japan Basin is an abyssal plain that is the result of turbidite deposition; the sediment thickness there reaches 2500 m. In contrast, the western part of the Japan Basin has shallower water depth and slightly rougher seafloor topography (Fig. 11.2); the sediment thickness there is accordingly thinner, 1000–1500 m. The contrast between the western and eastern parts of the Japan Basin is significant to our understanding of its origin.

There are many topographic rises in the southern part of the Japan Sea. Larger ones are the Yamato Rise in the center of the Japan Sea and the Korea Plateau at the east of the Korea peninsula. Abundant granitic rocks recovered from these and surrounding topographic highs suggest that these rises comprise continental crust (e.g., Lelikov and Bersenev, 1973; Tamaki, 1988). The oldest rock is the Archean gneiss recovered from the Korea Plateau with the absolute age of 2729 Ma (Tamaki, 1988). Seismic refraction results also suggest the existence of a layer with seismic velocity of 6.0 km/s, suggesting continental crust (Ludwig *et al.*, 1975). Granitic rocks were recovered from most other topographic highs in the southern part of the Japan Sea.

Two sedimentary basins are distributed between these topographic highs. The Yamato Basin is between Honshu Island and the Yamato Rise. The Tsushima (Ullung) Basin is at the southwestern corner and surrounded by the Korea Plateau, the Korea Peninsula, and Honshu Island. The water depth of the Yamato Basin is 3000 m, which is 500 m shallower than that of the Japan Basin. The sediment thickness of the Yamato Basin is 500 m thinner than that of the Japan Basin. Accordingly, the basement depth of the Yamato Basin is 1000–

1500 m shallower than that of the Japan Basin. Although the data are not sufficient in the Tsushima Basin, the sediment thickness of the Tshushima Basin is estimated to be over 2000 m and the basement depth appears to be comparable with that of the Yamato Basin.

3. THE JAPAN AND YAMATO BASINS

The topography is closely related to the crustal structure of the Japan Sea. The crustal structures present information critical to understanding the basin formation tectonics. The crust of the Japan Sea has been intensively studied by seismic refraction methods since the 1950s. The volume of crustal data is the largest for any of the backarc basins of the world; hence the Japan Sea presents us the best opportunity to study basin formation tectonics. In the last several years, techniques for using long arrays of ocean bottom seismometers were introduced for studies of detailed crustal structure in the basin areas (e.g., Hirata *et al.*, 1987, 1992). During ODP Leg 128, a borehole seismometer was used for the first time for actual crustal study of an ocean bottom hole (Shinohara *et al.*, 1992).

Figure 11.3 summarizes principal results of these studies in the Japan Sea. The figure clearly displays the differences between the crustal structure of the eastern part of the Japan Basin and that of the Yamato Basin. The crustal thickness of the Japan Basin is 6–6.5 km, while that of the Yamato Basin is 10–13 km, twice as thick as the Japan Basin. Regardless of the difference of crustal thickness, the seismic propagation velocity of each layer is identical between the two basins. Below the sedimentary layer the crustal structure with seismic velocity shows that the crust of the eastern part of the Japan Basin is a typical oceanic crust. Magnetic anomaly lineations are also identifiable in the eastern part of the Japan Sea, and the tentatively identified age is consistent with the age of the crust estimated on the basis of heat flow data (Tamaki *et al.*, 1990). Thus, it is concluded that the eastern part of the Japan Basin was formed by the normal seafloor spreading system.

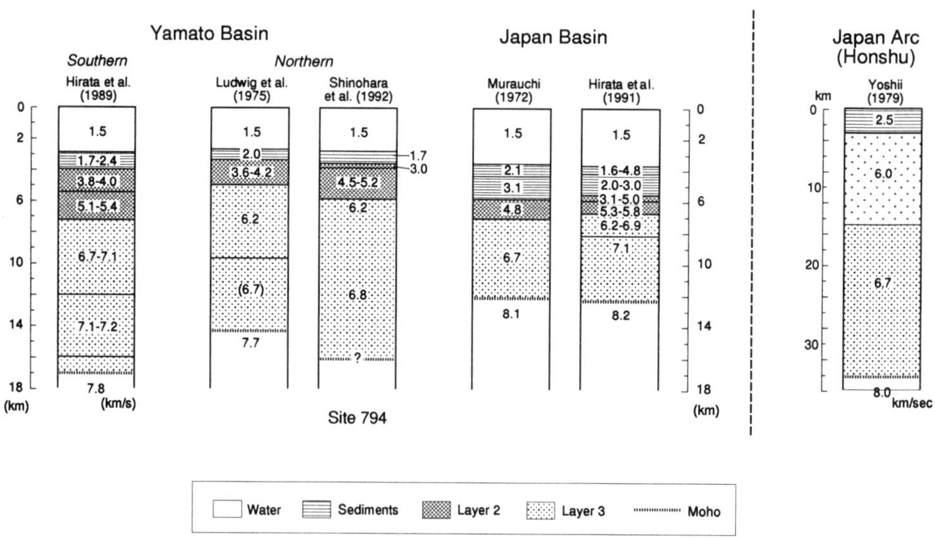

FIGURE 11.3. Comparison of the crustal structure of the eastern Japan Basin and the Yamato Basin.

In contrast, the origin of the thicker crust of the Yamato Basin is not simple. One possible explanation is that it is oceanic crust with thickness twice that of normal oceanic crust (Hirata *et al.*, 1989). The thickness of oceanic crust positively correlates to the temperature of upwelling upper mantle beneath the spreading center (McKenzie and Bickle, 1988). Thick oceanic crust is usually caused by the contamination of the spreading center and hot spots associated with high-temperature mantle. However, it is difficult to assume hot-spot interaction for the formation of the Japan Sea for the following reasons.

The oceanic crust, observed at the Yamato Basin, is twice as thick as the normal oceanic crust and is comparable to those of oceanic plateaus (Schubert and Sandwell, 1989) and suggests activity of a large hot spot. Present heat flow in the Yamato Basin, however, does not show any appreciable anomaly indicating previous hot-spot activity (Langseth and Tamaki, 1992). The average heat flow value of the Yamato Basin is actually 10 mW/m^2 lower than that of the Japan Basin, which has normal oceanic crust. It is implausible to introduce hot-spot interaction for the formation of the Yamato Basin in this regard.

The other idea for the origin of the Yamato Basin is that it is extended continental crust. The seismic velocity, 6.2–7.2 km/s, of the lower crust of the Yamato Basin appears to be identical to that of the lower crust of Honshu Island of the Japanese island arc in Fig. 11.3. The thickness of the lower crust of Honshu Island is twice that of the Yamato Basin. The hypothesis of extended continental crusts as the origin of the Yamato Basin crust assumes that the lower crust of the Yamato Basin is identical to that of the Japanese island arcs. According to the hypothesis, the lower crust of Honshu Island, with a thickness of 20 km, was extended and thinned to 10 km during the opening process of the Japan Sea to form the present Yamato Basin. The initial thickness of the lower crust of the proto-Japanese island arcs may have been thicker than the present 20 km, and then, in this case, the ratio of extension needs to be increased.

In contrast to the case of the lower crust, the seismic velocities of the upper crust are quite different for the Yamato Basin and for Honshu Island. The upper crust velocity of Honshu Island is 6.0 km/s, a typical seismic velocity of the upper crust of a continent, while that of the Yamato Basin is less than 6.0 km/s. The upper crust of the Yamato Basin has a velocity of 3.6–5.4 km/s, nearly identical to the upper crust of the Japan Basin. If we assume the extensional origin of the Yamato Basin, the above problem on the origin of upper crust of the Yamato Basin should be answered consistently with the extension tectonics of the lower crust. To approach this problem, we introduce a whole Japan Sea crustal distribution map in the next section.

4. EXTENSION OF ISLAND-ARC CRUST

Fig. 11.4 shows a compilation of crustal structures in the Japan Sea based on seismic refraction studies, basement topography, and bottom rock sampling (Tamaki *et al.*, 1992). Abundant results of seismic refraction studies by Russian scientists were included in the compilation. Crusts of the Japan Sea are classified into four groups in Fig. 11.4. They are continental crust, rifted continental crust, extended continental crust, and oceanic crust. The Yamato Basin–type crust was denoted as extended continental crust in the compiled map. Topographic highs such as the Yamato Rise, which were not appreciably thinned by extension, are classified as continental crusts. Rifted continental crust represents the intermediate condition between extended crust and nonextended crust. Rifted continental

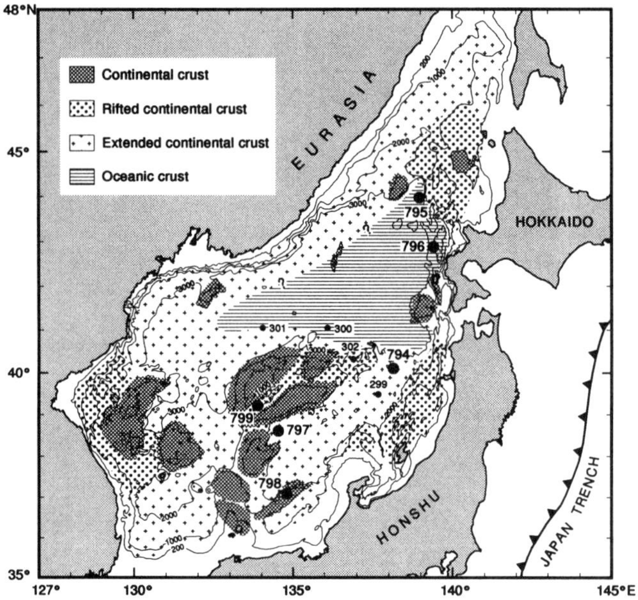

FIGURE 11.4. Crustal structure of the Japan Sea. The discrimination of the crustal structure is based on seismic reflection/refraction survey data, bottom sampling data, geomagnetic data, and basement depth and topography.

crust, which is distributed surrounding the continental crusts, is characterized by the occurrence of many normal faults. The region of rifted continental crust represents deeper water depths than that of continental crust, suggesting submergence by some extension and thinning of the crust.

Characteristics of the distribution of the variable crusts in the Japan Sea are summarized as follows.

1. Distribution of the oceanic crusts is restricted to the eastern part of the Japan Basin. The region of the oceanic crust widens to the east and narrows to the west in the western Japan Basin.
2. Extended continental crust is widely distributed in the Japan Sea, in contrast to the limited distribution of the oceanic crust. Distribution in the southern part of the Japan Sea is predominant.
3. Blocks of continental crust are distributed mostly in the southern part of the Japan Sea.

Restricted distribution of the oceanic crust in the Japan Sea is one of the most outstanding features of Fig. 11.4, which is different from the previously accepted idea that the oceanic crust is widely distributed in the Japan Sea. The predominant distribution of the extended continental crust suggests that the extension and thinning of the continental crust have an important role for the formation of the Japan Sea.

The southern part of the Japan Sea is characterized by the basins that are formed by the extended continental crust and by the rises that are the continental blocks. These characteristics introduce the following hypothesis: proto-Japanese island arcs, which were at the continental margin of the Eurasian continent, suffered a differential deformation between the upper crust and the lower crust under extensional tectonics. The lower crust was extended and thinned by ductile deformation, while the upper crust was collapsed by brittle deformation. The fragmented upper crusts were left as rises in the basins, while the extended lower crust formed basins and lacked upper crusts. The upper mantle was upheaved by the thinning of the crust and produced magma by partial melting of the upper mantle. Seismic refraction studies show that the upper crust of the Yamato Basin has a velocity of oceanic layer 2 (basaltic effusive and intrusive rocks) and basinal ubiquitous distribution of this layer. The upper crust of the Yamato Basin is hypothesized to be formed by basinal volcanism during extension tectonics. The extended lower crust in the basins that lack upper crust was thus covered by volcanic rocks. A cross-sectional reconstruction of the southern Japan Sea in Fig. 11.5 shows this model. We need more detailed crustal data, especially beneath the rises, to improve the above hypothesis.

5. FORMATION AGE

ODP Legs 127 and 128 successfully recovered basaltic rocks beneath the sediments and provided reliable information on the formation age of the Japan Sea (Tamaki *et al.*, 1992). The basement volcanic rocks were recovered at ODP Site 795 in the northern Japan Basin, at Site 794 in the northern Yamato Basin, and at Site 797 in the southern part of the Yamato Basin (Fig. 11.1). Two sets of data are available on age information; one is ^{40}Ar-^{39}Ar radiometric age determination measured by Kaneoka *et al.* (1992), and the other is the age of sediments overlying the uppermost part of the basement volcanic rocks. The latter age is determined by using microfossils such as foraminifera, diatoms, and pollen. Tada and

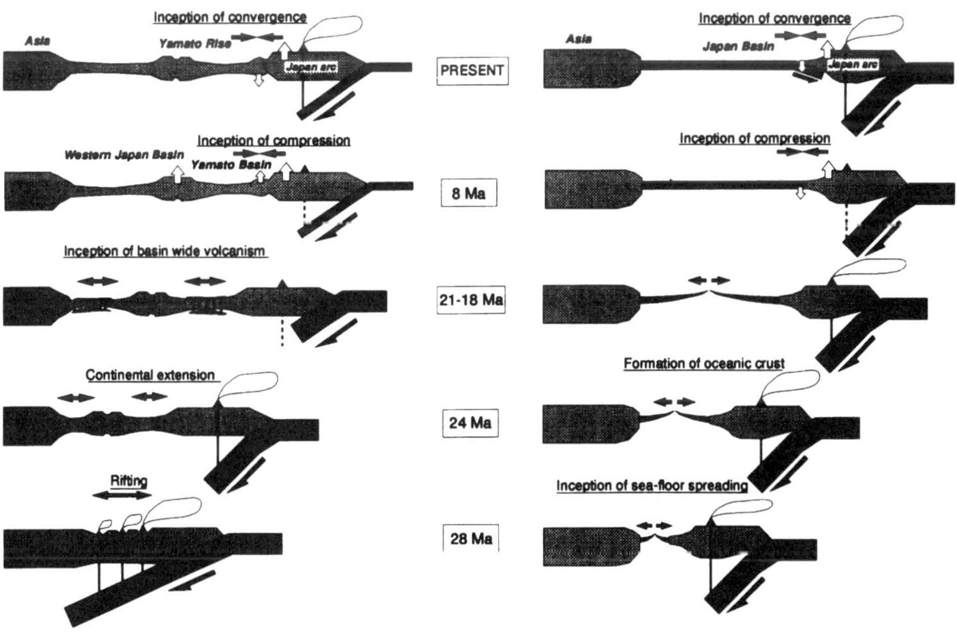

FIGURE 11.5. Cross section of the evolution of the southern Japan Sea (left), and northern Japan Sea (right).

Tamaki (1992) compiled and evaluated the analyzed results of microfossils and determined the fossil age of the uppermost volcanic rocks. Table I summarizes the radiometric ages and fossil ages of the uppermost volcanic rocks at the three drilling sites.

Table I shows that the fossil age is always younger than the radiometric age. The difference between the radiometric age and fossil age at Site 795 is interpreted as due to an unconformity between the sediments and underlying volcanic basement that is observed in seismic reflection profiles (Tamaki *et al.*, 1990). The difference between radiometric and fossil ages at site 794, which forms a slightly shallower topography than the major part of the Yamato Basin, is also explained as an unconformity (Tamaki *et al.*, 1992). The radiometric and fossil ages are consistent at Site 797 where most volcanic layers are sills intruded into the sediments of concurrent deposition (Tamaki *et al.*, 1990). In conclusion, radiometric ages of volcanic rocks are considered to represent the age of eruption of basalts in the basins.

The oldest age among three sites is 24 Ma in the Japan Basin. The age of 24 Ma is consistent with the preliminary identification of magnetic anomalies in the eastern part of the Japan Basin. An older radiometric age of 32 Ma is also reported from the basaltic rocks recovered at the outcrops along the Okushiri Ridge in the eastern margin of the Japan Basin (Kaneoka, personal comm.); this information suggests that the age of Japan Basin volcanism is older than that of Yamato Basin volcanism. The radiometric age at Site 797 in the western Yamato Basin is 1 to 3 m.y. younger than that at Site 794 in the eastern Yamato Basin.

6. MODEL OF OPENING OF THE JAPAN SEA

The existence of oceanic crust in the eastern Japan Basin indicates that the mid-oceanic-type seafloor spreading tectonics actually happened and that the history of seafloor spreading tectonics should have been preserved and observed as magnetic anomaly lineations. Preliminary identification of magnetic anomalies in the eastern Japan Basin proposes the age range of the crust of the Japan Basin as 28–18 Ma (Tamaki *et al.*, 1992). The age range of 28–18 Ma includes a part that is older than that of the Yamato Basin.

Critical information comes from the configuration of magnetic anomaly lineations in the eastern Japan Basin. A clear oblique displacement is observed in the magnetic anomaly lineations (Tamaki *et al.*, 1992). The trend of displacement is NW-SE, and the overall feature suggests WSW propagation of the spreading system in the eastern Japan Basin.

TABLE I
Age of Basement Volcanic Rocks Recovered by ODP

ODP Sites	Radiometric age[a] (Ma)	Fossil age[b] (Ma)
Site 795 (Japan Basin)	17–24	14–15.5
Site 794 (Yamato Basin)	20–21	16–18
Site 797 (Yamato Basin)	18–19	17–19

[a]Kaneoka *et al.* (1992).
[b]Tada and Tamaki (1992).

The WSW propagation of the spreading system is consistent with the westward narrowing of the oceanic crust in the Japan Basin as shown in Fig. 11.4.

Jolivet and Tamaki (1992) stressed the importance of the described propagation tectonics and introduced an idea that the oceanic crust of the Japan Basin was initiated from the eastern margin of the Japan Basin and propagated to the westward (Fig. 11.6). A N-S trending large strike-slip fault was assumed at the eastern margin of the Japan Basin. The strike-slip fault was generated in the margin of the basin to accommodate the strike-slip movement in accordance with the pull-apart-type opening of the Japan Sea. Because strike-slip faults form the weakest line of the lithosphere (Mueller and Phillips, 1991), the initial split of lithosphere was assumed to be formed along the strike-slip fault. The split of lithosphere evolved to seafloor spreading, and the seafloor spreading system propagated into the extended and thinned arc crusts of the southwestern Japan Sea.

The hypothesized opening process of the Japan Sea is summarized as follows.

1. The Japan Sea began forming by the extension and thinning of the proto-Japan arc that was at the margin of the Eurasian continent. The timing of initiation of extension is estimated to be before 30 Ma, based on the reconstruction of the basin margin's subsidence history of the Japan Sea (Ingle, 1992).
2. A split of the lithosphere was initiated along the strike-slip fault at the eastern margin of the proto-Japan Sea. Seafloor spreading was initiated from the split and started propagating to the west into the thinned arc crustal region. The initiation of the seafloor spreading is estimated to be at 28 Ma, based on the oldest magnetic anomaly lineations in the Japan Basin.
3. The propagation of the seafloor spreading system lasted till 18 Ma and formed the oceanic crust in the western Japan Basin. In contrast, the southwestern part of the Japan Sea was opened in association with the extension and thinning of the arc crust, and the resultant topography with rises and basins topography was formed during this period.

Thus, the contrasting topography in the northern and southern Japan Sea and the characteristic crustal distribution of the Japan Sea are well explained by our opening model

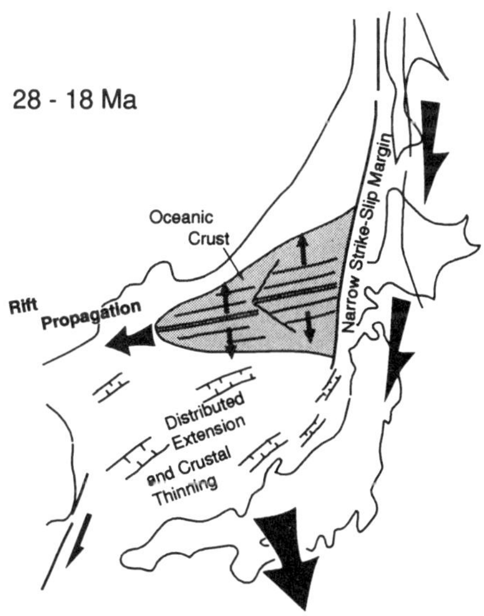

FIGURE 11.6. Simplified cartoon of opening of the Japan Sea (modified from Jolivet and Tamaki, 1992). Crustal thinning and extension prevailed at the initial stage. Seafloor spreading was triggered by the breakup of the lithosphere along the strike-slip margin at the eastern side of the Japan Sea. The spreading center propagated southwestward to increase the area of the oceanic crust. In the meantime the crust of the southern Japan Sea was being extended and thinned.

of the Japan Sea. Fig. 11.5 shows contrasting tectonics in the southern and northern Japan Sea. The principal processes of this model are the subsequent development of the extension and thinning of the arc crust, a split of the lithosphere along the strike-slip fault in the basin margin, initiation of seafloor spreading, and propagation of the spreading system into the thinned crustal region. This process appears to be fundamental for the backarc basin formation as discussed in the next section.

7. COMMON PROCESS OF BACKARC BASIN OPENING

The opening process of the Japan Sea introduced in the previous chapters appears to be applicable to the Lau, North Fiji, South China Sea, and Shikoku basins, too (Fig. 11.7).

ODP Leg 135 drilled seven sites along a transect in the Lau Basin (Leg 135 Scientific Party, 1992). The Lau Basin was considered, before ODP drilling, to be formed by a simple seafloor spreading system. Drilling results, however, showed that the western part of the basin is composed not of oceanic crust but of extended and thinned island-arc crust. A detailed geophysical survey for ODP drilling revealed a southward propagating spreading center in the eastern part of the Lau Basin (Fig. 11.7a). There is a large strike-slip fault at the northern margin of the Lau Basin. The fault is a boundary between the Pacific plate and the Lau Basin (the Australia plate). The area of the oceanic crust widens to the north, approaching the strike-slip fault as shown in Fig. 11.7a. The distribution of the oceanic crust and the extended crust shows a pattern similar to that of the Japan Sea. It is assumed that

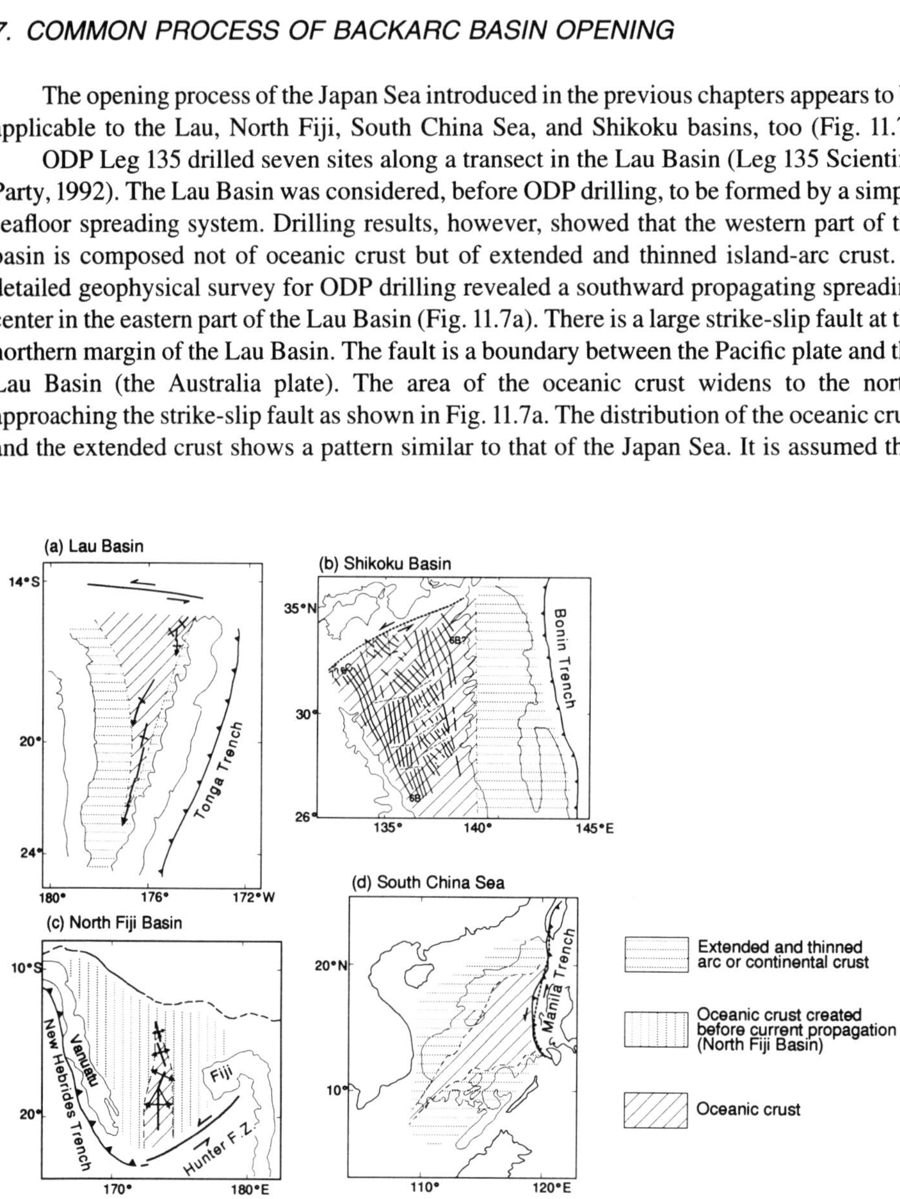

FIGURE 11.7. Opening process of the Lau Basin (a), Shikoku Basin (b), North Fiji Basin (c), and South China Sea (d).

the Lau Basin was initially opened by extension of island-arc crust, and seafloor spreading was triggered at the strike-slip fault in the northern basin margin. The seafloor spreading subsequently propagated southward into the thinned arc crust region.

The present configuration of the Shikoku Basin widens to the north (Fig. 11.7b). Magnetic anomaly lineations are clearly identified in the Shikoku Basin (Okino *et al.*, 1994). The oldest magnetic anomalies along the western margin of the Shikoku Basin show a systematic variation, older to the north and younger to the south. This variation is the unequivocal result of the southward propagation of the spreading system of the Shikoku Basin. The Nankai Trough subduction zone bounds the northern margin of the Shikoku Basin. As the Nankai Trough is supposed to have been originally a strike-slip fault, initiation of a spreading system at the strike-slip fault is again introduced. The extension of the Bonin arc during the opening of the Shikoku Basin is well documented by seismic reflection profiling and ODP drilling (Taylor, 1992). The Bonin arc widens and deepens to the south, suggesting more intensive extension in the southern Bonin arc than in the northern Bonin arc. This variation is consistent with the southward propagation of the Shikoku Basin seafloor spreading system. When the seafloor spreading intruded into the area, the extension of arc crust is degraded and ceases. The southern part of the Bonin arc underwent a longer period of extension tectonics than the northern part and resulted in the present along-axis variation of water depths and widths. Thus, the Shikoku Basin and the Bonin arc also show a combination of propagation of seafloor spreading apart from a strike-slip fault, and extension of island-arc crust.

The accumulation of marine geophysical data is outstanding in the North Fiji Basin during the past several years, especially by the cooperative France–Japan project (e.g., Tanahashi *et al.*, 1991). Tanahashi *et al.* (1991) detected northward propagation of the currently active spreading center by a swath mapping survey (Fig. 11.7c). The magnetic anomaly lineations associated with the spreading system widen to the south, suggesting broader distribution of oceanic crust generated by the spreading activity. A large strike-slip fault, the Hunter fracture zone, bounds the southern margin of the spreading system. The configuration above also suggests an initiation of seafloor spreading at the strike-slip fault zone and subsequent propagation to the north. The crustal structure of the area surrounding the presently active spreading system, however, appears to have been formed by another event of seafloor spreading (Auzende *et al.*, 1988). Then, in the case of the North Fiji Basin, the current propagation is considered to be propagating into the previously formed oceanic crust (Fig. 11.7c).

The last example is the South China Sea. The South China Sea shows a configuration similar to the Japan Sea. The topography and the distribution of magnetic anomaly lineations suggest that the zone of oceanic crust widens to the east (Fig. 11.7d). The Manila Trench bounds the eastern margin of the zone of oceanic crust. As the trench subduction zone is preferentially formed along the strike-slip fault (Mueller and Phillips, 1991), it is assumed that the eastern margin of the South China Sea was initially bounded by a strike-slip fault. Detailed magnetic anomaly data show a westward propagation of a past spreading system (Ishihara *et al.*, 1994). Abundant banks are distributed in the southwestern part of the South China Sea. Since the South China Sea is an intracontinental extensional basin, the zone of banks is supposed to be the result of the extension of continental crust; the mechanism is similar to that of the southern Japan Sea. This situation, again, begins the basin opening by thinning of continental crust and initiation and propagation of the spreading system at a strike-slip margin.

Among the foregoing cases, the Lau Basin, the Shikoku Basin, the South China Sea, and the Japan Sea show a common process of opening. Opening of the backarc basin is started by the extension and thinning of island-arc or continental crust. A strike-slip fault is generated at a basin margin during the extension tectonics. The lithosphere splits at the strike-slip fault and a seafloor spreading system appears, which propagates into the thinned crustal region to generate oceanic crustal region. Starting the propagation at a margin strike-slip fault is also well documented in the North Fiji Basin, suggesting the common setting of the initiation of propagation in backarc basins.

The same scenario may be applicable to other backarc basins. The configuration of the Mariana Trough suggests an ongoing northward propagation. The Okinawa Trough appears to have started splitting the thinned lithosphere at its southwestern margin (Oshida *et al.*, 1992).

8. CONCLUSIONS

The opening of the Japan Sea and the other backarc basins is described in the following simplified model (Fig. 11.8).

1. Backarc basins commence opening by an extension and thinning of island arc crust (Fig. 11.8a).
2. A large strike-slip fault develops at the margin of the basin and triggers a split of the lithosphere and the subsequent initiation of seafloor spreading (Fig. 11.8b).
3. Seafloor spreading propagates into the thinned crustal region and forms an oceanic crustal region (Fig. 11.8c).

This model recognizes that the extension and thinning of island-arc crust are ubiquitous phenomena during the opening of backarc basins. This means that the island-arc crust is easily extended and thinned because of its hot and weak condition. The extension, however, does not develop to seafloor spreading without the help of a strike-slip fault. The strike-slip fault forms the weakest lithospheric zone and initiates the seafloor spreading process by a lithosphere split. Once the seafloor spreading process commences, the spreading system propagates easily into the thinned and weakened crusts. The resultant configuration of backarc basins is a composite structure of oceanic crust, extended arc crust, and continental crust. The Japan Sea is most typically formed by these processes, and its extensive data base of crustal structure and geology will provide an opportunity to further improve the proposed model.

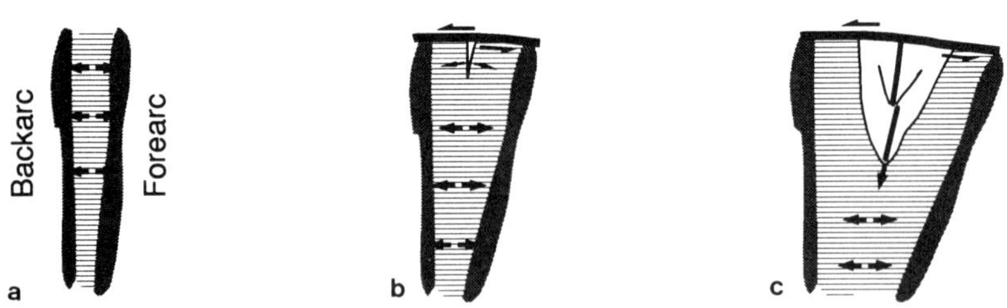

FIGURE 11.8. A common process of opening of backarc basins: (a) crustal extension, (b) breakup of lithospere at margin strike-slip fault, (c) propagation of seafloor spreading.

Acknowledgments

I wish to thank B. Taylor of SOEST, University of Hawaii, for his encouragement and review of this work and P. Jarvis of Ocean Research Institute, University of Tokyo, for his comments on the manuscript.

REFERENCES

Auzende, J., Lafoy, Y., and Marsset, B. 1988. Recent geodynamic evolution of the north Fiji basin (southwest Pacific), *Geology* **16:**925–929.

Hirata, N., Karp, B. Ya., Yamaguchi, T., Kanazawa, T., Suyehiro, K., Kasahara, J., Shiobara, H., Shinohara, M., and Kinoshita, H. 1991. Seismic structure of the Japan Basin in the Japan Sea by the 1990 Japan–USSR expedition, Abstract, IUGG 20th Symp. U11.

Hirata, N., Karp, B. Ya., Yamaguchi, T., Kanazawa, T., Suyehiro, K., Kasahara, J., Shiobara, H., Shinohara, M., and Kinoshita, H. 1992. Oceanic crust in the Japan Basin of the Japan Sea by the 1990 Japan-USSR expedition, *Geophys. Res. Lett.* **19:**2027–2030.

Hirata, N., Kinoshita, H., Suyehiro, K., Suyemasu, M., Matsuda, N., Ouchi, T., Katao, H., Koresawa, S., and Nagumo, S. 1987. Report on DELP 1985 Cruises in the Japan Sea, Part II: seismic refraction experiment conducted in the Yamato Basin, Southeast Japan Sea, *Bull. Earthquake Res. Inst. Univ. Tokyo* **62:**347–365.

Hirata, N., Tokuyama, H., and Chung, T. W. 1989. An anomalously thick layering of the crust of the Yamato Basin, southeastern Sea of Japan: the final stage of backarc spreading, *Tectonophysics* **165:**303–314.

Ingle, J. C. Jr. 1992. Subsidence of the Japan Sea: evidence from ODP sites and onshore sequence, in *Proc. ODP, Sci. Results*, 127/128, (K. Tamaki, K. Suyehiro, J. Allan, M. McWilliams *et al.*, eds.), pp. 1197–1218, Ocean Drilling Program, College Station, TX.

Ishihara, T., *et al.* 1994. Magnetic anomaly map of East Asia, *Magnetic Anomaly Map*, Geological Survey of Japan, 2.

Jolivet, L., and Tamaki, K. 1992. Neogene kinematics in the Japan Sea region and volcanic activity of the NE-Japan arc, in *Proc. ODP, Sci. Results*, 127/128, Pt. 2 (K. Tamaki, K. Suyehiro, J. Allan, M. McWilliams *et al.*, eds.), pp. 1311–1331, Ocean Drilling Program, College Station, TX.

Kaneoka, I., Takigami, Y., Takaoka, N., Yamashita, S., and Tamaki, K. 1992. ^{40}Ar-^{39}Ar analysis of volcanic rocks recovered from the Japan Sea floor: Constraints on the formation age of the Japan Sea, in *Proc. ODP, Sci. Results*, 127/128, Pt. 2 (K. Tamaki, K. Suyehiro, J. Allan, M. McWilliams *et al.*, eds.), pp. 819–836, Ocean Drilling Program, College Station, TX.

Langseth, M., and Tamaki, K. 1992. Geothermal measurements: Thermal evolution of the Japan Sea basins and sediments, in *Proc. ODP, Sci. Results*, 127/128, Pt. 2 (K. Tamaki, K. Suyehiro, J. Allan, M. McWilliams *et al.*, eds.), pp. 1297–1309, Ocean Drilling Program, College Station, TX.

Leg 135 Scientific Party. 1992. Evolution of backarc basins: ODP Leg 135, Lau Basin, *EOS, Trans. AGU.* **73:** 241–247.

Lelikov, Ye. P., and Bersenev, I. I. 1973. Early Proterozoic gneiss-magmatite complex in the southwestern part of the Sea of Japan, *Dokl. Akad. Nauk SSSR* **223:**74–76.

Ludwig, W. L., Murauchi, S., and Houtz, R. E. 1975. Sediments and structure of the Japan Sea, *Geol. Soc. Am. Bull.* **86:**651–664.

McKenzie, D., and Bickle, M. J. 1988. The volume and composition of melt generated by extension of the lithosphere, *J. Petrol.* **29:**625–679.

Mueller, S., and Phillips, R. J. 1991. On the initiation of subduction, *J. Geophys. Res.* **96:**651–665.

Murauchi, S. 1972. Crustal structure of the Japan Sea, *Kagaku (Science)* **38:**367–375. (in Japanese)

Okino, K., Shimakawa, Y., and Nagaoka, S. 1994. Evolution of the Shikoku Basin, *J. Geomagn. Geoelectr.* **46:**463–479.

Oshida, A., Tamaki, K., and Kimura, M. 1992. Origin of the magnetic anomalies in the southern Okinawa Trough, *J. Geomagn. Geoelectr.* **44:**345–359.

Schubert, G., and Sandwell, D. 1989. Crustal volumes of the continents and of oceanic and continental submarine plateaus, *Earth Planet. Sci. Lett.* **92:**234–246.

Shinohara, M., Hirata, N., Nambu, H., Suyehiro, K., Kanazawa, T., and Kinoshita, H. 1992. Detailed crustal structure of northern Yamato Basin, in *Proc. ODP, Sci. Results*, 127/128, Pt. 2 (K. Tamaki, K. Suyehiro, J. Allan, M. McWilliams *et al.*, eds.), pp. 1075–1106, Ocean Drilling Program, College Station, TX.

Tada, R., and Tamaki, K. 1992. Scientific results of ODP Japan Sea legs, and their implications for stratigraphy, *J. Jpn. Assoc. Petrol. Technol.* **57:**103–111.

Tamaki, K. 1988. Geological structure of the Japan Sea and its tectonic implications, *Bull. Geol. Surv. Jpn* **39:** 269–365.

Tamaki, K., Pisciotto, K., Allan, J., *et al.* 1990. *Proc. ODP, Init. Repts.*, 127, Ocean Drilling Program, College Station, TX.

Tamaki, K., Suyehiro, K., Allan, J., Ingle, J. C., and Pisciotto, K. A. 1992. Tectonic synthesis and implications of Japan Sea ODP drilling, in *Proc. ODP, Sci. Results*, 127/128, Pt. 2 (K. Tamaki, K. Suyehiro, J. Allan, M. McWilliams *et al.*, eds.), pp. 1333–1348, Ocean Drilling Program, College Station, TX.

Tanahashi, M., Kisimoto, K., Joshima, M., Lafoy, Y., Honza, E., and Auzende, J. M. 1991. Geological structure of the central spreading system, North Fiji Basin, *Mar. Geol.* **98:**187–200.

Taylor, B. 1992. Rifting and the volcanic-tectonic evolution of the Izu-Bonin-Mariana arc, in *Proc. ODP, Sci. Results*, 126, (B. Taylor, K. Fujioka *et al.*, eds.), pp. 627–651, Ocean Drilling Program, College Station, TX.

Yoshii, T. 1979. *Crustal Structure of the Japan Sea*, Tokyo, Univ. Tokyo Press. (in Japanese)

12

Kuril (South Okhotsk) Backarc Basin

Helios S. Gnibidenko, Thomas W. C. Hilde, Elena V. Gretskaya, and Andrey A. Andreyev

The researches of many commentators have already thrown much darkness on this subject and it is possible that, if they continue, we shall soon know nothing at all about it.

Mark Twain

ABSTRACT

The Kuril backarc basin is a 3300-m-deep basin located behind the Kuril island arc and subduction zone. It is underlain by oceanic crust, the top of which is 7–8 km deep, and it contains 4 to 5+ km of tectonically undisturbed sedimentary fill. The oldest sedimentary rocks sampled from the basin margin are Miocene/Oligocene. Although no seafloor spreading magnetic lineations exist, we conclude that the basin has formed by backarc extension and spreading. Elongated in the NE-SW direction, parallel to the Kuril arc, the basin is ~250 km wide at its SW end, narrowing to closure at its NE termination at the southern tip of Kamchatka. Opening of the basin was apparently related to a relative motion pole at or near the south end of Kamchatka, consistent with Miocene/Oligocene lateral fault trends between Sakhalin and Hokkaido and the strike of transverse basement ridges extending across the basin. The age of the basin may be as young as Miocene or as old as late Cretaceous. Ocean drilling is needed to establish its precise age.

1. INTRODUCTION

The Kuril (South Okhotsk) backarc basin lies behind the Kuril island arc in the position of a typical subduction related backarc basin (Fig. 12.1). However, the origin of this basin, long inactive, has remained unclear. The Kuril backarc basin structural evolution has been interpreted (Belousov, 1982; Belousov and Pavlenkova, 1986; Ermakov, 1991) as a result of alteration of continental crust to oceanic crust by a process of basification or oceanization, and, alternatively (Karig, 1971; Matsuda and Uyeda, 1971; Hilde *et al.*, 1977; Bogdanov, 1988) as a result of crustal extension and seafloor spreading behind the Kuril island arc, related to subduction of the Pacific oceanic plate. Taking into account data that

---

Helios S. Gnibidenko • Shirshov Institute of Oceanology, Russian Academy of Sciences, Moscow 117851, Russia. *Present address*: Geodynamics Research Institute, Texas A&M University, College Station, Texas 77843. *Thomas W.C. Hilde* • Geodynamics Research Institute, Texas A&M University, College Station, Texas 77843. *Elena V. Gretskaya and Andrey A. Andreyev* • Institute of Marine Geology and Geophysics, Russian Academy of Sciences, Yuzhno-Sakhalinsk, Sakhalin 693002, Russia.

Backarc Basins: Tectonics and Magmatism, edited by Brian Taylor, Plenum Press, New York, 1995.

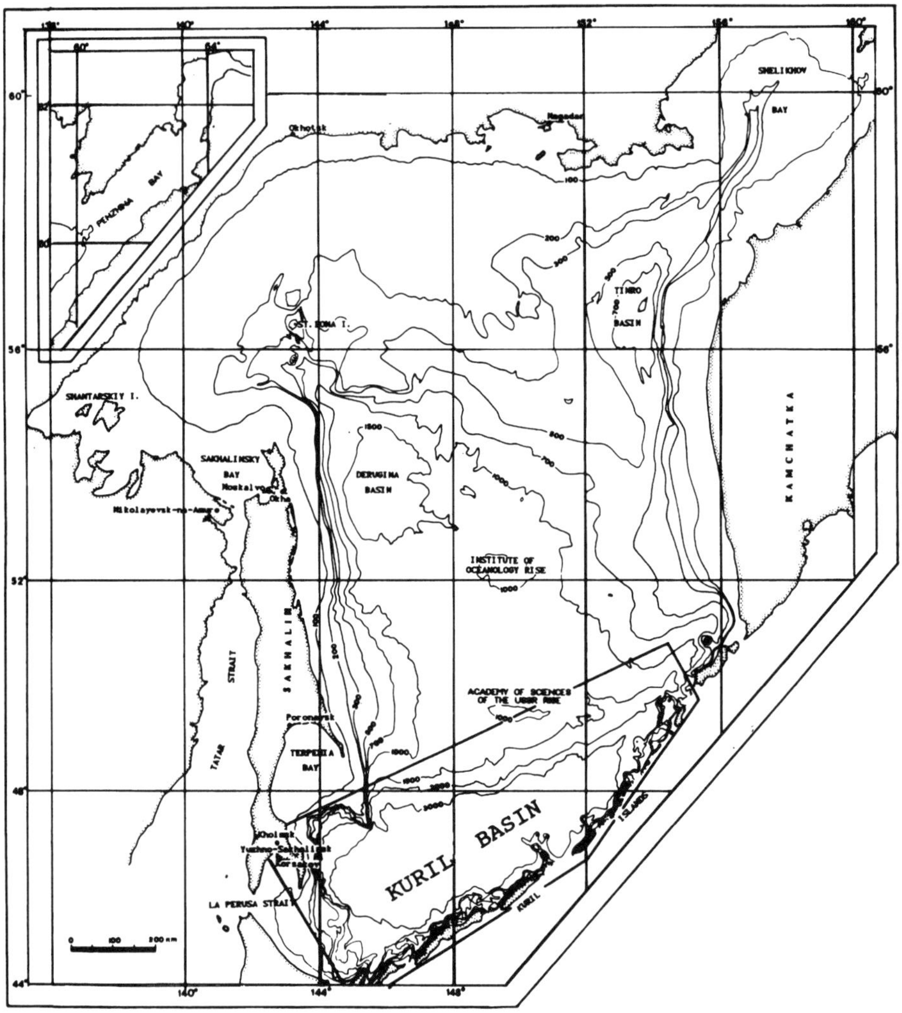

FIGURE 12.1. General bathymetry of the Okhotsk Sea. Framed area: the Kuril (South Okhotsk) backarc basin.

indicate the basin is inactive and possibly has existed since Cretaceous time, it has also been suggested to be a relict structural element (Vassilkovsky, 1967; Snegovskoy, 1974; Sychev and Snegovskoy, 1976; Gnibidenko and Svarichevsky, 1984) entrapped by Kuril island-arc formation.

Early models of the geodynamic evolution of the Okhotsk sea plate (Chapman and Solomon, 1976; Savostin *et al.*, 1983) invoked counterclockwise rotation of the Okhotsk plate converging with the Eurasian and/or Amurian plates along Sakhalin with subsequent subduction in the Kuril–Kamchatka Trench. Two recent geodynamics models for the Okhotsk plate, including the Kuril backarc basin, suggest additional recent relative plate motions:

a. The Okhotsk plate lithospheric block is under compression between the North American and Eurasian plates and is being extruded to the southwest (Riegel *et al.*, 1993). In such a case recent extension would be unlikely in the Kuril (South Okhotsk) backarc basin.

b. There is no separate Okhotsk plate. It is part of the North American plate with southerly motion (DeMets, 1992a,b). Again, no recent extension would be likely in the Kuril backarc basin.

We examine the bathymetric, structural, and geophysical characteristics of the Okhotsk Sea and particularly the Kuril (South Okhotsk) backarc basin in an attempt to better understand the origin of this backarc basin.

2. BATHYMETRY

The bathymetry of the Kuril backarc basin (Fig. 12.2) reveals an unbroken, relatively flat deep-sea plain bounded by the Okhotsk continental slope on the northwest and the steep Kuril backarc slope on the southeast. The basin extends NE-SW, parallel to the Kuril arc, being 200–250 km wide at its southwest end and narrowing toward Kamchatka. The nearly horizontal floor of the basin lies mainly at a depth of 3200–3300 m with isolated 3360+-m-deep northwest trending subbasins. The basin floor morphology suggests depositional origin with some underlying structural control. Seamounts, rising up to 1000–1500 m above the basin floor, are found only along the slope of the Kuril island arc. No axial rift or structural trends are observed.

The high density of bathymetric data allows three subbasins to be distinguished (Fig. 12.2). From the northeast, these are Atlasov, Urup, and Iturup basins, with depths greater than 3360 m. They are separated by subtle northwest trending swells, in line with the Bussol

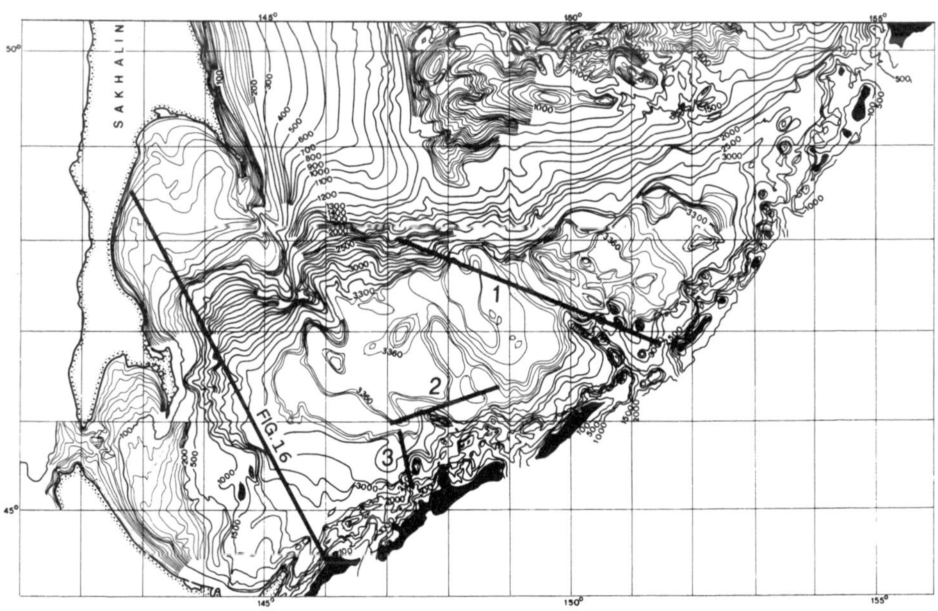

FIGURE 12.2. Bathymetric map of the Kuril backarc basin based on an echo-sounding profiling grid of 1.5–3.0 miles between profiles (Institute of Marine Geology & Geophysics, Russian Academy of Sciences). Location of refraction (Fig. 12.16) and reflection (Figs. 12.17 and 12.18) profiles are shown.

and Friza straits of the Kuril arc, and are underlain by basement structures of the same trend.

All the slopes bounding the basin are relatively steep with submarine canyons, slumps, and slides. The Kuril arc slope of the basin is steepest, ranging from 10° to 25° (Fig. 12.3). This slope exhibits a pattern of en echelon ridges striking into the basin at angles of about 25° to perpendicular to the general trend of the Kuril arc. Some of these ridges are volcanic edifices, others are faulted arc basement. Flat terraces are found at various depths: from shelf level at about 100–170 m and at 950–1200 m (Svarichevsky and Svarichevskaya, 1982). These terraces indicate substantial Pliocene-Holocene subsidence of the northwestern inner slope of the Kuril arc.

The Sakhalin–Hokkaido slope of the basin has a well–defined shelf break between 140 m in the north and 190 m in the south (Figs. 12.4–12.6). The slopes dip basinward at 10°–15°, steepening to the plane of the basin. Canyons, slumps, and deep sea terraces of probable structure origin characterize the western slope of the basin. Clearly, Terpeniya Ridge and the steep slopes below 500 m in this region (Figs. 12.4–12.6) are southward extensions of structural features of Sakhalin.

The northern slope of the basin is distinctly different, with a slope break near 1400 to 1600 m and constituting the structural boundary with the Academy of Sciences Rise. It dips from this depth to the basin plain at angles between 3° and 10° (Fig. 12.7). It is also dissected by canyons, exhibits slides and a sharp angle boundary with the basin caused by basin turbidite sediment onlap.

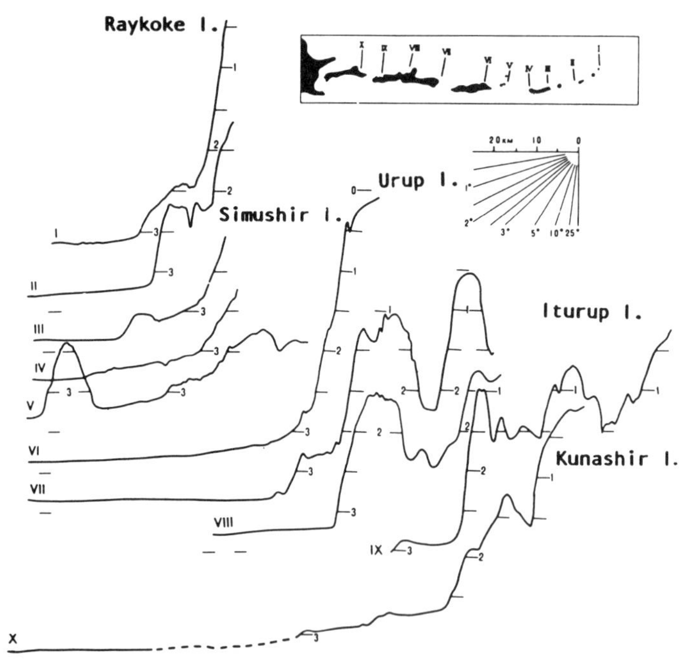

FIGURE 12.3. Bathymetry profiles across the Kuril arc slope of the Kuril backarc basin.

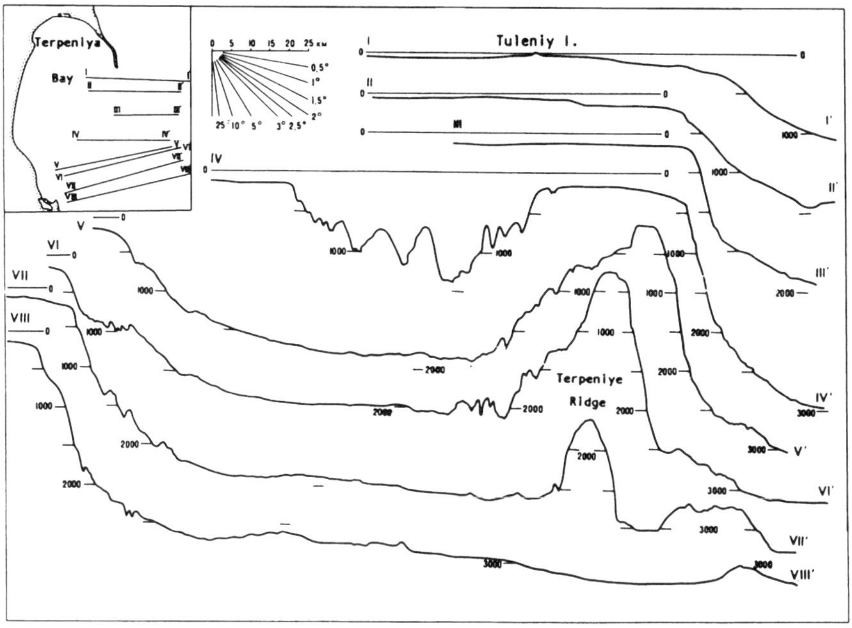

FIGURE 12.4. Bathymetry profiles across the northern (Terpenyia bay area) Sakhalin–Hokkaido slope of the Kuril backarc basin.

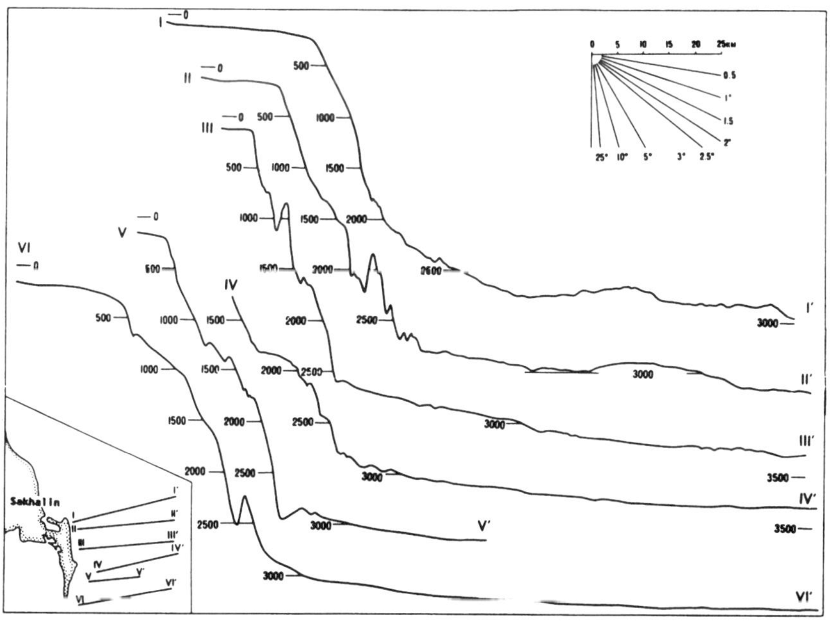

FIGURE 12.5. Bathymetry profiles across the central (offshore Tonino-Aniva peninsula) Sakhalin–Hokkaido slope of the Kuril backarc basin.

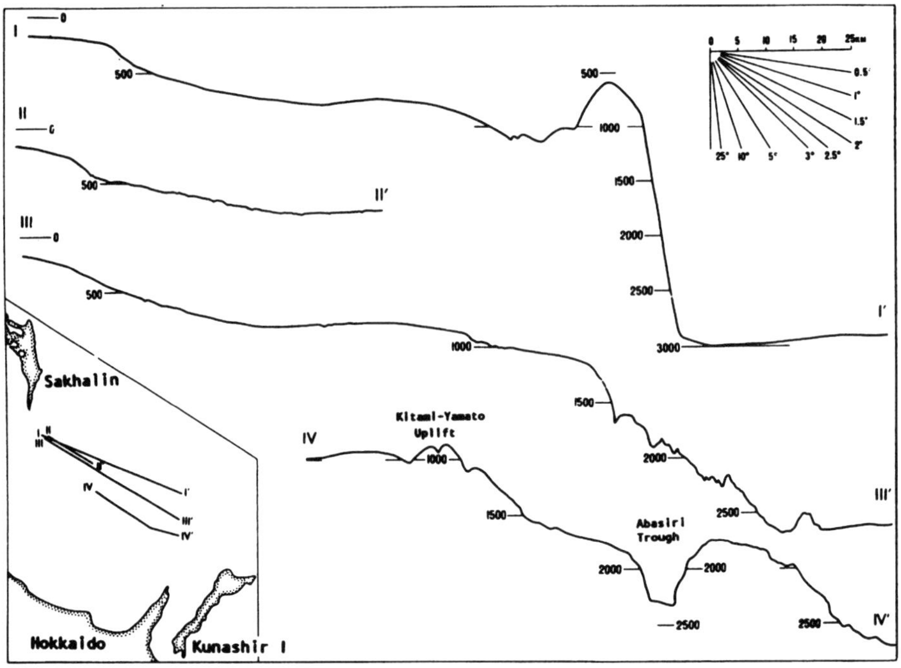

FIGURE 12.6. Bathymetry profiles across the southern Sakhalin-Hokkaido slope of the Kuril backarc basin.

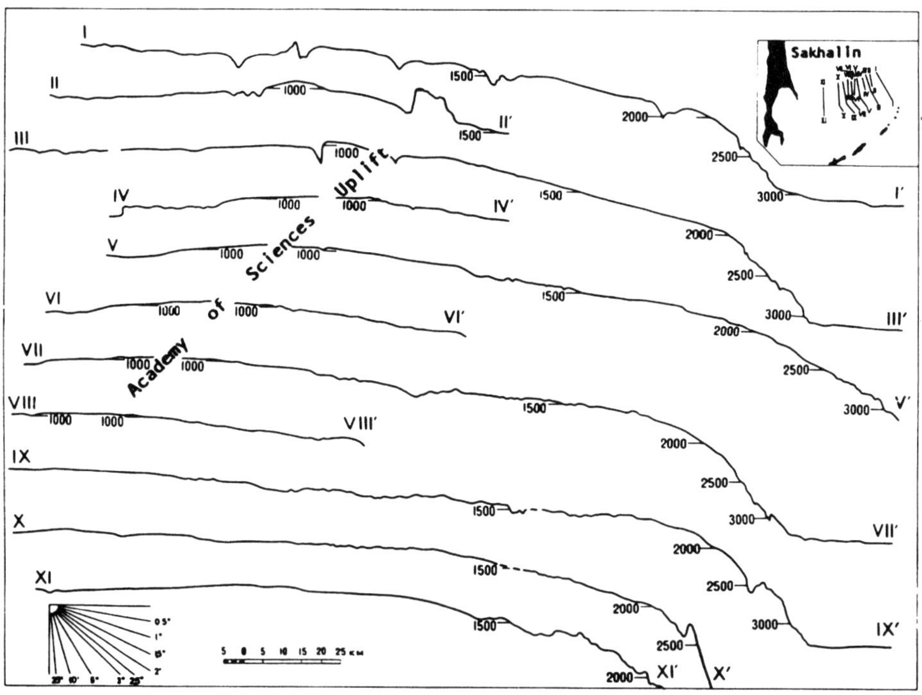

FIGURE 12.7. Bathymetry profiles across the northern slope (south slope of the Academy of Sciences of the USSR) of the Kuril backarc basin.

3. GEOPHYSICAL FIELDS

3.1. Magnetic Anomalies

The magnetic anomaly map presented here is based on combining previously compiled magnetic anomalies of the Okhotsk Sea (Tereschenkov *et al.*, 1988) with the latest marine magnetic surveys (Fig. 12.8). It is necessary to point out that much of the available data for the Okhotsk Sea backarc basin were obtained along NE-SW ship tracks. Nevertheless, the data is sufficient to allow us to conclude that there are no well-pronounced NE-SW linear magnetic anomalies oriented along the strike of the long axis of the basin (Krasny, 1990; Andreyev and Vorobyev, 1991).

Generally, negative magnetic values are associated with the SW portion of the basin and the southward offshore extension of Sakhalin, while more positive values are associated with the remaining area of the Okhotsk Sea. A rather subtle approximately N-S set of broad anomalies is associated with the SW portion of the South Okhotsk Basin that appear to be marine extensions of Kuril arc, Hokkaido, and Sakhalin structural trends. An abrupt change in magnetic character of the basin is observed along a NW-SE strike between 150° and 151°E.

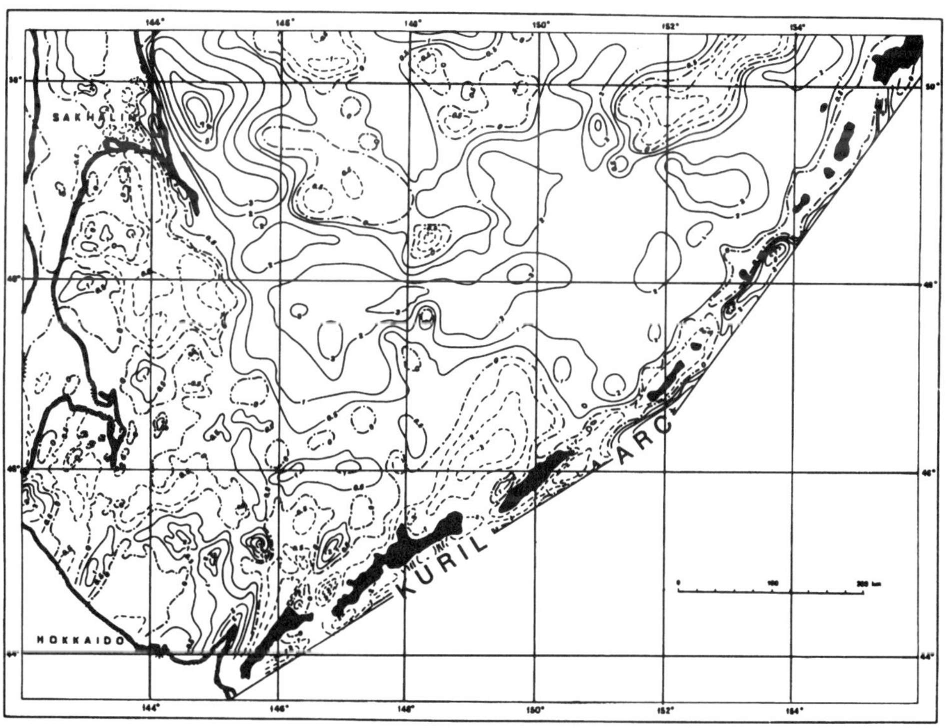

FIGURE 12.8. Magnetic anomalies map of the Kuril backarc basin. Positive anomalies are shown by solid lines, negative by dashed lines, dash-dotted lines mark zero anomaly contour; the unit of measurement is ×100 nT.

3.2. Gravimetry

A 5 × 5 minute interpolated grid of satellite- (SEASAT) derived gravity data were used for preparation of the free-air gravity anomaly map of the Okhotsk Sea including the South Okhotsk backarc basin (Fig. 12.9). The contours generally follow structural/bathymetric boundaries. Positive anomalies are associated with basement highs, while negative values outline deep sedimentary basins. A broad negative anomaly along the northern and western parts of the backarc basin suggests a thick sediment fill in this portion of the basin between the Mesozoic (and probably Paleozoic) Academy of Sciences Rise and the Kuril arc, and along the Sakhalin–Hokkaido island-arc system. The anomaly pattern indicates that the greatest sediment thicknesses beneath the basin are along the base of the slope of the Academy of Sciences Rise. On the basis of marine free-air anomalies, Kogan (1975) has proposed a reduced upper mantle density beneath the Kuril backarc basin due to elevated upper mantle temperatures.

3.3. Heat Flow

The Kuril backarc basin is characterized by high heat flow (Fig. 12.10; Sychev *et al.*, 1983; Tuezov *et al.*, 1984) of up to twice the global average. According to models of U, Th, and K content in the crust (Veselov and Volkova, 1981; Volkova, 1982) the main part of heat flow is contributed from the upper mantle (80–90%) in the Kuril backarc basin, whereas in thick crustal continental margins this is not more than 30%. The sources of such anomalies are located at depths of 15–20 km.

Heat flow distribution in the Kuril backarc basin (Fig. 12.11) as a whole is charac-

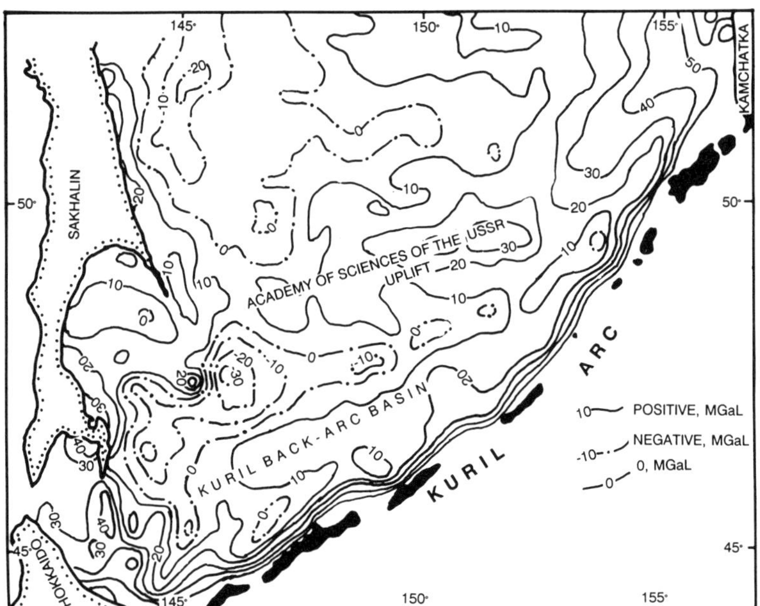

FIGURE 12.9. Free-air gravity anomalies map of the Kuril backarc basin based on the satellite altimetry data.

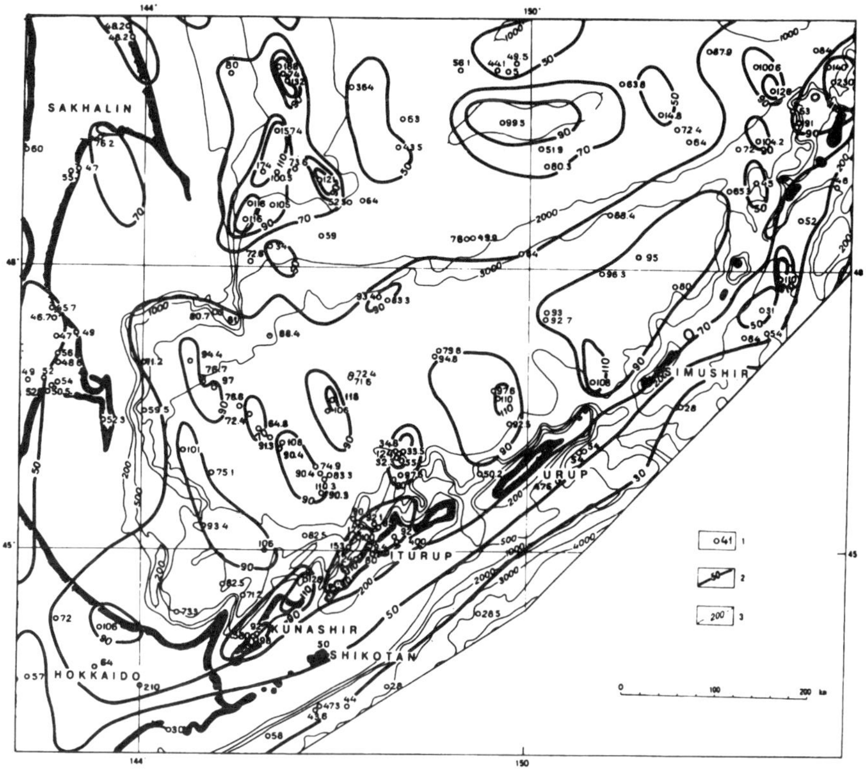

FIGURE 12.10. Heat flow map of the Kuril backarc basin. 1—measured heat flow value (mW/m^2); 2—heat flow isolines in mW/m^2; 3—isobaths in m.

terized by a heat flow of greater than 70 mW/m^2. All three subbasins (Atlasov, Urup, and Iturup) have high heat flow values of 90–110 mW/m^2. Heat flow is also particularly high along the southern Kuril arc.

Estimations of the crustal temperature distribution (Fig. 12.11) gives temperatures of about 300–500°C for layers 2 and 3, respectively, and upper mantle temperatures beneath the basin of 1100–1200°C at depths of 20–40 km. Given these heat flow values and temperature/crustal relationships, it is possible that any formerly magnetized oceanic crust beneath the basin could be partially demagnetized, especially in the lower crust.

3.4. Seismicity and Stress Field

The Kuril arc–Kuril backarc basin region has intensive earthquake seismic activity (Tarakanov and Kim, 1983). Twelve earthquakes with $M>5.5$ and more than 100 with $M>4.5$ have occurred in the basin area since 1986 (Far East, 1986–1989; Seismological Bull., 1990–93). A few earthquakes have occurred at relatively shallow depths directly beneath the basin (0–90 km, $M<4.5$), but most are beneath the Kuril arc and associated with the subducting Pacific plate (Figs. 12.12 and 12.13).

Two zones of seismic activity are defined by the distribution of seismicity of the subduction zone (Fig. 12.13): a more compact upper zone extending to a depth of 170 km

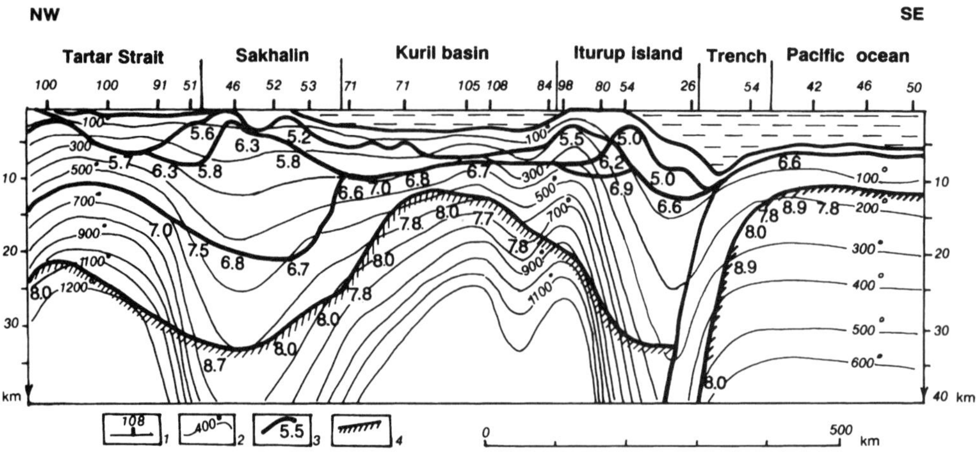

FIGURE 12.11. Temperature distribution pattern along the crustal–upper mantle section crossing Kuril backarc basin (Modified from Sychev *et al.*, 1983). 1—heat flow stations in mW/m²; 2—isotherms in °C; 3—generalized boundaries according to refraction studies and P-wave velocities in km/s; 4—upper mantle-crust boundary (Moho).

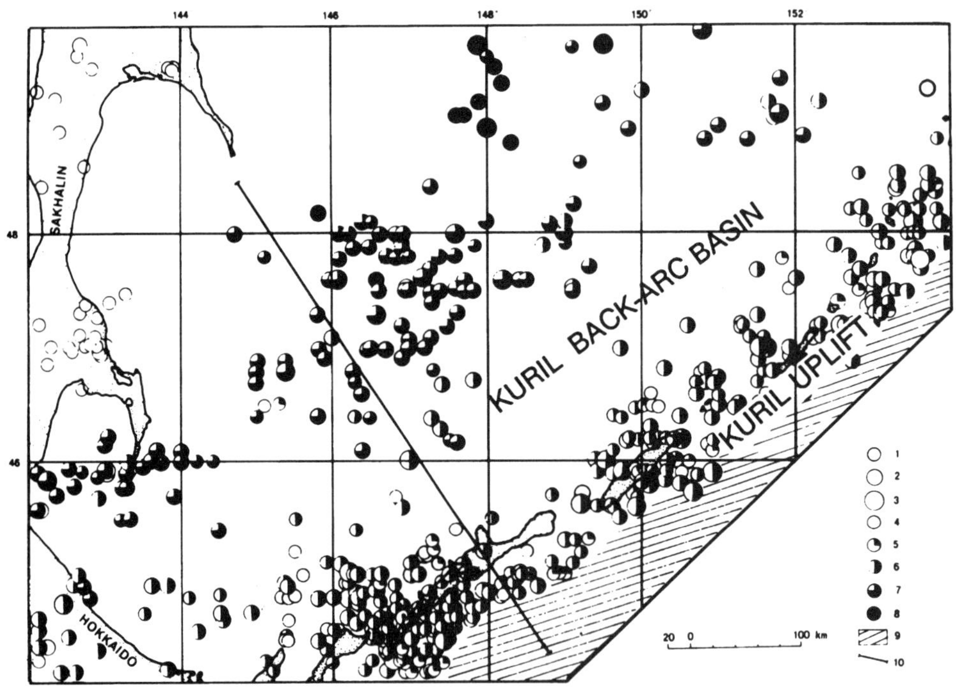

FIGURE 12.12. Earthquake locations from 1986 to February 1993 in the Kuril backarc basin. Earthquake magnitudes: 1: <4.55, 2: 4.5<5.5, 3: 5.5<6.5. Focal depths: 4: <30 km, 5: 30<90 km, 6: 90<300 km, 7: 300<500 km, 8: >500 km. 10: cross-sectional line (see Fig. 12.13).

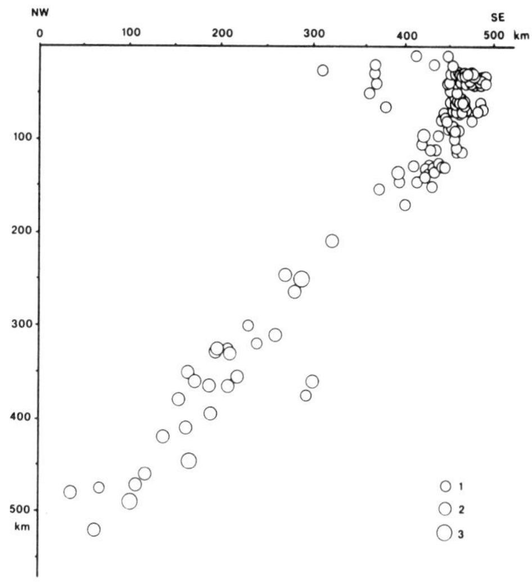

FIGURE 12.13. Cross section of hypocenters from 1986 to February 1993 events. The cross section scatter is 100 km wide (see location in Fig. 12.12). Earthquakes magnitudes: 1: <4.5, 2: 4.5<5.5, 3: 5.5<6.5.

and a lower more diffuse zone from 240 km to about 500 km. A 100 km zone of relative seismic quiescence separates these zones (Tarakanov and Leviy, 1968).

It has been shown (Simbireva *et al.*, 1976) that there is a remarkable change in the stress field of the subduction zone at the depths between 100 and 300 km, beneath the basin. Subhorizontal tension characterizes this zone (north of Bussol Strait) beneath the Atlasov subbasin, while subhorizontal compression is dominant for this zone beneath the southwest part of the basin (Urup and Iturup subbasins).

4. TECTONICS

4.1. Crustal and Lithospheric Structure

The morphostructural depression of the Kuril backarc basin coincides with a decreased crustal thickness, generally outlined by the 20-km crustal isopach (Figs. 12.14 and 12.15).

Three principal layers are distinguished in the crustal structure of the basin (Figs. 12.15–12.18): layer 1 (V = 1.7–4.3 km/s) is a sedimentary cover that consists of mainly layers of clay, silt, claystone, and siltstone with an average thickness up to 5 km (Fig. 12.19); layer 2 (V = 4.8–5.2-km/s) is volcanic and volcanoclastic in composition with a thickness of 0.5–2.0 km in the basin; and layer 3 (V = 6.4–7.2 km/s) is probably gabbro-basaltic in composition with average thickness about 5–7 km.

Starshinova (1980) suggests that the upper mantle beneath the basin down to 30 km deep has a sandwich-like structure consisting of alternating 8- and 7-km/s layers. Deeper discontinuities in the upper mantle are noted at depths about 80–85, 330–335, and 400 km (Zhang and Lay, 1993) with the first probably being the regional base of the lithosphere.

The roof of the crustal layer 2 serves as a basement (Fig. 12.20) for the sedimentary cover (layer 1) in the basin and it can be traced from the basin to its slopes through the seismic profiling (Figs. 12.17 and 12.18).

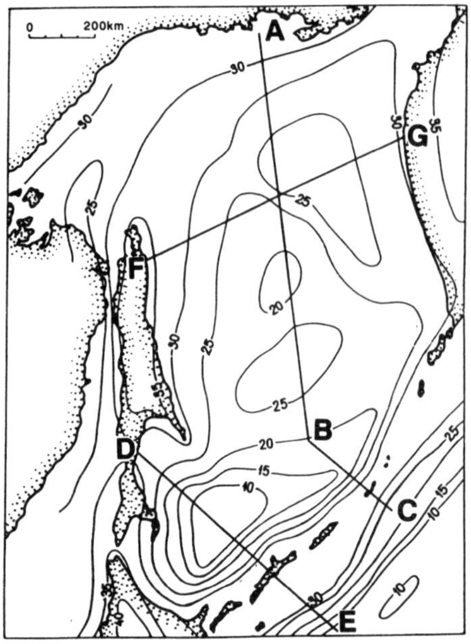

FIGURE 12.14. Crustal thickness map of the Okhotsk Sea plate including Kuril backarc basin based on the data (Galperin and Kosminskaya, 1964; Zverev and Tulina, 1971; Popov and Anosov, 1978; Bikkenina *et al.*, 1987; and Zlobin and Zlobina, 1991). Crustal cross sections shown in Fig. 15.

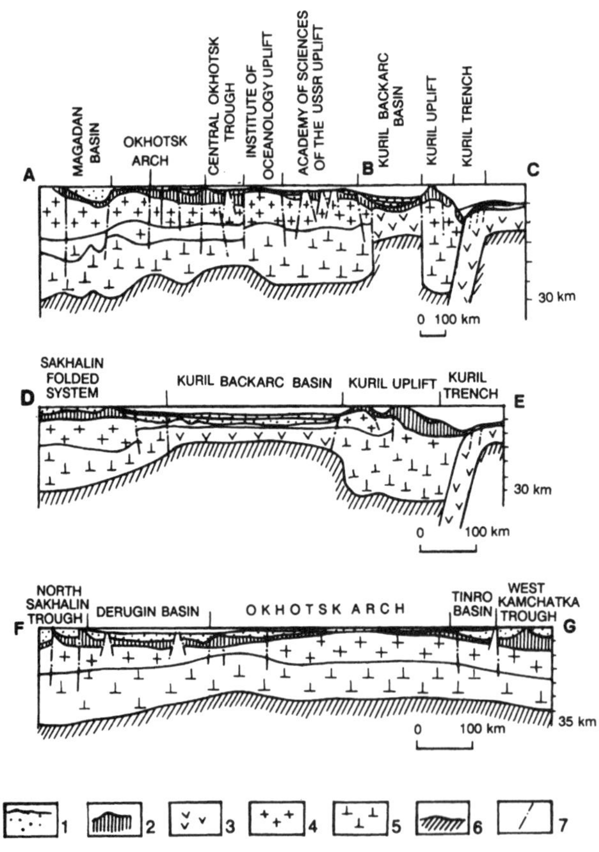

FIGURE 12.15. Crustal cross sections of the Okhotsk Sea plate including the Kuril backarc basin. Locations in Fig. 15. 1—terrigenous sedimentary layer; 2—volcanic sedimentary complex; 3—oceanic crust; 4—upper continental crust; 5—lower continental crust; 6—upper mantle; 7—faults.

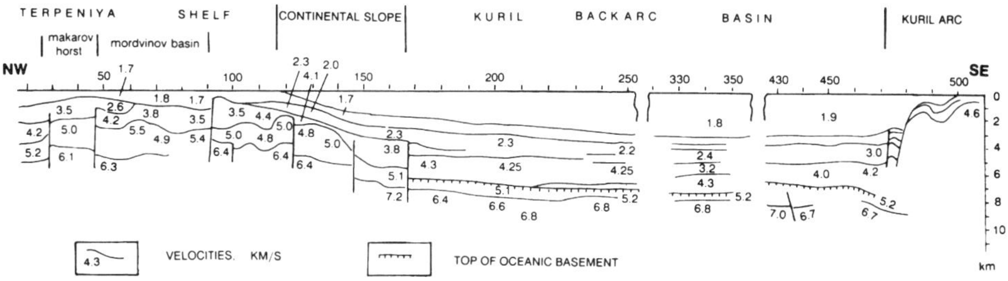

FIGURE 12.16. Upper crustal cross section of the Kuril backarc basin based on refraction studies (Bikkenina *et al.*, 1987). Location in Fig. 12.2.

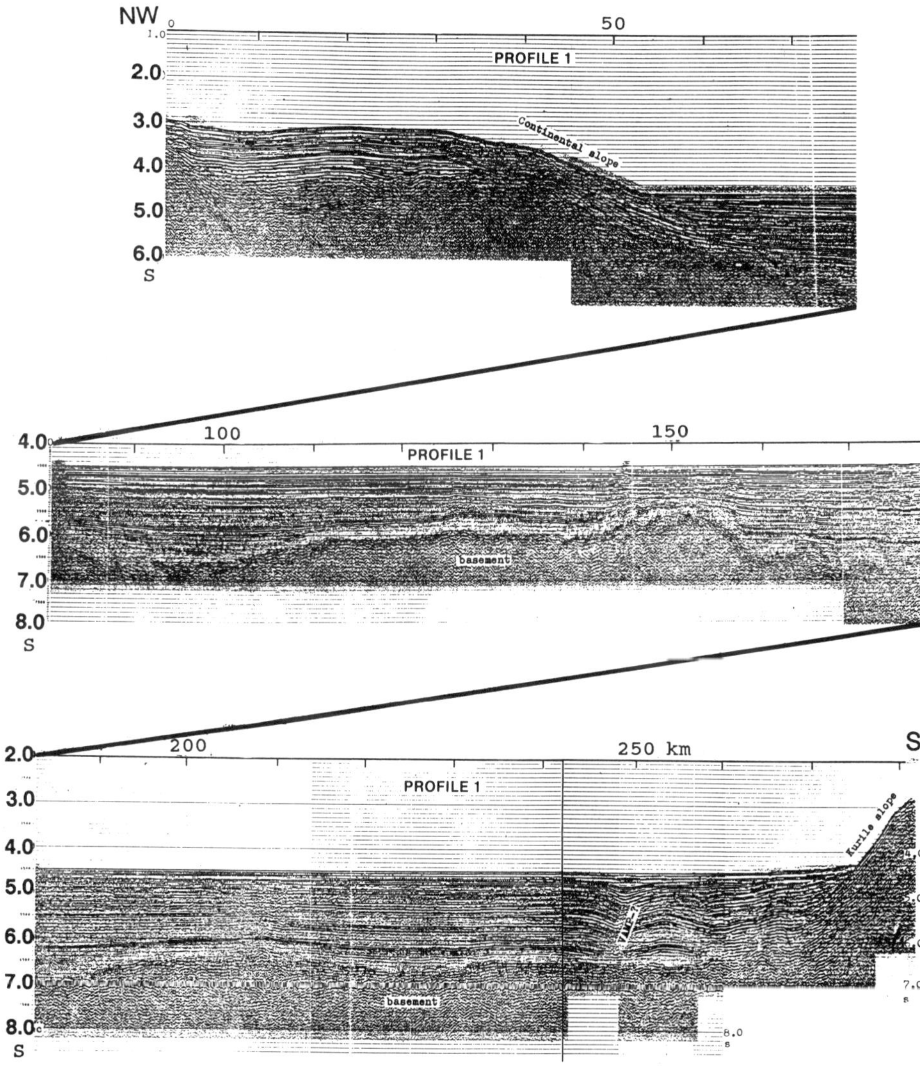

FIGURE 12.17. Multichannel seismic profile illustrating the structure of the sedimentary cover and its relationship with basement. Location in Fig. 12.2: profile 1.

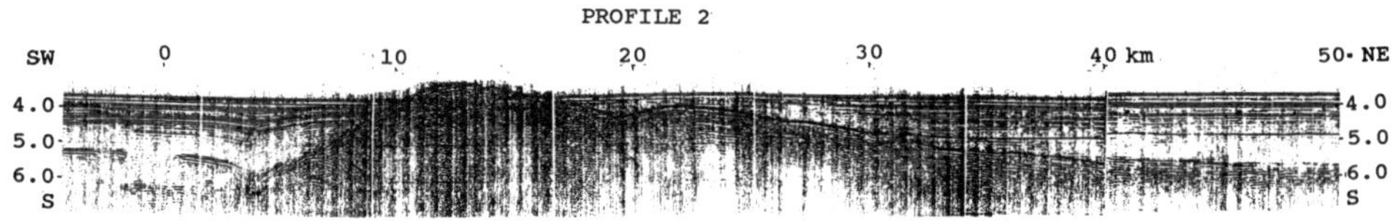

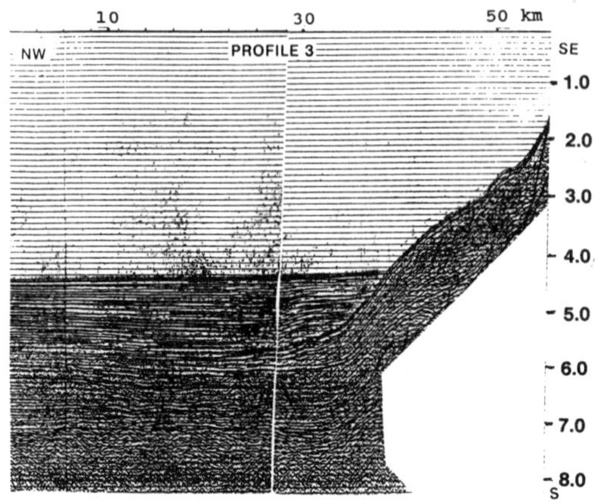

FIGURE 12.18. Single-channel (profile 2) and multichannel (profile 3) seismic profiles (profile 2) illustrating the structure of the sedimentary cover and its relationship with basement. Location in Fig. 12.2.

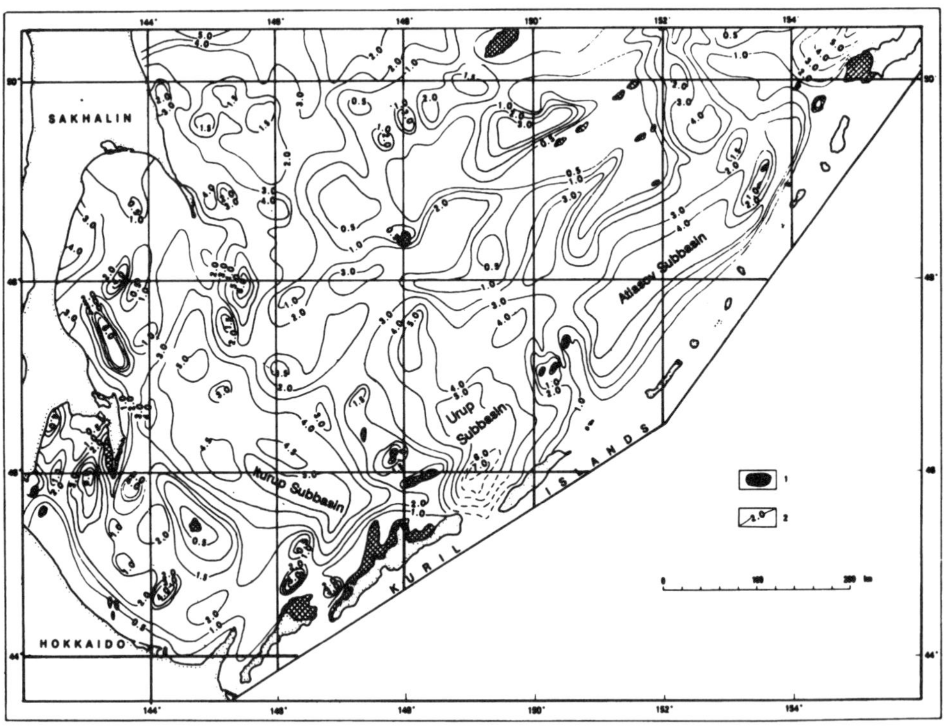

FIGURE 12.19. Isopach map of the Kuril backarc basin. 1—basement outcrops according to seismic data; 2—isopachs of sedimentary cover.

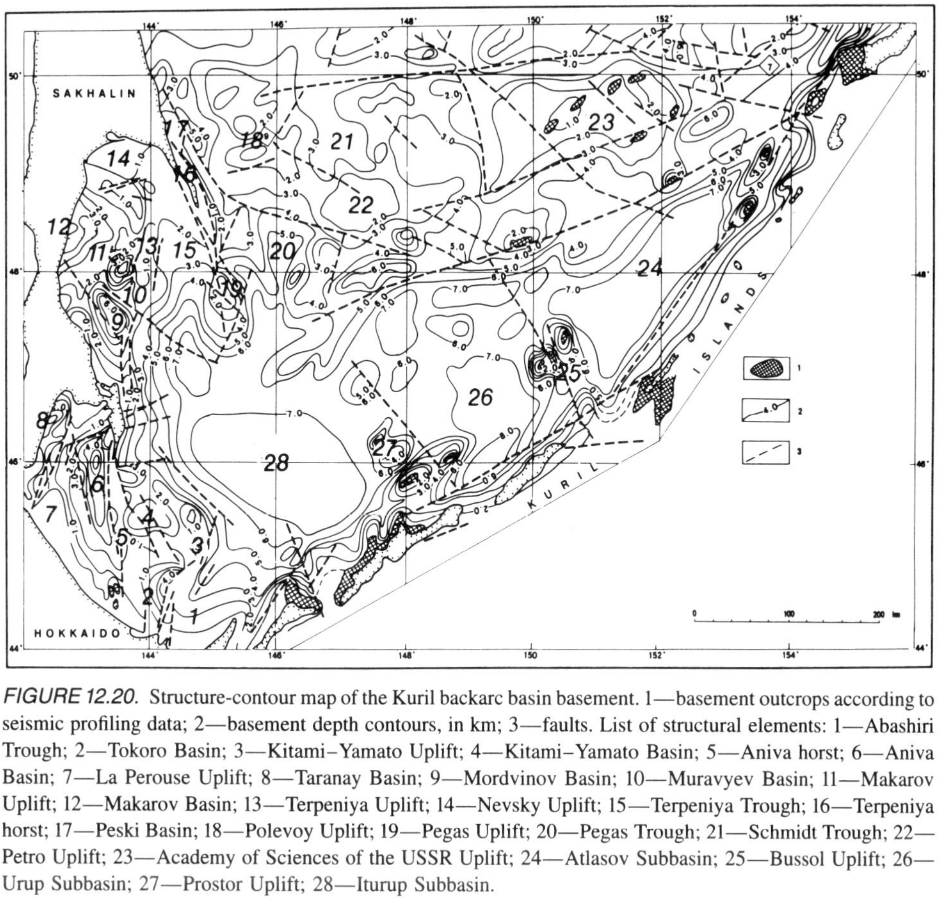

FIGURE 12.20. Structure-contour map of the Kuril backarc basin basement. 1—basement outcrops according to seismic profiling data; 2—basement depth contours, in km; 3—faults. List of structural elements: 1—Abashiri Trough; 2—Tokoro Basin; 3—Kitami–Yamato Uplift; 4—Kitami–Yamato Basin; 5—Aniva horst; 6—Aniva Basin; 7—La Perouse Uplift; 8—Taranay Basin; 9—Mordvinov Basin; 10—Muravyev Basin; 11—Makarov Uplift; 12—Makarov Basin; 13—Terpeniya Uplift; 14—Nevsky Uplift; 15—Terpeniya Trough; 16—Terpeniya horst; 17—Peski Basin; 18—Polevoy Uplift; 19—Pegas Uplift; 20—Pegas Trough; 21—Schmidt Trough; 22—Petro Uplift; 23—Academy of Sciences of the USSR Uplift; 24—Atlasov Subbasin; 25—Bussol Uplift; 26—Urup Subbasin; 27—Prostor Uplift; 28—Iturup Subbasin.

4.2. *Basement*

The data from dredging (Gnibidenko and Ilyev, 1992) of basement outcrops on the slopes around the Kuril backarc basin are combined here (Fig. 12.21) for only those stations for which either radiometric or fossils ages have been determined.

4.3. *Composition*

Among the dredged samples are extrusive, intrusive, and metamorphic rocks. Extrusive rocks consist of basalts to rhyolites with an abundance of widespread outcrops of intermediate and basic composition. Many of the extrusive rocks exhibit varying degrees of greenstone alteration (Vasilyev *et al.*, 1990), metabasalts, and meta-andesites. There are also relatively fresh olivine basalts, bipyroxene basalts, and andesites. The presence of transparent plagioclases and nonaltered volcanic glass characterize these rocks. Fresh hyaloclastites are also found. The nonaltered rocks were dredged from the Kuril arc slope of the basin only.

Extrusive rocks with moderate-to-weak greenstone alteration are the most widespread

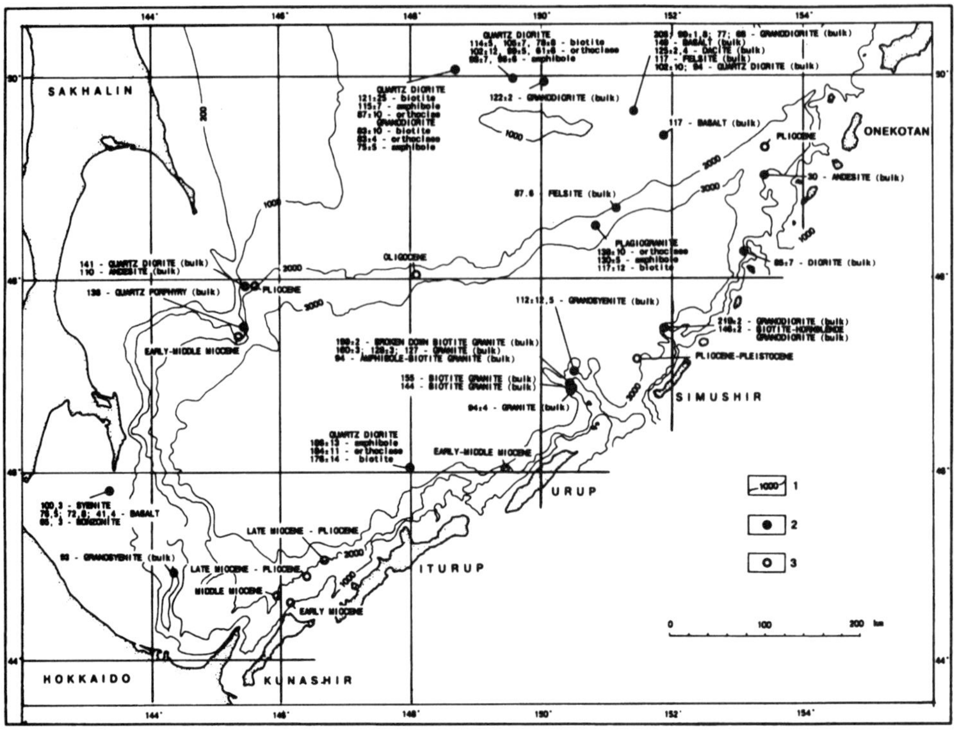

FIGURE 12.21. Dredge sites (only sites with ages determinated are presented) around the Kuril backarc basin. 1—isobaths, 2—magmatic rocks (K-Ar), 3—sedimentary rocks (biostratigraphy).

on the South Okhotsk backarc basin slopes. This group includes basalts, andesite-basalts, andesites, andesite-dacites, and rhyolites. Fresh nonaltered minerals exist within the matrix of these rocks in spite of clear indications of autometamorphic processes.

Intrusive rocks are mainly represented by granitoids, diorites, monzonites, and syenites. Gabbroids are found rarely as single blocks. A fragment of lherzolite has been exposed eastward of the Academy of Sciences Rise.

Granodiorites and quartz-diorites are the most abundant of all basement magmatic rocks. These rocks are light gray and gray, usually medium and coarse grain, sometimes porphyry rocks composed of acidic and medium plagioclase, quartz, potassic feldspar, and biotite. Autometamorphic alteration is slightly developed in these rocks.

Fine-grain, medium-grain, and coarse-grain, sometimes porphyry biotite-hornblende granites, plagiogranites, and quartz monzonites have been dredged from the Academy of Sciences Rise and on the lower slope of the central part of Kuril arc. Cataclastic texture is sometimes found.

Metamorphic rocks are mainly represented by hornfels. Relict porphyroblast textures are typical for many of them, which testifies to a volcanogenic origin. Angular blocks of granite-gneisses were also dredged from the Academy of the Sciences Rise. These leucocratic rocks consist of acidic plagioclase, quartz, muscovite, biotite, and garnet.

The chemical composition of the dredged rocks shows that the extrusives belong mainly to the calc-alkali series or close to it (Fig. 12.22). The swarm of intrusive rock

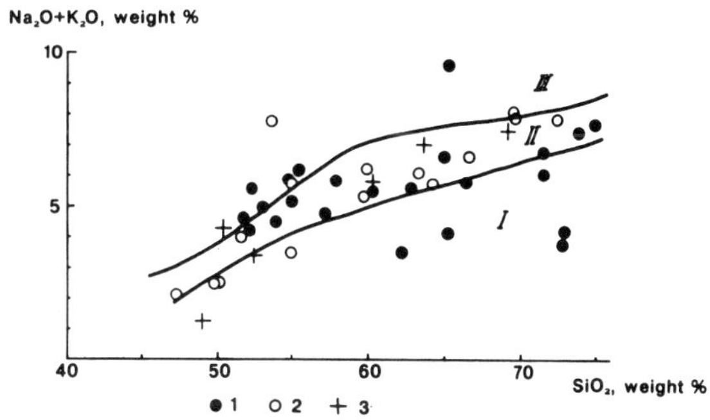

FIGURE 12.22. Relationship between alkali and SiO_2 (Kuno diagram) in effusive rocks from basement outcrops around the Kuril backarc basin. I—tholeiite (alkalic basalt) series area; II—calc-alkalic series area; III—calc-basaltic series area. 1–3 = extrusive rocks of: 1—Academy of Sciences Rise, 2—Kuril arc slope, 3—Hokkaido–Sakhalin slope.

compositions coincide with the calc-alkali series field (Fig. 12.23) and testifies to the close chemical relationship of these rock groups. All magmatic rocks of the Kuril backarc basin slopes belong to the sodium type where $Na_2O > K_2O$.

4.4. Age

K-Ar ages of the magmatic rocks from 40 samples of basement outcrops along the basin margins vary between 219 Ma and 30 Ma (Figs. 12.21 and 12.24). Most of the dated magmatic rocks (75%) are Cretaceous. Keeping in mind that all radiometric ages are of magmatic rocks, the host strata should be older, likely Jurassic in age. Granodiorites of ages 146 Ma and 219 Ma from basement outcrops on the Kuril arc slope of the basin, north of Simushir Island, and other similar age samples from the Kuril arc slope generally match the ages of rocks from the north side of the basin. Since the Kuril backarc basin clearly has a thin crustal structure and velocities characteristic of oceanic crust, it is likely that its development postdates and separates the Cretaceous and Jurassic basement rocks described above.

Because of the high latitudes for the Okhotsk Sea and the impact of Neogene and Pleistocene glaciation on the region, ice rafting must be considered as a process for distributing rocks over broad areas of the seafloor and especially for those rocks that seem

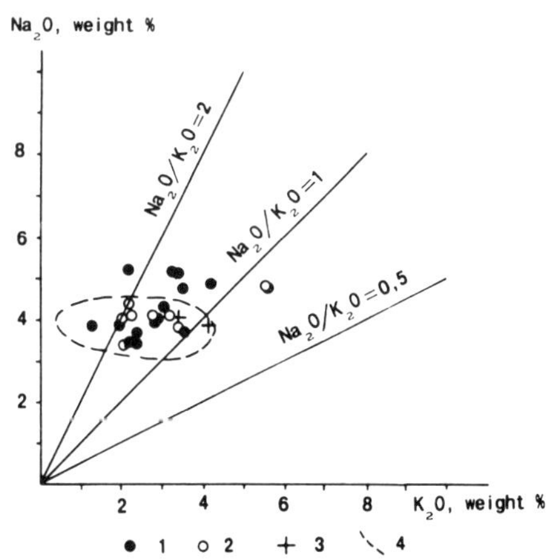

FIGURE 12.23. A diagram of alkalinity of intrusive rocks around the Kuril backarc basin. 1–3 = intrusive rocks of: 1—Academy of Science Uplift, 2—Kuril arc slope, 3—Hokkaido–Sakhalin slope, 4—contour of extrusive rocks of differentiated calc-alkalic series.

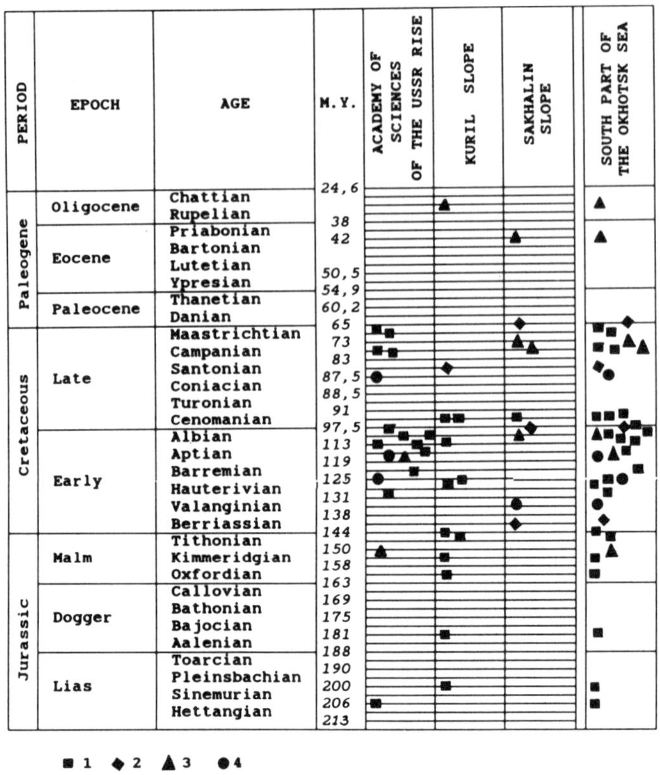

FIGURE 12.24. Radiometric age of the basement outcrops rocks around the Kuril backarc basin. The geochronology scale is according to Harland *et al.* (1985). 1—granitoids, 2—intrusive rocks of intermediate and basic composition, 3—rocks of intermediate and basic composition, 4—extrusive rocks of acid composition.

to be exotic to the tectonic setting from which they were dredged. This was considered in the reporting of the dredge samples described in this report. The dredge sites for these samples were all from topographic highs that were mostly clear basement exposures. The analyzed and reported samples were all angular, and several exhibited fresh surfaces. Any rounded or striated samples were excluded from the study. Further, those dredge samples from the Okhotsk Sea that have been identified as ice-rafted debris are typically Upper Cretaceous to Paleocene acidic volcanic and granitic rocks and were also excluded. Their source area is from onshore exposures north of the Okhotsk Sea. These rocks are distinctly different from the rocks reported here from around the Kuril backarc basin. While the possibility remains that ice rafting could be responsible for distributing the rocks described in the report, we feel that it is more likely that they represent the basement of the southern Okhotsk Sea region.

4.5. Structural Features

The acoustic basement of the basin is characterized by seismic velocities of 5.1–5.2 km/s (Fig. 12.16) and is overlain by 4 to 5 km of sediment throughout most of the basin (Fig. 12.19). Basement highs are observed which appear to be associated with either arc structures or basement at the edge of the basin (Figs. 12.17 and 12.18). They outcrop at the sea bottom only near and along the slopes of the basin. These features have previously been described as elevated portions of basement, volcanoes, or outcrops of layer 2 (Soloviev *et al.*, 1977). We interpret them as a parts of the arc basement or the basement of the northern margin of the basin.

Distinct differences exist in the velocity structure beneath the Kuril backarc basin and shallower portions of the Okhotsk Sea to the north and the northwest (Fig. 12.16). Beneath the basin there is a well-defined layered structure with the top of the 5.1–5.2 km/s layer correlating with acoustic basement of the seismic reflection data. The boundary of the basin and the continental slope (Fig. 12.16) marks a dramatic change in the velocity structure.

To the north of the basin, this boundary is characterized by greater structural variability in relief and velocities. These structures are representative of the Academy of Sciences Uplift and the Sakhalin–Hokkaido fold belt and can be correlated with Mesozoic geosynclinal and ophiolitic (Cretaceous and older) formations. The overlying sedimentary cover displays less folding and faulting.

Samples of sedimentary outcrops from around the margin of the Kuril backarc basin (Fig. 12.21) are all Miocene in age except for one Oligocene sample. They appear to correlate with the lower or middle parts of the Kuril basin sedimentary section as revealed in the seismic reflection profile (Fig. 12.17). Assuming that these rocks represent early deposition into the Kuril backarc basin, their age then places a minimum age on the formation of the basin.

5. BASIN SEDIMENTARY FILL

5.1. Seismic Stratigraphy

The most distinctive feature of the sedimentary section of the Kuril backarc basin is that from seafloor to basement, or near basement, it is an undisturbed, horizontally stratified section (Figs. 12.17 and 12.18). No evidence for extension exists within the sedimentary section. Gentle folds are observed at the base of the Kuril arc (Fig. 12.17) which could be evidence of relatively recent minor compression at the base of the arc. A minor unconformity exists between the uppermost sediments and the gentle folds at 4.7-s depth near the SE end of profile 1. These features are, however, indications of only very minor deformation.

Basement is clearly distinguished all across the basin (Fig. 12.17) as a rough, high-amplitude, highly reflective surface, typical of oceanic basement. This surface corresponds to the top of the 5.2-km/s layer (Fig. 12.16) detected in seismic refraction studies (Popov and Anosov, 1978; Bikkenina *et al.*, 1987). The sediments immediately overlying basement display the greatest variability, in places horizontally onlapping basement (SE portion of profile 1 and profile 3) and displaying dip toward the basin beneath both margins.

The relatively transparent lowest deposits overlying basement in the central portion of profile 1, displaying apparent drape, are possibly associated with the basement high that extends perpendicular to the arc, across the basin (Fig. 12.2). Onlap onto the basement high displayed in profile 2 (Fig. 12.18) also shows tectonic quiescence during deposition and another basement feature that is perpendicular to the arc (Fig. 12.2). In general, the sedimentary section and its basement relationships indicate that the Kuril backarc basin has been inactive throughout the basin's depositional history.

Two additional general features of the basin sedimentary section are worth noting. As a whole, approximately the upper half of the section is more reflective than the deeper sediments. On the basis of this and and other characteristics, we divide the basin fill into upper and lower sedimentary sequences. The boundary between these sequences corresponds to the boundary between the 2.4–3.0-km/s unit and the underlying 4.0–4.3-km/s unit (Fig. 12.16). This boundary is observed in the reflection profiles as a high-amplitude reflector and subtle unconformity at about 5.5-s depth across profile 1 (Fig. 12.17), 4.6–4.9 s in profile 2 and at 5.9–6.0 s in profile 3 (Fig. 12.18). The upper sequence clearly onlaps the margin slopes of the basin, while the lower sequence dips basinward at the margins and is exposed on the margin slopes (Fig. 12.20). These relationships suggest whole basin subsidence during deposition of the lower sequence.

5.2. Thickness and Distribution

A generally greater than 4-km-thick sedimentary section fills the Kuril backarc basin, with an axial extent closely matching the deep topographic basin defined by the 3300-m isobath (Figs. 12.2 and 12.19). While the basin as a whole has an arc-parallel long axis, it is divided by transverse basement highs or ridges into three subbasins: Atlasov, Urup, and Iturup. Sediment thicknesses exceed 5 km in the central Urup and the SW Iturup subbasins (Fig. 12.19). Basement relief has clearly controlled the distribution of the primarily turbidite deposits.

5.3. Composition and Age

Seismic reflection profiles and cores (see following section) indicate that the upper sedimentary sequence of the Kuril backarc basin is primarily of turbidite origin, consisting of turbidites, biogenic, and ash layers. Likewise the lower sedimentary sequence, while somewhat less reflective, also generally displays numerous closely spaced horizontal reflectors in the seismic reflection profiles that characterize turbidites (Figs. 12.17 and 12.18). The reflection profiles and dredge samples along the margin slopes (Gnibidenko and Ilyev, 1992; Fig. 12.21) indicate that the lower sequence is also primarily turbidites with alternating ash layers. These are the deposits that are exposed at the basin margins, that dip toward the basin, and correlate with the deeper sediments in the basin.

The dredged samples along the basin margin are Miocene and Oligocene in age (Fig. 12.21). Assuming that these samples represent the lower sequence within the basin as we have correlated them to the reflection profiles, the basement age of basin could then be as young as Miocene/Oligocene in age.

6. HOLOCENE SEDIMENTATION

Data obtained for the uppermost part of the sedimentary cover of the South Okhotsk backarc basin by coring (Bezrukov, 1955, 1960; Zhuze, 1957; Saidova, 1960; Ilyev *et al.*, 1979, and Gretskaya, 1990) provide us with Holocene sedimentation characteristics which may be used to consider limits on the depositional history for the older deposits.

Bottom sediments of the basin are the mainly fine-grain greenish gray, soft and plastic sediments up to a depth of 2.2 m. They are dominantly silts and clays and are generally poorly sorted with concentrations of silt 25–40%; clay 50–75%; water content 60–70%; and bulk density 1.2–14 g/cm^3. In the cores located in the central part of the basin (sites 62 and 63, Fig. 12.25) the fine clay fraction (0.001 mm) is considerably increased (70–80% Fig. 12.26) with water content up to 80% (Fig. 12.27).

A fine ooze is abundant in the central part of the basin. Silts and sands exist closer to the margins. Distribution of the terrigenous material can be traced throughout the basin in the upper horizon (0–15 cm) and deeper layers, suggesting powerful turbidity flows.

The distance of transportation of terrigenous material traced according to variations in grainsize was studied from the cores of Sites 60–63. In Core 61 (Fig. 12.26) silt-rich sediments form the upper 0–15 cm and three layers within the interval of 100–180 cm. The saturation with silt in this part of the basin reflects both the periodicity of the turbidity flows (four cycles) and the area of their unloading. Two turbidity cycles are distinguished in Core 62, while in core 63 in the central part of the basin no turbidites are detected (Fig. 12.26).

All Holocene basin sediments are rich in biogenic silica, 23–35% weight, and have a pronounced biogenic-detrital texture. Diatom plankton is the source of silica. The areal zoning of biogenic silica corresponds, in general, to the circumbasin pattern of the sedimentary material distribution. All sediments deposited closer to the base of the Kuril arc slope contain greater quantities of silica.

Thus, there are two main sources of Holocene sedimentary material: terrigenous and biogenic. The terrigenous material is mainly supplied by the turbidity flows, whereas the

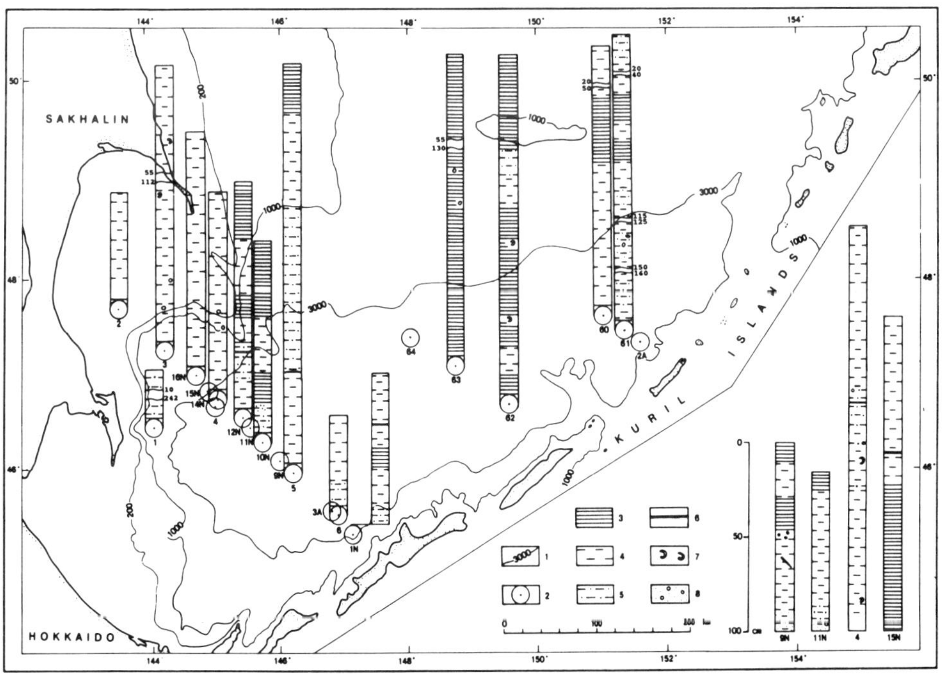

FIGURE 12.25. Core locations and composition in the Kuril backarc basin. 1—isobaths, 2—core sites, 3—ooze, 4—silty clay, 5—silty ooze, 6—ash layer, 7—faunal residuals, 8—dispersed sand and gravel material.

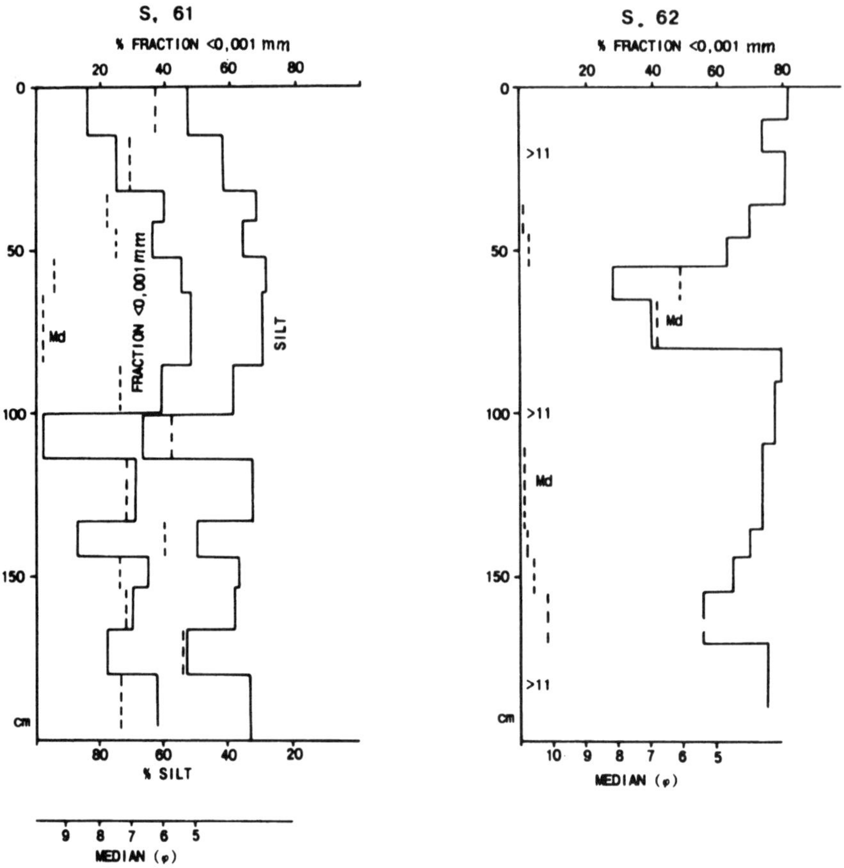

FIGURE 12.26. Grain size distribution of sediments for the central part of the Kuril backarc basin. For location of sites see Fig. 12.25.

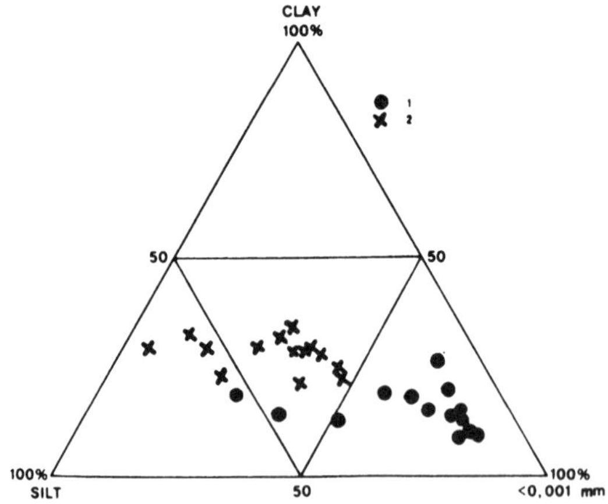

FIGURE 12.27. Log and measured grain-size parameters of sediments at Sites 61 and 62.

biogenic component is provided by high bioproductivity, resulting in a relatively rich organic content in the sediments.

Age and Sedimentation Rate

The upper 2.0–2.5 m of sediments are dated as Holocene. Sedimentation rates have been calculated as 20–25 cm/1000 year (200–250 m/Ma; Zhuze, 1957; Bezrukov, 1960; Saidova, 1960). More recent studies (Gretskaya, 1990) estimate 30–40 cm/1000 year (300–400 m/Ma). A high degree of water content and abundant <0.001-mm-size fraction in the sediments are cited as causing slow consolidation and preservation of initial thicknesses for long periods of time.

On the whole, Holocene sedimentation of the South Okhotsk backarc basin occurs as a result of the turbidity flows into a closed basin along with high diatom productivity. If we assume a rate of turbidite sedimentation of about 300 m/m.y. has existed for the late Cenozoic, the onset of intense turbidite deposition would have been in the middle Miocene.

7. DISCUSSION: STRUCTURAL EVOLUTION AND GENESIS

During the last decade the geological and geophysical characteristics of Okhotsk Sea plate, including the Kuril backarc basin and the Sakhalin–Hokkaido–Kuril–Kamchatka island-arc system, have been extensively studied. These studies allow us to distinguish the main features of the basement structure and age, and the distribution, age, and nature of the sedimentary basins. They reveal a Sakhalin–Hokkaido–Kuril–Kamchatka island-arc system which began its development in early Mesozoic time surrounding a cratonic Okhotsk Sea Plate. Extensive fold and thrust deformation began in late Cretaceous time and continues to present in the current island-arc system. The Kuril backarc basin developed within this complex sometime between late Cretaceous and Miocene time.

While the Kuril backarc basin is underlain by oceanic crust the remainder of the Okhotsk Sea region is underlain by continental crust of predominantly Mesozoic age (Gnibidenko, 1985). Crustal thickness varies between 25 km in the central Okhotsk plate up to 35 km in the surrounding folded zones. The crust of the Okhotsk plate is highly block-faulted with Cretaceous and Cenozoic sediment-filled basins organized in NW-SE and E-W tectonic patterns. Normal-faulted Cenozoic sedimentary basin development is inherited from these Mesozoic structural trends.

The Okhotsk Sea is bounded on the west and north by the Hokkaido–Sakhalin and Okhotsk–Chukotka Cretaceous and Cenozoic volcanic-plutonic belts, which are products of convergence and subduction of the Okhotsk plate beneath the Eurasian plate during Mesozoic time (Jolivet *et al.*, 1992). Collision of the continental crust of the Okhotsk Plate along this zone is evident by the late Mesozoic highly deformed subduction complexes of Sakhalin. This zone includes ophiolite and molasse formations of Mesozoic and Cenozoic age. Paleozoic rocks have been found in the lower sections (Melnikov, 1988). The zone is composed of en echelon anticlinoria, structurally connected between Sakhalin and Hokkaido across La Perouse Strait (Gnibidenko and Snegovskoy, 1975). All geological formations are highly folded and thrust faulted with vergence mainly to the west. During the Cenozoic a dextral wrench-faulting system was superimposed. This zone is still tectonically active with a N-S zone of seismicity all along Sakhalin that extends southward and links with the seismicity of Japan (Okada, 1982; Kiminami and Kontani, 1983; Rozhdestvensky, 1986; Jolivet *et al.*, 1992).

The southern and eastern boundary of the Okhotsk Sea consists of the Kuril–Kamchatka subduction zone/arc system where the Pacific plate is subducting northwestward beneath the Okhotsk Sea plate. Here, also, along the entire Kuril–Kamchatka zone, geological evidence exists for a late Mesozoic convergence/subduction history. The Kuril–Kamchatka fold system is composed of horst-anticlinorial uplifts of Mesozoic formations, separated by Cenozoic sediment-filled graben-synclinorial troughs which strike northeastward, parallel with the general trend of the arc. The average crustal thickness of the region is about 30 km. Crustal and upper mantle inhomogeneities are recognized with a resolution of a few kilometers (Averynova, 1975; Balakina, 1979, 1981; Gnibidenko *et al.*, 1983).

To the southwest, the Kuril fold system is structurally connected with eastern Hokkaido structural elements (Geological Map, 1978; Honza, 1978; Gnibidenko *et al.*, 1983). To the northeast this zone extends into Kamchatka, where the axial parts of uplifts are traced by outcrops of Mesozoic ophiolites (Gnibidenko *et al.*, 1974) within horst-anticlinorial uplifts on the continental slope. These formations consist of deformed geosynclinal volcanogenic and sedimentary rocks which are intruded by gabbroides, granitoides, and granodiorites. Graben-synclinal troughs are filled with Neogene sedimentary deposits, which in some basins exceed 3 km in thickness.

The crustal and tectonic evolution of the Okhotsk Sea region has involved a Mesozoic (and possible older) cratonic block colliding with the eastern Eurasian plate margin during late Mesozoic to early Cenozoic time (Den and Hotta, 1973; Savostin *et al.*, 1983; Takanashi, 1983; Watson and Fujita, 1985; Kimura, 1986; Jolivet *et al.*, 1989; Maeda, 1990). This was preceded and accompanied by construction of the Hokkaido–Sakhalin–Chutkotka volcanic/deformation belts discussed above.

Recent and present-day plate configurations within the region are deduced from zones of seismicity and focal mechanism solutions (Chapman and Solomon, 1976; Zonenshain and Savostin, 1979; Tarakanov and Kim 1983; Lundgren and Giardini, 1990; DeMets, 1992a,b; Riegel *et al.*, 1993). Bounded along the Kuril–Kamchatka Trench by the subducting Pacific plate and by a shallow zone of seismicity extending from Japan through Sakhalin into the northeast of Russia, the Okhotsk Sea region may presently be part of the North American plate.

Placing the origin and development of the Kuril backarc basin within the broader geological framework and tectonic evolution of the Okhotsk Sea region must take into account both the data from the basin and the surrounding structures. The basin is appropriately located behind the Kuril island arc where a backarc spreading produced basin may be expected, and it has oceanic crustal thicknesses. However, the magnetic anomalies are not seafloor spreading-type anomalies, so it is not possible to date the basement on the basis of reversal time-scale modeling. To the extent that the anomalies are linear, they strike transverse to the NE-SW long axis of the basin and are associated with basement highs. Unfortunately, there are no samples of the deep oceanic crust for direct dating.

The thick sedimentary section filling the Kuril backarc basin is essentially tectonically undisturbed, exhibiting no evidence for either significant extension or compression. Formation of the basin was apparently completed before most or all of the sedimentary fill was deposited. Only vertical motion is suggested (subsidence) based on sedimentary structural features at the basin margins.

Three lines of evidence are available to use in attempting to determine the age of formation for the Kuril backarc basin: the ages of the dredged rock, sedimentation rates, and age/depth values for oceanic crust. As we have noted, the Miocene/Oligocene age dredge samples place a likely minimum age on the basin. Likewise, the Mesozoic age dredge samples from the arc and the northwestern margin of the basin place a maximum possible age on the basin of late Cretaceous. The Mesozoic rock samples represent the older basement of the arc and the cratonic portion of the Okhotsk Sea, as found by many investigators and discussed above, and the Kuril backarc basin certainly postdates the age of these rocks. Based solely on the dredge samples, these are, however, rather broad limits for the basin's age.

Sedimentation rates in the Kuril backarc basin are known only from the sampling of the Holocene deposits, and estimates range from 200 to 400 m/Ma from these studies. Because these rates are only for the uppermost few meters of the basin deposits and have a great range, it is highly speculative to use them alone for estimating rates for the entire sediment section. In nearby backarc settings, the Japan Sea Basin and the Kamchatka–Komandorsky Basin, where there has been ODP and DSDP drilling, deposition rates are known for most or all of the sediment sections there and may be representative of sedimentation rates for the Kuril backarc basin. Rates in the Japan Sea Basin rates range from 9 to 77 m/m.y. (Burckle *et al.*, 1992). In the Kamchatka–Komandorsky Basin, the range is 70 to 200 m/m.y. (Scholl and Creager, 1973). Therefore, if we use a minimum of 30 m/m.y. and a maximum of 400 m/m.y., these values encompass those reported for Miocene to present rates from the nearby backarc basins for Holocene sedimentation in the Kuril basin. At 30 m/m.y. for an average 4500-m thickness of sediments the basement age would be 150 Ma. At 400 m/m.y. the basement age would be 11.25 Ma. Again, these extreme values are not very instructive. We note that for all except one of the Japan Sea and Kamchatka–Komandorsky basin sites the average range of depositional rates is between 30 and 90 m/Ma. If, on this basis, an average rate of 60 m/Ma is applied to the Kuril Basin, the basement age would be 75 Ma. The objective conclusion of this depositional rates exercise is that ocean drilling is needed in the Kuril backarc basin.

Basement depth for the deeper portion of the Kuril backarc basin is between 7 and 8 km with 4 to 5 km of overlying sediments. Taking into account that backarc basins are generally on the order of 1 km deeper than normal ocean crust for a given age and the loading of the sedimentary deposits, these basement depths are still deeper than that expected for a young backarc basin. Basement depths and sediment thicknesses more closely match those of the Aleutian Basin known to be underlain by Mesozoic crust (Marlow *et al.*, 1990) than those for the Japan Sea Basin where the sediment thickness is generally not as great and basement is 1 to 2 s shallower. These observations and comparisons would suggest the Kuril Basin is older than the minimum Miocene–Oligocene age limit based on the dredged sedimentary rocks.

The lack of reversal-type magnetic lineations in the Kuril backarc basin can be explained in two ways: The crust has been demagnetized by heating, or the basin was completely formed during a period of single magnetic polarity. The only reasonable time period for long single-magnetic polarities is the late Cretaceous.

We conclude that the Kuril backarc basin is floored by oceanic crust and that it was formed by backarc spreading within the southwestern margin of the Mesozoic continental structure of the Okhotsk Sea plate, behind the Kuril arc in response to Pacific plate subduction. Based on the broadly distributed, rather uniformly thick and undisturbed sediment fill, it apparently opened rapidly and has since remained inactive. The greater width of the basin at its SW end, narrowing to closure at the south tip of Kamchatka, suggests that the basin opened about a rotation pole very near the south end of Kamchatka. The transverse shallow basement ridges which strike across the basin and the NNW striking basement structures which extend between Sakhalin and Hokkaido at the SW end of the basin are believed to be shear/lateral fault zones that defined the opening direction. Displacement on the onland portion of these faults is documented during Oliogene(?)–Miocene age (Jolivet *et al.*, 1992). The Mesozoic age of the transverse basement ridges indicates that these features are shear/rafted remnants of the prebasin Kuril arc–Okhotsk Sea plate basement.

While we would like to be more definitive about the age of opening for the basin, the only objective conclusion we can reach is that the basin opened over a short period, sometime between late Cretaceous and Miocene time. The higher deposition rates and the known age of lateral faulting from Sakhalin to Hokkaido support a Miocene age, while the lower deposition rates, the depth of basement, and the lack of magnetic reversal lineations support a late Cretaceous age.

Acknowledgments

The authors thank Dave Scholl for his many useful comments, Larisa Khankishieva for her help in drafting illustrations, and Sandy Dunham for typing the manuscript. This is Geodynamics Research Institute contribution no. 98.

REFERENCES

Andreyev, A. A., and Vorobyev, V. M. 1991. On the tectonics of the Okhotsk Sea region in terms of geomagnetic data, *Pac. Geol.* **1**:27–33. (in Russian)

Averynova, V. N. 1975. *Deep Seismotectonics of Island Arcs (Northwest Pacific)*, Moscow, Nauka. (in Russian)

Balakina, L. M. 1979. Orientation and ruptures and movements in the sources of strong earthquakes of the north and northwest parts of the Pacific ocean, *Phys. Earth* **4**:43–52. (in Russian)

Balakina, L. M. 1981. Mechanism of intermediate earthquakes in the Kuril–Kamchatka focal zone, *Phys. Earth* **8**:3–24. (in Russian)

Belousov, V. V. 1982. *Transition Zones between Continent and Ocean*, Nedra, Moscow. (in Russian)

Belousov, V. V., and Pavlenkova, N. I. 1986. Interaction of crust and upper mantle, *Geotectonics* **6**:8–20. (in Russian)

Bezrukov, P. L. 1955. On the distribution and accumulation rate of siliceous sediments in the Okhotsk Sea, *Proc. Acad. Sci. USSR* **103**:433–476. (in Russian)

Bezrakov, P. L. 1960. Bottom sediments of the Okhotsk Sea, *Proc. Inst. Oceanol. USSR* **32**:15–95. (in Russian)

Bikkenina, S. K., Anosov, G. I., Argentov, V. V., and Sergeyev, K. F. 1987. *Crustal Structure of the Okhotsk Sea Southern Part According to Seismic Data*, Moscow, Nauka. (in Russian)

Bogdanov, N. A. 1988. *Tectonics of Marginal Sea Deep-Sea Basins*, Moscow, Nedra. (in Russian)

Burckle, L. H., Brunner, C. A., Alexandrovich, J., DeMenocal, P., Briscoe, J., Hamano, Y., Heusser, L., Ingle, J. C., Jr., Kheradyar, T., Koizumi, I., Krumsdiek, K. A. O., Ling, H.-Y., Muza, J. P., Rahamn, A., Sturz, A., Vigliotti, L., White, L. D., Wippern, J. J. M., and Yamanoi, T. 1992. Biostratigraphic and biochronologic synthesis of Legs 127 and 128: Sea of Japan, in *Proc. ODP, Sci. Results*, 127/128, Pt. 2 (K. Tamaki, K. Suyehiro, J. Allan, M. McWilliams *et al.*, eds.), p. 1228, Ocean Drilling Program, College Station, TX.

Chapman, M. E., and Solomon, S. C. 1976. North American–Eurasian plate boundary in north-east Asia, *J. Geophys. Res.* **81**:921–930.

DeMets, C. 1992a. Oblique convergence and deformation along the Kuril and Japan trenches, *J. Geophys. Res.* **97**:17,615–17,617, 17,625.

DeMets, C. 1992b. A test of present-day plate geometries for north-east Asia and Japan, *J. Geophys. Res.* **97**:17,627–17,635.

Den, N., and Hotta, H. 1973. Seismic refraction and reflection evidence supporting plate tectonics in Hokkaido, *Pap. Meteorol. Geophys.* **24**(1):31–54.

Ermakov, V. A. 1991. The origin of the Kuril deep-sea basin, *Pac. Geol.* **1**:34–49. (in Russian)

Far East Seismological Bulletin. 1986–1989. Institute of Marine Geology & Geophysics Publisher, Russian Academy of Sciences, Yuzhno-Sakhalinsk. (in Russian)

Galperin, E. I., and Kosminskaya, I. P. (Eds.). 1964. *Structure of the Earth's Crust in the Transition Zone from the Asian Continent to the Pacific Ocean*, Moscow, Nauka. (in Russian)

Geological Map of the Japan and Kuril Trenches and the Adjacent Areas. 1978. Marine Geology Map Ser., 11, Geol. Soc. Japan.

Gnibidenko, H. S., Gorbachev, S.Z., Lebedev, M. M., and Marakhanov, V. I. 1974. Geology and deep structure of Kamchatka peninsula, *Pac. Geol.* **7**:1–32.

Gnibidenko, H. S., and Snegovskoy, S. S. 1975. Structural relations between Sakhalin and Hokkaido, *Proc. Acad. Sci. USSR* **224**:1391–1394. (in Russian)

Gnibidenko, H. S., Bykova, T. G., Veselov, O. V., Vorobiev, V. M., and Svarichevsky, A. S. 1983. The tectonics of the Kuril-Kamchatka deep-sea trench, in *Geodynamics of the Western Pacific–Indonesian Region* (T. W. C. Hilde and S. Uyeda, eds.), pp. 249–285, Geodyn. Ser., Vol. 10, American Geophysical Union, Washington, DC.

Gnibidenko, H. S. 1985. The Sea of Okhotsk-Kuril islands ridge and Kuril–Kamchatka trench, in *The Ocean Basins and Margins*, Vol. 7A, *The Pacific Ocean* (A. E. S. Nairn, F. G. Stehli, and S. Uyeda, eds.), pp. 377–418, Plenum Press, New York.

Gnibidenko, H. S., and Ilyev, A. Ya. (Eds.). 1992. *Catalogue of Dredging Stations in the Okhotsk Sea*, Yuzhno-Sakhalinsk, Institute of Marine Geology and Geophysics. (in Russian)

Gnibidenko, H. S., and Svarichevsky, A. S. 1984. Tectonics of the South Okhotsk deep-sea basin, *Tectonophysics* **102**:225–244.

Gretskaya, E. V. 1990. *Initial Oil and Gas Parent Potential of Organic Matter of Sediments with Reference to the Okhotsk Sea Basins*, Vladivostok, Nauka. (in Russian)

Harland, W. B. (Ed.). 1985. *A Geological Time Scale*, Moscow, Mir. (in Russian)

Hilde, T. W. C., Uyeda, S., and Kroenke, L. W. 1977. Evolution of the western Pacific and its margins, *Tectonophysics* **38**:145–165.

Honza, E. (Ed.). 1978. *Geological Investigation of the Okhotsk and Japan Sea off Hokkaido June–July 1978 (GH77-3 Cruise)*, Hisamoto Geological Survey of Japan.

Ilyev, A. Ya., Voronova, V. A., Zakharova, M. A., Nesterova, O. N., Tarakanova, L. I., Sheremetieva, G. N., and Shustov, L. N. 1979. *Bottom Sediments of the Okhotsk Sea Southern Part*, Moscow, Nauka. (in Russian)

Jolivet, L., Fournier, M., Huchon, P., Pozhdestvenskiy, V., Sergeyev, K. F., and Oscorbin, L. S. 1992. Cenozoic intracontinental dextral motion in the Okhotsk–Japan Sea region, *Tectonics* **11**:968–977.

Jolivet, L., Huchon, F., and Rangin, C. 1989. Tectonic setting of Western Pacific marginal basins, *Tectonophysics* **160**:23–47.

Karig, D. E. 1971. Origin and developments of marginal basins in the Western Pacific., *J. Geophys. Res.* **76**:2542–2561.

Kiminami, K., and Kontani, Y. 1983. Mesozoic arc trench system in Hokkaido, Japan, in *Accretion Tectonics in the Circum-Pacific Regions* (M. Hashimoto and S. Uyeda, eds.), pp. 107–122, Terrapub, Tokyo.

Kimura, G. 1986. Oblique subduction and collision: forearc tectonics of the Kuril arc, *Geology* **14**:404–407.

Kogan, M. G. 1975. Gravity field of the Kuril-Kamchatka arc its relation to the thermal regime of the lithosphere, *J. Geophys. Res.* **80**:1381–1390.

Krasny, M. L. 1990. *Geophysical Fields and Deep Structure of the Okhotsk-Kuril Region*, Vladivostok, Nauka. (in Russian)

Lundgren, P. R., and Giardini, D. 1990. Lateral structure of the subducting Pacific plate beneath the Hokkaido corner from intermediate and deep earthquakes, *Pure Appl. Geophys.* **134**:385–404.

Maeda, J. 1990. Opening of the Kuril basin deduced from magmatic history of central Hokkaido, North Japan, *Tectonophysics* **174**:235–255.

Marlow, M. S., Cooper, H. K., Dodisman, S. V., Geist, E. L., and Carlson, P. R. 1990. Bowers swell: Evidence for a zone of compressive deformation concentric with Bowers Ridge, Bering Sea, *Mar. Petrol. Geol.* **7**:398–409.

Matsuda, T., and Uyeda, S. 1971. On the Pacific-type orogeny and its model-extension of the paired belts concept and possible origin of marginal sea, *Tectonophysics* **11**:5–27.

Melnikov, O. A. 1988. *Geology of Sakhalin–Hokkaido Folded Belt*, Vladivostok, Nauka. (in Russian)

Popov, A. A., and Anosov, G. I. 1978. New data on crustal structure of the Kuril Basin, *Proc. Acad. Sci. USSR* **240**:166–168. (in Russian)

Okada, H. 1982. Geological evolution of Hokkaido, Japan: An example of collision orogenesis, *Proc. Geol. Assoc.* **93**:201–212.

Riegel, S. A., Fujita, K., Kozmin, B. M., Imaev, V. S., and Cook, D. B. 1993. Extrusion tectonics of the Okhotsk plate, north-east Asia, *Geophys. Res. Lett.* **20**:607–610.

Rozhdestvensky, V. S. 1986. Evolution of the Sakhalin fold system, *Tectonophysics* **127**:331–339.

Saidova, H. M. 1960. Regulations of foraminiferal distribution in the Okhotsk Sea bottom sediments, *Proc. Inst. Oceanol. USSR* **32**:96–159. (in Russian)

Savostin, L., Zonenshain, L., and Baranov, B. 1983 Geology and plate tectonics of the Sea of Okhotsk, in *Geodynamics of the Western Pacific–Indonesian Region* (T. W. C. Hilde and S. Uyeda, eds.), pp. 189–221, Geodyn. Ser., Vol. 11, American Geophysical Union, Washington, DC.

Scholl, D. W., and Creager, J. S. 1973. Geologic synthesis of Leg 19 (DSDP) results. Far North Pacific, and Aleutian Ridge, and Bering Sea, in *Init. Repts. DSDP*, 19 (J. S. Creager, D. W. Scholl *et al.*, eds.), pp. 897–913, U.S. Government Printing Office, Washington, DC.

Seismological Bulletin. 1990–1993. Russian Academy of Sciences, Obninsk. (in Russian)

Simbireva, I. G., Fedotov, S. A., and Feofilaktov, V. D. 1976. Heterogeneities of the stress field in the Kuril-Kamchatka arc as derived from seismological data, *Geophys. Geol.* **1**:70–86. (in Russian)

Snegvskoy, S. S. 1974. *Seismic Reflection Studies and Tectonics of the Okhotsk Sea Southern Part and the Adjacent Pacific Margin*, Nauka, Novosibirsk. (in Russian)

Soloviev, V. V., I. K. Tuezov, I. K., and Vasilyev, B. I. 1977. The structure and origin of the Okhotsk and Japan Sea abyssal depressions according to new geophysical and geological data, *Tectonophysics* **37**:153–166.

Starshinova, E. A. 1980. Crustal and upper mantle structure inhomogeneity of the Okhotsk Sea, *Proc. Acad. Sci. USSR* **255**:1349–1343.

Svarichevsky, A. S., and Svarichevskaya, L. V. 1982. The Okhotsk Sea slope relief in the southern part of the Main Kuril Ridge, in *Relief and Volcanism of Kuril Island Arc System* (B. V. Ezhov, ed.), pp. 54–62, Vladivostok, Nauka. (in Russian)

Sychev, P. M., and Snegovskoy, S. S. 1976. Abyssal depression of the Okhotsk, Japan and Bering Seas, *Pac. Geol.* **11**:57–80.

Sychev, P. M., Soinov, V. V., Veselov, O. V., and Volkova, N. A. 1983. Heat flow and geodynamics of the transition zone from Asia to the North Pacific, in *Geodynamics of the Western Pacific–Indonesian Region* (T. W. C. Hilde and S. Uyeda, eds.), pp. 237–247, Geodyn. Ser., Vol. 11, American Geophysical Union, Washington, DC.

Takanashi, M. 1983. Space-time distribution of late Mesozoic to early Cenozoic magmatism in East Asia and its tectonic implications, in *Accretion Tectonics in the Circum-Pacific Region*. (M. Hashimoto and S. Uyeda, eds.), pp. 69–87, Terrapub, Tokyo.

Tarakanov, R. Z. and Kim, Chung Un. 1983. Seismofocal zones and geodynamics of the Kuril-Japan region, in *Geodynamics of the Western Pacific–Indonesian Region* (T. W. C. Hilde and S. Uyeda, eds.), pp. 223–236, Geodyn. Ser., Vol. 11, American Geophysical Union, Washington, DC.

Tarakanov, R. Z., and Leviy, N. V. 1968. A model for the upper mantle with several channels of lower velocity and strength, in *The Crust and Upper Mantle of the Pacific Area* (G. H. Sutton, M. H. Manghnani, and R. Moberly, eds.), pp. 43–50, Geophys. Monogr., American Geophysical Union, Washington, DC.

Tereschenkov, A. A., Baboshina, V. A., Tuezov, I. K., and Kharahinov, V. V. 1988. Structure of the Okhotsk Sea region anomalous magnetic fields, *Geophys. Geodyn. Res.* **10**:10–19. (in Russian)

Tuezov, I. K., Veselov, O. V., and Lipina, E. N. 1984. *Heat Flow of Asia, Australia and Western Pacific*, Vladivostok, Nauka. (in Russian)

Vassilkovsky, N. P. 1967. On the geological nature of the Pacific mobile belts, *Tectonophysics* **4**:583–593.

Veselov, O. V., and Volkova, N. A. 1981. Radioactivity of rocks in the Okhotsk Sea region, in *Geophysical Fields of the Transition Zone of Pacific Type* (M. L. Krasny, ed.), pp. 51–709, Vladivostok, Nauka. (in Russian)

Volkova, N. A. 1982. The thermoconductivity model of the crust of the Okhotsk Sea region, *Geol. Geophys.* **5**: 92–97. (in Russian)

Watson, B. F., and Fujita, K. 1985. Tectonic evolution of Kamchatka and the Sea of Okhotsk and implications for the Pacific basin, in *Tectonostratigraphic Terranes of the Circum-Pacific Region* (D. G. Howell, ed.), pp. 333–336, Circum-Pacific Council Energy and Mineral Resources, Houston, TX.

Zhang Zhi, and Lay, T. 1993. Investigation of upper mantle discontinuities near northwestern Pacific subduction zones using precursors to SSH, *J. Geophys. Res.* **98:**4389–4405.

Zhuze, A. P. 1957. Diatomic in the surface layer of the Okhotsk Sea sediments, *Proc. Inst. Oceanol. USSR* **22:**164–220. (in Russian)

Zlobin, T. K., and Zlobina, L. M. 1991. Crustal structure of the Kuril Island arc system, *Pac. Geol.* **6:**24–35. (in Russian)

Zonenshain, L. P., and Savostin, L. A. 1979. *Introduction to Geodynamics*, Moscow, Nedra. (in Russian)

Zverev, S. M., and Tulina, Yu. V. (Eds.). 1971. *Deep Seismic Sounding of the Earth's Crust in the Sakhalin-Hokkaido-Primorye Zone*, Moscow, Nauka. (in Russian)

13

Hydrothermal Activity Related to Arc-Backarc Magmatism in the Western Pacific

Jun-ichiro Ishibashi and Tetsuro Urabe

ABSTRACT

A compilation of 27 sites of hydrothermal mineralization in the western Pacific was made in terms of their tectonic settings, mineral commodity, and fluid chemistry. These sites constitute about 20% of the known seafloor hydrothermal sites, which are dominated by those occurring at mid-ocean ridges (MORs). High-temperature hydrothermal activities in the western Pacific exclusively occur in association with submarine volcanism found either on backarc spreading centers, backarc rifts, or volcanic fronts in arc-backarc systems. Magmatic contribution to the hydrothermal systems in arc-backarc settings is more obvious than those in MORs, and several lines of evidence suggest direct interactions between the magma chamber and hydrothermal fluids.

Hydrothermal activity at backarc spreading centers shares several common chemical, mineralogical, and morphological features with those in MORs. This is understandable because of small differences in parameters, such as magma signature and composition of oceanic crust, between those two systems. Frontal-arc volcanoes are proved to be the sites of vigorous high-temperature hydrothermal activity. Several investigated submarine calderas in the Izu–Bonin (Ogasawara) and Mariana arcs have signs of present or recent activity, indicating ubiquitous hydrothermal mineralization regardless of their magma composition. Backarc rifting centers are particularly interesting because of their analogous tectonic settings with the Japan arc of Miocene age where many Kuroko-type volcanic massive sulfide deposits (VMSDs) formed in association with bimodal volcanisms. Some hydrothermal sites in the Okinawa Trough backarc basin represent modern examples of such mineralization.

The wide variety in modes of occurrence of hydrothermal systems in the western Pacific region, in contrast to those in the MOR setting, reflects diversity in magma composition, contribution from heterogeneous island-arc crust rather than homogeneous oceanic crust, and the unique tectonic setting of the arc-backarc system. These systems

---

Jun-ichiro Ishibashi • Laboratory for Earthquake Chemistry, Faculty of Science, University of Tokyo, Bunkyo-ku, Tokyo 113, Japan. *Tetsuro Urabe* • Geological Survey of Japan, Tsukuba 305, Japan.

Backarc Basins: Tectonics and Magmatism, edited by Brian Taylor, Plenum Press, New York, 1995.

provide the best field to investigate analogs to VMSDs, because most ancient VMSDs are hosted by the rock types in a subduction-related volcanic suite or an extension-related bimodal suite representative of island-arc- or continental-rift-related tectonic settings.

1. INTRODUCTION

Five or six years after the discovery of hydrothermal mineralization at 21°N East Pacific Rise (EPR; e.g., Francheteau *et al.*, 1979; Spiess *et al.*, 1980), reports of similar seafloor hydrothermal activity were noted in the western Pacific (e.g., Both *et al.*, 1986; Hawkins, 1986; Hawkins and Helu, 1986; Craig *et al.*, 1987a,b; Urabe *et al.*, 1987). Continued exploration efforts in the last decade have shown that the western Pacific region has as much seafloor hydrothermal mineralization associated with it as with the mid-ocean ridges (MORs). Hydrothermal mineralization in the western Pacific appears to have wider variations in tectonic setting, mode of occurrence, and chemistry of fluid and resultant deposits than those of MORs.

Prior to 1970, the number of known seafloor hydrothermal mineral deposits was only one, that discovered in the Red Sea in 1966 (Miller *et al.*, 1966). But it had increased to 50 by 1983 (Rona, 1983), 80 by 1988 (Rona, 1988), and to about 140 in the latest compilation (Rona and Scott, 1993). By contrast, the number of the arc-backarc-related hydrothermal occurrences in the western Pacific compiled in this chapter is 27, thus comprising about 20% of all known seafloor hydrothermal occurrences. It is notable that this estimation would be biased from the actual distribution, because only a small portion of the seafloor has been investigated. Indeed, in recent years, the Japanese deep submergence research vessel (DSRV) *Shinkai 2000* and *Shinkai 6500*, the American DSRV *Alvin*, the French DSRV *Nautile*, and the Russian DSRV *Mir* have been deployed energetically and widely in the western Pacific, away from their home countries, and have contributed to successive discoveries of hydrothermal occurrence.

This chapter summarizes contemporary hydrothermal mineralization in arc-backarc settings in the western Pacific region. Several occurrences of hot-spring activity around shallow submarine volcanoes and volcanic islands, such as the Esmeralda Bank of the Mariana arc (Stüben *et al.*, 1992) and the Wakamiko caldera in the Kagoshima Bay in southwest Japan (Nedachi *et al.*, 1991), low-temperature hydrothermal manganese oxides on frontal volcanoes (e.g., Moorby *et al.*, 1984; Usui *et al.*, 1986), and cold seepage from accretionary prism of the Japan Trench and Nankai Trough (e.g., Taira and Pickering, 1991) are excluded because of their less important nature in terms of metal concentrations. However, cold venting found at serpentinite seamounts of the Mariana forearc (Fryer *et al.*, 1987b, 1990) is discussed because of the formation of aragonite chimneys.

Relevant comparative data from MOR hydrothermal systems are not cited here; readers are referred to excellent review papers by Rona (1983, 1984, 1988), Bäcker and Lange (1987), von Damm (1990), and Rona and Scott (1993).

2. TECTONIC SETTING AND ASSOCIATED VOLCANISM

Seafloor hydrothermal mineralization has occurred throughout the history of the Earth. The oldest known example is at Isua, Greenland, where copper sulfide layers occur concordantly within a banded iron formation (Appel, 1979) in a supracrustal belt of about 3.8 Ga (Compston *et al.*, 1986). More than 500 VMSDs are known on land. The VMSDs are found in association with volcanism of virtually every lithologic type: felsic to ultramafic rocks in predominantly volcanic or mixed volcanic-sedimentary environments.

Since the 1970s, many attempts have been made to relate the geologic setting of VMSDs to certain tectonic environments such as MORs, volcanic arc-subduction zones, and marginal basins (e.g., Sawkins, 1972; Sillitoe, 1972; Mitchell and Bell, 1973; Baranov and Levin, 1993). There is unanimous agreement that most of the major Phanerozoic VMSDs formed in island-arc and/or backarc settings rather than at MORs (e.g., Sawkins, 1972; Franklin *et al.*, 1981; Scott, 1985). Rona (1988) indicated that basalt-hosted VMSDs that might occur at spreading centers of MORs or backarc basins constitute only 17% of the 508 deposits of his compilation, and that the majority of deposits (56%) are hosted by rhyolite, which is considered to form in a subduction-related volcanic suite or to an extension-related bimodal basalt-rhyolite suite within an island arc or continental rift. Because of the range of acidic to basic submarine volcanism existing in diverse tectonic settings, the western Pacific provides the best field test as to whether or not present-day VMSDs are forming in similar settings to those formed during the Phanerozoic.

In the western Pacific, many backarc basins have been accreted and preserved along the margin of the Pacific plate since late Cretaceous time (Kroenke, 1984; Honza, 1991; Tamaki and Honza, 1992). About 75% of backarc basins worldwide occur in the western Pacific region and six of them have active spreading centers: Mariana Trough, Andaman Sea, Manus Basin, Woodlark Basin, North Fiji Basin, and Lau–Havre Basin (cf. Tamaki and Honza, 1992). These backarc spreading centers are the most important locations of submarine volcanism comparable to that at MORs. In contrast, the majority of present-day arc-trench systems in the western Pacific lack an associated backarc basin (Taylor and Karner, 1983). Volcanism in such arc-trench systems peaks on the volcanic front, which is located on the trench-side border of the active volcanic sites. Other tectonic units where sporadic volcanoes are observed include the forearcs, intra-arc rifts, and backarc rifts. Since the mechanism of this rifting activity is one of the least understood aspects of the evolution of arc-backarc systems, it is in practice difficult to categorize without ambiguity every known volcano into a simple tectonic regime such as forearc, volcanic front, backarc rift, or backarc spreading center.

Regardless of their tectonic category, it is quite reasonable to expect vigorous hydrothermal activity at most of the active volcanic centers in arc-backarc systems. Rifting of an island-arc or continental crust is likely to provide a favorable setting for the circulation of seawater eventually evolving as a hydrothermal fluid. This, combined with the fact that a submarine volcano has a shallow magma chamber inside the edifice will provide an ideal structure for the hydrothermal system to focus high-temperature venting at the summit (Scott, 1985, 1987). Backarc spreading centers show several similarities to MOR systems where hydrothermal activity is widely distributed. This variability in possible location makes investigation of hydrothermal activity in the western Pacific more complex but potentially more interesting than in those arguably simple tectonic settings, such as MORs and hot spots.

This chapter reviews some major sites of hydrothermal mineralization associated with three major tectonic settings: backarc spreading center, backarc rift, and volcanic front. These, and other miscellaneous examples, are briefly summarized in Table I together with

TABLE I
Hydrothermal Mineral Occurrences in Arc-Backarc Systems in the Western Pacific

Location No. location (Fig. 13.1)	Water depth (m)	Geologic structure	Type of deposit	Mineralogy
1. Backarc spreading center				
1 North Fiji Basin "Station 4" (16°59′S, 173°55′E)	1980	Axial graben at topographic high of N.-central segment near triple junction. Sheet lava floor.	Active (T = 290°C) anhydrite chimneys standing on dead sulfide mound. Forest of dead sulfide chimneys.	Anhydrite, amorphous silica in dead chimneys; pyrite, marcasite chalcopy., sphal., wurtz., goethite.
2 North Fiji Basin "Station 14" (18°50′S, 173°30′E)	2720	Collapsed lava lake on flat rise crest of fast spreading S-central segment. No sediment cover.	Warm (T = 5.2°C) fluid discharge through mussel bed. No hydrothermal minerals. Site of megaplume (Nojiri *et al.*, 1989).	None.
3 Fiji transform fault "Extensional Relay Zone A" (16°10′S, 177°′25E)	1860–2335	Short spreading ridge axis which displaces Fiji transform fault as interpreted by Jarvis *et al.* (1994)	Hydrothermal sulfide impregnation in MORB-like basalt dredged from axial valley.	Magnetite, pyrrhotite, chalcopyrite and opal on fracture surface.
4 Central Manus Basin "Vienna Woods" (3°10′S, 150°17′E)	2500	2-km-wide axial rift graben of the northeast spreading center. Mostly massive pillow lava floor.	Sulfide chimneys up to 20 m high are venting clear, milky and black fluids. Sulfate smokers are also present.	Sphalerite, wurtzite, pyrite, marcas, chalcopy, galena, am. silica, barite. Sulfate chimney; anhy., silica, barite.
5 Eastern Manus Basin "DESMOS cauldron" (3°42′S, 151°52′E)	2000	Caldera of basalt/basaltic andesite at an intersection of a spreading center and a transform fault?	Sulfide ores were not recovered. Megaplume-like methane anomalies in water column over the caldera.	Ferruginous oxide deposits. Pyrite and native sulfur disseminated in basaltic andesite.
6 Eastern Manus Basin "PACMANUS field" (3°42′S, 151°42.6′E)	1650	Crest of a prominent ridge of dacite flows and domes called Pual Ridge which exists in pull-apart basin.	4-m-high sulfide chimney venting "smoke" (no temperature information). Only one ore chip (1 cm) was recovered.	Cohesive anhydrite, chalcopy., bornite, tennantite, and sphalerite.

7 Valu Fa Ridge, Lau Basin "Vai Lili site" (22°18′S, 176°35′W)	1700	400-m-long zone along a normal fault on a ridge crest. Basaltic andesite to rhyodacite lava.	Black and white smokers venting up to 400°C fluid. Height up to 17 m. Massive Zn- and Cu-sulfide ores are present.	Pyrite, chalcopy., marcasite, sphal., barite, tennantite, galena, gratonite, native gold.
8 Valu Fa Ridge, Lau Basin "Hine Hina site" (22°33′S, 176°43′W)	1900	On south Valu Fa Ridge crest dominated by andesite and dacite. Strongly bleached volcanic rocks.	Low-temperature Mn-oxide crust is covering high-temperature fossil sulfide deposits. Diffusive fluid (T = 40°C).	d-MnO_2, birnessite, todorokite, amorphous Fe-oxide, goethite, sphalerite, barite, pyrite, chalcopy.
9 Valu Fa Ridge, Lau Basin "White Church site" (21°55′S, 176°32′W)	1946–1966	At the top and along the flanks of Northern Valu Fa Ridge. Brecciated and pillow lava, and pumice field.	10–15-m-high inactive barite-sulfide chimney field extends over 300 m along normal faults.	Barite and sphalerite are dominant with lesser galena, tennantite, chalcopy, gold, and pyrrhotite.
10 Northeastern Lau Basin "Papatua expedition site" (15°17′S, 174°45′W)	2100	Axial region of northeasterly trending active spreading ridge of the northern Lau Basin.	Dredged black smoker chimney samples.	Wurtzite, pyrite, chalcopyrite, barite, and amorphous silica. Thin film of Mn-oxyhydroxide.
11 Peggy Ridge, Lau Basin (16°55′S, 176°49.5′W)	1664–1900	At a fracture zone offsetting northwesterly trending active spreading center of the northern Lau Basin.	Dredged hyaloclastite basalt fragments which are cemented by opal and euhedral barite crystals.	Opal, barite, montmorillonite, phillipsite (oxygen-isotope study suggests temp. of 117–134°C).
12 Western Woodlark Basin "Franklin Seamount" (9°55′S, 151°50′E)	2143–2366	Westernmost propagating tip of spreading center. Basaltic andesite and inferred sodic rhyolite.	Spires and mounds of Fe-Mn-Si oxide up to several meters thick and 200 m in extent. Venting 20–30°C clear solution.	Inactive barite silica chimneys contain up to 21 ppm Au. Si-bearing Fe oxyhydroxide.
13 Central Mariana Trough "Alice Spring field" and other (18°12 E, 143°30′E)	3600–3700	Flank of an axial volcano of basaltic andesite on backarc spreading center. Volcanics are fractionated.	Active barite-rich sulfate chimneys, 1 m high, venting clear fluid (T = 287°C).	Barite chimneys. Sulfides; abundant sphal with lesser amounts of galena, chalcopy, pyrite.
14 Central Mariana Trough (18°02 N, 144°45′E)	3675	Crest of an axial ridge where the relief is greatest (ca. 800 m). A low mound, 20–30 m in diameter.	Several active chimneys surrounded by a low mound of hydrothermal precipitates.	Sphalerite, barite, amorphous silica.

(continued)

TABLE I
(Continued)

Location No. location (Fig. 13.1)	Water depth (m)	Geologic structure	Type of deposit	Mineralogy
2 Backarc rift				
15 Sumisu rift Izu–Ogasawara (Bonin) arc (31°06′N, 139°54′E)	1530–1600	On the flank of a rhyolite dome of en echelon ridge of bimodal (rhyolite-BABB) volcanism.	Several tens of fossil barite-silica chimneys with thin manganese oxyhydroxide coating.	Barite, amorphous silica, opal-CT (oxygen isotope study suggests temp. of formation <150°C).
16 Northern Okinawa Trough "Minami-Ensei Knoll" (28°23.5′N, 127°38.5′E)	690–705	Volcanic depression and hills where pumice, diorite and granite fragments are recovered.	Several anhydrite chimneys venting clear fluid (T = 267–278°C) are observed on barite-sulfide mound.	Anhydrite chimneys. Sphal., wurtz., tetrahedrite, chalcopy., galena, pyrite, barite, bornite, covellite
17 Middle Okinawa Trough "Iheya Ridge (CLAM site)" (27°33′N, 126°58′E)	1400	Vesicular pillow basalt ridge located on the axial part of backarc rift of the Okinawa Trough.	Carbonate-rich hydrothermal precipitates with disseminated sulfides. Fluid emanation (T = 100–220°C) from fissures.	Calcite, rhodocrosite, anhydrite, wurtzite, pyrrhotite, galena, isocubanite, chalcopyrite, argentite.
3 Volcanic front				
18 Middle Okinawa Trough "Izena Cauldron (JADE site)" (27°16′N, 127°05′E)	1300–1550	Northeastern slope of a sediment-rich rectangular depression with rhyolitic lava domes.	Active black smoker (T = 320°C), white smokers and sulfide crusts. Liquid-Co_2-hydrate bubbling are also observed.	Sphalerite, tetrahedrite, galena, barite, chalcopyrite, pyrite, anhydrite, stibnite and other sulfides.
19 Izu–Ogasawara (Bonin) arc "Kita-Bayonnaise Caldera" (32°06′N, 139°51′E)	1300–1400	A submarine caldera with diameter of 5–7 km and has dacitic central cone.	Submersible dive failed to locate active discharge but box coring recovered barite-sulfide fragments from the floor.	Barite, pyrite, chalcopyrite, sphalerite, tetrahedrite-tennantite, epidote, carbonate, smectite.
20 Izu–Ogasawara (Bonin) arc "Myojinsho Caldera" (31°53′N, 139°59′E)	1000–1100	Partly emerged submarine caldera with flat-topped central cone of dacite.	Massive barite and sulfide samples were recovered from the caldera floor sediments.	Barite, pyrite, and other minor sulfides.
21 Izu–Ogasawara (Bonin) arc "Suiyo seamount" (28°34′N, 140°39′E)	1370	Calc-alkaline dacite seamount of low-K series. Dimension of the caldera is 1.5 × 0.8 km.	High-temperature (T = 230–311°C) fluid discharged from chimneys (<1m) in a field of 150 × 300 m on the caldera floor.	Chalcopyrite, sphalerite, anhydrite, barite. High Au content (up to 71 ppm).

22 Izu–Ogasawa (Bonin) arc "Kaikata seamount" (26°42′N, 141°05′E)	930	Andesitic caldera with basaltic central cone called KC caldera. Low-temp. activity on the cone.	Veinlets of sulfides in altered andesite. Water in caldera is contaminated with iron hydroxide.	Pyrite, sphalerite, Cu-Fe-sulfide (filling temp. of fluid inclusions are 290°C).
23 Northern Mariana arc "Kasuga 2 seamount" (21°35′N, 143°37′E)	402	Basalt and basaltic andesite volcano at the intersection of Mariana volcanic arc and a fracture.	Warm fluid (T = 39°C) rich in carbon dioxide and sulfur dioxide is venting from sulfur-encrusted volcanic sand.	Elemental sulfur, Fe- and Mn-oxide, nontronite.
24 Southern Mariana Ridge "Forecast Vent Field" (13°24′N, 143°55′E)	1470	At the summit of a seamount of island-arc basalt located about 17 km east of spreading axis.	White smoker chimneys emanating clear fluid (T = 202°C). Height of chimneys are about 1 m.	Anhydrite, barite, with lesser amounts of sulfide minerals.
4 Other seamounts				
25 Derugin Depression, east of Sakhalin Island, Okhotsk Sea (54°02′N, 146°16′E)	1460–1480	A knoll in volcano-tectonic depression. Submarine caldera? Strong hydrogen sulfide odor on sample.	Fragments of travertine-like barite aggregates were dredged. No temperature measurements.	Barite ($\delta^{34}S$ = +20 per mil, which suggests its seawater sulfate origin).
26 Komandorsky Basin "Piip volcano" (55°23′N, 167°27′E)		A volcanic cone of fresh dacite to the rear of a transform fault of Aleutian arc. Backarc rift?	Emission of volcanic gases (methane, hydrogen) and precipitation of nontronite and ferro- and manganese hydroxide.	Nontronite, Fe- and Mn-hydroxide found in dredged sample.
5 Forearc Area				
27 Mariana forearc region Conical seamount (19°33′N, 146°39′E)	1500	Mud volcano or solid intrusion of serpentinite in forearc region.	Cold seepage of dilute fluid from Mg-silicate chimneys. Aragonite chimneys are also common.	Aragonite, calcite and amorphous Mg-silicate, Mg-silicate, ferromanganese oxide

Major reference of each site: 1, 2. Bendel *et al.* (1993); Gracia *et al.* (1994); 3. von Stackelberg *et al.* (1990); 4. Tufar (1990); 5. Gamo *et al.* (1993c); 6. Binns and Scott (1993); 7–9. Fouquet *et al.* (1993); Lisitsyn *et al.* (1992); 10. Hawkins and Helu (1986); 11. Bertine and Keene (1975); 12. Lisitsyn *et al.* (1991); Binns *et al.* (1993); 13. Craig *et al.* (1987b); 14. Hawkins *et al.* (1990); 15. Urabe and Kusakabe (1990); 16. Nedachi *et al.* (1992); 17. Gamo *et al.* (1991b); 18. Halbach *et al.* (1989); 19. Iizasa *et al.* (1993); 20. Iizasa *et al.* (1992); 21. Watanabe and Kajimura (1993); 22. Urabe *et al.* (1987); 23. McMurtry *et al.* (1993); 24 Gamo *et al.* (1993a); 25. Astakhova (1992); 26 Baranov *et al.* (1991); 27. Fryer *et al.* (1990).

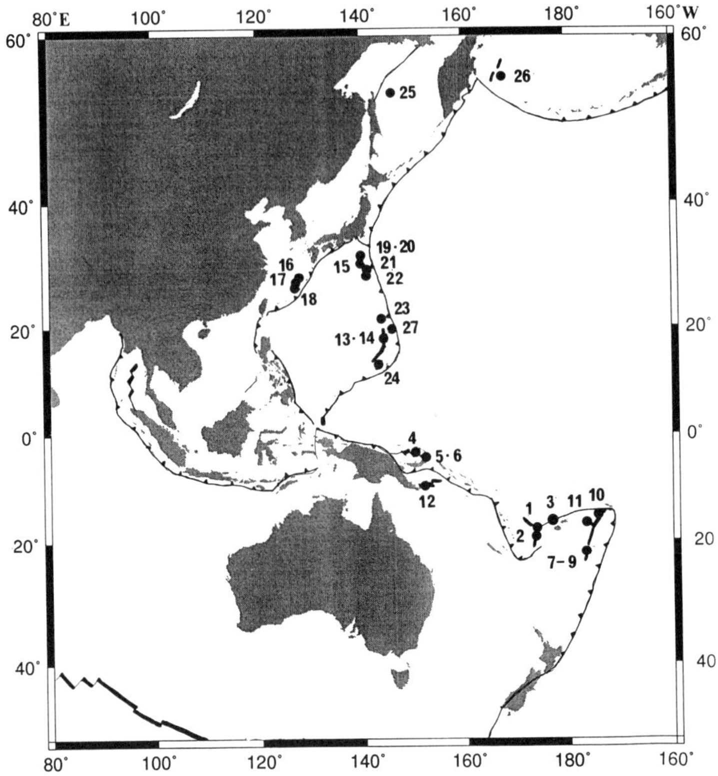

FIGURE 13.1. Hydrothermal mineral occurrences related to arc-backarc systems of the western Pacific. Numbers are listed in Table I.

their mineralogy and mode of occurrence. The locations of the study fields are plotted in Fig. 13.1.

3. CASE STUDIES

3.1. North Fiji Basin

The North Fiji Basin is an active, mature marginal basin in the southwestern Pacific (Fig. 13.2), which was investigated during the Japan–France–SOPAC cooperative project STARMER (e.g., Auzende *et al.*, 1989, 1990, 1991, 1992; Tanahashi *et al.*, 1991; Auzende and Urabe, 1994; Auzende *et al.*, Chapter 4 this volume). The 800-km-long linear backarc spreading center is comparable in size to a MOR spreading system (Tanahashi *et al.*, 1994). The Wadati–Benioff zone does not extend beneath the spreading center. Petrological studies of basalt samples have revealed features similar to N-MORB with some influence of coexisting oceanic island basalts (OIB) components (Eissen *et al.*, 1991, 1994; Nohara *et al.*, 1994). This spreading ridge is divided into four first-order segments: northern (N160°E), northern-central (N15°E), southern-central (N5°E), and southern offset (NS) segments (Tanahashi *et al.*, 1994).

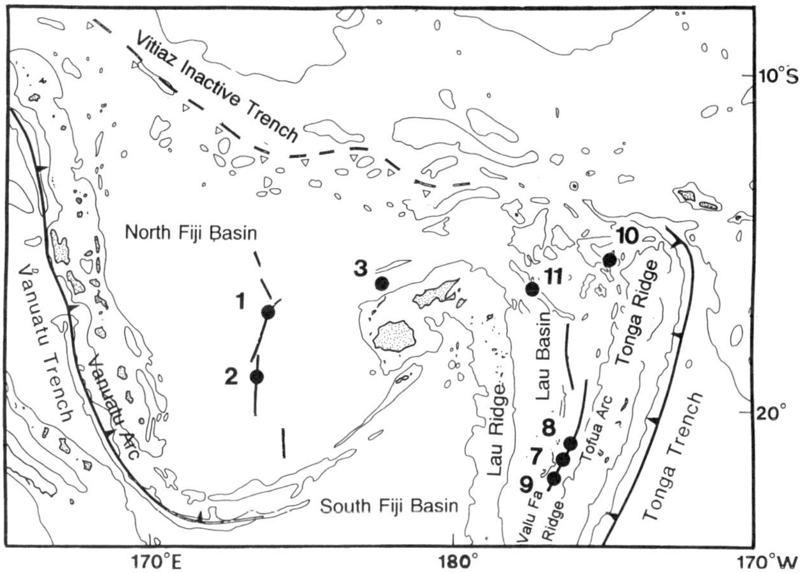

FIGURE 13.2. Location and tectonic setting of North Fiji and Lau basins. The solid circles mark the hydrothermal fields listed in Table I. The thick lines indicate backarc spreading centers, wheras the thin lines delinate contours of 2000 m water depth.

Hydrothermal activity along the North Fiji Basin spreading ridge has been denoted by chemical anomalies detected in hydrothermal plumes during the PAPATUA expedition of 1986 (Craig *et al.*, 1987a), the SEAPSO III cruise of 1986 (Auzende *et al.*, 1988), and the R/V *Moana Wave* cruise of 1987 (Sedwick *et al.*, 1990). Systematic investigation of the spreading ridge sites during the STARMER project, using a deep-tow TV camera system in conjunction with dives made by the DSRV *Nautile* and DSRV *Shinkai 6500*, confirmed two active hydrothermal sites (Honza *et al.*, 1989; Auzende *et al.*, 1991).

An active hydrothermal chimney known as "White Lady" occurs in the 2-km-wide axial graben on a bathymetric rise (location 1 on Fig. 13.2). This site is located at the northern tip of the northern-central (N15E) segment and adjacent to a ridge-ridge-fracture zone triple junction (Lafoy *et al.*, 1991). The White Lady consists exclusively of anhydrite and is developed on a 2-m-high sulfide mound. Other anhydrite chimneys were found at the STARMER II site about 150 m southwest of the White Lady chimney. The broad hydrothermal field named "Station 4" is distributed over a thinly sedimented fresh sheet lava flow (Bendel *et al.*, 1993). Dense hydrothermally supported fauna of balanoid, black gastropods, hairy gastropods, and mussels thrive around the chimneys (Desbruyeres *et al.*, 1994). These anhydrite chimneys were venting clear hydrothermal fluid of 230 to 290°C. Their fluid chemistry is characterized by low chlorinity (47–49% of seawater) and depletion in metal elements such as iron and manganese. This chemistry indicates segregation between brine and vapor beneath the seafloor and that the latter phase has condensed to form the hydrothermal fluid that was sampled (Grimaud *et al.*, 1991, Ishibashi *et al.*, 1994a). The hydrothermal activity at White Lady persisted for two years between the two diving cruises from 1989 to 1991 (Auzende *et al.*, 1992), with little change in the fluid chemistry (Ishibashi *et al.*, 1994a). In contrast to active sulfate mineral occurrences, no active sulfide chimneys have been observed. At the northern side of the White Lady site, a 2-km-wide, fossil hydrothermal area was discovered. The height of the sulfide chimneys in this "Pere Lachaise" area reaches 15 m, implying that intense black smoker hydrothermal activity previously existed in this same graben (Auzende *et al.*, 1991; Bendel *et al.*, 1993). The mineral assemblages of the sulfide mounds and of the fossil chimneys are very simple and resemble those found at mid-ocean ridges.

In the southern-central (N5°E) segment, diving revealed collapsed lava lakes on sediment-free axial depressions a few tens of meters wide and a few meters deep (Auzende *et al.*, 1992). Biological communities of mussels, galatheids, and synaptic holothurians exist in the collapsed lava lake at a depth about 2720 m (Auzende *et al.*, 1991). At this site, known as "Station 14" (location 2 on Fig. 13.2), a massive release of fluid as a megaplume was identified from a water column anomaly during the KAIYO 87 cruise (Nojiri *et al.*, 1989). However, only warm fluid emanations (5.2°C) were observed during the *Shinkai 6500* dives of 1993 (Gracia *et al.*, 1994). The low-temperature fluid samples showed slight chemical anomalies in SiO_2, Mn, Li, and CH_4 contents and helium isotopic composition, suggesting that the high-temperature fluid was diluted by a high degree of mixing with ambient seawater (Ishibashi *et al.*, 1994c).

3.2. Manus Basin

The Manus Basin, located in the Bismarck Sea, is a backarc basin situated to the north of the New Britain island arc–trench system (Fig. 13.3). Backarc spreading in this area is characterized by an exceptionally high spreading rate, greater than 10 cm/yr full rate (Taylor, 1979). Spreading centers in the western, central, and eastern basins are connected by transform faults.

The first discovery of hydrothermal deposits in the Manus Basin was made in 1986 via seafloor camera photographs (Both *et al.*, 1986). The photographs showed several inactive

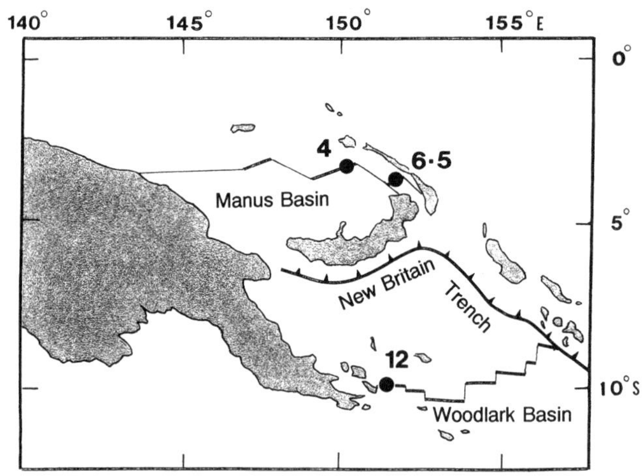

FIGURE 13.3. Location and tectonic setting of Manus and Woodlark basins. The solid circles mark the hydrothermal fields listed in Table I. The solid lines indicate backarc spreading centers and associated transform faults.

chimneys about 2 m in height with associated fauna composed of gastropods, galatheid crabs, and horny corals. This area, subsequently named "Red Star," is located on a small axial volcanic swell at a depth of about 2500 m inside the rift valley of the spreading center in the central basin. In 1990, the OLGA II expedition located a number of hydrothermal fields including four major fields, in this spreading ridge and retrieved several mineral samples (Tufar, 1990). At the same time as the OLGA II cruise, the DSRV *Mir* investigated some of these hydrothermal fields (Lisitsyn *et al.*, 1993; Shadlun *et al.*, 1993).

The active hydrothermal site named "Vienna Woods" (3°9.9′S, 150°16.8′E, depth 2490–2500 m; location 4 on Fig 13.3), extending for about 1000 m along strike of a N60°E direction on the Red Star swell, rests on pillow basalt lava without any sediment cover. Active and inactive massive sulfide chimneys occur close to one another, resembling tree trunks (Lisitsyn *et al.*, 1993). Several active fluid emissions were observed as colorless, milky, and, in places, black smoke. The highest-fluid-temperature measurement at one of the vent orifices was 276°C (Lisitsyn *et al.*, 1993). A fluid sample was substantially diluted with ambient bottom seawater (14% fluid vs. 86% ambient seawater), although the chemical composition was considered to be comparable to hydrothermal fluids from the hydrothermal systems in MORs. Near the area of high-temperature activity, sulfide precipitates and stockwork mineralization into oxides and hydroxides were noted. Mineral samples from the Vienna Woods and adjacent fields are characterized by sulfides dominantly rich in zinc, with lesser copper and locally lead, and by the occurrence of barite. Along the same spreading ridge where Vienna Woods site, two other active hydrothermal fields (3°6.7′S, 150°21.8′E, depth 2600 m and 3°22.2′S, 150°2.2′E, depth 2185 m) were located. These hydrothermal fields are restricted to the central graben of the ridge, and their linear distribution along a N60°E direction is attributed to fissures paralleling the spreading axis (Tufar, 1990).

Hydrothermal activity was also observed in some sites along the spreading ridge in the eastern basin. Water-column anomalies of CH_4 and helium-isotopic composition attributed to hydrothermal activity were first detected during the PAPATUA Expedition of 1986 (Craig *et al.*, 1987a). Gamo *et al.* (1993c) confirmed a large hydrothermal plume from CH_4, Mn, and Al anomalies during the AQUARIUS expedition in 1990. The plume was centered over an elongated caldera structure, named DESMOS cauldron, which has a relative depth of 200 m and a diameter of 1–2 km (location 5 on Fig. 13.3). A deep-tow camera and TV survey identified vent-associated biological colonies of squat lobsters, clams, and tube worms and bottom-temperature anomalies up to 0.26°C at the northern terrace. The hydrothermal dredged minerals were only native sulfur and disseminated pyrite attached to basaltic rock samples.

Another active hydrothermal site (PACMANUS field; location 6 on Fig. 13.3) in the eastern basin was identified about 20 km west of the DESMOS cauldron by dredging, sediment coring, and bottom photography during the PACMANUS I expedition of 1991 (Scott and Binns, 1992; Wheller *et al.*, 1992; Binns and Scott, 1993). Discontinuous hydrothermal deposits were observed throughout a 3-km by 800-m zone on the crest of Pual Ridge, where lavas range in chemical composition from andesite through dacite to rhyodacite. Ore fragments were recovered by the camera video sled, which had collided with a chimney. These fragments included anhydrite, chalcopyrite, and other sulfide minerals.

The spreading center of the western Manus Basin was surveyed during the R/V *Hakurei-Maru* No. 2 cruise in 1992. Large amounts of iron oxyhydroxide were dredged from the axial part of the ridge, but no high-temperature vent fields were discovered

(J. Date, pers. comm.). Recently, discovery of submarine hydrothermal venting was reported in the New Ireland forearc basin, which is in a rifted forearc setting behind Manus–Kilinailau trench (Herzig *et al.*, 1994). At volcanic cones on the southern flank of Lihir Island, distinct hydrothermal fields of vein mineralization contain clays, silica and disseminated sulfides, and clam colonies were identified.

3.3. Lau Basin

The Lau Basin is a typical example of an active backarc basin accreted between a remnant arc of Lau Ridge and the volcanic front of Tofua arc (Fig. 13.2). Its backarc spreading center is propagating southward as suggested by the triangular shape of the basin (von Stackelberg *et al.*, 1985, 1988; Hawkins, Chapter 3 this volume). The Valu Fa Ridge is 150 km long in the southern Lau Basin and considered to have started its spreading activity at 1 Ma with 6 cm/yr full spreading rate. This ridge is oriented asymmetrically within the basin, situated only 20 to 40 km west of the active Tofua volcanic arc. Small nontransform offsets and an overlapping spreading center divide the ridge into three major segments: the southern, central, and northern Valu Fa ridges (von Stackelberg *et al.*, 1988). Petrological studies show highly fractionated, diverse volcanic rock suites of backarc basin lava, i.e., basaltic andesite, andesite, and rhyodacite (e.g., Jenner *et al.*, 1987). These studies also show an increasing effect of the subducted slab component from north to south as the distance between the backarc spreading center and the Tofua volcanic arc narrows. Highly vesicular and autobrecciated lava dominate samples from this area, strongly suggesting that there is a high volatile content in the backarc magma.

Indications of hydrothermal activity were observed at the Valu Fa Ridge during the SO-35 cruise of 1984 and the SO-48 cruise of 1987 (von Stackelberg *et al.*, 1985,1988; Herzig *et al.*, 1990). Subsequent dives with the DSRV *Nautile* in 1989 located four hydrothermal sites (Fouquet *et al.*, 1990, 1991a,b, 1993). Fouquet *et al.* (1993) discussed a range of hydrothermal activity in these sites, which are controlled by the relative interplay of volcanism and tectonism. Early, low-temperature activity forms manganese crusts (the Hine Hina site) in the volcanic stage, while sulfide deposition starts to develop as the influence of tectonism improves fluid pathways (the Vai Lili site), and finally hydrothermal activity is completely controlled by fault systems (the White Church site).

The high-temperature fluid (up to 342°C) venting in the Vai Lili site (location 7 on Fig. 13.2) on the central ridge was the first discovery of black smokers in backarc basins of the western Pacific (Fouquet *et al.*, 1990). Besides numerous black smokers and white smokers both venting high-temperature fluid, diffuse discharges from the highly porous brecciated andesite causes temperature anomalies within the surrounding bottom seawater up to 30°C. This vigorously active hydrothermal field has dimensions of 400 m by 100 m and lies along a normal fault at a depth of 1720 m (Fouquet *et al.*, 1993). Compared to MOR sulfide deposits, the surface deposits of the Vai Lili site are characterized by abundant barite, sphalerite, tennantite, and galena with rare pyrrhotite and no pyrite. The chemical composition of sulfides samples shows an enrichment in Ba, Zn, As, Pb, Ag, Au, and Hg and depletion in Mo, Se, and Co. Primary grains of native gold were identified in the outer layer of Ba-Zn chimneys, which was the first documented occurrence of gold in seafloor sulfides (Herzig *et al.*, 1993). The fluid chemistry of the Vai Lili vents is characterized by very low pH and high dissolved base metals contents (Zn, Pb, Cu, and Cd). Herzig *et al.* (1993) pointed out that the chemistry of the hydrothermal fluid is significantly modified by subseafloor mixing and cooling processes that probably induced sulfide precipitation and oxidation.

Inactive barite-sulfide chimneys at the White Church site (location 8 on Fig. 13.2) were observed along the top and on the flanks of the northern Valu Fa Ridge segment, at depths between 1960 and 1800 m (Fouquet *et al.*, 1993). This hydrothermal field consists of chimneys up to 15 m in height and barite boulders mixed with massive sulfides. Hundreds of small manganese oxide chimneys up to 50 cm high and low-temperature hydrothermal discharges of 25°C were found along the faults.

On the southern Valu Fa Ridge segment (location 9 on Fig. 13.2), occurrences of manganese oxide crusts covering sulfide deposits, strongly altered and bleached surface rocks, and widespread discharge of low-temperature (40°C) fluid characterize a third site called Hine Hina which located at a depth of 1900 m (Fouquet *et al.*, 1993).

In the northern Lau Basin, the occurrence of hydrothermal deposits at some sites have been reported with less information: sulfides from the northeastern ridge (Hawkins and Helu, 1986; location 10 on Fig. 13.2) and barite from the Peggy Ridge (Bertine and Keene, 1975; location 11 on Fig. 13.2). In addition, extensive inactive sulfide deposits with andesitic lava at 15°23′S, and with basalt at 18°36′S, were recently discovered by DSRV *Mir* and are cited in Fouquet *et al.* (1993).

3.4. Mariana Trough

The Mariana Trough is an actively spreading backarc basin that accreted between the remnant volcanic arc of the West Mariana Ridge and the volcanic front of the Mariana arc (Fig. 13.4). Its spreading rate is considered to have been 3.0–3.4 cm/yr as a full rate for the past 3 m.y. (Hussong and Uyeda, 1981). The age of the Mariana Trough is less than 7 Ma, and the Wadati–Benioff zone does not lie beneath the backarc spreading axis (Hawkins *et al.*, 1990; Fryer, Chapter 6 this volume).

In 1987, DSRV *Alvin* dive programs confirmed present-day hydrothermal activity in the Mariana Trough (Craig *et al.*, 1987b), which had been evidenced by hydrothermal plume identification during the CEPHEUS cruise in 1982 (Horibe *et al.*, 1986). Together with observations by the DSRV *Shinkai 6500* in 1992 and 1993, hydrothermal occurrences are known in various tectonic settings across the trench–arc-backarc systems—that is, (1) axial volcanoes of backarc spreading center between 18°13′N and 18°02′N, (2) frontal volcanoes at 21°N in the northern Mariana arc (McMurtry *et al.*, 1993) and at 13°24′N in the southern Mariana Trough (Gamo *et al.*, 1993a; Johnson *et al.*, 1993), (3) a forearc serpentinite mud volcano (Fryer *et al.*, 1987b, 1990), and (4) off-axis hydrothermal mounds at 18°02′N (Leinen *et al.*, 1987; Wheat and McDuff, 1987,1994).

Active hydrothermal vents were first discovered on the flanks of axial volcanoes on the spreading center at the depth of 3600–3700 m (Craig *et al.*, 1987b; location 13 on Fig. 13.4). During the DSRV *Alvin* dive programs of 1987, four separate hydrothermal fields were discovered along the crestline of the trough (Hessler and Lonsdale, 1991). The Alice Spring site (18°12.6′N, 144°42.4′E, depth 3640 m) and another high-temperature venting site (18°12.8′N, 144°42.4′E, depth 3595 m) are located on the south flank of the central peak. Many sulfide-sulfate chimneys rise from large mounds of amorphous iron oxy-hydroxide and amorphous silica (Hawkins *et al.*, 1990). Chimney samples collected from this area are characterized by barite deposition with the absence of anhydrite as the matrix to the chimney walls (Kastner *et al.*, 1987; Stüben *et al.*, 1994). The mineral composition of

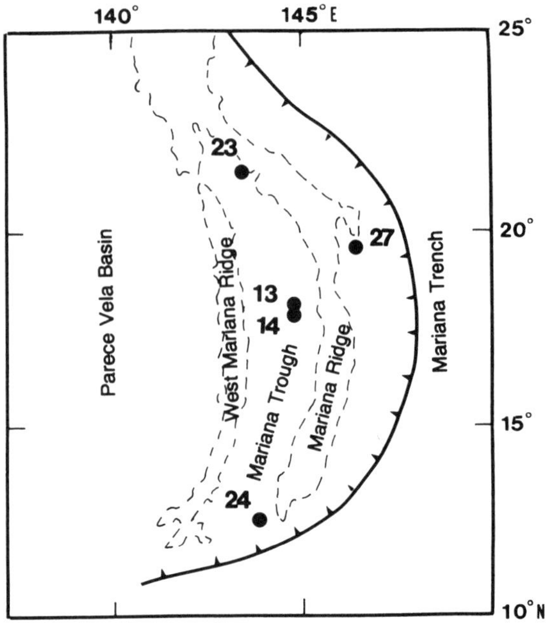

FIGURE 13.4. Location and tectonic setting of Mariana Trough. The solid circles mark the hydrothermal fields listed in Table I. The broken lines delinate contours of 3000-m water depth.

the sulfides is dominantly sphalerite, abundant galena, and lesser chalcopyrite, which clearly contrasts with those from MOR hydrothermal systems (Kastner *et al.*, 1987; Kusakabe *et al.*, 1990; Moore and Stakes, 1990). Some chimneys are venting a clear fluid with temperatures up to 287°C (Craig *et al.*, 1987b). The fluid chemistry, especially the chemical and isotopic composition of boron, indicated evidence for fluid interaction with oceanic crust in a backarc setting that is enriched in slab-derived boron (Campbell *et al.*, 1987; Palmer, 1991). Hydrothermal vent communities are represented by hairy gastropods, bresilid shrimp, and brachyuran crabs (Hessler *et al.*, 1987, 1988; Okutani and Ohta, 1988). In 1992, the Alice Spring site was revisited by the DSRV *Shinkai 6500* (Gamo *et al.*, 1993a,b). Negligible changes occurred in the five years since the DSRV *Alvin* dives, both in the fluid chemistry and biological communities. On the southeastern flank of the south peak, at 18°10′N, "Anemone Heaven" site is a milky, low-temperature (23°C) fluid discharging from an open crack. This field has been colonized by large anemones and galatheid crabs, although little mineralization was observed (Hessler *et al.*, 1987; Hawkins *et al.*, 1990). A high-temperature fluid (250–285°C) venting field, the "Snail Pits" site (18°10.95′N, 144°43.2′E, depth 3660 m), was also found on the south peak. At 20 km south of these sites, another hydrothermal field was located at 18°02.8′N, 144°45.2′E in a water depth of 3676 m (location No. 14 on Fig. 13.4), where sulfide-sulfate chimneys, including active high-temperature fluid discharges and large irregularly shaped mounds of soft ochreous oxide mud composed mainly of amorphous iron oxyhydroxide and some amorphous silica, were observed (Hawkins *et al.*, 1990).

The Kasuga seamounts in the northern Mariana arc are located along a N-S trending cross-arc fault zone (location 23 on Fig. 13.4) which probably represents the tectonic region of egress of arc magma leaking into the actively spreading Mariana backarc (Fryer *et al.*, 1987a). At the summit areas of seamounts Kasuga 2 (depth 400 m) and Kasuga 3 (depth 1140 m), emanations of warm fluid and hydrothermal deposits were confirmed by DSRV *Alvin* dives (McMurtry *et al.*, 1993). A sample of 39°C fluid collected from the Kasuga 2 seamount has very unusual characteristics, which are attributed to significant subseafloor addition of volatile components (CO_2 and SO_2) directly from magma into the hydrothermal fluids (McMurtry *et al.*, 1993). Enrichment in CO_2 results in "chemical weathering" reactions, whereby igneous minerals and/or alteration phases are strongly altered, adding magnesium and other cations to the hydrothermal fluid. Hydrolysis of SO_2 to sulfate and native sulfur is responsible for the high sulfate content, with a corresponding light $\delta^{34}S$ signature. Hydrothermal deposits consisting of elemental sulfur, Fe- and Mn-oxides, and nontronite are consistent with this fluid chemistry. The magmatic contribution caused by the reduced solubility of the volatile components at a shallow depth controls both the major elements composition of the fluid and the hydrothermal mineralization (McMurtry *et al.*, 1993).

On the southern Mariana Ridge, Gamo *et al.* (1993a) discovered the hydrothermal site named "Forecast Vent Field" (location 24 on Fig. 13.4) at the summit of the seamount Peak B (13°24′N, 143°55′E, depth 1470 m). Around this region, the distance between the Mariana Ridge and the Mariana Trough narrows and both island-arc volcanism and backarc spreading activity appear along the fault zone, which crosscuts the arc-backarc system (Fryer, 1993; Masuda *et al.*, 1993). Basalt samples recovered from the Peak B seamount have low K_2O content consistent with the Mariana backarc basalts, but their MgO-TiO_2 relationship would classify them as an island-arc basalt (Masuda *et al.*, 1993). Hydrothermal fluid up to 202°C emanates from anhydrite chimneys (Johnson *et al.*, 1993).

The Conical seamount is a conical edifice 20 km in diameter and 1500 m in height, located in the Mariana forearc, 80 km west of the Mariana Trench and 120 km east of the Mariana Ridge (location 27 on Fig. 13.4). The seamount is composed of unconsolidated clay to sand-sized serpentinite with suspended clasts of variously serpentinized harzburgite, dunite, and rare metabasalt (Fryer *et al.*, 1990). DSRV *Alvin* dives found white chimneys of two types, carbonate and silicate, near the summit knoll at a depth of about 3150 m (Fryer *et al.*, 1987b). The carbonate chimneys consist of aragonite, calcite, and minor amounts of amorphous gel-like Mg silicate. The other chimneys are Mg silicate with trace amounts of carbonate and a thin coating of ferromanganese oxide (Haggerty, 1987a; Fryer *et al.*, 1990). Age determinations by the ^{210}Pb method show that the chimneys are very young, <150 years old. A combined carbon- and oxygen-isotopic study suggests that the carbonates originated from the oxidation of methane of probable thermogenic origin and from ambient, low-temperature seawater (Haggerty, 1987b; Fryer *et al.*, 1990). Cold fluid was noted to seep from orifices of the silicate chimneys when they were scraped by the *Alvin* manipulator. Fluid temperatures were slightly lower (0.03°C) than that of ambient seawater, and the chemistry has several unique features, such as high pH and enrichments in silica, carbonate, sulfate, and H_2S. Pore fluids sampled by ODP drilling at the Conical seamount (Sites 778–780) confirmed the existence of this peculiar fluid (Mottl, 1992) with chemical characteristics (e.g., pH = 12) less mixed with seawater than the chimney fluid. Mottl (1992) attributed the fluid chemistry to a combination of deep-seated serpentinization processes and seawater interactions with crustal rocks including the surficial serpentinite. He proposed that the serpentinite mud volcanoes such as the Conical seamount in forearc regions act as important conduits through which volatile components (H_2O, CO_2, and other S and C species) are released from the subducting slab and return to the ocean.

3.5. Izu–Bonin (Ogasawara) Arc

The Izu–Bonin (Ogasawara) arc is a well-developed intraoceanic island arc associated with the Izu–Ogasawara Trench that includes a volcanic chain called Sichito–Iwojima Ridge and discontinuous backarc depressions (Fig. 13.5). The backarc depressions just behind the volcanic front are considered to be fault-bounded backarc rift grabens (Honza and Tamaki, 1985). The chain of active volcanic islands and submarine volcanoes about 1200 km long extends from the Izu Peninsula of Japan to the northern end of the Mariana arc (Yuasa, 1985). This trench–arc–backarc system is situated on the eastern rim of the Philippine Sea plate, where the Pacific plate is being subducted toward the northwest.

In 1984–1988, the Geological Survey of Japan conducted a research program to evaluate the possibility of submarine hydrothermal mineralization in this region (Nakao *et al.*, 1990). Several hydrothermal occurrences were found within submarine calderas of the frontal volcanoes. Subsequent dive programs by the DSRV *Shinkai 2000* confirmed the present-day hydrothermal activity. Moreover, in 1990, the first example of active high-temperature venting (up to 311°C) and accompanying mineralization located on the submarine frontal volcanoes was discovered (Kasuga and Kato, 1992; Watanabe and Kajimura, 1993, 1994). Together with frontal volcanoes, the rift grabens were considered to be an ideal tectonic setting for hydrothermal activity. Fujioka (1983) pointed out several topographic

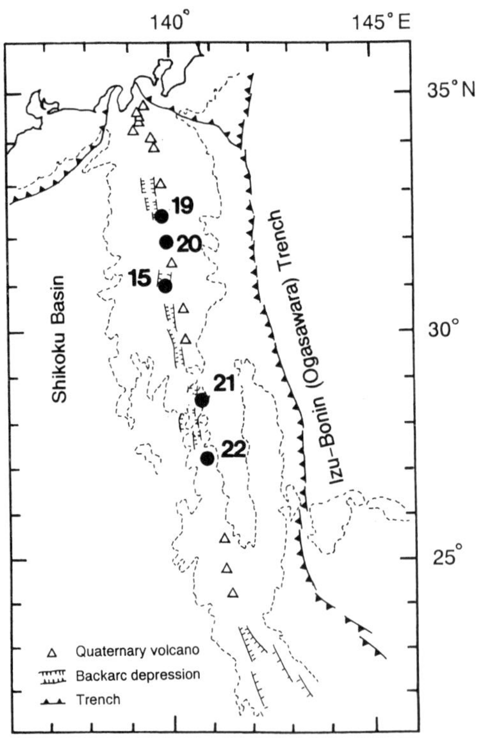

FIGURE 13.5. Location and tectonic setting of Izu–Bonin arc. The solid circles mark the hydrothermal fields listed in Table I. The broken lines delinate contours of 3000-m water depth.

similarities between the Izu–Bonin arc system and the restored Japan arc of Miocene age where major Kuroko-type VMSDs are known to exist.

The Kaikata seamount is one of the active frontal volcanoes and has four peaks: KM, KN, KS, and KC (location 22 on Fig. 13.5). The KC peak has a submarine caldera of 2-km radius in which the central cone is composed of lava and volcanic sand with a rim of andesitic lava (Yuasa *et al.*, 1988; Naka *et al.*, 1989a). From the caldera floor, altered andesite, intensely disseminated and veined with sulfide, was dredged during the GH85-1 cruise of 1985 (Urabe *et al.*, 1987). Sulfide minerals present are dominated by pyrite with minor amounts of sphalerite. High homogenization temperatures of fluid inclusions in vein quartz of 290°C and enrichment of gold in vein materials are indicative of vigorous fossil hydrothermal activity within this caldera (Urabe *et al.*, 1987). During a subsequent *Shinkai 2000* dive, Yuasa *et al.* (1988) noted an alteration zone along the lower part of the caldera wall more than 150 m high. Warm, shimmering fluid about 10°C higher than ambient seawater, associated biological communities dominated by white blind crab, and volcanic sands cemented by marcasite and minor amounts of pyrite at the shimmering site were confirmed in both the central cone at a depth about 900 m and the caldera wall (Naka *et al.*, 1989a; Fujikura *et al.*, 1993). Contrary to this evidence, present-day high-temperature hydrothermal activity was not seen.

On the northwestern slope of the KM peak of the Kaikata seamount, at depths between 900 and 1200 m, hydrothermal manganese oxide deposits cover the volcanic sands in the NE-SW trending zone over an area 10 km by 2 km (Usui *et al.*, 1986, 1989; Usui and Nishimura, 1992). The manganese deposits are characterized by high Mn/Fe ratios and very low transition element contents. Usui and Nishimura (1992) concluded that low-temperature hydrothermal solutions ascending through fractures or faults in permeable volcanic sands are responsible for precipitation of the several-centimeters-thick manganese layers.

The Suiyo seamount is one of the Sichito seamounts chain that is a part of the volcanic front in the middle part of the Izu–Bonin arc (location 21 on Fig. 13.5). This seamount is composed of dacitic rocks that belong to the low-potassium calc-alkaline rock series (Nagaoka *et al.*, 1991, 1992; Watanabe and Kajimura, 1993). In 1991, dives of the DSRV *Shinkai 2000* discovered sulfide chimneys venting clear, high-temperature fluid (230°C) at the slope of the caldera (Kasuga and Kato, 1992). During the next year, vigorous hydrothermal activity comparable to that of MORs with respect to their temperature (260–311°C) was located on the caldera floor at a depth of 1370 m (Watanabe and Kajimura, 1993, 1994; Watanabe *et al.*, 1994). Active smokers and a hydrothermal alteration zone are distributed within an area of about 300 m by 150 m in a NNW-SSE direction. Mineralogical studies show abundant chalcopyrite, especially inside chimneys walls. Columnar structures of chalcopyrite crystals are enclosed by colloform sphalerite, anhydrite, and barite. High Au contents (71 ppm maximum) and high Au/Ag ratios (average = 0.142) in the samples were also noted (Watanabe and Kajimura, 1994). The fluid chemistry of the Suiyo seamount is characterized by a higher Cl content than seawater, coupled with Ca enrichment and Na depletion (Tsunogai *et al.*, 1994; Ishibashi *et al.*, 1994b).

The Mokuyo seamount (28°14′N, 140°34′E) is located about 30 km south of the Suiyo seamount and is composed of basaltic rocks (Nagaoka *et al.*, 1991). Within the caldera wall, at a depth of 1210 m, low-temperature fluid emanation (40°C maximum) is associated with clam-dominated biological communities (Nagaoka *et al.*, 1992). Although mineralization was not observed, helium isotopic compositions of fluid samples suggest the existence of a hydrothermal end-member fluid originating from magmatic activity (Tsunogai *et al.*, 1993).

The Kita–Bayonnaise caldera (Myojin Knoll) is located in the northern Izu–Bonin arc, which has an outer rim of 5–7 km (location 19 on Fig. 13.5). The caldera rim and a central cone are composed of dacitic rock, while significant amounts of pumice are observed in the upper caldera wall (Yuasa *et al.*, 1991; Iizasa *et al.*, 1993). From the caldera floor at depths around 1300–1400 m, abundant sulfides and barite, comprising about 1.0–3.5 wt% of the sediment, were recovered by box-and-gravity coring (Iizasa, 1993a). The mineral assemblage of barite-pyrite-chalcopyrite-sphalerite, with or without galena and tetrahedrite, was identified in a single grain sample, indicating hydrothermal mineralization took place and was subsequently eroded by mass wasting (Iizasa, 1993a). During a 1992 diving expedition with *Shinkai 2000*, mineralized tuff consisting of sulfides and barite was sampled from the eastern part of the caldera floor (Iizasa *et al.*, 1993a). The chemical composition of this tuff is characterized by enrichments in Zn, Fe, Cu, Ba, Au (up to 19 ppm), and Ag (up to 272 ppm). Fluid inclusions in barite indicating formation temperature of 190–215°C and those in sphalerite indicating 205–225°C support the hydrothermal mineralization (Iizasa, 1993a).

The Myojinsho submarine caldera is located about 30 km south of the Kita–Bayonnaise caldera (location 20 on Fig. 13.5). It has an outer rim of 5–6 km and a flat-topped central cone composed of dacitic rocks. From the caldera floor at depths of 1000–1100 m, several massive hydrothermal barite and sulfide samples, together with rocks containing disseminated sulfides, were recovered. On the basis of a textural study, Iizasa *et al.* (1992) showed there are three sulfide mineralization sequences at the Myojinsho caldera: pyritization, quartz and sulfide veinlet-type, and massive sulfides. They also reported chemical compositions of sulfide samples characterized by enrichments in Zn, Pb, Cu, Au (up to 2.7 ppm), and Ag (up to 350 ppm). Factor analysis on the chemical compositions of sediment samples collected from the whole caldera floor suggests that the mineralization took place in the western part of the Myojinsho caldera (Iizasa, 1993b).

The Sumisu rift is one of the rift grabens in the northern Izu–Bonin arc (location 15 of Fig. 13.5). Taylor *et al.* (1990) conducted a DSRV *Alvin*–Sea Beam study and found a fossil hydrothermal site on an axial volcano. Chimneys, veins, and crusts are composed of silica, barite, and iron oxide. They occur primarily on the crown of a rhyolite lava dome at a depth of 1600 to 1530 m (Urabe and Kusakabe, 1990). The $\delta^{18}O$ values of barite and the occurrence of amorphous silica suggest low-temperature (<150°C) formation of the deposits, a possible analog to the hydrothermal hematic chert associated with the Kuroko deposition (Urabe and Kusakabe, 1990). Surface sediments of the Sumisu rift basin show manganese contents commonly in excess of 1%, suggesting a hydrothermal input (Nishimura and Murakami, 1988), although lower manganese content (<0.3%) of the pelagic and hemipelagic sediments sampled by ODP Leg 126 indicate this hydrothermal influence has been negligible back to 1.1 Ma (Nishimura *et al.*, 1992a,b). Combined with high heat flow values in the basin (Yamazaki, 1988), the evidence given above would suggest that the recent hydrothermal activity that formed the dead barite-silica chimneys could have been a source of the surficial manganese, derived from manganese-rich plumes spread all over the rift basin (Urabe and Kusakabe, 1990).

3.6. Okinawa Trough

The Okinawa Trough is considered to be a nascent backarc basin within the continental margin (Japanese DELP Research Group on backarc basins, 1986; 1991). Several geophysical and geological investigations (e.g., Sibuet *et al.*, 1987, Chapter 9 this volume) have confirmed rifting activities in the middle Okinawa Trough, which is characterized by the development of normal faulting in brittle continental crust (Fig. 13.6). Kimura *et al.* (1986) have demonstrated recent bimodal volcanism in the middle Okinawa Trough. Ishizuka *et al.* (1990) concluded that the magmatic sources in this area are in transition from island-arc type to backarc type, based on petrological studies. Extremely high and localized heat flow anomalies support very recent magmatic intrusions in this region (Yamano *et al.*, 1989).

In 1984 and 1986, dive programs with the DSRV *Shinkai 2000* first located low-temperature hydrothermal activity in the middle Okinawa Trough (Kimura *et al.*, 1988). A large hydrothermal field with extensive occurrence of sulfides was discovered during the SO-56 cruise of 1988 (Halback *et al.*, 1989). Subsequent *Shinkai 2000* dives confirmed the presence of high-temperature fluid venting up to 320°C (Nakamura *et al.*, 1990; Sakai *et al.*, 1990a). The middle Okinawa Trough mineralization is considered to be a modern analog for Kuroko-type mineralization (Halbach *et al.*, 1993). In 1988, a deep-tow TV camera survey conducted by R/V *Kaiyo* located hydrothermal activity at other sites in the Okinawa Trough (Momma *et al.*, 1989). These hydrothermal fields have been visited every year since their discovery (recently, however, the Safety Evaluation Committee of JAMSTEC prohibited *Shinkai 2000* diving at one of these sites, because the installation of submerged buoys by fisheries groups makes further dive studies at this site unsafe). All of the hydrothermal

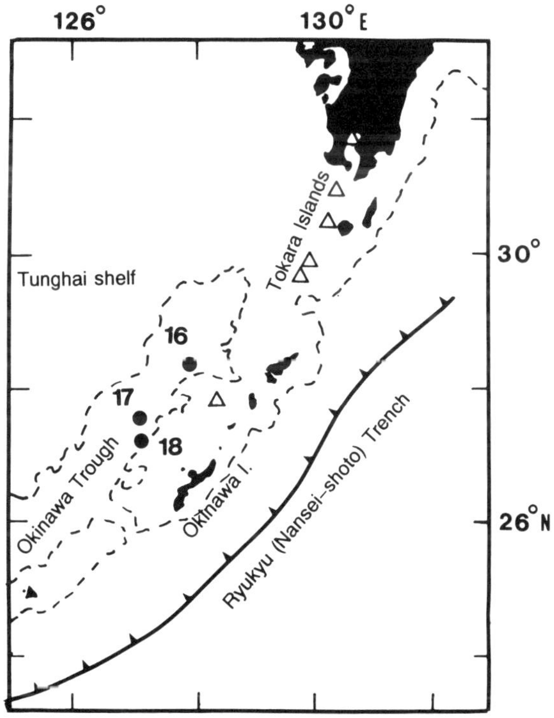

FIGURE 13.6. Location and tectonic setting of Okinawa Trough. The solid circles mark the hydrothermal fields listed in Table I. The broken lines show contours of 1000-m water depth. The triangles indicate Quaternary volcanoes.

systems in the Okinawa Trough formed in environments with extensive terrigenous sediment, supplied from the adjacent Asian continent.

The Izena cauldron is a rectangular (6 km by 3 km) depression formed by strike-slip strain accompanying the rifting (Halbach *et al.*, 1989, 1993; location 18 on Fig. 13.6). It is situated approximately in the southwestern continuation of a chain of Quaternary volcanoes on the Ryukyu arc; however, this location is also just in front of the eastern end of the Aguni rift graben segment. It is difficult to judge whether the Izena magmatic activity should be ascribed to volcanic front or backarc rifting. From the outer wall of the cauldron, occurrence of rhyolite, dacite, poorly vesciculated pumice, mudstone pumice, and tuff breccia have been reported, with twin rhyolite lava domes located in its center part (Nakamura *et al.*, 1989; Tanaka *et al.*, 1990). The hydrothermal field known as "Jade site" lies within an elongated region of 1000 m by 200 m on the northeastern slope at depths between 1300 and 1550 m. At the center of the hydrothermal field, the "Black smoker" chimney is vigorously discharging 320°C hydrothermal fluid. Surrounding this chimney, several groups of clear smoker chimneys are emitting transparent fluids with temperatures up to 220°C (Nakamura *et al.*, 1990). The fluid chemistry can be explained by the interaction with acidic to intermediate volcanic rocks and by fluid-sediment interaction (Sakai *et al.*, 1990a). One characteristic of the Izena hydrothermal activity is the abundance of CO_2, reflecting a magmatic source (Ishibashi *et al.*, 1990). At the outskirts of the hydrothermal field, liquid-CO_2 bubble emanation was observed (Sakai *et al.*, 1990b) Mineral samples from this area contain sphalerite, tetrahedrite, galena, chalcopyrite, pyrite, barite, and anhydrite (Aoki and Nakamura, 1989; Urabe, 1989; Halbach *et al.*, 1993). High Au (up to 24 ppm) and Ag (up to 1.1%) contents have also been noted.

At a small depression along the Iheya Ridge, which lies in the axial part of the Iheya rift graben, another active hydrothermal field has been discovered (location 17 on Fig. 13.6). This field is called the "Clam site," after an associated biological community that consists dominantly of clams, scalpellids, and slender vestimentiferan tube worms (Ohta, 1990). The Iheya Ridge consists of highly vesicular pillow basalt; the gassy lava surfaces are suggestive of recent eruptions (Naka *et al.*, 1989b). Hydrothermal activity is limited to narrow area at a depth of 1400 m. Mineralization at the Iheya site is characterized by carbonate precipitates, which occur in a form of cemented multilayered eaves covering large portions of the seafloor (Gamo *et al.*, 1991b). The sampled carbonates consist of manganoan calcite, rhodochrosite, anhydrite, and amorphous silica, with only trace amounts of sulfides including wurtzite, pyrrhotite, and galena present as disseminations (Tanaka *et al.*, 1989; Izawa *et al.*, 1991). Native sulfur was also recovered from the outer region of the field. Moderate emanations of fluid from edges and fissures of the cemented carbonates were observed in several places (Sakai *et al.*, 1990a; Gamo *et al.*, 1991a,b). Fluid temperatures showed both temporal and spatial variation, with temperature ranging from 100 to 220°C. Nakashima *et al.* (1993) estimated the fluid-rock interaction temperature at between 250 and 300°C, based on fluid inclusion homogenization temperatures in calcite. Carbon-isotopic compositions for carbonates are very close to those of the venting fluids, providing evidence that the carbonate precipitated directly from the CO_2-rich fluids (Kimura *et al.*, 1989). Gamo *et al.* (1991a) demonstrated some unique features of the fluid chemistry, such as anomalously high ammonium content, high alkalinity, and ^{34}S-enriched sulfate, all interpreted as a result of organic matter decomposition during subseafloor mixing with penetrated seawater. Gamo *et al.* (1991a) proposed a model of multiple secondary hydrothermal circulation cells within sediment layers which are in turn associated with high-temperature (around 300°C) primary circulation.

At the summit area of the Natsushima 84-1 knoll (water depth of 1540 m), which is located about 15 km east of the Iheya Ridge, low-temperature hydrothermal activity was first discovered in 1986 in the middle Okinawa Trough (Kimura *et al.*, 1988). This small knoll is composed of dacite and rhyolitic rocks (Kimura *et al.*, 1986) and situated on the south side of a small depression, where Yamano *et al.* (1986) reported extremely high-heat-flow measurements. Shimmering water from one of the mounds has a temperature 2–3°C higher than that of the ambient seawater, while methane and manganese anomalies were detected (Gamo *et al.*, 1987). The mound was covered with a yellowish brown deposit of Fe-rich smectite and Fe-Mn oxyhydroxide (Masuda *et al.*, 1987). Oxygen isotopes of these deposits indicate a temperature of precipitation of about 50°C, which confirms the low-temperature activity at this site (Masuda *et al.*, 1987).

The Minami–Ensei knoll is located at a rifting center in the northern Okinawa Trough (location 16 on Fig. 13.6). Its topography is characterized by small hills and depressions within a rim of 2-km radius and is attributed to a volcanic complex (Aoki *et al.*, 1993). Poorly vesiculated pumice, diorite, and granite samples were recovered from this knoll. Living and/or dead biological communities dominated by mussels were observed in several places in the small depressions (Hashimoto *et al.*, 1990, 1993). The strongest hydrothermal activity is in "C depression," located on the western slope of the volcanic complex. Several chimneys are aligned along a N-S direction at depths between 690 and 705 m. Most of the chimneys are discharging transparent fluids with temperatures of 267–278°C (Chiba *et al.*, 1993). Nedachi *et al.* (1992) described the hydrothermal samples as siliceous ores, sulfate chimneys, massive sulfide ores, and clastic ores. Siliceous samples contain high SiO_2 contents (up to 83%) and some sulfides and barite. All of the active hydrothermal chimney samples are composed mainly of anhydrite, with only fine-grain sulfide minerals precipitated on the anhydrite surfaces. Some Zn-rich massive sulfides were found at the foot of the chimneys. The clastic ore samples are characterized by enrichment in both Zn and Pb and a high average Au/Ag ratio (around 0.08). Moreover, components of the clastic material do not coincide with the samples observed on the seafloor. Nedachi *et al.* (1992) suggest that the clastic ore samples were formed beneath the seafloor and ejected on the seafloor during an explosive boiling stage. A similar scenario has been proposed for the sulfides from the Jade site (Urabe, 1987). The chemical composition of the sampled hydrothermal fluid is characterized by a depletion in both metal elements and H_2S and by a chlorinity only 10% less than that of seawater, which can be explained by a model of phase separation just below the seafloor which induces subseafloor sulfide precipitation (Chiba *et al.*, 1993). Nakashima *et al.* (1993) reported fluid inclusion homogenization temperatures are similar to the measured venting fluid temperature. Hydrothermal activity at the Minami–Ensei knoll has been shown to be very close to *P-T* conditions for the phase separation.

4. DISCUSSION

4.1. Volcanism and Hydrothermal Activity

About three quarters of terrestrial volcanic rocks have been produced at divergent plate boundaries during the Cenozoic era (McBirney, 1984). The oceanic hot spots and arc-backarc systems contributed 7–10% and 10–15% of the total volume of volcanic rocks, respectively. Although submarine arc-backarc volcanism is less important volumetrically, it is very significant for its lithologic diversity.

All the known high-temperature hydrothermal activity in the western Pacific region is located at the loci of such submarine volcanism. This is understandable because the volcanism provides favorable crustal environments for hydrothermal mineralization: extremely high heat flow and fracture- or rift-induced permeability to facilitate voluminous hydrothermal circulation, magma chambers of favorable geometry (e.g., Rona, 1988), supply of volatile and other components from the subduction zone (e.g., Stolper and Newman, 1992), the volatile-rich nature of arc and backarc magmas (e.g., Hochstaedter *et al.*, 1990), and tectonic events conducive to intense hydrothermal activity. The hydrothermal activity at hot spots seems to be the exception for these favorable features. Fluids from some hot spots in the central Pacific, the Loihi seamount in the Hawaii volcanic chain (Sedwick *et al.*, 1992,1994), the Teahitia seamount at the southern extension of the Society Islands (Michard *et al.*, 1993), and the Macdonald seamount in the Austral Island volcanic chain (Chemineé *et al.*, 1991) are highly enriched in volatile components of magmatic origin such as CO_2 and He. However, the hydrothermal deposits in these submarine volcanoes are mainly composed of iron hydroxides, and no sulfide mineralization has been identified (De Carlo *et al.*, 1983; Puteanus *et al.*, 1991).

Rona (1988) pointed out, on the basis of a compilation of ancient massive sulfide deposits, that basalt-hosted massive sulfide deposits that may have formed at spreading centers constitute 17% of the known 508 deposits. Most of them are less than 30×10^6 tonnes in size. On the other hand, the most numerous (56%) and largest (up to 231×10^6 tonnes) massive sulfide deposits are hosted in rhyolite. Hydrothermal activity associated with dacitic to rhyolitic volcanism is known at ten sites (6, 7, 8, 12, 15, 16, 18, 19, 20, and 21, Table I) in the western Pacific, and eight (except 12 and 15) of them have sulfide mineralization. This number is comparable to the nine sulfide mineralized sites (1, 4, 9, 10, 13, 14, 17, 22, and 24, Table I) that are associated with basaltic to andesitic volcanism. Since rhyodacite is extremely rare in MOR systems (Christie and Sinton, 1981), this statistic supports the idea that the ancient arc-backarc setting is the exclusive site for the formation of rhyolite-hosted massive sulfide deposits. It is still premature to discuss the size and tonnage of the present-day mineralization in the western Pacific, but relatively large deposits (e.g., 7, 8, 18, and 21) tend to be associated with acidic volcanism.

4.2. Time Scale of Volcanic and Hydrothermal Events

Oshima island, the northernmost frontal volcano of the Izu–Bonin arc, displays a remarkably regular periodic volcanic activity. The stratovolcano of arc tholeiite basalt erupted at average time intervals of 135 ± 50 years in the past 1330 years and of 100 years in the past 10,000 years (Aramaki *et al.*, 1992). Although there are insufficient data to prove such periodicity in other frontal volcanoes of the arc, it is reasonable to deduce that the replenishing of arc magma happens periodically at interval of order 100 years. The associated hydrothermal activity is, therefore, likely to periodically reactivate and cease in accordance with the volcanic activity.

Estimation of the time scale of hydrothermal activity at submarine arc volcanoes seems to be even more difficult. However, repeated *Shinkai 2000* dives on the Kaikata Seamount provide some clues. In November 1986, Yuasa *et al.* (1988) observed reddish brown flocculates of Fe-oxyhydroxide drifting within the caldera and found weak fluid shimmering from hydrothermally altered caldera wall. A revisit in May, 1988 (1.5 years later) by one of the authors (T.U.) showed that all the Fe-oxyhydroxide had been completely swept away.

An intensive search for the shimmering failed, probably because hydrothermal activity had ceased. In contrast, activity at the summit of the central cone was confirmed by later dives in the same month. Shimmering of warm fluid (max. temperature 22°C) marcasite-cemented sediments (T. Urabe; unpublished data) and associated biological community of mussel, deep-sea sole, and brachyuran crab characterize the latter site (Mitsuzawa *et al.*, 1989). Furthermore, the shimmering became hardly identifiable in 1990. These lines of evidence suggest that the lifetime of some hydrothermal activity at frontal volcanoes is likely to be short, on the order of years, and possibly much shorter than that of MORs.

4.3. Hydrothermal Mineralization in Relation to Ancient Analogs

Sawkins (1976) proposed a fourfold classification of massive sulfide deposits based on their composition and tectonic settings: Kuroko, Cyprus, Besshi, and Sullivan types. The hydrothermal mineralization in the western Pacific provides present-day examples of the first three deposit types, but there are no observed present-day examples for the sediment-hosted massive sulfide deposits of the last type. The Kuroko deposits in Japan are located in a fault-bounded backarc trough of the northeast Japan arc of Miocene age (Ohmoto *et al.*, 1983). Rifting with 1–3 km of vertical displacement, intense bimodal volcanism, and associated Kuroko mineralization occurred between 16 and 15 Ma. Several lines of evidence, including volcano-tectonic setting, geology, and isotopic composition, all suggest their magmatic hydrothermal origin (Urabe, 1987; Urabe and Marumo, 1991). A Besshi type deposit, named after the Besshi deposit in Japan, is defined as a cupriferous-pyrite massive sulfide deposit hosted in metamorphosed, submarine tholeiitic basalt (Kanehira and Tatsumi, 1970). These deposits differ significantly from the Cyprus-type by an abundance of epiclastic rocks and the lack of an ophiolite suite. The Kuroko and Besshi types are considered to have formed in arc-backarc environments rather than MORs, which may have produced the Cyprus type deposits (Sawkins, 1976).

Hydrothermal mineralization in the Okinawa Trough, the Sumisu rift, and the Lau Basin occurs in association with rhyodacite volcanism at backarc rifting-spreading centers, which develop by splitting preexisting island-arc crust and are regarded as the best examples of contemporaneous Kuroko-type mineralization (Halbach *et al.*, 1989, Urabe and Kusakabe, 1990, Fouquet *et al.*, 1993). The basalt/andesite-hosted deposits such as those in the central Mariana Trough and the Vienna Woods site of the Manus Basin seem to provide the modern analog of the Besshi-type deposits. The present-day deposit will be covered by epiclastic sediment like the latter deposits after the termination of the mineralization: Recent submersible dives at various volcanic centers in the western Pacific provided unequivocal evidence that most of the flanks of the volcanic edifices are thickly covered by sand to pebble-size sediments of volcanic rock and/or pumice origin. This is in contrast with sediment-starved MORs, where we observed almost no coarse-grain sediments except scarce glassy fragments. Such volcaniclastic sediments may be derived either from eruption of arc-backarc volcanism or by mass wasting on the flank of the volcano. Later accretion of the deposits and their host rocks during subduction may give rise to metamorphism and intense deformation (Karig, 1985) which are commonly observed in the Besshi-type deposits in the world (e.g., Franklin *et al.*, 1981).

The deposit in the North Fiji Basin that is located at the spreading center of a mature backarc basin may be a candidate for the Cyprus-type deposit. This type of deposit may also be formed at the MORs.

4.4. Mineralogy, Geochemistry, and Sulfur Isotopes of the Deposits

The hydrothermal deposits of the Western Pacific (Table I) are characterized by their low pyrite/marcasite content and absence of pyrrhotite (except at a few sites such as Lau Basin and the Fiji transform fault). This is partly a result of the low iron content in hydrothermal fluids in the western Pacific and the higher fO_2 condition in the backarc settings, since pyrite is more stable than pyrrhotite at such conditions, as discussed by Herzig *et al.* (1993). They attributed high gold content of the western Pacific deposits to higher fO_2 and fS_2 conditions.

Mineralogical zoning observed in chimneys and mounds is essentially the same as that of MORs; Cu-rich sulfides in the center and Zn-rich sulfides in the outer edges (e.g., Bendel *et al.*, 1993). Galena and tetrahedrite-tennantite (and realger, orpiment, and sulfosalts minerals in the case of the Okinawa Trough), which are unique to the arc-backarc systems, are enriched in the outer zone, sometimes together with barite.

The chemical analyses on the sulfide samples from the western Pacific region commonly show a wide range of values even in a single hydrothermal field. This is due to the gross mineral zoning observed in these ores and to their complex growth history. However, the average chemical composition of the sulfides shows certain characteristics that differ from those of the MORs (Table II).

Figure 13.7 shows the variation in representative element contents, which are normalized against average EPR sulfide deposits. The average composition of sulfide samples of the North Fiji Basin spreading center (Fig. 13.7A), which is in the most advanced stage of backarc spreading, strongly resembles that of the EPR. It is also obvious that the composition is almost identical to the average Besshi-type ores, which are believed to have formed in backarc spreading centers of the Mesozoic to Paleozoic eras. On the other hand, samples of the Manus Basin (PACMANUS site) show slight enrichment in gold and arsenic, which are characteristic elements in subaerial epithermal gold deposits (e.g. Hedenquist *et al.*, 1990) and high-temperature hot springs observed on the acidic to intermediate volcanoes (Karpov and Naboko, 1990). This is rather remarkable because the crustal abundance of these elements is quite small and the enrichment factors should be around several thousand. Such an extreme enrichment is unlikely to occur simply by leaching from the host rocks.

One of the most striking occurrences of present-day gold mineralization is observed at the shore of a crater lake of Osorezan volcano, northern Japan (Aoki and Thompson, 1990; Izawa and Aoki, 1991). Extremely enriched gold (6510 ppm Au) and arsenic (3650 ppm) are precipitating near the surface together with Hg, Pb, Zn, and Sb from neutral pH, chloride-rich hot-spring water with high H_2S content (Aoki, 1992a). Oxygen- and hydrogen-isotope data suggested a significant contribution of the fluid from cooling dacite magma of the Osorezan (Aoki, 1992b). Dacitic volcanism observed at the PACMANUS site is the likely source of such "epithermal elements" in the seafloor deposits, by an analogy of the terrestrial example.

The composition of sulfide samples from the Lau Basin, Mariana Trough (Fig. 13.7B), and Minami–Ensei (Fig. 13.7C) matches closely that of the average black ore (Zn-Pb-Ba massive ore) of the Kuroko deposits in Japan, which strongly indicates their common mechanism of formation. In fact, the Kuroko and these three present-day examples all formed at centers of rifting and early stages of backarc spreading.

The composition of the sulfide samples from frontal volcanoes (Suiyo seamount, Okinawa Trough) shows an unusual variation (Fig. 13.7C): The iron-poor nature of the

Okinawa Trough ore makes all the elements more enriched than that of the EPR. The sulfide samples from the Suiyo seamount are enriched in copper but depleted in lead and barium, compared with other backarc examples.

The significant contribution of magmatic components to the hydrothermal fluids is manifested by the presence of volatile elements such as sulfur and carbon. Sulfur-isotopic values of the known sulfide samples are around 8‰ at the Izena cauldron (H. Sakai; pers. comm.), 2.1–3.1‰ at the Mariana Trough (Kusakabe *et al.*, 1990), 0.3–2.2‰ at the Minami–Ensei knoll (Nedachi *et al.*, 1992), 2.1–2.8‰ at the Kita–Bayonnaise caldera (Iizasa, 1993a), and 0.9–1.2‰ at the Kaikata caldera (vein part) (Urabe, unpubl. data). This value range from 0 and 8‰ corresponds to the compositional range of magmatic sulfur.

4.5. Chemical Composition of the Hydrothermal Fluid

Geochemical studies of venting hydrothermal fluid are considered to be important and instructive, because fluid chemistry evolves from original seawater and is affected by chemical interactions with various materials encountered during its circulation beneath the seafloor. Active venting of high-temperature fluid (temperature around 300°C) was observed in several sites in the western Pacific. At six sites, enough fluid samples were collected and analyzed to estimate chemical composition of each hydrothermal end member. Table III summarizes chemical data of these hydrothermal fluids together with those of the EPR 21°N, OBS vent for comparison with the MOR hydrothermal system. There are some reports of chemical data of fluid samples collected at other sites. Samples from Vienna Woods in the Manus Basin (Lisitsyn *et al.*, 1993) and Franklin seamount in the Woodlark basin (Binns *et al.*, 1993) were not listed in this table, because those samples were significantly diluted with ambient seawater, which made end-member estimation ambiguous. Several fluid samples were collected and studied from low-temperature vents: the Clam site (Iheya site) in the Okinawa Trough (Gamo *et al.*, 1987, 1991a) and the Kasuga 2 seamount in the northern Mariana arc (McMurtry *et al.*, 1993). However, these data are excluded from the following discussion, because the significant temperature difference makes it difficult to compare the chemistry.

The major element composition of hydrothermal fluids is considered to be essentially controlled by chemical equilibria with various mineral phases. In the reaction zone just above the magma chamber, the hydrothermal fluid is saturated with respect to all of the primary and secondary (alteration) minerals at hydrothermal conditions. This model has been confirmed via geochemical studies by several works (e.g., Seyfried, 1987; Campbell *et al.*, 1988; Bowers, 1989). As compiled in Table III, the high-temperature fluids from the western Pacific have some characteristics in common, such as the depletion of Mg and SO_4, the enrichment in cations relative to seawater, and low pH and reducing conditions, regardless of the type of hydrothermal system or the tectonic setting, including MOR. The spectrum of major element compositions does not diverge more than an order of magnitude. This similarity of major element composition is understandable, because both mineral assemblages and physical conditions ($T > 300°C$, $P > 100$ bars) where hydrothermal fluid-mineral interactions occur are similar in these systems.

Among the chemical equilibria at hydrothermal conditions, the control of dissolved SiO_2 concentration, via quartz dissolution, has been well studied (e.g., Von Damm *et al.*, 1991). In Fig. 13.8, the SiO_2 concentration of some fluids are plotted together with quartz solubility data for isobars from 100 to 500 bars. Most data for the EPR systems (open

TABLE II

Chemical Compositions of Hydrothermal Sulfide Samples in the Western Pacific

Location no.	1 N. Fiji (Bendel *et al.*, 1993)	1 N. Fiji (T. Urabe, unpubl. data)	4 Manus (Binns and Scott, 1993)	7 Lau Vai Lili (Fouquet *et al.*, 1993)	8 Lau Hine Hina (Fouquet *et al.*, 1993)	9 Lau White Church (Fouquet *et al.*, 1993)	7–9 Lau (Fouquet *et al.*, 1993)	13 Mariana (Hannington, 1989)
Fe (%)	30.05	30.96	30	10.9	30.4	7.23	17.4	2.39
Cu (%)	7.45	2.76	30.3	8.18	2.77	4.19	4.56	1.15
Pb (%)	0.057	0.025	0.035	0.27	0.88	0.4	0.33	7.4
Zn (%)	6.64	5.49	1.4	26.4	15.4	12.2	16.1	9.96
Mn (ppm)	761		65	522		744	542	175
Co (ppm)	238	90	20		6.9		3	2
Ni (ppm)			0.45				6	
S (%)	36.67	38.28	30.2	29.05	42.4	20	30.12	17.8
As (ppm)	182	170	2440	1575	4629	445	2213	126
Sb (ppm)	30	21	320	35		89	51	190
Se (ppm)	168	230	56		21		8	10
Cd (ppm)	260	150	73	780		350	482	465
Sn (ppm)			13				4	
W (ppm)								
Mo (ppm)	270	80	360	27	20	56	32	5
Au (ppm)	1.077	1.2	10	1.21	2.51	2.42	1.4	0.784
Ag (ppm)	151	47.7	42	147	680	118	256	184
Hg (ppm)								22
Ba (%)	0.82	0.043	0.75	11.76	2.75	25.8	11.56	33.33

Location no.	13 Mariana (T. Urabe, unpubl. data)	16 Okinawa Minami–Ensei (Nedachi *et al.*, 1992)	18 Okinawa Izena (Halbach *et al.*, 1989)	18 Okinawa Izena (Urabe, 1989; Aoki and Nakamura, 1989)	21 Izu–Bonin Suiyo (Watanabe and Kajimura, 1994)	EPR average (Fouquet *et al.*, 1993)	Black Ore average ($n = 160$) (ANRE[a], 1993)	Besshi average ($n = 32$) (ANRE[a], 1993)
Fe (%)	4.57	3.29	7.33	4.84	16.7	25.93	9.62	36.38
Cu (%)	1.4	1.46	1.77	9.42	14.5	2.10	1.67	5.95
Pb (%)	0.29	5.01	14.27	15.8	0.15	0.06	6.76	0.09
Zn (%)	8.83	12.8	22	31.3	11.6	11.92	13.58	1.55
Mn (ppm)		300	1567	1236	165	961	301	348
Co (ppm)					13	336	23.19	1116
Ni (ppm)					2.5	16.94	12.50	27.93
S (%)						36.92	25.16	36.56
As (ppm)	110	417	27537	15300	892	168	873	112
Sb (ppm)					495	14.12	489	36.20
Se (ppm)					0.92	112	1.73	17.92
Cd (ppm)			950	925	938	434	860	48.33
Sn (ppm)						2.56	12.70	6.94
W (ppm)						15.00	178	67.52
Mo (ppm)					120	52.62	113	80.28
Au (ppm)	0.65	2.43	4.6	3.6	19.7	0.20	2.02	0.84
Ag (ppm)	55.5	90.3	2100	1445	113	51.07	285	33.92
Hg (ppm)						3.68	42.31	3.31
Ba (%)	18.1	1.11	2.76		0.089	0.08	10.33	0.03

[a]Agency of Natural Resources and Energy.

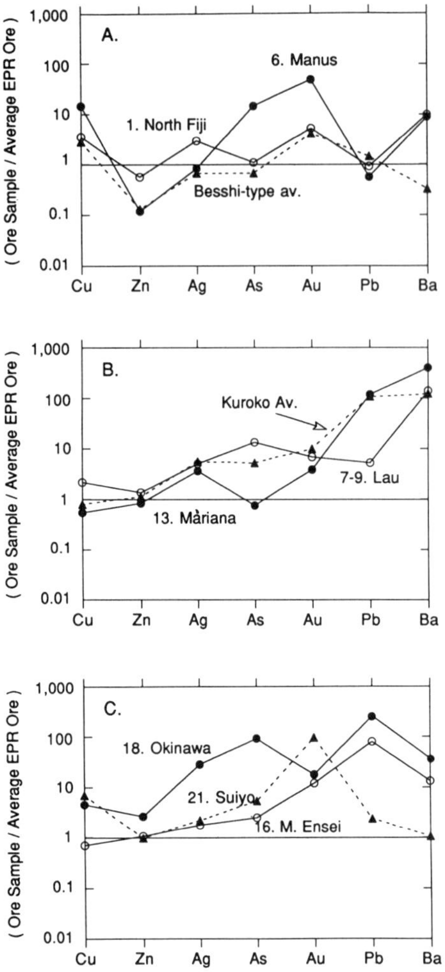

FIGURE 13.7. Variation in composition of hydrothermal mineral samples from the weatern Pacific. Metal contents are nomalized with the average East Pacific Rise ore (listed in Table II). Average assay of Besshi-type ore and Balk ore (sphalerite-galena-barite ore) of the Kuroko deposit are also superimopsed (data after Agency of Natural Resources and Energy, 1993). Numbers in the figure correspond to the location in Table I.

squares) are plotted on isobars deeper than the seafloor, which is considered evidence for deeper hydrothermal reaction zones. In contrast, SiO_2 data for arc-related fluids, such as those at Lau Basin, Suiyo seamount, Minami–Ensei knoll, and Izena cauldron, suggest hydrothermal interaction at pressures and temperatures very close to the seafloor. This could be attributed to very shallow hydrothermal circulation cells, caused probably by a near surface magmatic body. For example, at the Izena site in the Okinawa Trough, heat flow measurements suggest convection cells at about 1-km depth (Kinoshita, 1990). An alternative explanation is that hydrothermal fluid interacts with conduit rock during its ascent because quartz dissolution occurs very quickly. Highly permeable volcaniclastic rocks resulting from volatile-rich volcanic activity are commonly observed around the above four sites, which is consistent with both explanations. For the Mariana Trough and North Fiji Basin hydrothermal fluids, substantial cooling near the surface might explain the reason why silica dissolution equilibrium is not attained in these systems.

The process of subseafloor phase separation, which appears more commonly in arc-

TABLE III
Chemical Compositions of Hydrothermal Fluids in the Western Pacific

Location no.	1 N. Fiji White Lady	7 Lau Vai Lili	13 Mariana Alice Spring	16 Okinawa Minami–Ensei	18 Okinawa Izena	21 Izu–Bonin Suiyo	EPR OBS
Temperature (°C)	285	334	285	278	320	311	350
pH	4.7	2.0	4.4	5.0	4.7	3.7	3.4
Li (μM)	200	623	5800	1860	600	891	
Na (mM)	210	590	431	430	446	432	
K (mM)	10.5	79.0	31.0	50.9	73.7	29.7	23.2
Rb (μM)	8.8	68	30	360	28		
NH_4 (mM)	0			4.70	5.32	<0.1	
Mg (mM)	0	0	0	0	0	0	0
Ca (mM)	6.5	41.3	23.6	22.1	23.2	89.0	15.6
Sr (μM)	30	20	73	227	110	300	81
Ba (μM)	5.9	>39		55	60	100	8
Mn (μM)	12	7100		94	370	587	960
Fe (μM)	13	2500			21	435	1660
Cu (μM)		34			0.003		35
Zn (μM)		3000			7.6		106
Pb (nM)		3900			36		308
SO_4 (mM)	0	0	0	0	0	0	0
Cl (mM)	255	790	544	527	550	658	489
Br (μM)	306	1140	866		1045		802
Alkalinity (meq)	0.12		0.43	3.51	0.88	−0.20	−0.40
B (mM)	0.47	0.83	0.81	4.0	3.41	1.43	0.51
Al (μM)	6.0				4.9	17.0	5.2
Si (mM)	14.0	14.5	14.0	10.8	12.5	13.2	17.6
H_2S (mM)	2.0	<5	2.6	2.44	13.1	1.6	7.3
CO_2 (mM)	14.4	15		96	200	40	8
$^{87}Sr/^{86}Sr$	0.7046	0.7044	0.7033	0.7100	0.7089		0.7031
$\delta^{34}S$ (H_2S) (‰)			3.6–4.8	3.6	7.4–7.7		1.3–5.5

Data sources:
1 North Fiji Basin, White Lady (Grimaud *et al.*, 1991; Ishibashi *et al.*, 1994a).
7 Lau Basin, Vai Lili Field, VL-3 (Fouquet *et al.*, 1993)
13 Mariana Trough, Alice Spring (Campbell *et al.*, 1987; Campbell and Edmond, 1989; Palmer and Edmond, 1989; Palmer, 1991).
16 Minami–Ensei knoll (Chiba *et al.*, 1993).
18 Izena cauldron, (Sakai *et al.*, 1990a; Gamo *et al.*, 1991a; Ishibashi, 1991; Chiba *et al.*, 1993; You *et al.*, 1994; Shitashima, 1994).
21 Suiyo seamount (Ishibashi *et al.*, 1994b).
EPR East Pacific Rise 21°N, OBS vent (Von Damm *et al.*, 1985a).

related hydrothermal systems, is one of the most important factors that control fluid chemistry and mineral occurrence. Boiling drastically changes the chemical composition of the hydrothermal fluid during its ascent from the reaction zone. Phase segregation and condensation between the vapor and the brine components result in a wide divergence in the concentrations of the major elements (Butterfield *et al.*, 1990). Moreover, boiling induces subseafloor precipitation of ore-forming metal elements (e.g., Drummond and Ohmoto, 1985), because of the sudden increase in pH, mainly caused by CO_2 partitioning into the vapor phase. The White Lady site in the North Fiji Basin was the first example

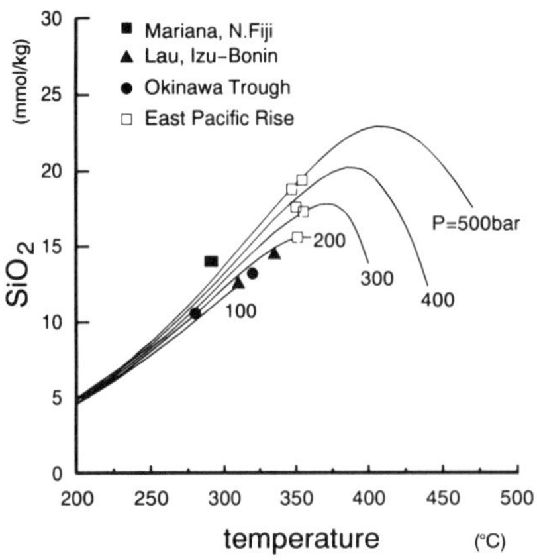

FIGURE 13.8. Relationship between SiO_2 concentration and equilibrium condition with quartz. Isobars are drawn according to the calculated data and plots of Von Damm *et al.* (1991). SiO_2 concentrations of EPR 21°N fluids are from Von Damm *et al.* (1985a) and Campbell et al. (1988), and those of others are from Table III.

described in the western Pacific (Grimaud *et al.*, 1991) of a phase-separated hydrothermal system and the second in the world after the ASHES vent field, Juan de Fuca Ridge, reported by Massoth *et al.* (1989). Since the venting fluids at the White Lady site are the vapor-rich fraction of the phase-separated hydrothermal fluid, they have low concentrations in all of the major elements (see Table III). The fluids discharging from these seafloor orifices do not form sulfide deposits, only sulfate chimneys. Figure 13.9 shows the relationship between temperature and depth (pressure) of the venting hydrothermal fluids in the western Pacific region. While the fluids at East Pacific Rise (EPR), Mid-Atlantic Ridge (MAR), and Mariana Trough plot in the one phase region, the Lau Basin, Suiyo seamount, Minami–Ensei knoll, and Izena cauldron plot close to the two-phase boundary. For North Fiji hydrothermal fluid, significant cooling of the condensed vapor-rich fluid near the surface of the seafloor is responsible for the shift of the plot far from the two-phase boundary region, although the fluid certainly experienced boiling (Ishibashi *et al.*, 1994a).

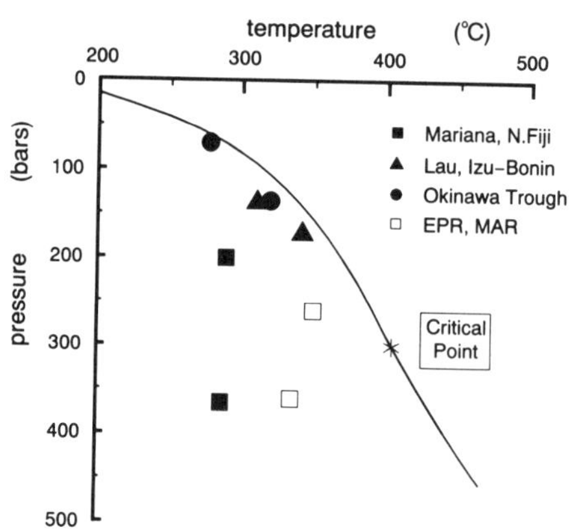

FIGURE 13.9. Relationship between temperatures and depth of hydrothermal fluids at the seafloor. Two-phase boundary curve for 3.2 wt% NaCl in H_2O system (equivalent to seawater) is drawn from the data of Bischoff and Rosenbauuer (1985).

Many examples of arc-related hydrothermal systems are located at shallower water depths than those at MOR, which gives rise to *P-T* conditions close to the two-phase boundary. Besides, the contribution of volatiles from arc magmas into the hydrothermal fluid would further promote the initiation of boiling. In several deposit samples from within the Okinawa Trough, there is evidence for explosions that are attributed to subseafloor boiling.

In contrast to the major elements, trace elements do not form separate secondary (alteration) minerals. Their distribution between the hydrothermal fluid and the alteration minerals is likely to be controlled by ion exchange with major elements. For example, the Sr/Ca ratio of hydrothermal fluid has common values among the various hydrothermal systems, which is attributed to a single-step control, probably by recrystallization of plagioclase (Berndt *et al.*, 1988). The "soluble elements," such as lithium and boron, are considered to be rarely involved in secondary (alteration) minerals during hydrothermal interaction (Seyfried *et al.*, 1984). In other words, they are only leached from primary minerals of rocks during early stages of fluid circulation. Thus, if we assume similar water/rock ratios during hydrothermal interactions among the various tectonic settings, lithium and boron concentrations of the hydrothermal fluid should reflect their contents in the underlying rocks. This relationship could be used as an indicator of the rock type that the fluid encountered. Palmer (1991) showed that the Mariana Trough hydrothermal fluid has a higher boron content and lower $\delta^{11}B$ isotopic composition than the EPR fluids, which is attributed to the abundance of boron in backarc basalt magma having been influenced by the subducting slab. The Lau Basin hydrothermal fluid has boron systematics quite similar to the Mariana Trough hydrothermal fluid. Together with enrichment in other trace elements compared to the MOR fluid, the trace element composition of the Lau hydrothermal fluid could be interpreted by a similar explanation (Fouquet *et al.*, 1991a,b). The high boron content of the Suiyo seamount fluid would be consistent with the underlying rhyolitic magma, since the latter is, in general, enriched in boron compared to basaltic magma. Palmer (1991) also showed that higher boron contents occur in hydrothermal fluids of sediment-hosted systems (e.g., 1.5–1.7 mM at Guaymas Basin and 1.7–2.1 mM at Escanaba Trough). The boron abundance in the middle Okinawa Trough fluids could be attributed to extensive fluid-sediment interaction (You *et al.*, 1994).

4.6. Gas Geochemistry of the Hydrothermal Fluid

Gas species are basically unrelated to fluid-mineral interactions, but they have been shown to be good tracers for the source of the hydrothermal fluid during subseafloor circulation. For sediment-starved hydrothermal systems on ridge crests, both carbon- and helium-isotopic compositions are very close to those of MORB, indicating that they are dominantly derived from a magma (e.g., Welhan and Craig, 1983). In contrast, in sediment-hosted environments, fluid interaction with organic matter during hydrothermal circulation results in the significant increase of CO_2 and CH_4 with their ^{13}C-depleted signature (e.g., Welhan and Lupton, 1987). Helium-isotopic compositions are known to be an unequivocal geochemical indicator of the mantle component in hydrothermal fluids, because helium has less fractionation (e.g., Lupton *et al.*, 1980). Compilation of the helium-isotopic composition of various fluids from the western Pacific (Fig. 13.10) provides information on the isotopic signature of their magmatic sources beneath the hydrothermal system. The Mariana Trough hydrothermal fluid has helium-isotopic composition encompassed by the MORB range, in spite of the subtle difference in fluid chemistry. The North Fiji hydrother-

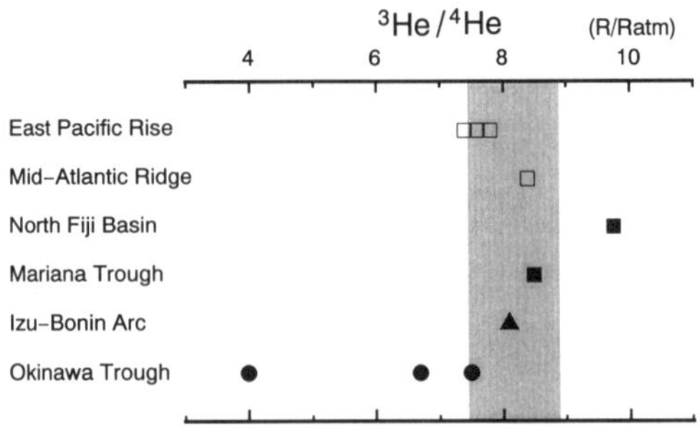

FIGURE 13.10. Comparison of helium-isotopic compositions of various hydrothermal fluids in the western Pacific. The hatched area represents the averaged helium-isotopic composition of MORB (Hilton *et al.*, 1993). Data sources: East Pacific Rise: 21°N (Welhan and Craig, 1983), 13°N (Merlivat *et al.*, 1987), Mid-Atlantic Ridge: Snake Pit site (Jean-Baptiste *et al.*, 1991), North Fiji Basin: Station 4 (Ishibashi *et al.*, 1994c), Mariana Trough: central peak (Craig *et al.*, 1987b), Izu–Bonin arc: Suiyo seamount (Tsunogai *et al.*, 1994), Okinawa Trough: Izena cauldron, Iheya Ridge (Sakai *et al.*, 1990a), Minami–Ensei knoll (Chiba *et al.*, 1992).

mal fluid has a higher $^3He/^4He$ ratio, which is likely to reflect the influence of the ocean island basalt (OIB) component into the magma source (Ishibashi *et al.*, 1994c). The Okinawa Trough hydrothermal fluids have lower helium-isotopic ratios than MORB; these ratios are commonly found in subaerial island-arc volcanic gases and geothermal systems. It is interesting to note, therefore, that the hydrothermal fluid of the Suiyo seamount has a MORB-like isotopic ratio in spite of its arc-related setting. Tsunogai *et al.* (1994) attributed this to minor contamination by crustal radiogenic helium from the thin continental crust of the Izu–Bonin intraoceanic island arc.

As shown in Table III, CO_2 concentrations in the hydrothermal fluids of arc-related systems range from 15 to 200 mM, spanning more than an order of magnitude. This range is significantly higher than that of the MOR hydrothermal fluids which are typically 10 mM. This abundance of CO_2 should be attributed to the volatile-rich nature of arc magmas compared to MORB magmas, since the gas species are dominantly derived from magmatic source. Although organic-derived CO_2 are also, at least partially, involved in the Okinawa Trough hydrothermal systems, it is worthy to note that the CO_2 concentrations of 200 mM correspond well with CO_2 contents observed in fluid inclusions (50–350 mM) contained within stockwork ores from Kuroko deposits (Pisutha–Arnond and Ohmoto, 1983). Moreover, the wide variance of CO_2 concentration in hydrothermal fluids in different tectonic settings is notable, since the major element compositions are rather restricted, being controlled by fluid-mineral equilibria, as discussed before. As an interesting demonstration of the abundance of CO_2 in the Okinawa Trough, the discharge of liquid-CO_2 bubbles forming CO_2 hydrate has been observed at two sites, the Izena hydrothermal field (Sakai *et al.*, 1990b) and the Minami–Ensei knoll (Chiba *et al.*, 1992). In both cases, the bubbles are dominated by 85–90% CO_2, and their carbon-isotopic compositions coincide with that of the dissolved CO_2 in high-temperature fluids. This suggests that these bubbles originated from a magmatic source, although the mechanism of transportation (direct degassing, separated fluids due to boiling) is not clear. Sakai *et al.* (1990b) proposed that the CO_2-rich fluids accumulate under a gas-hydrate seal barrier within a porous sediment layer and that the bubble emanation leaks through a crack of this barrier, probably caused by a change of hydrothermal activity.

4.7. Relationship of Fluid Chemistry to Seafloor Mineralization

The contents of dissolved ore-forming metal elements are controlled by various parameters, such as temperature, pH, redox condition, and chlorine concentration. For hydrothermal systems on MORs, the pH and redox condition of the hydrothermal fluid is controlled only by fluid-rock mineral equilibria (e.g., Bowers *et al.*, 1988). In contrast, in sediment-hosted systems, organic-derived species play an important role in regulating these parameters. The Okinawa Trough hydrothermal systems are examples of sediment-hosted systems located in a continental margin, where hydrothermal fluid has interacted with volcanic, terrigenous, and organic material (Sakai *et al.*, 1990a; Gamo *et al.*, 1991a,b). Hydrothermal fluid interactions within the sediment layer would be another important source for the various fluid components, including gas species (e.g., Von Damm *et al.*, 1985b). In particular, the thermal decomposition of organic matter causes significant enrichment in CO_2, CH_4, and NH_3. Involvement of these species in the hydrothermal fluid would control various parameters of the fluid chemistry, in particular pH, as NH_3 has a high pH buffering potential because of its strong dissolution capacity at high temperatures (Ishibashi, 1991). Such pH regulation may be advantageous in continuous ore precipitation. Chlorine concentration is one of the important parameters, since many metal elements are transported as species of chloride complexes in the hydrothermal fluid. Although chlorine is not related to any major alteration mineral phases, many of the chlorine concentrations in hydrothermal fluids (Campbell and Edmond, 1989), including the Suiyo and Lau fluids, deviate from that of ambient seawater. To explain the chlorine enrichment, several mechanisms have been proposed including rock hydration, hydrothermal fluid phase separation, mixing of the fluid with deep brine formed during previous phase separation (Von Damm and Bischoff, 1987), dissolution of a retrograde soluble phase (Campbell and Edmond, 1989), and trapping of mantle-derived chlorine-rich fluids, but they are still controversial.

In summary, the fluid chemistry of hydrothermal systems located in the western Pacific region is variable, reflecting their diverse geological environment and tectonic setting. Among them, a magmatic contribution of volatile components plays an important role in regulating some parameters of the fluid chemistry, such as the boiling point, pH, and redox condition, which in turn control the solubility of ore-forming metal elements. The variety of fluid chemistries provides a good opportunity to examine the relationship of fluid chemistry and mode of occurrence of seafloor mineralization.

5. SUMMARY

A compilation of 27 sites of hydrothermal mineralization in the western Pacific is given in terms of their tectonic settings, mode of mineral occurrence, and fluid chemistry. These sites comprise about 20% of the known seafloor hydrothermal sites which are dominated by those occurring at MORs. All the high-temperature hydrothermal sites discussed above occur in association with submarine volcanism found either in backarc spreading centers (14 sites), backarc rifts (3 sites), or volcanic fronts (7 sites) within arc-backarc systems in the western Pacific region. No sulfide mineralization has been identified in hot-spot volcanoes.

The lavas of the backarc spreading and backarc rift centers span a compositional range that include slightly to moderately fractionated basalt, andesite, and rhyodacite. Hydrothermal activity associated with dacitic to rhyolitic volcanism is known at ten sites in the western Pacific and eight of them accompany sulfide mineralization. Those include the PACMANUS site in eastern Manus Basin, the southern Valu Fa Ridge sites in the Lau Basin, three sites in the Okinawa Trough, and Suiyo seamount in the Izu–Ogasawara arc. Since rhyodacite is extremely rare in MOR systems, ancient rhyolite-hosted deposits (Kuroko type) that constitute 56% of the known massive sulfide deposits now on land were likely to have formed in arc/backarc setting.

There are nine high-temperature hydrothermal sites with sulfide mineralization in association with basaltic to andesitic volcanism, such as central Manus Basin, central Mariana Trough, and northern Valu Fa Ridge. These deposits seem to provide the modern analog of the Besshi-type deposits, which is defined as a cupriferous-pyrite massive sulfide deposit hosted in metamorphosed, submarine tholeiitic basalt. The deposits differ significantly from the Cyprus type by an abundance of epiclastic rocks and the lack of ophiolite suite. The ubiquitous occurrence of sand- to pebble-size epiclastic sediments, which are absent in MOR, on the flank of arc/backarc volcanoes suggests that these volcanoes were the sites for the Besshi-type mineralization. The deposit in the North Fiji Basin, located at a spreading center of the matured backarc basin, may be a candidate for the Cyprus type.

It is reasonable to deduce that the replenishing of arc magma happens periodically at an interval of 10^2–10^3 years. The associated hydrothermal activity is, therefore, likely to periodically reactivate and cease in accordance with the volcanic activity. There is not enough data to estimate the time scale of hydrothermal activity at submarine arc volcanoes. However, repeated *Shinkai 2000* dives on the Kaikata seamount suggest that the lifetime of the hydrothermal activity at frontal volcano is likely to be less than 10 years.

The hydrothermal deposits of the western Pacific are characterized by their high gold content, low pyrite/marcasite content, and lack of pyrrhotite except at a few sites. This is due to low iron content in hydrothermal fluids and higher fO_2 and fS_2 conditions. Average chemical composition of the sulfides shows certain characteristics compared with those from the MORs. The average composition of sulfide samples of the North Fiji Basin, which is in the most advanced stage of backarc spreading, closely resembles that of the EPR. On the other hand, that of the Manus Basin (PACMANUS site) shows slight enrichment in gold and arsenic. Such "epithermal elements" were likely derived from underlying dacite magma, by an analogy of the terrestrial gold mineralization at Osorezan volcano, northern Japan. The composition of sulfide samples from the Lau Basin, Mariana Trough, and Minami–Ensei matches closely that of the average black ore (Zn-Pb-Ba massive ore) of the Kuroko deposits in Japan, a strong indication of their common mechanism of formation.

Significant contribution of magmatic components to the hydrothermal fluids is also manifested by volatile elements such as sulfur, carbon, and helium. Sulfur-isotopic values of the known sulfide samples fall in the range 0‰ to 8‰, which corresponds to the compositional range of magmatic sulfur.

Enough samples of high-temperature fluid (around 300°C) have been collected and analyzed at six sites to estimate the chemical composition of end-member fluid. The fluids have some characteristics in common, such as the depletion of Mg and SO_4, the enrichment in cations relative to seawater, low pH, and reduced conditions, regardless of the type of hydrothermal systems or the tectonic setting. The spectrum of major element compositions does not diverge more than an order of magnitude from those of the MOR sites. The SiO_2 concentration of some arc-related hydrothermal systems, such as Lau Basin, Suiyo seamount, Minami–Ensei knoll, and Izena cauldron sites, lies along quartz solubility isobars very close to the pressure of the seafloor, which suggests the existence of very shallow hydrothermal circulation cells driven by a near-surface magmatic body.

The process of subseafloor phase separation appears to be more common in arc-related hydrothermal systems. This process drastically changes the chemical composition of the hydrothermal fluid and induces subseafloor sulfide precipitation. The White Lady site in the North Fiji Basin was the first example of a phase-separated hydrothermal system described in the western Pacific. Fluid chemistry is characterized by low concentrations in all of the major elements and depletion in metal elements, which corresponds to sulfate chimney formation. Several examples of arc-related hydrothermal systems are located at shallower water depths than those at MOR system, which gives rise to *P-T* conditions close to the two-phase boundary. Also, the added contribution of volatiles from arc magmas into the hydrothermal fluid would promote the initiation of boiling. In several deposit samples within the Okinawa Trough, there is evidence for explosions that are attributed to sub-seafloor boiling.

Gas species are shown to be a good tracer for the source of the hydrothermal fluid during subseafloor circulation. Compilation of the helium isotopic composition of various fluids provides information on the isotopic signature of their magmatic sources beneath the hydrothermal system. The CO_2 concentration (15–200 mM) in the hydrothermal fluids of arc-related systems is significantly higher than that of the MOR hydrothermal fluids (ave. 10 mM) and corresponds well with that (50–350 mM) observed in fluid inclusions from the stockwork part of Kuroko deposits. This is attributed to the volatile-rich nature of arc magmas, since the gas species are dominantly derived from a magmatic source. The hydrothermal systems located in the western Pacific region are variable, reflecting their diverse geologic environment and tectonic setting. Nevertheless, a magmatic contribution to the systems is more obvious than those in MOR, where several lines of evidence also suggest direct interactions between the magma chamber and the hydrothermal fluids. This is consistent with a hypothesis of magmatic hydrothermal origin of Kuroko-type deposits.

Acknowledgments

We are grateful to Brian Taylor, Steven Scott, Toshitaka Gamo, and Cornel deRonde for their critical comments and improvement of the English in an early version of the manuscript. Discussion with Hitoshi Sakai, Gary Massoth, Masahiro Aoki, Manabu Tanahashi, Hitoshi Chiba, and Urumu Tsunogai was quite helpful during the preparation of this text. Ms. Hinako Shimizu helped in the word processing of the text. We particularly wish to acknowledge the captains and crews of the various cruises and operational teams of DSRVs whose cooperation enabled us to explore the wonder of seafloor hydrothermal activity.

REFERENCES

Agency of Natural Resources and Energy. 1993. *Report of the Survey on the Occurrence of Rare Metal Resources; Evaluation of Its Potential*, Ministry of International Trade and Industry, Tokyo. (in Japanese).

Aoki, M. 1992a. Active gold mineralization in the Osorezan caldera, in *29th IGC Field Trip Guide Book*, Vol. 6, pp. 69–75, Society of Resource Geology, Tokyo.

Aoki, M. 1992b. Magmatic fluid discharging to the surface from the Osorezan geothermal system, Northern

Honshu, Japan, in *Magmatic Contributions to Hydrothermal Systems* (J. W. Hedenquist, ed.), pp. 16–21, Geological Survey of Japan Report No. 279.
Aoki, M., Matsumoto, T., Kimura, M., and Fujioka, K. 1993. Hydrothermal activity and topographic features in the western part of the Minami–Ensei Knoll, Northern Okinawa Trough, in *Proc. JAMSTEC Symp. Deep Sea Res.*, **9**:309–320. (in Japanese with English abstract).
Aoki, M., and Nakamura, K. 1989. The occurrence of chimneys in Izena Hole No. 2 ore body and texture and mineral composition of the sulfide chimneys, in *Proc. JAMSTEC Symp. Deep Sea Res.*, **5**:197–210. (in Japanese with English abstract).
Aoki, M., and Thompson, M. 1990. The Osorezan hydrothermal system, Japan: Gold bearing hot springs, *Geotherm. Resour. Council, Trans.* **14**(2):1365–1369.
Appel, P. W. U. 1979. Stratabound copper sulfides in a banded iron-formation and in basaltic tuffs in the early Precambrian Isua supracrustal belt, west Greenland, *Econ. Geol.* **74**:45–52.
Aramaki, S., Oshima, O., Uto, K., and Kawanabe, Y. 1992. Izu–Oshima and Hakone volcanoes, in *29th IGC Field Trip Guide Book*, Vol. 4., pp. 111–148, Geological Survey of Japan.
Astakhova, N. V. 1992. Hydrothermal barite in the Okhotsk Sea, in *Proc. 29th IGC 1992* (S. Ishihara, T. Urabe, and H. Ohmoto eds.), *Resource Geol.*, Special Issue No. 17, pp. 169–172.
Auzende, J.-M., Bendel, V., Fujikura, K., Geistdoerfer, P., Gracia-Mont, E., Joshima, M., Kisimoto, K., Mitsuzawa, K., Murai, M., Nojiri, Y., Ondreas, H., Pratt, C., and Ruellan, E. 1992. Resultats preliminaires des plongees du Shinkai 6500 sur la dorsale du Bassin Nord-Fidjien (SW Pacifique)-Programme STARMER, *C.R. Acad. Sci. Paris* **314**(II):491–498. (in French with English abstract).
Auzende, J.-M., Eissen, J. P., Lafoy, Y., Gente, P., and Charlou, J. L. 1988. Seafloor spreading in the North Fiji Basin (South Pacific), *Tectonophysics* **146**:317–351.
Auzende, J.-M., Honza, E., Boespflug, X., Deo, S., Eissen, J.-P., Hashimoto, J., Huchon, P., Ishibashi, J., Iwabuchi, Y., Jarvis, P., Joshima, M., Kisimoto, K., Kuwahara, Y., Lafoy, Y., Matsumoto, T., Maze, J.-P., Mitsuzawa, K., Monma, H., Naganuma, T., Nojiri, Y., Ohta, S., Otsuka, K., Okuda, Y., Ondreas, H., Otsuki, A., Ruellan, E., Sibuet, M., Tanahashi, M., Tanaka, T., and Urabe, T. 1990. Active spreading and hydrothermalism in North Fiji Basin (SW Pacific). Results of Japanese French cruise Kaiyo 87, *Mar. Geophys. Res.* **12**:269–283.
Auzende, J.-M., and Urabe, T. 1994. The STARMER French–Japanese joint project, 1987–1992, *Mar. Geol.* **116**:1–3.
Auzende, J.-M., Urabe, T., Bendel, V., Deplus, C., Eissen, J.-P., Grimaud, D., Huchon, P., Ishibashi, J., Joshima, M., Lagabrielle, Y., Mevel, C., Naka, J., Ruellan, E., Tanaka, T., and Tanahashi, M. 1991. In situ geological and geochemical study of an active hydrothermal site on the North Fiji Basin ridge, *Mar. Geol.* **98**:259–269.
Auzende, J.-M., Urabe, T., Deplus, C., Eissen, J. P., Grimaud, D., Huchon, P., Ishibashi, J., Joshima, M., Lagabrielle, Y., Mevel, C., Naka, J., Ruellan, E., Tanaka, T., and Tanahashi, M. 1989. Geological setting of an active hydrothermal site: preliminary results of the STARMER 1 cruise of the submersible Nautile in the North Fiji Basin, *C.R. Acad. Sci. Paris* **309**(II):1247–1254. (in French with English abstract).
Bäcker, H., and Lange, J. 1987. Recent hydrothermal metal accumulation: Products and conditions of formation, *NATO ASI Series C* **194**:317–337.
Baranov, E., and Levin, L. E. 1993. Geotectonic nature of localization of the Besshi and Kuroko type deposits in the mobile belts of the continents and in island arcs, in *Proc. 29th IGC 1992* (S. Ishihara, T. Urabe, and H. Ohmoto, eds.), *Resource Geol.*, Special Issue No. 17, pp. 331–335.
Baranov, B. V., Seliverstov, N. I., Murav'ev, A. V., and Muzurov, E. L. 1991. The Komandorsky Basin as a product of spreading behind a transform plate boundary, *Tectonophysics* **199**:237–269.
Bendel, V., Auzende, J.-M., Lagabrielle, Y., Grimaud, D., Fouquet, Y., and Urabe, T. 1993. The White Lady hydrothermal site in the north Fiji Basin (STARMER), *Econ. Geol.* **88**:2237–2249.
Berndt, M. E., Seyfried, W. E., Jr., and Beck, W. 1988. Hydrothermal alteration processes at midocean ridges: Experimental and theoretical constraints from Ca and Sr exchange reactions and Sr isotope ratios, *J. Geophys. Res.* **93**:4573–4583.
Bertine, K. K., and Keene, J. B. 1975. Submarine barite-opal rocks of hydrothermal origin, *Science* **188**:150–152.
Binns, R. A., and Scott, S. D. 1993. Actively forming polymetallic sulfide deposits associated with felsic volcanic rocks in the eastern Manus back-arc basin, Papua New Guinea, *Econ. Geol.* **88**:2226–2236.
Binns, R. A., Scott, S. D., Bogdanov, Y. A., Lisizin, A. P., Gordeev, V. V., Gurvich, E. G., Finlayson, E. J., Boyd, T., Dotter, L. E., Wheller, G. E., and Muravyev, K. G. 1993. Hydrothermal oxide and gold-rich sulfate deposits of Franklin Seamount, Western Woodlark Basin, Papua New Guinea, *Econ. Geol.* **88**:2122–2153.
Bischoff, J. L., and Rosenbauer, R. J. 1984. The critical point and two-phase boundary of seawater, 200–500°C, *Earth Planet. Sci. Lett.* **68**:172–180.

Bischoff, J. L., and Rosenbauer, R. J. 1985. An empirical equation of state for hydrothermal seawater, (3.2 percent NaCl), *Am. J. Sci.* **285:**725–763.

Both, R. A., Crook, K., Taylor, B., Brogan, S., Chappell, B., Frankel, E., Liu, L., Sinton, J., and Tiffin, D. 1986. Hydrothermal chimneys and associated fauna in the Manus back-arc basin, Papua New Guinea, *EOS, Trans. AGU* **67:**489–490.

Bowers, T. S. 1989. Stable isotope signatures of water-rock interaction in mid-ocean ridge hydrothermal systems: Sulfur, oxygen, and hydrogen, *J. Geophys. Res.* **94:**5775–5786.

Bowers, T. S., Campbell, A. C., Measures, C. I., Spivack, A. J., Khadem, M. and Edmond, J. M. 1988. Chemical controls on the composition of vent fluids at 13°–11°N and 21°N, East Pacific Rise, *J. Geophys. Res.* **93:**4522–4536.

Butterfield, D. A., Massoth, G. J., McDuff, R. E., Lupton, J. E., and Lilley, D. 1990. Geochemistry of hydrothermal fluids from Axial Seamount Hydrothermal Emissions Study vent field, Juan de Fuca Ridge: Subseafloor boiling and subsequent fluid-rock interaction, *J. Geophys. Res.* **95:**12895–12921.

Campbell, A. C., Bowers, T. S., Measures, C. I., Falkner, K. K., Khadem, M., and Edmond, J. M. 1988. A time series of vent fluid compositions from 21°N, East Pacific Rise (1979, 1981, 1985), and the Guaymas Basin, Gulf of California (1982, 1985), *J. Geophys. Res.* **93:**4537–4549.

Campbell, A. C., and Edmond, J. M. 1989. Halide systematics of submarine hydrothermal vents, *Nature* **342:** 168–180.

Campbell, A. C., Edmond, J. M., Colodner, D., Palmer, M. R., Falkner, K. K. 1987. Chemistry of hydrothermal fluids from the Mariana Trough back arc basin in comparison to mid-ocean ridge fluids., *EOS, Trans. AGU* **68:**1531 (abstract).

Chemineé, J. L., Stoffers, P., McMurtry, G. M., Richnow, H., Puteanus, D., and Sedwick, P. 1991. Gas-rich submarine exhalations during the 1989 eruption of Macdonald Seamount, *Earth Planet. Sci. Lett.* **107:** 318–327.

Chiba, H., Nakashima, K., Gamo, T., Ishibashi, J., Tsunogai, U., and Sakai, H. 1993. Hydrothermal activity at the Minami–Ensei Knoll, Okinawa Troguh: Chemical characteristics of hydrothermal solutions, in *Proc. JAMSTEC Symp. Deep Sea Res.*, **9:**271–282. (in Japanese with English abstract)

Chiba, H., Sakai, H., Gamo, T., Ishibashi, J., Nakashima, K., Minami, H., and Dobashi, F. 1992. Chemistry and isotopic composition of bubbles emerging from the seafloor at the Minami–Ensei Knoll, Okinawa Trough, in *Proc. JAMSTEC Symp. Deep Sea Res.*, **8:**81–87. (in Japanese with English abstract).

Christie, D. M., and Sinton, J. M. 1981. Evolution of abyssal lavas along propagating segments of the Galapagos spreading center, *Earth Planet. Sci. Lett.* **56:**321–355.

Compston, W., Kinny, P. D., Williams, I. S., and Foster, J. J. 1986. The age and Pb loss behavior of zircons from Isua supracrustal belt as determined by ion microprobe, *Earth Planet. Sci. Lett.* **80:**71–81.

Craig, H., Craig, V. K., and Kim, K. R. 1987a. PAPATUA Expedition 1: Hydrothermal vent surveys in back-arc Basins: The Lau, N. Fiji, Woodlark, and Manus basins and Havre Trough, *EOS, Trans. AGU* **68:**100 (abstract).

Craig, H., Horibe, Y., Farley, K. A., Welhan, J. A., Kim, K.-R. and Hey, R. N. 1987b. Hydrothermal vents in the Mariana Trough; Results of the first Alvin dives, *EOS, Trans. AGU* **68:**1531 (abstract).

De Carlo, E. H., McMurtry, G. M., and Yeh, H.-W. 1983. Geochemistry of hydrothermal deposits from Loihi submarine volcano, Hawaii, *Earth Planet. Sci. Lett.* **66:**483–499.

Desbruyeres, D., Alayse-Danet, A.-M., Ohta, S., and Scientific Parties of BIOLAU and STARMER Cruises. 1994. Deep-sea hydrothermal communities in southwestern Pacific back-arc basins (the North Fiji and Lau basins): Composition, microdistribution and food web, *Mar. Geol.* **116:**227–242.

Drummond, S. E., and Ohmoto, H. 1985. Chemical evolution and mineral deposition in boiling hydrothermal system, *Econ. Geol.* **80:**126–147.

Eissen, J.-P., Lefevere, P., Maillet, P., Morvant, G., and Nohara, M. 1991. Petrology and geochemistry of the central North Fiji Basin spreading center (SW Pacific) between 16°S and 22°S, *Mar. Geol.* **98:**201–239.

Eissen, J.-P., Nohara, M., Cotten, J., and Hirose, K. 1994. North Fiji Basin basalts and their magma sources. Part 1: Incompatible element constraints, *Mar. Geol.* **116:**153–178.

Fouquet, Y., and the NAUTILAU Group. 1990. Hydrothermal activity in the Lau Basin, *EOS, Trans. AGU* **71:** 678–679.

Fouquet, Y., von Stackelberg, U., Charlou, J. L., Donval, J. P., Erzinger, J., Foucher, J. P., Herzig, P., Muhe, R., Soakai, S., Wiedicke, M., and Whitechurch, H. 1991a. Hydrothermal activity and metallogenesis in the Lau back-arc basin, *Nature* **349:**778–781.

Fouquet, Y., von Stackelberg, U., Charlou, J.-L., Donval, J.-P., Foucher, J.-P., Erzinger, J., Herzig, P., Muhe, R.,

Wiedicke, M., Soakai, S., and Whitechurch, H. 1991b. Hydrothermal activity in the Lau back-arc basin: Sulfides and water chemistry, *Geology* **19:**303–306.

Fouquet, Y., von Stackelberg, U., Charlou, J.-L., Erzinger, J., Herzig, P. M., Muhe, R., and Wiedicke, M. 1993. Metallogenesis in back-arc environments: The Lau Basin example, *Econ. Geol.* **88:**2154–2181.

Francheteau, J., Needham, H. D., Choukroune, P., Juteau, T., Ségret, M., Ballard, R. D., Fox, P. J., Normark, W., Carranza, A., Cordova, D., Guerrero, J., Ranjin, C., Bougault, H., Gambon, P., and Hekinian, R. 1979. Massive deep-sea sulfide ore deposits discovered on the East Pacific Rise, *Nature* **277:**523–528.

Franklin, J. M., Lydon, J. W., and Sangster, D. F. 1981. Volcanic-associated massive sulfide deposits, *Econ. Geol.*, Special Issue, 75th Anniversary Volume: 485–627.

Fryer, P. 1993. The relationship between tectonic deformation, volcanism, and fluid venting in the southeastern Mariana convergent plate margin, in *Proc. JAMSTEC Symp., Deep Sea Res.*, **9:**161–179.

Fryer, P., Gill, J., Jackson, M., Fiske, R., Hochstaedter, A., McMurtry, G., Sedwick, P., Malahaoff, A., Mouginins-Mark, P., and Horikoshi, E. 1987a. Results of recent *Alvin* studies of the Kasuga volcanoes, northern Mariana Arc, *EOS, Trans. AGU* **68:**1531 (abstract).

Fryer, P., Haggerty, J., Tilbrook, B., Sedwick, P., Johnson, L., Saboda, K., Newsom, S., Karig, D., Uyeda, S., and Ishii, T. 1987b. Results of *Alvin* studies of Mariana forearc serpentinite diapirism, *EOS, Trans. AGU* **68:**1534 (abstract).

Fryer, P., Saboda, K. L., Johnson, L. E., Mackay, M. E., Moore, G. F., and Stoffers, P. 1990. Conical seamount: SeaMARCII, *Alvin* submersible, and seismic-reflection studies, *Proc. ODP, Init. Repts.*, 125, (P. Fryer, J. A. Pearce, L. B. Stokking *et al.*, eds.), pp. 69–80, Ocean Drilling Program, College Station, TX.

Fujikura, K., Hashimoto, J., Segawa, S., and Fujiwara, Y. 1993. Thermal tolerance of White Blind Crab, Bythograeidea, inhabited at hydrothermal vents, in *Proc. JAMSTEC Symp. Deep Sea Res.*, **9:**383–391. (in Japanese with English abstract)

Fujioka, K. 1983. Where were the "Kuroko deposits" formed, looking for the present day analogy, *Mining Geol.*, Special Issue, No. 11:55–68. (in Japanese with English abstract)

Gamo, T., Chiba, H., Fujioka, K., Ishibashi, J., Kelly, K., Masuda, H., Ohta, S., Reysenbach, A.-L., Rona, P., Shibata, T., Tamaoka, J., Tanaka, H., and Yamaguchi, T. 1993b. Alice springs hydrothermal field, mid-Mariana Trough 1992 revisit, *EOS, Trans. AGU* **43:**361 (abstract).

Gamo, T., Ishibashi, J., Sakai, H., Kodera, M., Igarashi, G., Ozima, M., Akagi, T., and Masuda, A. 1987. Geochemistry of hydrothermal solutions in the Okinawa Trough: Report on Dive 235 of the SHINKAI2000, in *Proc. JAMSTEC Symp. Deep Sea Res.*, **3:**213–224. (in Japanese with English abstract).

Gamo, T., Sakai, H., Ishibashi, J., Oomori, T., Chiba, H., Shitashima, K., Nakashima, K., Tanaka, Y., and Masuda, H. 1991b. Growth mechanism of the hydrothermal mounds at the CLAM site, Mid Okinawa Trough, inferred from their morphological, mineralogical, and chemical characteristics, in *Proc. JAMSTEC Symp. Deep Sea Res.* **7:**163–184. (in Japanese with English abstract)

Gamo, T., Sakai, H., Ishibashi, J., Nakayama, E., Isshiki, K., Matsuura, H., Shitatshima, K., Takeuchi, K., and Ohta, S. 1993c. Hydrothermal plumes in the eastern Masnus Basin, Bismarck Sea: CH_4, Mn, Al and pH anomalies, *Deep-Sea Res.* **40:**2335–2349.

Gamo, T., Sakai, H., Kim, E. S., Shitashima, K., and Ishibashi, J. 1991a. High alkalinity due to sulfate reduction in the CLAM hydrothermal field, Okinawa Trough, *Earth Planet. Sci. Lett.* **107:**328–338.

Gamo, T., and Shipboard Scientific Party. 1993a. Revisits to the mid-Mariana Trough hydrothermal site and discovery of new venting in the southern Mariana region by the Japanese Submersible Shinkai 6500, *InterRidge News* **2:**11–14.

Gracia, E., Ondreas, H., Bendel, V., and STARMER Group. 1994. Multi-scale morphologic variability of the North Fiji Basin ridge (southwest Pacific), *Mar. Geol.* **116:**133–151.

Grimaud, D., Ishibashi, J., Lagabrielle, Y., Auzende, J. M., and Urabe, T. 1991. Chemistry of hydrothermal fluids from the 17°S active site on the North Fiji Basin ridge (SW Pacific), *Chem. Geol.* **93:**209–218.

Haggerty, J. A. 1987a. Cold-water, Deep-sea chimneys from the Mariana forearc serpentinite seamounts, *EOS, Trans. AGU* **68:**1534 (abstract).

Haggerty, J. A. 1987b. Petrology and geochemistry of Neogene sedimentary rocks from Mariana forearc seamounts, in *Seamounts, Islands and Atolls* (B. H. Keating, P. Fryer, R. Batiza, and G. W. Boehlert, eds.), Geophys. Monogr. Ser., Vol. 43, pp. 175–186, American Geophysical Union, Washington, DC.

Halbach, P., Nakamura, K. I., Wahsner, M., Lange, J., Sakai, H., Kaeselitz, L., Hansen, R. D., Yamano, M., Post, J., Prause, B., Seifert, R., Michaelis, W., Teichmann, F., Kinoshita, M., Maerten, A., Ishibashi, J., Czervinski, S., and Bulm, N. 1989. Probable modern analogue of Kuroko-type massive sulphide deposits in the Okinawa Trough back-arc basin, *Nature* **338:**496–499.

Halbach, P., Pracejus, B., and Andreas M. 1993. Geology and mineralogy of massive sulfide ores from the central Okinawa Trough, Japan, *Econ. Geol.* **88:**2210–2225.

Hannington, M. D. 1989. The geochemistry of gold in modern sea-floor hydrothermal systems and implications for gold mineralization in ancient volcanogenic massive sulfides, Ph.D thesis, Univ. of Toronto, 554 p.

Hashimoto, J., Fujikura, K., and Hotta, H. 1990. Observation of deep sea biological communities at the Minami–Ensei knoll, in *Proc. JAMSTEC Symp. Deep Sea Res.* **6:**167–180. (in Japanese with English abstract)

Hashimoto, J., Fujikura, K., Ohta, S., and Miura, T. 1993. Observations of hydrothermal vent communities at the Minami–Ensei Knoll-2, in *Proc. JAMSTEC Symp. Deep Sea Res.* **9:**327–336. (in Japanese with English abstract)

Hawkins, J. 1986. "Black smoker" vent chimneys, *EOS, Trans. AGU* **67:**430.

Hawkins, J., and Helu, S. 1986. Polymetallic sulphide deposits from "black smoker" chimney, Lau Basin, *EOS, Trans. AGU* **67:**378.

Hawkins, J. W., Lonsdale, P. F., Macdougall, J. D., and Vope, A. M. 1990. Petrology of the axial ridge of the Mariana Trough back arc spreading center, *Earth Planet. Sci. Lett.* **100:**226–250.

Hedenquist, J. W., White, N. C., and Siddeley, G. (eds.). 1990. *Epithermal Gold Mineralization of the Circum-Pacific*, Association of Explorative Geochemistry, Special Publication No. 16ab.

Herzig, P. M., Hannington, M. D., Fouquet, Y., von Stackelberg, U., and Petersen, S. 1993. Gold-rich polymetallic sulfides from the Lau back arc and implications for the geochemistry of gold in sea-floor hydrothermal systems of the Southwest Pacific, *Econ. Geol.* **88:**2182–2225.

Herzig, P., Hannington, M., McInnes, B., Stoffers, P., Villinger, H., Seifert, R., Binns, R., and Liebe, T. 1994. Submarine volcanism and hydrothermal venting studied in Papua, New Guinea, *EOS, Trans. AGU* **75:** 513–516.

Herzig, P., von Stackelberg, U., and Petersen, S. 1990. Hydrothermal mineralization from the Valu Fa Ridge, Lau backarc basin (SW Pacific), *Mar. Mining* **9:**271–301.

Hessler, R. R., France, S. C., and Boudrias, M. A. 1987. Hydrothermal vent communities of the Mariana back-arc basin, *EOS, Trans. AGU* **68:**1531 (abstract).

Hessler, R. R., and Lonsdale, P. F. 1991. The biogeography of the Mariana Trough hydrothermal vents, in: *Marine Biology—Its Accomplishment and Future Prospect* (J. Mauchline and T. Nemoto, eds.), pp. 165–182, Hokusensha, Tokyo.

Hessler, R. R., Lonsdale, P., and Hawkins, J. 1988. Patterns on the ocean floor, *New Sci.* **117:**47–51.

Hilton, D. R., Hammerschimdt, K., Loock, G., and Friederichsen, H. 1993. Helium and argon isotope systematics of the central Lau Basin and Valu Fa Ridge: Evidence of crust/mantle interactions in a back-arc basin, *Geochim. Cosmochim. Acta* **57:**2819–2841.

Hochstaedter, A. G., Gill, J. B., Kusakabe, M., Newman, S., Pringle, M., Taylor, B., and Fryer, P. 1990. Volcanism in the Sumisu rift. I: Major element volatile, and stable isotope geochemistry, *Earth Planet. Sci. Lett.* **100:**179–194.

Honza, E. 1991. The Tertiary arc chain in the western Pacific, *Tectonophysics* **187:**285–303.

Honza, E., Auzende, J. M., and KAIYO88 Shipboard Party. 1989. Geology of the rift system in the North Fiji Basin: Results of Japan–France cooperative research on board KAIYO88, *La Mer* **27:**53–61.

Honza, E., and Tamaki, K. 1985. Bonin arc, in *The Ocean Basins and Margins, 7A, The Pacific Ocean* (A. E. M. Nairn, F. G. Stehli, and S. Uyeda, eds.), pp. 459–502, Plenum Press, New York.

Horibe, Y., Kim, K. R., and Craig, H. 1986. Hydrothermal methane plumes in the Mariana back-arc spreading centre, *Nature* **324:**131–133.

Hussong, D. M., and Uyeda, S. 1981. Tectonic processes and the history of the Mariana arc: A systhesis of the results of Deep Sea Drilling Project Leg 60, in *Init. Repts. DSDP*, 60 (D. M. Hussong and S. Uyeda, *et al.*, eds.), pp. 909–929, Deep Sea Drilling Project, Washington, DC.

Iizasa, K. 1993a. Petrographic investigations of seafloor sedeiments from the Kita–Bayonnaise submarine caldera, Shichito–Iwo Jima Ridge, Izu–Ogasawara arc, northwestern Pacific, *Mar. Geol.* **112:**271–290.

Iizasa, K. 1993b. Assessment of the hydrothermal contribution to seafloor sediements in the Myojinsho submarine caldera, Shichito–Iwo Jima Ridge, Izu–Ogasawara arc, Japan, *Mar. Geol.* **114:**119–132.

Iizasa, K., Terashima, S., Sasaki, M., and Marumo, K. 1993. Hydrothermal mineralization in the Kita–Bayonnaise submarine caldera, Izu–Ogasawara arc, in *Proc. JAMSTEC Symp. Deep Sea Res.* **9:**105–115. (in Japanese with English abstract)

Iizasa, K., Yuasa, M., and Yokota, S. 1992. Mineralogy and geochemistry of volcanogenic sulfides from the Myojinsho submarine caldera, the Shichito–Iwo Jima Ridge, Izu–Ogasawara arc, northwestern Pacific. *Mar. Geol.* **108:**39–58.

Ishibashi, J. 1991. Geochemical studies of hydrothermal fluids in the middle Okinawa Trough back arc basin, Ph.D Thesis, Univ. Tokyo, 131p.

Ishibashi, J., Grimaud, D., Nojiri, Y., Auzende, J. M., and Urabe, T. 1994a. Fluctuation of chemical compositions of the phase-separated hydrothermal fluid from the North Fiji Basin Ridge, *Mar. Geol.* **116:**215–226.

Ishibashi, J., Sano, Y., Wakita, H., Gamo, T., Tsutsumi, M., and Sakai, H. 1990. Geochemical studies on the hydrothermal activity in the mid-Okinawa Trough: Characterization of hydrothermal fluids from chemical and isotopical composition of the gas components, in *Proc. JAMSTEC Symp. Deep Sea Res.* **6:**63–68. (in Japanese with English abstract)

Ishibashi, J., Tsunogai, U., Wakita, H., Watanabe, K., Kajimura, T., Shibata, A., Fujiwara, Y., and Hashimoto, J. 1994b. Chemical composition of hydrothermal fluids from the Suiyo and the Mokuyo Seamounts, Izu–Bonin Arc, in *JAMSTEC Journal of Deep Sea Res. 10*, pp. 89–97. (in Japanese with English abstract)

Ishibashi, J., Wakita, H., Nojiri, Y., Grimaud, D., Jean-Baptiste, P., Gamo, T., Auzende, J. M., and Urabe, T. 1994c. Helium and carbon geochemistry of hydrothermal fluids from the North Fiji Basin spreading ridge, Southern Pacific, *Earth Planet. Sci. Lett.* **128:**183–197.

Ishizuka, H., Kawanobe, Y., and Sakai, H. 1990. Petrology and geochemistry of volcanic rocks dredged from the Okinawa Trough, an active back-arc basin, *Geochem. J.* **24:**75–92.

Izawa, E., and Aoki, M. 1991. Geothermal activity and epithermal gold mineralization in Japan, *Episodes* **14:** 269–273.

Izawa, E., Motomura, Y., Tanaka, T., and Kimura, M. 1991. Hydrothermal carbonate chimneys in the Iheya Ridge of the Okinawa Trough, in *Proc. JAMSTEC Symp. Deep Sea Res.* **7:**185–192. (in Japanese with English abstract)

Japanese DELP Research Group on Backarc Basins. 1986. Report on DELP 1984 cruises in the middle Okinawa Trough. Part 1: General outline, *Bull. Earthquake Res. Inst. Univ. Tokyo* **61:**159–165.

Japanese DELP Research Group on Backarc Basins. 1991. Report on DELP 1988 cruises in the southern Okinawa Trough. Part 1: General outline, *Bull. Earthquake Res. Inst. Univ. Tokyo* **66:**1–16.

Jarvis, P., Hughes-Clarke, J., Tiffin, D., Tanahashi, M., and Kroenke, L. 1994. The western Fiji transform fault and its role in the dismemberment of the Fiji Platform, *Mar. Geol.* **116:**57–68.

Jean-Baptiste, P., Charlou, J. L., Stievenard, M., Donval, J. P., Bougault, H., and Mevel, C. 1991. Helium and methane measurements in hydrothermal fluids from the mid-Atlantic Ridge: The Snake Pit site at 23°N, *Earth Planet. Sci. Lett.* **106:**17–28.

Jenner, G. A., Cawood, P. A., Rautenschlein, M., and White, W. M. 1987. Composition of back-arc basin volcanics, Valu Fa Ridge, Lau Basin: Evidence for slab-derived component in their mantle source, *J. Volcanol. Geotherm. Res.* **32:**209–222.

Johnson, L. E., Fryer, P., Masuda, H., Ishii, T., Gamo, T. 1993. Hydrothermal vent deposits and two magam sources for volcanoes near 13°20′N in the Mariana backarc: A view from Shinkai 6500, *EOS, Trans. AGU* **74:**681 (abstract).

Kanehira, K., and Tatsumi, T. 1970. Bedded cupriferous iron sulphide deposits in Japan, a review, in *Volcanism and Ore Genesis*, pp. 51–76, Tokyo Press, Tokyo.

Karig, D. E. 1985. Kinematics and mechanism of deformation across some accreting forearcs, in *Formation of Active Ocean Margins* (N. Nasu *et al.*, eds.), pp. 155–177, Terra Scientific, Tokyo.

Karpov, G. A., and Naboko, S. I. 1990. Metal contents of recent thermal waters, mineral precipitates and hydrothermal alteration in active geothermal fields, Kamchatka, *J. Geochem. Explor.* **36:**57–71.

Kastner, M., Craig, H., and Sturz, A. 1987. Hydrothermal deposition in the Mariana Trough; Preliminary mineralogical investigations, *EOS, Trans. AGU* **68:**1531 (abstract).

Kasuga, S., and Kato, Y. 1992. Discovery of hydrothermal ore deposits in the crater of the Suiyo Smt. on the Izu–Ogasawara arc, in *Proc. JAMSTEC Symp. Deep Sea Res.* **8:**249–255. (in Japanese with English abstract)

Kimura, M., Kaneoka, I., Kato, Y., Yamamoto, S., Kushiro, I., Tokuyama, H., Kinoshita, H., Isezaki, N., Masaki, H., Ohsida, A., Uyeda, S., and Hilde, T. W. C. 1986. Report on DELP 1984 cruise in the middle Okinawa Trough. 5: Topography and geology of the central grabens and their vicinity, *Bull. Earthquake Res. Inst. Univ. Tokyo* **61:**269–310.

Kimura, M., Tanaka, T., Kyo, M., Ando, M., Oomori, T., Izawa, E., and Yoshikawa, I. 1989. Study of topography, hydrothermal deposits and animal colonies in the middle Okinawa Trough hydrothermal areas using the submersible "SHINKAI2000" system, in *Proc. JAMSTEC Symp. Deep Sea Res.* **5:**223–244. (in Japanese with English abstract)

Kimura, M., Uyeda, S., Kato, Y., Yamano, M., Gamo, T., Sakai, H., Kato, S., Izawa, E., and Oomori, T. 1988. Active hydrothermal mounds in the Okinawa Trough backarc basin, Japan, *Tectonophysics* **145:**319–324.

Kinoshita, M. 1990. Heat flow anomaly in some western Pacific trench–arc-backarc systems associated with interstitial water circulation, Ph.D. thesis, Univ. Tokyo, 74p.

Kroenke, L. W. 1984. Cenozoic tectonic development of the Southwest Pacific, U.N. ESCAP, CCOP/SOPAC Tech. Bull., vol. 6.

Kusakabe, M., Mayeda, S., and Nakamura, E. 1990. S, O and Sr isotope systematics of active vent materials from the Mariana backarc basin spreading axis at 18°N, *Earth Planet. Sci. Lett.* **100:**275–282.

Lafoy, Y., Auzende, J.-M., Ruellan, E., Huchon, P. and Honza, E. 1991. The 16°40′S triple junction in the North Fiji Basin (SW Pacific), *Mar. Geopys. Res.* **12:**285–296.

Leinen, M., McDuff, R., Delaney, J., Becker, K., and Schultheiss, P. 1987. Off-axis hydrothermal activity in the Mariana mound field, *EOS, Trans. AGU* **68:**1531 (abstract).

Lisitsyn, A. P., Binns, R. A., Boganov, Yu. A., Scott, S., Zonenshayn, L. P., Gordeyev, V. V., Gurvich, Ye. G., Murav'yev, K. G., and Serova, V. V. 1991. Active hydrothermal activity at Franklin seamount, western Woodlark Sea (Papua New Guinea), *Int. Geol. Rev.* **33:**914–929.

Lisitsyn, A. P., Crook, K. A. W., Bogdanov, Yu. A., Zonenshayn, L. P., Murav'yev, K. G., Tufar, W., Gurvich, Ye. G., Gordeyev, V. V., and Ivanov, G. V. 1993. A hydrothermal field in the rift zone of the Manus Basin, Bismark Sea, *Inter. Geol. Rev.* **35:**05–126.

Lisitsyn, A. P., Malahoff, A. R., Bogdanov, Yu. A., Sione Soakai, Zonenshayn, L. P., Gurvich, Ye. G., Murav'yev, K. G., and Ivanov, G. V. 1992. Hydrothermal formations in the northern part of the Lau Basin, Pacific Ocean, *Inter. Geol. Rev.* **34:**828–847.

Lupton, J. E., Klinkhammer, G. P., Normark, W. R., Haymon, R., MacDonald, K. C., Weiss, R. F., and Craig, H. 1980. Helium-3 and manganese at the 21°N East Pacific Rise hydrothermal site, *Earth Planet. Sci. Lett.* **50:**115–127.

Massoth, G. J., Butterfield, D. A., Lupton, J. E., McDuff, R. E., Lilley, M. D., and Jonasson, I. R. 1989. Submarine venting of phase-separated hydrothermal fluids at Axial volcano, Juan de Fuca Ridge, *Nature* **340:**702–705.

Masuda, H., Gamo, T., Freyer, P., Ishii, T., Jonson, L.E., Tanaka, H., Tsunogai, U., Matsumoto, S., Masumoto, S., and Fujioka, K. 1993. The major element chemistry of submarine volcanic rocks from the Southern Mariana Trough and its relation to the topography, in *Proc. JAMSTEC Symp. Deep Sea Res.* **9:**181–189. (in Japanese with English abstract)

Masuda, H., Ishibashi, J., Kato, Y., Gamo, T., and Sakai, H. 1987. Oxygen isotope ratio and trace element composition of hydrothermal sediments from Okinawa Trough, collected with SHINKAI2000, dive 231, in *Proc. JAMSTEC Symp. Deep Sea Res.* **3:**225–231. (in Japanese with English abstract)

McBirney, A. R. 1984. *Igneous Petrology*, Freeman & Cooper, San Francisco.

McMurtry, G. M., Sedwick, P., Fryer, P., Vonder Haar, D. L., and Yeh, H. W. 1993. Unusual geochemistry of hydrothermal vents in submarine arc volcanoes: Kasuga seamounts, northern Mariana arc, *Earth Planet. Sci. Lett.* **114:**517–528.

Merlivat, L., Pineau, F., and Javoy, M. 1987. Hydrothermal vent waters at 13°N on the East Pacific Rise: Isotopic composition and gas concentration, *Earth Planet. Sci. Lett.* **84:**100–108.

Michard, A., Michard, G., Stüben, D., Stoffers, P., Cheminee, J.- L., and Binard, N. 1993. Submarine thermal springs associated with young volcanoes: The Teahitia vents, Society Islands, Pacific Ocean, *Geochim. Cosmochim. Acta* **57:**4977–4986.

Miller, A. R., Densmore, C. D., Degens, E. T., Pocklington, R., and Jokela, A. 1966. Hot brines and recent iron deposits in deeps of the Red Sea, *Geochim. Cosmochim. Acta* **30:**341–359.

Mitchell, A. H. G., and Bell, J. D. 1973. Island-arc evolution and related mineral deposits, *J. Geol.* **81:**381–405.

Mitsuzawa, K., Momma, H., Hotta, H., and Deep Sea Research Group. 1989. Measurements of water temperature and current around the submarine caldera in the KC Kaikata seamount, in *Proc. JAMSTEC, Symp. Deep Sea Res.* **5:**47–55. (in Japanese with English abstract)

Momma, H., Hashimoto, J., Tanaka, T., and Deep Sea Research Group. 1989. Preliminary report of deep tow surveys in the Okinawa Trough, DK88-2-OKN-LEG1,2, JAMSTEC Technical Report, Vol. 21, pp. 203–221. (in Japanese with English abstract)

Moorby, S. A., Cronan, D. S., and Glasby, G. P. 1984. Geochemistry of hydrothermal Mn-oxide deposits from the SW Pacific island arc, *Geochim. Cosmochim. Acta* **48:**433–441.

Moore, W. S., and Stakes, D. 1990. Ages of barite-sulfide chimneys from the Mariana Trough, *Earth Planet. Sci. Lett.* **100:**265–274.

Mottl, M. J. 1992. Pore waters from serpentinite seamounts in the Mariana and Izu–Bonin forearcs, Leg 125: Evidence for volatiles from the subducting slab, in *Proc. ODP, Sci. Results*, 125 (P. Fryer, J. A. Pearce, L. B. Stokking *et al.*, eds.), pp. 373–385, Ocean Drilling Program, College Station, TX.

Nagaoka, N., Kasuga, S., and Kato, Y. 1992. Geology of Mokuyo smt. Doyo smt., and Suiyo smt. in the Sichiyo seamounts on the Ogasawara arc, in *Proc. JAMSTEC Symp. Deep Sea Res.* **8**:237–248. (in Japanese with English abstract)

Nagaoka, N., Okino, K., and Kato, S. 1991. Landforms of submarine volcanoes in central part of the Izu–Ogasawara arc, by multi-beam sounding system, *Rep. Hydrogr. Res.* **27**:145–172. (in Japanese with English abstract)

Naka, J., and Deep Sea Research Group. 1989a. Sea bottm observation around the KC peak of the Kaikata seamount, Bonin Islands, in *Proc. JAMSTEC Symp. Deep Sea Res.* **5**:57–65. (in Japanese with English abstract)

Naka, J., and Deep Sea Research Group. 1989b. Volcanic products of the Iheya Ridge, central Okinawa Trough, in *Proc. JAMSTEC Symp. Deep Sea Res.* **5**:245–257. (in Japanese with English abstract)

Nakamura, K., Kato, Y., Kimura, M., Ando, M., and Kyo, M. 1989. Occurrence and distribution of the hydrothermal ore deposits at the Izena Hole in the Okinawa Trough—Summary of the knowledge in 1988, in *Proc. JAMSTEC Symp. Deep Sea Res.*, pp. 183–189. (in Japanese with English abstract)

Nakamura, K., Marumo, K., and Aoki, M. 1990. Discovery of a black smoker vent and a pockmark emitting CO_2-rich fluid on the seafloor hydrothermal mineralization field at the Izena cauldron in the Okinawa Trough, in *Proc. JAMSTEC Symp. Deep Sea Res.* **6**:33–50. (in Japanese with English abstract)

Nakao, S., Yuasa, M., Nohara, M., and Usui, A. 1990. Submarine hydrotheraml activity in the Izu–Ogasawara arc, western Pacific, *Rev. Aqua. Sci.* **3**:95–115.

Nakashima, K., Sakai, H., Yoshida, H., Chiba, H., Tanaka, Y., Gamo, T., Ishibashi, J., and Tsunogai, U. 1993. Mineralogical and fluid inclusion studies on some hydrothermal deposits at the Iheya Ridge and Minami–Ensei Knoll, Okinawa Trough: Report of research dive 621 and hydrothermal precipitates collected by dives 487 and 622 of "Shinkai2000," in *Proc. JAMSTEC Symp. Deep Sea Res.* **9**:253–269. (in Japanese with English abstract)

Nedachi, M., Ueno, H., Oki, K., Shiga, Y., Hayasaka, S., Ossaka, J., Nogami, K., Ito, N., and Hashimoto, J. 1991. Sulfide veinlets and the surrounding marine sediments in the fumarole area in the Wakamiko caldera, northern Kagoshima Bay, in *Proc. JAMSTEC Symp. Deep Sea Res.* **7**:235–243. (in Japanese with English abstract)

Nedachi, M., Ueno, H., Ossaka, J., Nogami, K., Hashimoto, J., Fujikura, K., and Miura, T. 1992. Hydrothermal ore deposits on the Minami–Ensei Knoll of the Okinawa Trough—Mineral Assemblages, in *Proc. JAMSTEC Symp. Deep Sea Res.* **8**:95–106. (in Japanese with English abstract)

Nishimura, A., Mita, N., and Nohara, M. 1992b. Pelagic and hemipelagic sediments of the Izu-Bonin region, Leg 126: Geochemical and compositional features, in *Proc. ODP, Sci. Results*, 126 (B. Taylor, K. Fujioka *et al.*, eds.), pp. 487–503, Ocean Drilling Program, College Station, TX.

Nishimura, A., and Murakami, F. 1988. Sedimentation of the Sumisu rift, Izu–Ogasawara arc, *Bull. Geol. Surv. Jpn.* **39**:39–61.

Nishimura, A., Rodolfo, K. S., Koizumi, A., Gill, J., and Fujioka, K. 1992a. Episodic deposition of Pliocene–Pleistocene pumice from the Izu–Bonin arc, Leg 126, in *Proc. ODP, Sci. Results*, 126 (B. Taylor, K. Fujioka *et al.*, eds.), pp. 3–21, Ocean Drilling Program, College Station, TX.

Nohara, M., Hirose, K., Eissen, J.-P., Urabe, T., and Joshima, M. 1994. The North Fiji Basin basalts and their magma sources. Part 2: Sr-Nd isotopic and trace element constraints, *Mar. Geol.* **116**:179–195.

Nojiri, Y., Ishibashi, J., Kawai, T., Otsuki, A., and Sakai, H. 1989. Hydrothermal plumes along the North Fiji Basin spreading axis, *Nature* **342**:667–670.

Ohmoto, H., Tanimura, S., Date, J., and Takahashi, T. 1983. Geologic setting of the Kuroko deposits, Japan, *Econ. Geol. Monogr.* **5**:9–54.

Ohta, S. 1990. Deep-sea submersible survey of the hydrothermal vent community on the northeastern slope of the Iheya Ridge, the Okinawa Trough, in *Proc. JAMSTEC Symp. Deep Sea Res.* **6**:145–156. (in Japanese with English abstract)

Okutani, T., and Ohta, S. 1988. A new gastropod mollusk assciated with hydrothermal vents in the Mariana back-arc basin, western Pacific, *Venus (Jpn. J. Malacology)* **47**:1–9.

Palmer, M. R. 1991. Boron isotope systematics of hydrotherm fluids and tourmalines: a synthesis, *Chem. Geol. (Isotope Geosci. Sect.)* **94**:111–121.

Palmer, M. R., and Edmond, J. M. 1989. The strontium isotope budget of the modern ocean, *Earth Planet. Sci. Lett.* **92**:11–26.

Pisutha-Arnond, V., and Ohmoto, H. 1983. Thermal history, and chemical and isotopic compositions of the ore-forming fluids responsible for the Kuroko massive sulfide deposits in the Hokuroku district of Japan, *Econ. Geol. Monogr.* **5**:523–558.

Puteanus, D., Glasby, G. P., Stoffers, P., and Kuznendorf, H. 1991. Hydrothermal iron-rich deposits from the Teahitia–Mehetia and Macdonald hot spot areas, Southwest Pacific, *Mar. Geol.*, **98**:389–409.

Rona, P. A. 1983. Exploration for hydrothermal mineral deposits at seafloor spreading centers, *Mar. Mining* **4**:7–38.

Rona, P. A. 1984. Hydrothermal mineralization at seafloor spreading centers, *Earth Sci. Review* **20**:1–104.

Rona, P. A. 1988. Hydrothermal mineralization at oceanic ridges, *Can. Mineral.* **26**:431–465.

Rona, P A., and Scott, S. D. 1993. Preface for a special issue on sea-floor hydrothermal mineralization: New perspectives, *Econ. Geol.* **88**:1935–1976.

Sakai, H., Gamo, T., Kim, E.-S., Tsutsumi, M., Tanaka, T., Ishibashi, J., Wakita, H., Yamano, M., and Oomori, T. 1990b. Venting of carbon dioxide-rich fluid and hydrate formation in mid-Okinawa Trough backarc basin, *Science* **248**:1093–1096.

Sakai, H., Gamo, T., Ishibashi, J., Shitashima, K., Kim., E. S., Yanagisawa, F., Tsutsumi, M., Sano, Y., Wakita, H., Tanaka, T., Matsumoto, T., Naganuma, T., and Mitsuzawa, K. 1990a. Unique chemistry of the hydrothermal solutions in the mid-Okinawa Trough backarc basin, *Geophys. Res. Lett.* **17**:2133–2136.

Sawkins, F. J. 1972. Sulfide ore deposits in relation to plate tectonics, *J. Geol.* **80**:377–397.

Sawkins, F. J. 1976. Massive sulphide deposits in relation to geotectonics, *Geol. Assoc. Can.* Special Paper **14**:221–240.

Scott, S. D. 1985. Seafloor polymetallic sulfide deposits: Modern and ancient, *Mar. Mining* **5**:191–212.

Scott, S. D. 1987. Seafloor polymetallic sulfides: Scientific curiosities or mines of the future?, in *Marine Minerals* (P. G. Teleki *et al.*, eds.), NATO ASI Series C, Vol. 194, pp. 277–300.

Scott, S.D., and Binns, R. A. 1992. An actively-forming, flesic volcanic-hosted polymetallic sulfide deposit in the southeast Manus back-arc basin of Papua New Guinea, *EOS, Trans. AGU* **73**:626 (Abstract).

Sedwick, P. N., Gamo, T., and McMurtry, G. M. 1990. Manganese and methane anomalies in the North Fiji Basin, *Deep-Sea Res.* **37**:891–896.

Sedwick, P. N., McMurtry, G. M., Hilton, D. R., and Goff, F. 1994. Carbon dioxide and helium in hydrothrmal fluids from Loihi seamount, Hawaii, USA: Temporal variability and implications for the release of mantle volatiles, *Geochim. Cosmochim. Acta* **58**:1219–1227.

Sedwick, P. N., McMurtry, G. M., and Macdougall, J. D. 1992. Chemistry of hydrothermal solutions from Pele's vent, Loihi seamount, Hawaii, *Geochim. Cosmochim. Acta* **56**:3643–3667.

Seyfried, W.E. Jr. 1987. Experimental and theoretical constraints on hydrothermal alteration processes at mid-ocean ridges, *Ann. Rev. Earth Planet. Sci.* **15**:317–335.

Seyfried, W. E. Jr., Janeckey, D. R., and Mottl, M. J. 1984. Alteration of the oceanic crust: Implications for geochemical cycles of lithium and boron, *Geochim. Cosmochim. Acta* **48**:557–569.

Shadlum, T. N., Bortnikov, N. S., Bogdanov, Yu. A., Tufar, W., Murav'yev, K. G., Gurvich, Ye. G., Muravitskaya, G. N., Korina, Ye. A., and Topa, T. 1993. Mineralogy, textures, and formation conditions of modern sulfide ores, Manus Basin rift zone, *Int. Geol. Rev.* **35**(2):127–145.

Shitashima, K. 1994. Distribution and behavior of trace metals on the Okinawa Trough, in *JAMSTEC Journal of Deep Sea Res.* 10, pp. 291–298. (in Japanese with English abstract)

Sibuet, J. C., Letouzey, J., Barbier, F., Charvet, J., Foucher, J. P., Hilde, T. W. C., Kimura, M., Chiao, L. Y., Marsset, B., Muller, C., and Stephan, J. F. 1987. Back arc extension in the Okinawa Trough, *J. Geophys. Res.* **92**:14,041–14,063.

Sillitoe, R. H. 1972. Formation of certain massive sulphide deposits at sites of sea-floor spreading, *Inst. Mining Metall. Trans.* **81**:B141–148.

Spiess, F. N., and RISE Group. 1980. East Pacific Rise; hot springs and geophysical experiments, *Science* **297**:1421–1433.

Stolper, E. M., and Newman, S. 1992. Fluids in the source regions of subduction zone magmas: Clue from the study of volatiles in Mariana Trough magmas, in *Magmatic Contributions to Hydrothermal Systems* (J. W. Hedenquist, ed.), pp. 161–169, Geol. Surv. Jpn Rep. No. 279.

Stüben, D., Bloomer, S. H., Taïbi, N. E., Neumann, Th., Bendel, V., Püschel, U., Barone, A., Lange, A., Shiying, W., Cuizhong, L., and Deyu, Z. 1992. First results of study of sulphur-rich hydrothermal activity from an island-arc environment: Esmeralda Bank in the Mariana arc, *Mar. Geol.* **103**:521–528.

Stüben, D., Taibi, N. S., McMurtry, G. M., Scholten, J., Stoffers, P., and Zhang, D. 1994. Growth history of a hydrothermal silica chimney from the Mariana backarc spreading center (southeast Pacific, 18°13′N), *Mar. Geol.* **113**:273–296.

Taira, A., and Pickering, K. T. 1991. Sediment deformation and fluid activity in the Nankai, Izu–Bonin and Japan forearc slopes and trenches, in *The Behaviour and Influence of Fluids in Subduction Zones* (J. Tarney, K. T. Pickering, R. J. Knipe, and J. F. Gewey, eds.), pp. 63–88, The Royal Society, London.

Tamaki, K., and Honza, E. 1992. Global tectonics and formation of marginal basins: Role of the western Pacific, *Episodes* **14:**224–230.

Tanahashi, M., Kisimoto, K., Joshima, Jarvis. P., Iwabuchi, Y., Ruellann, E., and Auzende, J. M. 1994. 800 km long N-S spreading system of the North Fiji Basin, *Mar. Geol.* **116:**5–24.

Tanahashi, M., Kisimoto, K., Joshima, M., Lafoy, Y., Honza, E., and Auzende, J. M. 1991. Geological structure of the central spreading system, North Fiji Basin, *Mar. Geol.* **98:**187–200.

Tanaka, T., Hotta, H., Sakai, H., Ishibashi, J., Oomori, T., Izawa, E., and Oda, N. 1990. Occurrence and distribution of hydrothermal deposits in the Izena Hole, central Okinawa Trough, in *Proc. JAMSTEC Symp. Deep Sea Res.* **6:**11–26. (in Japanese with English abstract)

Tanaka, T., Mitsuzawa, K., and Hotta, H. 1989. "SHINKAI2000" diving surveys in the east of Iheya Small ridge in the central Okinawa Trough, in *Proc. JAMSTEC Symp. Deep Sea Res.* **5:**267–281. (in Japanese with English abstract)

Taylor, B. 1979. Bismark Sea; evolution of a back arc basin, *Geology* **7:**171.

Taylor, B., Brown, G., Fryer, P., Gill, J., Hochstaedter, A., Hotta, H., Langmuir, C., Leinen, M., Nishimura, A., and Urabe, T. 1990. ALVIN-SeaBeam studies of the Sumisu rift, Izu–Bonin arc, *Earth Planet. Sci. Lett.* **100:** 127–147.

Taylor, B., and Karner, G. D. 1983. On the evolution of marginal basins, *Rev. Geophys. Space Phys.* **21:**1727–1741.

Tsunogai, U., Ishibashi, J., Wakita, H., Gamo, T., Watanabe, K., Kajimura, T., Kanayama, S., and Sakai, H. 1994. Peculiar features of Suiyo seamount hydrothermal fluids, Izu–Bonin arc: Differences from subaerial volcanism, *Earth Planet. Sci. Lett.* **126:**289–301.

Tsunogai, U., Ishibashi, J., Wakita, H., Watanabe, K., Kazimura, T., and Hashimoto, J. 1993. Dissolved gas geochemistry of submarine hydrothermal fluids from the Suiyo and the Mokuyo seamount, Izu–Bonin, Ogasawara arc, in *Proc. JAMSTEC Symp. Deep Sea Res.* **9:**92–103. (in Japanese with English abstract)

Tufar, W. 1990. Modern hydrothermal activity, formation of complex massive sulfide deposits and associated vent communities in the Manus back-arc basin, Bismark Sea (Papua New Guinea), *Mitt. Österr. geol. Ges.* **82:**183–210.

Urabe, T. 1987. Kuroko deposit modeling based on magmatic hydrothermal theory, *Mining Geol.* **37:**159–176.

Urabe, T. 1989. Mineralogical characteristics of the hydrothermal ore body at Izena Hole No.1 in comparison with Kuroko deposits, in *Proc. JAMSTEC Symp. Deep Sea Res.* **5:**191–196. (in Japanese with English abstract)

Urabe, T., and Kusakabe, M. 1990. Barite silica chimneys from the Sumisu rift, Izu–Bonin arc: Possible analog tohematitic chert associated with Kuroko deposits, *Earth Planet. Sci. Lett.* **100:**283–290.

Urabe, T., and Marumo, K. 1991. A new model for Kuroko-type deposits of Japan, *Episodes* **14:**246–251.

Urabe, T., Yuasa, M., Nakao, S., and On-board Scientists. 1987. Hydrothermal sulfides from a submarine caldera in the Shichito–Iwo Jima Ridge, northwestern Pacific, *Mar. Geol.* **74:**295–299.

Usui, A., Mellin, T.A., Nohara, M., and Yuasa, M. 1989. Structural stability of marine 10A manganates from the Ogasawara, Bonin, arc: Implication for low-temperature hydrothermal activity, *Mar. Geol.* **86:**41–56.

Usui, A., and Nishimura, A. 1992. Submersible observations of hydrothermal manganese deposits on the Kaikata seamount, Izu–Ogasawara, Bonin arc, *Mar. Geol.* **106:**203–216.

Usui, A., Yuasa, M., Yokota, S., Nohara, M., Nishimura, A., and Murakami, F. 1986. Submarine hydrothermal manganese deposits from the Ogasawara (Bonin) arc, off the Japan islands, *Mar. Geol.* **73:**311–322.

Von Damm, K. L. 1990. Seafloor hydrothermal activity: Black smoker chemistry and chimneys, *Ann. Rev. Earth Planet. Sci.* **18:**173–204.

Von Damm, K. L., and Bischoff, J. L. 1987. Chemistry of hydrothermal solutions from the southern Juan de Fuca Ridge, *J. Geophys. Res.* **92:**11,334–11,346.

Von Damm, K. L., Bischoff, J. L., and Rosenbauer, R. J. 1991. Quartz solubility in hydrothermal seawater: an experimental study and equation describing quartz solubility for up to 0.5 M NaCl solutions, *Am. J. Sci.* **291:**977–1007.

Von Damm, K. L., Edmond, J. M., Grant, B., Measures, C. I., Walden, B., and Weiss, R. F. 1985a. Chemistry of submarine hydrothermal solutions at 21°N, East Pacific Rise, *Geochim. Cosmochim. Acta* **49:**2197–2220.

Von Damm, K. L., Edmond, J. M., Measures, C. I., and Grant, B. 1985b. Chemistry of submarine hydrothermal solutions at Guaymas Basin, Gulf of California, *Geochim. Cosmochim. Acta* **49:**2221–2237.

von Stackelberg, U., Marchig, V., Muller, P., and Weiser, T. 1990. Hydrothermal mineralization in the North Fiji basins, *Geologisch. Jahrbuch* **D92:**547–613.

von Stackelberg, U., and Shipboard Scientific Party. 1985. Hydrothermal sulfide deposits in back-arc spreading centers in the Southwest Pacific, BGR circ., Vol. 2, pp. 3–14, Bundesanst. Geowiss. Rohstoff., Hannover, FRG.

von Stackelberg, U., and Shipboard Scientific Party. 1988. Active hydrothermalism in the Lau back-arc basin (SW Pacific): First results of the SONNE48 Cruise, 1987, *Mar. Mining* **7**:431–442.

Watanabe, K., and Kajimura, T. 1993. Topography, geology and hydrothermal deposits at Suiyo seamount, in *Proc. JAMSTEC Symp. Deep Sea Res.* **9**:77–89. (in Japanese with English abstract)

Watanabe, K., and Kajimura, T. 1994. The hydrothermal mineralization at Suiyo seamount, in the Izu–Ogasawara arc, *Resource Geol.* **44**:133–140. (in Japanese with English abstract)

Watanabe, K., Shibata, A., Kajimura, T., Ishibashi, J., Tsunogai, U., Aoki, M., and Nakamura, K. 1994. Survey method about the submarine volcano and its sea-floor hydrothermal ore deposit-A example of Suiyo seamount in the Izu–Ogasawara arc with the submersible "Shinkai2000," *J. Jpn. Soc. Mar. Surv. Tech.* **6**: 29–44. (in Japanese with English abstract)

Welhan, J. A., and Craig, H. 1983. Methane, hydrogen and helium in hydrothermal fluids at 21°N on the East Pacific Rise, in *Hydrothermal Processes at Seafloor Spreading Venters* (P. A. Rona, K. Bostrom, L. Laubier, and K. L. Smith, eds.), pp. 391–409, Plenum Press, New York.

Welhan, J. A., and Lupton, J. E. 1987. Light hydrocarbon gases in Guaymas Basin hydrothermal fluids: Thermogenic versus abiogenic origin, *Am. Assoc. Petrol. Geol. Bull.* **71**:215–223.

Wheat, C. G., and McDuff, R. E. 1987. Advection of pore waters in the Marianas mounds hydrothermal region as determined from nutrient profiles, *EOS, Trans. AGU* **68**:1531 (abstract).

Wheat, C. G., and McDuff, R. E. 1994. Hydrothermal flow through the Mariana Mounds: Dissolution of amorphous silica and degradation of organic matter on a mid-ocean ridge flank, *Geochim. Cosmochim. Acta* **58**:2461–2475.

Wheller, G., Binns, R., and Scott, S. 1992. The PACMANUS/PACLARK program: Search for modern "Kuroko-type" analogues in the SW Pacific, *InterRidge News* **1**:7–9.

Yamano, M., Uyeda, S., Foucher, J. P., and Sibuet, J. C. 1989. Heat flow anomaly in the middle Okinawa Trough, *Tectonophysics* **159**:307–318.

Yamano, M., Uyeda, S., Kinoshita, H., and Hilde, T. W. C. 1986. Report on DELP1984 cruises in the middle Okinawa Trough. 4. Heat flow measurements, *Bull. Earthquake Res. Inst. Univ. Tokyo* **61**:251–267.

Yamazaki, T. 1988. Heat flow in the Sumisu rift, Izu–Ogasawara (Bonin) arc, *Bull. Geol. Surv. Jpn.* **39**:63–70.

You, C.-F., Butterfield, D. A., Spivack, A. J., Gieskes, J. M., Gamo, T., and Campbell, A. J. 1994. Boron and halide systematics in submarine hydrothermal systems: Effects of phase separation and sedimentary contributions, *Earth Planet. Sci. Lett.* **123**:227–238.

Yuasa, M. 1985. Sofugan tectonic line, a new tectonic bounary separating northern and southern parts of the Ogasawara, Bonin arc, northwest Pacific, in *Formation of Active Ocean Margins* (N. Nasu *et al.*, eds.), pp. 483–496, Terra Scientific, Tokyo.

Yuasa, M., Murakami, F., Saito, E., and Watanabe, K. 1991. Submarine topography of seamounts on the volcanic front of the Izu–Ogasawara, Bonin arc, *Bull. Geol. Surv. Jpn* **42**:703–743.

Yuasa, M., Urabe, T., and Murakami, F. 1988. A submersible study of hydrothermal fields at the Kaikata seamount, Izu–Ogasawara arc, in *Proc. JAMSTEC Symp. Deep Sea Res.* **4**:129–139. (in Japanese with English abstract)

14

Tectonic and Magmatic Controls on Backarc Basin Sedimentation
The Mariana Region Reexamined

Kathleen M. Marsaglia and Kathleen A. Devaney

ABSTRACT

Previous models for intraoceanic backarc basin sedimentation are static in that they focus on a long-lived, stationary, magmatic source along the arc axis. These models are adequate in the case of backarc rifting, but a more dynamic model is needed in the case of forearc rifting. We present a revised model for backarc basin sedimentation that reflects progressive shifts in volcanism and associated depocenters through the history of a backarc basin developed by forearc rifting. This model is based on recent Ocean Drilling Program results from the Lau Basin and a reexamination of the sedimentary sequences recovered on Deep Sea Drilling Project legs across the Mariana arc and backarc system. We report new petrographic data for volcaniclastic sand from the Mariana forearc and backarc regions and compare them to data collected from other backarc basins. The proportions of colorless, brown and black glassy volcanic fragments within coarse, sandy intervals associated with these arc systems suggest that the early rift phase of backarc basin formation is characterized by felsic (bimodal) volcanism and that volcanic centers become progressively more intermediate with time. This trend is consistent with the geochemical evolution of arc lavas. In the case of forearc rifting, this compositional trend may coincide with a shift in volcanism from the protoremnant arc, across the nascent backarc basin, toward the frontal arc.

1. INTRODUCTION

The depositional history of backarc basins is closely tied to regional tectonics and volcanism; hence it is appropriate that they be discussed in this volume. Proposed models for the development of, and sediment distribution within, backarc basins have primarily

Kathleen M. Marsaglia and Kathleen A. Devaney • Department of Geological Sciences, University of Texas at El Paso, El Paso, Texas 79968.

Backarc Basins: Tectonics and Magmatism, edited by Brian Taylor, Plenum Press, New York, 1995.

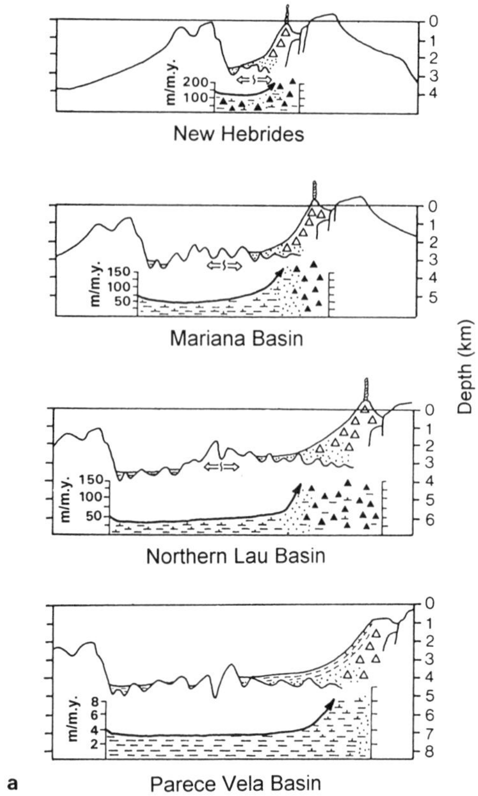

FIGURE 14.1. Models of backarc basin evolution and sedimentation from (a) Karig and Moore (1975) and (b) Carey and Sigurdsson (1984). (a) The distribution, thickness and type of surficial sediments are indicated on a series of four idealized cross sections based on various backarc basins. Sedimentation rates (m/m.y.) and distribution of surficial sediment types are illustrated below each cross section. Symbols are as follows: dashes = clay, barbed dashes = nannofossil ooze, dots = clastics, and triangles = coarse-grained volcaniclastic material. (b) Facies distribution during various stages of backarc basin formation: early rifting (Stage 1), backarc spreading (Stage 2), basin maturity (Stage 3), and cessation of backarc spreading and initiation of a new cycle of backarc rifting (Stage 4). Bold curved arrows indicate extension due to rifting or seafloor spreading.

stemmed from Deep Sea Drilling Project (DSDP) drilling results (Karig and Moore, 1975; Klein, 1975a,b, 1985a,b; Carey and Sigurdsson, 1984). The most detailed volcano-tectonic model, by Carey and Sigurdsson (1984; Fig. 14.1) is largely based on drilling results from DSDP Legs 59 and 60; during these legs, a series of sites was drilled on a transect across the Parece Vela Basin and Mariana Trough. In this chapter we present new petrographic data on the proportions of varieties of volcaniclastic lithic fragments in samples from the Mariana backarc and forearc and use these data to help unravel the history of backarc basin evolution. We then reevaluate the history of sedimentation across the Mariana "model" region in light of recent Ocean Drilling Program (ODP) drilling results in other backarc basins and propose a revised model for sedimentation in backarc basins formed by forearc rifting.

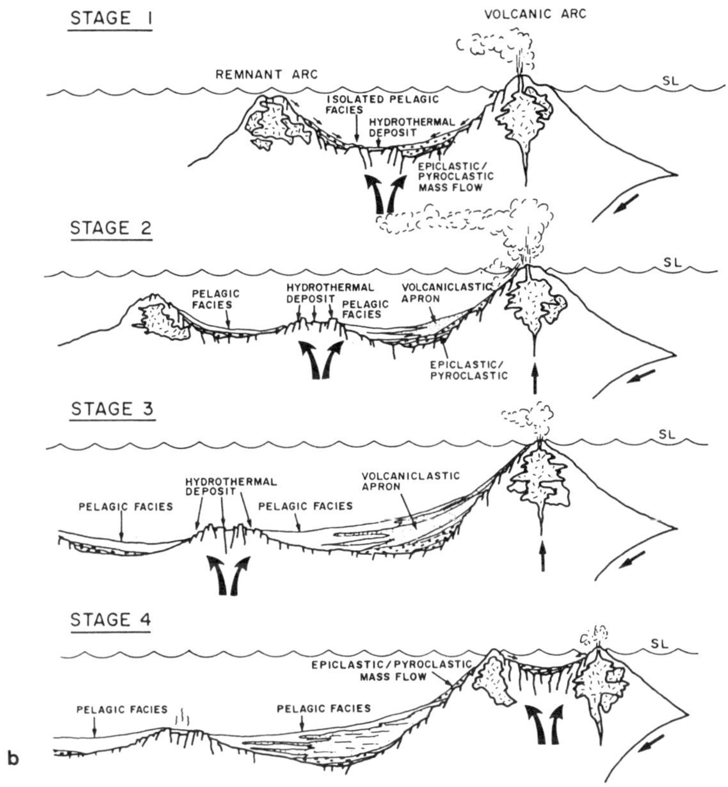

FIGURE 14.1. (*Continued*)

2. CHARACTERISTICS OF AND CONTROLS ON INTRAOCEANIC BACKARC BASIN SEDIMENTATION

The following summary of intraoceanic backarc basin sedimentation is largely based on previous syntheses (Karig and Moore, 1975; Carey and Sigurdsson, 1984; Klein 1985a,b; Marsaglia, in press). Additional information comes from recent drilling in the western Pacific during ODP Legs 124 (Sulu Sea; Rangin *et al.*, 1990; Silver and Rangin, 1991), 126 (Izu–Bonin arc; see Klaus *et al.*, 1992; Taylor, 1992), and 135 (Lau Basin; see Hawkins, Chapter 3 this volume; Parson *et al.*, 1992).

Sedimentary lithofacies in intraoceanic backarc basins are quite variable and include pyroclastic and volcaniclastic deposits, pelagic and resedimented carbonates, biogenic siliceous sediments, sandy to silty turbidites, and debris flows (Klein, 1985a,b). The distribution of these lithofacies is partly a function of latitude (carbonate vs. silica productivity), ocean current circulation, and proximity to subaerial land sources of hemipelagic clay and turbidites (Klein, 1985a). Eustacy may also affect sediment supply into backarc basins (e.g., Shipboard Scientific Party, 1990; Betzler *et al.*, 1991).

Tectonism and arc magmatism play the most important roles in intraoceanic backarc basin sedimentation. Tectonism dictates basin depth (subsidence) and hence the localization of depocenters. In incipient backarc basins such as the Sumisu rift, subsidence is localized between arc volcanoes, and basins have half- to full-graben configurations (Taylor *et al.*, 1991; Klaus *et al.*, 1992). Rapid uplift of the Sumisu rift-flank margins coincides with differential subsidence in the basin (Taylor *et al.*, 1990a; Taylor, 1992). Subsidence also controls water depth, which directly affects the preservation of pelagic carbonate facies (Klein, 1985a). The rate of tectonic uplift related to volcano building and changes in subduction and plate boundary tectonics may also affect sediment input into backarc basins. Klein (1985b) found that uplift rates of 400 m/m.y. or more along the arc were required to support canyon-fed submarine fan development in the adjacent backarc basin. In contrast, rapidly building volcanic edifices along a dominantly submerged arc are characterized by the development of submarine volcanic aprons. Sedimentation on these aprons is a function of multiple-sourced, coalesced depositional systems and so is characterized by rare systematic facies transitions (Carey and Sigurdsson, 1984). The loose pyroclastic materials that make up these aprons accumulate rapidly during periodic volcanic eruptions and are subject to downslope redistribution into backarc and interarc areas as mass-flow deposits (Nishimura *et al.*, 1991).

Possible sources of volcaniclastic material in nascent backarc basins include the protoremnant arc, intrarift volcanic centers and the active arc. During the early rift phase of backarc basin formation intrarift volcanism may be concentrated along fault-controlled transfer zones such as those documented in the Izu–Bonin arc system (Taylor *et al.*, 1990b). The central ridge in the Sumisu rift is one such transfer zone; it is composed primarily of basaltic lavas with lesser rhyolite and dacite (Fryer *et al.*, 1990; Hochstaedter *et al.*, 1990a). Volcanic vents are also present on the flanks of the volcanoes that compose the Sumisu rift protoremnant arc, but their relative sedimentary contribution is unknown. Overall, the main source of volcanic material in the Sumisu rift appears to be the active magmatic arc (Marsaglia, 1992).

As a backarc basin matures, dike-fed intrabasinal volcanic ridges may develop into seafloor spreading centers that supply minor amounts of pyroclastic material. In mature backarc basins, the arc axis, rather than the inactive remnant arc, is the dominant source of pyroclastic debris. Other minor sources of volcanic material include chains of small volcanoes that develop on fractures extending from the arc into the backarc basin, such as those found in the Mariana backarc basin (Hussong and Fryer, 1983; Jackson and Fryer, 1986; Bloomer *et al.*, 1989).

The nature and volume of volcaniclastic debris supplied to a backarc basin, therefore, are controlled by the distribution, composition, eruption rate, and mode of eruption of volcanoes along the arc axis (Fisher, 1984). This volcaniclastic material ranges in size from fine ash and silt to coarse lapilli and gravel, and from basaltic to rhyolitic in composition. Mafic eruptions likely produce minor fragmental material, whereas intermediate volcanoes and felsic calderas may produce large volumes of pyroclastic debris. The pyroclastic debris are variably vesicular and crystalline, depending on the eruption mode. For example, submarine eruptions may produce thick pumice deposits, especially during the early phases of rift development (Cashman and Fiske, 1991; Nishimura *et al.*, 1991). The localization of coarse sandy material is also a function of water and atmospheric currents; Sigurdsson *et al.* (1980) have shown that along the modern Lesser Antilles arc the fine ash is carried to the east, whereas coarse clastic materials are preferentially carried to the west and deposited in the backarc basin. The amount of epiclastic debris is a function of the degree of volcano emergence, which is related to rates of tectonic uplift and volcanism (volcano size).

Backarc spreading has been tied to both minima and maxima in volcanism (Karig, 1975, 1983; Kroenke *et al.*, 1980; Scott and Kroenke, 1980; Taylor, 1992). According to

Taylor (1992), many arc segments (excluding the Mariana arc) show cyclic volcanism with maxima before rifting, during early rifting, and middle to late backarc spreading, and show minima during latest rifting and early backarc spreading. In addition, along-strike transport of pyroclastic debris can camouflage local lulls in volcanism (Taylor, 1992). Temporal variations are affected by the locus of rifting, forearc versus backarc; in the case of forearc rifting, a temporal gap in volcanic activity may result from the shift in volcanism from the arc toward the rifted frontal arc. The effects of forearc versus backarc rifting on the localization of volcaniclastic debris are discussed below.

Karig and Moore (1975) constructed the first tectonic and volcanic model for backarc basin sedimentation (Fig. 14.1). It is a simple model that shows the distribution of sediment types and range of sedimentation rates in various backarc basins, from young (e.g., New Hebrides), to mature (Mariana and Lau), to inactive (Parece Vela Basin) stages. In this model, volcaniclastic sediments and higher sediment accumulation rates are localized near the frontal arc. Carey and Sigurdsson (1984) modified the Karig and Moore (1975) tectonic model to better reflect subsequent DSDP drilling results from the western Pacific (Legs 59 and 60) and their research on volcaniclastic sedimentation in the Grenada backarc basin. This model shows more detailed facies distributions within a hypothetical, evolving backarc basin (Fig. 14.1). These models indicate that proximal coarse-grain sedimentary packages, as opposed to more distal facies, provide the most detailed information on the tectonic and volcanic evolution of the arc during backarc basin formation.

3. MARIANA DRILLING RESULTS

During Legs 59 and 60, a series of sites were drilled along a transect across the West Mariana Ridge, the Mariana Trough, and the Mariana arc. Significant coarse volcaniclastic sections were recovered at a number of these sites, and the stratigraphic relationship and depositional history of these sandy units are summarized below. Much of this information has been taken from Kroenke *et al.* (1980) and Hussong *et al.* (1981).

Site 450 is located on the eastern side of the Parece Vela Basin, west of the West Mariana Ridge (Fig. 14.2). Basaltic basement at this site is overlain by Middle to Upper Miocene vitric tuffs and minor volcaniclastic conglomerates that pass upward in an interbedded fashion into an Upper Miocene to Pleistocene pelagic clay sequence (Kroenke *et al.*, 1980). The volcaniclastic units at Site 450 are thought to have been derived from the West Mariana Ridge, prior to an eastward shift of volcanism during extension and formation of the Mariana Trough (Scott *et al.*, 1980). An upsection decrease in carbonate content at this site is attributed to thermal subsidence through the CCD (Klein, 1985a).

Site 451 lies on the eastern edge of the Western Mariana Ridge (Fig. 14.2). The stratigraphic sequence at this site consists of approximately 900 m of Upper Miocene vitric tuff and andesitic to basaltic breccia and conglomerate, gradationally overlain by a thin (36 m) Plio-Pleistocene cover of calcareous ooze (Kroenke *et al.*, 1980). Geochemical analysis of basalt clasts from this site indicates that they are evolved island-arc tholeiites with strong calc-alkalic affinities (Mattey *et al.*, 1980; Wood *et al.*, 1980). The degree of clast rounding and the presence of shallow-water bioclasts and lignite fragments suggest that the source of the Upper Miocene section was an at-least-partly emergent volcanic arc along the West Mariana Ridge (Kroenke *et al.*, 1980). Klein (1985a) interpreted this sequence as volcaniclastic apron deposits.

Site 453 was drilled just east of Site 451 in a sediment pond, in what is considered the

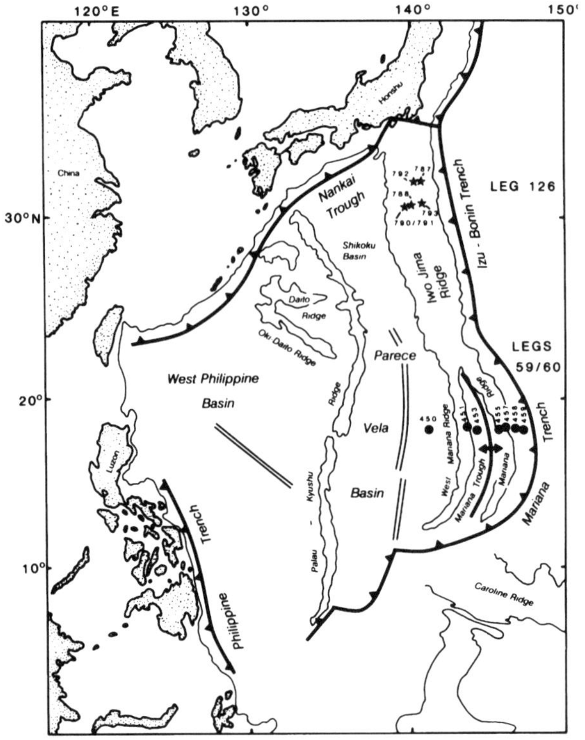

FIGURE 14.2. Location map of DSDP and ODP sites for which data are presented in this study (from Marsaglia, 1992).

oldest part of the Mariana Trough (Fig. 14.2). The stratigraphic sequence begins with approximately 150 m of gabbro-metabasalt polymict breccia of unknown age derived from the rifted roots of a calc-alkaline magmatic arc (Natland and Tarney, 1981). The breccia is in turn overlain by Pliocene to Pleistocene volcaniclastic turbidites and pelagic to hemipelagic muds. Klein (1985a) interpreted this volcaniclastic sequence as submarine fan deposits. The age of the sediments and basement rocks recovered at this site constrain the time of initial rifting in the Mariana Trough to approximately 5 Ma. The frequency and grain size of sandy units decrease from the Lower to the Upper Pliocene sections; associated changes in sedimentary structures within these turbidites indicate a shift from a more proximal to a more distal setting. Given the paucity of pyroclastic debris in equivalent sections to the west at Site 451, the source of volcaniclastic material at Site 453 was either the Mariana arc (Hussong *et al.*, 1981) or, perhaps, intrabasinal volcanoes. According to Hussong *et al.* (1981), the progressive decrease in arc input upsection could be a function of sequential westward displacement of the site with respect to the volcanic source due to backarc spreading.

The presence of unconsolidated coarse volcaniclastic sediment at arc-proximal sites (455 and 457; Fig. 14.2) resulted in poor recovery (30–37%), unstable hole conditions, and early abandonment of the sites. Site 455 is located near the eastern margin of the Mariana Trough on the sediment apron of the Mariana arc. The 104 m of sedimentary section penetrated at this site consists of Pleistocene vitric mud and mudstone, vitric nannofossil ooze and chalk, vitric ash and tuff, and unconsolidated volcanic gravel and sand. Site 457 is located on the arc axis, near Alamagan Island. A 61-m sequence of coarse volcanic sediment and volcanic rock (drilling breccia) was penetrated at this site; the volcaniclastic units contain red scoria fragments suggesting a subaerial source (Alamagan Island).

Site 458 is located in the Mariana forearc region approximately 130 km east of the active arc. At this site a 256-m, Oligocene to Pleistocene sedimentary section rests on pillowed to massive flows of high-MgO bronzite andesite and basalt. The sedimentary sequence is composed of Oligocene and Miocene chalks, tuffs, and vitric siltstones and sandstones (turbidites), overlain by Neogene to Pleistocene siliceous and calcareous oozes with variable ash content. Ash is present throughout the sequence, with maxima in the Lower Oligocene and Upper Pliocene to Pleistocene sections. The Lower Oligocene to Miocene volcanic input was likely derived from both the Palau–Kyushu and West Mariana ridges prior to rifting, whereas the upper Pliocene to Pleistocene sections were likely derived from the modern Mariana arc. A number of hiatuses occur in the section, the longest from 7 to 3 Ma.

At Site 459, 31 km to the east of Site 458, a similar section was drilled. The stratigraphy at this site begins with basalt, overlain by a thin sequence of Upper Eocene claystone and chert, followed by a thick sequence of Oligocene to Miocene sandy to silty tuffaceous turbidites, chalk and ooze and mudstone, and then a thin (50 m) Pliocene to Pleistocene section of vitric and siliceous mud, ash, and calcareous ooze. As at Site 458, this sequence is characterized by a series of hiatuses, the longest in duration from 10 to 3 Ma. The major Mio-Pliocene hiatuses at Sites 458 and 459 likely correspond to the advent of rifting that formed the Mariana Trough.

From west to east, the volcaniclastic units cored during DSDP Legs 59 and 60 on the Mariana transect record volcanic activity associated with the prerift to seafloor spreading stages of Mariana Trough formation. Coarse volcaniclastic sediment packages at these sites young from west to east, recording an eastward shift of volcanism associated with rifting processes in the Mariana arc: Site 450 (Middle to Upper Miocene), Site 451 (Upper Miocene), Site 453 (Pliocene), Site 455 (Pleistocene), and Site 457 (Pleistocene). The forearc sections at Sites 458 and 459 provide discontinuous records, with significant coarse volcaniclastic sediments restricted to the Miocene and Oligocene sections. In combination with lava geochemistry, the volcaniclastic record at these sites should provide a means of determining evolutionary trends in the composition and distribution of magmatism during backarc basin formation.

4. PREVIOUS PETROGRAPHIC STUDIES OF BACKARC BASIN SAND

Packer and Ingersoll (1986) determined detrital modes for 60 volcaniclastic sand and sandstone samples from DSDP Sites 450–459 (Fig. 14.2). They found sands from these sites were relatively homogeneous in comparison to sands derived from the Japan arc and did not document any trends along this transect. Marsaglia and Ingersoll (1992) combined the Packer and Ingersoll (1986) data set with data from other circum-Pacific intraoceanic-arc systems and found that active intraoceanic and remnant arcs have very similar compositional ranges. Petrographic data for these systems tend to fall along the base of a QFL ternary diagram and cluster at the plagioclase and volcanic apices of QmKP and LmLvLs ternary diagrams, respectively (Packer and Ingersoll, 1986; Marsaglia, 1992; Marsaglia and Ingersoll, 1992).

Studies of the glass fractions of these sediments provide additional information on the provenance of backarc sediment. Rodolfo and Warner (1981) examined the coarse fraction of 25 ash-bearing samples from Site 450 (predominantly very fine sand), and Schmincke (1981) examined the very fine sand-sized (65–125 μm) glass populations of 32 samples from Sites 453, 454, 455, 458, and 459. Both studies found a wide range of glass colors and refractive indices in these samples, and both demonstrate a link between glass color, refractive index, and silica content. Glass fragments with refractive indices of <1.53 are colorless and have high silica contents, whereas those with refractive indices of >1.56 are dark brown and have low silica contents. Light brown to tan to green glass fragments exhibit intermediate refractive indices and silica contents. According to Schmincke (1981), grain shape and vesicularity are not diagnostic of composition, but of eruption mode or mechanism.

Because standard detrital modes of intraoceanic-arc volcaniclastic sands are very similar and offer little insight into variations in sand provenance, provenance information might best be determined by subdividing the volcanic lithic populations according to texture (crystallinity) and color (composition). Glassy volcanic lithic fragments can be categorized as vitric (holohyaline), microlitic (silt-sized crystals), or lathwork (sand-sized crystals), depending on the presence or absence and size of crystal inclusions. The glassy groundmass may range in color from colorless to brown to black. Marsaglia (1992) developed this classification system in order to maximize petrographic information for Leg 126 sand and sandstone samples from the Izu–Bonin arc. Based on the studies by Rodolfo and Warner (1981) and Schmincke (1981), glassy fragments are categorized as either colorless or brown (includes tan, brown, and green varieties), providing information on the proportion of felsic and intermediate to mafic components respectively. Black tachylitic glassy fragments ("lithics") are also differentiated. Tachylitic glass is black (opaque) in plane-polarized light due to the presence of fine Fe-Ti opaque minerals. This texture is thought to be produced by lower cooling rates of mafic lava in both subaerial and subaqueous conditions (Macdonald, 1972; Fisher and Schmincke, 1984; Cas and Wright, 1987). In the Leg 126 example (Marsaglia, 1992), this petrographic method provided information on source lava composition and the effects of mixing and segregation during sediment transport.

Standard modal analysis of sand from Leg 126 and Mariana transect (Legs 59 and 60) sites shows them to be very similar and relatively uniform in composition (Fig. 14.3). In this chapter, we present a petrographic data set acquired using the scheme outlined in Marsaglia (1992), which allows us to compare and contrast the Izu–Bonin and Mariana arc systems. The Leg 126 samples represent an early rift phase of backarc basin formation, whereas the Legs 59 and 60 samples represent prerift to mature stages of backarc basin formation. When combined, these data provide a detailed picture of changes in volcaniclastic provenance during backarc basin evolution.

5. METHODS

Thin sections examined in this study were originally prepared and analyzed by Packer and Ingersoll (1986). Their preparation techniques included sieving unconsolidated samples for the sand fraction (0.0625–2 mm), epoxying the sand concentrates to glass slides, and then grinding them to 30-μm thickness. Thin sections of sandstone and unconsolidated

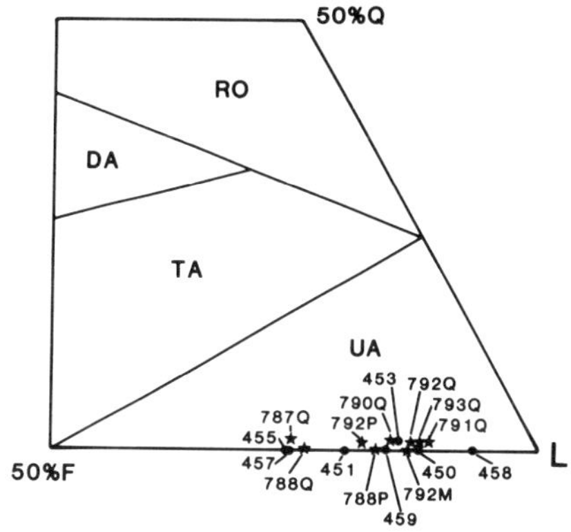

FIGURE 14.3. Right-lower portion of QFL ternary plot of Mariana (Legs 59 and 60) site means (filled circles) and Izu–Bonin (Leg 126) site means (stars) by age from Marsaglia (1992). Apices of ternary plot defined as Q = total quartz, F = total feldspar, and L = total lithic fragments except polycrystalline quartz. Compositional fields taken from Dickinson *et al.* (1983): RO = recycled orogenic, DA = dissected arc, TA = transitional arc, and UA = undissected arc. Age designations follow site numbers: Q = Quaternary, P = Pliocene, and M = Miocene.

sand were subsequently stained for both calcium- and potassium-rich feldspars according to the method outlined in Marsaglia and Tazaki (1992). Petrographic data for these samples can be found in Packer and Ingersoll (1986). They used the Gazzi–Dickinson method of point counting (Dickinson, 1970; Ingersoll *et al.*, 1984) and divided volcanic lithic clasts into various textural types, including vitric, microlitic, lathwork, and felsitic categories (Dickinson, 1970). The volcanic fraction of these thin sections was recounted in this study, further separating volcanic lithic grains according to glass color under plane-polarized light (i.e., colorless, tan or brown, and black). All nonopaque "colored" glassy fragments were grouped into the brown glass category, in part, because color is a function of grain thickness. The Gazzi–Dickinson method was used and 300 volcanic lithic grains were counted per thin section. The categories and recalculated parameters used in this study are defined in Table I and illustrated in Fig. 14.4. Note that in accordance with the Gazzi–Dickinson method, sand-sized (0.0625–2 mm) phenocrysts in volcanic lithic fragments were skipped and not included in the lithic categories. Recalculated data are presented in Table II.

6. PETROGRAPHIC RESULTS

Petrographic data were obtained from sand-rich sections recovered at Sites 450, 451, 453, 455, 457, 458, and 459. The age of the main volcaniclastic sand and sandstone sections encountered at each site varied from Oligocene (459) to Miocene (450, 451, and 458), to Pliocene (453) to Pleistocene (455, 457). The proportions of colorless, brown, and black glassy fragments for samples from these sites are shown in Fig. 14.5. Overall, these volcaniclastic sand samples are dominated by brown and colorless glassy fragments. Black glassy fragments are most prevalent in the Upper Miocene section at Site 451 (Fig. 14.5). The distributions of vitric, microlitic, and lathwork textures for colorless, brown, and black glassy populations by site are shown in Figs. 14.6 and 14.7. Sites 450 and 453 show similar vitric- and microlitic-dominated colorless, brown, and black glass populations, with a lesser lathwork component. At each of these sites, mean values for brown glassy fractions are slightly enriched in microlitic component with respect to the colorless and black glassy fractions, and the black glassy fraction shows a higher range in percentage of lathwork

TABLE I
Counted and Recalculated Parameters

Counted Parameters
Lvv: Vitric volcanic lithic fragments
brgl = brown glass
clgl = colorless glass
blgl = black glass
Lvml: Volcanic lithic fragments with microlitic texture
brgl = brown glass
clgl = colorless glass
blgl = black glass
Lvl: Volcanic lithic fragments with lathwork texture
brgl = brown glass
clgl = colorless glass
blgl = black glass
(altered glass counted but not included in categories)
Recalculated Parameters
Total colorless glass = ColBlkBrwn%col = [Lvv(clgl) + Lvml(clgl) + Lvl(clgl)]/[Lvv(clgl) + Lvml(clgl) + Lvl(clgl) + Lvv(brgl) + Lvml(brgl) + Lvl(brgl) + Lvv(blgl) + Lvml(blgl) + Lvl(blgl)]
Total brown glass = ColBlkBrwn%brwn = [Lvv(brgl) + Lvml(brgl) + Lvl(brgl)]/[Lvv(clgl) + Lvml(clgl) + Lvl(clgl) + Lvv(brgl) + Lvml(brgl) + Lvl(brgl) + Lvv(blgl) + Lvml(blgl) + Lvl(blgl)]
Total black glass = ColBlkBrwn%blk = [Lvv(blgl) + Lvml(blgl) + Lvl(blgl)]/[Lvv(clgl) + Lvml(clgl) + Lvl(clgl) + Lvv(brgl) + Lvml(brgl) + Lvl(brgl) + Lvv(blgl) + Lvml(blgl) + Lvl(blgl)]
LvvLvmLvl%Lvv = 100*Lvv/(Lvv + Lvml + Lvl)
LvvLvmlLvl%Lvml = 100*Lvml/(Lvv + Lvml + Lvl)
LvvLvmlLvl%Lvl = 100*Lvl/(Lvv + Lvml + Lvl)
(LvvLvmlLvl percentages were calculated separately for Colorless/Pumice, Brown and Black glassy grains)

grains. A second group of samples from Sites 451, 455, and 457 are enriched in grains exhibiting microlitic and lathwork textures. Sites 458 and 459B distributions are intermediate between these two groups.

7. DISCUSSION

7.1. Spatial and Temporal Trends in Sand Composition across the Mariana Trough

There appear to be both spatial and temporal shifts in the locus of arc volcanism across the Mariana arc and backarc region, and, as discussed below, these shifts are reflected in the nature of associated volcaniclastic sediment. Sites that penetrated significant coarse clastic sedimentary sequences, or arc proximal deposits, are limited to (1) the forearc (Palau–Kyushu and West Mariana ridges) in the Oligocene and early Miocene, (2) the proto-remnant arc (West Mariana Ridge) in the Middle to Upper Miocene, (3) within the Mariana Trough in the Pliocene, and (4) along the modern arc axis (Mariana Ridge) in the Pleistocene. This shift in volcanism could be explained as an artifact of drill site location and degree of penetration, or it could be due to a physical shift in the locus of volcanism from the now "remnant" arc, across the Mariana Trough to the present arc axis. Models based on backarc spreading (i.e., Fig. 14.1) imply that the locus of arc volcanism remains

TABLE II

Recalculated Petrographic Data

Site[a]		Core	Section	Interval (cm)	Age	Colorless LvvLvmlLvl%			Brown LvvLvmlLvl%			Black LvvLvmlLvl%			% Total ColBlkBrwn%		
						Lvv	Lvml	Lvl	Lvv	Lvml	Lvl	Lvv	Lvml	Lvl	col	blk	brwn
450	59.11	12	1	25–30	Mio	68	29	2	5	91	5	11	79	10	30	32	38
450	59.12	13	1	136–140	Mio	94	6	0	72	27	1	72	28	0	38	11	51
450	59.13	14	6	12–14	Mio	89	10	1	52	41	7	43	48	9	27	31	42
450	59.14	16	1	129–131	Mio	84	16	0	32	66	2	42	47	11	40	25	35
450	59.15	18	3	141–143	Mio	83	17	0	58	37	5	63	33	4	36	27	37
450	59.15a	18	3	143–145	Mio	94	6	0	88	12	0	64	25	11	42	9	48
450	59.15b	20	1	54–56	Mio	94	6	0	89	11	0	86	14	0	53	7	40
450	59.17	28	1	29–31	Mio	94	6	0	85	14	1	89	5	5	36	19	45
450	59.18	30	2	130–134	Mio	94	6	0	35	60	4	71	20	9	42	27	31
450	59.19	35	3	11–14	Mio	89	10	1	32	66	2	27	52	22	47	22	32
451	59.20	32	1	24–27	Mio	4	93	4	1	88	11	11	59	30	23	32	45
451	59.21	34	1	90–94	Mio	38	58	4	17	74	9	32	57	11	19	24	57
451	59.22	35	1	17–21	Mio	36	64	0	3	86	10	20	48	32	21	35	44
451	59.23	38	1	18–22	Mio	43	50	7	15	69	16	28	52	20	6	38	57
451	59.24	40	2	20–25	Mio	40	60	0	4	94	2	33	57	10	20	42	37
451	59.25	41	1	63–67	Mio	46	54	0	12	81	6	32	46	21	23	29	48
451	59.27	46	1	88–92	Mio	35	65	0	15	75	10	28	61	11	11	34	55
451	59.28	48	CC		Mio	6	94	0	7	84	9	33	45	22	12	26	61
451	59.35	66	1	79–83	Mio	83	17	0	2	84	13	32	54	13	19	39	42
451	59.36	68	1	10–14	Mio	18	79	4	6	63	32	25	49	26	11	36	53
451	59.37	69	1	7–11	Mio	80	18	2	7	83	10	30	45	25	40	29	31
451	59.37a	70	2	140–143	Mio	91	8	1	35	55	9	47	31	23	35	26	40
451	59.33	72	1	134–140	Mio	100	0	0	39	56	6	37	28	35	9	43	48
451	59.41	79	2	24–27	Mio	92	6	2	11	74	15	22	45	33	32	36	32
451	59.42	80	2	57–60	Mio	94	6	0	14	64	22	26	33	41	18	32	50
451	59.43	83	1	62–65	Mio	84	16	0	14	78	9	62	22	16	34	38	28
451	59.44	84	1	21–26	Mio	93	6	1	24	69	7	53	25	22	25	48	27
451	59.45	86	2	57–61	Mio	57	43	0	13	72	15	15	54	31	17	43	40

451	59.45a	86	3	75–78	Mio	86	14	0	21	66	14	52	23	25	5	44	51
451	59.45b	87	2	139–142	Mio	73	27	0	20	55	25	39	27	35	4	47	50
451	59.46	89	1	144–148	Mio	74	26	0	22	73	5	52	37	11	28	29	43
451	59.48	93	1	14–20	Mio	100	0	0	14	80	6	35	50	15	2	53	46
451	59.49	94	1	144–150	Mio	71	29	0	11	86	3	25	66	9	2	56	42
453	60.01	29	1	32–33	Plio	38	58	4	27	70	4	62	33	5	18	33	48
453	60.02	32	2	23–25	Plio	22	76	2	12	74	14	27	54	19	51	34	15
453	60.03	33	2	72–74	Plio	69	31	0	53	46	1	71	21	7	54	5	41
453	60.04	34	2	40–43	Plio	79	21	0	85	15	1	78	22	0	49	3	48
453	60.06a	39	3	108–112	Plio	94	6	0	80	18	2	68	21	11	75	6	19
453	60.07	41	3	31–34	Plio	75	21	4	52	39	10	78	15	7	80	9	11
453	60.08	42	2	20–22	Plio	43	54	3	45	51	4	44	24	32	37	24	39
455	60.13	1	6	35–37	Pleist	25	71	4	39	59	2	47	39	14	39	20	40
455	60.14	3	3	55–58	Pleist	38	54	8	34	64	2	66	32	2	59	20	21
455	60.15	4	CC		Pleist	26	43	31	16	68	16	16	51	33	20	28	51
455	60.16	5	1	95–98	Pleist	18	55	27	12	66	22	26	57	17	47	26	27
455	60.17	7	1	7–10	Pleist	19	79	2	10	63	27	22	44	33	41	7	53
455	60.18	8	CC		Pleist	14	80	6	14	57	29	19	52	29	43	13	44
455	60.19	9	CC		Pleist	4	71	25	6	52	42	0	64	36	18	10	73
455	60.20	10	2	66–68	Pleist	10	84	5	6	58	36	4	68	29	21	10	70
457	60.21	2	3	18–20	Pleist	10	68	23	43	27	30	20	43	37	11	32	57
457	60.22	6	2	64–67	Pleist	11	47	42	16	58	26	20	56	24	7	32	61
458	60.23	5	3	88–90	Mio	63	37	1	28	60	12	35	59	6	55	6	39
459b	60.25	12	CC		Mio	66	34	0	26	69	5	22	56	22	58	6	36
459b	60.26	15	1	59–61	Mio	56	42	2	39	58	3	40	55	5	31	29	40
459b	60.27	19	1	56–58	Mio	100	0	0	93	6	1	43	48	9	2	8	90
459b	60.28	21	1	74–76	Mio	74	26	0	37	57	6	24	76	0	27	23	49
459b	60.30	25	2	98–100	Mio	50	47	3	29	62	10	12	65	24	76	6	18
459b	60.33	32	1	59–61	Mio	79	19	2	32	62	6	29	65	6	39	6	55
459b	60.34	36	1	71–73	Mio	95	4	1	40	57	4	25	71	4	55	8	37
459b	60.35	37	1	126–128	Mio	88	11	1	28	65	7	12	58	31	63	9	28
459b	60.36	39	2	100–102	Mio	93	7	0	18	76	6	0	100	0	82	1	17
459b	60.37	49	1	39–41	Oligo	88	12	0	26	71	3	32	58	11	74	6	20
459b	60.38	50	2	10–12	Oligo	50	45	6	44	40	17	45	48	7	61	15	24

[a]Sample numbers from Packer and Ingersoll (1986).

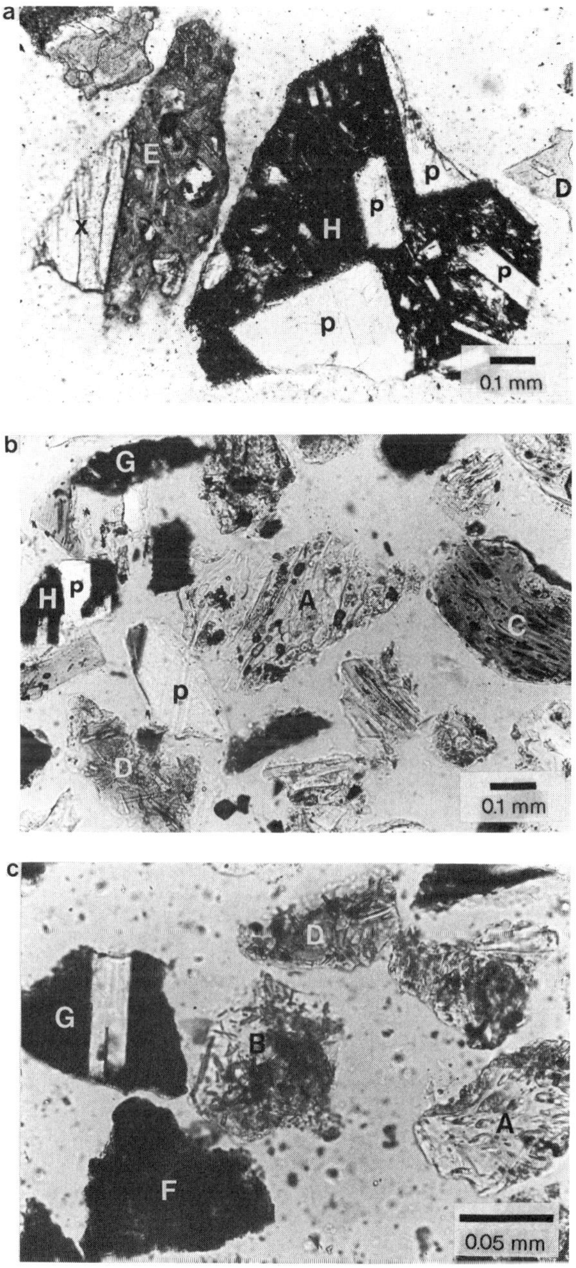

FIGURE 14.4. Photomicrographs of various volcanic lithic fragments, all plane-polarized light in unstained portion of slide. (a) Sample 60–20, (b) Sample 59–19, (c) Sample 60–20. See Table II for sample information. Lithic fragment types are keyed as follows: A = colorless vitric (Lvv), B = colorless microlitic (Lvml), C = brown vitric (Lvv), D = brown microlitic (Lvml), E = brown lathwork (Lvl), F = black vitric (Lvv), G = black microlitic (Lvml), H = black lathwork (Lvl). Colorless and brown vitric fragments are commonly vesicular. Sand-sized phenocrysts in lathwork fragments include plagioclase (p) and pyroxene (x); the groundmass of lathwork fragments shown here are microlitic, but may also be glassy (vitric) in other instances. Edges of colorless grains may appear dark due to adhesion of matrix material. Note that in stained portion of slide, the stain often masks nonopaque glass color.

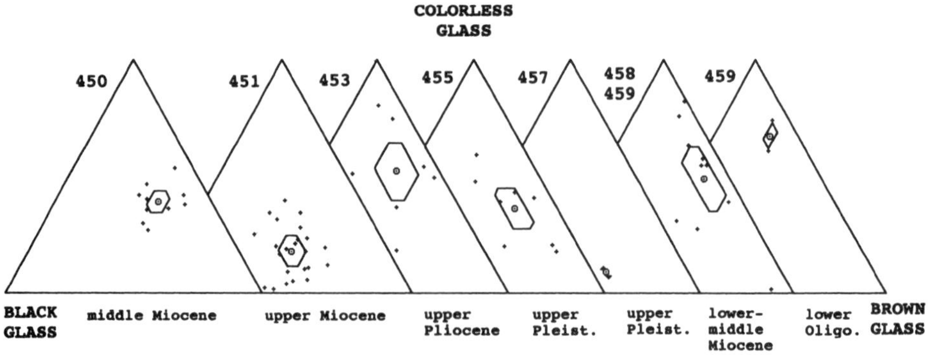

FIGURE 14.5. Ternary plots of colorless, brown, and black glassy lithic proportions by site and age. Polygons represent fields of variation defined by one standard deviation on either side of mean ("circle"). See Table I for definition of recalculated parameters and Table II for recalculated data.

fixed along the frontal arc. Therefore a temporal or spatial shift might be consistent with forearc rifting rather than backarc rifting along this arc segment. Sedimentary sequences cored in the Mariana forearc also present an incomplete history of arc volcanism. Nascent Mariana arc volcanoes appear to have developed at the boundary between backarc and old frontal arc crust, or on downfaulted frontal arc crust, to the west of and structurally lower than the frontal arc scarp (Hussong and Uyeda, 1981; Bloomer *et al.*, 1989). The active arc is largely submerged and located to the west of the frontal arc scarp, which acts as a topographic barrier; hence coarse volcaniclastic materials associated with these volcanoes are probably preferentially distributed in the backarc basin. A successful deep penetration of Mariana arc apron deposits in the backarc region would help document the longevity of the modern Mariana arc and test the theory of spatial shifts in volcanism through time across the Mariana Trough.

Temporal trends in volcaniclastic composition across the Mariana Trough are portrayed in Fig. 14.7, in which site means are plotted by age. These data indicate a gradual shift from more felsic populations in the Oligocene (Site 459) and early Miocene (Sites 458 and 459) to more intermediate to mafic populations (brown glass) in the middle (Site 450) and late Miocene (Site 451). A subsequent shift to more felsic components in the Pliocene (Site 453) is coincident with rifting and the inception of seafloor spreading in the Mariana Trough. Samples from Site 453 are also slightly enriched in quartz with respect to the other sites, suggesting more felsic input (Fig. 14.3). Schmincke (1981) noted the dominance of intermediate to silica-rich volcaniclastic material in the Pliocene sections at Site 453 and found Pleistocene vitric muds at Sites 454, 456, 458, and 459B to be rich in glass shards of basaltic andesitic composition. A corresponding trend from more felsic Pliocene compositions toward more intermediate to mafic compositions (brown and black glass) in the Pleistocene is also seen in the glassy proportions determined in this study (Fig. 14.7).

By focussing on coarser-grain samples from more volcaniclastic-rich sections, we hope we have circumvented provenance problems related to widespread distribution of airborne ash. Note that paleoceanographic currents may have favored accumulation of coarse clastic debris in the backarc region, similar to the modern Lesser Antilles arc as described by Carey and Sigurdsson (1984). Schmincke (1981) discusses the likely input at Leg 60 sites from distant versus proximal volcanoes and suggests that finer-grain felsic components produced by highly explosive volcanic eruptions are more apt to be far traveled. Klein (1985a) indicates that wind patterns in the Mariana region would likely result in along-strike transport of airborne ash, from the north or south, depending on the season. Some of the apparent trends in provenance that Schmincke (1981) outlines in the younger sections at Site 453 may reflect the interplay of distal versus proximal sources, as opposed to actual changes in volcanic provenance from calc-alkaline to bimodal rhyolite-basalt. Whereas Schmincke (1981) sees a change from more mafic to felsic compositions from Cores 36 to 19 at Site 453, data presented in Table II indicate a shift from more colorless (felsic) to more brown (intermediate to mafic) glass from Cores 48 to 29. Discrepancies between our data sets are understandable given the differences in depth range, grain size, and number of samples.

The volcaniclastic history of Site 453 is key to understanding the evolution of the Mariana Trough. The thick Pliocene volcaniclastic section present at Site 453 is represented by only traces of coarse pyroclastic debris in the Pliocene sections at Site 451. The fact that Site 451 is more proximal to West Mariana Ridge than Site 453 suggests that the likely source of the Pliocene volcaniclastic units was to the east, derived from the Mariana arc during the early phases of seafloor spreading (Hussong *et al.*, 1981). Thin layers of tachylitic glass present in this section (Hussong *et al.*, 1981) may have been derived from the nascent spreading center or intrabasinal volcanoes, but apparently these layers were not sampled by Packer and Ingersoll (1986), and so were not counted in this study. Hussong *et al.* (1981) believed the primary source of the volcanic material at Site 453 to be the Mariana arc, but thought that this material had probably been reworked from the highs flanking the small basin in which Site 453 was drilled. The possibility of intrabasinal volcanic sources for the main portion of the volcaniclastic section was not discussed by Hussong *et al.* (1981). However, in light of recent drilling in the Lau Basin (Parson *et al.*, 1992; Hawkins, Chapter 3 this volume), the possibility of an intrabasinal volcanic source for volcaniclastic materials at Site 453 should not be ruled out (see discussion below).

7.2. Relationship between Lava Geochemistry and Detrital Modes

The petrologic evolution of Mariana arc and backarc igneous rocks is reviewed by Natland and Tarney (1981). They outline the dominant lava compositions as follows: Oligocene—arc tholeiite; Miocene—calc-alkalic; Pliocene—arc tholeiite; and Quaternary—arc tholeiite to calc-alkalic. Tholeiitic volcanic rocks characterize the early rift and spreading phases of backarc basin formation, whereas calc-alkalic volcanism associated with fractional crystallization characterizes the more mature arc and backarc systems (Hussong and Uyeda, 1981; Natland and Tarney, 1981). Thus these compositional shifts are related to backarc basin evolution (Natland and Tarney, 1981).

There is a general correlation between the volcanic lithic proportions outlined in Fig. 14.5 and the temporal trends outlined in Natland and Tarney (1981): periods of calc-alkaline volcanism (Miocene and Quaternary–Recent) are characterized by glass populations dominated by brown glassy fragments, whereas periods of arc tholeiite volcanism (Oligocene–early Miocene and Pliocene–Pleistocene) are characterized by glass populations dominated by colorless glassy fragments. The high felsic and minor mafic components associated with periods of tholeiitic volcanism can be best explained by bimodal volcanism during the early rift phases of backarc basin formation. After the onset of seafloor spreading and concurrent with arc evolution, volcanism becomes more intermediate and brown glass dominates.

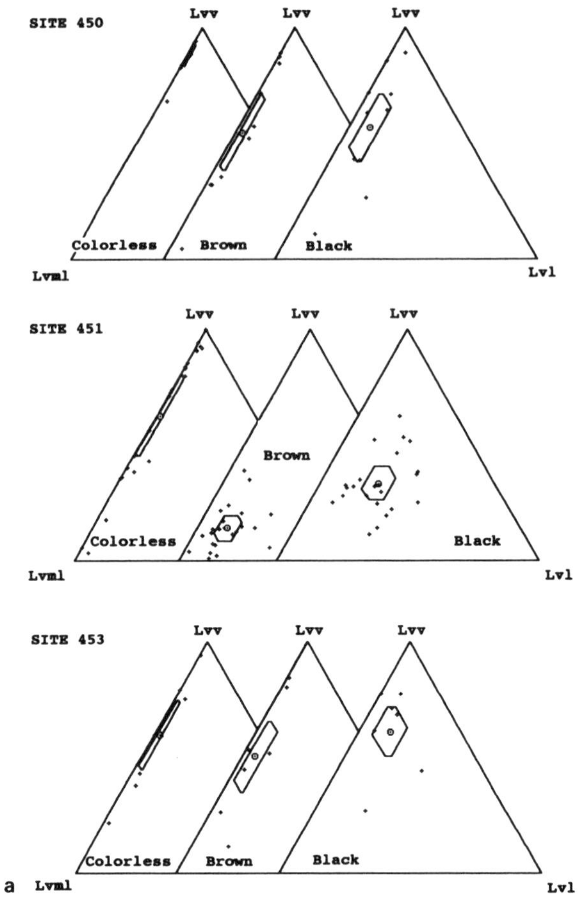

FIGURE 14.6. Ternary plots of vitric (Lvv), microlitic (Lvml), and lathwork (Lvl) proportions in glassy fractions (colorless, brown, and black) by site. (a) Sites 450, 451, and 453. (b) Sites 455, 457, 458, and 459b. Means and polygons as in Fig. 14.5.

Electron microprobe analyses of glass and mineral fragments from Site 453 by Packham and Williams (1981) demonstrate the Pliocene to Recent trend from arc-tholeiite to calc-alkaline volcanism. Glass compositions in the Lower Pliocene volcaniclastic units indicate a broad spectrum of silica contents (49–73%), with a bimodal distribution in some intervals (Packham and Williams, 1981). The dominance of felsic components in these sediments has been ascribed to preferential eruption of differentiated felsic magmas or explosive volcanism (Packham and Williams, 1981; Schmincke, 1981). The glass proportions presented in Fig. 14.5 suggest that bimodal volcanism, such as that of the incipiently rifting Izu–Bonin arc, may have characterized the early history of both the Mariana Trough and the Parece Vela Basin.

7.3. Comparison of Mariana Sand Composition to That of Other Backarc Basins

The Sumisu rift is a well-studied example of a nascent backarc basin within the Izu–Bonin arc system. It is an asymmetrical structural graben located between two submarine

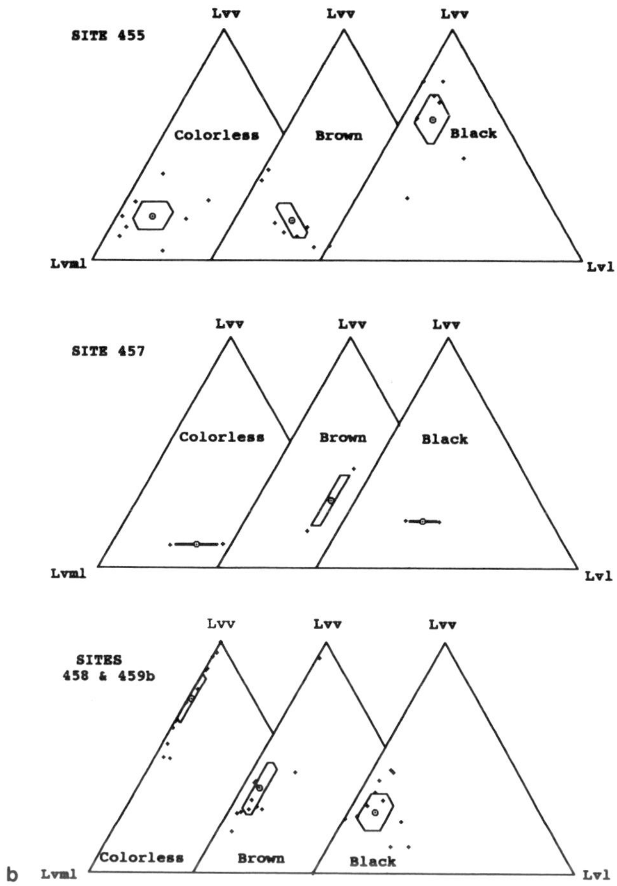

FIGURE 14.6. (*Continued*)

arc calderas (Taylor *et al.*, 1990b, 1991; Klaus *et al.*, 1992). Arc volcanism prior to and during the present rifting of the Izu–Bonin arc has been dominantly silicic (subalkaline andesite-rhyolite) with tholeiitic basalt limited to the backarc region (Fryer *et al.*, 1985; Fujioka *et al.*, 1992; Gill *et al.*, 1992; Rodolfo *et al.*, 1992; Taylor, 1992). Pliocene and Pleistocene sand detrital modes across the Izu–Bonin arc are consistent with this picture of volcanism. Pliocene volcaniclastic deposits from the rift-flank uplift (Site 788) are dominantly composed of colorless glassy fragments, whereas samples from intrarift sites (790 and 791) are extremely heterogeneous, showing wide ranges of basaltic and intermediate brown and black glassy components (Fig. 14.8; Marsaglia, 1992). Their heterogeneous nature can be explained by either syneruptive mixing or mixing during mass transport within the basin. The proximity of Sites 790 and 791 to intrarift mafic volcanic centers suggests that sand compositions could have been produced by mixing of intrabasinal basaltic materials during mass flow of siliceous debris from nearby calderas or by reworking off rift flank uplifts (Marsaglia, 1992). A similar compositional range is also observed in Quaternary deposits cored at Izu–Bonin forearc sites, but mean forearc-site compositions are much more enriched in intermediate to mafic materials (Fig. 14.9; Marsaglia, 1992). These sites are structurally isolated from backarc basaltic volcanoes, and therefore only

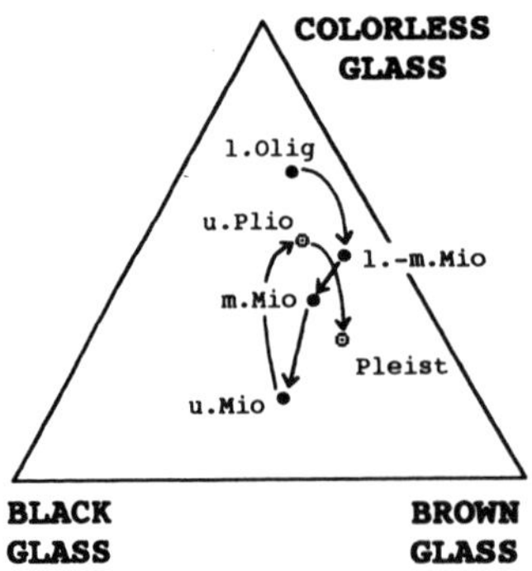

FIGURE 14.7. Ternary plot of mean proportions of colorless, black, and brown glassy fragments for Mariana sites. Temporal trends indicated by arrows connecting Oligocene to Miocene means (filled circles), and Pliocene to Pleistocene means (open circles). Lower Oligocene = Site 459; Lower to Middle Miocene = Sites 458/459; Middle Miocene = Site 450; Upper Miocene = Site 451; Upper Pliocene = Site 453; Pleistocene = Sites 455 and 457.

syneruptive mixing of felsic and mafic components could be responsible for these mixtures. In a syneruptive mixing scenario, the tachylitic fraction might represent arc lavas fragmented during explosive caldera eruptions. According to Marsaglia (1992), the tachylitic sand fragments at Site 787 (Fig. 14.8) may be epiclastic detritus eroded from an emergent arc volcano and concentrated during transport down the submarine canyon in which Site 787 is located. Detailed geochemical analyses of tachylite grains that might constrain their provenance are not available.

The Lower Pliocene volcaniclastic units at Site 453 within the Mariana Trough could represent ancient analogs to Quaternary sediments deposited at Sites 790 and 791 in the Sumisu rift. Sediments from all three sites are enriched in felsic components such as quartz (Fig. 14.3) and colorless glassy fragments (Figs. 14.5 and 14.8). Their mean glassy proportions are almost identical (Figs. 14.5 and 14.8).

Sequences drilled at a number of sites in the Lau Basin on Leg 135 are remarkably similar to sequences recovered at Site 453 in the Mariana Trough and, in some respects, to those in the Sumisu rift. Sites 834 through 839 drilled a series of fault-bounded subbasins, initially filled by Pliocene to Pleistocene coarse volcaniclastic turbidites and debris flows, which are overlain by fine clayey pelagic units (Shipboard Scientific Party, 1992). Se-

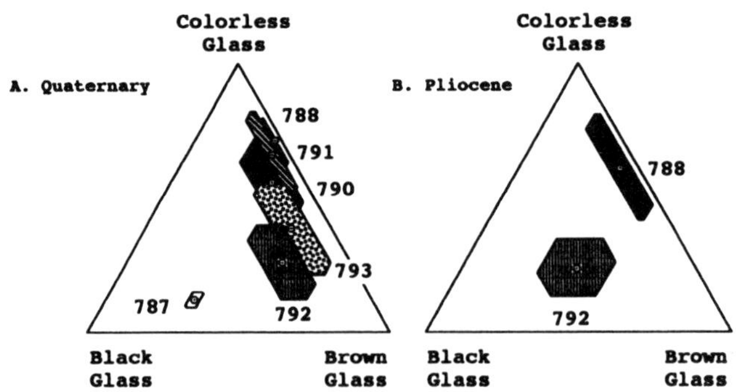

FIGURE 14.8. Ternary plots of colorless, brown and black glassy lithic proportions by site and age, for samples from ODP Leg 126, Izu–Bonin arc Sites 787, 789, 790–793. Polygons represent fields of variation defined by one standard deviation on either side of mean (circle). See Table I for definition of recalculated parameters and Marsaglia (1992) for recalculated data.

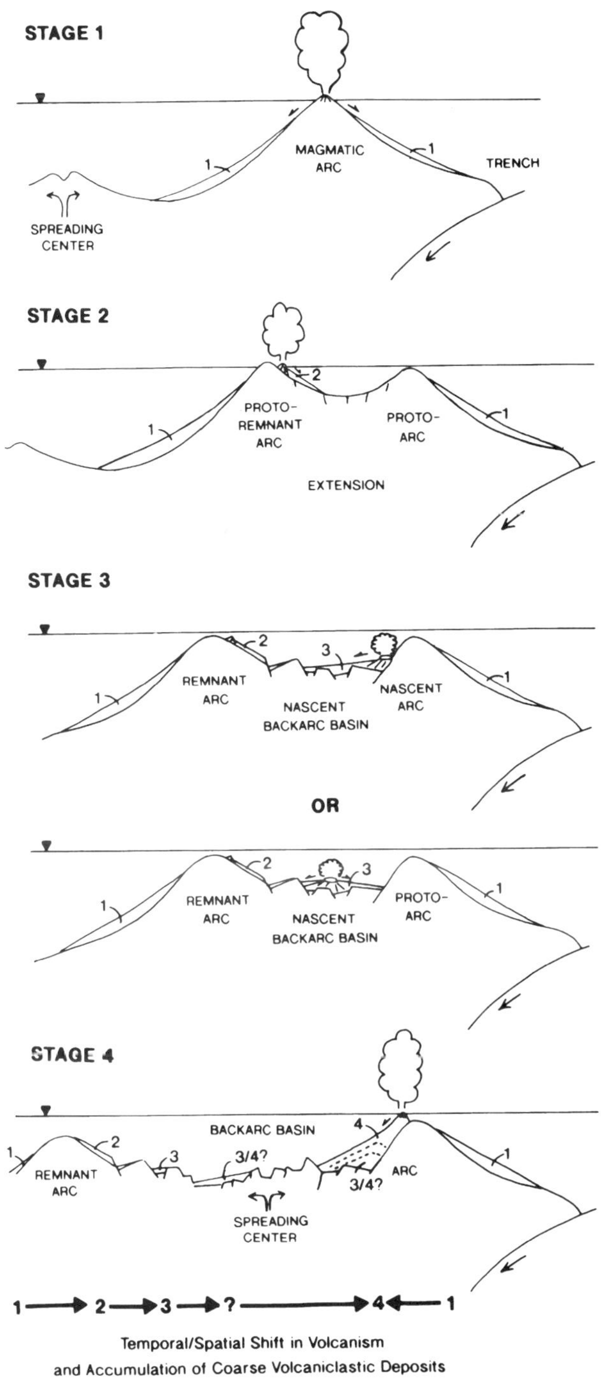

FIGURE 14.9. Model of backarc basin evolution (Stage 1 [oldest] to Stage 4 [youngest]) due to forearc rifting. Main volcanic centers depicted by erupting subaerial (Stages 1, 2, and 4) and submarine (Stage 3) volcanoes. Associated coarse volcaniclastic apron or fan deposits are numbered according to the stage in which they are produced. Temporal and spatial shifts are summarized at base of diagram. Model concept modified from Karig and Moore (1975) and Carey and Sigurdsson (1984). Filled triangle denotes sea level.

quences within these basins are similar to those at Sites 453, 790, and 791 not only in terms of their stratigraphy but also in terms of their apparent volcaniclastic populations. No detrital modes for these sediments have been reported, but general descriptions from the initial scientific reports (Shipboard Scientific Party, 1992) indicate that the sandy turbidites are predominantly composed of colorless shards and pumice fragments, dominantly rhyodacitic (70–71% SiO_2) in composition. Less common layers of scoriaceous basaltic sand and hyaloclastite layers of brown to brownish green glass ranging from basaltic andesite to andesite (52–57% SiO_2) in composition are also present. The majority of these materials appear to have been derived from (an) unidentified intrabasinal or arc-related silicic volcanic center(s) (Shipboard Scientific Party, 1992). Deposition occurred during a geographically broad extensional event, coeval with magmatism and eastward migration of volcanic centers. This migration is indicated by eastward younging of both volcanic basement and coarse clastic packages at Leg 135 sites (Shipboard Scientific Party, 1992).

7.4. Implications of Forearc versus Backarc Rifting on Sedimentary Models

The Lau and Mariana backarc basins have been proposed as possible examples of "backarc" basins initiated by forearc spreading (Hawkins *et al.*, 1984). The forearc spreading model of Hawkins *et al.* (1984) predicts a gap in volcanism during the time of backarc basin initiation. Volcanic activity along the main arc axis is thought to wane during forearc rifting, and then the center of volcanism shifts to the forearc "remnant" ridge. In contrast, Leg 135 documented a record of continuous volcanism across the Lau Basin, but with an eastward migration of the locus of volcanism (Shipboard Scientific Party, 1992). We suggest that such a shift in volcanism may also be recorded in the temporal and spatial distribution of coarse proximal volcaniclastic facies across the Mariana Trough.

In both the Karig and Moore (1975) and Carey and Sigurdsson (1984) models, the effects of forearc rifting are not considered. According to these models (Fig. 14.1), the history of arc magmatism is best recorded in the proximal deposits of the arc apron. The fine ash component is not as useful because of probable transport from along strike or from nearby arc volcanoes. To date, arc aprons, particularly those along the Mariana arc system, have proved unsuitable for drilling, and only minimal penetration of this facies has been achieved at any given site. The complete history of Mariana volcanism is not preserved at any one site drilled during Legs 59 and 60; in part because of unsuccessful drilling, unconformities in the forearc sections, and an apparent shift in the locus of magmatism, this history is only preserved piecemeal at a combination of sites.

The backarc basin sedimentation models presented in Fig. 14.1 can be modified to represent the possible effects of forearc rifting on the localization of coarse volcaniclastic sedimentation. If rifting begins in the backarc region, the arc is static and semicontinuously supplies volcaniclastic material into the forearc and backarc basins. If rifting initiates just in front of the arc axis, in the forearc region, then the locus of volcanism and volcaniclastic sedimentation shifts temporally and spatially from the protoremnant arc, across the basin concurrent with extension, rifting, and seafloor spreading. A simplified model depicting such a scenario is shown in Fig. 14.9. It is loosely based on the Mariana and Lau examples discussed in this chapter and shows the development of coarse, volcaniclastic-apron facies across the region with time. Stage 1 represents the prerift phase, in which volcaniclastic aprons flank a mature intraoceanic arc. Stage 2 indicates a shift in the locus of volcanism and volcaniclastic sedimentation away from the arc axis associated with early extension of the forearc region. Two scenarios are presented in Stage 3, one with volcanism shifting directly from the protoremnant arc to the protoarc, and another with volcanism moving across the nascent backarc basin. In Stage 4, the volcanic arc is reestablished and seafloor spreading has commenced. The likely spatial distribution of coarse volcaniclastic sediments produced during forearc rifting is depicted for each stage (sequentially numbered 1 through 4).

In the Mariana example, Stage 1 would represent the middle Miocene, Stage 2 would correspond to the late Miocene, Stage 3 would represent the Pliocene, and Stage 4 would depict the Pleistocene to Recent setting. According to the petrographic data presented in Figs. 14.5 and 14.7, volcaniclastic apron deposits in Stages 1, 2, and 4 are characterized by intermediate volcaniclastic debris, whereas Stage 3 deposits are characterized by felsic to bimodal volcaniclastic debris. Felsic to bimodal volcaniclastic sequences cored on Leg 135 in the western Lau Basin also represent Stage 3 sedimentation and, as discussed previously, appear to be compositionally similar to Site 453 (Stage 3) volcaniclastic debris. Sumisu rift sediments cored on Leg 126 correspond to intra-arc sedimentation during Stages 1 and 4 of the Carey and Sigurdsson (1984) backarc-rifting model shown in Fig. 14.1.

8. CONCLUSIONS

This chapter presents a nonclassical view of intraoceanic volcaniclastic sedimentation focusing on the populations of sand-sized volcanic lithic fragments rather than on bulk and individual grain geochemistries. The proportion of volcaniclastic components yields information on magma composition and is also influenced by posteruption depositional processes. By focusing on the sand-size components of intraoceanic backarc sequences, we have decreased the effects of long-distance transport of fine ash in the atmosphere and have emphasized proximal volcanism. Compositional variations between and across these regions may be linked to the stage (incipient vs. mature) of backarc basin formation. The proportions of colorless, brown, and black glassy sand fragments within coarse volcaniclastic intervals suggest that the early rift phase of backarc basin formation is characterized by felsic (bimodal) volcanism and that volcanic centers become progressively more intermediate with time. This trend is consistent with the geochemical evolution of arc lavas. In the case of forearc spreading, these igneous and volcaniclastic compositional trends coincide with a shift in volcanism from the backarc, across the nascent backarc basin, toward the frontal arc.

Acknowledgments

We thank Bonnie Packer and Raymond Ingersoll for use of their thin sections and Loren Kroenke and Neil Lundberg for their thoughtful reviews.

REFERENCES

Betzler, C., Nederbragt, A. J., and Nichols, G. J. 1991. Significance of turbidites at Site 767 (Celebes Sea) and Site 768 (Sulu Sea), in *Proc. ODP, Sci. Results*, 124 (A. E. Silver, C. Rangin, T. von Breymann *et al.*, eds.), pp. 431–446, Ocean Drilling Program, College Station, TX.

Bloomer, S. H., Stern, R. J., and Smoot, N. C. 1989. Physical volcanology of the submarine Mariana and Volcano arcs, *Bull. Volcanol.* **51**:210–224.

Carey, S., and Sigurdsson, H. 1984. A model of volcanogenic sedimentation in marginal basins, in *Marginal Basin Geology* (B. P. Kokelaar and M. F. Howells, eds.), pp. 37–58, Geological Society Special Publ., Vol. 16, Oxford.

Cas, R. A., and Wright, J. V. 1987. *Volcanic Successions*, Allen and Unwin, London.

Cashman, K. V., and Fiske, R. S. 1991. Fallout of pyroclastic debris from submarine volcanic eruptions, *Science* **253**:275–280.

Dickinson, W. R. 1970. Interpreting detrital modes of graywacke and arkose, *J. Sediment. Petrol.* **40**:695–707.

Dickinson, W. R., Beard, L. S., Brakenridge, G. R., Erjavec, J. L., Ferguson, R. C., Inman, K. F., Knepp, R. A., Lindberg, F. A., and Ryberg, P. T. 1983. Provenance of North American Phanerozoic sandstones in relation to tectonic setting, *Geol. Soc. Am. Bull.* **94**:222–235.

Fisher, R. V. 1984. Submarine volcaniclastic rocks, in *Marginal Basin Geology* (B. P. Kokelaar and M. F. Howells, eds.), pp. 5–27, Geological Society Special Publ., Vol. 16, Oxford.

Fisher, R. V., and Schmincke, H.-U. 1984. *Pyroclastic Rocks*, Springer-Verlag, New York.

Fryer, P., Langmuir, C., Taylor, B., Zhang, Y., and Hussong, D. 1985. Rifting of the Izu arc. III: Relationship of chemistry to tectonics, *Eos* **66**:421.

Fryer, P., Taylor, B., Langmuir, C. H., and Hochstaedter, A. G. 1990. Petrology and geochemistry of lavas from the Sumisu and Torishima backarc rifts, *Earth Planet. Sci. Lett.* **100**:161–178.

Fujioka, K., Nishimura, A., Matsuo, Y., and Rodolfo, K. S. 1992. Correlation of Quaternary tephras throughout the Izu–Bonin areas, in *Proc. ODP, Sci. Results*, 126 (B. Taylor, K. Fujioka *et al.*, eds.), pp. 23–46, Ocean Drilling Program, College Station, TX.

Gill, J. B, Seales, C., Thompson, P., Hochstaedter, A. G., and Dunlap, C. 1992. Petrology and geochemistry of Pliocene–Pleistocene volcanic rocks from the Izu arc, Leg 126, in *Proc. ODP, Sci. Results*, 126 (B. Taylor, K. Fujioka *et al.*, eds.), pp. 383–404, Ocean Drilling Program, College Station, TX.

Hawkins, J. W., Bloomer, S. H., Evans, C. A., and Melchior, J. T. 1984. Evolution of intra-oceanic arc-trench systems, *Tectonophysics* **102**:175–205.

Hochstaedter, A. F., Gill., J. B., Kusakabe, M., Newman, S., Pringle, M., Taylor, B., and Fryer, P. 1990a. Volcanism in the Sumisu Rift. I: Major element, volatiles and stable isotope geochemistry, *Earth Planet. Sci. Lett.* **100**:179–194.

Hussong, D. M., and Fryer, P. 1983. Backarc seamounts and the SeaMARC II seafloor mapping system, *Eos* **64**:627–632.

Hussong, D. M., and Uyeda, S. 1981. Tectonic processes and the history of the Mariana arc: A synthesis of the results of Deep Sea Drilling Project Leg 60, in *Init. Repts, DSDP*, 60 (D. M. Hussong, S. Uyeda *et al.*, eds.), pp. 909–929, Deep Sea Drilling Project, U.S. Govt. Printing Office, Washington, DC.

Hussong, D. M., Uyeda, S. *et al.* 1981. *Init. Repts, DSDP*, 60, Deep Sea Drilling Project, U.S. Govt. Printing Office, Washington, DC.

Ingersoll, R. V., Bullard, T. F., Ford, R. L., Grimm, J. P., Pickle, J. D., and Sares, S. W. 1984. The effect of grain size on detrital modes: A test of the Gazzi–Dickinson point-counting method, *J. Sediment. Petrol.* **54**: 103–116.

Jackson, M. C., and Fryer, P. 1986. The Kasuga volcanic cross-chain: A calc-alkaline basalt-dacite suite in the northern Mariana island arc, *Eos* **67**:276.

Karig, D. E. 1975. Basin genesis in the Philippine Sea, in *Init. Repts, DSDP*, 31 (D. E. Karig, J. C. Ingle, Jr. *et al.*, eds.), pp. 857–879, Deep Sea Drilling Project, U.S. Govt. Printing Office, Washington, DC.

Karig, D. E. 1983. Temporal relationships between back arc basin formation and arc volcanism with special reference to the Philippine Sea, in *The Tectonic and Geologic Evolution of Southeast Asian Seas and Islands*. Part II (D. E. Hayes, ed.), pp. 318–325, Geophys. Monogr. Ser, Vol. 27, American Geophysical Union, Washington, DC.

Karig, D. E., and Moore, G. F. 1975. Tectonically controlled sedimentation in marginal basins, *Earth Planet. Sci. Lett.* **26**:233–238.

Klaus, A., Taylor, B., Moore, G. F., MacKay, M. E., Brown, G. R., Okamura, Y., and Murakami, F. 1992. Structural and stratigraphic evolution of Sumisu rift, Izu–Bonin arc, in *Proc. ODP, Sci. Results*, 126 (B. Taylor, K. Fujioka *et al.*, eds.), pp. 555–574, Ocean Drilling Program, College Station, TX.

Klein, G. deV. 1975a. Sedimentary tectonics in southwest Pacific marginal basins based on Leg 30 Deep Sea Drilling Project cores from the South Fiji, Hebrides, and Coral Sea basins, *Geol. Soc. Am. Bull.* **86**: 1012–1018.

Klein, G. deV. 1975b. Depositional facies of leg 30 DSDP sediment cores, in *Init. Repts, DSDP*, 30 (J. E. Andrews, G. Packham, G. *et al.*, eds.), pp. 423–442, Deep Sea Drilling Project, U.S. Govt. Printing Office, Washington, DC.

Klein, G. deV. 1985a. The control of depositional depth, tectonic uplift, and volcanism on sedimentation processes in the backarc basins of the western Pacific Ocean, *J. Geol.* **93**:1–25.

Klein, G. deV. 1985b. The frequency and periodicity of preserved turbidites in submarine fans as a quantitative record of tectonic uplift in collision zones, *Tectonophysics* **119**:181–193.

Kroenke, L., Scott, R. *et al.* 1980. *Init. Repts., DSDP*, 59, U.S. Govt. Printing Office, Washington, DC.

Macdonald, G. A. 1972. *Volcanoes*, Prentice-Hall, Englewood Cliffs, NJ.

Marsaglia, K. M. 1992. Petrography and provenance of volcaniclastic sands recovered from the Izu–Bonin arc, Leg 126, in *Proc. ODP, Sci. Results*, 126 (B. Taylor, K. Fujioka *et al.*, eds.), pp. 139–154, Ocean Drilling Program, College Station, TX.

Marsaglia, K. M., in press, Interarc and backarc basins, in *Tectonics of Sedimentary Basins* (R. V. Ingersoll, and C. J. Busby, eds.), Blackwell Scientific.

Marsaglia, K. M., and Ingersoll, R. V. 1992. Compositional trends in arc-related, deep-marine sand and sandstone: A reassessment of magmatic-arc provenance, *Geol. Soc. Am. Bull.* **104**:1637–1649.

Marsaglia, K. M., and Tazaki, K. 1992. Diagenetic trends in ODP Leg 126 sandstones, in *Proc. ODP, Sci. Results*, 126 (B. Taylor, K. Fujioka *et al.*, eds.) pp. 125–138, Ocean Drilling Program, College Station, TX.

Mattey, D. P., Marsh, N. G., and Tarney, J. 1980. The geochemistry, mineralogy, and petrology of basalts from the West Philippine and Parece Vela basins and from the Palau–Kyushu and West Mariana ridges, Deep Sea Drilling Project Leg 59, in *Init. Repts, DSDP*, 59 (L. Kroenke, R. Scott *et al.*, eds.), pp. 753–800, Deep Sea Drilling Project, U.S. Govt. Printing Office, Washington, DC.

Natland, J. H., and Tarney, J. 1981. Petrologic evolution of the Mariana arc and backarc system—a synthesis of drilling results in the south Philippine Sea, in *Init. Repts, DSDP*, 60 (D. M. Hussong, S. Uyeda *et al.*, eds.), pp. 877–908, Deep Sea Drilling Project, U.S. Govt. Printing Office, Washington, DC.

Nishimura, A., Marsaglia, K. M., Rodolfo, K. S., Colella, A., Hiscott, R. N., Tazaki, K., Janecek, T., Firth, J., Isiminger-Kelso, M., Herman, Y., Gill, J. B., Taylor, R. N., Taylor, B., Fujioka, K., and Leg 126 Shipboard Scientific Party. 1991. Pliocene–Quaternary submarine pumice deposits in the Sumisu rift area, Izu–Bonin Arc, in *Sedimentation in Volcanic Settings* (R. V. Fisher and G. A. Smith, eds.), pp. 201–208, SEPM Special Publication 45.

Packham, G. H., and Williams, K. L. 1981. Volcanic glasses from sediments from Sites 453 and 454 in the Mariana Trough, in *Init. Repts., DSDP*, 60 (D. M. Hussong, S. Uyeda *et al.*, eds.), pp. 483–495, Deep Sea Drilling Project, U.S. Govt. Printing Office, Washington, DC.

Packer, B. M., and Ingersoll, R. V. 1986. Provenance and petrology of Deep Sea Drilling Project sands and sandstones from the Japan and Mariana forearc and backarc regions, *Sediment. Geol.* **51**:5–28.

Parson, L., Hawkins, J., Allan, J., *et al.* 1992. *Proc. ODP, Init. Repts.*, 135, Ocean Drilling Program, College Station, TX.

Rangin, C., Silver, E. A., von Breyman, M. T. *et al.* 1990. *Proc. ODP, Init. Repts.*, 124, Ocean Drilling Program, College Station, TX.

Rodolfo, K. S., Solidum, R. U., Nishimura, A., Matsuo, Y., and Fujioka, K. 1992. Major oxide stratigraphy of glass shards in volcanic ash layers of the Izu–Bonin arc-backarc sites (Sites 788/788 and 790/791), *Proc. ODP, Sci. Results*, 126 (B. Taylor, K. Fujioka *et al.*, eds.), pp. 505–517, Ocean Drilling Program, College Station, TX.

Rodolfo, K. S., and Warner, R. J. 1981. Tectonic, volcanic, and sedimentologic significance of volcanic glasses from Site 450 in the eastern Parece Vela basin, in *Init. Repts., DSDP*, 59 (L. Kroenke, R. Scott *et al.*, eds.), pp. 603–607, Deep Sea Drilling Project, U.S. Govt. Printing Office, Washington, DC.

Schmincke, H.-U. 1981. Ash from vitric muds in deep sea cores from the Mariana Trough and fore-arc region (South Philippine Sea)(Sites 453, 454, 455, 458, 459), in *Init. Repts., DSDP*, 60 (D. M. Hussong, S. Uyeda *et al.*, eds.), pp. 473–481, Deep Sea Drilling Project, U.S. Govt. Printing Office, Washington, DC.

Scott, R., and Kroenke, L. 1980. Evolution of back arc spreading and arc volcanism in the Philippine Sea: Interpretation of Leg 59 DSDP results, in *The Tectonic and Geologic Evolution of Southeast Asian Seas and Islands* (D. E. Hayes, ed.), pp. 283–291, Geophys. Monogr. Ser., Vol. 23, American Geophysical Union, Washington, DC.

Scott, R. B., Kroenke, L., and Zakariadze, G. 1980. Evolution of the South Philippine Sea: Deep Sea Drilling Project Leg 59 drilling results, in *Init. Repts., DSDP*, 59 (L. Kroenke, R. Scott *et al.*, eds.), pp. 803–815, Deep Sea Drilling Project, U.S. Govt. Printing Office, Washington, DC.

Shipboard Scientific Party. 1990. Introduction, background, and principal results of Leg 128 of the Ocean Drilling

Program, Japan Sea, in *Proc. ODP, Init. Repts.*, 128 (J. C. Ingle, Jr., K. Suyehiro, T. von Breymann *et al.*, eds.), pp. 5–38, Ocean Drilling Program, College Station, TX.

Shipboard Scientific Party. 1992. Introduction, background, and principal results of Leg 135, Lau Basin, in *Proc. ODP, Init. Repts.*, 135 (L. Parson, J. Hawkins, J. Allan *et al.*, eds.), pp. 5–47, Ocean Drilling Program, College Station, TX.

Sigurdsson, H., Sparks, R. S., Carey, S. N., and Huang, T. C. 1980. Volcanogenic sedimentation in the Lesser Antilles arc, *J. Geol.* **88:**523–540.

Silver, E. A., and Rangin, C. 1991. Leg 124 tectonic synthesis, in *Proc. ODP, Sci. Results*, 124 (A. E. Silver, C. Rangin, T. von Breymann *et al.*, eds.), pp. 3–9, Ocean Drilling Program, College Station, TX.

Taylor, B. 1992. Rifting and the volcanic-tectonic evolution of the Izu–Bonin–Mariana arc, in *Proc. ODP, Sci. Results*, 126 (B. Taylor, K. Fujioka *et al.*, eds.), pp. 627–651, Ocean Drilling Program, College Station, TX.

Taylor, B., Brown, G., Fryer, P., Gill, J. B., Hochstaedter, A. G., Hotta, H., Langmuir, C. H., Leinen, M., Nishimura, A., and Urabe, T. 1990b. ALVIN-Sea Beam studies of the Sumisu rift, Izu–Bonin arc, *Earth Planet. Sci. Lett.* **100:**127–147.

Taylor, B., Klaus, A., Brown, G. R., and Moore, G. F. 1991. Structural development of Sumisu rift, Izu–Bonin arc, *J. Geophys. Res.* **96:**16,113–16,129.

Taylor, B., Fujioka, K., *et al.* 1990a. *Proc. ODP, Init. Repts.*, 126, Ocean Drilling Program, College Station, TX.

Wood, D. A., Mattey, D. P., Joron, J. L., Marsh, N. G., Tarney, J., and Treuil, M. 1980. A geochemical study of 17 selected samples from basement cores recovered at Sites 447, 448, 449, and 451, in *Init. Repts., DSDP*, 59 (L. Kroenke, R. Scott *et al.*, eds.), pp. 743–752, Deep Sea Drilling Project, U.S. Govt. Printing Office, Washington, DC.

Index